The Retina in Health and Disease

The Retina in Health and Disease

Editors

Maurice Ptito
Joseph Bouskila

MDPI • Basel • Beijing • Wuhan • Barcelona • Belgrade • Manchester • Tokyo • Cluj • Tianjin

Editors
Maurice Ptito
School of Optometry
University of Montreal
Montreal
Canada

Joseph Bouskila
School of Optometry
University of Montreal
Montreal
Canada

Editorial Office
MDPI
St. Alban-Anlage 66
4052 Basel, Switzerland

This is a reprint of articles from the Special Issue published online in the open access journal *Cells* (ISSN 2073-4409) (available at: www.mdpi.com/journal/cells/special_issues/Retina_Disease).

For citation purposes, cite each article independently as indicated on the article page online and as indicated below:

LastName, A.A.; LastName, B.B.; LastName, C.C. Article Title. *Journal Name* **Year**, *Volume Number*, Page Range.

ISBN 978-3-0365-2653-9 (Hbk)
ISBN 978-3-0365-2652-2 (PDF)

Contents

About the Editors

Maurice Ptito

Professor Maurice Ptito, Dr.med., Ph.D. is the director of the Laboratory on Development and Plasticity of the Visual System and Holder of the Harland Sanders Research Chair in Visual Science at the School of Optometry (University of Montreal, Montreal, Canada). He is also a Guest Researcher at the Department of Neuroscience at the University of Copenhagen (Copenhagen, Denmark) and Adjunct Professor at the Department of Neurology and Neurosurgery at the Montreal Neurological Institute, McGill University (Montreal, Canada). His laboratory was the first to thoroughly examine the localization and function of the endocannabinoid system in monkey retinas. Prof Ptito's interests also include the study of sensory substitution and cross-modal plasticity in humans who are born blind and the mechanisms of prenatal alcohol and cannabis exposure in vervet monkeys. His work has led to several rewards (The Sir John William Dawson Medal from the Royal Society of Canada and the Hensen Prize, Denmark) and distinctions: Fellow of the Royal Society of Canada, Fellow of the Royal Society of Denmark, and Knight of the National Order Of Quebec. He has authored or co-authored more than 200 publications.

Joseph Bouskila

Dr. Joseph Bouskila, Ph.D., is a Research Associate in the Laboratory on Development and Plasticity of the Visual System at the School of Optometry (University of Montreal, Montreal, Canada). He is particularly interested in the role of the endocannabinoid system in vision in the vervet monkey. He has characterized the endocannabinoid signaling system in the central nervous system with an emphasis on the visual system, particularly the retina, lateral geniculate nucleus, and primary visual cortex. His work has contributed significantly to the field of cannabinoid research through its originality and the impact of the results on the development of pharmacological targets to treat visual diseases using the primate model. He has received numerous scholarships and rewards during his studies and has published in high-impact journals.

Preface to "The Retina in Health and Disease"

Vision is the most important sense in higher mammals. The retina is the first step in visual processing and the window to the brain. It is not surprising that problems arising in the retina would lead to moderate to severe visual impairments. According to the World Health Organization, 1.2 billion people suffer from visual impairments and blindness. Moreover, people live longer, and many retinal diseases become more frequent. New advances in retina research offer hope to cure many of these illnesses. In this Special Issue, we were fortunate enough to gather a number of prominent scientists working with various retinal pathologies in humans and animal models. We present in this volume up-to-date advances on the healthy retina (i.e. endocannabinoid system), as well as the most common retinal diseases and potential treatments. The book is divided into five topics: neurovisual development, diabetic retinopathy, neurodegenerative diseases, the role of glial cells, and the molecular and physiological aspects of visual function. This collection of articles (reviews as well as original articles) will be of interest to readers interested in vision and offers comprehensive up-to-date information on healthy and diseased retinas.

Maurice Ptito, Joseph Bouskila

Editors

cells

Editorial

The Retina: A Window into the Brain

Maurice Ptito [1,2,3,*], **Maxime Bleau** [1] and **Joseph Bouskila** [1]

1   School of Optometry, University of Montreal, Montreal, QC H3T 1P1, Canada;
      maxime.bleau.1@umontreal.ca (M.B.); joseph.bouskila@umontreal.ca (J.B.)
2   Department of Neuroscience, Copenhagen University, 2200 Copenhagen, Denmark
3   Department of Neurology and Neurosurgery, Montreal Neurological Institute, McGill University,
      Montreal, QC H3A 2B4, Canada
*   Correspondence: maurice.ptito@umontreal.ca

Citation: Ptito, M.; Bleau, M.; Bouskila, J. The Retina: A Window into the Brain. *Cells* **2021**, *10*, 3269. https://doi.org/10.3390/cells10123269

Received: 14 November 2021
Accepted: 18 November 2021
Published: 23 November 2021

Publisher's Note: MDPI stays neutral with regard to jurisdictional claims in published maps and institutional affiliations.

In the course of evolution, animals have obtained the capacity to perceive and encode their environment via the development of sensory systems such as touch, olfaction, audition, and vision. In many vertebrate species, vision is the most predominant sense for accumulating and perceiving environmental information. Vision plays a pivotal role in many interactions with the environment and with other living organisms. The eyes have adapted to complex environments, enabling animals to effectively navigate, procreate, forage for food, hunt for prey, or shelter from predators (reviewed in [1,2]).

The mammalian eye is composed of three concentric and distinct layers of tissue: the sclera, the uvea, and the retina. It is a laminar tissue containing neurons and glial cells that plays the crucial role of phototransduction, the process of converting light energy into encoded neural signals delivered to the brain [3,4]. The retina has a complex laminar organization and a cellular composition that are similar in all vertebrates. However, the layout of the retina varies between species to meet their specific needs, behaviors, and habitat (Figure 1). Humans and other primates need to identify food and predators against cluttered environments, and consequently evolved a zone of acute vision located at the center of the retina, namely, the macula. This area (5.5 mm in diameter) is specialized for color and detailed vision enabled by its higher density of cones and ganglion cells [5]. At the macular center is located the fovea, a zone with a 700 μm deep focus approximately 1.5 mm in diameter [2], composed exclusively of cones, thus providing the highest resolution for daytime vision. Nasally to the macula is located the optic disc; in this zone, all ganglion cell axons converge to exit the eye *en route* to the visual brain via the retinofugal pathways [6]. Due to this retinal configuration with a central zone of higher visual acuity extending over a limited field of view, humans and other higher primates have developed highly motile eyes [7–9]. This retinal organization and associated visual behavior had significant consequences for the evolutionary development of the occipital, frontal and prefrontal cortical areas in the simian lineage, as vision became the most important sensory system.

The human retina is a complex mosaic (Figure 2B,C) comprised of five classes of neurons specialised in processing the visual information received by the eye. These neuron types are the photoreceptors, bipolar cells, ganglion cells, horizontal cells and amacrine cells [4]. These neurons have a characteristic distribution in the ten distinct layers of the retina, and have interconnections via two different pathways, namely, the vertical and horizontal pathways. Their extensive interconnections enable the processing of the visual image projected on the retina and its transmission to the brain through the optic nerve. The vertical pathway begins with the transduction of light signals by the photoreceptors and ends with transmission to the brain via the axons of retinal ganglion cells (RGCs). The horizontal pathway is comprised of horizontal cells and amacrine cells that connect laterally to provide feedback and feedforward signals between photoreceptors and bipolar cells (for horizontal cells) and between bipolar cells and RGCs (for the amacrine cells) [10]. The horizontal pathway mediates photoreceptor convergence, motion processing and contextual modulation [11]. Therefore, these two pathways embody the basic characteristics of retinal architecture and, hence, visual processing through their connections within the retinal layers.

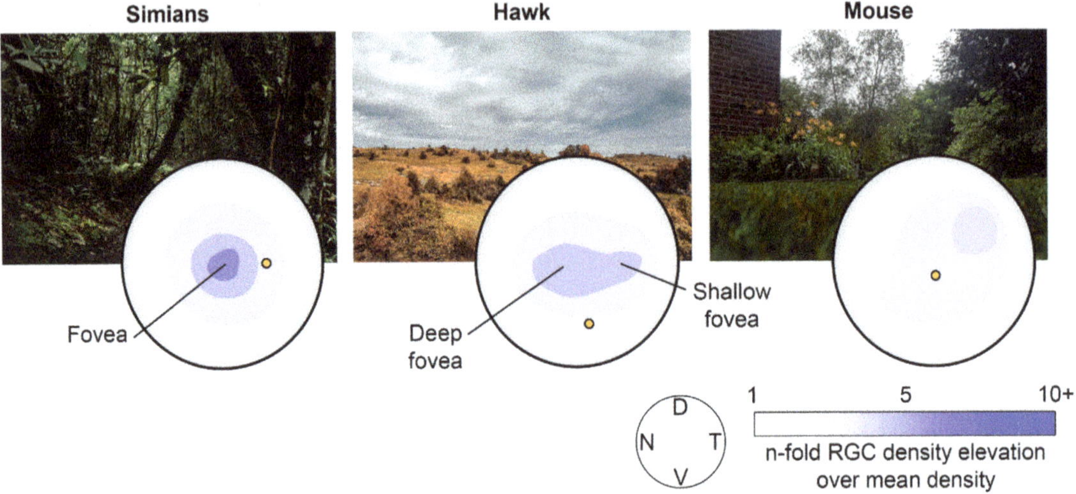

Figure 1. Snapshots of the habitat of different species along with their representative RGC density distribution across their retinal surfaces. (**Left**) Habitat of simians (jungle) and the RGC density distribution in the human retina. (**Center**) example of a typical habitat of the hawk (sky) and its retina RGC density distribution. (**Right**) example of a typical habitat of the mouse (ground) and its retina RGC density distribution. D, dorsal; N, nasal; T, temporal; V, ventral. Redrawn from [1].

The ten layers of the retina (Figure 2B,C), proceeding from the innermost to the outermost, are traditionally named as follows: the inner limiting membrane, nerve fibre layer (NFL), ganglion cell layer (GCL), inner plexiform layer (IPL), inner nuclear layer (INL), outer plexiform layer (OPL), outer nuclear layer (ONL), external limiting membrane, photoreceptors layer (PL), and the retinal pigment epithelium (RPE). The RPE is composed of epithelial cells with a rich content of melanosomes and melanin granules. RPE cells support the metabolic activity of the retina and supply the photoreceptors with nutrients and oxygen [12–14]. The photoreceptor layer is composed of the photosensitive outer segments and inner segments of the photoreceptors, the rods and cones, which are specialized for capturing and transducing light energy into electrochemical signals. The next layer, the outer limiting membrane, is formed by the extensions of large glial cells, known as the retinal Müller cells. The fourth layer is the outer nuclear layer, which contains the photoreceptors somata and nuclei. The photoreceptors then make synaptic contact with the bipolar and horizontal cells in the fifth layer, forming the outer plexiform layer. The sixth layer, the inner nuclear layer, is composed of cell bodies of horizontal, amacrine and bipolar cells. The latter cell type makes synapses with amacrine and ganglion cells in the inner plexiform layer. Ganglion cells bodies then form the ganglion cell layer, where their long axons run horizontally along the nerve fiber layer towards the optic disk. The tenth and final layer, the inner limiting membrane, is formed by extensions of Müller cells.

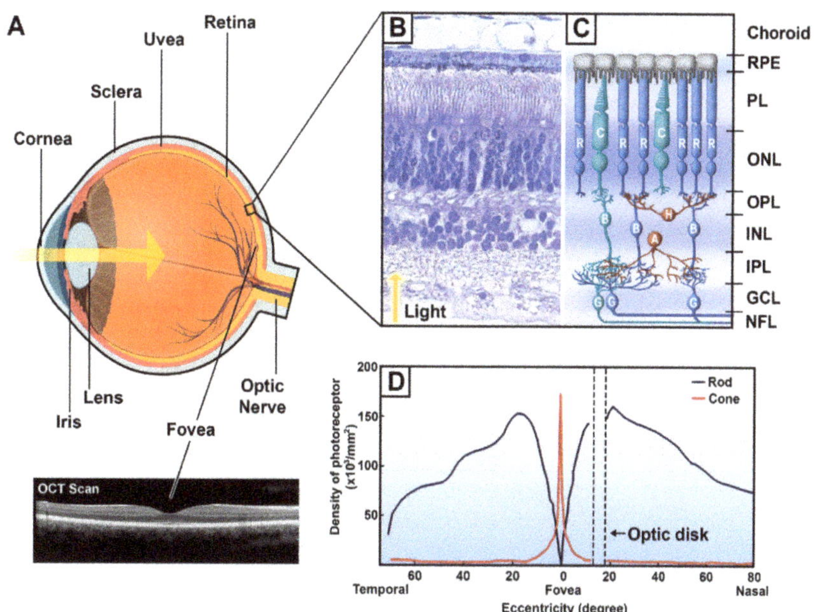

Figure 2. The eye and the retinal mosaic. (**A**) A sagittal section of the eye (created with Biorender.com) and an Optical Coherence Tomography (OCT) scan of the central retina. The yellow arrow represents the path of the light rays entering the eye. (**B**) An H&E-stained transverse section of a human retina. Adapted from [15]. (**C**) Schematic organization of retinal cells from the retinal pigment epithelium to the nerve fiber layer. (**D**) Density of cones and rods throughout the retinal surface. Adapted from [16]. OCT, optical coherence tomography; R, rod; C, cone; H, horizontal cells; B, bipolar cells; A, amacrine cells; G, retinal ganglion cells; RPE, retinal pigment epithelium; PL, photoreceptor layer; ONL, outer nuclear layer; OPL, outer plexiform layer; INL, inner nuclear layer; IPL, inner plexiform layer; GCL, ganglion cell layer; NFL, nerve fiber layer.

The vertical pathway (Figure 3) is defined by the capture and transduction of photons by the photoreceptors and the transmission of the resulting electrical signal to RGCs via their connections with bipolar cells. The main neurochemical involved in this pathway is glutamate, an excitatory neurotransmitter. There are two types of photoreceptors, the rods and cones, which both possess outer segments that are composed of stacked disks of infolded membranes containing the visual photopigments (opsin or rhodopsin coupled to a chromophore) [17]. Rods and cones differ not only in shape but also with respect to the composition of their outer segment disks, light and spectral sensitivity, and convergence towards RGCs. Rods are thinner (averaging 2 μm) and longer (averaging 50 μm) than cones. The functional particularity of rod outer segments derives from their abundant photopigment disks [18]. This property imparts greater light absorption capacity and, thus, higher light sensitivity compared to cones [13]. Moreover, through the retina, rods have a greater degree of convergence towards RGCs through their connections with bipolar cells, which serves to provide greater signal amplification, but with lower visual acuity. As such, rods are responsible for scotopic vision, but are saturated during the day or in other situations of high luminosity [19].

Cones, on the other hand, are generally thicker (3 to 5 μm) and shorter (40 μm) than rods and have a lower degree of convergence; this ratio even reaches 1 RGC for 1 cone in the fovea. The less numerous photopigment disks of the cones float freely in their outer segments [20,21]. Being less sensitive to low levels of light and only activated in situations of high luminosity, cones mediate photopic vision, operating when rods are saturated. However, both cones and rods are active in intermediate lighting conditions, such as daybreak and twilight [22]. Cones are also responsible for high-acuity vision, based on their 1-to-1 ratio of convergence in the central retina. Furthermore, three types of cones (S, L and M cones) are present in the human and nonhuman primate retina and contain the three types of opsins with different spectral sensitivity [3,23,24].

Populations of cones and rods have different distributions throughout the retina (Figure 2D). Typically, cones are outnumbered by rods by a ratio of 20- or 30-to-1 [3]. The human retina contains approximately 90 to 120 millions rods and 5 to 7 millions cones [16,25]. Throughout the peripheral retina, the density of rods greatly exceeds that of cones. However, this ratio is shifted in the fovea, where the density of cones increases almost 200-fold [25]. In that region, the individual cones are thinner (Figure 3B), allowing for the highest photoreceptor density recorded in the retina [13,25]. As a result, the central retina mediates photopic vision with a high degree of resolution (or visual acuity) and color perception, while the periphery is responsible for scotopic vision and motion sensitivity, albeit with lower visual acuity. Both types of photoreceptors send signals to the parallel pathways of bipolar cells (reviewed in [3]) via the release of glutamate.

Bipolar cells (Figure 3C) are the interneurons linking photoreceptors to RGCs and amacrine cells. Morphologically, bipolar cells are recognized by their two protrusions, one extending in the outer retina and making synaptic contact with photoreceptors and horizontal cells, and the other protrusion extending in the inner retina as the axon that relays signals from the photoreceptors to RGCs and amacrine cells. There are diverse morphological types of bipolar cells, of which the axons terminate different levels, or strata, of the inner plexiform layer. Bipolar cells thus contact different types and sets of RGCs and amacrine cells (reviewed in [26]). There are 12 types of cone bipolar cell types, and only one type of rod bipolar cell that relay the signals from rods at low light intensities [27–30]. Bipolar cells are divided into ON and OFF types, thus subserving the first step in encoding visual information according to the intensity of light received by photoreceptors [26,31,32].

In the second synaptic layer of the retina, the inner plexiform layer, bipolar cells make synaptic contact with the dendritic arborizations of the third-order neurons, namely, the RGCs (Figure 3D). These cells have relatively large somata located in the ganglion cells layer that give rise to long axons extending horizontally in the nerve fiber layer and exiting the eye through the optic nerve. As such, the RGCs are the direct portal from the retina to the brain. Numbering around 0.7 to 1.5 million in the human retina [33–35], RGCs are separated from each other by glial processes of Müller cells. They are arranged in a single cell layer, except at the macula, where the ganglion cell layer is about 8 to 10 cells thick and contains 50% of all RGCs due to the far lesser photoreceptor convergence [2,36]. RGCs were traditionally classified as ON- and OFF-center RGCs, responding to increases or decreases in light intensity presented at the center of their receptive field [37]. However, there are many subtypes of RGCs differentiated by morphological criteria (e.g., dendritic arborisation), functional criteria (response to different stimuli) and molecular criteria [38–40]. There are also five types of intrinsically photosensitive RGCs (ipRGCs), which were discovered only recently [41]. These light-sensitive cells express melanopsin, which, upon stimulation by light, activates a signaling cascade that hyperpolarizes the neuron. ipRGCs mostly project to the suprachiasmatic nucleus and thus contribute to the synchronization of the circadian oscillator [40,42,43]. There are also three types of alpha RGCs [44], three types of Local Edge Detectors RGCs [45] and three types of J-RGCs expressing junctional adhesion molecule B [46]. Because of this high diversity of ganglion cell types, each sensitive to different visual features, the retina is not a simple relay structure but the first center of complex visual processing, sending preprocessed images of the external world to the brain.

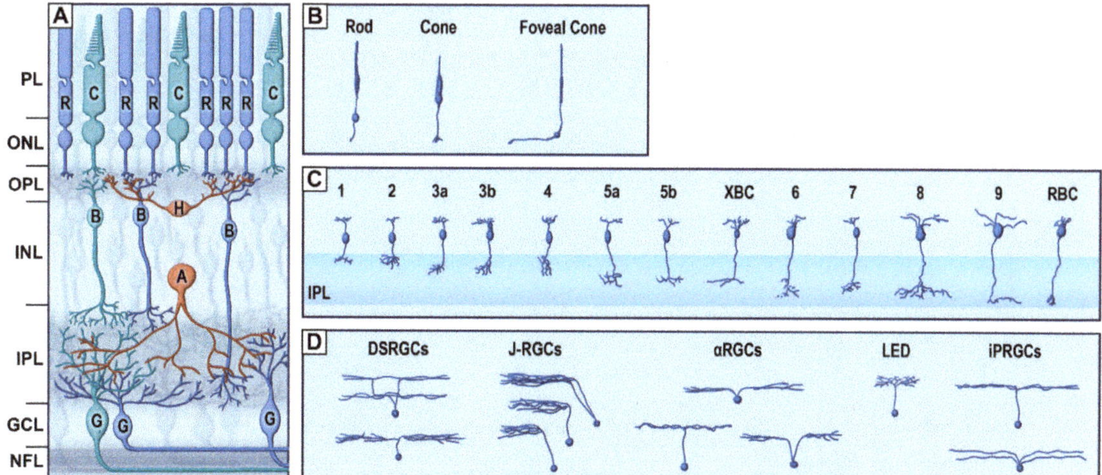

Figure 3. The vertical pathway of the retina. (**A**) Schematic organization of retinal cells. (**B**) Subtypes of photoreceptors (redrawn from von Greef, 1900). (**C**) Subtypes of bipolar cells. Redrawn from [26]. (**D**) Subtypes of retinal ganglion cells (RGCs) (redrawn from [38]). PL, photoreceptor layer; ONL, outer nuclear layer; OPL, outer plexiform layer; INL, inner nuclear layer; IPL, inner plexiform layer; GCL, ganglion cell layer; NFL, nerve fiber layer; R, rod; C, cone; H, horizontal cells; B, bipolar cells; A, amacrine cells; G, RGCs; DSRGCs, direction selective retinal ganglion cells; J-RGCs, junctional adhesion molecule B-positive retinal ganglion cells; αRGCs, alpha RGCs; LED, local edge detector; iPRGCs, intrinsically photosensitive RGCs.

The horizontal pathway (Figure 4) is comprised of two types of interneurons, horizontal and amacrine cells, which together subserve the connectivity and complementarity of subparts of the vertical circuitry at two different levels of the retina (OPL and IPL). This cellular architecture establishes the interconnected mosaic that defines the vast array of retinal image-processing functions. Indeed, horizontal and amacrine cells are essential for the creation of the center-surround properties of RGC receptive fields, along with many other visual functions observed in the retina, such as surround inhibition of photoreceptor cells [3,10,33]. The main chemical involved in this pathway is gamma aminobutyric acid (GABA), an inhibitory neurotransmitter. Horizontal cell types (Figure 4B) are distinguishable by morphological criteria and molecular markers [47]. The main classification is into A-type, or axon-bearing horizontal cells, and B-type, or axon-less horizontal cells [47]. Both subtypes are GABAergic interneurons, providing inhibitory feedback to cones or rods.

Amacrine cells mainly extend laterally (but some vertically) in the inner plexiform layer and receive inputs from bipolar cells (Figure 4C). Amacrine cells are astonishingly specialized interneurons that send feedforward signals to RGCs, feedback signals to bipolar cells and even inhibitory signals to other nearby amacrine cells (reviewed in [10,33]). Indeed, amacrine cells are inhibitory interneurons that mediate the spatial and temporal characteristics of RGCs' receptive fields and light responses [48,49], refine the center-surround receptive fields of bipolar cells [50], sharpen the bipolar cell responses timing [51] and can regulate the gain of feedforward signals [52]. Amacrine cells, similar to horizontal cells, express the atypical endocannabinoid receptor TRPV1 [53].

The numerous subtypes of amacrine cells are all inhibitory (GABAergic or glycinergic, along with a secondary neurotransmitter such as acetylcholine) and are mostly categorized functionally into narrow-, medium- and wide-field amacrine cells, according to the size of their dendritic arborization [10,33]. Wide-field amacrine cells are mostly GABAergic interneurons with arborizations that extend in diameter from 100 μm (average 350 μm) to the millimeter scale [54], whereby they mediate long-range interactions as well as inhibitory surrounds in RGCs [3,55,56]. Starburst amacrine cells are a subtype of GABAergic/cholinergic wide-field amacrine cells that impart direction of movement selectivity in some RGCs [33,57]. Medium-field amacrine cells, of which there are at least eight types, including spiny AC, secretoneurin AC are GABAergic/glycinergic, have stratified dendritic arborizations extending between 100 and 500 µm. These cells gather and distribute signals across multiple levels of the IPL [33]. Finally, narrow-field amacrine cells, of which there are at least nine types, have dendritic arborization less than 100 µm wide and are commonly glycinergic interneurons [58,59].

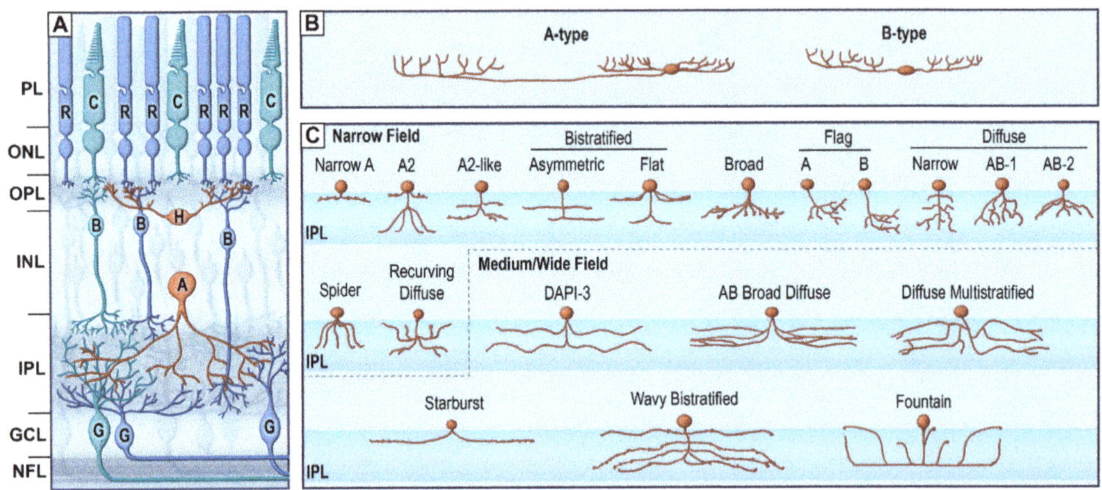

Figure 4. The retina horizontal pathway. (**A**) schematic organization of retinal cells. (**B**) Subtypes of horizontal cells. Redrawn from [60]. (**C**) Subtypes of narrow-field, medium-field and wide-field amacrine cells. Redrawn from [60,61]. PL, photoreceptor layer; ONL, outer nuclear layer; OPL, outer plexiform layer; INL, inner nuclear layer; IPL, inner plexiform layer; GCL, ganglion cell layer; NFL, nerve fiber layer; R, rod; C, cone; H, horizontal cells; B, bipolar cells; A, amacrine cells; G, retinal ganglion cells.

The intricate organization of the retina makes it the perfect window into the brain. Unsurprisingly, vision is the dominant sense in many species, such that 30 to 55% of the brain is devoted to vision, depending on the species [62]. As a result, any loss of retinal function will have dire consequences, extending from various perceptual impairments to blindness (see Table 1). Indeed, many visual problems diagnosed in patients arise due to a disorder of the retinal mosaic. Most human retinal diseases mainly affect one of two sites in the human retina: the photoreceptor layer and the RPE or RGC and nerve fibers layers. The RPE and photoreceptors layers are the main locus of serious pathologies such as Age-Related Macular Degeneration (AMD) and autosomal disorders such as Retinitis Pigmentosa (RP), Juvenile Macular Degeneration (Stargardt's disease and Best's disease), achromatopsia and retinal detachment. AMD is the most prevalent cause of visual loss in older adults around the globe. It is characterized by the deposition of lipids and proteins (Drusen) in RPE cells, along with the degeneration of photoreceptors, especially in the macula region (reviewed in [63]). The onset of AMD is marked by progressive loss of acuity in the central vision (scotoma), which impedes important visual functions such as reading and face recognition. RP is an autosomal-dominant retinal dystrophy marked by the progressive loss of RPE cells and apoptosis of photoreceptor cells, causing progressive vision loss, extending from the peripheral to central vision (i.e., ring scotoma or tunnel vision). Peripheral vision loss often leads to night blindness and challenges during locomotion, such as collisions with unseen obstacles. Stargardt's disease is an autosomal recessive disorder affecting photoreceptor cells and causing vision loss at an early age, whereas Best's disease (or vitelliform macular dystrophy) is an autosomal dominant disorder the leads to protein/lipid deposits between the RPE and photoreceptor layers [64]. Furthermore, achromatopsia, or color blindness, is an X-linked form of congenital cone dystrophy, leading to improper light transduction and reduced perception [65]. The junction between the RPE and photoreceptor cell layers is also a common site of retinal detachment. RGCs and their axons (GCL and NFL) are also commonly affected in glaucoma and other retinopathies (i.e., diabetic, hypertensive and retinopathy of prematurity) associated with micro-haemorrhages in the NFL. Glaucoma, which usually results from increased intraocular pressure, is marked by the degeneration of RGCs, leading to significant losses in visual field and contrast sensitivity [66].

Table 1. The most common retinal diseases, from hereditary disorders to age- or trauma-related conditions, linked to their functional repercussions.

Pathology	Glare	VF Loss	Scotoma	Night Blindness	Light Adaptation	Nystagmus	Fluctuating Vision	Depth Perception
Achromatopsia	x				x	x		
Albinism	x				x	x		
Coloboma	x	x						
Diabetes	x	x	x		x		x	x
Glaucoma	x	x		x	x		x	x
Macular degeneration	x		x				x	x
Optic atrophy	x	x	x			x	x	x
Retinal detachment	x	x					x	
Retinopathy of prematurity	x	x	x				x	x
Retinitis pigmentosa	x	x		x	x		x	x

Vision is undoubtedly the most important sense for humans, and visual impairments have dire consequeces for quality of life and productive activities [67–70]. The retina contains a surprising complexity in its cellular architecture, and literally presents a window to the brain; no other part of the central nervous system is amenable to direct observation. Because of the importance of human vision, researchers are making a broad effort to understand better the complex details of retinal function, even to the extent of developing highly invasive cybernetic inputs replacing the retina, or advanced methods for behavioral sensory substitution [71–73].

Therefore, this Special Issue of *Cells* is timely and brings a collection of groundbreaking novel research on the cellular and molecular aspects of healthy and diseased retinas. We present here 20 original articles and six reviews from the international research community, all with the goal of advancing knowledge towards the prevention and cure of visual pathologies, mainly AMD, RP and diabetic retinopathy. Papers in this Special Issue can be regrouped into five major themes.

The first group of articles focuses on the neurovisual development of the retina in rodents and birds. Laroche et al. bring interesting results on the role of L-lactate and its receptor GPR81 on the growth of RGC axons during development in mice. They show that treatment of axon growth cones with L-Lactate or GPR81 agonist increased the cones size and number of filopodia. The following two papers study the role of cell death (autophagy and apoptosis) in the postnatal development of the retina. Pesce et al. show increased autophagy levels in the early retinal hypoxic phase and normalized levels in the mature, fully vascularized retina, and Álvarez-Hernán et al. propose that programmed cell death markers during early retinal development in hatched altricial birds is a potential model to study post-natal retinal development.

The second group of articles deals with diabetic retinopathy, its ischemic and hemorrhagic mechanisms, related markers and possible therapies in many species, including man. Shaw et al. investigate the use of quantitative Optical Coherence Tomography Angiography (OCTA) to track systemic vascular functioning and microvascular complications in black patients with diabetes mellitus. Adeghate et al. examine the effects of early onset diabetes on the retinal ultrastructure and cellular bioenergetics in rats. They found an increase in incretins and antioxidant levels as well as increased oxidative phosphorylation, which are events that may transiently preserve visual function. Kim et al. investigate the inhibition of Drp1, its diabetes-induced overexpression being linked to mitochondrial dysfunction and apoptosis of retinal endothelial cells, with the goal to protect the retina against vascular lesions in diabetic retinopathy. Musayeva et al. studied the neuroprotective effects of betulinic acid against ischemia and reperfusion injuries, often related to pathologies such as diabetic retinopathy and glaucoma, in the mouse retina. They showed that the injection of betulinic acid improved endothelial functioning and caused a reduction in reactive oxygen species (ROS) levels following ischemia. Middel et al. propose a review about studies on the zebrafish neurovascular unit in the context of diabetic retinopathy and discuss the advantages of the zebrafish model in studying this specific pathology. As for ischemic injuries, Agrawal et al. tested iSMC as a potential therapy for reducing inflammation and retinal degeneration.

The third group of articles focuses on changes and markers associated with retinal neurodegenerative diseases and on potential therapies. Hamid et al. explore the effects of anti-VEGF drugs on different AMD genetic markers in retinal epithelial cells. Holan et al. review the literature on mesenchymal stem cell (MSC) therapy for many retinal degenerative diseases, such as AMD, RP, glaucoma and diabetic retinopathy. They provide insights on MSCs' therapeutic properties, such as the production of growth/neurotrophic factors. Othman et al. propose an interesting review on the kallikrein-kinin system (KKS), its important role in neurodegenerative diseases such as AMD and diabetic retinopathy, and how it provides promising therapeutic targets against these diseases. The last two original articles in this section explore changes and markers in neurodegeneration induced by light in rodents. Riccitelli et al. investigate the time course of retinal changes (neurodegeneration and recovery) following light exposure in rats and suggest the development of diagnostic tools to monitor the progression of pathologies and efficacy of treatments. Park et al. investigate the expression of stress markers in retinal degeneration induced by blue LED stimuli. They show that these markers may serve an important role in the activation of glial cells during degeneration of photoreceptors. Rajendran et al. underline the therapeutic potential of RPE grafts during long-term observations in a rat model of RPE. Candadai et al. conclude on the neuroprotective effects of Fingolimod in an in vitro model of optic neuritis. Girol et al. demonstrate that in a case of endotoxin-induced uveitis, the inflammation can be treated with piperlongumine and/or Annexin A1 mimetic peptide (Ac2-26).

The fourth group of articles explores the role of glial cells in retinal function. Lee et al. investigate the response of Müller cells and microglia in retinal detachment. They found that glial activation markers were differently expressed in intact and detached regions. Choi, Guo and Cordeiro propose a review exploring the role and morphology of retinal microglia as well as their activation by stress stimuli in multiple sclerosis and other neurodegenerative diseases such as Alzheimer's disease, Parkinson's disease, glaucoma and RP. Finally, Yoo et al. present an interesting review on the role of retinal astrocytes and Müller cells and the potential molecular pathways that can induce these cells to become growth-supportive and thus promote the survival and regeneration of RGCs in neurodegenerative diseases.

The fifth and final group of articles brings novel data on the molecular and physiological aspects of visual function. Solanki et al. investigate the role of the motor protein MYO1C, an unconventional myosin, and how its loss causes the mis-localization of rhodopsin in photoreceptors, leading to impaired visual function in mice. Dao et al. bring novel data on high-fat diets associated with the development and progression of many retinal diseases, such as AMD and diabetic retinopathy. They show that such diets can alter the retinal transcriptome independently of gut microbiota. Fusz et al. investigate the variation in gap junction connections across the mammalian (mouse, rat and cat) retina via the topographical distribution of connexin-36 (associated with electrical synapses) in ON and OFF amacrine cells. Bouskila et al. present a review on the expression, localization and function of endocannabinoids in the horizontal and vertical pathways of the monkey's retina and discuss their potential as therapeutic targets for visual pathologies such as AMD, glaucoma and diabetic retinopathy. Piarulli et al. report a case of a patient with Charles Bonnet syndrome, a health condition characterized by vivid visual hallucinations in individuals with retinal vision loss. Dumitrascu et al. showed that retinal curcumin-fluorescence imaging is a good predictor of verbal memory loss linked to abnormal retinal vasculature and amyloid count. Finally, long exposure to short wavelength LED light causes retinal damage that can be reduced by the use of short wavelength selective filters apposed on the LED screens, as demonstrated by Sanchez-Ramos et al.

We sincerely hope that the papers presented in this Special Issue of *Cells* will contribute to the understanding of retinal diseases and their underlying mechanisms and will lead to the development of new pharmaceutical tools to treat visual disorders.

Funding: This research received no external funding.

Acknowledgments: The Editors are grateful to the contributors to this Special Issue and also the numerous referees who generously gave their time and effort to review the papers. The editorial staff at *Cells*, particularly Sybil Miao, deserves a special mention for their professionalism, availability and patience.

Conflicts of Interest: The author declares no conflict of interest.

References

1. Baden, T.; Euler, T.; Berens, P. Understanding the retinal basis of vision across species. *Nat. Rev. Neurosci.* **2019**, *21*, 5–20. [CrossRef] [PubMed]
2. Mashige, K.P.; Oduntan, O.A. A review of the human retina with emphasis on nerve fibre layer and macula thicknesses. *Afr. Vis. Eye Health* **2016**, *75*, 1–8. [CrossRef]
3. Demb, J.B.; Singer, J.H. Functional Circuitry of the Retina. *Annu. Rev. Vis. Sci.* **2015**, *1*, 263–289. [CrossRef] [PubMed]
4. Cajal, S.R. La retine des vertebres. *Cellule* **1893**, *9*, 119–255.
5. Ishikawa, H.; Stein, D.M.; Wollstein, G.; Beaton, S.; Fujimoto, J.G.; Schuman, J. Macular Segmentation with Optical Coherence Tomography. *Investig. Opthalmology Vis. Sci.* **2005**, *46*, 2012–2017. [CrossRef]
6. Gilbert, C.D. The Constructive Nature of Visual Processing. In *Principles of Neural Science*, 6th ed.; Kandel, E.R., Schwartz, J.H., Jessell, T.M., Eds.; McGraw-Hill: NewYork, NY, USA, 2021; pp. 556–576.
7. Wu, C.-C.; Wick, F.A.; Pomplun, M. Guidance of visual attention by semantic information in real-world scenes. *Front. Psychol.* **2014**, *5*, 54. [CrossRef] [PubMed]
8. Carpenter, R.H. *Movements of the Eyes, 2nd Rev*; Pion Limited: London, UK, 1988.
9. Fuchs, A.F. Saccadic and smooth pursuit eye movements in the monkey. *J. Physiol.* **1967**, *191*, 609–631. [CrossRef]
10. Diamond, J.S. Inhibitory Interneurons in the Retina: Types, Circuitry, and Function. *Annu. Rev. Vis. Sci.* **2017**, *3*, 1–24. [CrossRef]
11. Franke, K.; Baden, T. General features of inhibition in the inner retina. *J. Physiol.* **2017**, *595*, 5507–5515. [CrossRef]
12. Steinberg, R.H. Interactions between the retinal pigment epithelium and the neural retina. *Doc. Ophthalmol.* **1985**, *60*, 327–346. [CrossRef]
13. Nag, T.C.; Wadhwa, S. Ultrastructure of the human retina in aging and various pathological states. *Micron* **2012**, *43*, 759–781. [CrossRef] [PubMed]
14. Strauss, O. The Retinal Pigment Epithelium in Visual Function. *Physiol. Rev.* **2005**, *85*, 845–881. [CrossRef]
15. Swaroop, A.; Zack, D.J. Transcriptome analysis of the retina. *Genome Biol.* **2002**, *3*, 1–4. [CrossRef] [PubMed]
16. Osterberg, G.A. Topography of the layer of the rods and cones in the human retima. *Acta Ophthalmol* **1935**, *13*, 1–102.
17. Kottke, T.; Xie, A.; Larsen, D.S.; Hoff, W.D. Photoreceptors Take Charge: Emerging Principles for Light Sensing. *Annu. Rev. Biophys.* **2018**, *47*, 291–313. [CrossRef] [PubMed]
18. Borwein, B. The Retinal Receptor: A Description. *Vertebr. Photoreceptor Opt.* **1981**, *23*, 11–81. [CrossRef]
19. Turner, P.L.; Mainster, M.A. Circadian photoreception: Ageing and the eye's important role in systemic health. *Br. J. Ophthalmol.* **2008**, *92*, 1439–1444. [CrossRef]
20. Johnson, L.V.; Hageman, G.S.; Blanks, J.C. Interphotoreceptor matrix domains ensheath vertebrate cone photoreceptor cells. *Investig. Ophthalmol. Vis. Sci.* **1986**, *27*, 129–135.
21. Blanks, J.C.; Hageman, G.S.; Johnson, L.V.; Spee, C. Ultrastructural visualization of primate cone photoreceptor matrix sheaths. *J. Comp. Neurol.* **1988**, *270*, 288–300. [CrossRef]

22. Yin, L.; Smith, R.G.; Sterling, P.; Brainard, D.H. Chromatic Properties of Horizontal and Ganglion Cell Responses Follow a Dual Gradient in Cone Opsin Expression. *J. Neurosci.* **2006**, *26*, 12351–12361. [CrossRef]

23. Jacobs, G.H.; Rowe, M.P. Evolution of vertebrate colour vision. *Clin. Exp. Optom.* **2004**, *87*, 206–216. [CrossRef] [PubMed]

24. Williams, D.R. Imaging single cells in the living retina. *Vis. Res.* **2011**, *51*, 1379–1396. [CrossRef] [PubMed]

25. Curcio, C.A.; Sloan, K.R.; Kalina, R.E.; Hendrickson, A.E. Human photoreceptor topography. *J. Comp. Neurol.* **1990**, *292*, 497–523. [CrossRef]

26. Euler, T.; Haverkamp, S.; Schubert, T.; Baden, T. Retinal bipolar cells: Elementary building blocks of vision. *Nat. Rev. Neurosci.* **2014**, *15*, 507–519. [CrossRef]

27. Tsukamoto, Y.; Omi, N. Functional allocation of synaptic contacts in microcircuits from rods via rod bipolar to AII amacrine cells in the mouse retina. *J. Comp. Neurol.* **2013**, *521*, 3541–3555. [CrossRef]

28. Wässle, H.; Puller, C.; Müller, F.; Haverkamp, S.; Mueller, F. Cone Contacts, Mosaics, and Territories of Bipolar Cells in the Mouse Retina. *J. Neurosci.* **2009**, *29*, 106–117. [CrossRef]

29. Ghosh, K.K.; Bujan, S.; Haverkamp, S.; Feigenspan, A.; Wässle, H. Types of bipolar cells in the mouse retina. *J. Comp. Neurol.* **2004**, *469*, 70–82. [CrossRef]

30. Breuninger, T.; Puller, C.; Haverkamp, S.; Euler, T. Chromatic Bipolar Cell Pathways in the Mouse Retina. *J. Neurosci.* **2011**, *31*, 6504–6517. [CrossRef] [PubMed]

31. Euler, T.; Masland, R.H. Light-evoked responses of bipolar cells in a mammalian retina. *J. Neurophysiol.* **2000**, *83*, 1817–1829. [CrossRef]

32. Hartveit, E. Functional Organization of Cone Bipolar Cells in the Rat Retina. *J. Neurophysiol.* **1997**, *77*, 1716–1730. [CrossRef]

33. Grünert, U.; Martin, P.R. Cell types and cell circuits in human and non-human primate retina. *Prog. Retin. Eye Res.* **2020**, *78*, 100844. [CrossRef]

34. Watson, A.B. A formula for human retinal ganglion cell receptive field density as a function of visual field location. *J. Vis.* **2014**, *14*, 15. [CrossRef]

35. Herbin, M.; Boire, D.; Ptito, M. Size and distribution of retinal ganglion cells in the St. Kitts green monkey (Cercopithecus aethiops sabeus). *J. Comp. Neurol.* **1997**, *383*, 459–472. [CrossRef]

36. Tan, O.; Li, G.; Lu, A.T.-H.; Varma, R.; Huang, D. Mapping of Macular Substructures with Optical Coherence Tomography for Glaucoma Diagnosis. *Ophthalmology* **2008**, *115*, 949–956. [CrossRef]

37. Kuffler, S.W. Discharge patterns and functional organization of mammalian retina. *J. Neurophysiol.* **1953**, *16*, 37–68. [CrossRef]

38. Sanes, J.R.; Masland, R.H. The Types of Retinal Ganglion Cells: Current Status and Implications for Neuronal Classification. *Annu. Rev. Neurosci.* **2015**, *38*, 221–246. [CrossRef] [PubMed]

39. Baden, T.; Berens, P.; Franke, K.J.; Rosón, M.R.; Bethge, M.; Euler, T. The functional diversity of retinal ganglion cells in the mouse. *Nature* **2016**, *529*, 345–350. [CrossRef] [PubMed]

40. Nguyen-Ba-Charvet, K.T.; Rebsam, A. Neurogenesis and Specification of Retinal Ganglion Cells. *Int. J. Mol. Sci.* **2020**, *21*, 451. [CrossRef] [PubMed]

41. Hattar, S.; Liao, H.-W.; Takao, M.; Berson, D.M.; Yau, K.-W. Melanopsin-containing retinal ganglion cells: Architecture, projections, and intrinsic photosensitivity. *Science* **2002**, *295*, 1065–1070. [CrossRef]

42. Do, M.T.H.; Yau, K.-W. Intrinsically Photosensitive Retinal Ganglion Cells. *Physiol. Rev.* **2010**, *90*, 1547–1581. [CrossRef]

43. Chew, K.S.; Renna, J.M.; McNeill, D.S.; Fernandez, D.C.; Keenan, W.T.; Thomsen, M.B.; Ecker, J.L.; Loevinsohn, G.S.; Van Dunk, C.; Vicarel, D.C.; et al. A subset of ipRGCs regulates both maturation of the circadian clock and segregation of retinogeniculate projections in mice. *eLife* **2017**, *6*, e22861. [CrossRef]

44. Pang, J.-J.; Gao, F.; Wu, S.M. Light-Evoked Excitatory and Inhibitory Synaptic Inputs to ON and OFF α Ganglion Cells in the Mouse Retina. *J. Neurosci.* **2003**, *23*, 6063–6073. [CrossRef] [PubMed]

45. Zhang, Y.; Kim, I.-J.; Sanes, J.R.; Meister, M. The most numerous ganglion cell type of the mouse retina is a selective feature detector. *Proc. Natl. Acad. Sci. USA* **2012**, *109*, E2391–E2398. [CrossRef] [PubMed]

46. Kim, I.-J.; Zhang, Y.; Yamagata, M.; Meister, M.; Sanes, J.R. Molecular identification of a retinal cell type that responds to upward motion. *Nature* **2008**, *452*, 478–482. [CrossRef]

47. Boije, H.; Shirazi Fard, S.; Edqvist, P.-H.; Hallböök, F. Horizontal Cells, the Odd Ones Out in the Retina, Give Insights into Development and Disease. *Front. Neuroanat.* **2016**, *10*, 77. [CrossRef]

48. Lee, S.; Zhang, Y.; Chen, M.; Zhou, Z.J. Segregated glycine-glutamate co-transmission from vGluT3 amacrine cells to contrast-suppressed and contrast-enhanced retinal circuits. *Neuron* **2016**, *90*, 27–34. [CrossRef] [PubMed]

49. Tien, N.-W.; Kim, T.; Kerschensteiner, D. Target-Specific Glycinergic Transmission from VGluT3-Expressing Amacrine Cells Shapes Suppressive Contrast Responses in the Retina. *Cell Rep.* **2016**, *15*, 1369–1375. [CrossRef]

50. Flores-Herr, N.; Protti, D.A.; Wässle, H. Synaptic Currents Generating the Inhibitory Surround of Ganglion Cells in the Mammalian Retina. *J. Neurosci.* **2001**, *21*, 4852–4863. [CrossRef]

51. Dong, C.-J.; Hare, W.A. Temporal Modulation of Scotopic Visual Signals by A17 Amacrine Cells in Mammalian Retina In Vivo. *J. Neurophysiol.* **2003**, *89*, 2159–2166. [CrossRef]

52. Grimes, W.N.; Zhang, J.; Tian, H.; Graydon, C.W.; Hoon, M.; Rieke, F.; Diamond, J.S. Complex inhibitory microcircuitry regulates retinal signaling near visual threshold. *J. Neurophysiol.* **2015**, *114*, 341–353. [CrossRef]

53. Bouskila, J.; Micaelo-Fernandes, C.; Palmour, R.M.; Bouchard, J.-F.; Ptito, M. Transient receptor potential vanilloid type 1 is expressed in the horizontal pathway of the vervet monkey retina. *Sci. Rep.* **2020**, *10*, 1–10. [CrossRef]
54. Balasubramanian, R.; Gan, L. Development of Retinal Amacrine Cells and Their Dendritic Stratification. *Curr. Ophthalmol. Rep.* **2014**, *2*, 100–106. [CrossRef]
55. Zaghloul, K.A.; Manookin, M.B.; Borghuis, B.G.; Boahen, K.; Demb, J.B. Functional Circuitry for Peripheral Suppression in Mammalian Y-Type Retinal Ganglion Cells. *J. Neurophysiol.* **2007**, *97*, 4327–4340. [CrossRef] [PubMed]
56. Baccus, S.A.; Ölveczky, B.P.; Manu, M.; Meister, M. A Retinal Circuit That Computes Object Motion. *J. Neurosci.* **2008**, *28*, 6807–6817. [CrossRef]
57. Vaney, D.I.; Sivyer, B.; Taylor, W.R. Direction selectivity in the retina: Symmetry and asymmetry in structure and function. *Nat. Rev. Neurosci.* **2012**, *13*, 194–208. [CrossRef] [PubMed]
58. Roska, B.; Werblin, F. Vertical interactions across ten parallel, stacked representations in the mammalian retina. *Nature* **2001**, *410*, 583–587. [CrossRef] [PubMed]
59. Zhang, C.; McCall, M.A. Receptor targets of amacrine cells. *Vis. Neurosci.* **2012**, *29*, 11–29. [CrossRef] [PubMed]
60. Masland, R.H. The fundamental plan of the retina. *Nat. Neurosci.* **2001**, *4*, 877–886. [CrossRef] [PubMed]
61. Masland, R.H. The tasks of amacrine cells. *Vis. Neurosci.* **2012**, *29*, 3–9. [CrossRef] [PubMed]
62. Sheth, B.R.; Young, R. Two Visual Pathways in Primates Based on Sampling of Space: Exploitation and Exploration of Visual Information. *Front. Integr. Neurosci.* **2016**, *10*, 37. [CrossRef]
63. Fleckenstein, M.; Keenan, T.D.L.; Guymer, R.H.; Chakravarthy, U.; Schmitz-Valckenberg, S.; Klaver, C.C.; Wong, W.T.; Chew, E.Y. Age-related macular degeneration. *Nat. Rev. Dis. Prim.* **2021**, *7*, 1–25. [CrossRef]
64. Altschwager, P.; Ambrosio, L.; Swanson, E.A.; Moskowitz, A.; Fulton, A.B. Juvenile Macular Degenerations. *Semin. Pediatr. Neurol.* **2017**, *24*, 104–109. [CrossRef] [PubMed]
65. Remmer, M.H.; Rastogi, N.; Ranka, M.P.; Ceisler, E.J. Achromatopsia. *Curr. Opin. Ophthalmol.* **2015**, *26*, 333–340. [CrossRef] [PubMed]
66. Weinreb, R.N.; Aung, T.; Medeiros, F.A. The Pathophysiology and Treatment of Glaucoma. *JAMA* **2014**, *311*, 1901–1911. [CrossRef]
67. Eckert, K.A.; Carter, M.J.; Lansingh, V.C.; Wilson, D.A.; Furtado, J.M.; Frick, K.D.; Resnikoff, S. A Simple Method for Estimating the Economic Cost of Productivity Loss Due to Blindness and Moderate to Severe Visual Impairment. *Ophthalmic Epidemiol.* **2015**, *22*, 349–355. [CrossRef]
68. Ramrattan, R.S.; Wolfs, R.C.; Panda-Jonas, S.; Jonas, J.B.; Bakker, D.; Pols, H.A.; de Jong, P.T. Prevalence and causes of visual field loss in the elderly and associations with impairment in daily functioning: The Rotterdam Study. *Arch. Ophthalmol.* **2001**, *119*, 1788–1794. [CrossRef] [PubMed]
69. Palmer, C. Educating learners with vision impairment in inclusive settings. *Int. Congr. Ser.* **2005**, *1282*, 922–926. [CrossRef]
70. Schinazi, V.R.; Thrash, T.; Chebat, D. Spatial navigation by congenitally blind individuals. *WIREs Cogn. Sci.* **2016**, *7*, 37–58. [CrossRef]
71. Ptito, M.; Bleau, M.; Djerourou, I.; Paré, S.; Schneider, F.C.; Chebat, D.-R. Brain-Machine Interfaces to Assist the Blind. *Front. Hum. Neurosci.* **2021**, *15*, 46. [CrossRef] [PubMed]
72. Fernandez, E. Development of visual Neuroprostheses: Trends and challenges. *Bioelectron. Med.* **2018**, *4*, 1–8. [CrossRef]
73. Chebat, D.-R.; Schneider, F.C.; Ptito, M. Spatial Competence and Brain Plasticity in Congenital Blindness via Sensory Substitution Devices. *Front. Neurosci.* **2020**, *14*, 815. [CrossRef] [PubMed]

Article

Participation of L-Lactate and Its Receptor HCAR1/GPR81 in Neurovisual Development

Samuel Laroche [1], **Aurélie Stil** [1], **Philippe Germain** [1], **Hosni Cherif** [1], **Sylvain Chemtob** [2,3] **and Jean-François Bouchard** [1,*]

1   Neuropharmacology Laboratory, School of Optometry, Université de Montréal, Montreal, QC H3T 1P1, Canada; samuel.laroche.1@umontreal.ca (S.L.); aurelie.stil@umontreal.ca (A.S.); philippe.germain.1@umontreal.ca (P.G.); hosni.cherif@hotmail.com (H.C.)
2   Department of Pediatrics, Research Center-CHU Sainte-Justine, Montreal, QC H3T 1C5, Canada; sylvain.chemtob@umontreal.ca
3   Department of Ophtalmology, Faculty of Medicine, Université de Montréal, Montreal, QC H3T 1J4, Canada
*   Correspondence: jean-francois.bouchard@umontreal.ca

Abstract: During the development of the retina and the nervous system, high levels of energy are required by the axons of retinal ganglion cells (RGCs) to grow towards their brain targets. This energy demand leads to an increase of glycolysis and L-lactate concentrations in the retina. L-lactate is known to be the endogenous ligand of the GPR81 receptor. However, the role of L-lactate and its receptor in the development of the nervous system has not been studied in depth. In the present study, we used immunohistochemistry to show that GPR81 is localized in different retinal layers during development, but is predominantly expressed in the RGC of the adult rodent. Treatment of retinal explants with L-lactate or the exogenous GPR81 agonist 3,5-DHBA altered RGC growth cone (GC) morphology (increasing in size and number of filopodia) and promoted RGC axon growth. These GPR81-mediated modifications of GC morphology and axon growth were mediated by protein kinases A and C, but were absent in explants from $gpr81^{-/-}$ transgenic mice. Living $gpr81^{-/-}$ mice showed a decrease in ipsilateral projections of RGCs to the dorsal lateral geniculate nucleus (dLGN). In conclusion, present results suggest that L-lactate and its receptor GPR81 play an important role in the development of the visual nervous system.

Keywords: lactate; GPR81; HCAR1; retinal ganglion cells; growth cone; dLGN; retina; axon; 3,5-DHBA

Citation: Laroche, S.; Stil, A.; Germain, P.; Cherif, H.; Chemtob, S.; Bouchard, J.-F. Participation of L-Lactate and Its Receptor HCAR1/GPR81 in Neurovisual Development. *Cells* **2021**, *10*, 1640. https://doi.org/10.3390/cells10071640

Academic Editor: Stefan Liebau

Received: 22 May 2021
Accepted: 22 June 2021
Published: 30 June 2021

Publisher's Note: MDPI stays neutral with regard to jurisdictional claims in published maps and institutional affiliations.

1. Introduction

The physiological significance of lactic acid and its conjugate base lactate have been a major source of controversy since their discovery in biological tissues. Lactic acid was long considered to be simply the waste product of anaerobic glycolysis. Mainly occurring as the L-enantiomer in physiological conditions, lactate is now known to have multiple effects on cell homeostasis, serving as a metabolic fuel and buffering agent, while also acting as a signaling molecule, also known as "Lactormone". This signaling action is obtained via the hydroxycarboxylic acid receptor 1 (HCAR1) [1,2]. Also known as GPR81, this is a G-protein-coupled receptor activated by L-lactate and the exogenous agonist 3,5-dihydroxybenzoic acid (3,5-DHBA) [3]. GPR81 is expressed in diverse organs, including adipose tissues, skeletal muscle, liver, kidney, brain, and retina [4,5].

The retina and central nervous system have an inherently high energy demand due to the continuous depolarization of neuronal membranes [6]. During the development of the visual system, the axons of the retinal ganglion cells (RGCs) that form the optic nerves follow chemotropic molecules to grow toward and across the optic chiasma [7]. In mammals, most of the axons cross at the optic chiasm to reach the contralateral side of the optic tract while only axons from RGCs on the ventro-temporal side of the retina do not cross at the optic chiasm. Ventro-temporal RGCs specifically express the ephrin receptor (EphB1) on their GC. When the GCs come into contact with the Ephrin-B2 ligand secreted by radial glial cells in the midline of the optic chiasm, a chemorepulsive effect redirects them and the axons project to the ipsilateral optic tract [8]. Following the optic tract, some of RGC axons join the dorso-lateral geniculate nucleus (dLGN) in the thalamus, making synapses with dLGN neurons. These neurons project their axons to the layer 4 of the primary visual cortex [9].

The consequent high metabolic demand for ATP production is met by glycolysis, resulting in L-lactate generation, despite a rich endowment of mitochondria for aerobic respiration. ATP generation by oxidative phosphorylation is limited and high energy demand leads to an increase of glycolysis filling the ATP equilibrium [2,10]. It has been previously reported that the lactate dehydrogenase (LDH-1) subunit, which preferentially catalyzes the conversion of L-lactate to pyruvate, is found in neurons and astrocytes [11,12]. However, LDH-5, which is more highly expressed in astrocytes, preferentially favors the conversion of pyruvate back to L-lactate [11]; LDH-5 is also present in glycolytic tissues such as skeletal muscles [13,14]. This cell-type distribution of LDH subunits indicates a close link with characteristic features of L-lactate metabolism, which has implications for neuronal development, growth, and survival [6,11].

GPR81 has been reported to mediate effects of L-lactate in diverse processes, including wound healing, angiogenesis, neuroprotection, cancer cell survival, attenuation of inflammation, and antilipolytic effects [5,15–20]. Recent studies have revealed the presence of GPR81 in Müller cells and retinal ganglion cells (RGCs) [21]. Its activation in these cells regulates angiogenic Wnt ligands and Norrin, which together participate in intra-retinal vascularization. Receptors specific for other metabolic intermediates have also been shown to govern angiogenesis and neuronal growth in the visual system. Thus, receptors for the Krebs cycle intermediates succinate (GPR91) and α-ketoglutarate (GPR99) control vascular and axonal growth [22,23]. Hence, there is a close link between metabolism and cell signaling in regulating the development of vascular and neuronal systems. However, the role of L-lactate and its cell membrane target GPR81 during the development of the central nervous system (CNS) is yet to be explored. In this study, we built upon established findings in the CNS to investigate the role of GPR81 during development of the visual system. More specifically, we investigated the effects of L-lactate/GPR81 on axonal growth guidance in the developing mouse neuro-visual system.

2. Materials and Methods

2.1. Ethics Statement and Animals

All animal procedures were performed in accordance with the Animal Care Committee of the University of Montreal following the guidelines from the Canadian Council on Animal Care. The *gpr81*$^{-/-}$ mice were purchased from Lexicon Pharmaceuticals (The Woodlands, TX, USA). These mice were developed by the insertion of a 4-kb IRES-lacZ-neo cassette in the trans-membrane domain 2 coding region (100 base pairs) of the *gpr81* gene on C57BL/6J mice [20], with control studies in C57BL/6J WT control mice. Other experiments were undertaken in golden Syrian hamsters (Charles River Laboratories, Saint-Constant, QC, Canada). All animals were maintained in an environmentally controlled room held at $21 \pm 2\,^{\circ}$C, and with 12 h dark/light circle. Mice and hamsters of both sexes were used in this study. Food and water were provided ad libitum.

2.2. Genotypic Screening

Mice were genotyped by PCR using the amputated tail tip for DNA extraction. Tail samples were immersed in 50 mM NaOH, incubated for 20 min at 95 $^{\circ}$C, vortexed, and neutralized with 1 M Tris-HCl, pH 8.0. The samples were then revortexed and centrifuged for eight minutes at 13,000 rpm in a Fisher Scientific^{TM} accuSpin^{TM} Micro R centrifuge. The supernatant was taken for DNA amplification and added to the PCR reagent mix containing PCR Buffer, MgCl$_2$, dNTP mix, Taq DNA polymerase, forward and reverse primers. PCR cycle conditions were: 5 min at 95.0 °C, 30 cycles of three steps (1 min at 50.0 °C, 1 min at 72.0 °C and 1 min at 95.0 °C). Genotype of the $Gpr81^{-/-}$ mice was confirmed using specific primers for wild-type (WT), with normal mouse DNA as a control. The primer pairs w (forward: 5′-CATCTTGTTCTGCTCGGTCA–3′ and reverse: 5′-GAGGAAGTAGAGCCTAGCCA-3′) were used to amplify a 160 bp fragment present in the $gpr81^{+/+}$ genome but absent in the $gpr81^{-/-}$ mice.

2.3. Reagents

L-(+)-lactic acid, bovine serum albumin (BSA), ciliary neurotrophic factor (CNTF), DNase, forskolin, insulin, laminin, poly-D-lysine, progesterone, selenium, putrescine, gelatin from porcine skin, chromium(III) potassium sulfate dodecahydrate, sucrose, sodium chloride (NaCl), PCR reagent mix, potassium chloride (KCl), hydrochloric acid (HCl), disodium hydrogen phosphate (Na_2HPO_4), potassium phosphate monobasic (KH_2PO_4), *gpr81* primers, mouse anti-Brn-3a (MAB1585), rabbit anti-GPR81-S296 (SAB1300790), mouse anti-MAP2 (M9942), transferrin, trypsin, and triiodothyronine were purchased from Sigma Aldrich (Oakville, ON, Canada). B27, N2, fetal bovine serum (FBS), neurobasal medium (NB), penicillin-streptomycin, Minimum Essential Medium Eagle medium, Spinner Modification (S-MEM), and sodium pyruvate were purchased from Life Technologies (Burlington, ON, Canada). The standard goat serum, Peroxidase-AffiniPure Donkey anti-rabbit IgG, and Peroxidase-AffiniPure Donkey anti-mouse IgG were from Jackson ImmunoResearch (West Grove, PA, USA). N-acetyl cysteine (NAC) was acquired from EMD technologies (Saint-Eustache, QC, Canada). 3,5-DHBA was purchased from Tocris (Oakville, ON, Canada). Cholera toxin subunit B (CTb) recombinant conjugate with Alexa Fluor 555 (C34776) and 647 (C34778), GlutaMAX™ Supplement, green Neg-50 Frozen section medium, Hank's 1X Balanced Salt Solutions (HBSS), Tween 20, Triton X-100, paraformaldehyde (PFA), sucrose, TEMED, sodium hydroxide (NaOH), Tris base, Taq DNA Polymerase, Shandon Immu-Mount, AlexaFluor donkey anti-mouse 488, Alexa Fluor 488 goat anti-rabbit IgG (H+L), AlexaFluor goat anti-rabbit 546, Alexa Fluor 546 goat anti-mouse, and Alexa Fluor 546 phalloidin were obtained from Fisher Scientific (Ottawa, ON, Canada). Rabbit anti-PKA C-α (#4782), rabbit anti-Phospho-PKA C (Thr197) (#4781), rabbit anti-PKCα (#2056), and rabbit anti-Phospho-PKC (pan) (βII Ser660) (#9371) were acquired from Cell Signalling Technology (Whitby, ON, Canada). Heparin and sterile saline solution (0.9%) were purchased from CDMV (St-Hyacinthe, QC, Canada).

2.4. Tissue Preparation for Immunohistochemistry

Adult mice and postnatal day 5 (P5) golden Syrian hamsters were euthanized by an isoflurane overdose. A transcardial perfusion was conducted with 10 U/mL of heparin in 60 mL of phosphate-buffered 0.9% saline (PBS; 0.1 M, 4 °C, pH 7.4), followed by 60 mL phosphate-buffered 4% paraformaldehyde 4 °C (PFA). Following the harvesting of mouse embryos E14-16 extraction by Cesarian, they were deeply anesthetized by hypothermia. The orbits were removed and two small holes were made in the cornea prior to immersion fixation in 4% PFA 4 °C for 60 min. The eyecups were washed in PBS, cryoprotected in 30% sucrose overnight, embedded in Neg 50 medium, flash-frozen, and kept at −80 °C until processing. Sections 14 μm-thick were cut with a cryostat (Leica Microsystems, Exton, PA, USA) and mounted on slides coated with gelatin/chromium (double-frosted microscope slides, Fisher Scientific, Ottawa, ON, Canada).

2.5. Immunohistochemistry

Frozen sections were thawed, washed three times for five minutes each time in 10 mM PBS with 0.05% Tween 20(PBST), and then blocked in 1% BSA, gelatin, and 0.5% Triton X-100 in PBS for one hour. The sections were then co-incubated overnight with rabbit anti-GPR81 and mouse anti-Brn3a (a specific marker for RGCs). The next morning, sections were washed three times for five each time minutes in PBST and incubated for one h with secondary antibodies: AlexaFluor donkey anti-mouse 488 and AlexaFluor goat anti-rabbit 546. After three washes in PBST, the sections were slide mounted with Shandon Immu-Mount.

2.6. L-Lactate Solution

Daily fresh solution of 100 mM L-lactate was made using 0.1 g of L-(+)-lactic acid (MW = 90.08 g/mol) dissolved in 10 mL of NB (vehicle). The pH was adjusted to 7.4 ± 0.1 with NaOH, the volume reached 11 mL with NB, and the solution was filtered with Corning™ PES 0.20 µm pore Syringe Filters (09-754-29). The prepared L-lactate solution was equilibrated for at least one h at 37 °C in a 5% CO_2 incubator before use in experiments in vitro.

2.7. Retinal Explant Culture

The retina were isolated from E14-E16 mouse embryos, dissected into small segments in HBSS, and plated on 12 mm diameter glass coverslips previously coated with poly-D-Lysine (20 µg/mL) and laminin (5 µg/mL) placed in 24-well plates. The explants were cultured in NB medium supplemented with 100 U/mL penicillin/streptomycin, 5 µg/mL NAC, 1% B27, 40 ng/mL selenium, 16 µg/mL putrescine, 0.04 ng/mL triiodothyronine, 100 µg/mL transferrin, 60 ng/mL progesterone, 100 µg/mL BSA, 1 mM sodium pyruvate, 2 mM glutamine (glutaMAX™), 10 ng/mL CNTF, 5 µg/mL insulin, and 10 µM forskolin at 37 °C and 5% CO_2. At 0 DIV, starting 1 h following plating, the explants were treated for 15 h for projection analysis or for 1 h at one day in vitro (DIV) for the growth cone morphology analysis. Photomicrographs were taken with a Olympus IX71 microscope (Olympus, Markham, ON, Canada) and the axonal projection and growth cone measurement analysis were made using ImageJ software. These analyses were performed by operators blind to the experimental condition.

2.8. Growth Cone Behavior Assay

Similarly, embryonic retinal explants were cultured in Thermo Scientific™ Nunc™ Lab-Tek™ Chambered Coverglass with a borosilicate glass bottom (Lab-Tek; Rochester, NY, USA). After one or two DIV, explants were transferred to an incubator (Live cell chamber) at 37 °C and 5% CO_2, mounted on an inverted Olympus IX71 microscope (Olympus, Markham, ON, Canada). A micropipette was positioned at a 45° angle about 100 µm from the growth cone of interest, as previously described [24–26]. A micro-injector (Harvard Apparatus, St-Laurent, QC, Canada) was used to deliver NB vehicle or L-lactate 20 mM (pH 7.4) at a rate of 0.1 µL/min of in the NB. Measurements were performed with ImageJ software at baseline and after 60 min of these treatments.

2.9. Primary Neuron Culture

Primary cortical neurons were used in this study because of the ease in culturing and harvesting sufficient numbers for biochemical assays, which is technically difficult for RGCs. C57BL/6J WT pregnant mice were used to obtain E14-16 embryo brains. The superior layer of each cerebral cortex was isolated and transferred to 2 mL S-MEM containing 2.5% trypsin and 2 mg/mL DNase and incubated at 37 °C for 15 min. After trituration, the pellet was transferred to 10 mL S-MEM with 10% FBS and stored at 4 °C. The pellet was again transferred in 2 mL S-MEM supplemented with 10% FBS and triturated three or four times. The supernatant was transferred to 10 mL NB medium. Dissociated neurons were counted under a microscope and plated at a density of 100,000 cells per well on 12 mm glass coverslips previously coated with poly-D-lysine (20 µg/mL) for immunocytochemistry, or at 250,000 cells per 35 mm Petri dish for western blot analysis. Neurons were cultured for two-four days in NB medium supplemented with 1% B-27, 100 U/mL penicillin/streptomycin, 0.25% N2, and 0.5 mM of glutaMAX™. They were then treated with the GPR81 agonist L-lactate for 1 h to study acute effects on growth cone morphology, or for various other intervals for identifying the activation of signaling pathways by western blot analysis.

2.10. Immunocytochemistry

After treatment, retinal explants and primary cortical neuron cultures were washed with PBS (pH 7.4), fixed in 4% PFA (pH 7.4), and blocked with 2% normal goat serum (NGS) and 2% BSA in PBS containing 0.1% Tween 20 (pH 7.4) for 30 min at room temperature. The samples were then incubated overnight at 4 °C in a blocking solution containing anti-GPR81, anti-MAP2. The following day, the samples were washed and labeled with Alexa Fluor 488 and 546 secondary antibodies against the host species of the primary antibodies, Hoechst 33258, or AlexaFluor Phalloidin 546. The coverslips were mounted with Shandon Immu-Mount and photomicrographs were taken with an Olympus IX71 microscope (Olympus, Markham, ON, Canada) for quantitative analysis on ImageJ software.

2.11. Western Blot Analysis

Following L-lactate treatment at various time points, the primary cortical neurons were washed with cold 4 °C PBS (pH 7.4) and then lysed with 125 μL of laemmli sample buffer pre-warmed to 100 °C. The samples were then frozen at −20 °C until the day of Western blot analysis. On that day, the samples were thawed at 4 °C, placed in a 100 °C dry bath for 10 minutes, quickly vortexed and then centrifuged at 13,000 rpm at 4 °C. Twenty microliter samples were then resolved on a 10% SDS-polyacrylamide gel along with the trihalo compound from TGX Stain-Free Technology (Bio Rad, Mississauga, ON, Canada). During the electrophoresis, the trihalo compound covalently modifies tryptophan residues in the proteins to impart a fluorescence signal. Visualization of this signal was obtained by UV excitation in a Chemidoc imaging system (Bio-Rad). After this activation, gels were transferred onto a PVDF membrane with a TransBlot Turbo (Bio-Rad), and blots were imaged on the Chemidoc to reveal the total protein transferred on the membrane, which was used for later normalization of antibody signals to total protein. The blots were then blocked with 2% BSA in TBST (Tris 10 mM and NaCl 150 mM saline with 0.1% Tween 20) for 1 h and incubated overnight with antibodies against Phospho-PKA, PKA, Phospho-PKC, and PKC. They were then exposed to the species-appropriate HRP-coupled secondary antibodies for two h in blocking buffer, and dected using the Chemidoc with Clarity Max ECL substrate from Bio-Rad. The target protein expressions were then analyzed on the Image Lab v.6.0.1 software. All procedures were completed according to Bio-Rad protocols [27].

2.12. Eye Specific Segregation

In this in vivo experiment, *gpr81*$^{+/+}$ and *gpr81*$^{-/-}$ adult mice under anesthesia received an intraocular injection of 2 μL of 1% (mg/mL) CTb in 0.9% sterile saline conjugated to AlexaFluor 555 into the left vitrious humor eye, and with CTb conjugated to AlexaFluor 647 into the right vitrious humor eye. Four days after the injections, the animals were anesthetized and perfused transcardially, as described in the tissue preparation section above. The brains were removed, post-fixed in PFA 4% overnight, and then cryoprotected by immersion in gradient sucrose solutions of 10%, 20%, and 30% until the brains sank, followed by storage at −80 °C. Coronal sections 40 μm thick were cut on glass slides in a cryostat, air-dried, and mounted with Shandon Immu-Mount. Fluorescent images of entire brain sections were taken using 561 and 640 nm laser emissions and a 4× objective on an Fluoview FV3000 Olympus Confocal microscope to identify the three sections containing the largest ipsilateral projection. The dorsal lateral geniculate nucleus (dLGN) was scanned in these sections with a 20× objective and 2× amplification (total magnification of 40×). Z-stacks of 19 images with both lasers were taken from top to bottom of the signal emissions. The colocalization of both channels with a multi-threshold analysis was performed on CellSens Dimension software. The percentage of ipsilateral signal overlapping with the contralateral signal was measured for each stack. The mean for the 19 stacks in each section was reported for each threshold, and differences evaluated by two-way ANOVA with Tukey post hoc testing [28,29].

2.13. Statistical Analysis

Data were imported in GraphPad Prism 8 software. Tests for normal distribution were performed by an Anderson–Darling test (α = 0.05). Depending on parametric or nonparametric distribution, the appropriate statistical analysis was computed. Values were reported as the mean $\pm$ SEM.

3. Results

3.1. GPR81 Is Expressed in the Retina

We used rabbit anti-GPR81-S296 (SAB1300790) immunohistochemistry to identify the retinal laminar distribution of GPR81. Results in retina from P5 Syrian gold hamster pups, E16 mouse embryos, and adult mice all showed retinal GPR81 expression. GPR81 protein immunoreactivity was consistently detected in the Outer Nuclear Layer (ONL), Inner Nuclear Layer (INL), Inner Plexiform Layer (IPL), RGCs, and the RGC fiber layer of hamsters (Figure 1A–C). GPR81 immunoreactivity was detected in all layers of the embryonic and adult mouse retina (Figure 1D-I). The colocalization of GPR81 with the specific RGC marker Brn-3a showed that GPR81 is predominantly expressed in RGCs and in the RGC fiber layer. From this result, we inferred possible involvement of GPR81 in retinal development and in projection navigation towards brain innervation targets.

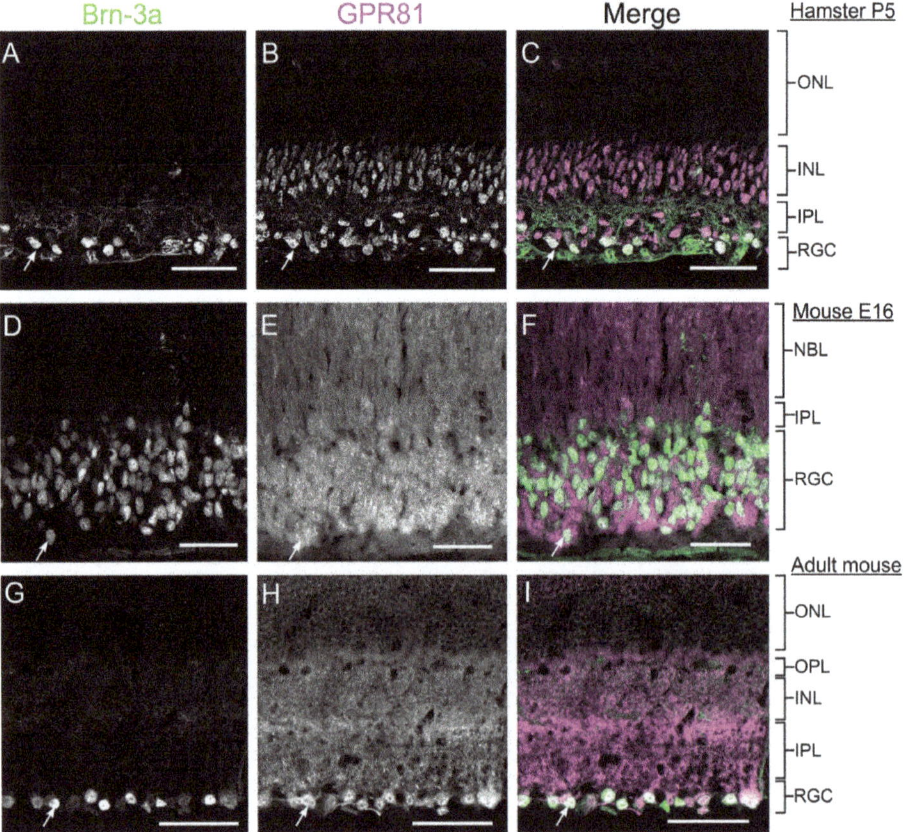

Figure 1. GPR81 is Expressed in the Retina. Expression of GPR81 in RGCs of retinal sections from P5 hamster pups (**A–C**) E16 WT mouse embryo (**D–F**), and adult mouse (**G–I**). ONL: outer nuclear layer, RGC: retinal ganglion cells, OPL: outer plexiform layer, INL: inner nuclear layer, NBL: neuroblast layer, IPL: inner plexiform layer. The white arrows indicate colocalization of GPR81 and Brn-3a. Scale bars: 50 μm.

Primary cortical neurons and retinal explant cultures in vitro also expressed the lactate receptor; GPR81 fluorescence signal was visualized on the neuronal soma, neurites, GC, and filopodia (Figure 2A, Figure S1).

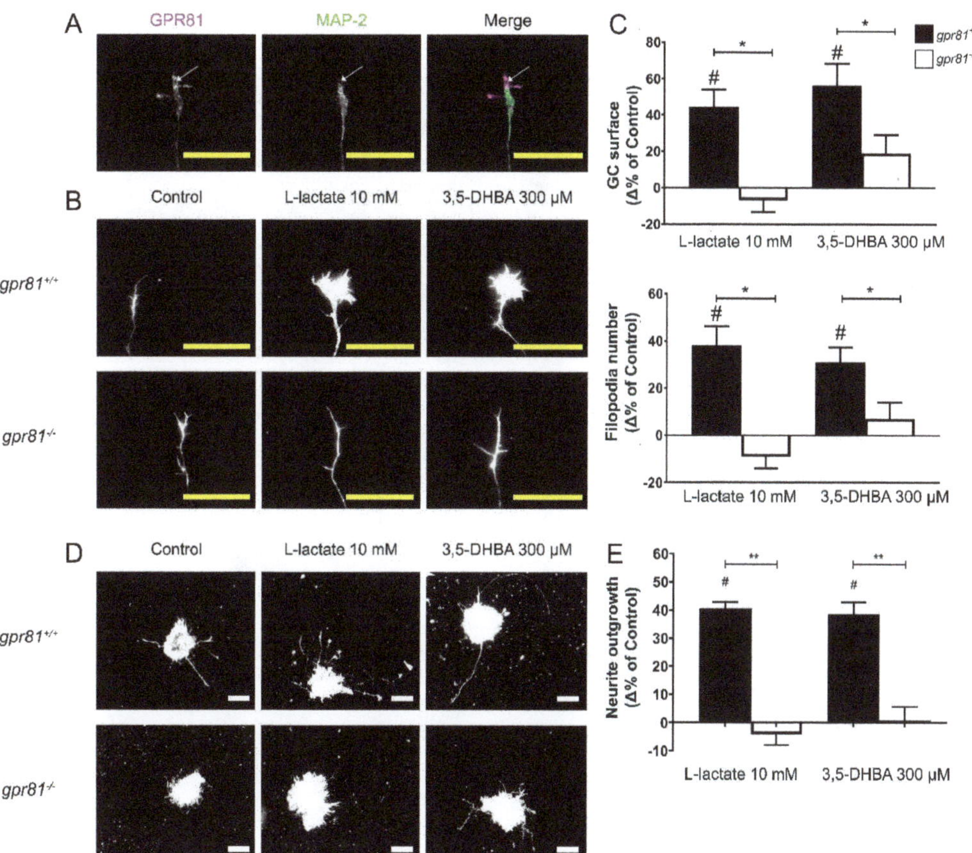

Figure 2. L-lactate and 3-5-DHBA influence GC dynamics and increase axon growth via GPR81 agonism. (**A**) GPR81 is expressed in GCs in vitro. (**B**) Microscopic photography of GCs following 1 h treatment with GPR81 agonists. Explants were marked with phalloidin conjugated with Alexa fluor 546. (**C**) Analysis of GC surface area and filopodia number (n = 48–147 GCs per condition). Values are presented as the means ± SEM. (**D**) Microscopic photography of retinal explants following 15 h treatment with the GPR81 agonists. Explants were marked with phalloidin conjugated with Alexa fluor 546. (**E**) Analysis of RGC axon projections(n = 278–2572 axons per condition). Values are presented as the means ± SEM. # Indicates significant changes compared to the control group. * Indicates significant changes $p < 0.05$ between *gpr81+/+* and *gpr81−/−* genotype. ** Indicates significant changes $p < 0.01$ between *gpr81+/+* and *gpr81−/−* genotype. Statistical test used is the Krustal–Wallis nonparametric ANOVA. White scale bar: 100 µm. Yellow scale bar 25 µm.

3.2. GPR81 Influences GC Morphology and Axon Growth

A number of GPCRs exert effects on axon guidance [22,24–26,30,31]. To determine the implication of GPR81 in regulating GC morphology, we cultivated embryonic retinal explants during 1 DIV. Afterwards, treatment for one h with the GPR81 agonist L-lactate (10 mM) significantly increased the RGC GC surface area by 44.2 ± 9.6%, whereas 3,5-DHBA (300 µM) increased surface area by 56.1 ± 12.0% compared to the vehicle control. These same treatments increased filopodia numbers of GC by 38.1 ± 8.1% and 31.0 ± 6.5%, respectively on retinal explants obtained from *gpr81+/+* mice. The corresponding changes induced by GPR81 agonists were abrogated in similar retinal explants obtained from *gpr81⁻/⁻* mouse embryos (Figure 2B–C).

We next measured the effect of GPR81 agonists on axon growth of RGCs on retinal explants. Interestingly, RGCs from *gpr81⁺/⁺* mice incubated with GPR81 agonists for 15 h increased axon outgrowth by 40.7 ± 2.2% (10 mM L-lactate) and 38.7 ± 4.2% (300 μM 3,5-DHBA) (Figure 2D–E). As expected, RGCs obtained from *gpr81⁻/⁻* mice did not respond to the agonist treatment (−4.2 ± 3.7% and 4.8 ± 5.0%, respectively) (Figure 2D–E). This impact of L-lactate on GC morphology and axon growth was confirmed with time-lapse live cell imaging in the GC behavior assay. Here, we exposed a GC of a retinal explant to a microgradient of L-lactate concentration to visualize any axon growth or guidance effects. In this experiment, we measured the GC surface area, filopodia numbers, axon length, and the angle between the GC and the micropipette at baseline (T0) and at 60 min of drug exposure (T60). The time-dependent differences were obtained by subtraction. At 60 min, the vehicle-treated explants (NB medium) showed negligible changes in the GC (−0.42 ± 3.03 μm^2 area), 0.5 ± 0.5 filopodia, and −1.39 ± 2.78 μm axon growth; in contrast, L-lactate-treated explants displayed mean increases of 22.8 ± 11.1 μm^2 GC area, 4.0 ± 1.8 in number of filopodia, and 4.31 ± 2.14 μm in axon growth. However, we did not observe any significant changes in the angle of orientation after the addition of lactate (−2.1 ± 2.0° for NB medium versus 2.5 ± 3.5° for 20 mM L-lactate gradient; Figure 3).

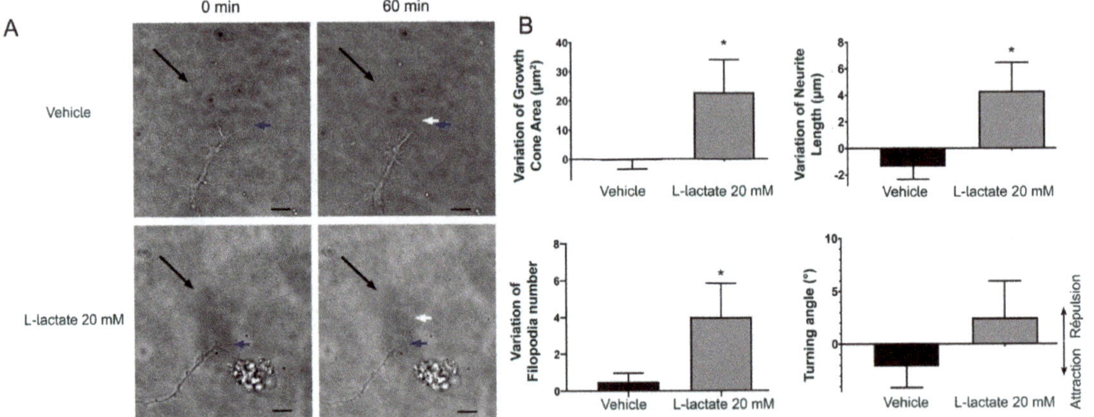

Figure 3. L-lactate influences GC morphology and axon growth, but not turning. (**A**) Time-lapse microscopy time of a DIV1-2 mouse RGC growth cone before and at 60 min after the addition of L-lactate. The black arrows indicate the microgradient direction created by a micropipette with a tip of 2 μm diameter and positioned at a 45° angle and around 100 μm distal to the growth cone. The blue arrows indicate the tips of the growth cone before the application of L-lactate or vehicle (T0). The white arrows indicate the position of the growth cone 60 min later (T60). (**B**) Analysis of growth cone dynamics. The values of the growth cone area, filopodia number on the growth cone, axon length, and the angle of the growth within the L-lactate microgradient were measured before (T0) and at 60 min (T60) after the addition of L-lactate or vehicle. Bar graphs show the means ± SEM of the effects induced by the treatments (*n* = 6–8 cells per condition). * Indicates significant changes compared to the vehicle group, compared by parametric unpaired *t*-test (*p* < 0.05). Scale bar = 15 μm.

3.3. Lactate Increases PKC and PKA Phosphorylation

GPR81 is coupled to G$_i$, such that agonism induces Ca^{2+} release in CHO-K1 adipocytes, resulting in decreased lipolysis [32]. To determine the signaling pathway resulting in GPR81 activation, we treated primary neurons obtained from mouse embryos with a GPR81 agonist (10 mM L-lactate). Interestingly, there was an increase of PKC phosphorylation (2.1 ± 0.4 fold) after five minutes and an increase of PKA phosphorylation (1.9 ± 0.2 fold) at 15 min after the addition of 10 mM L-lactate (Figure 4A). These data show that GPR81 agonism stimulates the PKC and the PKA signaling pathways in primary neurons.

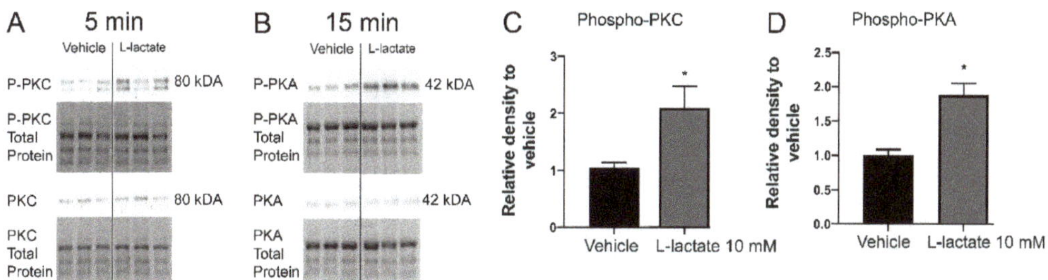

Figure 4. L-lactate increases PKC and PKA phosphorylation (**A,B**) Protein expression levels of P-PKC, PKC, P-PKA, PKA, and to-tal proteins in primary neuron cultures incubated with L-lactate (10 mM) at 37 °C for five min (for PKC) and 15 min (for PKA). The total protein Western blots confirmed equal loading in all lanes with TGX Stain-Free Technology. Values of relative density to vehicle are presented as the means ± SEM (**C**) for P-PKC and (**D**) for P-PKA. * Indicates significant changes compared to the control group, as compared by the nonparametric ANOVA Krustal Wallis ($p < 0.05$) with Dunn's post hoc correction.

3.4. GPR81 Affects Retinothalamic Projections In Vivo

We next studied the contribution of GPR81 to the development of retinothalamic projections in living transgenic *gpr81*$^{-/-}$ mice. Left and right RGC projections were visualized with the neuron tracer CTb conjugated to Alexa fluorescent molecules. RGC projections normally segregate through various brain structures to join dLGN [33]. Interestingly, adult *gpr81*$^{-/-}$ mice had significantly fewer ipsilateral projections in the dLGN, irrespective of the intensity threshold used for the analysis (Figure 5). This observation demonstrates, for the first time, an essential role of GPR81 in the normal development of the rodent retinothalamic pathway.

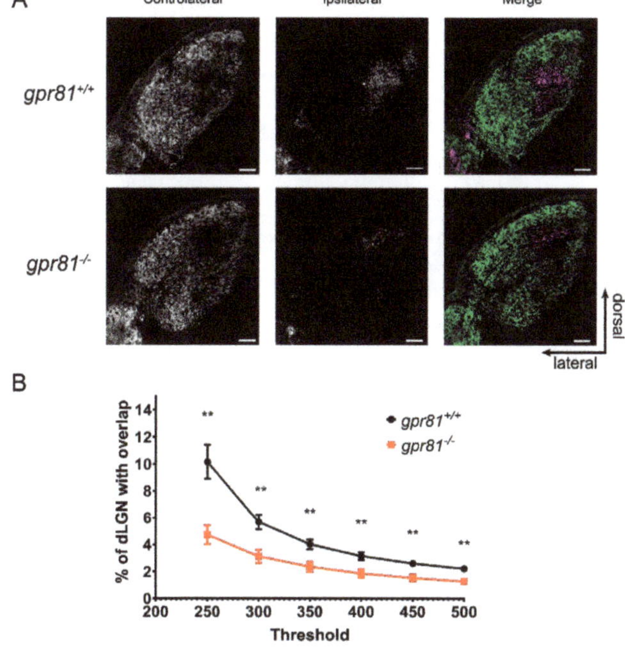

Figure 5. GPR81 Affects Retinothalamic Projections in vivo. (**A**) Micrographs of the dorsal lateral geniculate nucleus (dLGN) following the injection of CTb-Alexa 555 in the left eye and CTb-Alexa 647 in the right eye of *gpr81*$^{+/+}$ and *gpr81*$^{-/-}$ mice. Controlateral and ipsilateral projections from both

eyes RGC segregate in the dLGN. RGC projections from both eyes are shown in the merged images. (**B**) Line graph shows the percentage of ipsilateral projections overlapping in the contralateral projections expressed as the mean ± SEM. Percentage of overlapping signals is significantly lower in *gpr81*$^{-/-}$ mice irrespective of the intensity analysis threshold. ** Indicates significant changes between *gpr81*$^{+/+}$ (n = 18) and *gpr81*$^{-/-}$ (n = 12) genotypes, according to two-way ANOVA with Tukey post hoc test (p < 0.01). Scale bars: 100 μm.

4. Discussion

Lactic acid and L-lactate have drawn considerable attention since the early days of metabolic research. Indeed, the indisputable importance of lactate in the glycolytic pathway may have led to persisting misconceptions about the bodily functions of L-lactate [1]. The introduction of the astrocyte-neuron Lactate shuttle hypothesis (ANLSH) [34] led to a more mature understanding of the importance of L-lactate metabolism in the CNS. According to this model, glucose is predominantly taken up by astrocytes during neuronal activation. The astrocytes produce and release L-lactate, which serves as the primary metabolic fuel for activated neurons [35]. Furthermore, glucose consumption by the outer retina forms L-lactate by aerobic glycolysis [36,37], in metabolic support of RGCs. The recent discovery of receptors for metabolic intermediates has brought new physiologic perspectives on the relevance of L-lactate concentrations in tissue. We undertook in this study to explore the localization and neural functions of GPR81 in the developing retina and its CNS projections. We show for the first time expression of the lactate receptor GPR81 in the retina of hamsters and mice, noting its high abundance in the RGC layer and in the RGC nerve fiber layer. This result was confirmed by studies in vitro using retinal explants and cortical neurons of embryonic mice, which both proved to express GPR81 on their soma, axons, GCs, and GC filopodia. We demonstrated that specific activation of the GPR81 receptor by its agonists modulates GC size, filopodia numbers, and axon growth of RGCs. In vitro, treatment with L-lactate at a physiological concentration increased phosphorylation of PKC and PKA, which are key mediators of the pathways that likely promote neurite growth effects [38]. Importantly, genetic absence of GPR81 curtailed formation of RGC projections to the dLGN, as evident in the *gpr81*$^{-/-}$ mouse in vivo. Altogether, these results suggest an essential involvement of L-lactate and its receptor GPR81 in the development of RGCs and their axon projections in the CNS.

Previous findings for GPR81 impart an extra layer of complexity with regard to its involvement in CNS development, especially for retinal projections. The colocalization of GPR81 with glutamine synthetase, a specific marker of Müller cells, was previously established in early postnatal mice (P8-17). GPR81 activation in Müller cells regulates Norrin and Wnt ligands, which in turn promote angiogenesis and produce stimulate production of neuroprotective growth factors in the retina [21]. In our present embryonic model (E15), Müller cells have not yet been differentiated in the developing retina. Accordingly, the mechanism responsible for axon growth at this development stage likely differs from the retinal angiogenic effects produced by Müller cells in postnatal rodents [39].

L-lactate was long considered to be a metabolic waste of anaerobic glycolysis, but it is now known to be a metabolic fuel of aerobic glycolysis, especially in neurons [1,11,40]. In this capacity, L-lactate is converted by LDH to pyruvate, which then enters into the Krebs cycle for ATP generation in mitochondria [41]. In order to confirm that the present findings of L-lactate effect are not due trivially to an increase in the energy metabolite pool, we used 3,5-DHBA as a non-metabolized specific GPR81 agonist [3]. Activation of GPR81 by L-lactate and 3,5-DHBA of RGCs in retinal explants from embryonic mice positively modulated GC morphology and increased axon projection length. Interestingly, these effects of GPR81 agonist treatment were absent in explants from *gpr81*$^{-/-}$ mice. These findings support our claim that L-lactate and 3,5-DHBA act on RGCs through GPR81 agonism.

RGCs in the intact organism project their axons from retina to transmit visual information to CNS targets. GCs must detect and respond to a complex combination of chemotactic signals to direct their axons through the visual pathway. Our time-lapse microscopy experiments confirmed that L-lactate concentration gradients modify GC morphology and increase axon growth without having a significant effect on axon guidance *per se*. However, L-lactate was without effect in explants from retinal mice genetically depleted of GPR81, thus suggesting a crucial role for GPR81 in the retinothalamic pathway formation.

We show in this study that L-lactate activates PKC and PKA through agonism at GPR81, as was reported previously. Indeed, PKC activation is well-known to participate in synaptic remodeling, presynaptic plasticity, and neuronal repair [38,42,43]. Some PKC isoforms are known to influence neurite adhesion and outgrowth in various neuronal cell types, including RGCs [40,42–44]. Activation of these second messenger pathways results in a reorganization of the growth cone cytoskeleton, which manifest to microscopic examination of the filopodia, lamellipodia, and axonal growth in vitro [45–47]. Our present results are thus also in accordance with studies showing that PKA activation increases GC area and filopodia numbers to promote RGC axon extension during development [48].

In conclusion, as demonstrated in this study for the first time, the GPR81 L-lactate receptors contribute critically to neuro-visual development in the rodent retina and CNS. This study thus provides a foundation for the ongoing investigation of these actors in broader aspects of central and peripheral nervous system development. Present results support the formulation of new hypotheses for elaborating effective therapies targeting the development and regeneration of the nervous system. In fact, our results suggest a possible therapeutic role of L-lactate and other GPR81 agonists in the neuroregenerative and neuroreparative field. First of all, L-lactate can act by itself as an efficient energy substrate [49]. It can also protect and attenuate neuronal death induces by oxygen and glucose deprivation following cerebral ischemia [50,51]. In traumatic brain injury model, preconditioning with L-lactate promotes plasticity-related expression helping to reduce neurological deficits effect via the GPR81 signaling pathway [52]. In accordance with our findings, the utilization of L-lactate or a GPR81 agonist could stimulate axon regeneration and prevent neuronal degeneration.

Supplementary Materials: The following are available online at https://www.mdpi.com/article/10.3390/cells10071640/s1, Figure S1: GPR81 is expressed in DIV2 E15 cortical neuron.

Author Contributions: Conceptualization, S.L., A.S., and J.-F.B.; methodology, S.L., A.S., H.C., and J.-F.B.; software S.L. and P.G.; validation, A.S. and J.-F.B.; formal analysis, S.L.; investigation, S.L. and J.-F.B.; resources, J.-F.B. and S.C.; writing—original draft preparation, Samuel Laroche; writing—review and editing, S.L., S.C., H.C., and J.-F.B.; supervision, J.-F.B.; project administration, J.-F.B.; funding acquisition, J.-F.B. and S.C. All authors have read and agreed to the published version of the manuscript.

Funding: This research was funded by the National Sciences and Engineering Research Council of Canada (NSERC, RGPIN-2020-05739 and the Canadian Institutes of Health Research (CIHR, PJT-156029) grants to Jean-François Bouchard.

Institutional Review Board Statement: The study was conducted according to the guidelines of the Institutional Review Board of Comité de déontologie de l'expérimentation sur les animaux ((CDEA) (20-052) date of approval: May 2020).

Informed Consent Statement: Not applicable.

Data Availability Statement: All relevant data are within the paper and its Supporting Information files.

Acknowledgments: We thank the Sylvain Chemtob laboratory members Xiaojuan Yang, Xin Hou, Sonja L'Espérance, and Christiane Quiniou for their precious help to acquire *gpr81*$^{-/-}$ mice. We thank Bruno Cécyre for his technical support and Paul Cumming for critical reading of the manuscript.

Conflicts of Interest: The authors declare no conflict of interest.

References

1. Ferguson, B.S.; Rogatzki, M.J.; Goodwin, M.L.; Kane, D.A.; Rightmire, Z.; Gladden, L.B. Lactate metabolism: Historical context, prior misinterpretations, and current understanding. *Eur. J. Appl. Physiol.* **2018**, *118*, 691–728. [CrossRef]
2. Sun, S.; Li, H.; Chen, J.; Qian, Q. Lactic Acid: No Longer an Inert and End-Product of Glycolysis. *Physiology* **2017**, *32*, 453–463. [CrossRef]
3. Liu, C.; Kuei, C.; Zhu, J.; Yu, J.; Zhang, L.; Shih, A.; Mirzadegan, T.; Shelton, J.; Sutton, S.; Connelly, M.A.; et al. 3,5-Dihydroxybenzoic Acid, a Specific Agonist for Hydroxycarboxylic Acid 1, Inhibits Lipolysis in Adipocytes. *J. Pharmacol. Exp. Ther.* **2012**, *341*, 794. [CrossRef]
4. Madaan, A.; Chaudhari, P.; Nadeau-Vallee, M.; Hamel, D.; Zhu, T.; Mitchell, G.; Samuels, M.; Pundir, S.; Dabouz, R.; Howe Cheng, C.W.; et al. Muller Cell-Localized G-Protein-Coupled Receptor 81 (Hydroxycarboxylic Acid Receptor 1) Regulates Inner Retinal Vasculature via Norrin/Wnt Pathways. (1525-2191 (Electronic)). *Am. J. Pathol.* **2019**, *189*, 1878–1896. [CrossRef]
5. Liu, C.; Wu, J.; Zhu, J.; Kuei, C.; Yu, J.; Shelton, J.; Sutton, S.W.; Li, X.; Yun, S.J.; Mirzadegan, T.; et al. Lactate inhibits lipolysis in fat cells through activation of an orphan G-protein-coupled receptor, GPR81. *J. Biol. Chem.* **2009**, *284*, 2811–2822. [CrossRef]
6. Wong-Riley, M.T.T. Energy metabolism of the visual system. *Eye Brain* **2010**, *2*, 99–116. [CrossRef]
7. Plump, A.S.; Erskine, L.; Sabatier, C.; Brose, K.; Epstein, C.J.; Goodman, C.S.; Mason, C.A.; Tessier-Lavigne, M. Slit1 and Slit2 Cooperate to Prevent Premature Midline Crossing of Retinal Axons in the Mouse Visual System. *Neuron* **2002**, *33*, 219–232. [CrossRef]
8. Nakagawa, S.; Brennan, C.; Johnson, K.G.; Shewan, D.; Harris, W.A.; Holt, C.E. Ephrin-B regulates the Ipsilateral routing of retinal axons at the optic chiasm. *Neuron* **2000**, *25*, 599–610. [CrossRef]
9. Reese, B.E. Development of the retina and optic pathway. *Vis. Res.* **2011**, *51*, 613–632. [CrossRef] [PubMed]
10. Rogatzki, M.J.; Ferguson, B.S.; Goodwin, M.L.; Gladden, L.B. Lactate is always the end product of glycolysis. *Front. Neurosci.* **2015**, *9*, 22. [CrossRef]
11. Bittar, P.G.; Charnay, Y.; Pellerin, L.; Bouras, C.; Magistretti, P.J. Selective Distribution of Lactate Dehydrogenase Isoenzymes in Neurons and Astrocytes of Human Brain. *J. Cereb. Blood Flow Metab.* **1996**, *16*, 1079–1089. [CrossRef]
12. Friede, R.L.; Fleming, L.M. A mapping of the distribution of lactic dehydrogenase in the brain of the rhesus monkey. *Am. J. Anat.* **1963**, *113*, 215–234. [CrossRef]
13. Karlsson, J.; Frith, K.; Sjödin, B.; Gollnick, P.D.; Saltin, B. Distribution of LDH Isozymes in Human Skeletal Muscle. *Scand. J. Clin. Lab. Investig.* **1974**, *33*, 307–312. [CrossRef]
14. Sjödin, B.; Thorstensson, A.; Frith, K.; Karlsson, J. Effect of Physical Training on LDH Activity and LDH Isozyme Pattern in Human Skeletal Muscle. *Acta Physiol. Scand.* **1976**, *97*, 150–157. [CrossRef]
15. Porporato, P.E.; Payen, V.L.; De Saedeleer, C.J.; Préat, V.; Thissen, J.-P.; Feron, O.; Sonveaux, P. Lactate stimulates angiogenesis and accelerates the healing of superficial and ischemic wounds in mice. *Angiogenesis* **2012**, *15*, 581–592. [CrossRef] [PubMed]
16. Schurr, A.; Payne, R.S.; Miller, J.J.; Tseng, M.T.; Rigor, B.M. Blockade of lactate transport exacerbates delayed neuronal damage in a rat model of cerebral ischemia. *Brain Res.* **2001**, *895*, 268–272. [CrossRef]
17. Shen, Z.; Jiang, L.; Yuan, Y.; Deng, T.; Zheng, Y.-R.; Zhao, Y.-Y.; Li, W.-L.; Wu, J.-Y.; Gao, J.-Q.; Hu, W.-W.; et al. Inhibition of G Protein-Coupled Receptor 81 (GPR81) Protects Against Ischemic Brain Injury. *CNS Neurosci. Ther.* **2015**, *21*, 271–279. [CrossRef] [PubMed]
18. Roland, C.L.; Arumugam, T.; Deng, D.; Liu, S.H.; Philip, B.; Gomez, S.; Burns, W.R.; Ramachandran, V.; Wang, H.; Cruz-Monserrate, Z.; et al. Cell Surface Lactate Receptor GPR81 Is Crucial for Cancer Cell Survival. *Cancer Res.* **2014**, *74*, 5301. [CrossRef] [PubMed]
19. Hoque, R.; Farooq, A.; Ghani, A.; Gorelick, F.; Mehal, W.Z. Lactate Reduces Liver and Pancreatic Injury in Toll-Like Receptor- and Inflammasome-Mediated Inflammation via GPR81-Mediated Suppression of Innate Immunity. *Gastroenterology* **2014**, *146*, 1763–1774. [CrossRef]
20. Madaan, A.; Nadeau-Vallée, M.; Rivera, J.C.; Obari, D.; Hou, X.; Sierra, E.M.; Girard, S.; Olson, D.M.; Chemtob, S. Lactate produced during labor modulates uterine inflammation via GPR81 (HCA1). *Am. J. Obstet. Gynecol.* **2017**, *216*, 60.e1–60.e17. [CrossRef] [PubMed]
21. Kolko, M.; Vosborg, F.; Henriksen, U.L.; Hasan-Olive, M.M.; Diget, E.H.; Vohra, R.; Gurubaran, I.R.S.; Gjedde, A.; Mariga, S.T.; Skytt, D.M.; et al. Erratum to: Lactate Transport and Receptor Actions in Retina: Potential Roles in Retinal Function and Disease. *Neurochem. Res.* **2016**, *41*, 1237. [CrossRef]
22. Cherif, H.; Duhamel, F.; Cécyre, B.; Bouchard, A.; Quintal, A.; Chemtob, S.; Bouchard, J.-F. Receptors of intermediates of carbohydrate metabolism, GPR91 and GPR99, mediate axon growth. *PLoS Biol.* **2018**, *16*, e2003619. [CrossRef]
23. Sapieha, P.; Sirinyan, M.; Hamel, D.; Zaniolo, K.; Joyal, J.S.; Cho, J.H.; Honoré, J.C.; Kermorvant-Duchemin, E.; Varma, D.R.; Tremblay, S.; et al. The succinate receptor GPR91 in neurons has a major role in retinal angiogenesis. (1546-170X (Electronic)). *Nat. Med.* **2008**, *14*, 1067–1076. [CrossRef]
24. Argaw, A.; Duff, G.; Zabouri, N.; Cécyre, B.; Chainé, N.; Cherif, H.; Tea, N.; Lutz, B.; Ptito, M.; Bouchard, J.-F. Concerted Action of CB1 Cannabinoid Receptor and Deleted in Colorectal Cancer in Axon Guidance. *J. Neurosci.* **2011**, *31*, 1489. [CrossRef]
25. Cherif, H.; Argaw, A.; Cécyre, B.; Bouchard, A.; Gagnon, J.; Javadi, P.; Desgent, S.; Mackie, K.; Bouchard, J.-F. Role of GPR55 during Axon Growth and Target Innervation. *eNeuro* **2015**, *2*. [CrossRef] [PubMed]

26. Duff, G.; Argaw, A.; Cecyre, B.; Cherif, H.; Tea, N.; Zabouri, N.; Casanova, C.; Ptito, M.; Bouchard, J.-F. Cannabinoid Receptor CB2 Modulates Axon Guidance. *PLoS ONE* **2013**, *8*, e70849. [CrossRef] [PubMed]

27. Taylor, S.C.; Posch, A. The Design of a Quantitative Western Blot Experiment. *BioMed Res. Int.* **2014**, *2014*, 361590. [CrossRef]

28. Torborg, C.L.; Feller, M.B. Unbiased analysis of bulk axonal segregation patterns. *J. Neurosci. Methods* **2004**, *135*, 17–26. [CrossRef]

29. Muir-Robinson, G.; Hwang, B.J.; Feller, M.B. Retinogeniculate Axons Undergo Eye-Specific Segregation in the Absence of Eye-Specific Layers. *J. Neurosci.* **2002**, *22*, 5259. [CrossRef] [PubMed]

30. Jassen, A.K.; Yang, H.; Miller, G.M.; Calder, E.; Madras, B.K. Receptor Regulation of Gene Expression of Axon Guidance Molecules: Implications for Adaptation. *Mol. Pharmacol.* **2006**, *70*, 71. [CrossRef] [PubMed]

31. Xiang, Y.; Li, Y.; Zhang, Z.; Cui, K.; Wang, S.; Yuan, X.-B.; Wu, C.-P.; Poo, M.-M.; Duan, S. Nerve growth cone guidance mediated by G protein–coupled receptors. *Nat. Neurosci.* **2002**, *5*, 843–848. [CrossRef] [PubMed]

32. Cai, T.-Q.; Ren, N.; Jin, L.; Cheng, K.; Kash, S.; Chen, R.; Wright, S.D.; Taggart, A.K.P.; Waters, M.G. Role of GPR81 in lactate-mediated reduction of adipose lipolysis. *Biochem. Biophys. Res. Commun.* **2008**, *377*, 987–991. [CrossRef]

33. Shanks, J.A.; Ito, S.; Schaevitz, L.; Yamada, J.; Chen, B.; Litke, A.; Feldheim, D.A. Corticothalamic Axons Are Essential for Retinal Ganglion Cell Axon Targeting to the Mouse Dorsal Lateral Geniculate Nucleus. *J. Neurosci.* **2016**, *36*, 5252–5263. [CrossRef]

34. Chih, C.-P.; Roberts, E.L. Energy Substrates for Neurons during Neural Activity: A Critical Review of the Astrocyte-Neuron Lactate Shuttle Hypothesis. *J. Cereb. Blood Flow Metab.* **2003**, *23*, 1263–1281. [CrossRef] [PubMed]

35. Pellerin, L.; Pellegri, G.; Bittar, P.G.; Charnay, Y.; Bouras, C.; Martin, J.L.; Stella, N.; Magistretti, P.J. Evidence Supporting the Existence of an Activity-Dependent Astrocyte-Neuron Lactate Shuttle. *Dev. Neurosci.* **1998**, *20*, 291–299. [CrossRef] [PubMed]

36. Wang, L.; TÖRnquist, P.; Bill, A. Glucose metabolism of the inner retina in pigs in darkness and light. *Acta Physiol. Scand.* **1997**, *160*, 71–74. [CrossRef]

37. Hurley, J.B.; Lindsay, K.J.; Du, J. Glucose, lactate, and shuttling of metabolites in vertebrate retinas. *J. Neurosci Res.* **2015**, *93*, 1079–1092. [CrossRef]

38. Bouchard, J.-F.; Horn, K.E.; Stroh, T.; Kennedy, T.E. Depolarization recruits DCC to the plasma membrane of embryonic cortical neurons and enhances axon extension in response to netrin-1. *J. Neurochem.* **2008**, *107*, 398–417. [CrossRef]

39. Sanes, D.H.; Reh, T.A.; Harris, W.A. 4-Determination and differentiation. In *Development of the Nervous System*, 3rd ed.; Sanes, D.H., Reh, T.A., Harris, W.A., Eds.; Academic Press: London, UK, 2012; pp. 77–104. [CrossRef]

40. Hsu, L.; Jeng, A.Y.; Chen, K.Y. Induction of neurite outgrowth from chick embryonic ganglia explants by activators of protein kinase C. *Neurosci. Lett.* **1989**, *99*, 257–262. [CrossRef]

41. Mot, A.I.; Liddell, J.R.; White, A.R.; Crouch, P.J. Circumventing the Crabtree Effect: A method to induce lactate consumption and increase oxidative phosphorylation in cell culture. *Int. J. Biochem. Cell Biol.* **2016**, *79*, 128–138. [CrossRef]

42. Kolkova, K.; Novitskaya, V.; Pedersen, N.; Berezin, V.; Bock, E. Neural Cell Adhesion Molecule-Stimulated Neurite Outgrowth Depends on Activation of Protein Kinase C and the Ras–Mitogen-Activated Protein Kinase Pathway. *J. Neurosci.* **2000**, *20*, 2238. [CrossRef]

43. Rosdahl, J.A.; Mourton, T.L.; Brady-Kalnay, S.M. Protein Kinase C δ (PKCδ) Is Required for Protein Tyrosine Phosphatase μ (PTPμ)-Dependent Neurite Outgrowth. *Mol. Cell. Neurosci.* **2002**, *19*, 292–306. [CrossRef] [PubMed]

44. Heacock, A.M.; Agranoff, B.W. Protein Kinase Inhibitors Block Neurite Outgrowth from Explants of Goldfish Retina. *Neurochem. Res.* **1997**, *22*, 1179–1185. [CrossRef] [PubMed]

45. Cirulli, V.; Yebra, M. Netrins: Beyond the brain. *Nat. Rev. Mol. Cell Biol.* **2007**, *8*, 296–306. [CrossRef] [PubMed]

46. Bouchard, J.-F.; Moore, S.W.; Tritsch, N.X.; Roux, P.P.; Shekarabi, M.; Barker, P.A.; Kennedy, T.E. Protein Kinase A Activation Promotes Plasma Membrane Insertion of DCC from an Intracellular Pool: A Novel Mechanism Regulating Commissural Axon Extension. *J. Neurosci.* **2004**, *24*, 3040. [CrossRef] [PubMed]

47. Barallobre, M.J.; Pascual, M.; Del Río, J.A.; Soriano, E. The Netrin family of guidance factors: Emphasis on Netrin-1 signalling. *Brain Res. Rev.* **2005**, *49*, 22–47. [CrossRef]

48. Argaw, A.; Duff, G.; Boire, D.; Ptito, M.; Bouchard, J.-F. Protein kinase A modulates retinal ganglion cell growth during development. *Exp. Neurol.* **2008**, *211*, 494–502. [CrossRef]

49. Pellerin, L.; Magistretti, P.J. Glutamate uptake into astrocytes stimulates aerobic glycolysis: A mechanism coupling neuronal activity to glucose utilization. *Proc. Natl. Acad. Sci. USA* **1994**, *91*, 10625. [CrossRef]

50. Berthet, C.; Lei, H.; Thevenet, J.; Gruetter, R.; Magistretti, P.J.; Hirt, L. Neuroprotective Role of Lactate after Cerebral Ischemia. *J. Cereb. Blood Flow Metab.* **2009**, *29*, 1780–1789. [CrossRef]

51. Berthet, C.; Castillo, X.; Magistretti, P.J.; Hirt, L. New Evidence of Neuroprotection by Lactate after Transient Focal Cerebral Ischaemia: Extended Benefit after Intracerebroventricular Injection and Efficacy of Intravenous Administration. *Cerebrovasc. Dis.* **2012**, *34*, 329–335. [CrossRef]

52. Zhai, X.; Li, J.; Li, L.; Sun, Y.; Zhang, X.; Xue, Y.; Gao, Y.; Li, S.; Yan, W.; Yin, S.; et al. L-lactate preconditioning promotes plasticity-related proteins expression and reduces neurological deficits by potentiating GPR81 signaling in rat traumatic brain injury model. *Brain Res.* **2020**, *1746*, 146945. [CrossRef] [PubMed]

 *cells*

MDPI

Article

Autophagy Involvement in the Postnatal Development of the Rat Retina

Noemi Anna Pesce [1,2], Alessio Canovai [2], Emma Lardner [1], Maurizio Cammalleri [2], Anders Kvanta [1], Helder André [1,*,†] and Massimo Dal Monte [2,†]

[1] Department of Clinical Neuroscience, Division of Eye and Vision, St. Erik Eye Hospital, Karolinska Institutet, Eugeniavägen 12, 17164 Solna, Sweden; n.pesce@student.unisi.it (N.A.P.); emma.lardner@sll.se (E.L.); anders.kvanta@ki.se (A.K.)

[2] Department of Biology, University of Pisa, Via San Zeno 31, 56127 Pisa, Italy; a.canovai@studenti.unipi.it (A.C.); maurizio.cammalleri@unipi.it (M.C.); massimo.dalmonte@unipi.it (M.D.M.)

* Correspondence: helder.andre@ki.se; Tel.: +46-700-923-479

† These authors contributed equally to the present work.

Abstract: During retinal development, a physiologic hypoxia stimulates endothelial cell proliferation. The hypoxic milieu warrants retina vascularization and promotes the activation of several mechanisms aimed to ensure homeostasis and energy balance of both endothelial and retinal cells. Autophagy is an evolutionarily conserved catabolic system that contributes to cellular adaptation to a variety of environmental changes and stresses. In association with the physiologic hypoxia, autophagy plays a crucial role during development. Autophagy expression profile was evaluated in the developing retina from birth to post-natal day 18 of rat pups, using qPCR, western blotting and immunostaining methodologies. The rat post-partum developing retina displayed increased active autophagy during the first postnatal days, correlating to the hypoxic phase. In latter stages of development, rat retinal autophagy decreases, reaching a normalization between post-natal days 14-18, when the retina is fully vascularized and mature. Collectively, the present study elaborates on the link between hypoxia and autophagy, and contributes to further elucidate the role of autophagy during retinal development.

Keywords: eye; retina; development; vascularization; hypoxia; autophagy

Citation: Pesce, N.A.; Canovai, A.; Lardner, E.; Cammalleri, M.; Kvanta, A.; André, H.; Dal Monte, M. Autophagy Involvement in the Postnatal Development of the Rat Retina. *Cells* **2021**, *10*, 177. https://doi.org/10.3390/cells10010177

Received: 18 December 2020
Accepted: 14 January 2021
Published: 17 January 2021

Publisher's Note: MDPI stays neutral with regard to jurisdictional claims in published maps and institutional affiliations.

1. Introduction

The mature retina is considered one of the highest oxygen-demanding tissues in the body, with a considerable metabolic activity [1,2]. The heightened metabolic demand of the retina is supplied by a structured vascular systems, including retinal vessels and the choriocapillaris, which provide nutrients and oxygen to the inner and the outer layers of the retina respectively [3,4]. During development of the mammalian eye, the retinal vasculature undergoes considerable changes and reorganization [5]. In the early stages of embryogenesis, the interior of the eye is metabolically supplied by a transient embryonic circulatory network in the vitreous, referred to as the hyaloid system [6]. In the latter stages of development, the hyaloid vasculature regresses and concurrently is replaced by the retinal vasculature [7]. The physiologic hypoxia in uterus (O_2 levels < 5%) drives the proliferation of retinal blood vessels from the optic nerve to the periphery [8], through vascular endothelial growth factor (VEGF)-mediated angiogenesis [9,10].

At this level, the developing retinal vasculature lacks a functional barrier, necessary to maintain homeostasis into the retina and controlling vascular permeability [11,12]. Thus, the retinal capillary endothelial cells interact with each other to create a complex network, composed of tight junctions between transmembrane and peripheral membrane proteins. In this manner, the retinal endothelial cells form an inner blood retinal barrier (BRB), which contributes to preserve neuronal environment regulating the entry of molecules from the blood into the retina [13,14]. At these critical times in developmental events, both retinal and endothelial cells (ECs) endure morphological changes and reorganization; as consequence, they require mechanisms for the degradation and recycling of obsolete cellular components [15].

Autophagy is an essential process in maintaining the normal cellular homeostasis under physiological conditions [16]; and it plays an important role in the turnover of damaged organelles, such as peroxisomes and endoplasmic reticulum, as well as in removing unnecessary aggregated or misfolded proteins [17,18]. Previous findings indicate that vascular remodeling in ocular development can be regulated by autophagy [19]; the blood vessels need autophagy to balance their bioenergetic dynamic mechanisms [20]. Moreover, autophagic mechanisms seem to have a critical role in anatomical involution of the hyaloid blood vessels [19]. Several studies have demonstrated that autophagy can be induced by physiological hypoxia, the key stimulus for retinal angiogenesis [21–23]. At the molecular level, hypoxia stimulates several molecules involved in different signaling pathways, including hypoxia-inducible factors (HIFs) that induce angiogenesis through VEGF; and adenosine monophosphate-activated protein kinase (AMPK), a positive regulator of autophagy [24,25]. In conditions where nutrients are scarce, such as during development, AMPK is activated by a decreased ATP/AMP ratio and leads to phosphorylation of several molecules, including Unc-51-like autophagy activating kinase (ULK)1 [26,27]. Activated ULK1 is involved in the formation of multiple protein complexes that are responsible for the initiation of autophagic mechanisms that lead to the formation of autophagosome [28,29]. Elongation and maturation of the autophagosome involve the microtubule-associated protein I light chain 3 (LC3 I) system. LC3 I is conjugated to phosphatidylethanolamine, converted to LC3 II and inserted into the autophagosome membrane [30]. The synthesis and processing of LC3 II is increased during autophagy, thus acting as a key marker of levels of autophagy in cells [31]. The cargo is selected by targeted ubiquitination and carried to the autophagosome through the binding of LC3 II with sequestosome-1 (SQSTM-1), also known as ubiquitin-binding protein p62 [32,33]. The p62 is degraded by autophagy and a decrease in its protein levels correlates with an active autophagic flux [34]. Autophagy ends with the fusion of the autophagosome with the lysosome, where the inner cargo is degraded by lysosomal hydrolases.

Considering the myriad of autophagic mechanisms, the aim of the present study was to examine the changes of expression of autophagy markers in the developing retina in postnatal rats. Due to its postnatal development and accessibility, the rat retinal vasculature warrants a bonafide model to assess vascular developmental autophagy mechanisms from birth through postnatal day (P) 18 when retinal vasculature has attained its adult pattern.

2. Material and Methods

2.1. Animals and Ethics Statements

After birth, 84 Wistar rat pups were maintained with their nursing mothers through the experimental times P7, P14 and P18 in a regulated environment (24 ± 1°C, 50 ± 5% humidity), with a 12 h light/dark cycle and provided with food and water. Rat pups were euthanized with an intraperitoneal injection of 30 mg/kg of pentobarbital. All animal protocols were in accordance with the Statement for the Use of Animals in Ophthalmic and Vision Research (ARVO), the Italian regulation for animal care (DL 116/92), and the European Communities Council Directive (86/609/EEC). Animal procedures were authorized by the Ethical Committee in Animal Experiments of the University of Pisa (permit number: 133/2019-PR, 14 February 2019).

2.2. Vascular Labeling

A total of 24 rat pups of different ages (birth, P7, P14 and P18; six rats for each time point) were used to prepare whole-mount and retina sections. Isolated retinas were fixed in 4% paraformaldehyde in 0.1 M phosphate buffer, pH 7.4 (PB), at room temperature for

3 h. Subsequently, retinas were washed three times (5 min per wash) in PB and incubated for 1 h at room temperature (RT) in blocking buffer (PB containing 10% donkey serum and 0.5% Triton X-100; Sigma-Aldrich, St. Louis, MO, USA) to prevent non-specific labeling. Sequentially, retinas were incubated with fluorescein-labelled isolectin B4 (1:200; Vector Laboratories, Burlingame, CA, USA) in blocking solution, at 4 °C overnight (ON). Finally, after three washes with PB, retinas were placed onto a slide, mounted and covered with a coverslip. Immunostaining was observed by a digital fluorescence microscope (Ni-E; Nikon-Europe, Amsterdam, The Netherlands) and immunofluorescent images of the retinal vasculature were acquired using a digital camera (DS-Fi1c; Nikon-Europe). The vascular area and the total area were measured using ImageJ freeware. Vascular area was reported as percentage of the total area.

2.3. Western Blot Analysis

Proteins were extracted from retinas of 24 rat pups at different ages (birth, P7, P14 and P18; six rats for each time point) using RIPA lysis buffer (Santa Cruz Biotechnology, Dallas, TX, USA), supplemented with phosphatase inhibitor (Sigma-Aldrich) and protease inhibitor (Roche, Mannheim, Germany) cocktails. Protein extracts were quantified by the microBCA method (Thermo Fisher Scientific, Waltam, MA, USA) and 15 µg of total proteins were separated by SDS-PAGE and transferred onto polyvinylidene difluoride (PVDF) membranes (Bio-Rad Laboratories, Hercules, CA, United States). Membranes were blocked either with 5% of skim milk in Tris-buffered saline (TBS-T; Bio-Rad Laboratories, containing 0.05% Tween-20; Sigma-Aldrich) or with 4% of Bovine Serum Albumin (BSA; Sigma-Aldrich) in TBS-T, for 1 hour at RT. Subsequently, the membranes were incubated at 4 °C ON with primary antibodies: anti-HIF-1α (1:500, rabbit polyclonal, cat. no. NB100479; Novus Biologicals, Centennial, Colorado, USA); anti-pAMPKα (Thr172, 1:500, rabbit monoclonal, cat. no. 2535S; Cell Signaling Technology); anti-AMPKα (1:1000, rabbit monoclonal, cat. no. 5832S; Cell Signaling Technology); anti-pULK1 (Ser^{555}; 1:500, rabbit monoclonal, cat. no. 5869S; Cell Signaling Technology, Danvers, MA, USA); anti-ULK1 (1:1000, rabbit monoclonal, cat. no. 8054S; Cell Signaling Technology); anti-LC3 I and II (1:1000, rabbit polyclonal, cat. no. 4108S; Cell Signaling Technology); anti-p62 (1:1000, rabbit polyclonal, cat. no. ab-91526; Abcam, Cambridge, UK); and anti-β-actin (1:5000, rabbit monoclonal, cat. no. SAB5600204; Sigma-Aldrich). Secondary antibody anti-rabbit-IgG conjugated to horseradish peroxidase (1:10,000, cat. no. P044801-2; Dako, Carpinteria, CA, USA) was incubated for 1 h at RT. Following the incubation with both primary and secondary antibodies the membranes were washed with TBS-T, three times for 5 min. Finally, the protein of interest was visualized using the Clarity Western ECL substrate with a ChemiDoc XPS$^+$ imaging system (Bio-Rad Laboratories, Hercules CA, USA). The optical density (OD) of the bands was determined with the Image Lab 3.0 software (Bio-Rad Laboratories). Protein levels were corrected to the β-actin loading control or non-phosphorylated proteins.

2.4. Quantitative PCR

Total RNA was extracted and purified from retinas of 24 rat pups at different ages (birth, P7, P14 and P18; six rats for each time point), using the Trizol$^®$ reagent (Invitrogen, Waltam MA, USA), resuspended in RNAse-free water, and quantified by spectrophotometry (BioSpectrometer basic; Eppendorf AG, Hamburg, Germany). First-strand cDNA was generated from 1 µg of total RNA (QuantiTect Reverse Transcription Kit; Qiagen, Hilden, Germany). Quantitative PCR was performed with a kit (SsoAdvanced Universal SYBR Green Supermix; Bio-Rad Laboratories) on a CFX96 Real Time PCR Detection System (equipped with the CFX manager software (Bio-Rad Laboratories). Forward and reverse sequence of primers were chosen to hybridize to unique region of the appropriate gene sequence: occludin-1 (Forward: 5'-TTTCATGCCTTGGGGATTGAG-3'/Reverse: 5'-GACTTCCCAGAGTGCAGAGT-3'; Invitrogen); zonula-occludens-1 (ZO-1; Forward: 5'-AGTCTCGGAAAAGTGCCAGG-3'/Reverse: 5'-GGGCACCATACCAACCATCA-3'; Invitrogen); VEGF-A (Forward: 5′-CAATGATGAAGCCCTGGAGTG-3′/Reverse: 5′-AGGTTTG ATCCGCATGATCTG-3; Invitrogen) and Ribosomal Protein L13a (Rpl13a; Forward: 5′-GGATCCCTCCACCCTATGACA-3′/Reverse: 5′-CTGGTACTTCCACCCGACCTC-3′; Invitrogen). Samples were compared by the threshold cycle analysis (Ct) and absolute expression values were calculated using the $2^{-\Delta\Delta Ct}$ formula, with Rpl13a as the housekeeping gene.

2.5. Immunohistofluorescence

In total, 12 rat pups at different ages (birth, P7, P14 and P18; three rats for each time point) were fixed in 4% paraformaldehyde in PB at RT for 5 days prior to immunofluorescence staining. Sections of 4 micrometers were processed for immunohistofluorescence in a Bond III robotic system (Leica Biosystems, Newcastle, UK), as previously described [35]. According to the manufacturer's instructions, antigen retrieval was performed in 10 mM citrate buffer pH6 and sections were incubated with primary antibody step with anti-LC3 I and II (1:100) and isolectin-biotin (1:1000; cat. no. I21414; Thermo Fisher Scientific), while secondary antibody steps included Streptavidin-Alexa 488 (1:500; cat. no. S11223, Thermo Fisher Scientific), anti-rat-Alexa 546 (1:500; cat. no. A11081, Invitrogen) and anti-goat-Alexa 647 (1:500; cat. no. SAB4600175, Sigma-Aldrich). Sections were mounted by using a vector Vectashield with DAPI mounting medium (Vector Laboratories, CA, Burlingame, USA), and visualized with an Axioscope 2 plus with the AxioVision software (Zeiss, Gottingen, Germany).

2.6. Statistical Analysis

The statistical analyses were performed using Prism software (GraphPad software, Inc., San Diego, CA, USA), applying one-way ANOVA with Bonferroni's multiple comparisons posttest. Data were presented as mean ± standard error of mean (SEM) of $n = 6$; p values < 0.05 were considered statistically significant.

3. Results

3.1. The Rat Retina Is Partially Avascular at Birth

Fluorescein-labeled isolectin B4 was used to assess the progress of retinal vascularization in the rat. As previously described, rat retinal vascularization is almost completed around P13-P16 [36]. As depicted in Figure 1A, at birth the hyaloid vasculature was still present and the retina remained partially avascular, as determined by measuring the vascular area as less than 50% of total area of the retina (Figure 1B). Quantitative analysis demonstrated a significant difference at P7, where the retina was 90% vascularized ($p < 0.001$) and in the late stages of ocular blood vessel development, P14-P18 ($p < 0.001$), where the retinas were fully vascularized (100%) as compared to birth.

3.2. Different Expression of Blood-Retina Barrier Genes during Rat Retina Development

In retinal blood vessels, ECs present tight junctions that function as a part of the BRB, fundamental to maintain retinal homeostasis and to mediate selective diffusion of molecules from the circulation to the retinal tissue [14,37]. Consequently, analysis of the expression of BRB genes in the rat retinas at birth, P7, P14 and P18 was performed by qPCR (Figure 2). Transcript expression levels of occludin-1 and ZO-1 were upregulated at P14 and P18, compared to birth and P7 ($p < 0.01$ both). These results confirmed that at birth and P7 the rat retinal vasculature was still under development, presenting an incomplete vascular network.

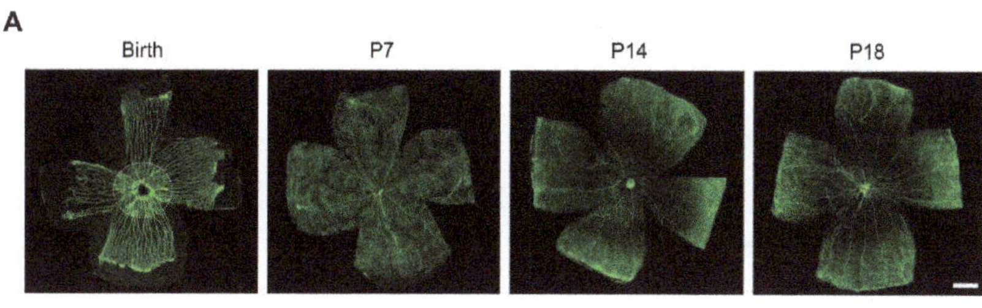

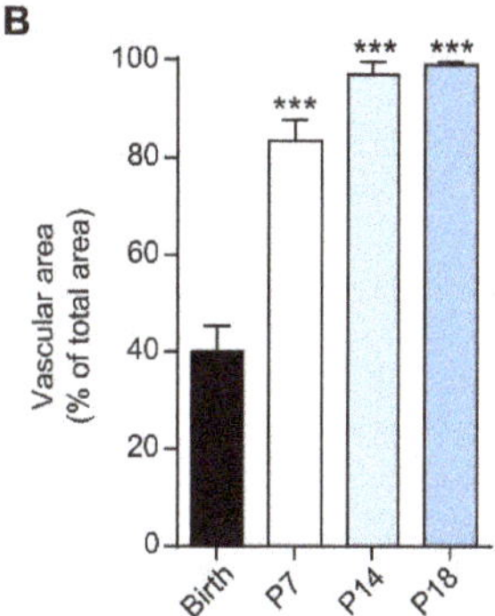

Figure 1. Development of the vascular network in rat retina. (**A**) Visualization of blood vessels by isolectin B4 staining of rat pup retinas at birth, postnatal day (P)7, P14 and P18. Dashed circle delineates the developing retinal vasculature (inner of the dashed circle) with the hyaloid vasculature (outer of the dashed circle). Scale bar = 500 μm. (**B**) Quantitative analysis of vascular area of the retina. Data is presented as mean ± SEM. One-way ANOVA was used as statistical analysis, followed by Bonferroni's multiple comparisons test ($n = 6$; *** $p < 0.001$ vs. birth).

3.3. At P7 the Rat Retina Is Hypoxic

During retinal development, physiologic hypoxia induces the activation of HIF-1α, which promotes the transcription of *VEGF-A* gene. This process is pivotal to induce endothelial cell proliferation and migration to form the vasculature network [38]. In this context, Western blotting was performed to analyze HIF-1α protein levels in the rat retina (Figure 3A) at birth, P7, P14 and P18. Densitometric analysis showed an increase of HIF-1α protein levels at P7 ($p < 0.001$) compared to birth, followed by a decrease at P14 ($p < 0.01$ vs. birth; $p < 0.01$ vs. P7) and at P18 ($p < 0.01$ vs. P7; Figure 3B).

Since HIF-1α promotes the transcription of *VEGF-A* gene, expression analysis was performed by qPCR in rat retinas at birth, P7, P14 and P18. As depicted in Figure 3C, VEGF-A was upregulated at P7, ($p < 0.001$ vs. birth) while at P14 and P18 its levels were comparable to those at birth, in agreement with the reduced levels of HIF-1α at the latter postnatal days.

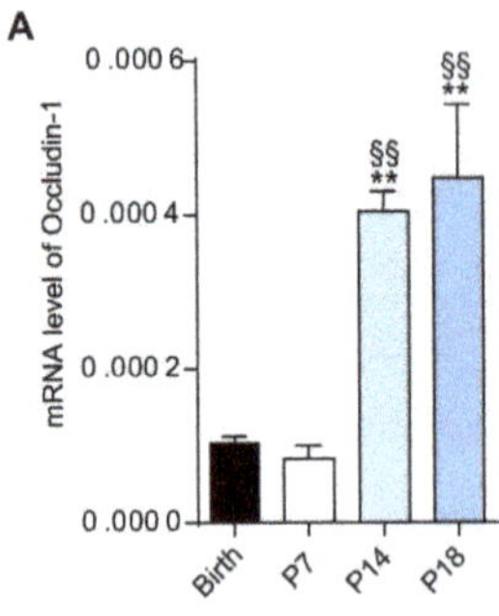

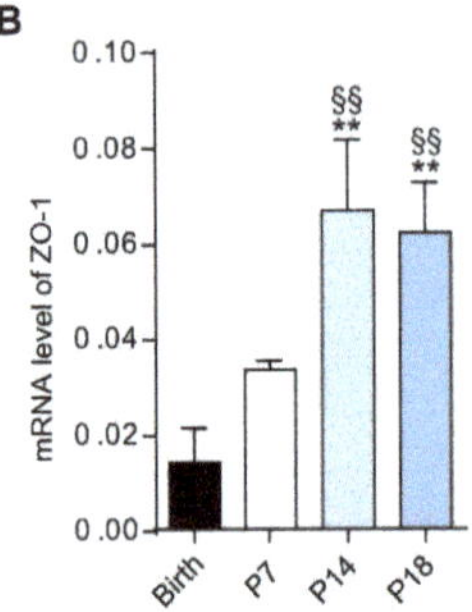

Figure 2. Expression of blood-retina barrier genes during retina rat development. mRNA expression of occludin-1 (**A**) and zonula occludens (ZO-1) (**B**) genes was evaluated by qPCR in rat retinas at birth, P7, P14 and P18. Data is presented as mean $\pm$ SEM. One-way ANOVA was used as statistical analysis, followed by Bonferroni's multiple comparisons test ($n = 6$; ** $p < 0.01$ vs. birth, §§ $p < 0.01$ vs. P7).

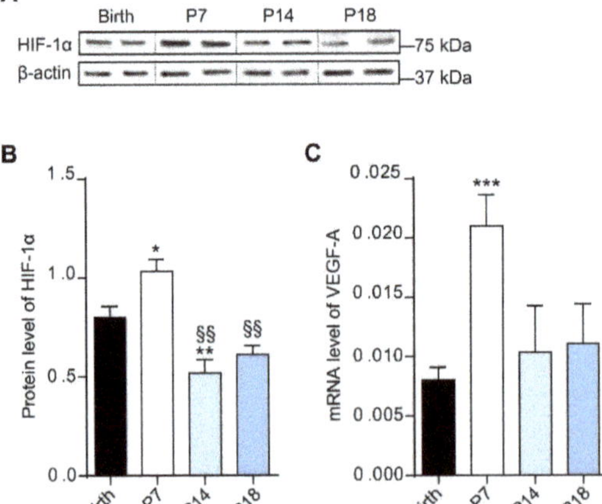

Figure 3. Protein levels of hypoxia-inducible factor (HIF) change during rat retina development. (**A**) Western blots illustrate representative immunoreactive bands of HIF-1α and β-actin (loading control) in the retina of rat pups from birth to P18. (**B**) Quantitative analysis of optical density of the immunoreactive bands of HIF-1α. (**C**) mRNA expression of vascular endothelial growth factor (*VEGF*)-*A* gene evaluated with qPCR in rat retinas at birth, P7, P14 and P18. One-way ANOVA followed by Bonferroni's multiple comparisons test was used as statistical analysis of mean $\pm$ SEM datasets ($n = 6$; * $p < 0.05$, ** $p < 0.01$, *** $p < 0.001$ vs. birth, §§ $p < 0.01$ vs. P7).

3.4. Autophagic Mechanisms Increased during the Hypoxic Phase

During development, autophagic mechanisms support cells to adapt and respond to several processes, including proliferation, differentiation and migration [39]. In this context, Western blotting analysis was performed to evaluate the levels of proteins involved in autophagy during retina development (Figure 4A). Densitometric analysis demonstrated a significant increase of phosphorylated levels of AMPKα ($p < 0.001$ vs. birth; Figure 4C) and ULK1 ($p < 0.01$ vs. birth; Figure 4B) as well as LC3 II ($p < 0.01$ vs. birth; Figure 4D) at P7. At P14 and P18, the levels of these autophagic markers decreased then at levels comparable to those at birth.

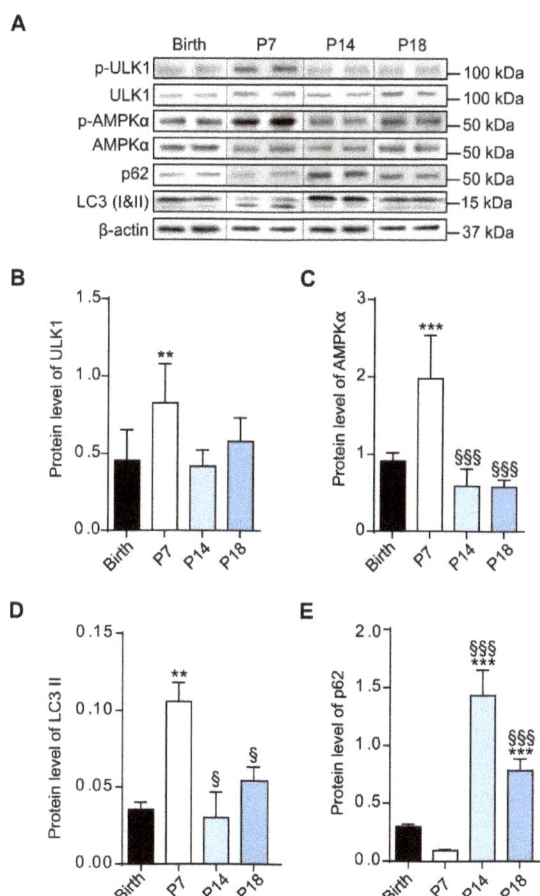

Figure 4. Protein levels of autophagic markers during rat retinal development. (**A**) Western blots depict representative immunoreactive bands of proteins involved in autophagic mechanisms in rat retinas, from birth to P18. Quantitative analysis of optical density of the ratio of immunoreactive bands between pSer^{555}-Ulk1 / Ulk1 (**B**), p-AMPKα / AMPKα (**C**); LC3 II / β-actin (**D**) and p62 / β-actin (**E**). One-way ANOVA followed by Bonferroni's multiple comparisons test was used as statistical analysis of mean ± SEM datasets ($n = 6$; ** $p < 0.01$, *** $p < 0.001$ vs. birth, § $p < 0.05$, §§§ $p < 0.001$ vs. P7).

On the contrary, a trend to a decrease of SQSTM1/p62 protein levels, yet not statistically significative, was observed at P7, followed by a substantial increase at both P14 ($p < 0.001$ vs. birth; $p < 0.001$ vs. P7) and P18 ($p < 0.001$ vs. birth; $p < 0.001$ vs. p7).

3.5. High Expression of Autophagic Marker at P7 in Rat Retina

In rat, developing retinal cells are already organized into layers from P7, giving the tissue its stratified feature [40,41]. The variation of autophagic flux was evaluated relative to the retinal layers using immunostained retinal sections to visualize autophagy and vasculature, with LC3 as an autophagic marker and isolectin B4 as a marker of endothelial cells. The immunofluorescence demonstrated a clear variation of autophagy during retinal development, indicating a predominant expression of the autophagic marker LC3 at P7 (Figure 5). At this specific time point, a substantial expression of LC3 was observed in both the inner plexiform and outer plexiform layers (IPL; OPL). At birth and on latter stages of retinal developmental, LC3 was predominantly expressed in the IPL and almost undetectable in the OPL. In addition, heightened colocalized expression of the autophagy and vascular markers was observed at P7, as compared to birth, P14 and P18, which could be related to the hypoxia associated with the involution of the hyaloid blood vessels. Albeit a positive staining for LC3 was detected in ganglion cell layer (GCL) and the photoreceptor outer segments (OS), no changes were observed in the studied times of development.

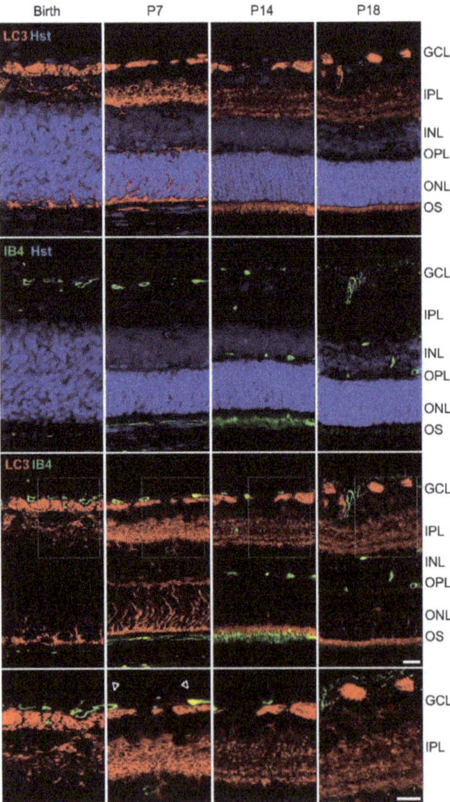

Figure 5. Expression pattern of LC3 in the developing rat retina. Representative immunohistofluorescence analysis of LC3 (red) and isolectin B4 (IB4; green) and Hoechst (Hst; blue) in retina sections of rat pups at birth, P7, P14 and P18. Dashed squares indicate the magnification area of GCL and IPL layers. Arrows represent colocalization of LC3 with IB4 in retinal vasculature. GCL, ganglion cell layer; IPL, inner plexiform layer; INL, inner nuclear layer; OPL, outer plexiform layer; ONL, outer nuclear layer; OS, outer segments of photoreceptors. Scale bar = 50 μm.

4. Discussion

During mammalian development, cells go through proliferation, cell death and differentiation, culminating in an adult organism formation. During these stages, autophagy assures cell adaptation, by promoting rapid changes in cytosolic composition and accelerating organelle and protein turnover [39,42]. The present study elaborates on the influence of autophagy mechanisms in the development of rat's retina from birth to P18.

In rats, the retinal vasculature develops postnatally. At birth, the retina surface is still covered by the hyaloid vessels, and the retinal vessels have merely begun to raise from the optic disc [43]. Previous studies have shown that the hyaloid network persists until P7, by which time the retinal vessels have almost propagated to the periphery of the retina [44,45]. Interestingly, in newborn rat retinas a large area lacking retinal capillary coverage is observed, indicating the presence of hyaloid vessels, while at P7 the rat retinas display nearly full vascularization. At this time point, the developing retina vascular network is still immature and the BRB is not formed [11]. To form the BRB, endothelial cells require tight junctions, essential to regulate the movement of solutes and nutrients from the outer to the inner retinal layers [46]. Tight junctions comprise several proteins, including occludin-1 and ZO-1, responsible for anchoring the junctional complex to the cytoskeleton [47]. As demonstrated here, at birth and P7 the retina of rat pups presents low transcript levels of both occludin-1 and ZO-1 genes. Reversely, mRNA levels of these genes are increased at P14 and P18, when the retinal vasculature is covering the total surface of retina. This suggests that retina vascularization is completed and mature around P14-P18, in alignment with the presence of tight junctions in retinal endothelial cells.

From embryonic stages to birth, a physiological hypoxia is paramount to drive retinal neovascularization, through the upregulation of HIF-1α and subsequently VEGF-A [9,48]. During the early postpartum retinal developmental stage in rats, the involution of the hyaloid vasculature to the retinal vascular network results in a partially avascular and ischemic retina, correlating to increased oxygen demand and resulting in a hypoxic stimulus [10]. In the present study, a peak of HIF-1α and VEGF-A expression is determined at P7, which decreases during latter stages when the retina is fully vascularized. The hypoxic environment contributes to activate essential mechanisms in adaptation and survival, ensuring cellular homeostasis during angiogenesis [49,50]. In fact, the developing retina exposed to changing environment and metabolic stress requires autophagy to adjust its bioenergetic and biosynthetic demands [20,49]. In this respect, both hypoxia and energy deprivation can promote AMPK activation, a known inducer of autophagy [22,25]. In agreement, an increase of p-AMPKα, p-ULK1 and LC3 II protein levels are demonstrated at P7, with a trend of decreased p62 protein levels, indicating an active autophagic flux at this developmental stage. At P14 and P18, with the presence of a fully vascularized retina, a decrease in p-ULK1 and LC3 II protein levels is determined concomitantly with an increase of p62. The observed accumulation of p62 with a decrease of LC3 II protein levels at P14 could be associated with a transition from autophagy-dependent to -independent mechanisms of retinal homeostasis, as previously suggested in retinal pigment epithelium cells [51,52].

To elaborate on the role of autophagy in the different cell layers during rat retinal development, a predominant expression of LC3 is confirmed in the GCL, IPL, OPL and OS, in agreement with previous studies in rodent retinas [53–55]. At P7, an increase of LC3 staining is denoted in the IPL and OPL, with a noticeable reduction in the autophagy marker at P14 and P18. During the early postpartum developmental stage, the rat retinal cells are affected by ischemia, which is correlated to the peak of expression of HIF-1α and VEGF-A, concomitantly with LC3. Moreover, an expression of LC3 is observed in endothelial cells at P7, suggesting that autophagy may contribute directly to the formation of the retinal vascular network. These findings indicate that during the physiologic hypoxia in the rat retina, HIF-mediated signaling induces the increase in VEGF-A to promote endothelial cell proliferation, and an upregulation of autophagy markers to sustain cellular homeostasis and cellular quality control in retinal cells.

5. Conclusions

The present study indicates that increased autophagy is intrinsically associated with the hypoxic phase of retinal development and critically contributes to the physiologic development of the different cell layers of the retina during the transition from the hyaloid to the retinal vasculature, thus allowing the normal development of the retina.

Author Contributions: Conceptualization, H.A. and M.D.M.; formal analysis, N.A.P.; investigation, N.A.P., A.C., H.A. and M.D.M.; methodology, N.A.P., A.C. and E.L.; resources, M.C. and A.K.; supervision, H.A. and M.D.M.; writing—original draft, N.A.P., H.A. and M.D.M.; writing—review and editing, N.A.P., H.A. and M.D.M. All authors have read and agreed to the published version of the manuscript.

Funding: This research was funded by the grants from the Karolinska Institutet Foundations and The Swedish Eye Foundation.

Institutional Review Board Statement: Animal protocols were conducted in accordance with the Statement for the Use of Animals in Ophthalmic and Vision Research (ARVO), the Italian regulation for animal care (DL 116/92), and the European Communities Council Directive (86/609/EEC). Animal procedures were authorized by the Ethical Committee in Animal Experiments of the University of Pisa (permit number: 133/2019-PR, 14 February 2019).

Informed Consent Statement: Not applicable.

Data Availability Statement: The data presented in this study are available on request from the corresponding author.

Acknowledgments: The authors thank Flavia Plastino, Maria Grazia Rossino and Filippo Locri for technical support.

Conflicts of Interest: The authors declare no conflict of interest.

References

1. Wong-Riley, M. Energy metabolism of the visual system. *Eye Brain* **2010**, *2*, 99. [CrossRef]
2. Joyal, J.S.; Gantner, M.L.; Smith, L.E.H. Retinal energy demands control vascular supply of the retina in development and disease: The role of neuronal lipid and glucose metabolism. *Prog. Retin. Eye Res.* **2018**, *64*, 131–156. [CrossRef]
3. Archer, D.B.; Gardiner, T.A.; Stitt, A.W. Functional anatomy, fine structure and basic pathology of the retinal vasculature. In *Retinal Vascular Disease*; Springer: Berlin/Heidelberg, Germany, 2007; pp. 3–23.
4. Sun, Y.; Smith, L.E.H. Notice of withdrawal: Retinal vasculature in development and diseases. *Annu. Rev. Vis. Sci.* **2020**. [CrossRef]
5. Rust, R.; Grönnert, L.; Dogançay, B.; Schwab, M.E. A revised view on growth and remodeling in the retinal vasculature. *Sci. Rep.* **2019**, *9*, 3263. [CrossRef] [PubMed]
6. Fruttiger, M. Development of the retinal vasculature. *Angiogenesis* **2007**, *10*, 77–88. [CrossRef] [PubMed]
7. Liu, C.; Wang, Z.; Sun, Y.; Chen, J. Animal models of ocular angiogenesis: From development to pathologies. *FASEB J.* **2017**, *31*, 4665–4681. [CrossRef]
8. Selvam, S.; Kumar, T.; Fruttiger, M. Retinal vasculature development in health and disease. *Prog. Retin. Eye Res.* **2018**, *63*, 1–19. [CrossRef] [PubMed]
9. Krock, B.L.; Skuli, N.; Simon, M.C. Hypoxia-induced angiogenesis: Good and evil. *Genes Cancer* **2011**, *2*, 1117–1133. [CrossRef]
10. Stone, J.; Itin, A.; Alon, T.; Pe'er, J.; Gnessin, H.; Chan-Ling, T.; Keshet, E. Development of retinal vasculature is mediated by hypoxia-induced vascular endothelial growth factor (VEGF) expression by neuroglia. *J. Neurosci.* **1995**, *15*, 4738–4747. [CrossRef]
11. Díaz-Coránguez, M.; Ramos, C.; Antonetti, D.A. The inner blood-retinal barrier: Cellular basis and development. *Vision Res.* **2017**, *139*, 123–137. [CrossRef]
12. Chow, B.W.; Gu, C. Gradual suppression of transcytosis governs functional blood-retinal barrier formation. *Neuron* **2017**, *93*, 1325–1333. [CrossRef] [PubMed]
13. van der Wijk, A.E.; Vogels, I.M.C.; van Veen, H.A.; van Noorden, C.J.F.; Schlingemann, R.O.; Klaassen, I. Spatial and temporal recruitment of the neurovascular unit during development of the mouse blood-retinal barrier. *Tissue Cell* **2018**, *52*, 42–50. [CrossRef] [PubMed]
14. Hosoya, K.I.; Tachikawa, M. The inner blood-retinal barrier molecular structure and transport biology. *Adv. Exp. Med. Biol.* **2013**, *763*, 85–104. [CrossRef]
15. Dejana, E.; Hirschi, K.K.; Simons, M. The molecular basis of endothelial cell plasticity. *Nat. Commun.* **2017**, *8*, 1–11. [CrossRef]
16. Chun, Y.; Kim, J. Autophagy: An essential degradation program for cellular homeostasis and life. *Cells* **2018**, *7*, 278. [CrossRef]
17. Glick, D.; Barth, S.; Macleod, K.F. Autophagy: Cellular and molecular mechanisms. *J. Pathol.* **2010**, *221*, 3–12. [CrossRef]

18. Levine, B.; Kroemer, G. Autophagy in the pathogenesis of disease. *Cell* **2008**, *132*, 27–42. [CrossRef]
19. Kim, J.H.; Kim, J.H.; Yu, Y.S.; Mun, J.Y.; Kim, K.W. Autophagy-induced regression of hyaloid vessels in early ocular development. *Autophagy* **2010**, *6*, 922–928. [CrossRef]
20. Schaaf, M.B.; Houbaert, D.; Meçe, O.; Agostinis, P. Autophagy in endothelial cells and tumor angiogenesis. *Cell Death Differ.* **2019**, *26*, 665–679. [CrossRef]
21. Bellot, G.; Garcia-Medina, R.; Gounon, P.; Chiche, J.; Roux, D.; Pouysségur, J.; Mazure, N.M. Hypoxia-induced autophagy is mediated through hypoxia-inducible factor induction of BNIP3 and BNIP3L via their BH3 domains. *Mol. Cell. Biol.* **2009**, *29*, 2570–2581. [CrossRef]
22. Mazure, N.M.; Pouysségur, J. Hypoxia-induced autophagy: Cell death or cell survival? *Curr. Opin. Cell Biol.* **2010**, *22*, 177–180. [CrossRef] [PubMed]
23. Blagosklonny, M.V. Hypoxia, MTOR and autophagy. *Autophagy* **2013**, *9*, 260–262. [CrossRef] [PubMed]
24. Zhang, J.; Zhang, C.; Jiang, X.; Li, L.; Zhang, D.; Tang, D.; Yan, T.; Zhang, Q.; Yuan, H.; Jia, J.; et al. Involvement of autophagy in hypoxia-BNIP3 signaling to promote epidermal keratinocyte migration. *Cell Death Dis.* **2019**, *10*, 1–15. [CrossRef] [PubMed]
25. Dengler, F. Activation of ampk under hypoxia: Many roads leading to rome. *Int. J. Mol. Sci.* **2020**, *21*, 2428. [CrossRef] [PubMed]
26. Mihaylova, M.M.; Shaw, R.J. The AMPK signalling pathway coordinates cell growth, autophagy and metabolism. *Nat. Cell Biol.* **2011**, *13*, 1016–1023. [CrossRef]
27. Egan, D.F.; Shackelford, D.B.; Mihaylova, M.M.; Gelino, S.; Kohnz, R.A.; Mair, W.; Vasquez, D.S.; Joshi, A.; Gwinn, D.M.; Taylor, R.; et al. Phosphorylation of ULK1 (hATG1) by AMP-activated protein kinase connects energy sensing to mitophagy. *Science* **2011**, *331*, 456–461. [CrossRef]
28. Grasso, D.; Renna, F.J.; Vaccaro, M.I. Initial steps in mammalian autophagosome biogenesis. *Front. Cell Dev. Biol.* **2018**, *6*, 146. [CrossRef]
29. Pyo, J.O.; Nah, J.; Jung, Y.K. Molecules and their functions in autophagy. *Exp. Mol. Med.* **2012**, *44*, 73–80. [CrossRef]
30. Metlagel, Z.; Otomo, C.; Takaesu, G.; Otomo, T. Structural basis of ATG3 recognition by the autophagic ubiquitin-like protein ATG12. *Proc. Natl. Acad. Sci. USA* **2013**, *110*, 18844–18849. [CrossRef]
31. Tanida, I.; Ueno, T.; Kominami, E. LC3 and autophagy. *Methods Mol. Biol.* **2008**, *445*, 77–88. [CrossRef]
32. Itakura, E.; Mizushima, N. p62 targeting to the autophagosome formation site requires self-oligomerization but not LC3 binding. *J. Cell Biol.* **2011**, *192*, 17–27. [CrossRef] [PubMed]
33. Wurzer, B.; Zaffagnini, G.; Fracchiolla, D.; Turco, E.; Abert, C.; Romanov, J.; Martens, S. Oligomerization of p62 allows for selection of ubiquitinated cargo and isolation membrane during selective autophagy. *eLife* **2015**, *4*, e08941. [CrossRef] [PubMed]
34. Yoshii, S.R.; Mizushima, N. Monitoring and measuring autophagy. *Int. J. Mol. Sci.* **2017**, *18*, 1865. [CrossRef] [PubMed]
35. André, H.; Tunik, S.; Aronsson, M.; Kvanta, A. Hypoxia-inducible factor-1α Is associated with sprouting angiogenesis in the murine laser-induced Choroidal neovascularization model. *Investig. Ophthalmol. Vis. Sci.* **2015**, *56*, 6591–6604. [CrossRef]
36. Inagaki, K.; Koga, H.; Inoue, K.; Suzuki, K.; Suzuki, H. Spontaneous intraocular hemorrhage in rats during postnatal ocular development. *Comp. Med.* **2014**, *64*, 34–43.
37. Naylor, A.; Hopkins, A.; Hudson, N.; Campbell, M. Tight junctions of the outer blood retina barrier. *Int. J. Mol. Sci.* **2020**, *21*, 211. [CrossRef]
38. Oladipupo, S.; Hu, S.; Kovalski, J.; Yao, J.; Santeford, A.; Sohn, R.E.; Shohet, R.; Maslov, K.; Wang, L.V.; Arbeit, J.M. VEGF is essential for hypoxia-inducible factor-mediated neovascularization but dispensable for endothelial sprouting. *Proc. Natl. Acad. Sci. USA* **2011**, *108*, 13264–13269. [CrossRef]
39. Cecconi, F.; Levine, B. The role of autophagy in mammalian development: Cell makeover rather than cell death. *Dev. Cell* **2008**, *15*, 344–357. [CrossRef]
40. Amini, R.; Rocha-Martins, M.; Norden, C. Neuronal migration and lamination in the vertebrate retina. *Front. Neurosci.* **2018**, *11*, 742. [CrossRef]
41. Kuwabara, T.; Weidman, T. Development of the prenatal rat retina. *Invest. Ophthalmol.* **1974**, *13*, 725–739.
42. Di Bartolomeo, S.; Nazio, F.; Cecconi, F. The role of autophagy during development in higher eukaryotes. *Traffic* **2010**, *11*, 1280–1289. [CrossRef] [PubMed]
43. Engerman, R.L.; Meyer, R.K. Development of retinal vasculature in rats. *Am. J. Ophthalmol.* **1965**, *60*, 628–641. [CrossRef]
44. Kim, Y.; Park, J.R.; Hong, H.K.; Han, M.; Lee, J.; Kim, P.; Woo, S.J.; Park, K.H.; Oh, W.Y. In vivo imaging of the hyaloid vascular regression and retinal and choroidal vascular development in rat eyes using optical coherence tomography angiography. *Sci. Rep.* **2020**, *10*, 12901. [CrossRef]
45. Cairns, J.E. Normal development of the hyaloid and retinal vessels in the rat. *Br. J. Ophthalmol.* **1959**, *43*, 385–393. [CrossRef] [PubMed]
46. Campbell, M.; Humphries, P. The blood-retina barrier tight junctions and barrier modulation. *Adv. Exp. Med. Biol.* **2013**, *763*, 70–84. [CrossRef]
47. Bazzoni, G.; Dejana, E. Endothelial cell-to-cell junctions: Molecular organization and role in vascular homeostasis. *Physiol. Rev.* **2004**, *84*, 869–901. [CrossRef]
48. Rattner, A.; Williams, J.; Nathans, J. Roles of HIFs and VEGF in angiogenesis in the retina and brain. *J. Clin. Invest.* **2019**, *129*, 3807–3820. [CrossRef]
49. Nussenzweig, S.C.; Verma, S.; Finkel, T. The role of autophagy in vascular biology. *Circ. Res.* **2015**, *116*, 480–488. [CrossRef]

50. Oeste, C.L.; Seco, E.; Patton, W.F.; Boya, P.; Pérez-Sala, D. Interactions between autophagic and endo-lysosomal markers in endothelial cells. *Histochem. Cell Biol.* **2013**, *139*, 659–670. [CrossRef]
51. Ryhänen, T.; Hyttinen, J.M.T.; Kopitz, J.; Rilla, K.; Kuusisto, E.; Mannermaa, E.; Viiri, J.; Holmberg, C.I.; Immonen, I.; Meri, S.; et al. Crosstalk between Hsp70 molecular chaperone, lysosomes and proteasomes in autophagy-mediated proteolysis in human retinal pigment epithelial cells. *J. Cell. Mol. Med.* **2009**, *13*, 3616–3631. [CrossRef]
52. Viiri, J.; Hyttinen, J.M.T.; Ryhänen, T.; Rilla, K.; Paimela, T.; Kuusisto, E.; Siitonen, A.; Urtti, A.; Salminen, A.; Kaarniranta, K. p62/sequestosome 1 as a regulator of proteasome inhibitor-induced autophagy in human retinal pigment epithelial cells. *Mol. Vis.* **2010**, *16*, 1399–1414. [CrossRef] [PubMed]
53. Cammalleri, M.; Locri, F.; Catalani, E.; Filippi, L.; Cervia, D.; Dal Monte, M.; Bagnoli, P. The beta-adrenergic receptor blocker propranolol counteracts retinal dysfunction in a mouse model of oxygen induced retinopathy: Restoring the balance between apoptosis and autopha. *Front. Cell. Neurosci.* **2017**, *11*, 395. [CrossRef] [PubMed]
54. Boya, P.; Esteban-Martínez, L.; Serrano-Puebla, A.; Gómez-Sintes, R.; Villarejo-Zori, B. Autophagy in the eye: Development, degeneration, and aging. *Prog. Retin. Eye Res.* **2016**, *55*, 206–245. [CrossRef] [PubMed]
55. Boya, P. Why autophagy is good for retinal ganglion cells? *Eye* **2017**, *31*, 185–190. [CrossRef]

cells

MDPI

Article

Analysis of Programmed Cell Death and Senescence Markers in the Developing Retina of an Altricial Bird Species

Guadalupe Álvarez-Hernán [1], José Antonio de Mera-Rodríguez [1], Ismael Hernández-Núñez [1], Alfonso Marzal [2], Yolanda Gañán [3], Gervasio Martín-Partido [1], Joaquín Rodríguez-León [3,*,†] and Javier Francisco-Morcillo [1,*]

1   Área de Biología Celular Departamento de Anatomía Biología Celular y Zoología, Facultad de Ciencias, Universidad de Extremadura, 06006 Badajoz, Spain; galvarezt@unex.es (G.Á.-H.); merarodja@unex.es (J.A.d.M.-R.); ihernandezn7694@gmail.com (I.H.-N.); gmartin@unex.es (G.M.-P.)
2   Área de Zoología, Departamento de Anatomía, Biología Celular y Zoología, Facultad de Ciencias, Universidad de Extremadura, 06006 Badajoz, Spain; amarzal@unex.es
3   Área de Anatomía y Embriología Humana, Departamento de Anatomía, Biología Celular y Zoología, Facultad de Medicina, Universidad de Extremadura, 06006 Badajoz, Spain; yolandag@unex.es
*   Correspondence: jrleon@unex.es (J.R.-L.); morcillo@unex.es (J.F.-M.)
†   Current address: Área de Anatomía Humana, Departamento de Anatomía, Biología Celular y Zoología, Facultad de Medicina, Universidad de Extremadura, 06006 Badajoz, Spain.

check for updates

Citation: Álvarez-Hernán, G.; de Mera-Rodríguez, J.A.; Hernández-Núñez, I.; Marzal, A.; Gañán, Y.; Martín-Partido, G.; Rodríguez-León, J.; Francisco-Morcillo, J. Analysis of Programmed Cell Death and Senescence Markers in the Developing Retina of an Altricial Bird Species. *Cells* **2021**, *10*, 504. https://doi.org/10.3390/cells10030504

Academic Editor: Maurizio Sorice

Received: 29 January 2021
Accepted: 23 February 2021
Published: 26 February 2021

Publisher's Note: MDPI stays neutral with regard to jurisdictional claims in published maps and institutional affiliations.

Abstract: This study shows the distribution patterns of apoptotic cells and biomarkers of cellular senescence during the ontogeny of the retina in the zebra finch (*T. guttata*). Neurogenesis in this altricial bird species is intense in the retina at perinatal and post-hatching stages, as opposed to precocial bird species in which retinogenesis occurs entirely during the embryonic period. Various phases of programmed cell death (PCD) were distinguishable in the *T. guttata* visual system. These included areas of PCD in the central region of the neuroretina at the stages of optic cup morphogenesis, and in the sub-optic necrotic centers (St15–St20). A small focus of early neural PCD was detected in the neuroblastic layer, dorsal to the optic nerve head, coinciding with the appearance of the first differentiated neuroblasts (St24–St25). There were sparse pyknotic bodies in the non-laminated retina between St26 and St37. An intense wave of neurotrophic PCD was detected in the laminated retina between St42 and P8, the last post-hatching stage included in the present study. PCD was absent from the photoreceptor layer. Phagocytic activity was also detected in Müller cells during the wave of neurotrophic PCD. With regard to the chronotopographical staining patterns of senescence biomarkers, there was strong parallelism between the SA-β-GAL signal and p21 immunoreactivity in both the undifferentiated and the laminated retina, coinciding in the cell body of differentiated neurons. In contrast, no correlation was found between SA-β-GAL activity and the distribution of TUNEL-positive cells in the developing tissue.

Keywords: programmed cell death; cellular senescence; retinogenesis; altricial bird species; precocial bird species; senescence-associated galactosidase activity

1. Introduction

Programmed cell death (PCD) and cellular senescence during vertebrate embryogenesis are transient phenomena that contribute mainly to tissue remodeling [1–3] through the degeneration of temporary structures in the embryo. Indeed, it has been described that PCD processes are accompanied by cell senescence in interdigital regression [4–6], heart morphogenesis [7], pronephros and mesonephros degeneration [8–11], and degeneration of structures in the developing otic vesicle [12–14].

The vertebrate visual system constitutes an excellent model for investigating the mechanisms involved in cell degeneration and the phases of PCD that affect different structures (for a review, see [3]). Areas of intense PCD have been described in the developing visual system in fish [15–19], amphibians [20–22], reptiles [23–25], and mammals [26–31]. With respect to birds, similar studies have been conducted in the chicken [32–38] and in the quail [39], two precocial bird species. In these species, PCD during visual system morphogenesis and retinogenesis is completely restricted to the embryonic period. During early stages of avian eye morphogenesis, two pyknotic zones have been described in the central region of the retinal neuroepithelium and in the dorsal rim of the optic cup [35]. Furthermore, two areas of intense PCD appear in the neuroepithelium located laterally to the optic chiasm, in the so-called sub-optic necrotic centers (SONCs) [40,41]. With the onset of neurogenesis in the neural retina, PCD also affects neuroepithelial cells and newborn ganglion cell neuroblasts [33,36–38]. At later stages, coinciding with the synaptogenesis between retinal neurons, PCD affects those neurons that are unable to successfully innervate their targets [34,37,39]. The last wave of cell death follows different gradients that resemble the spatiotemporal patterns of cell differentiation [34,39].

With regard to developmental cellular senescence, several markers are currently employed to identify the distribution of senescent cells in vertebrate embryos. One of the most commonly used is the histochemical technique that detects the presence of β-galactosidase enzymatic activity at pH 6.0 (senescence-associated β-galactosidase, SA-β-GAL), different from that normally observed at pH 4.0 within lysosomes [42]. Increased expression of intracellular proteins such as p21, p16, p63, and p73 and the Btg/Tob tumor suppressor gene family also identifies cell senescence in several regions of the developing embryo [6]. These markers have been described in different embryonic tissues, but little is known about their distribution in the developing visual system. In this sense, we have recently described that some of these senescence markers are detected not only in several subpopulations of neurons in the developing retina, but also in the retinal pigment epithelium [43,44].

Although the ontogenetic mechanisms involved in visual system development and the basic structure of the retina are similar across bird species, the developmental rate and the acquisition of retinal structures are highly variable. Visual system morphogenesis and retinogenesis occur early in embryogenesis in precocial bird species [45,46], while these ontogenetic processes are delayed in altricial birds [47–49]. This delay can reach the stage of hatching and the first week of life, in which intense postnatal neurogenesis has been detected in the altricial retina [50]. The timing of histogenesis and cell differentiation and the state of retinal maturation at hatching thus differ significantly between precocial and altricial bird species.

All these data suggest that it is necessary to study visual system development across a broad range of avian species to conduct interspecific comparisons that can clarify the ontogenetic patterns. In the present study, we use classical histological, histochemical, and immunohistochemical methods (i) to describe the chronotopographical patterns of cell death and cell senescence markers in the developing visual system of an altricial bird species, the zebra finch (*Taeniopygia guttata*, Vieillot 1817), (ii) to study whether the distribution of senescence markers correlates with the progression of cell death in the *Taeniopygia guttata* retinal tissue, and (iii) to compare these results with those described in other precocial bird species, such as *Gallus gallus* or *Coturnix japonica*, and in the rest of the vertebrates.

2. Materials and Methods

2.1. Animal and Tissue Processing

All animals were treated according to the regulations and laws of the European Union (EU Directive 2010/63/EU) and Spain (Royal Decree 53/2013). A total of twenty-seven *T. guttata* embryos and twelve hatchlings were used in the present study (Table 1). Embryos were obtained by incubating eggs in a rotating egg incubator (Masallés S.A., Spain) that was maintained at 37.5 ± 1 °C, 80–90% humidity. The degree of development of the embryos and hatchlings (Figure 1) was determined in accordance with the stages (St) established by by [51]. Embryos and hatchlings were fixed with paraformaldehyde (PFA)

4% in phosphate-buffered solution (PBS) (0.1 M, pH 7.4) overnight at 4 °C. For histological analysis with toluidine blue staining, some fixed embryos were dehydrated in a graded series of acetone and propylene oxide and embedded in Spurr's resin. Serial frontal 3 µm sections were cut in a Reichert Jung microtome.

For the histochemical and immunohistochemical procedures, embryos and hatchlings were immersed overnight in a cryoprotective solution (15% sucrose in PBS) at 4 °C, soaked in embedding medium, and frozen. Cryosections of 20 µm were obtained in a cryostat microtome (Leica CM 1900, Charleston, SC, USA), thaw mounted on SuperFrost Plus slides, air dried, and stored at 20 °C.

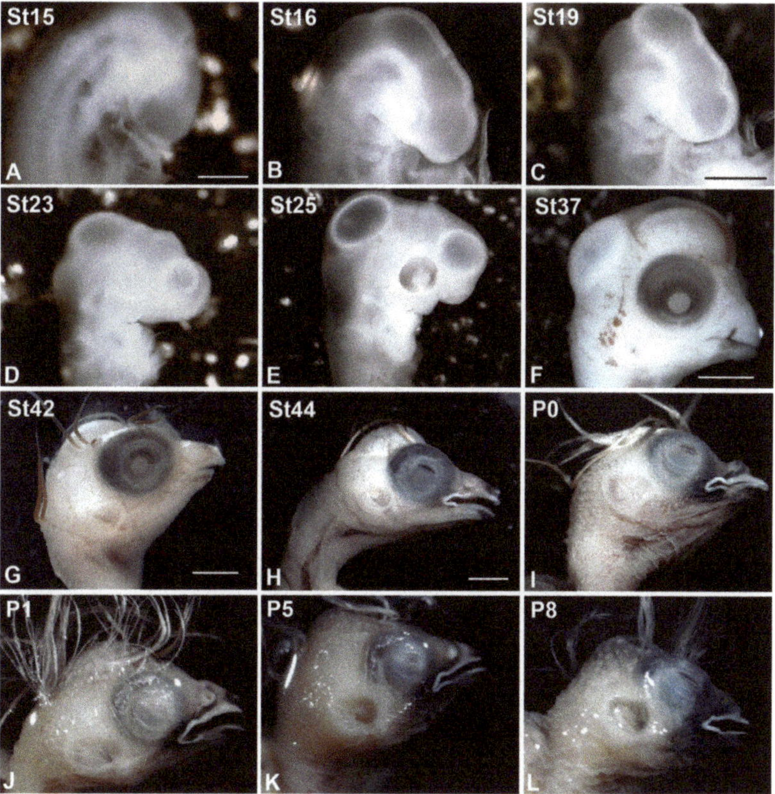

Figure 1. Stereomicroscope images of some embryos and postnatal specimens of *Taeniopygia guttata* showing the external morphological changes of the eye. The embryos were staged in accordance with the developmental stages (St) established by [51]. The optic cup was distinguishable between St15 and St23 (**A–D**). Pigmentation in the RPE was observed at St25 (**E**). At St37, the eye was completely pigmented (**F**). From St42 until perinatal stages, the eyelids progressively covered the eye (**G–J**). Eyelids were closed at P5 (**K**), but slightly open at P8 (**L**). Scale bars: 2 mm (**A,B**); 3 mm (**C–E**); 6 mm (**F**); 7 mm (**G**); 10 mm (**H–L**).

Table 1. *T. guttata* embryos and hatchlings used in the present study.

Stage	*n*	Incubation Time (Approximate)
St11	2	54 h
St15	3	66 h
St16	3	3 days
St19	3	3.5 days
St20	3	3.5 days
St24	3	4.5 days
St25	3	5 days
St37	3	8 days
St42	3	11 days
St44	3	13 days
P0	3	14 days
P1	3	15 days
P5	3	19 days
P8	3	22 days

2.2. Toluidine Blue Staining

Morphological analysis of development of cell death was conducted on resin sections stained with toluidine blue 0.5% and sodium tetraborate 0.5% solution. For this purpose, slides were put in the colorant at 90 °C for 45 s and then rinsed with distilled water. Sections were mounted with Eukitt (Kindler, Freiburg, Germany).

2.3. Detection of β-Galactosidase Activity

We followed the protocol described by [52]. Cryosections were incubated in 450 µL of chromogenic SA-β-GAL substrate X-gal (5-bromo-4-chloro-3-indolyl-β-D-galactopyranoside) in $PBS-MgCl2$ at pH 6.0 at 37.5 °C for 24 h. A blue-green precipitate was developed by $SA-\beta- GAL$-positive cells. Then, sections were washed in $PBS -MgCl2$ acid buffer for 10 min. After histochemical reaction, some of the sections were counterstained with DAPI (Sigma-Aldrich, Madrid, Spain, Ref. D9542) and others were used to perform immunohistochemical analyses. Slides were rinsed in PBS and mounted with Mowiol (Polyvinyl alcohol 40–88, Fluka, Madrid, Spain, Ref. 81386).

2.4. Immunohistochemistry

After histochemical analyses to detect β-galactosidase activity, slides were subjected to an antigen retrieval process with citrate buffer (pH 6) at 90 °C or 30 min. Sections were chilled at RT for 20 min. Slides were washed several times in 0.1% Triton-X-100 in PBS (PBS-T) and pre-blocked in 0.2% gelatin, 0.25% Triton-X-100, and Lys 0.1M in PBS (PBS-G-T-L) for 1 h.

Sections were incubated with mouse anti-p21 monoclonal antibody (1:200, Abcam, Madrid, Spain, ab109199) overnight at RT in a humidified chamber. The day after, slides were washed several times in PBS-T and PBS-G-T and incubated with Alexa Fluor 488 goat anti-mouse IgG antibody (1:200, Molecular Probes, Eugene, OR, USA, A11029) for 2 h at RT in a humidified chamber in darkness. Sections were washed several times in PBS-T and PBS-G-T in darkness and incubated for 10 min with DAPI at RT, followed by two washes in PBS. Slides were mounted with Mowiol.

2.5. TUNEL Technique

The TUNEL technique (Tdt-mediated dUTP Nick End Labeling, Sigma-Aldrich, Madrid, Spain, Cat. No. 11 684 795 910), described by [53], is the histochemical technique commonly used to detect apoptotic nuclei. Cryosections were washed in PBS for 15 min at RT and incubated in 10 µg/mL of proteinase K in PBS for 10 min at 37 °C. The slides were then washed in PBS and incubated in blocking solution (3% H_2O_2 in PBS) for 15 min. Subsequently, sections were washed several times in PBS and then incubated for 60 min at 37 °C with TUNEL reaction mixture, consisting of the enzyme terminal deoxynucleotidyl transferase (TdT) and fluorescein-conjugated nucleotides in a reaction buffer. After rinsing in PBS, sections were incubated in blocking solution (PBS-G-T-L) and covered with the HRP-conjugated anti-fluorescein antibody solution. The apoptotic nuclei were visualized using DAB as a chromogen. The sections were then washed thrice in PBS, dehydrated, and mounted with Eukitt® (Kindler, Freiburg, Germany) for observation. In control sections in which the enzyme TdT was absent from the reaction solution, no stained nuclei were observed.

2.6. Quantification of TUNEL-Positive Nuclei

Quantification was performed by counting all TUNEL-positive nuclei in micrographs of the central region of the retina. The surface area of the retina in digital microphotographs was measured using the ImageJ free open-source software package (http://rsb.info.nih.gov/ij/ accessed on 28 January 2021). The density profiles were expressed as the mean $\pm$ sem of the number of apoptotic nuclei per square millimeter (an/mm^2). Similar procedures have been described in the literature [23,34,36]. Statistical analyses were performed using Student's two-tailed t-test. Differences between groups were considered as significant (*) when $p < 0.05$ and (**) when $p < 0.01$.

2.7. Image Acquisition and Processing

Toluidine blue-stained, TUNEL, and $SA-\beta-GAL$ and immunofluorescence sections were observed with a bright-field and epifluorescence Nikon Eclipse 80i microscope and photographed using an ultra-high definition Nikon DXM1200F digital camera. Images were processed with Adobe Photoshop CS4.

3. Results

3.1. Programmed Cell Death in the Developing T. guttata Visual System

In order to identify dying cells in the developing *T. guttata* visual system, we used some of the methods for detecting PCD in embryonic tissues [3]. Light microscopy observation of toluidine blue-stained semi-thin sections revealed pyknotic bodies in the ganglion cell layer (GCL) and in the inner nuclear layer (INL) of the retinal tissue at the hatching day (P0) (Figure 2A–D). Cryosections labeled with DAPI staining identified nuclear condensation in the laminated retina (Figure 2E,E'). Abundant TUNEL-positive nuclei were observed both in the GCL and in the INL (Figure 2F), but also in other eye tissues, such as the lens (Figure 2G) where DNA of cells of the equatorial zone breaks down due to nuclear endodeoxyribonuclease activity [54]. Therefore, PCD was intense and clearly detected in the developing *T. guttata* visual system.

The distribution of pyknotic nuclei and TUNEL-positive bodies was carefully examined from stage 11 (St11), coinciding with the formation of the optic vesicle [48,51], to postnatal day 8 (P8), the last postnatal stage considered in the present study. Pyknotic bodies were absent from the optic anlage from St11 to St14 (not shown). At St15, when the lateral wall of the optic vesicle invaginates to form the optic cup, abundant pyknotic bodies were found in the central undifferentiated neural retina (Figure 3A,B). Moreover, dead cell fragments were observed in two groups of neuroepithelial cells located on either side of the presumptive optic chiasm (Figure 3A,C). Similar areas of cell degeneration have been described in the chicken embryo, the so-called sub-optic necrotic centers (SONCs) [40,41]. The distribution of PCD was similar at St16 in the neuroretina (Figure 3D–G), but the presence of pyknotic bodies in the SONCs (Figure 3E–G) increased notably. Furthermore, pyknotic bodies were also detected in the anterior wall of the lens anlage (Figure 3D).

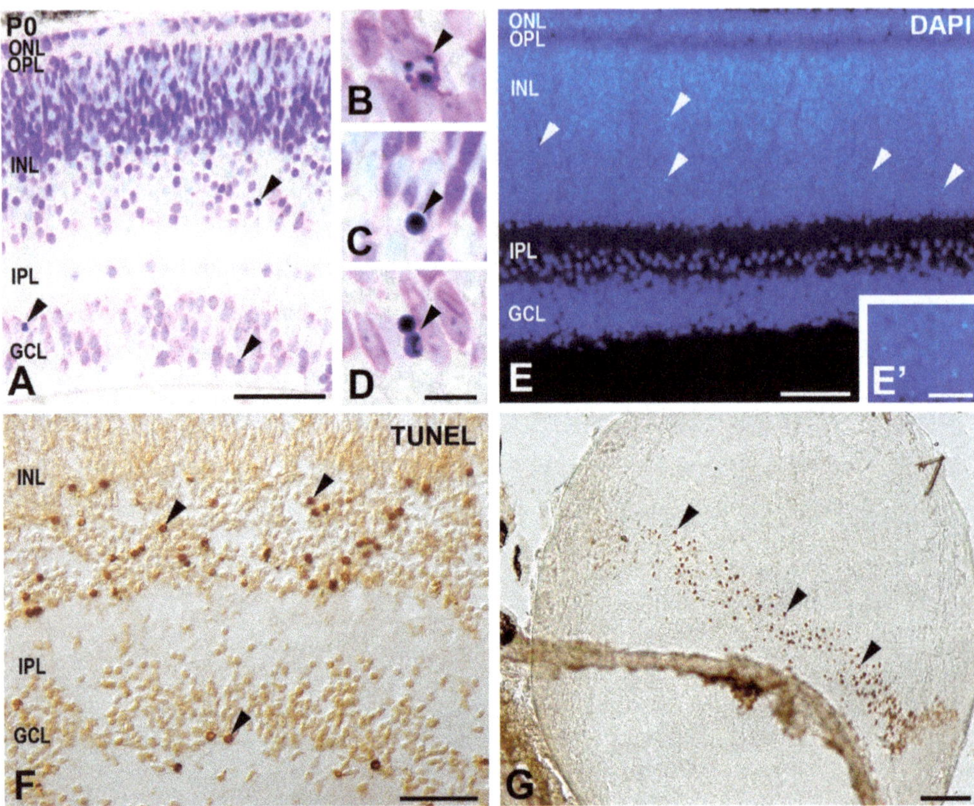

Figure 2. Programmed cell death in the *T. guttata* retina detected by using various sensitive methods. (**A–D**) Transversal semi-thin section of the P0 retina showing pyknotic bodies with morphological features typical of apoptosis after toluidine blue staining. (**E,E′**) Identification of neuronal cell death in the ganglion cell layer (GCL) and in the inner nuclear layer (INL) (arrowheads) in cryosections of *T. guttata* retinas at P0 stained with DAPI. (**F,G**) Eye cryosections of a P0 *T. guttata* hatchling showing intense abundant TUNEL-positive bodies in the GCL and INL (arrowheads in (**C**)) and in the equatorial region of the lens (arrowheads in (**D**)). Abbreviations: GCL, ganglion cell layer; INL, inner nuclear layer; IPL, inner plexiform layer; ONL, outer nuclear layer; OPL, outer plexiform layer. Scale bars: 50 μm (**A,E–G**), 7 μm (**B–D,E′**).

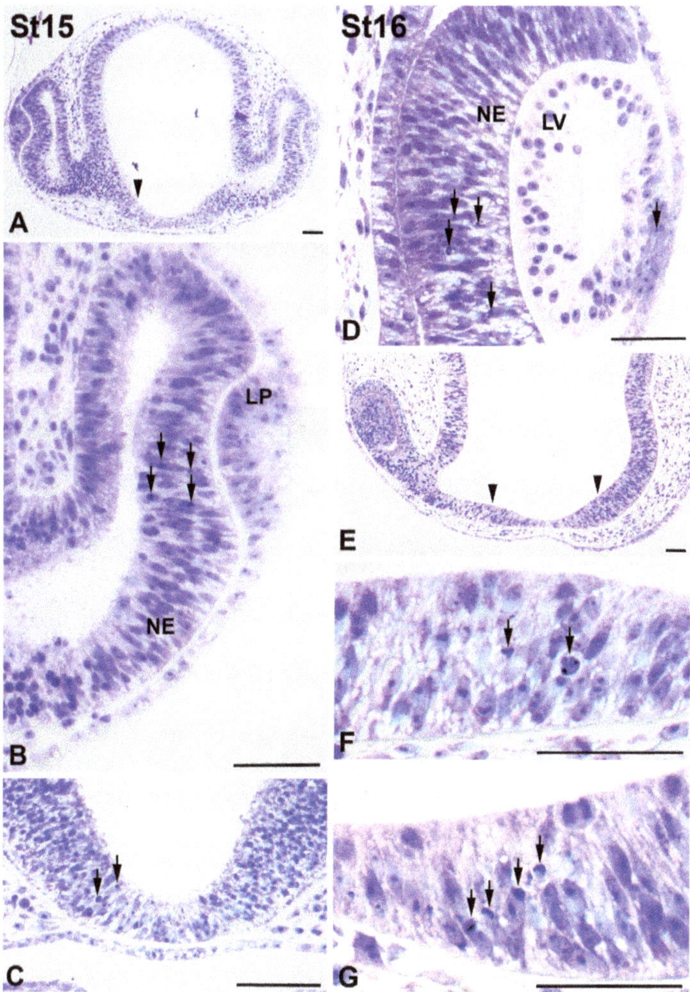

Figure 3. Pyknotic fragments during visual system development in *T. guttata*. Toluidine blue-stained semi-thin sections were obtained from the heads of embryos at different stages of development. Pyknotic bodies were mainly located in the central neural retina ((**A**), arrows in (**B**)) and in the sub-optic necrotic centers (SONCs) (arrowhead in (**A**), arrows in (**C**)) at St15 in the early optic cup. At St16, pyknotic fragments were restricted to the central neural retina, to the anterior wall of the lens vesicle (arrows in (**D**)), and to the SONCs (arrowheads in (**E**), arrows in (**F,G**)). Abbreviations: LP, lens placode; LV, lens vesicle; NE, neuroepithelium. Scale bars: 50 μm.

At St19, sparse pyknotic bodies were detected in the anterior wall of the lens vesicle (Figure 4A). Pyknotic bodies were still detected in the SONCs (Figure 4B,C). The first differentiating retinal neuroblasts in *T. guttata* appeared by St24 [48,49]. At this stage, pyknotic bodies were concentrated in the NbL in a region located dorsally to the optic nerve head (Figure 4D,E). PCD was also detected in the presumptive retinal pigment epithelium (pRPE), adjacent to the region of the distal optic nerve (Figure 4F,G). At St25, pyknotic bodies were concentrated at the level of the distal optic nerve (Figure 4H,I). From St26 (not shown) to St36, pyknotic bodies were sparsely observed, randomly localized throughout the NbL (Figure 4J–L).

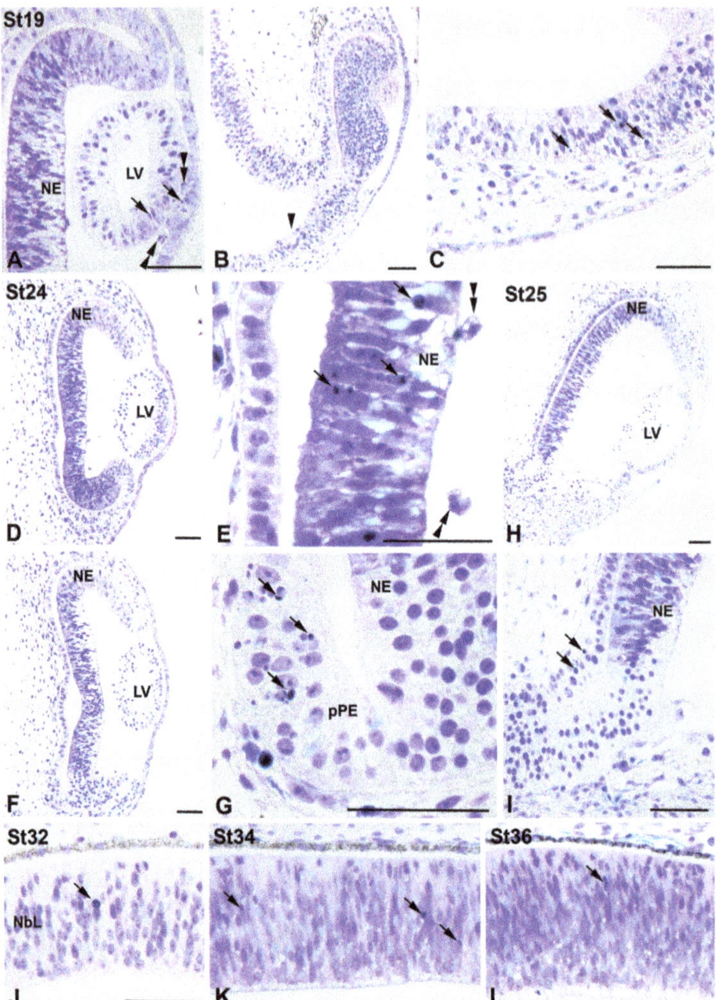

Figure 4. Pyknotic bodies during visual system development in *T. guttata*. Toluidine blue-stained semi-thin sections were obtained from the heads of embryos at different stages of development. At E19, pyknotic fragments were mainly detected in the anterior wall of the lens vesicle (arrows in (**A**)), but also in the SONCs (arrowhead in (**B**), arrows in (**C**)). At St24 (**D–G**), pyknotic bodies were concentrated in retinal regions located dorsally to the optic nerve head (arrows in (**E**)) and in the presumptive pigment epithelium located surrounding the optic nerve head (arrows in (**G**)). At St25 (**H,I**), abundant pyknotic fragments were detected in the dorsal region of the distal optic nerve (arrows in (**I**)). Pyknotic bodies were sparse and dispersed throughout the neuroblastic layer (NbL) by St32 (arrow in (**J**)), St34 (arrows in (**K**)), and St 36 (arrow in (**L**)). Abbreviations: LV, lens vesicle; NbL, neuroblastic layer; NE, neuroepithelium; pRPE, presumptive retinal pigment epithelium. Scale bars: 50 μm.

At St37, scattered TUNEL-positive nuclei were found dispersed throughout the NbL (Figure 5A), similar to the distribution of pyknotic nuclei described from St26 to St36. At St42, retinal stratification was evident, and a few TUNEL-positive nuclei were observed in the GCL and in the INL (Figures 5B and 6). The incidence of cell death rose significantly in the GCL between St42 and St44 (Figures 5C and 6) (2 days before hatching), reaching the highest values in this layer by this stage (Figure 6). At P0, the density of TUNEL-positive nuclei in the GCL diminished (Figures 5D and 6), but increased significantly in the INL, reaching a peak at P5 (Figures 5E and 6). At P8, the last stage analyzed, there was a high incidence of cell death in the INL (Figures 5F and 6), but TUNEL-positive nuclei almost disappeared from the GCL (Figure 5F), reaching values close to 0 in this layer (Figure 6).

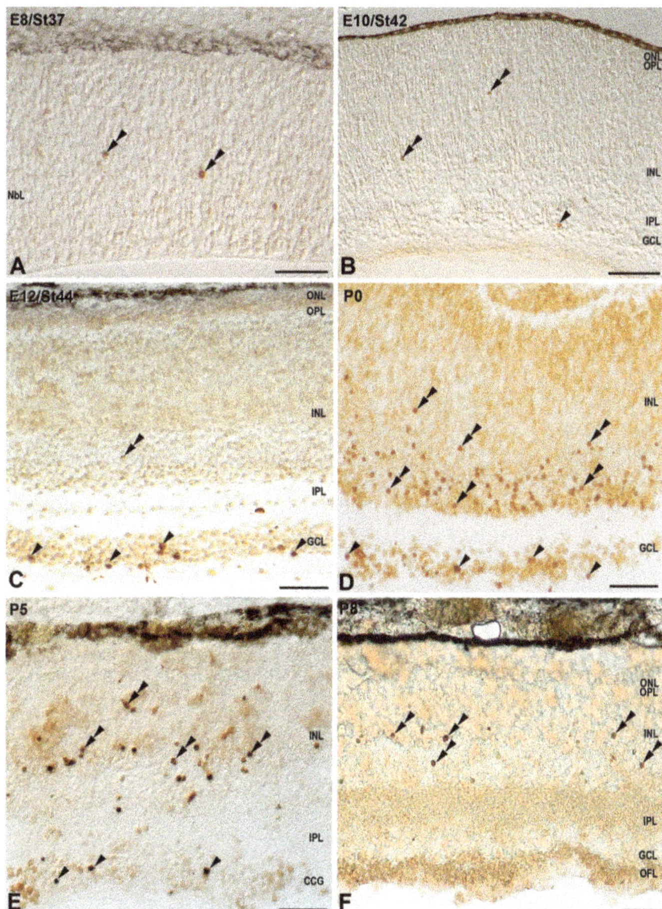

Figure 5. Spatial distribution of TUNEL-positive nuclei in the developing retina of *T. guttata*. Retinal cryosections of embryos and postnatal specimens were treated in accordance with this histochemical technique. Sparse randomly distributed TUNEL-positive nuclei were detected in the NbL at St37 (double arrowheads in (**A**)). At St42, sparse TUNEL-positive nuclei were detected both in the GCL (arrowhead in (**B**)) and in the INL (double arrowheads in (**B**)). TUNEL-positive nuclei were mainly detected in the GCL at St44 (arrowheads in (**B**)), but also in the INL (double arrowhead in (**C**)). TUNEL-positive nuclei progressively diminished from P0 to P8 in the GCL (arrowheads in (**D,E**)), but they increased markedly from P0 to P5 in the INL (double arrowheads in (**D,E**)). At P8, TUNEL-positive nuclei in the INL were less abundant than observed at previous stages (double arrowheads in (**F**)). Abbreviations: GCL, ganglion cell layer; INL, inner nuclear layer; IPL, inner plexiform layer; NbL, neuroblastic layer; ONL, outer nuclear layer; OPL, outer plexiform layer. Scale bars: 50 μm.

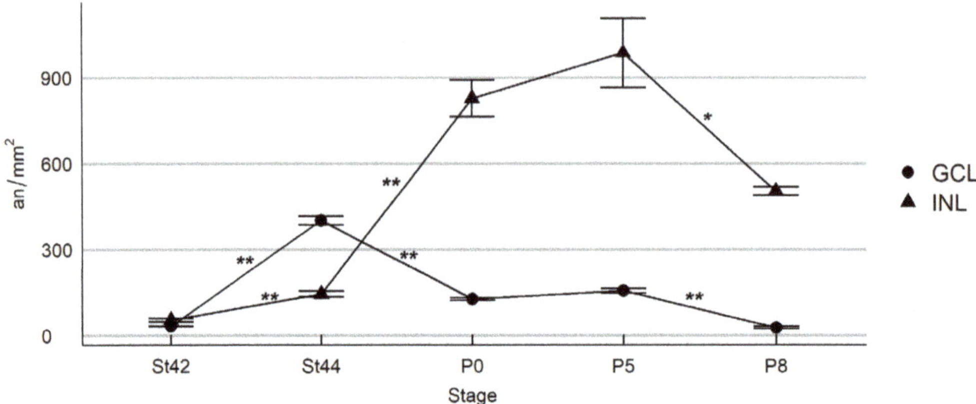

Figure 6. Quantitative analysis of the density of TUNEL-positive nuclei during *T. guttata* retinal histogenesis. The incidence of cell death rose significantly in the GCL between St42 and St44, reaching the highest values in this layer by this latter stage. From P0 onwards, the density of TUNEL-positive nuclei in the GCL diminished progressively. The INL contained a low density of TUNEL-positive nuclei between St42 and St44. A significant increase in the density of TUNEL-positive nuclei was observed at P0, reaching a maximum at P5. At P8, there was a high incidence of cell death in the INL. Abbreviations: an/mm², apoptotic nuclei per square millimeter; GCL, ganglion cell layer; INL, inner nuclear layer. Asterisks correspond to p values; * $p < 0.05$ and ** $p < 0.01$.

At late embryonic stages (St44) (Figure 7A,B) and at P0 (Figure 7C–F), TUNEL-labeling was occasionally detected in the cell somata and in fine processes of radially oriented cells with an apparent intact healthy morphology (Figure 7A–D). Some of the vitreal TUNEL-positive processes form endfeet that seemed to be anchored to the inner limiting membrane ILM (Figure 7A,B). In semi-thin sections, pyknotic bodies were found radially aligned in the cytoplasm of cell processes (Figure 7E,F).

Finally, it is important to note that cell death was completely absent from the ONL during all the embryonic stages and postnatal ages analyzed. Furthermore, the chrono-topographical distribution of TUNEL-positive nuclei in the developing *T. guttata* retinal tissue from St42 onwards followed central-to-peripheral and vitreal-to-scleral gradients, in concordance with the gradients of cell differentiation described in this altricial bird species [48,55].

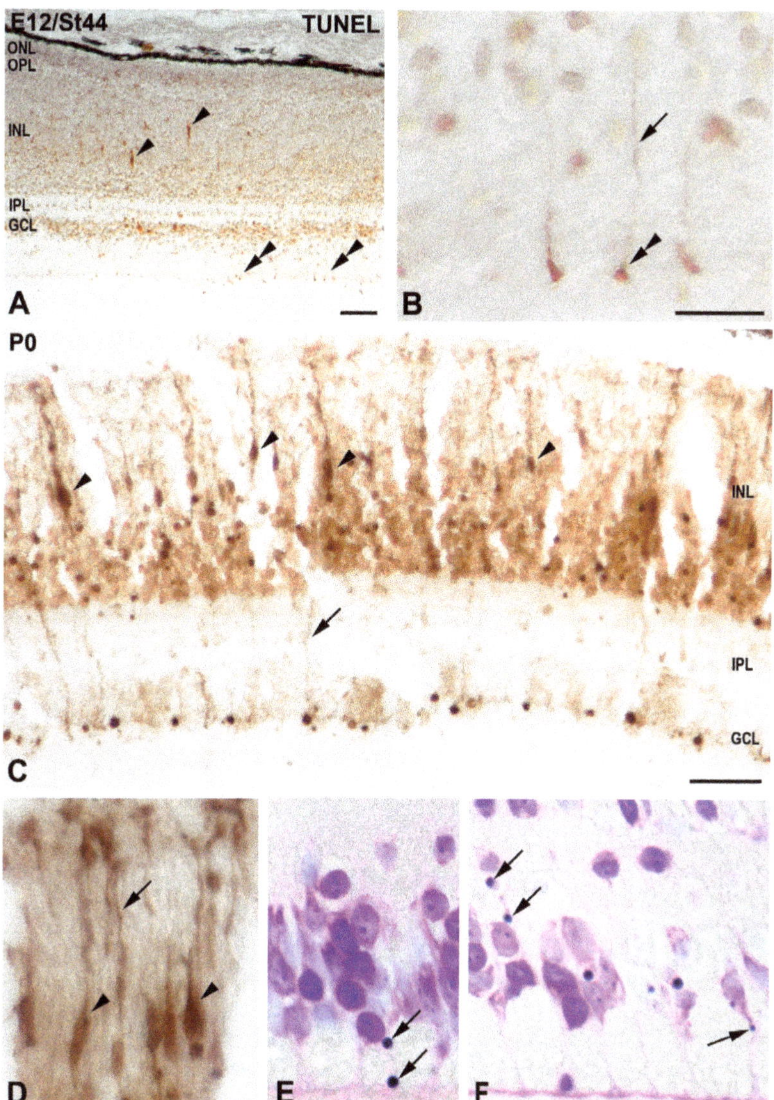

Figure 7. Spatial distribution of TUNEL-positive elements in the developing retina of *T. guttata*. Retinal cryosections of embryos (St44: (**A,B**)) and newly hatched chicks (P0: (**C,D**)) were treated in accordance with this histochemical technique. Elongated cell somata located in the INL (arrowheads in (**A,C,D**)) and fine processes (arrows in (**B–E**)) of radially oriented cells were diffusely labeled with this technique. Occasionally, TUNEL-positive Müller cell endfeet were labeled in the vitreal surface of the retina (double arrowheads in (**A,B**)). Semi-thin sections treated according to the toluidine blue technique revealed small pyknotic bodies within the cytoplasm of Müller cells (arrows in (**E,F**)). Abbreviations: GCL, ganglion cell layer; INL, inner nuclear layer; IPL, inner plexiform layer; ONL, outer nuclear layer; OPL, outer plexiform layer. Scale bars: 50 μm in (**A,C**); 10 μm in (**B,D–F**).

3.2. Senescence Markers in the Developing T. guttata Visual System

Retinal cryosections of zebra finch embryos and hatchlings were stained with SA-β-GAL histochemistry and examined for the appearance of positively stained cells. At St34, the vitreal-most region and the scleral surface of the central NbL appeared faintly stained with SA-β-GAL histochemistry (Figure 8A,B). In contrast, SA-β-GAL staining was mainly detected in the scleral region of the peripheral rim of the retina (Figure 8C,D). The staining pattern of SA-β-GAL changed with the appearance of plexiform layers. At St43, SA-β-GAL labeling was mainly detected in the GCL, amacrine cell layer, and horizontal cell layer (Figure 8E,F). Double labeling with antibodies against p21 (inhibitor of cyclin-dependent kinases), which has been demonstrated to be overexpressed in senescent cells during embryonic development [1,2,4], showed a strong parallelism between the SA-β-GAL signal and p21 immunoreactivity (Figure 8E–G). The same staining patterns were detected in the retina of *T. guttata* hatchlings (Figure 8H–J).

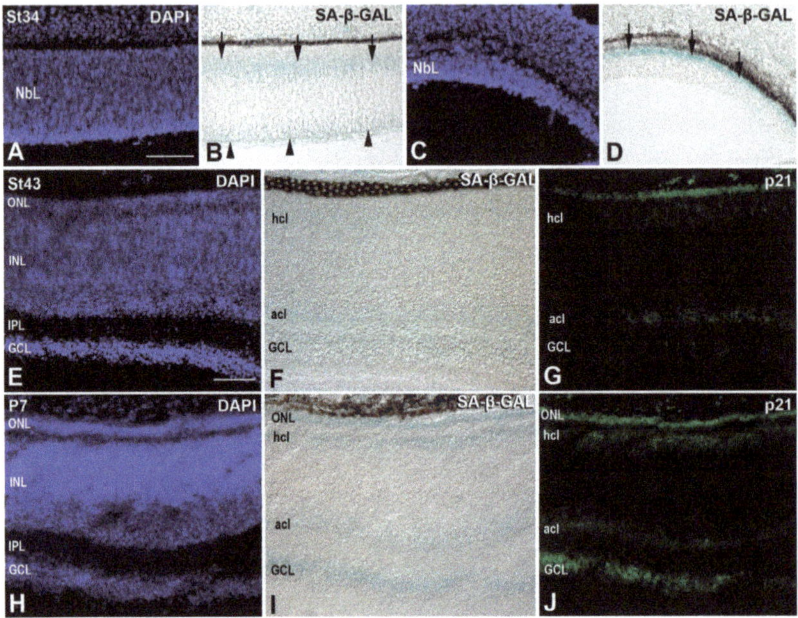

Figure 8. Distribution of SA-β-GAL labeling and p21 immunoreactivity in retinal cryosections of embryos (**A–G**) and post-hatched specimens (**H–J**) of *T. guttata*. All sections were counterstained with DAPI. DAPI staining showed that at St34, the retinal tissue comprised an NbL (**A,C**). SA-β-GAL activity presented two bands of labeling located in the vitreal and scleral regions of the NbL in the central retina (**B**), but in a single band located sclerally in the peripheral retina (**D**). At St43 and P7, DAPI staining revealed the central retina to present a multi-laminated structure (**E,H**). SA-β-GAL activity was mainly detected in the GCL, amacrine cell layer, horizontal cell layer, and photoreceptor cell layer (**F,I**). The p21 immunosignal was highly coincident with SA-β-GAL staining (**G,J**). Abbreviations: acl, amacrine cell layer; GCL, ganglion cell layer; hcl, horizontal cell layer; INL, inner nuclear layer; IPL, inner plexiform layer; NbL, neuroblastic layer; ONL, outer nuclear layer; OPL, outer plexiform layer. Scale bars: 50 μm.

These staining patterns of cell senescence markers were homogeneous throughout the GCL, amacrine, and horizontal cell layers. Furthermore, TUNEL-positive bodies in the horizontal cell layer were almost absent. Therefore, PCD and senescence markers did not correlate in the developing bird retina.

4. Discussion

We have presented details of the distribution of pyknotic bodies and TUNEL-positive nuclei during development of the visual system in the altricial bird species *T. guttata*. Previous work in our laboratory has shown that these are effective methods for the detection of dying cells in the developing visual system of vertebrates (for a review, see [3]).

To the best of our knowledge, the present study provides the first description of the spatiotemporal distribution of dying cells in an altricial bird species. Furthermore, in order to find any possible coincidence between apoptotic and senescent cells in the developing visual system, we also labeled retinal cryosections with SA-β-GAL histochemistry and p21 immunohistochemistry. All the results will be discussed below.

4.1. Cell Death during Early Visual System Morphogenesis in T. guttata

During optic cup stages, abundant pyknotic bodies were found in the central region of the neural retina, coinciding with previous results described in the chicken [32,33,35] and in the mouse [30,31,35]. This wave of PCD may be involved in shaping the optic cup [3].

With respect to the *T. guttata* lens vesicle, pyknotic bodies appeared during detachment of this structure from the head ectoderm, coinciding with results described in all vertebrates studied [19,27,31,56,57]. In this case, cell death seems to be involved in eliminating cells in the interface between the ectoderm and lens tissue, facilitating the separation of the lens vesicle.

Finally, abundant pyknotic nuclei were detected in the SONCs, areas of intense cell degeneration located laterally to the ventral midline of the diencephalon in the chicken [40,41] and in the mouse [31]. SONCs were detected between St15 and St20. This wave of cell death preceded the arrival of ganglion cell axons at the presumptive optic chiasm and therefore seems to be involved in the invasion of pioneer axons in this region of the visual system.

4.2. Cell Death during the Period of Cell Differentiation in the T. guttata Retina

At St24 (E4.5), coinciding with the appearance of the first differentiated neuroblasts in the *T. guttata* retinal tissue [49], pyknotic nuclei were found in the central retina, dorsally to the optic nerve head. At this stage, cell death affects mainly some proliferating neuroepithelial cells and recent newborn neuroblasts, coinciding with the emergence of the pioneer ganglion cell axons [33,36–38]. This wave of cell death (known as "early neural cell death") could be involved in the creation of extracellular channels that facilitate axonal guidance during early stages of ganglion cell differentiation (for reviews, see [3,58]).

An area of cell death was also detected by St25 in the distal optic nerve, at the junction of this structure with the rudiment of the eye. A similar area of degeneration has been described in the small-spotted catshark, *Scyliorhynus canicula* [19], at stages prior to the invasion of the ganglion cell axons. Neurotrophic cell death affected differentiated neurons in the layered *T. guttata* retina. The emergence of the plexiform layers occurred between St38 (E8.5) and St39 (E9) [48], but the presence of TUNEL-positive bodies was sparse until St42 (E11). At St44 (E13), the incidence of cell death in the GCL increased abruptly, reaching a peak by this stage. In contrast, the maximum of cell death density in the INL was reached at P5, indicating a vitreal-to-scleral progression of cell death, similar to the vitreal-to-scleral wave of cell differentiation described in this bird species [48,49].

These results also reveal marked differences in the timing of visual system maturation between altricial and precocial bird species (Figure 9). Neurotrophic cell death in the GCL occurs in the quail in the period E8–E14, peaking at E10 [39], while in the chicken, it takes place in the period E8–E15, also peaking at E10 [34]. In contrast, dying ganglion cells are detected in *T. guttata* from embryonic stages (St42–E10.5) to a post-hatching period (P8), peaking at St44 (E12). In the case of the INL, cell death extends from E8 to P1 in the quail, peaking at E12 [39], and from E8 to E19 in the chicken, peaking at E11 [34]. In the present study, we have shown that cell death in the INL is detected from St42 (E10.5) to at least P8, the last stage analyzed in the present study, peaking at P5. Therefore, the highest incidence of cell death in the *T. guttata* INL occurred in the post-hatching period, suggesting that most of the synapses established between retinal cells located in this nuclear layer occur during the first week of life. This is a very interesting finding which suggests that, during early post-hatching life, the retinal tissue is still immature and is unable to process the light information it receives.

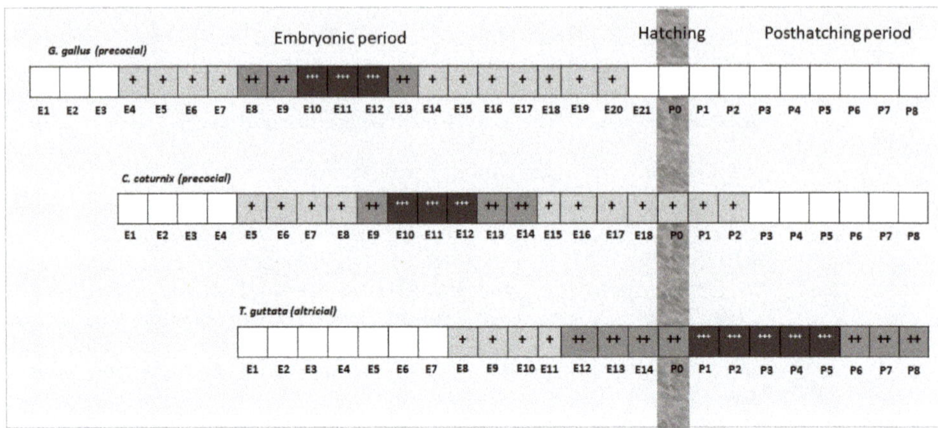

Figure 9. Schematic summary of the chronological patterns and the intensity of neurotrophic cell death in the developing retina of *G. gallus* [34], *C. coturnix* [39], and *T. guttata* (present study). Neurotrophic cell death occurred in the altricial bird at perinatal stages and extended through the first week of life. In contrast, it was restricted to the embryonic period in both of the precocial species. Color codes: white (absence of cell death); light gray (low levels of cell death density) (+); gray (moderate levels of cell death density) (++); dark gray (high levels of cell death density) (+++).

Previous studies in our laboratory have shown that mitotic activity is intense during the first postnatal week in the retina of this altricial species [50], reinforcing the idea of the immature state of this tissue during early life. Indeed, *T. guttata* hatchlings open their eyes at P7 [59], coinciding with a decrease in the incidence of cell death in the retina.

These differences in the timing of ontogenetic cell death between altricial and precocial species have been found in all vertebrates studied. The main waves of cell death occur during the embryonic period in precocial fish [16,19], reptiles [23–25], and birds [33,34,39,40]. In contrast, cell death takes place mainly after hatching/birth in altricial fish [18], birds (present study), and most of the mammals studied [26,29,31,55].

4.3. TUNEL Labeling in the Cytoplasm of Radially Oriented Cells

Diffuse TUNEL-labeling was also found in the cytoplasm of cells that have a bipolar morphology in the radial plane. Their somas were located at the center of the INL, from which radially oriented processes emerge to span the thickness of the neuroretina. Similar results have been described in the developing retina of fish [19], reptiles [23], birds [14], and mammals [60]. Similar staining following retinal injury has also been described in the retina of fish [61–63] and mammals [64,65]. These radially oriented TUNEL-positive cells were also GS-immunoreactive [62,63,65,66]. The morphology and immunochemical profiles of these labeled cells coincided with those described for Müller cells [67]. Müller glia possess phagocytic activity to remove degenerating cells during development or under experimental conditions (reviewed in [66]). This cytoplasmic labeling is due to the engulfment of TUNEL-positive cell debris by the phagocytic Müller cells.

4.4. Senescence Markers in the Developing Retina of T. guttata

Cellular senescence occurs in different embryonic tissues during restricted time windows, in most cases contributing to degeneration of the interdigital mesoderm [4,6], pronephros [9], mesonephros [1], and developing heart [7] or inner ear [12–14] structures. SA-β-GAL histochemistry is widely used as a biomarker of cellular senescence in vivo and in vitro [42], even in whole-mount embryos [1,2,4,7,9,11]. Most of these works report that SA-β-GAL labeling strongly correlates with areas of cell death. The developing visual system of vertebrates is also affected by several waves of cell death (for a review, see [3]), which we also detected in the *T. guttata* visual system (see above). However, we found no correlation of the labeling pattern of SA-β-GAL activity with the TUNEL-positive nuclei detected in the developing retina, in concordance with previous results obtained in the developing chicken retina [43,44]. In this sense, we clearly demonstrated that SA-β-GAL activity was restricted to several subpopulations of differentiated neurons (ganglion, amacrine, and horizontal cells) in the embryonic *T. guttata* retina.

Furthermore, the establishment of the state of cell senescence in embryos is associated with the expression of anti-proliferative mediators, such as p21 that seems to act independently of p53 [1,2]. It has been described that p21 expression in mouse embryos strongly correlates with known locations of developmental senescence [68]. In the present study, p21 immunoreactivity faithfully correlates with SA-β-GAL labeling, similar to results described in the developing chicken eye [43,44]. Therefore, the present work has clearly shown that the expression of typical senescence markers, including SA-β-GAL and p21, in the developing bird retina is up-regulated in subpopulations of differentiated neurons. Notably, both markers have been found to be highly expressed by the first differentiating retinal neurons in the chicken [43,44]. These data indicate that senescence is not the only developmental event that can increase SA-β-GAL activity and p21 expression in embryonic tissues. Senescent cells and differentiated retinal neurons share a common biological feature—they are in a characteristic non-proliferative state. Therefore, SA-β-GAL activity and p21 could be involved in distinct biological phenomena such as cell senescence and terminal cell differentiation of neurons. In this sense, typical senescence markers have been found to be associated with cell differentiation in the developing tendons [6] and the maturing ventricular myocardium of embryonic mice [7]. However, the possible relationship between the mechanistic events involved in cell senescence and terminal cell differentiation remains to be clarified.

5. Conclusions

Relative to precocial bird species, in altricial species, some aspects of brain maturation such as telencephalic neurogenesis are delayed into the post-hatching period [69–73]. Retinal neurogenesis is intense in altricial birds at hatching [48,49] and during the first week of life [50]. Furthermore, it has been demonstrated [74] that the formation of some retinal structures, the foveal pit in particular, is delayed until the second week of life (P10–P14). In the present study, we have demonstrated that there is intense ontogenetic cell death in the retina of the hatched animals. Thus, *T. guttata* constitutes an excellent model in which to study retinal development events during the first weeks of life.

Author Contributions: G.M.-P., J.R.-L. and J.F.-M. designed research; A.M. and Y.G. analyzed data; G.Á.-H., J.A.d.M.-R. and I.H.-N. performed research; J.R.-L. and J.F.-M. wrote the paper. All authors have read and agreed to the published version of the manuscript.

Funding: G.A.-H. was a recipient of a pre-doctoral studentship from the Universidad de Extremadura. This work was supported by grants from Dirección General de Investigación del Ministerio de Educación y Ciencia (BFU2017-85547-P), and Junta de Extremadura, Fondo Europeo de Desarrollo Regional, "Una manera de hacer Europa" (GR15158, GR18114, IB18113).

Institutional Review Board Statement: The study was conducted according to the guidelines of the Declaration of Helsinki, and approved by the Ethics Committee of the University of Extremadura (protocol code 264/2019, 29 June 2020).

Informed Consent Statement: Not applicable.

Data Availability Statement: Some or all data used during the study are available from the corresponding author by request.

Conflicts of Interest: The authors declare no conflict of interest.

References

1. Muñoz-Espín, D.; Cañamero, M.; Maraver, A.; Gómez-López, G.; Contreras, J.; Murillo-Cuesta, S.; Rodríguez-Baeza, A.; Varela-Nieto, I.; Ruberte, J.; Collado, M.; et al. Programmed Cell Senescence during Mammalian Embryonic Development. *Cell* **2013**, *155*, 1104–1118. [CrossRef]
2. Storer, M.; Mas, A.; Robert-Moreno, A.; Pecoraro, M.; Ortells, M.C.; Di Giacomo, V.; Yosef, R.; Pilpel, N.; Krizhanovsky, V.; Sharpe, J.; et al. Senescence Is a Developmental Mechanism that Contributes to Embryonic Growth and Patterning. *Cell* **2013**, *155*, 1119–1130. [CrossRef] [PubMed]
3. Francisco-Morcillo, J.; Bejarano-Escobar, R.; Rodríguez-León, J.; Navascués, J.; Martín-Partido, G. Ontogenetic Cell Death and Phagocytosis in the Visual System of Vertebrates: Cell Death and Phagocytosis during Ontogeny. *Dev. Dyn.* **2014**, *243*, 1203–1225. [CrossRef]
4. Lorda-Diez, C.I.; Montero, J.A.; Garcia-Porrero, J.A.; Hurle, J.M. Interdigital tissue regression in the developing limb of vertebrates. *Int. J. Dev. Biol.* **2015**, *59*, 55–62. [CrossRef]
5. Montero, J.A.; Sanchez-Fernandez, C.; Diez, C.I.L.; Garcia-Porrero, J.A.; Hurle, J.M. DNA damage precedes apoptosis during the regression of the interdigital tissue in vertebrate embryos. *Sci. Rep.* **2016**, *6*, 35478. [CrossRef] [PubMed]
6. Montero, J.A.; Lorda-Diez, C.I.; Hurle, J.M. Confluence of Cellular Degradation Pathways During Interdigital Tissue Remodeling in Embryonic Tetrapods. *Front. Cell Dev. Biol.* **2020**, *8*, 593761. [CrossRef]
7. Lorda-Diez, C.I.; Solis-Mancilla, M.E.; Sanchez-Fernandez, C.; Garcia-Porrero, J.A.; Hurle, J.M.; Montero, J.A. Cell senescence, apoptosis and DNA damage cooperate in the remodeling processes accounting for heart morphogenesis. *J. Anat.* **2019**, *234*, 815–829. [CrossRef] [PubMed]
8. Nacher, V.; Carretero, A.; Navarro, M.; Armengol, C.; Llombart, C.; Rodríguez, A.; Herrero-Fresneda, I.; Ayuso, E.; Ruberte, J. The Quail Mesonephros: A New Model for Renal Senescence? *J. Vasc. Res.* **2006**, *43*, 581–586. [CrossRef] [PubMed]
9. Davaapil, H.; Brockes, J.P.; Yun, M.H. Conserved and novel functions of programmed cellular senescence during vertebrate development. *Development* **2016**, *144*, 106–114. [CrossRef]
10. Villiard, É.; Denis, J.F.; Hashemi, F.S.; Igelmann, S.; Ferbeyre, G.; Roy, S. Senescence gives insights into the morphogenetic evolution of anamniotes. *Biol. Open* **2017**, *6*, 891–896. [CrossRef]
11. Da Silva-Álvarez, S.; Lamas-González, O.; Ferreirós, A.; González, P.; Gómez, M.; García-Caballero, T.; Barcia, M.G.; García-González, M.A.; Collado, M. Pkd2 deletion during embryo development does not alter mesonephric programmed cell senescence. *Int. J. Dev. Biol.* **2018**, *62*, 637–640. [CrossRef]
12. Gibaja, A.; Aburto, M.R.; Pulido, S.; Collado, M.; Hurle, J.M.; Varela-Nieto, I.; Magariños, M. TGFβ2-induced senescence during early inner ear development. *Sci. Rep.* **2019**, *9*, 1–13. [CrossRef]
13. Varela-Nieto, I.; Palmero, I.; Magariños, M. Complementary and distinct roles of autophagy, apoptosis and senescence during early inner ear development. *Hear. Res.* **2019**, *376*, 86–96. [CrossRef] [PubMed]
14. Magariños, M.; Barajas-Azpeleta, R.; Varela-Nieto, I.; Aburto, M.R. Otic Neurogenesis Is Regulated by TGFβ in a Senescence-Independent Manner. *Front. Cell. Neurosci.* **2020**, *14*, 217. [CrossRef]
15. Biehlmaier, O.; Neuhauss, S.C.; Kohler, K. Onset and time course of apoptosis in the developing zebrafish retina. *Cell Tissue Res.* **2001**, *306*, 199–207. [CrossRef]
16. Candal, E.; Anadón, R.; DeGrip, W.J.; Rodríguez-Moldes, I. Patterns of cell proliferation and cell death in the developing retina and optic tectum of the brown trout. *Brain Res. Dev. Brain Res.* **2005**, *154*, 101–119. [CrossRef]
17. Iijima, N.; Yokoyama, T. Apoptosis in the Medaka Embryo in the Early Developmental Stage. *Acta Histochem. Cytochem.* **2007**, *40*, 1–7. [CrossRef] [PubMed]
18. Bejarano-Escobar, R.; Blasco, M.; DeGrip, W.J.; Oyola-Velasco, J.A.; Martín-Partido, G.; Francisco-Morcillo, J. Eye development and retinal differentiation in an altricial fish species, the senegalese sole (*Solea senegalensis*, Kaup 1858). *J. Exp. Zoöl. Part B Mol. Dev. Evol.* **2010**, *314*, 580–605. [CrossRef] [PubMed]
19. Bejarano-Escobar, R.; Blasco, M.; Durán, A.C.; Martín-Partido, G.; Francisco-Morcillo, J. Chronotopographical distribution patterns of cell death and of lectin-positive macrophages/microglial cells during the visual system ontogeny of the small-spotted catshark Scyliorhinus canicula. *J. Anat.* **2013**, *223*, 171–184. [CrossRef] [PubMed]
20. Gaze, R.M.; Grant, P. Spatio-temporal patterns of retinal ganglion cell death duringXenopus development. *J. Comp. Neurol.* **1992**, *315*, 264–274. [CrossRef] [PubMed]
21. Hensey, C.; Gautier, J. Programmed Cell Death duringXenopusDevelopment: A Spatio-temporal Analysis. *Dev. Biol.* **1998**, *203*, 36–48. [CrossRef]
22. Hutson, L.D.; Bothwell, M. Expression and function of Xenopus laevis p75(NTR) suggest evolution of developmental regulatory mechanisms. *J. Neurobiol.* **2001**, *49*, 79–98. [CrossRef] [PubMed]
23. Francisco-Morcillo, J.; Hidalgo-Sánchez, M.; Martín-Partido, G. Spatial and temporal patterns of apoptosis during differentiation of the retina in the turtle. *Anat. Embryol.* **2004**, *208*, 289–299. [CrossRef] [PubMed]
24. Hidalgo-Sánchez, M.; Francisco-Morcillo, J.; Navascués, J.; Martín-Partido, G. Developmental changes in the fibre population of the optic nerve follow an avian/mammalian-like pattern in the turtle Mauremys leprosa. *Brain Res.* **2006**, *1113*, 74–85. [CrossRef]
25. Hidalgo-Sánchez, M.; Francisco-Morcillo, J.; Martín-Partido, G. Changes in fiber arrangement in the retinofugal pathway of the turtle Mauremys leprosa: An evolutionarily conserved mechanism. *Brain Res.* **2007**, *1186*, 124–128. [CrossRef]
26. Young, R.W. Cell death during differentiation of the retina in the mouse. *J. Comp. Neurol.* **1984**, *229*, 362–373. [CrossRef] [PubMed]

27. Knabe, W.; Kuhn, H.J. Pattern of cell death during optic cup formation in the tree shrew Tupaia belangeri. *J. Comp. Neurol.* **1998**, *401*, 352–366. [CrossRef]

28. Knabe, W.; Süss, M.; Kuhn, H.-J. The patterns of cell death and of macrophages in the developing forebrain of the tree shrew Tupaia belangeri. *Anat. Embryol.* **2000**, *201*, 157–168. [CrossRef] [PubMed]

29. Péquignot, M.; Provost, A.C.; Sallé, S.; Taupin, P.; Sainton, K.M.; Marchant, D.; Martinou, J.C.; Ameisen, J.C.; Jais, J.-P.; Abitbol, M. Major role of BAX in apoptosis during retinal development and in establishment of a functional postnatal retina. *Dev. Dyn.* **2003**, *228*, 231–238. [CrossRef] [PubMed]

30. Rodríguez-Gallardo, L.; Lineros-Domínguez, M.D.C.; Francisco-Morcillo, J.; Martín-Partido, G. Macrophages during retina and optic nerve development in the mouse embryo: Relationship to cell death and optic fibres. *Anat. Embryol.* **2005**, *210*, 303–316. [CrossRef]

31. Bejarano-Escobar, R.; Holguín-Arévalo, M.S.; Montero, J.A.; Francisco-Morcillo, J.; Martín-Partido, G. Macrophage and microglia ontogeny in the mouse visual system can be traced by the expression of Cathepsins B and D. *Dev. Dyn.* **2011**, *240*, 1841–1855. [CrossRef] [PubMed]

32. Cuadros, M.A.; Ríos, A. Spatial and temporal correlation between early nerve fiber growth and neuroepithelial cell death in the chick embryo retina. *Anat. Embryol.* **1988**, *178*, 543–551. [CrossRef]

33. Martín-Partido, G.; Rodríguez-Gallaro, L.; Alvarez, I.S.; Navascués, J. Cell death in the ventral region of the neural retina during the early development of the chick embryo eye. *Anat. Rec. Adv. Integr. Anat. Evol. Biol.* **1988**, *222*, 272–281. [CrossRef] [PubMed]

34. Cook, B.; Portera-Cailliau, C.; Adler, R. Developmental neuronal death is not a universal phenomenon among cell types in the chick embryo retina. *J. Comp. Neurol.* **1998**, *396*, 12–19. [CrossRef]

35. Trousse, F.; Esteve, P.; Bovolenta, P. BMP4 Mediates Apoptotic Cell Death in the Developing Chick Eye. *J. Neurosci.* **2001**, *21*, 1292–1301. [CrossRef]

36. Mayordomo, R.; Valenciano, A.I.; De La Rosa, E.J.; Hallböök, F. Generation of retinal ganglion cells is modulated by caspase-dependent programmed cell death. *Eur. J. Neurosci.* **2003**, *18*, 1744–1750. [CrossRef]

37. Chavarría, T.; Valenciano, A.I.; Mayordomo, R.; Egea, J.; Comella, J.X.; Hallböök, F.; De Pablo, F.; De La Rosa, E.J. Differential, age-dependent MEK-ERK and PI3K-Akt activation by insulin acting as a survival factor during embryonic retinal development. *Dev. Neurobiol.* **2007**, *67*, 1777–1788. [CrossRef]

38. Chavarría, T.; Baleriola, J.; Mayordomo, R.; De Pablo, F.; De La Rosa, E.J. Early Neural Cell Death Is an Extensive, Dynamic Process in the Embryonic Chick and Mouse Retina. *Sci. World J.* **2013**, *2013*, 627240. [CrossRef]

39. Marín-Teva, J.L.; Cuadros, M.A.; Calvente, R.; Almendros, A.; Navascués, J. Naturally Occurring Cell Death and Migration of Microglial Precursors in the Quail Retina during Normal Development. *J. Comp. Neurol.* **1999**, *412*, 255–275. [CrossRef]

40. Navascuéeas, J.; Martíian-Partido, G.; Alvarez, I.S.; Rodríiaguez-Gallardo, L. Cell death in suboptic necrotic centers of chick embryo diencephalon and their topographic relationship with the earliest optic fiber fascicles. *J. Comp. Neurol.* **1988**, *278*, 34–46. [CrossRef]

41. Navascués, J.; Martín-Partido, G. Glial cells in the optic chiasm arise from the suboptic necrotic centers of the diencephalon floor: Morphological evidence in the chick embryo. *Neurosci. Lett.* **1990**, *120*, 62–65. [CrossRef]

42. Dimri, G.P.; Lee, X.; Basile, G.; Acosta, M.; Scott, G.; Roskelley, C.; Medrano, E.E.; Linskens, M.; Rubelj, I.; Pereira-Smith, O.; et al. A biomarker that identifies senescent human cells in culture and in aging skin in vivo. *Proc. Natl. Acad. Sci. USA* **1995**, *92*, 9363–9367. [CrossRef] [PubMed]

43. De Mera-Rodríguez, J.A.; Álvarez-Hernán, G.; Gañán, Y.; Martín-Partido, G.; Rodríguez-León, J.; Francisco-Morcillo, J.; Mera-Rodríguez, J.A. Senescence-associated β-galactosidase activity in the developing avian retina. *Dev. Dyn.* **2019**, *248*, 850–865. [CrossRef]

44. De Mera-Rodríguez, J.A.; Álvarez-Hernán, G.; Gañán, Y.; Martín-Partido, G.; Rodríguez-León, J.; Francisco-Morcillo, J. Is Senescence-Associated β-Galactosidase a Reliable in vivo Marker of Cellular Senescence During Embryonic Development? *Front. Cell Dev. Biol.* **2021**, *9*, 623175. [CrossRef] [PubMed]

45. Francisco-Morcillo, J.; Sánchez-Calderón, H.; Kawakami, Y.; Belmonte, J.C.I.; Hidalgo-Sánchez, M.; Martín-Partido, G. Expression of Fgf19 in the developing chick eye. *Dev. Brain Res.* **2005**, *156*, 104–109. [CrossRef] [PubMed]

46. Calaza, K.D.C.; Gardino, P.F. Neurochemical phenotype and birthdating of specific cell populations in the chick retina. *Anais Acad. Bras. Ciências* **2010**, *82*, 595–608. [CrossRef] [PubMed]

47. Rojas, L.M.; Mitchell, M.A.; Ramírez, Y.M.; McNeil, R. Comparative Analysis of Retina Structure and Photopic Electro-retinograms in Developing Altricial Pigeons (Columba Livia) and Precocial Japanese Quails (Coturnix Coturnix Japonica). *Neotrop. Ornitholog. Soc.* **2007**, *18*, 503–518.

48. Álvarez-Hernán, G.; Sánchez-Resino, E.; Hernández-Núñez, I.; Marzal, A.; Rodriguez-Leon, J.; Martín-Partido, G.; Francisco-Morcillo, J. Retinal histogenesis in an altricial avian species, the zebra finch (Taeniopygia guttata, Vieillot 1817). *J. Anat.* **2018**, *233*, 106–120. [CrossRef] [PubMed]

49. Francisco-Morcillo, J.; Alvarez-Hernan, G.; De Mera-Rodríguez, J.A.; Gañán, Y.; Solana-Fajardo, J.; Martín-Partido, G.; Rodríguez-León, J. Development and postnatal neurogenesis in the retina: A comparison between altricial and precocial bird species. *Neural Regen. Res.* **2021**, *16*, 16–20. [CrossRef] [PubMed]

50. Álvarez-Hernán, G.; Hernández-Núñez, I.; Rico-Leo, E.M.; Marzal, A.; De Mera-Rodríguez, J.A.; Rodríguez-León, J.; Martín-Partido, G.; Francisco-Morcillo, J. Retinal differentiation in an altricial bird species, Taeniopygia guttata: An immunohistochemical study. *Exp. Eye Res.* **2020**, *190*, 107869. [CrossRef]

51. Murray, J.R.; Varian-Ramos, C.W.; Welch, Z.S.; Saha, M.S. Embryological staging of the Zebra Finch, Taeniopygia guttata. *J. Morphol.* **2013**, *274*, 1090–1110. [CrossRef]

52. Piechota, M.; Sunderland, P.; Wysocka, A.; Nalberczak, M.; Sliwinska, M.A.; Radwanska, K.; Sikora, E. Is senescence-associated β-galactosidase a marker of neuronal senescence? *Oncotarget* **2016**, *7*, 81099–81109. [CrossRef] [PubMed]

53. Gavrieli, Y.; Sherman, Y.; Ben-Sasson, S.A. Identification of programmed cell death in situ via specific labeling of nuclear DNA fragmentation. *J. Cell Biol.* **1992**, *119*, 493–501. [CrossRef] [PubMed]

54. Sanwal, M.; Muel, A.; Chaudun, E.; Courtois, Y.; Counis, M. Chromatin condensation and terminal differentiation process in embryonic chicken lens in vivo and in vitro. *Exp. Cell Res.* **1986**, *167*, 429–439. [CrossRef]

55. Sengelaub, D.R.; Dolan, R.P.; Finlay, B.L. Cell generation, death, and retinal growth in the development of the hamster retinal ganglion cell layer. *J. Comp. Neurol.* **1986**, *246*, 527–543. [CrossRef] [PubMed]

56. Cuadros, M.A.; Martin, C.; Ríos, A.; Martín-Partido, G.; Navascués, J. Macrophages of hemangioblastic lineage invade the lens vesicle-ectoderm interspace during closure and detachment of the avian embryonic lens. *Cell Tissue Res.* **1991**, *266*, 117–127. [CrossRef]

57. Nishitani, K.; Sasaki, K. Macrophage localization in the developing lens primordium of the mouse embryo—An immunohistochemical study. *Exp. Eye Res.* **2006**, *83*, 223–228. [CrossRef]

58. Valenciano, A.I.; Boya, P.; De La Rosa, E.J. Early neural cell death: Numbers and cues from the developing neuroretina. *Int. J. Dev. Biol.* **2009**, *53*, 1515–1528. [CrossRef]

59. Caspers, B.A.; Hagelin, J.C.; Paul, M.; Bock, S.; Willeke, S.; Krause, E.T. Zebra Finch chicks recognise parental scent, and retain chemosensory knowledge of their genetic mother, even after egg cross-fostering. *Sci. Rep.* **2017**, *7*, 1–8. [CrossRef]

60. Egensperger, R.; Maslim, J.; Bisti, S.; Holländer, H.; Stone, J. Fate of DNA from retinal cells dying during development: Uptake by microglia and macroglia (Müller cells). *Dev. Brain Res.* **1996**, *97*, 1–8. [CrossRef]

61. Thummel, R.; Kassen, S.C.; Enright, J.M.; Nelson, C.M.; Montgomery, J.E.; Hyde, D.R. Characterization of Müller glia and neuronal progenitors during adult zebrafish retinal regeneration. *Exp. Eye Res.* **2008**, *87*, 433–444. [CrossRef]

62. Bailey, T.J.; Fossum, S.L.; Fimbel, S.M.; Montgomery, J.E.; Hyde, D.R. The inhibitor of phagocytosis, O-phospho-l-serine, suppresses Müller glia proliferation and cone cell regeneration in the light-damaged zebrafish retina. *Exp. Eye Res.* **2010**, *91*, 601–612. [CrossRef]

63. Bejarano-Escobar, R.; Blasco, M.; Martín-Partido, G.; Francisco-Morcillo, J. Light-induced degeneration and microglial response in the retina of an epibenthonic pigmented teleost: Age-dependent photoreceptor susceptibility to cell death. *J. Exp. Biol.* **2012**, *215*, 3799–3812. [CrossRef]

64. Sakami, S.; Imanishi, Y.; Palczewski, K. Müller glia phagocytose dead photoreceptor cells in a mouse model of retinal degenerative disease. *FASEB J.* **2018**, *33*, 3680–3692. [CrossRef]

65. Nomura-Komoike, K.; Saitoh, F.; Fujieda, H. Phosphatidylserine recognition and Rac1 activation are required for Müller glia proliferation, gliosis and phagocytosis after retinal injury. *Sci. Rep.* **2020**, *10*, 1–11. [CrossRef]

66. Bejarano-Escobar, R.; Sánchez-Calderón, H.; Otero-Arenas, J.; Martín-Partido, G.; Francisco-Morcillo, J. Müller glia and phagocytosis of cell debris in retinal tissue. *J. Anat.* **2017**, *231*, 471–483. [CrossRef]

67. Charlton-Perkins, M.; Almeida, A.D.; Macdonald, R.B.; Harris, W.A. Genetic control of cellular morphogenesis in Müller glia. *Glia* **2019**, *67*, 1401–1411. [CrossRef]

68. Vasey, D.B.; Wolf, C.R.; Brown, K.; Whitelaw, C.B.A. Spatial p21 expression profile in the mid-term mouse embryo. *Transgenic Res.* **2011**, *20*, 23–28. [CrossRef]

69. Dewulf, V.; Bottjer, S.W. Age and Sex Differences in Mitotic Activity within the Zebra Finch Telencephalon. *J. Neurosci.* **2002**, *22*, 4080–4094. [CrossRef]

70. Dewulf, V.; Bottjer, S.W. Neurogenesis within the juvenile zebra finch telencephalic ventricular zone: A map of proliferative activity. *J. Comp. Neurol.* **2004**, *481*, 70–83. [CrossRef]

71. Striedter, G.F.; Charvet, C.J. Developmental origins of species differences in telencephalon and tectum size: Morphometric comparisons between a parakeet (*Melopsittacus undulatus*) and a quail (*Colinus virgianus*). *J. Comp. Neurol.* **2008**, *507*, 1663–1675. [CrossRef]

72. Striedter, G.F.; Charvet, C.J. Telencephalon enlargement by the convergent evolution of expanded subventricular zones. *Biol. Lett.* **2008**, *5*, 134–137. [CrossRef]

73. Charvet, C.J.; Striedter, G.F. Causes and consequences of expanded subventricular zones. *Eur. J. Neurosci.* **2011**, *34*, 988–993. [CrossRef]

74. Sugiyama, T.; Yamamoto, H.; Kon, T.; Chaya, T.; Omori, Y.; Suzuki, Y.; Abe, K.; Watanabe, D.; Furukawa, T. The potential role of Arhgef33 RhoGEF in foveal development in the zebra finch retina. *Sci. Rep.* **2020**, *10*, 1–11. [CrossRef]

Article

Quantitative Optical Coherence Tomography Angiography (OCTA) Parameters in a Black Diabetic Population and Correlations with Systemic Diseases

Lincoln T. Shaw [1], Saira Khanna [1], Lindsay Y. Chun [1], Rose C. Dimitroyannis [1,2], Sarah H. Rodriguez [1], Nathalie Massamba [1,3], Seenu M. Hariprasad [1] and Dimitra Skondra [1,3,*]

1   Department of Ophthalmology and Visual Science, University of Chicago Medical Center, Chicago, IL 60637, USA; Lincoln.Shaw@uchospitals.edu (L.T.S.); Saira.Khanna@uchospitals.edu (S.K.); Lindsay.Chun@uchospitals.edu (L.Y.C.); rose.dimitroyannis@gmail.com (R.C.D.); srodriguez5@bsd.uchicago.edu (S.H.R.); nmassamba@bsd.uchicago.edu (N.M.); retina@uchicago.edu (S.M.H.)
2   University of Chicago, Chicago, IL 60637, USA
3   J. Terry Ernest Ocular Imaging Center, University of Chicago Medical Center, Chicago, IL 60637, USA
*   Correspondence: dskondra@bsd.uchicago.edu; Tel.: 1-773-702-3937

Abstract: This is a cross-sectional, prospective study of a population of black diabetic participants without diabetic retinopathy aimed to investigate optical coherence tomography angiography (OCTA) characteristics and correlations with systemic diseases in this population. These parameters could serve as novel biomarkers for microvascular complications; especially in black populations which are more vulnerable to diabetic microvascular complications. Linear mixed models were used to obtain OCTA mean values $\pm$ standard deviation and analyze statistical correlations to systemic diseases. Variables showing significance on univariate mixed model analysis were further analyzed with multivariate mixed models. 92 eyes of 52 black adult subjects were included. After multivariate analysis; signal strength intensity (SSI) and heart disease had statistical correlations to superficial capillary plexus vessel density in our population. SSI and smoking status had statistical correlations to deep capillary plexus vessel density in a univariate analysis that persisted in part of the imaging subset in a multivariate analysis. Hyperlipidemia; hypertension; smoking status and pack-years; diabetes duration; creatinine; glomerular filtration rate; total cholesterol; hemoglobin A1C; and albumin-to-creatinine ratio were not significantly associated with any OCTA measurement in multivariate analysis. Our findings suggest that OCTA measures may serve as valuable biomarkers to track systemic vascular functioning in diabetes mellitus in black patients.

Keywords: diabetes mellitus; retinopathy; microvascular; complication; optical coherence tomography; angiography; black; African-American; systemic disease; biomarker

Citation: Shaw, L.T.; Khanna, S.; Chun, L.Y.; Dimitroyannis, R.C.; Rodriguez, S.H.; Massamba, N.; Hariprasad, S.M.; Skondra, D. Quantitative Optical Coherence Tomography Angiography (OCTA) Parameters in a Black Diabetic Population and Correlations with Systemic Diseases. *Cells* **2021**, *10*, 551. https://doi.org/10.3390/cells10030551

Academic Editor: Maurice Ptito

Received: 20 January 2021
Accepted: 2 March 2021
Published: 4 March 2021

Publisher's Note: MDPI stays neutral with regard to jurisdictional claims in published maps and institutional affiliations.

1. Introduction

Diabetes mellitus (DM) is the most common metabolic disorder worldwide and affects an estimated 463 million people worldwide and 34.2 million people in the United States (U.S.)—about 10.5% of the U.S. population [1,2]. Furthermore, epidemiological studies demonstrate higher rates of DM in non-Hispanic black populations, as well as a higher risk of microvascular complications of DM including nephropathy progressing to end-stage renal disease (ESRD) and diabetic retinopathy (DR) compared to white populations in the U.S [3–5].

DR is the leading cause of blindness between the ages of 20–74 in the U.S. [6]. Changes in retinal microvascular structure associated with DR include pericyte and endothelial cell loss, decreased perfusion, and ultimately ischemia which leads to upregulation of pro-angiogenic factors (such as vascular endothelial growth factor [VEGF]) and subsequent neovascularization [7]. These changes are reflected throughout the rest of the body as well, which explains the strong association between DR and other microvascular complications of diabetes such as nephropathy and neuropathy [8]. In addition, recent studies have correlated microvascular complications in diabetics (including DR) with generalized vascular dysfunction and an increased risk of cardiovascular disease and all-cause mortality [9], independent of other risk factors such as diabetes duration, glycemic control, smoking, and lipid profile [10].

Optical coherence tomography angiography (OCTA) has added more precise noninvasive methods of analyzing retinal vasculature in recent years. OCTA allows researchers to analyze quantitative parameters such as superficial capillary plexus vessel density (SCP VD), deep capillary plexus vessel density (DCP VD), foveal avascular zone (FAZ) area, acircularity index (AI), and vascular length density (VLD) [11]. Using OCTA analysis, diabetics without DR have been shown to have decreased superficial and deep retinal VD in the parafoveal area, increased nonperfusion, and increased FAZ area compared to healthy subjects [7,12–15]. Moreover, evaluation of these changes using OCTA analysis can help predict an increased risk for progression of DR and development of diabetic macular edema (DME) [16], and worsening of these changes ultimately leads to poorer visual outcomes [14,17–20].

Differences between races on OCTA analysis have also been identified, as prior studies from our group have suggested that black populations may have decreased macular capillary vasculature at baseline compared to white populations even in the absence of systemic disease [21]. To date, limited data has been published regarding changes in retinal microvasculature in black subjects with diabetes and how these changes may correlate with systemic biomarkers. In this study, OCTA was used to analyze retinal vasculature and study correlations between retinal microvascular environment characteristics and systemic diseases in black adults with DM but without DR to help establish microvascular changes that correlate with systemic diseases in this population. These parameters could serve as a novel biomarker for microvascular complications especially in the black populations that are more vulnerable to diabetic microvascular complications.

2. Materials and Methods

This prospective, single-center, cross-sectional study of participants was approved by the Institutional Review Board of the University of Chicago (IRB #17-0170). All study protocols adhered to the tenets of the Declaration of Helsinki. The study conformed to the Health Insurance Portability and Accountability Act of 1996 regulations. The study was conducted between 1 February 2017 and 23 January 2019. All subjects provided informed, written consent.

2.1. Participants

Subjects were recruited at a single center, large, academic retinal practice at the University of Chicago Eye Clinic. These patients were established clinic patients that had been diagnosed with diabetes mellitus and had a recent dilated fundus exam and evaluation by a retina specialist that confirmed no evidence of DR. Inclusion criteria included age over 18 years, history of diabetes mellitus (type 1 or 2), and self-identification of black/African-American ethnicity. Exclusion criteria included diabetic retinopathy, medical conditions outside the variables listed in the study that would compromise microvasculature, and ocular conditions or ocular surgical history that would compromise image quality or retinal microvasculature. A history of systemic disease was treated as a binary variable (i.e., the subject either had or did not have the disease). Furthermore, many patients were currently being treated with pharmacologic agents (e.g., insulin, oral hypoglycemics, statins, antihypertensives, etc.). Subjects were prospectively consented for additional OCTA imaging using Optovue RTVue XR Avanti 2017.1 (3 × 3 mm macula cube) for further evaluation of retinal vasculature. All subjects were properly informed of the risks and benefits of additional imaging procedures, as well as the use of subsequent data gained from these imaging procedures. Chart review for demographic information, clinical information (history of other diagnoses including presence of hypertension, hyperlipidemia, heart disease [defined as past medical history including heart failure, coronary artery disease, and/or myocardial infarction], and smoking status), ocular history, and relevant laboratory values (creatinine, estimated glomerular filtration rate [GFR], albumin-to-creatinine ratio [ACR], total cholesterol, hemoglobin A1c [HbA1c]) of subjects was performed. Only images with a signal strength intensity (SSI) $\geq$0.5 were used in the analysis.

2.2. OCTA Imaging

Images were obtained using the Optovue RTVue XR Avanti (Optovue Inc, Fremont, CA, USA Version 2017.1) with the AngioRetina mode (3 × 3 mm macular cube). Each image was made up of 304 clusters of repeated B-scans containing 304 A-scans, and the images were automatically segmented. AngioAnalytics software was used to analyze retinal vascular parameters after IRB approval. The software set the superficial capillary plexus (SCP) at the inner limiting membrane (ILM) and 9μm above the IPL (inner plexiform layer); the deep capillary plexus (DCP) was set between 9μm above the IPL and 9 μm below the outer plexiform layer (OPL).The superficial whole image was the entire 3 × 3 mm macular cube above the DCP. The parafoveal area was a 1–3 mm annulus surrounding the central fovea (see Figure 1). The FAZ (foveal avascular zone, in mm^2) at the level of the SCP and DCP was manually measured with the built-in software. The acircularity index (AI) was defined as the ratio of the perimeter of the FAZ to the perimeter of a circle with equal area [22,23]. The foveal vascular length density (FD-300 LD) was determined as vessel density within a ring of a width of 300 μm surrounding the FAZ. The foveal vascular area density in the 300-μm ring (FD-300 AD) was calculated by dividing the number of vessel pixels by the total number of pixels then multiplied by 100%. The threshold for signal strength intensity (SSI) was set at $\geq$50 based on previous studies [24–26].

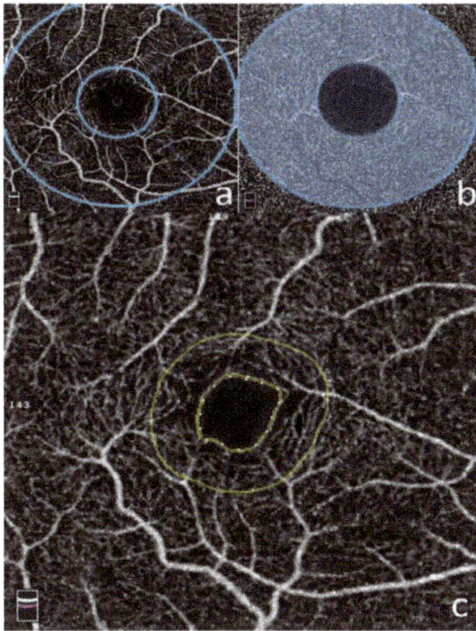

Figure 1. OCTA Software. Optovue RTVue XR Avanti 2017.1 (3 × 3 mm macula cube) software was used to analyze retinal microvasculature. Figure 1a represents superficial capillary plexus en-face image with 1mm and 3 mm rings surrounding fovea (blue). Figure 1b represents an image of the deep capillary plexus with 1–3 mm annulus representing the parafoveal region (transparent blue annulus). Figure 1c represents foveal avascular zone (FAZ) with marked perimeter (inner yellow circle) surrounded by 300 μm ring (outer yellow circle) used to calculate foveal vascular length density (FD300-LD) and foveal vascular area density (FD300-AD).

2.3. Vessel Density Analysis

Vessel Density (VD) measurements were automatically generated by the software. It represents the percentage of pixels in the 3 × 3 mm macular scan that correspond to blood vessels that were automatically calculated. Within the 3 × 3 mm macular scan, there was an annulus of concentric circles of 1mm and 3mm diameters, with the inner circle defined as the fovea and the parafovea defined as the ring between the two circles.

2.4. Statistical Analysis

Statistical analysis was performed using Stata13 (College Station, TX: StataCorp LP). To control for 2 eyes of the same patient, linear mixed models were used to obtain OCTA marginal means values ± standard deviation. Variables showing significant correlation with OCTA values on univariate mixed model analysis ($p < 0.05$) were analyzed with mixed-effects multivariate models, also controlling for 2 eyes of the same patient.

3. Results

92 eyes of 52 black adult subjects with DM were included in this study. 12 eyes were excluded from the study for meeting exclusion criteria for only one eye (signal strength intensity <0.5, ocular conditions that would affect microvasculature unilaterally but are unrelated to the study variables, and/or past ocular surgical history of the excluded eye). The mean age was 57.89 ± 15.85 (20.6–88.2 years, $p = 0.1537$). Most subjects were female (75%) and all the patients identified as black/African-American. Additional patient demographics and mean lab values are included in Table 1.

Table 1. Patient Demographics and Clinical Characteristics.

	Subjects (n = 52)
Baseline criteria	
Female gender, n (%)	39 (75)
DM, years, mean (SD)	8.79 (7.29)
Former or current smoker, n (%)	25 (48)
Pack-years, mean (SD)	6.11 (9.16)
Hyperlipidemia, n (%)	18 (35)
Hypertension, n (%)	31 (60)
Heart Disease, n (%)	11 (21)
Lab Values	
Creatinine [Cr] (SD)	0.89 (0.31)
Glomerular Filtration Rate [GFR] (SD)	78.15 (19.63)
Albumin-to-Cr Ratio [ACR] (SD)	186.33 (460.49)
Total Cholesterol (SD)	197.71 (157.06)
HbA1c (SD)	7.64 (1.98)

The mean SSI was 65.06 ± 8.17. The mean superficial vessel density was 43.41 ± 5.13% in the whole image and 46.27 ± 5.39% in the parafovea. The mean deep vessel density was 48.71 ± 4.64% in the whole image and 51.23 ± 4.66% in the parafovea. The mean FAZ area was 0.34 ± 0.13 mm^2. Additional measurements are included in Table 2.

Table 2. Vessel Density in Superficial and Deep Vascular Capillary Plexus and FAZ measurements.

	Vessel Density (*n* = 92 Eyes)
SSI	65.06 ± 8.17
Superficial vessel density (%)	
Whole image	43.41 ± 5.13
Parafovea	46.27 ± 5.39
Deep vessel density (%)	
Whole image	48.71 ± 4.64
Parafovea	51.23 ± 4.66
FAZ	
FAZ area (mm^2)	0.34 ± 0.13
Perimeter (mm)	2.35 ± 0.48
Acircularity index [AI]	1.16 ± 0.08
FD-300 area density (%)	48.41 ± 7.05
FD-300 length density (%)	15.38 ± 3.82

In the univariate mixed effect model for superficial whole image, age, SSI, heart disease, and Cr were statistically significant (age: -0.162, $p < 0.001$; SSI: 0.427, $p < 0.001$; heart disease: -3.536, $p = 0.022$; Cr: -5.198, $p = 0.017$) (Table 3a). The univariate analysis of the superficial parafovea had the same statistically significant variables (age: -0.15, $p < 0.001$, SSI: 0.45, $p < 0.001$, heart disease: -3.50, $p = 0.028$, Cr: -5.18, $p = 0.018$).

In the multivariate analyses for the superficial whole image VD and superficial parafovea, SSI and heart disease were statistically significantly correlated with VD. In the superficial parafovea, SSI had a positive association (0.40, $p < 0.001$), while having heart disease was negative (-2.82, $p = 0.019$) (Table 3b).

SSI and smoking status were significantly associated with the deep whole image VD (SSI: 0.182, $p = 0.002$, smoking status: -2.841, $p = 0.011$, respectively). In the multivariate analysis, only SSI remains statistically significant (0.154, $p = 0.011$).

SSI and smoking status were significantly associated with deep parafoveal VD (0.14, $p = 0.017$; -2.6, $p = 0.022$, respectively). Through the multivariate model, none of the clinical factors studied were significant in the deep parafoveal VD.

FAZ area was significantly associated with Cr (-0.131, $p = 0.015$), and FAZ perimeter with pack-years of smoking in univariate analysis (0.018, $p = 0.008$). Both FAZ acircularity index and FD300-AD were associated with SSI in multivariate analysis (-0.005, $p < 0.001$; 0.491, $p < 0.001$, respectively). SSI and heart disease were shown to be statistically associated in the multivariate analysis for FD300-LD (SSI: 0.293, $p < 0.001$; heart disease: -1.943, $p = 0.049$).

Hyperlipidemia, hypertension, smoking status, pack-years, DM duration, Cr, GFR, total cholesterol, HbA1C, and albumin-to-Cr ratio (ACR) were not significantly associated with any OCTA measurement in multivariate analysis.

Table 3. (a) Results of Univariate Analysis. Correlation coefficients are provided for all parameters and variables. Values that are bolded are statistically significant ($p < 0.05$) and the 95% confidence interval is provided. **(b)** Results of Multivariate Analysis. Correlation coefficients are provided for all parameters and variables. Values that are bolded are statistically significant ($p < 0.05$) and the 95% confidence interval is provided. Variables and parameters that did not have multiple statistically significant variables in the univariate analysis were not included in this table.

(a)

Variables	Superficial Whole Image	Superficial Parafovea	Deep Whole Image	Deep Parafovea	FAZ Area	Perimeter	Acircularity Index	FD300 Area Density	FD300 Length Density
Age	−0.162 (−0.238 to −0.086)	−0.153 (−0.23 to −0.07)	−0.073	−0.06	0.001	0.005	0.001 (0.000 to 0.002)	−0.213 (−0.311 to −0.116)	−0.146 (−0.199 to −0.092)
SSI	0.427 (0.332 to 0.523)	0.454 (0.350 to 0.560)	0.182 (0.068 to 0.297)	0.142 (0.030 to 0.260)	0.001	0.000	−0.003 (−0.005 to −0.002)	0.521 (0.376 to 0.666)	0.342 (0.273 to 0.410)
Hyperlipidemia	0.260	0.46	1.869	2.12	0.013	0.040	−0.014	1.297	0.326
Hypertension	1.017	1.33	−1.715	−1.72	0.022	0.068	−0.011	1.301	0.343
Heart disease	−3.536 (−6.563 to −0.509)	−3.498 (−6.563 to −0.509)	−1.392	−0.71	0.045	0.197	0.014	−4.365 (−8.350 to −0.379)	−3.264 (−5.514 to −1.015)
Current smoker	−1.859	−1.72	−2.841 (−5.021 to −0.660)	−2.600 (−4.830 to −0.370)	0.038	0.222	0.053 (−0.005 to −0.002)	−3.321(−6.641 to −0.002)	−2.500 (−4.409 to −0.591)
Pack years	−0.069	−0.07	−0.087	−0.06	0.004	0.018 (0.005 to 0.032)	0.003 (0.000 to 0.005)	−0.047	−0.071
DM duration	−0.057	−0.06	0.005	0.02	0.004	0.013	0.000	0.036	−0.015
Cr	−5.198 (−9.459 to −0.937)	−5.184 (−9.480 to −0.890)	−3.597	−3.76	−0.131 (−0.238 to −0.025)	−0.362	0.058 (0.001 to 0.115)	−6.371 (−12.178 to −0.565)	−3.750 (−7.255 to −0.244)
GFR	0.072 (0.004 to 0.140)	0.07	0.051	0.05	0.001	0.003	−0.001	0.101 (0.007 to 0.195)	0.064 (0.008 to 0.120)
Total cholesterol	−0.001	0.00	0.004	0.00	0.000	−0.001	0.001	−0.001	0.001
HbA1C	−0.142	−0.18	0.299	0.40	0.013	0.030	−0.007	0.716	0.342
ACR	0.002	0.00	0.002	0.00	0.000	0.000	0.000	0.001	0.002

(b)

Variables	Superficial Whole Image	Superficial Parafovea	Deep Whole Image	Deep Parafovea	Acircularity Index	FD300 Area Density	FD300 Length Density
Age	−0.001	0.052			−0.001	−0.016	−0.007
SSI	0.403 (0.278 to 0.527)	0.475 (0.350 to 0.60)	0.154 (0.036 to 0.272)	0.115	−0.005 (0.007 to −0.002)	0.491 (0.274 to 0.708)	0.293 (0.197 to 0.389)
Heart disease	−2.819 (−5.531 to −0.424)	−2.795(−5.200 to −0.390)				−3.176	−1.944 (−4.561 to −0.0535)
Current smoker			−2.118	−2.053	−0.059	−0.503	−0.823
Pack years					0.001		
Cr	−4.459	−2.577			0.012	−3.382	−0.472
GFR	−0.039					−0.025	−0.001

4. Discussion

In this study, we analyzed the retinal vasculature of black subjects with DM without DR and analyzed correlations to systemic diseases. After multivariate analysis, SSI and heart disease had statistically significant correlations to SCP VD in our population.

SSI and smoking status had statistical correlations to DCP VD in a univariate analysis that persisted in at least part of the imaging subset (e.g., deep whole image) in a multivariate analysis. None of the variables studied were associated with changes in deep parafoveal VD in a multivariate model of analysis.

In addition, none of the variables analyzed were statistically associated with FAZ area in a multivariate analysis, and SSI was the only variable that was significantly correlated with AI, FD300-AD, and FD300-LD.

Several of our results support previously published research, which is discussed below.

4.1. Age

Our study demonstrated a statistical correlation between age and decreased SCP VD on whole image and parafoveal imaging techniques in a univariate analysis, but these correlations were not shown to be statistically significant in a multivariate analysis. Additionally, only univariate analysis correlations were present between age and AI, FD300-AD, and FD300-LD, and none of these were statistically significant on multivariate analysis.

Prior studies have shown that age is correlated with decreased retinal capillary density in healthy subjects [27–30], including black populations [31]. Additionally, healthy black subjects have shown to have an increased FAZ area at baseline compared to white subjects [21,32], and diabetic patients have also been shown to have an increased FAZ area compared to non-diabetic patients even in the absence of diabetic retinopathy [13,15]. Based on this prior data, we know that black populations as well as diabetic populations have baseline structural differences in retinal microvasculature compared to subjects of other ethnicities and non-diabetics. Our data suggests that these structural differences in black diabetic populations may make the retinal microvasculature less susceptible to age-related changes. However, while our study included subjects with ages of 20–88 years, most of our population were middle-aged, and we may not have had statistical power to detect true correlations between age and OCTA changes as we had only a few individuals that fell towards the extremes of this age range.

4.2. Signal Strength Intensity

Signal strength was shown to be positively correlated with VD in our study in many subsections of OCTA analysis, including superficial whole image VD, superficial whole VD, superficial parafoveal VD, deep whole image VD, FD300-AD and FD300-LD in a multivariate analysis. Furthermore, our study showed a weak negative correlation between SSI and AI.

Yu et al. found on two OCTA platforms that VD measurements decreased linearly with decreasing signal strength with high statistical significance [30]. Similarly, Lim et al. found that VD, perfusion density (PD), and FAZ area significantly increased with increased signal strength [24]. In addition, Czakó et al. demonstrated that repeatability of OCTA metrics in diabetic populations are significantly affected by reduced signal strength [33].

The weak negative correlation with acircularity may have been related to processing of data, as vessels in images with less signal strength may have registered with the software as being more tortuous, and therefore deviating more from the circle of equal area that the software uses to calculate AI, due to projection artifacts associated with decreased signal strength.

Similar to conclusions suggested by prior authors [24,30,33,34], our results highlight the importance of using high-quality images with a high level of signal strength when interpreting OCTA metrics, including black diabetic populations.

4.3. Heart Disease

In a multivariate analysis, our study demonstrated that heart disease was negatively correlated with SCP VD in whole image and parafoveal regions, and was the strongest correlation we found of the variables studied. This supports the results of prior research regarding OCTA metrics and associations with heart disease in other populations.

Multiple prior studies have correlated retinal vessel atherosclerosis and/or reduced vessel diameter with increased incidence of coronary artery disease and risk of death secondary to coronary events and stroke [35–38]. The methods of analysis used in these studies was direct visualization of retinal vasculature by examiners or analysis of fundus photographs.

Limited data is available regarding the role of OCTA analysis in correlating quantitative changes in retinal microvasculature to heart disease; however, Wang et al. studied an Asian cohort of 316 subjects and compared coronary artery disease (CAD) patients to healthy controls. Their results showed that CAD patients had reduced retinal vessel density, choroidal vessel density, and flow area, and the authors concluded that OCTA may be a noninvasive strategy for identifying high-risk early stage CAD individuals that may benefit from further examination or cardiac procedures. Furthermore, they proposed that the mechanisms responsible for these microvascular changes are likely similar to those seen in larger vessels contributing to CAD, including atherosclerotic changes to the vessel walls that result in thickening and stenosis. Their study found that vessel density was directly correlated with Gensini score, which is a well-established weighted grading system of coronary artery stenosis that grades based on which coronary arteries are stenotic (left main coronary artery [LMCA] carrying the most weight, followed by left anterior descending branch [LAD], left circumflex coronary artery [LCX], etc.), and found that SCP VD and DCP VD, as well as choroidal capillary vessel density, was directly correlated to Gensini score. In other words, a greater degree of retinal and choroidal vessel density loss was present in those patients which had LMCA stenosis, followed by LAD and LCX stenosis, which correlates well to severity of coronary artery disease [39].

Our results support the results of Wang et al., and suggest that similar findings, at least in the superficial retinal vasculature layer, are found in not only an Asian population, but also the black diabetic population of our study. This provides further support that OCTA parameters may be an under-utilized tool in screening patients for early stage CAD that may warrant further evaluation.

4.4. Creatinine (Cr), Estimated Glomerular filtration Rate (eGFR), and Albumin-to-Creatinine Ratio

Creatinine was negatively associated with SCP VD, FAZ area, and FD300-AD, FD300-LD, and weakly positively associated with increased AI in a univariate analysis, but none of these associations were found to persist in a multivariate analysis. eGFR was positively associated with FD300-AD and FD300-LD in a univariate analysis but not in a multivariate analysis.

Some prior studies of Chinese populations have shown that higher creatinine level is associated with decreased retinal vessel density [34] and lower eGFR is associated with increased FAZ size in diabetic populations [15], while other studies with similar populations failed to show any similar correlations [14,40]. While microvascular complications of diabetes such as nephropathy are more common in black diabetic populations [3–5], our population likely included diabetics with relatively mild microvascular complications as demonstrated by their lack of retinopathy on exam and low average creatinine. Further studies of OCTA parameters including black diabetic patients with DR and more severe kidney dysfunction may better elucidate further correlations.

4.5. Smoking Status and Smoking Pack-Years

Current smoking status was negatively correlated with DCP VD in the whole image and parafoveal image analysis in our study in a univariate analysis but failed to show a correlation in a multivariate analysis. Additionally, our results showed a positive correlation with AI, and a negative correlation in FD300-AD and FD300-LD in a univariate analysis. Additionally, pack-years smoking was positively correlated with FAZ perimeter in a univariate analysis. However, none of these changes were statistically significant on multivariate analysis.

Lee et al. found that current smoking status was associated with decreased DCP VD but not SCP VD reduction in a diabetic population [15]. These results align with other studies that have shown decreased blood flow [41] and reduced retinal capillary density in smoking diabetic populations [42]. Our study population had a relatively low number of former or current smokers (25/52), which may be the reason that smoking status was not correlated to OCTA parameters analysis and we may not have had adequate statistical power to detect these correlations in the multivariate analysis and was not identified in the univariate.

4.6. HbA1C and Diabetes Duration

Our study did not show any statistical associations in OCTA parameters with HbA1c level and diabetes duration in either univariate or multivariate analysis. Prior studies in Chinese and African American diabetic populations also failed to show correlations with HbA1c (26, 36), although diabetes duration was correlated with reduced capillary vessel density on OCTA in a previous study of diabetic African-American subjects [31]. Our results, at least in part, highlight the multi-factorial nature of diabetes. While previous studies have shown that the risk of diabetic retinopathy increases with higher HbA1c levels and longer diabetes duration, especially in racial/ethnic minorities [3], our data may suggest that those eyes that do not exhibit clinical signs of disease also do not seem to have sub-clinical OCTA parameter changes correlated to glycemic control. It is possible that some eyes respond more robustly to changes in glycemic control (and therefore develop clinical retinopathy), while eyes such as the ones that were used in our study are less affected and may be more tolerant to these changes, even at the level of detail detectable by OCTA analysis. More data is needed to explain the mechanisms responsible for the differing response severity in these patients.

4.7. Other Variables (Hypertension, Hyperlipidemia, Total Cholesterol)

Our results did not show any correlations between OCTA parameters and hypertension, hyperlipidemia, or total cholesterol in either univariate or multivariate analysis. Limited previously published data is available for these variables with similar characteristics to our patient population, which makes these results difficult to compare to prior research. Changes in microvascular structure of the choriocapillaris on OCTA has been associated with hypertension in a Japanese study population [43], and hypertension has been shown to be associated with lower VD in diabetic patients [15] as well as otherwise healthy individuals [44] in Korean populations. Furthermore, lower VD has been associated with dyslipidemia in populations in South Korea [15] and Singapore [42]. These results highlight the need for analyzing OCTA parameters in diverse patient populations, as these changes may be due to ethnic differences among these populations or may simply be due to inter-study variability.

4.8. Limitations

Firstly, our population was a relatively small and homogenous population at a single-center academic retinal practice in Chicago, IL. The subject recruitment in the study was limited to available subjects that met inclusion and exclusion criteria in our single-center practice. Due to lack of normative OCTA vascular data for this patient population, upon which we could draw an estimated effect size, formal power calculations were not possible for this study. However, we believe our data serves as a starting point for future studies that investigate the establishment of normative OCTA databases in this population, and we acknowledge that we cannot claim with certainty that the differences found in this study would be applicable to a broader population. In the future, a larger study could potentially identify additional factors associated with OCTA parameters.

We decided to narrow our population to black diabetic subjects as the black population remains underrepresented in scientific literature [45] despite microvascular complications of diabetes being more prevalent in this population [3–5]. Previous studies have shown that young black patients without systemic disease have differences in OCTA parameters [21], which is why we focused on black patients with diabetes. Studying patients without retinopathy prevented retinal hemorrhage, cysts, or exudates from influencing OCTA parameters. This pilot study can help aid in the design for future studies in this patient population and ensure less bias is introduced in the interpretation of OCTA parameters and association in retinal disease burden. That being said, a larger data set from more subjects over a larger and more diverse geographical area would help add statistical power to the results found in this study. Similarly, future studies including both comparison groups of additional ethnicities and black patients without diabetes would further aid in establishing the effect of race/ethnicity in OCTA measurements.

Secondly, a potential confounder of any study comparing OCTA parameters to other previously published research is that OCTA analysis methods have been shown to lack interchangeability with one another [46]. This makes interpretation of data difficult, as inter-study variability exists in methods used for imaging and post-imaging processing and analysis. To an extent, this unfortunately also limits the uses of normative databases until a standard method of OCTA analysis is widely adopted.

Thirdly, a confounder that is inherent to studies using OCTA is the role of axial length in OCTA parameter analysis. Previous studies have shown that axial length has been shown to affect vessel density and FAZ area [14,28], but unfortunately not widely adopted and verified magnification correction factor has been developed to date, so this variable is difficult to correct. Unfortunately, axial length data was not available for analysis for our subset of patients, so conclusions drawn from our data could potentially be skewed by a non-gaussian distribution of axial lengths among our population.

Lastly, as described above, systemic diseases other than diabetes were treated as binary variables upon chart review, and we did not further stratify based on disease duration, severity of disease, and use of pharmacologic treatment agents. Hypertension, for example, has varying degrees of severity, and long-standing uncontrolled severe hypertension is likely much more damaging to microvasculature than milder cases of disease. Ultimately, more data is needed to further stratify whether these correlations persist if these diseases are diagnosed early and well-controlled.

5. Conclusions

Our findings suggest that OCTA measures may serve as valuable biomarkers in black patients to track systemic vascular functioning in DM and underscore the importance of working towards establishing normative databases that represent diverse populations. Clinically, these findings suggest OCTA may be helpful in identifying microvascular characteristics that may guide the physician to refer these patients for closer monitoring of other systemic diseases, such as heart disease. Furthermore, our data reflects that the presence of systemic diseases have correlations with baseline OCTA parameters in black diabetic populations. This highlights that baseline ethnic characteristics of study participants as well as the presence of systemic diseases need to be considered when analyzing OCTA imaging, especially when being used for research purposes. Further studies including larger sample size of patients with systemic diseases from diverse racial backgrounds are needed to further delineate the correlations of systemic vascular status with retinal microvascular environment and their involvement in retinal microvascular diseases.

Author Contributions: Conceptualization, L.Y.C. and D.S.; Data curation, L.Y.C., R.C.D.; N.M. and D.S.; Formal analysis, L.Y.C.; R.C.D. and D.S.; Funding acquisition, L.Y.C. and D.S.; Investigation, L.Y.C., R.C.D. and D.S.; Methodology, S.H.R. and D.S.; Project administration L.Y.C. and D.S.; Resources, L.Y.C. and D.S.; Software, S.H.R. and D.S.; Supervision, D.S.; Validation, L.T.S., S.K.,

Cells **2021**, 10, 551

L.Y.C., S.H.R. and D.S.; Visualization, L.T.S., S.K., S.M.H. and D.S.; Writing—original draft and revisions, L.T.S., S.K., S.M.H. and D.S. All authors have read and agreed to the published version of the manuscript.

Funding: This research was funded by the University of Chicago Diversity Grant awarded to Lindsay Chun and Dimitra Skondra.

Institutional Review Board Statement: This prospective, single-center, cross-sectional study of participants was approved by the Institutional Review Board of the University of Chicago (IRB #17-0170). All study protocols adhered to the tenets of the Declaration of Helsinki. The study conformed to the Health Insurance Portability and Accountability Act of 1996 regulations.

Informed Consent Statement: Informed consent was obtained from all subjects involved in the study.

Data Availability Statement: The data presented in this study are available on request from the corresponding author.

Acknowledgments: J. Terry Ernest Ocular Imaging Center Staff for assisting with imaging of the subjects used in this study.

Conflicts of Interest: The authors declare no conflict of interest. The funders had no role in the design of the study; in the collection, analyses, or interpretation of data; in the writing of the manuscript, or in the decision to publish the results.

References

1. Saeedi, P.; Petersohn, I.; Salpea, P.; Malanda, B.; Karuranga, S.; Unwin, N.; Colagiuri, S.; Guariguata, L.; Motala, A.A.; Ogurtsova, K.; et al. Global and regional diabetes prevalence estimates for 2019 and projections for 2030 and 2045: Results from the International Diabetes Federation Diabetes Atlas, 9th edition. *Diabetes Res. Clin. Pract.* **2019**, *157*, 107843. [CrossRef] [PubMed]
2. Centers for Disease Control and Prevention. *National Diabetes Statistics Report*; U.S. Dept of Heatlh and Human Services: Washington, DC, USA, 2020.
3. Zhang, X.; Saaddine, J.B.; Chou, C.-F.; Cotch, M.F.; Cheng, Y.J.; Geiss, L.S.; Gregg, E.W.; Albright, A.L.; Klein, B.K.; Klein, R. Prevalence of diabetic retinopathy in the United States, 2005–2008. *JAMA* **2010**, *304*, 649–656. [CrossRef] [PubMed]
4. Menke, A.; Casagrande, S.; Geiss, L.; Cowie, C.C. Prevalence of and Trends in Diabetes Among Adults in the United States, 1988-2012. *JAMA* **2015**, *314*, 1021–1029. [CrossRef] [PubMed]
5. Harris, M.I.; Klein, R.; Cowie, C.C.; Rowland, M.; Byrd-Holt, D.D. Is the risk of diabetic retinopathy greater in non-Hispanic blacks and Mexican Americans than in non-Hispanic whites with type 2 diabetes? A U.S. population study. *Diabetes Care* **1998**, *21*, 1230–1235. [CrossRef] [PubMed]
6. Ellis, M.P.; Lent-Schochet, D.; Lo, T.; Yiu, G. Emerging Concepts in the Treatment of Diabetic Retinopathy. *Curr. Diab. Rep.* **2019**, *19*, 137. [CrossRef]
7. Nesper, P.L.; Roberts, P.K.; Onishi, A.C.; Chai, H.; Liu, L.; Jampol, L.M.; Fawzi, A.A. Quantifying Microvascular Abnormalities With Increasing Severity of Diabetic Retinopathy Using Optical Coherence Tomography Angiography. *Invest Ophthalmol. Vis. Sci.* **2017**, *58*, BIO307–BIO315. [CrossRef] [PubMed]
8. Pontuch, P.; Vozár, J.; Potocký, M.; Krahulec, B. Relationship between nephropathy, retinopathy, and autonomic neuropathy in patients with type I diabetes. *J. Diabet. Complicat.* **1990**, *4*, 188–192. [CrossRef]
9. Garofolo, M.; Gualdani, E.; Giannarelli, R.; Aragona, M.; Campi, F.; Lucchesi, D.; Daniele, G.; Miccoli, R.; Francesconi, P.; Del Prato, S.; et al. Microvascular complications burden (nephropathy, retinopathy and peripheral polyneuropathy) affects risk of major vascular events and all-cause mortality in type 1 diabetes: A 10-year follow-up study. *Cardiovasc. Diabetol.* **2019**, *18*, 159. [CrossRef]
10. Cheung, N.; Wong, T.Y. Diabetic retinopathy and systemic vascular complications. *Prog. Retin Eye Res.* **2008**, *27*, 161–176. [CrossRef]
11. Tey, K.Y.; Teo, K.; Tan, A.C.S.; Devarajan, K.; Tan, B.; Tan, J.; Schmetterer, L.; Ang, M. Optical coherence tomography angiography in diabetic retinopathy: A review of current applications. *Eye Vis.* **2019**, *6*, 37. [CrossRef]
12. Dimitrova, G.; Chihara, E.; Takahashi, H.; Amano, H.; Okazaki, K. Quantitative Retinal Optical Coherence Tomography Angiography in Patients with Diabetes Without Diabetic Retinopathy. *Invest. Ophthalmol. Vis. Sci.* **2017**, *58*, 190–196. [CrossRef]
13. De Carlo, T.E.; Chin, A.T.; Bonini Fiho, M.A.; Adhi, M.; Branchini, L.; Salz, D.A.; Baumal, C.R.; Crawford, C.; Reichel, E.; Witkin, A.J.; et al. Detection of microvascular changes in eyes of patients with diabetes but not clinical diabetic retinopathy using optical coherence tomography angiography. *Retina Phila. Pa.* **2015**, *35*, 2364–2370. [CrossRef] [PubMed]
14. Tang, F.Y.; Ng, D.S.; Lam, A.; Luk, F.; Wong, R.; Chan, C.; Mohamed, S.; Fong, A.; Lok, J.; Tso, T.; et al. Determinants of Quantitative Optical Coherence Tomography Angiography Metrics in Patients with Diabetes. *Sci. Rep.* **2017**, *7*, 2575. [CrossRef] [PubMed]
15. Lee, D.-H.; Yi, H.C.; Bae, S.H.; Cho, J.H.; Choi, S.W.; Kim, H. Risk factors for retinal microvascular impairment in type 2 diabetic patients without diabetic retinopathy. *PLoS ONE* **2018**, *13*, e0202103. [CrossRef] [PubMed]

16. Sun, Z.; Tang, F.; Wong, R.; Lok, J.; Szeto, S.K.H.; Chan, J.C.K.; Chan, C.K.M.; Tham, C.C.; Ng, D.S.; Cheung, C.Y. OCT Angiography Metrics Predict Progression of Diabetic Retinopathy and Development of Diabetic Macular Edema: A Prospective Study. *Ophthalmology* **2019**, *126*, 1675–1684. [CrossRef] [PubMed]

17. Kim, A.Y.; Chu, Z.; Shahidzadeh, A.; Wang, R.K.; Puliafito, C.A.; Kashani, A.H. Quantifying Microvascular Density and Morphology in Diabetic Retinopathy Using Spectral-Domain Optical Coherence Tomography Angiography. *Invest. Ophthalmol. Vis. Sci.* **2016**, *57*, OCT362–OCT370. [CrossRef] [PubMed]

18. Balaratnasingam, C.; Inoue, M.; Ahn, S.; McCann, J.; Dhrami-Gavazi, E.; Yannuzzi, L.A.; Freund, K.B. Visual Acuity Is Correlated with the Area of the Foveal Avascular Zone in Diabetic Retinopathy and Retinal Vein Occlusion. *Ophthalmology* **2016**, *123*, 2352–2367. [CrossRef]

19. Dupas, B.; Minvielle, W.; Bonnin, S.; Couturier, A.; Erginay, A.; Massin, P.; Gaudric, A.; Tadayoni, R. Association Between Vessel Density and Visual Acuity in Patients With Diabetic Retinopathy and Poorly Controlled Type 1 Diabetes. *JAMA Ophthalmol.* **2018**, *136*, 721–728. [CrossRef]

20. Samara, W.A.; Shahlaee, A.; Adam, M.K.; Khan, M.A.; Chiang, A.; Maguire, J.I.; Hsu, J.; Ho, A.C. Quantification of Diabetic Macular Ischemia Using Optical Coherence Tomography Angiography and Its Relationship with Visual Acuity. *Ophthalmology* **2017**, *124*, 235–244. [CrossRef] [PubMed]

21. Chun, L.Y.; Silas, M.R.; Dimitroyannis, R.C.; Ho, K.; Skondra, D. Differences in macular capillary parameters between healthy black and white subjects with Optical Coherence Tomography Angiography (OCTA). *PLoS ONE* **2019**, *14*, e0223142. [CrossRef]

22. Tam, J.; Dhamdhere, K.P.; Tiruveedhula, P.; Manzanera, S.; Barez, S.; Bearse, M.A.; Adams, A.J.; Roorda, A. Disruption of the Retinal Parafoveal Capillary Network in Type 2 Diabetes before the Onset of Diabetic Retinopathy. Investig Opthalmology. *Vis. Sci.* **2011**, *52*, 9257. [CrossRef]

23. Krawitz, B.D.; Mo, S.; Geyman, L.S.; Agemy, S.A.; Scripsema, N.K.; Garcia, P.M.; Chui, T.Y.P.; Rosen, R.B. Acircularity index and axis ratio of the foveal avascular zone in diabetic eyes and healthy controls measured by optical coherence tomography angiography. *Vision Res.* **2017**, *139*, 177–186. [CrossRef] [PubMed]

24. Lim, H.B.; Kim, Y.W.; Kim, J.M.; Jo, Y.J.; Kim, J.Y. The Importance of Signal Strength in Quantitative Assessment of Retinal Vessel Density Using Optical Coherence Tomography Angiography. *Sci. Rep.* **2018**, *8*, 12897. [CrossRef] [PubMed]

25. Wang, X.; Kong, X.; Jiang, C.; Li, M.; Yu, J.; Sun, X. Is the peripapillary retinal perfusion related to myopia in healthy eyes? A prospective comparative study. *BMJ Open* **2016**, *6*, e010791. [CrossRef] [PubMed]

26. Garrity, S.T.; Iafe, N.A.; Phasukkijwatana, N.; Chen, X.; Sarraf, D. Quantitative Analysis of Three Distinct Retinal Capillary Plexuses in Healthy Eyes Using Optical Coherence Tomography Angiography. *Investig. Ophthalmol. Vis. Sci.* **2017**, *58*, 5548. [CrossRef]

27. Iafe, N.A.; Phasukkijwatana, N.; Chen, X.; Sarraf, D. Retinal Capillary Density and Foveal Avascular Zone Area Are Age-Dependent: Quantitative Analysis Using Optical Coherence Tomography Angiography. *Investig. Ophthalmol. Vis. Sci.* **2016**, *57*, 5780–5787. [CrossRef]

28. Leng, Y.; Tam, E.K.; Falavarjani, K.G.; Tsui, I. Effect of Age and Myopia on Retinal Microvasculature. *Ophthalmic Surg. Lasers Imaging Retina.* **2018**, *49*, 925–931. [CrossRef]

29. Jo, Y.H.; Sung, K.R.; Shin, J.W. Effects of Age on Peripapillary and Macular Vessel Density Determined Using Optical Coherence Tomography Angiography in Healthy Eyes. *Investig. Ophthalmol. Vis. Sci.* **2019**, *60*, 3492–3498. [CrossRef]

30. Yu, J.J.; Camino, A.; Liu, L.; Zhang, X.; Wang, J.; Gao, S.S.; Jia, Y.; Huang, D. Signal Strength Reduction Effects in OCT Angiography. *Ophthalmol. Retina.* **2019**, *3*, 835–842. [CrossRef]

31. Chang, R.; Nelson, A.J.; LeTran, V.; Vu, B.; Burkemper, B.; Chu, Z.; Fard, A.; Kashani, A.H.; Xu, B.Y.; Wang, R.K.; et al. Systemic Determinants of Peripapillary Vessel Density in Healthy African Americans: The African American Eye Disease Study. *Am. J. Ophthalmol.* **2019**, *207*, 240–247. [CrossRef]

32. Hsu, S.T.; Ngo, H.T.; Stinnett, S.S.; Cheung, N.L.; House, R.J.; Kelly, M.P.; Chen, X.; Enyedi, L.B.; Prakalpakorn, S.G.; Materin, M.M.; et al. Assessment of Macular Microvasculature in Healthy Eyes of Infants and Children Using OCT Angiography. *Ophthalmology* **2019**, *126*, 1703–1711. [CrossRef]

33. Czakó, C.; István, L.; Ecsedy, M.; Recsan, Z.; Sandor, G.; Benyo, F.; Horvach, H.; Papp, A.; Resch, M.; Borbandy, A.; et al. The effect of image quality on the reliability of OCT angiography measurements in patients with diabetes. *Int. J. Retina Vitr.* **2019**, *5*, 46. [CrossRef] [PubMed]

34. You, Q.S.; Chan, J.C.H.; Ng, A.L.K.; Choy, B.K.N.; Shih, K.C.; Cheung, J.J.C.; Wong, J.K.W.; Shum, J.W.H.; Ni, M.Y.; Lai, J.S.M.; et al. Macular Vessel Density Measured With Optical Coherence Tomography Angiography and Its Associations in a Large Population-Based Study. *Invest. Ophthalmol. Vis. Sci.* **2019**, *60*, 4830–4837. [CrossRef] [PubMed]

35. Tedeschi-Reiner, E.; Strozzi, M.; Skoric, B.; Reiner, Z. Relation of atherosclerotic changes in retinal arteries to the extent of coronary artery disease. *Am. J. Cardiol.* **2005**, *96*, 1107–1109. [CrossRef]

36. McGeechan, K.; Liew, G.; Macaskill, P.; Irwig, L.; Klein, R.; Klein, B.E.K.; Wang, J.J.; Mitchell, P.; Vingerling, J.R.; Dejong, P.T.V.M.; et al. Meta-analysis: Retinal vessel caliber and risk for coronary heart disease. *Ann. Intern. Med.* **2009**, *151*, 404–413. [CrossRef]

37. Wang, J.J.; Liew, G.; Klein, R.; Rochtchina, E.; Knudtson, M.D.; Klein, B.E.K.; Wong, T.Y.; Burlutsky, G.; Mitchell, P. Retinal vessel diameter and cardiovascular mortality: Pooled data analysis from two older populations. *Eur. Heart J.* **2007**, *28*, 1984–1992. [CrossRef] [PubMed]

38. Wang, J.J.; Liew, G.; Wong, T.Y.; Smith, W.; Klein, R.; Leeder, S.R.; Mitchell, P. Retinal vascular calibre and the risk of coronary heart disease-related death. *Heart Br. Card. Soc.* **2006**, *92*, 1583–1587. [CrossRef] [PubMed]
39. Wang, J.; Jiang, J.; Zhang, Y.; Qian, Y.W.; Zhang, J.F.; Wang, Z.L. Retinal and choroidal vascular changes in coronary heart disease: An optical coherence tomography angiography study. *Biomed. Opt Express* **2019**, *10*, 1532–1544. [CrossRef]
40. Cao, D.; Yang, D.; Huang, Z.; Zeng, Y.; Wang, J.; Hu, Y.; Zhang, L. Optical coherence tomography angiography discerns preclinical diabetic retinopathy in eyes of patients with type 2 diabetes without clinical diabetic retinopathy. *Acta. Diabetol.* **2018**, *55*, 469–477. [CrossRef]
41. Omae, T.; Nagaoka, T.; Yoshida, A. Effects of Habitual Cigarette Smoking on Retinal Circulation in Patients with Type 2 Diabetes. *Investig. Ophtalmol. Vis. Sci.* **2016**, *57*, 1345. [CrossRef] [PubMed]
42. Ting, D.S.W.; Tan, G.S.W.; Agrawal, R.; Yanagi, Y.; Sie, N.M.; Wong, C.W.; San Yeo, I.Y.; Lee, S.Y.; Cheung, C.M.G.; Wong, T.Y. Optical Coherence Tomographic Angiography in Type 2 Diabetes and Diabetic Retinopathy. *JAMA Ophthalmol.* **2017**, *135*, 306. [CrossRef] [PubMed]
43. Takayama, K.; Kaneko, H.; Ito, Y.; Kataoka, K.; Iwase, T.; Yasuma, T.; Matsuura, T.; Tsunekawa, T.; Shimizu, H.; Suzumura, A.; et al. Novel Classification of Early-stage Systemic Hypertensive Changes in Human Retina Based on OCTA Measurement of Choriocapillaris. *Sci. Rep.* **2018**, *8*, 15163. [CrossRef] [PubMed]
44. Lee, W.H.; Park, J.-H.; Won, Y.; Lee, M.-W.; Shin, Y.-I.; Jo, Y.-J.; Kim, J.-Y. Retinal Microvascular Change in Hypertension as measured by Optical Coherence Tomography Angiography. *Sci. Rep.* **2019**, *9*, 156. [CrossRef] [PubMed]
45. Clark, L.T.; Watkins, L.; Piña, I.L.; Elmer, M.; Akinboboye, O.; Gorham, M.; Jamerson, B.; McCullough, C.; Pierre, C.; Polis, A.B.; et al. Increasing Diversity in Clinical Trials: Overcoming Critical Barriers. *Curr. Probl. Cardiol.* **2019**, *44*, 148–172. [CrossRef] [PubMed]
46. Rabiolo, A.; Gelormini, F.; Sacconi, R.; Cicinelli, M.V.; Triolo, G.; Bettin, P.; Nouri-Mahdavi, K.; Bandello, F.; Querques, G. Comparison of methods to quantify macular and peripapillary vessel density in optical coherence tomography angiography. *PLoS ONE* **2018**, *13*, e0205773. [CrossRef] [PubMed]

Article

Early (5-Day) Onset of Diabetes Mellitus Causes Degeneration of Photoreceptor Cells, Overexpression of Incretins, and Increased Cellular Bioenergetics in Rat Retina

Jennifer O. Adeghate [1], Crystal D'Souza [2], Orsolya Kántor [3,4], Saeed Tariq [2], Abdul-Kader Souid [5] and Ernest Adeghate [2,*]

[1] Department of Ophthalmology, University of Pittsburgh School of Medicine, Pittsburgh, PA 15213, USA; jen.adeghate@gmail.com
[2] Department of Anatomy, College of Medicine & Health Sciences, United Arab Emirates University, Al Ain P.O. Box 17666, United Arab Emirates; crystal.dz@uaeu.ac.ae (C.D.); stariq@uaeu.ac.ae (S.T.)
[3] Department of Molecular Embryology, Institute of Anatomy and Cell Biology, Faculty of Medicine, University of Freiburg, D-79104 Freiburg, Germany; kantororsi@googlemail.com
[4] Department of Anatomy, Semmelweis University, Tűzoltó u. 58, H-1094 Budapest, Hungary
[5] Department of Pediatrics, College of Medicine & Health Sciences, United Arab Emirates University, Al Ain P.O. Box 17666, United Arab Emirates; asouid@uaeu.ac.ae
* Correspondence: eadeghate@uaeu.ac.ae; Tel.: +971-3-7137496

Abstract: The effects of early (5-day) onset of diabetes mellitus (DM) on retina ultrastructure and cellular bioenergetics were examined. The retinas of streptozotocin-induced diabetic rats were compared to those of non-diabetic rats using light and transmission electron microscopy. Tissue localization of glucagon-like-peptide-1 (GLP-1), exendin-4 (EXE-4), and catalase (CAT) in non-diabetic and diabetic rat retinas was conducted using immunohistochemistry, while the retinal and plasma concentration of GLP-1, EXE-4, and CAT were measured with ELISA. Lipid profiles and kidney and liver function markers were measured from the blood of non-diabetic and diabetic rats with an automated biochemical analyzer. Oxygen consumption was monitored using a phosphorescence analyzer, and the adenosine triphosphate (ATP) level was determined using the Enliten ATP assay kit. Blood glucose and cholesterol levels were significantly higher in diabetic rats compared to control. The number of degenerated photoreceptor cells was significantly higher in the diabetic rat retina. Tissue levels of EXE-4, GLP-1 and CAT were significantly ($p = 0.002$) higher in diabetic rat retina compared to non-diabetic controls. Retinal cellular respiration was 50% higher ($p = 0.004$) in diabetic ($0.53 \pm 0.16\ \mu M\ O_2\ min^{-1}\ mg^{-1}$, $n = 10$) than in non-diabetic rats ($0.35 \pm 0.07\ \mu M\ O_2\ min^{-1}\ mg^{-1}$, $n = 11$). Retinal cellular ATP was 76% higher ($p = 0.077$) in diabetic ($205 \pm 113\ pmol\ mg^{-1}$, $n = 10$) than in non-diabetic rats ($116 \pm 99\ pmol\ mg^{-1}$, $n = 12$). Thus, acute (5-day) or early onslaught of diabetes-induced hyperglycemia increased incretins and antioxidant levels and oxidative phosphorylation. All of these events could transiently preserve retinal function during the early phase of the progression of diabetes.

Keywords: diabetes; retinopathy; antioxidants; bioenergetics; respiration; ATP; glucagon-like peptide-1; exendin-4; catalase; immunohistochemistry; electron microscopy

Citation: Adeghate, J.O.; D'Souza, C.; Kántor, O.; Tariq, S.; Souid, A.-K.; Adeghate, E. Early (5-Day) Onset of Diabetes Mellitus Causes Degeneration of Photoreceptor Cells, Overexpression of Incretins, and Increased Cellular Bioenergetics in Rat Retina. *Cells* **2021**, *10*, 1981. https://doi.org/10.3390/cells10081981

Academic Editors: Maurice Ptito and Joseph Bouskila

Received: 22 June 2021
Accepted: 29 July 2021
Published: 4 August 2021

Publisher's Note: MDPI stays neutral with regard to jurisdictional claims in published maps and institutional affiliations.

1. Introduction

Diabetes mellitus (DM) is associated with impaired cellular bioenergetics, and its metabolic derangements are known to contribute to the development of retinopathy [1–3]. The degeneration of mitochondria in diabetic retina is evident in several studies, but measurements of retinal cellular bioenergetics have been limited [4,5].

Catalase (CAT) and other novel antioxidants, such as glucagon-like peptide-1 (GLP-1) and exendin-4 (EXE-4; a natural GLP-1 analog) are used as cytoprotective agents for the prevention and treatment of diabetic retinopathy [6,7]. The relevance of these incretins to cellular bioenergetics in diabetic retinal disease, however, remains unknown. In one study, EXE-4 was shown to improve oxygen consumption in adipocytes [8].

GLP-1 is a peptide produced by enteroendocrine L cells of the intestinal mucosa [9]. This hormone stimulates pancreatic beta cells and inhibits pancreatic alpha cells; thus, it lowers blood glucose by increasing insulin release and decreasing glucagon production. Although GLP-1 receptors are found predominantly in pancreatic islet cells, they are present in other organs as well, including the retina [10,11]. Importantly, the downregulation of GLP-1 receptors in the retinal pigment epithelium (RPE) causes cellular degeneration by the induction of apoptosis [12]. It remains to be seen, however, whether this cytoprotective activity of GLP-1 signaling could also improve mitochondrial function, especially in retinas affected by acute or early onset of DM.

EXE-4 occurs naturally and contains 39 amino acids. It was first isolated from the saliva of a giant lizard (gila monster), known as *Heloderma suspectum*. EXE-4 is an incretin that shares some similarities with GLP-1 [13]. EXE-4 has been shown to lower retinal inflammation by decreasing nuclear factor-κB (NF-κB) activation and leukocyte infiltration [14]. Other studies have also shown that EXE-4 lowers the intercellular adhesion molecule 1 (ICAM-1), the vascular cell adhesion protein 1 (VCAM-1), the receptor for advanced glycation end products (RAGE) and the placental growth factor (PGF) [15,16]. Hernández et al. have shown that local injections of GLP-1 or its analogs inhibit apoptosis and improve the function of retinal neurons [6]. Zhang et al. reported similar effects after injecting EXE-4 into the vitreous body of diabetic rats [17,18]. One study has shown a decrease in glucose-induced hydrogen peroxide in human and murine retinas following the administration of CAT [19]. In another study, oxidative stress-induced apoptosis was ameliorated by CAT [20].

A sudden increase in the blood glucose level, as seen in postprandial hyperglycemia has been shown to contribute to oxidative stress leading to many blood vessel-associated complications of DM, such as diabetic retinopathy [21]. These complications are even more pronounced in the presence of fluctuating blood levels of glucose [22]. The rodent model of experimental DM had a very high level of glucose in the first few weeks after the induction of DM [23].

Based on these results, we hypothesize that the retinal level of cytoprotective molecules such as incretins and CAT will increase in acute (early, 5-day) onset of DM to mitigate the sudden increase in hyperglycemia-induced oxidative stress. We also hypothesize that retinal bioenergetics will also increase.

In this study, we examined the distribution of incretins (GLP-1, EXE-4) and CAT (a representative member of endogenous antioxidants) in the retina of diabetic rats and examined whether incretins colocalize with endogenous antioxidants. In addition, we examined retinal ultrastructure and cellular bioenergetics after a sudden onset of hyperglycemia.

2. Materials and Methods

2.1. Experimental Animals

Twenty-four male Wistar rats (Harlan Laboratories, Oxon, UK) weighing approximately 272 g (3-month-old) were used for the study. The reason for including only male rats was to rule out the effects of hormones released during the estrous cycle. The experimental rats were bred and kept in plastic cages in a climate-controlled animal facility (23 ± 1 °C; 50 ± 4% humidity) with a 12 h day/12 h night daily cycle. Rodent chow and water were given ad libitum.

2.2. Induction of Diabetes

Diabetes was induced in twelve rats by streptozotocin (Sigma-Aldrich Chemical Co., St. Louis, MO, USA; administered intraperitoneally at 60 mg/kg body weight) and monitored with a One Touch Ultra 2 Glucometer® (Life scan Inc., Milpitas, CA, USA). Rats with a blood glucose ≥250 mg/dL 5 days after streptozotocin injection were used for the study. The weights were taken at the beginning and end (after 5 days) of the experiment using a laboratory scale (Sartorius, Goettingen, Germany).

2.3. Retina Retrieval for Immunohistochemistry and Transmission Electron Microscopy

The central part of the retina was rapidly excised from the left globe of 6 control and 6 diabetic rats (normal, $n = 6$; diabetic, $n = 6$), washed in phosphate-buffered saline (PBS), and fixed in McDowell's solution for transmission electron microscopy and Zamboni's solution for light microscopy as previously described [24,25]. Briefly, immediately following euthanasia (ether) and decapitation (guillotine), the eyes were removed in toto, placed on a paraffin wax-coated petri-dish filled with ice-cold PBS, and pinned down to paraffin wax via the surrounding connective tissue. The cornea was carefully excised at the corneo-scleral junction, and the lens and vitreous humor were removed to expose the retina. The trimming of the central portion of the retina was performed in the respective fixative (Zamboni or McDowell). The left eye was chosen for the light and electron microscopy methods while the right eye was set aside for the bioenergetics study. This method was also used to minimize the number of animals used for the different aspects of the study.

2.4. Retina Retrieval for Bioenergetics Experiments

Whole retinal extracts of the oculus dexter from twelve normal and twelve diabetic rats (normal $n = 12$, diabetic $n = 12$) were used for the bioenergetics experiments. In order to maintain the structure and intactness of the tissue, a retinal retrieval process was completed in less than 5 min. Briefly, immediately after ether anesthesia, the oculi were completely extracted and placed on a paraffin wax-filled petri-dish topped with ice-cold PBS. The eyes were pinned to the paraffin wax. The retinas were carefully extracted after the removal of the cornea, lens, and the vitreous humor. The entire retina from the right eye was then removed and transferred into a O_2-measuring glass vial containing 1 mL of Pd II phosphor solution (see section on cellular mitochondrial oxygen consumption and ATP content). The retinas were kept in the Pd II phosphor solution as soon as they were peeled off from the eye and were weighed using OHAUS microbalance (OHAUS Europe GmbH, Nänikon, Switzerland). The right eye was chosen to maintain consistency between animals and morphological and bioenergetics methods. All experiments including the bioenergetics experiments were conducted during the day.

2.5. Immunohistochemistry

Retina specimens were washed thoroughly in PBS (pH 7.2) and fixed at 4 °C overnight in Zamboni's solution [24]. The samples were then dehydrated in ascending concentrations of alcohol, cleared in xylene, and embedded in paraffin wax. Six μm-thick sections were cut and processed as previously described [24].

To determine whether EXE-4 or GLP-1 is co-localized with CAT, sections were incubated with antibodies against EXE-4 and CAT or GLP-1 and CAT before immune-labelling with tetramethylrhodamine isothiocyanate (TRITC)- or fluorescein isothiocyanate (FITC)-conjugated secondary antibodies [24,26]. Briefly, de-paraffinized retina sections were rinsed in PBS and immersed in 1% bovine serum albumin (BSA) for 30 min at 25 °C. Retina sections were immersed with either goat GLP-1 (Cat #: H-028-13) or EXE-4 polyclonal (Cat #: H-070-94; dilution, 1:200) antibodies (Phoenix Pharmaceuticals, 330 Beach Rd, Burlingame, CA, USA; 1:200) overnight at 4 °C. The sections were then incubated for 24 h at 4 °C with pre-diluted mouse anti-CAT antibodies (Abcam, 330 Cambridge Science Park, Cambridge, UK; Cat #: ab209211) after labeling GLP-1 and EXE-4 immunobinding sites with Goat Anti-Rabbit IgG H&L (TRITC) (Abcam Cat #: ab6718; Dilution: 1:100) in an overnight incubation at 4 °C. Immunoreactive tissue sites for CAT were identified with Fluorescein (FITC) AffiniPure Donkey Anti-Mouse IgG (H+L) (Jackson ImmunoResearch Laboratories, Inc., 872 West Baltimore Pike, West Grove, PA, USA; Dilution: 1:100,) (Table 1). The sections were cover-slipped using Immunomount® (Shandon, Pittsburgh, PA, USA), and were examined with a Nikon Eclipse Ni fluorescent microscope (Nikon Instruments Europe BV,

Tripolis 100, Burgerweeshuispad 101, Amsterdam, The Netherlands). Image acquisition parameters were identical in both non-diabetic and diabetic rats. The intensity of the fluorescence was quantified using Image J Software (NIH, Bethesda, MD, USA).

Table 1. List of primary and secondary antibodies used for immunohistochemistry.

#	Antibody	Source	Type	Cat No.	Dilution	Manufacturer
1	GLP-1	Rabbit	Polyclonal	H-028-13	1:200	Phoenix Pharmaceuticals Inc. Burlingame, CA, USA
2	EXE-4	Rabbit	Polyclonal	H-070-94	1:200	Phoenix Pharmaceuticals Inc. Burlingame, CA, USA
3	Catalase	Mouse	Monoclonal	ab209211	1:200	Abcam, Cambridge, MA, USA
4	TRITC-conjugated secondary antibody	Goat	Polyclonal	ab6718	1:100	Abcam, Cambridge, MA, USA
5	FITC-conjugated secondary antibody	Donkey	Monoclonal	715-095-150	1:100	Jackson ImmunoResearch Laboratories, Europe Ltd. (Ely, Cambridgeshire, UK)

2.6. Morphometric Analysis

The measurement of fluorescence intensity was performed using Image J software (NIH, Bethesda, MD, USA). The intensity of the signal was averaged over all the layers per field. Ten images were analyzed for each group (diabetic and non-diabetic rats).

2.7. Transmission Electron Microscopy

Retina samples (6 control and 6 diabetic retinas) taken from the central part of the left eye were dehydrated in ascending concentrations of ethanol, and were embedded in Epon resin (Agar Scientific, Essex, UK). Embedded capsules were polymerized with ultraviolet light (360–365 nm) overnight at 25 °C in a TAAB ultraviolet chamber [27]. Semi-thin sections were cut after the blocks were trimmed with a razor blade. Ultra-thin sections (95 nm each) were also made and were mounted onto carbon-formvar-coated 200-mesh Nickel grids using a Reichert Ultracuts ultramicrotome (Leica, Microsystems GmbH, Vienna, Austria). Contrast staining was performed with 2% uranyl acetate and lead citrate, for 15 and 7 min, respectively [27]. The sections were viewed with a Philips CM10 transmission electron microscope (Philips, Amsterdam, The Netherlands).

2.8. Blood Biochemistry

One week after streptozotocin injection, rat blood (~3 mL) was collected from the inferior vena cava into BD Vacutainer® Plus (Becton, Dickinson and Company, Franklin Lakes, NJ, USA) tubes (six rats per group). The samples were spun for 10 min at 3000 revolutions per minute, and the serum was stored at −80 °C. Glucose, triglycerides, total cholesterol, low-density lipoprotein (LDL), cholesterol, high-density lipoprotein (HDL), cholesterol, total protein, aspartate aminotransferase (AST), alanine aminotransferase (ALT), creatinine, and blood urea nitrogen (BUN) were measured using a COBAS III automated biochemical analyzer (Roche Diagnostics, Mannheim, Germany).

2.9. Measurements of Incretins and Catalase in the Retina and Plasma

For the enzyme-linked immunosorbent assay (ELISA) of EXE-4 and GLP-1, pre-weighed (using OHAUS microbalance), whole retina samples ($n = 6$ for diabetic; and $n = 6$ for non-diabetic rats) were homogenized in PBS, using a Janke & Kunkel Ultra Turrax t25 homogenizer (Freiburg, Baden-Württemberg, Germany). EXE-4 (*Heloderma suspectum*) was measured using the ELISA kit EK-070-94 (Phoenix Pharmaceuticals Inc., Burlingame, CA, USA), which had a sensitivity of 0.08 ng/mL (range, 0.08–0.86 ng/mL) and cross reacted only with EXE-4. GLP-1 measurements were performed as per the manufacturer's protocol (GLP-1—Sigma Aldrich RAB0201). The sensitivity of the assay was 1.17 pg/mL, the intra-assay coefficient of variation (CV) was <10%, and the inter-assay CV was <15%. The reagents did not cross-react with similar cytokines. The plasma of the corresponding rats was used for plasma EXE-4 and GLP-1 ELISA measurements. The measurement of CAT activity was performed using a colorimetric method. The measurement was performed according to the protocol provided with the commercial kit (Cayman Chemical, Ann Arbor, MI, USA).

2.10. Cellular Mitochondrial Oxygen Consumption and ATP Content

The rate of cellular respiration in the whole retina was determined using a phosphorescence O_2 analyzer as previously described [28]. The Pd (II) derivative of meso-tetra-(4-sulfonatophenyl)-tetrabenzoporphyrin was used, with an absorption maximum at 625 nm and a phosphorescence emission maximum at 800 nm (Porphyrin Products, Logan, UT, USA). The measurements were performed at 37 °C in closed 1 mL vials containing the Pd phosphor solution. The Pd phosphor solution was prepared daily, kept on ice, and was warmed to 37 °C prior to use. Briefly, whole retinas were rapidly excised and immediately immersed in an oxygen-measuring glass vial containing 1 mL PBS (with or without 5 mM glucose), 3 μM Pd phosphor, and 0.5% fat-free albumin. The vials were sealed with crimp-top aluminum seals. The mixing was performed with the aid of parylene-coated stirring bars. The samples were exposed to a diode array emitting pulse light (OTL630A-5-10- 66-E, Opto Technology, Inc., Wheeling, IL, USA). Captured emissions were recorded using a Hamamatsu photomultiplier tube after passing through a wide-band interference filter centered at 800 nm. The amplified phosphorescence was digitized at 1–2 MHz using an analog/digital converter (PCI-DAS 4020/12 I/O Board) with 1 to 20 MHz outputs (Computer Boards, Mansfield, MA, USA). The instrument was calibrated using the glucose oxidase system, as previously described [28]. Cellular respiration was inhibited with the addition of 10 mM sodium cyanide (CN), ensuring that oxidation occurred in the respiratory chain only. The addition of glucose oxidase (which catalyzes the reaction D-glucose + $O_2 \rightarrow$ D-glucono-δ-lactone + H_2O_2) depleted residual O_2 in the solution. The phosphorescence oxygen analyzer was used instead of other technologies, such as microelectrodes and the Seahorse XFe24 Analyzer, because of the ease of the runs, the constancy of the probe, and the low-cost.

To measure cellular ATP, small retina fragments were rapidly collected and immediately immersed in 1.0 mL ice-cold 2% trichloroacetic acid (prepared daily). The samples were homogenized by vigorous vortexing, and were centrifuged at −8 °C at 16,873 g for 10 min. The supernatants were stored at −80 °C in small aliquots until the analysis. Fifty μL of the cellular acid extracts were then thawed on ice and neutralized with 50 μL of 100 mM Tris-acetate and 2 mM ethylenediaminetetraacetic acid (pH 7.75). The measurements used the Enliten ATP Assay System (Bioluminescence Detection Kit, Promega, Madison, WI, USA). The luminescence reaction contained 5 μL of the neutralized acid-soluble supernatant and 45 μL of the luciferin/luciferase reagent. The luminescence intensity was measured at 25 °C using the Glomax Luminometer (Promega, Madison, WI, USA). The standard ATP concentration curve was linear (10 pM to 100 nM, $R^2 > 0.9999$).

2.11. Statistical Analysis

The analyses (including means and standard deviations) were performed using the SPSS statistical package (Version 20). The Mann–Whitney (non-parametric, 2 independent variables) test was used to compare the two groups of samples. A *p*-value of <0.05 was considered statistically significant.

3. Results

3.1. Metabolic Parameters and Blood Chemistry

The weight of the animals decreased significantly after the onset of acute or early 5-day) diabetes; however, the blood glucose level rose markedly to a high level of 26.6 ± 2.2 mmol/L. This is a significant rise of blood glucose compared to that of the normal control (4.0 ± 0.4 mmol/L). In addition, the blood levels of triglycerides and total cholesterol were significantly higher compared to the control at this stage of early hyperglycemia-associated DM. The kidney lesion markers also rose markedly in diabetic rats compared to the controls (Table 2).

Table 2. Body weight and serum biomarkers in non-diabetic and diabetic rats.

	Non-Diabetic	Diabetic	p
Weight, g	272 ± 11	221 ± 11	**<0.0002**
Glucose, mmol/L	4.0 ± 0.4	26.6 ± 2.2	**<0.0001**
Triglycerides, mmol/L	0.6 ± 0.2	1.7 ± 0.6	**<0.001**
Total cholesterol, mmol/L	0.9 ± 0.1	1.2 ± 0.3	**<0.03**
LDL, mmol/L	0.1 ± 0.0	0.1 ± 0.1	<0.5
HDL, mmol/L	0.7 ± 0.1	0.9 ± 0.5	<0.42
Total protein, g/L	60 ± 5	55 ± 11	<0.43
AST, U/L	86 ± 20	98 ± 20	<0.34
ALT, U/L	48 ± 11	63 ± 19	<0.11
Creatinine, µmol/L	27 ± 3	41 ± 15	**<0.04**
BUN, mmol/L	9 ± 1	12 ± 2	**<0.04**

Values are mean $\pm$ SD (n = 6 for each group). The measurements were performed five days after streptozotocin injection. LDL, low-density lipoprotein; HDL, high-density lipoprotein; AST, aspartate aminotransferase; ALT, alanine aminotransferase; BUN, blood urea nitrogen. Statistically significant p-values are in bold.

3.2. Transmission Electron Microscopy of the Retina (TEM)

The TEM showed degeneration of the photoreceptor layer of the retina 5 days after the induction of DM. The degeneration was prominent in the outer segments of the photoreceptor layer where the discs and photoreceptor proteins are located. In addition, the degeneration of the cells in the outer nuclear layer, containing the cell bodies of photoreceptor cells, was also conspicuous in the diabetic rat retina compared to those of the non-diabetic rats (Figure 1). The degeneration of photoreceptor mitochondria was also observed (Figure S1).

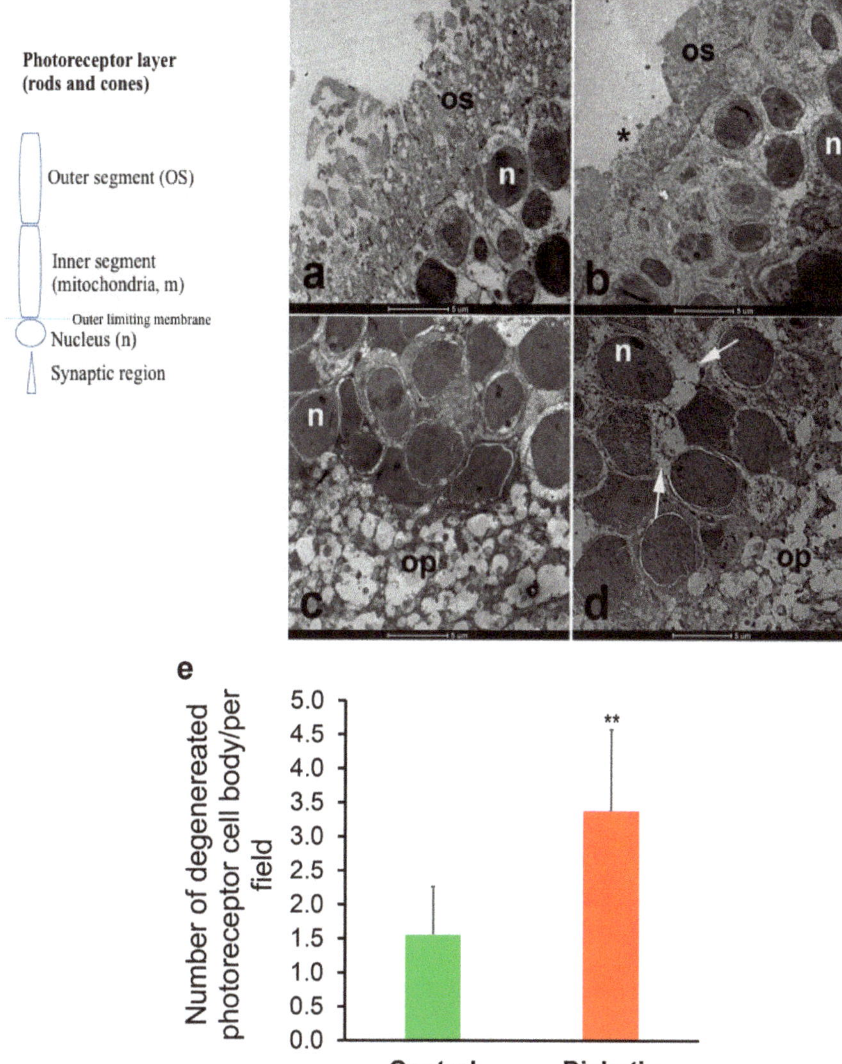

Figure 1. Representative transmission electron micrographs of the retina of non-diabetic (**a,c**) and diabetic (**b,d**) rats. Micrographs (**a,b**) are representatives of the outer segment (OS), the external limiting membrane and part of the outer nuclear layer of the photoreceptor layer. Micrographs (**c,d**) are representatives of the outer nuclear layer of the retina. The outer segment of the photoreceptor layer (OS) is significantly thinner in the diabetic rat retina (**b**). The degeneration of cells of the outer nuclear layer is also conspicuous in the diabetic rat retina ((**d**), arrows); $n = 6$ independent experiments. n, nucleus in the outer nuclear layer; asterisk = outer segment, op = outer plexiform layer. Scale bar = 5 µm. (**e**) shows the morphometric analysis of the number of degenerated photoreceptor cell bodies in the control versus in the diabetic rat retina. ** $p < 0.05$ (control versus diabetic).

3.3. Expression of Incretins and Catalase in the Diabetic Rat Retina

Immunofluorescence studies (Figure 2) and enzyme-linked immunosorbent assays were then performed to investigate the EXE-4, GLP-1, and CAT contents in the retinas of both normal and diabetic rats. These molecules have previously been shown to protect against the development of diabetic retinopathy. Their proposed mode of action is to intercept reactive oxygen species, which are increased in diabetes. As shown, both EXE-4 and GLP-1 were slightly, but not significantly, overexpressed in the retina of diabetic rats, most notably in the outer segment of photoreceptor cells (Figure 2A,B). These incretins co-localized with CAT in the rat retina layers (Figure 2A–C).

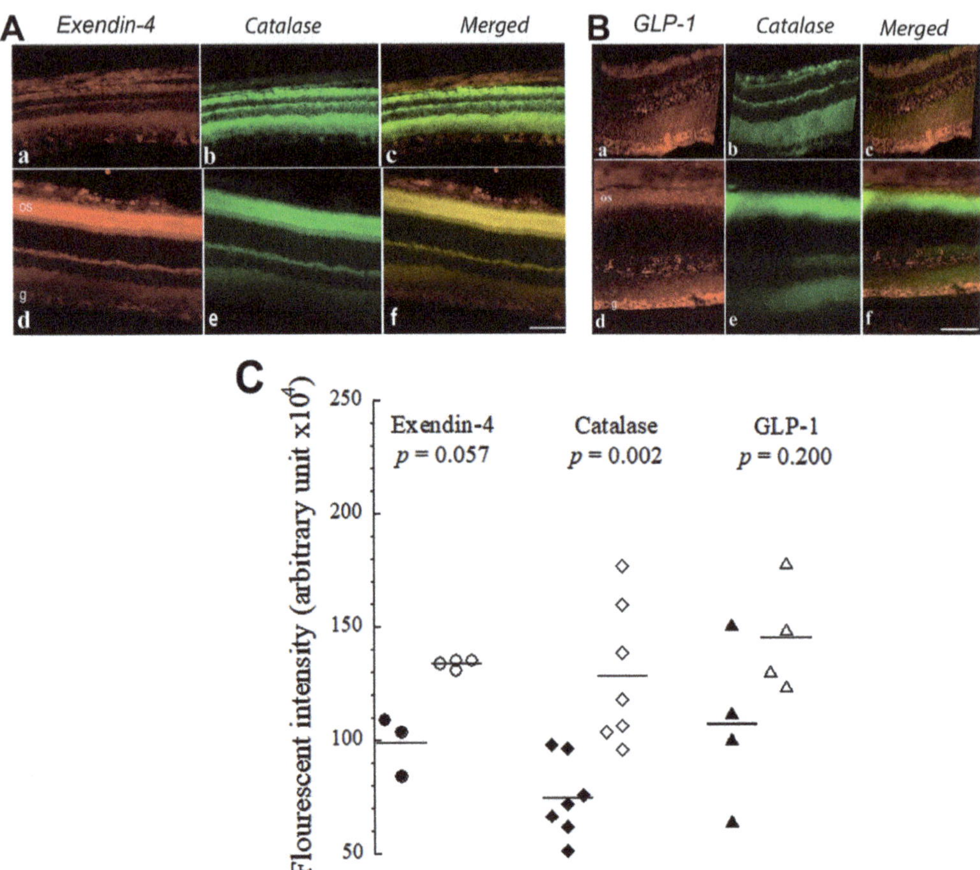

Figure 2. (**A**) Immunofluorescence images of Extendin (EXE-4) (**a,d**), catalase (CAT) (**b,e**) and merged signals (**c,f**) in the retina of non-diabetic (**a–c**) and diabetic (**d–f**) rats. (**B**) Immunofluorescence images of GLP-1 (**a,d**), CAT (**b,e**) and merged signals (**c,f**) in the retina of non-diabetic (**a–c**) and diabetic (**d–f**) rats. Scale bar = 25 µm. Note that EXE-4 and GLP-1 colocalize with CAT (merged signals (**c,f**)). (**C**) Dot plot of the total fluorescent intensities of experiments (**A,B**), respectively, using Image J software (NIH, Bethesda, MA, USA). Filled symbols are non-diabetic retinas and unfilled symbols are diabetic retinas (number of independent measurements = 4–7). The value of *p* is based on non-parametric, 2 independent samples (Mann–Whitney) test. The horizontal lines represent the mean. os = outer segment of photoreceptor cells; g = ganglion cell layer of retina.

3.4. Retinal and Plasma Levels of Incretins and Catalase in Normal and Diabetic Rats

Using the ELISA method, we measured the retina and plasma concentrations of EXE-4 and GLP-1 to determine whether the levels of these molecules changed after the onset of acute (early) diabetes. Consistent with these immunofluorescence findings, the protein levels in the retinas of diabetic rats were also significantly higher than in the retinas of non-diabetic rats (*p* = 0.004, Figure 3A). Their plasma protein levels were slightly, but not significantly higher in diabetic rats compared to the control rats (Figure 3B). The colorimetric method showed that the retinal and plasma levels of CAT increased significantly after an early (5-day) onset of diabetes when compared to normal controls (Figure 3C).

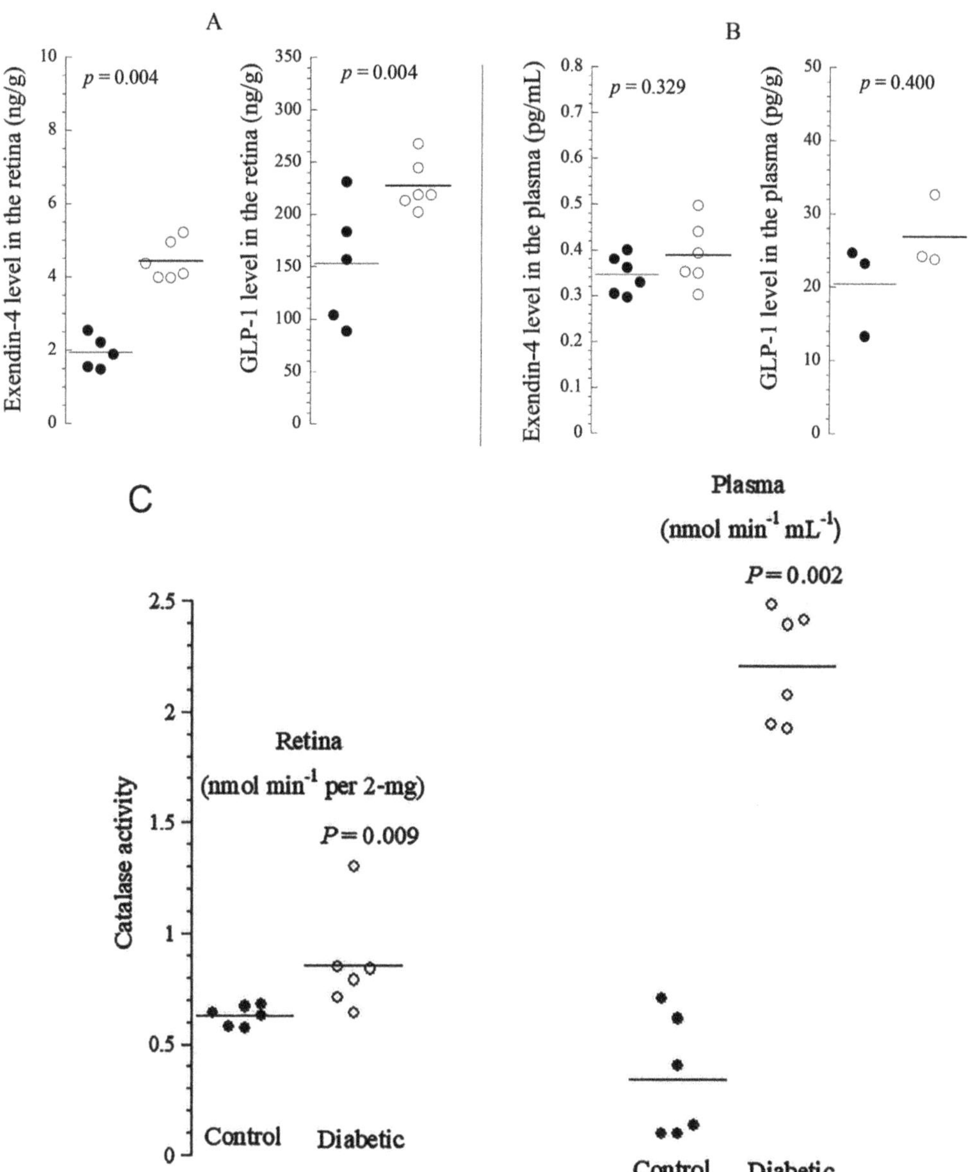

Figure 3. Enzyme-linked immunosorbent (ELISA) assays of EXE-4 and GLP-1 in the retina (**A**) and plasma (**B**) of non-diabetic retinas (filled circles) and diabetic retinas (unfilled circles); number of independent measurements = 3–6). (**C**) Retinal and plasma concentration of catalase (CAT). The value of p is based on non-parametric, 2 independent samples (Mann–Whitney) test. The horizontal lines represent the mean. Thus, the levels of incretins are significantly higher in the retinas of diabetic rats. Note that the retinal and plasma concentration of CAT is significantly higher in the plasma and diabetic rat retinas compared to controls.

3.5. Retinal Cellular Respiration and ATP Content

The observed ultrastructural changes in the photoreceptor cell layer (Figure 1) and the overexpression of incretins and catalase in diabetic rat retinas prompted the following functional oxidative phosphorylation studies. The oxygen consumption in the fragments of the retina was investigated as shown in Figure 4A–E. Briefly, the reaction mixture contained retinal fragments suspended in 1.0 mL of phosphate-buffered saline (PBS) in a sealed glass vial with and without added glucose, as described in the Methods section. The dissolved oxygen in the vial decreased linearly with time, indicating a zero-order kinetic process (Figure 4A). The zero-order rate constant (k_c, in μM O_2 min^{-1} mg^{-1}) in 'PBS + glucose' was 0.58 and in 'PBS + glucose oxidase' (which depletes extracellular glucose) was 0.43 (25% lower). Thus, the bulk of the metabolic fuel that supplied retinal cellular respiration was endogenous (Figure 4A). The addition of sodium cyanide completely inhibited O_2 consumption, confirming that the oxidation occurred only in the mitochondrial respiratory chain (Figure 4B–E). It is important to note that respiration was observed in the cornea fragment (Figure 4E), demonstrating that cellular respiration could also be elicited in other ocular tissues.

We then examined the rate of cellular respiration in non-diabetic and diabetic rat retinas (Figure 5A,B). As shown, cellular respiration was 50% higher ($p = 0.004$) in the diabetic rat retinas (0.53 ± 0.16 μM O_2 min^{-1} mg^{-1}, $n = 10$) compared to non-diabetic rat retinas (0.35 ± 0.07 μM O_2 min^{-1} mg^{-1}, $n = 11$). Cellular ATP content was also 76% higher ($p = 0.077$) in diabetic rat retinas (205 ± 113 pmol mg^{-1}, $n = 10$) compared to non-diabetic rat retinas (116 ± 99 pmol mg^{-1}, $n = 12$), as can be seen in Figure 5B. Thus, the cellular bioenergetics were increased but not at statistically significant levels in diabetic rat retinas compared to non-diabetic controls.

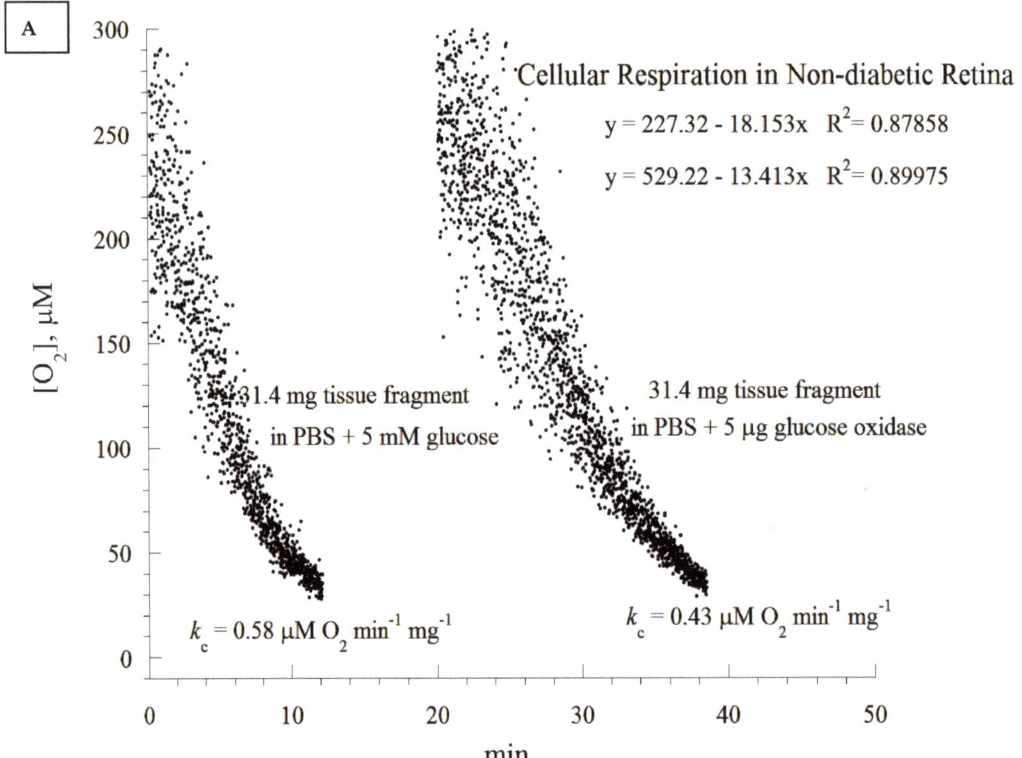

Figure 4. *Cont.*

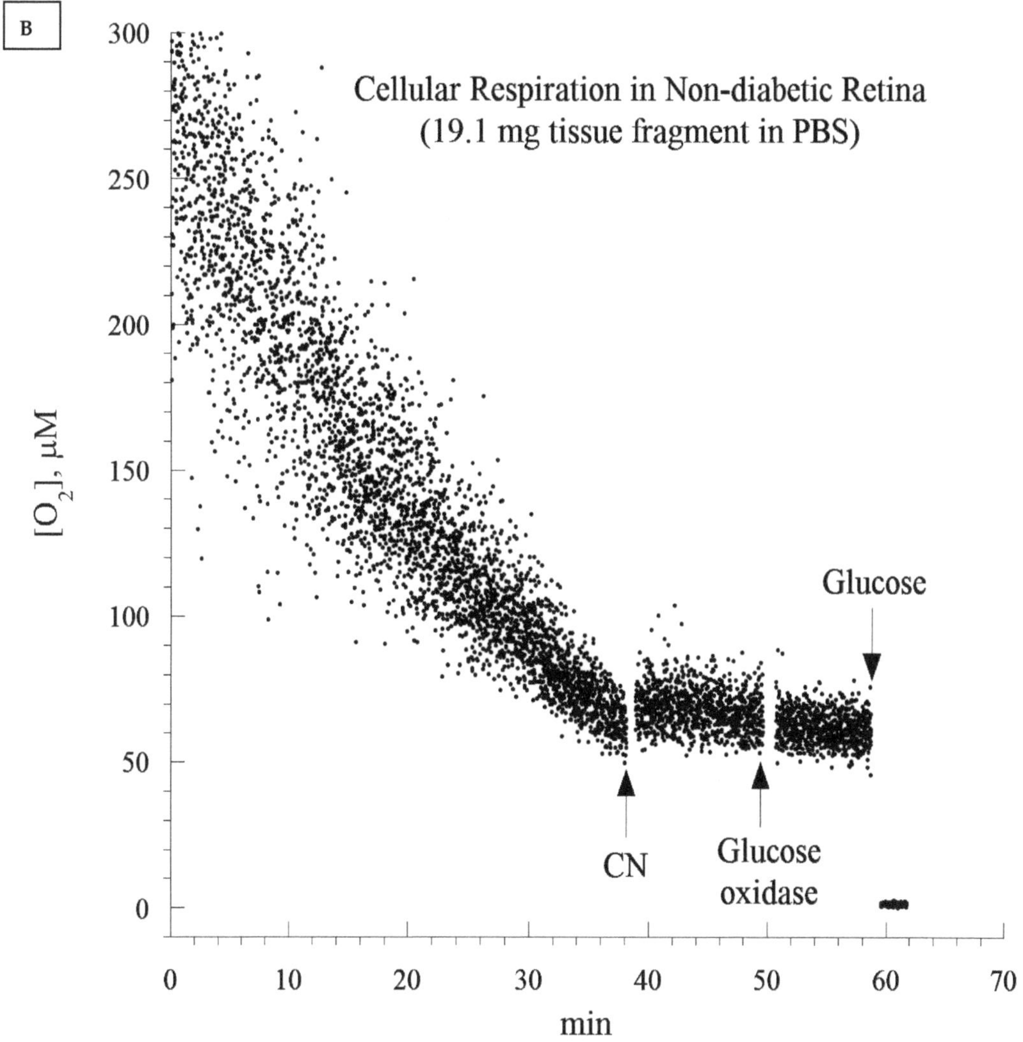

Figure 4. *Cont.*

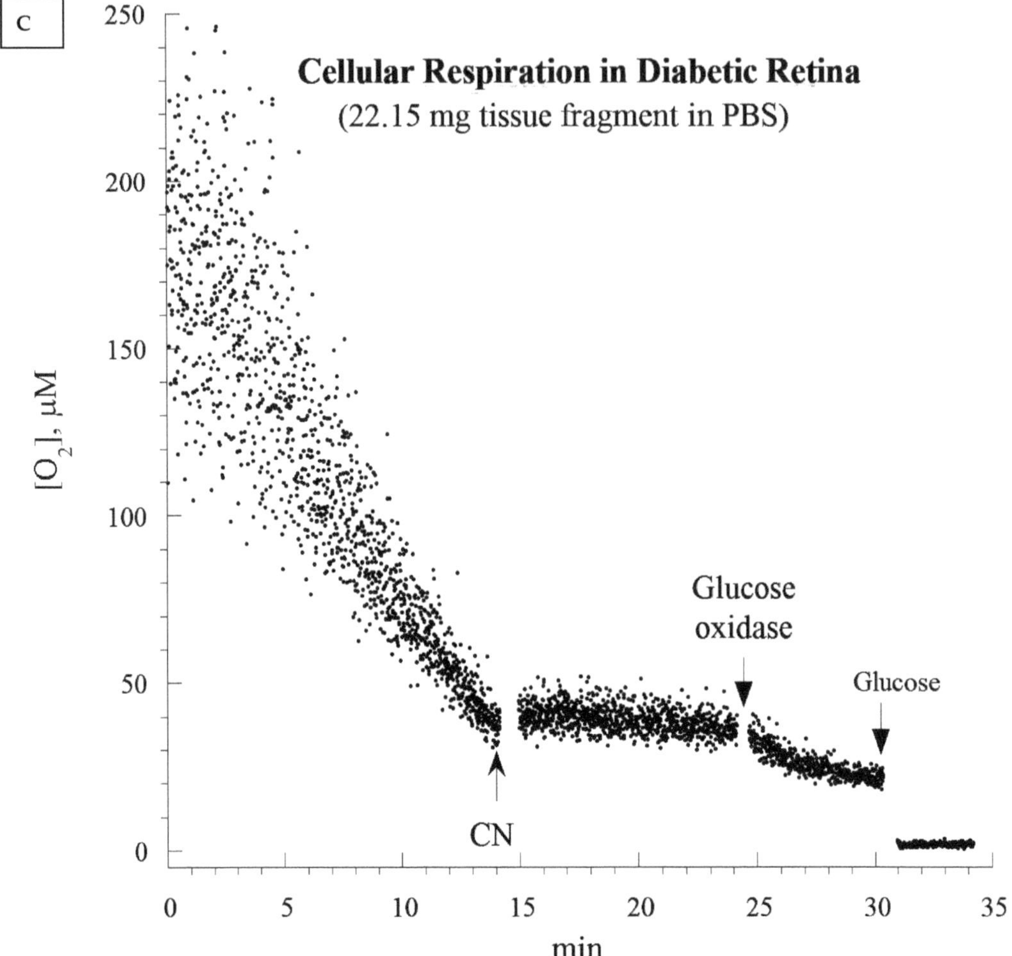

Figure 4. *Cont.*

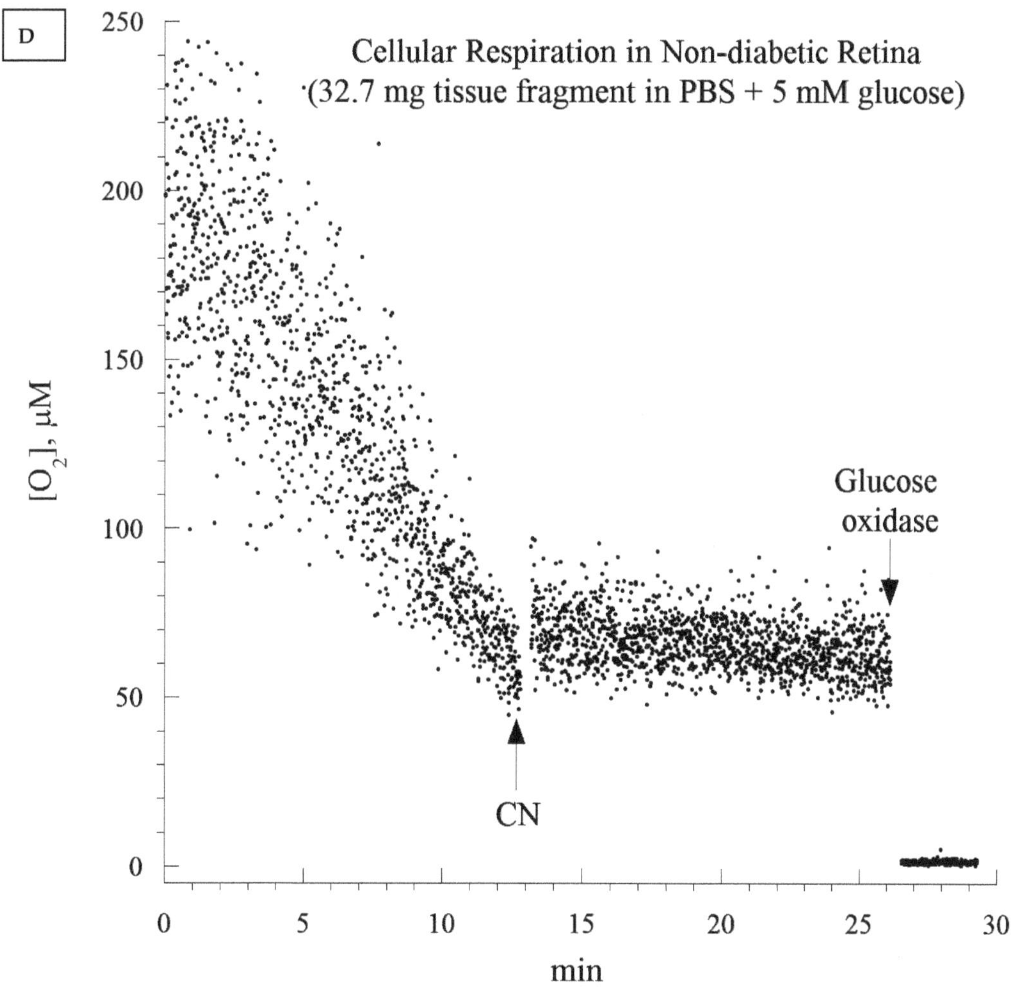

Figure 4. *Cont.*

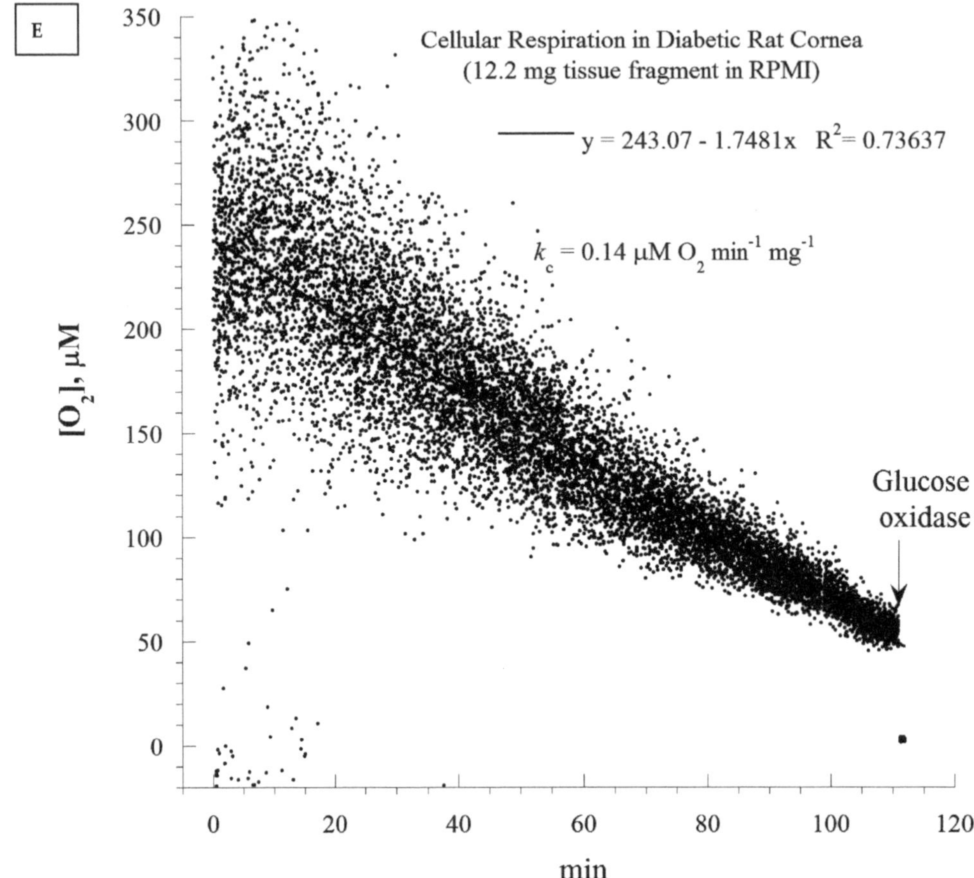

Figure 4. Retina specimens were processed for measurements of O_2 consumption in sealed vials at 37 °C, as described in the Methods section. Representative experimental runs are shown. The rate of cellular respiration (k, μM O_2 min^{-1}) was set as the negative of the slope of $[O_2]$ vs. time. The values of k_c (μM O_2 min^{-1} mg^{-1}) are shown at the bottom of each run. (**A**) Retinal cellular respiration in phosphate-buffered saline (PBS) supplemented with 5.0 mM glucose or 5.0 μg glucose oxidase (depletes extracellular glucose). Glucose or glucose oxidase were added on two separate samples (two separate conditions). (**B–D**) Retinal cellular respiration showing a complete inhibition of O_2 consumption with the addition of sodium cyanide (CN), confirming that oxidation occurred only in the mitochondrial respiratory chain. The addition of 'glucose + glucose oxidase' resulted in a rapid depletion of glucose in the media. In (**B**), the run was in PBS alone; i.e., cellular respiration was driven by endogenous nutrients. At minute 39, CN was added, halting cellular respiration by inhibiting complex IV of the respiratory chain, cytochrome oxidase. To validate this conclusion, glucose oxidase and glucose were then added successively, as shown. As a result, the enzyme glucose oxidase catalyzed the reaction: D-glucose + O_2 $\rightarrow$ D-glucono-δ-lactone + H_2O_2; thus, the remaining oxygen was depleted in the solution. In (**C**), the addition of glucose oxidase followed by glucose near the end of the run demonstrated that the halt of oxygen consumption after the addition of CN occurred in the presence of oxygen. The presence of both glucose oxidase and glucose rapidly depleted the remaining oxygen in the solution. (**E**) Corneal cellular respiration in Roswell Park Memorial Institute (RPMI) cell growth media with the addition of 5.0 μg glucose oxidase is shown. (**E**) shows the outcome of a control experiment, validating viable tissues such as the cornea, immediately following an eye removal surgery (i.e., before the removal of the retina). As cellular respiration was measured immediately after tissue extraction, the remaining experiments were confined to the retina.

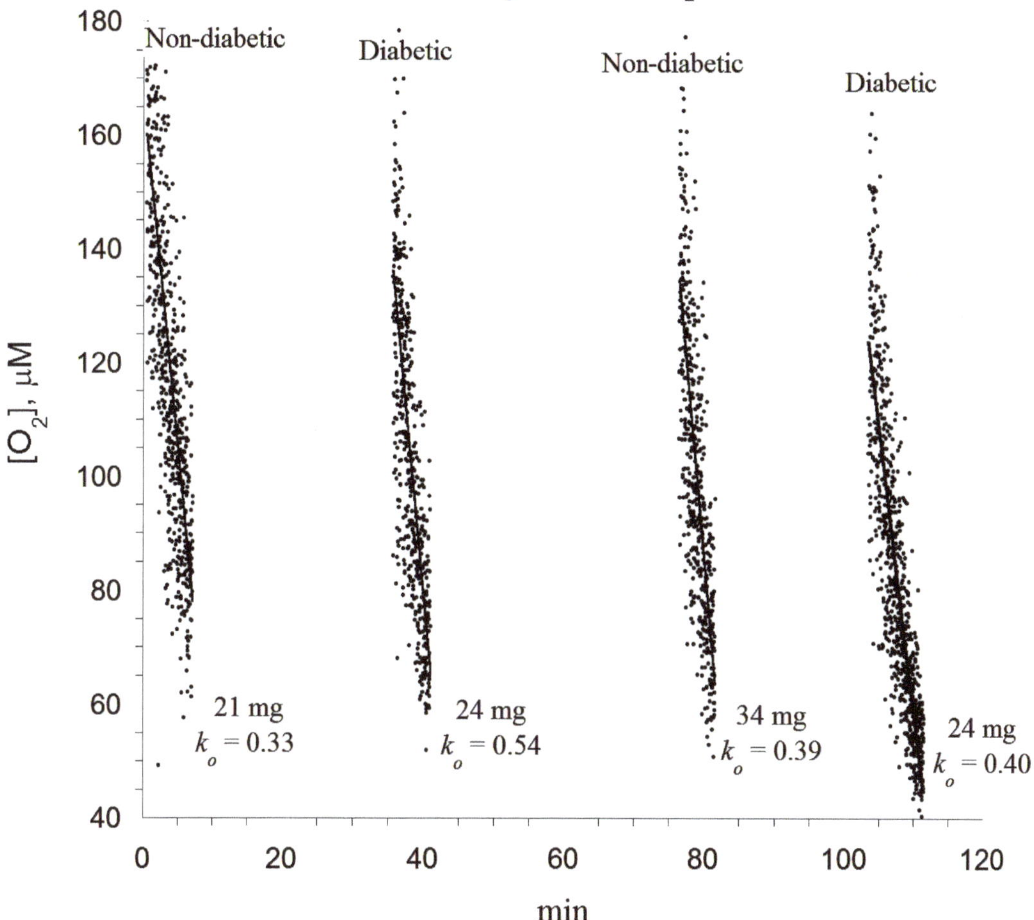

Figure 5. *Cont.*

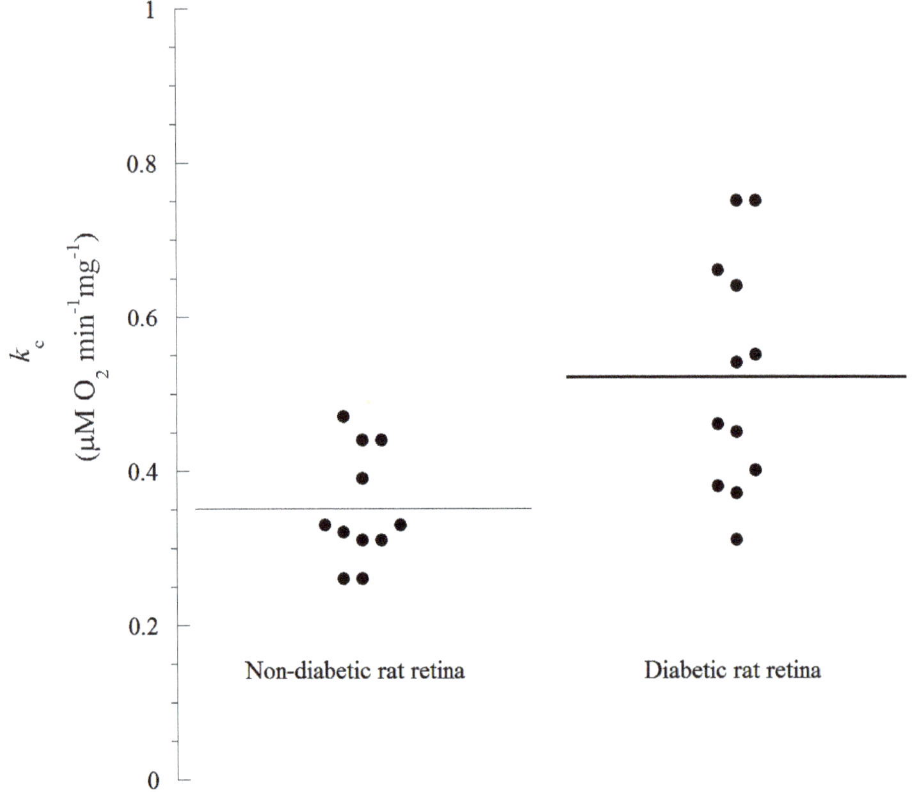

Figure 5. Representative, independent experimental runs of cellular respiration performed on individual non-diabetic and diabetic rat retinas followed by a summary of all results. (**A**) The rate of cellular respiration (k, μM O$_2$ min^{-1}) was set as the negative of the slope of [O$_2$] vs. time. The values of k_c (μM O$_2$ min^{-1} mg^{-1}) are shown at the bottom of each run. These runs (in PBS + 5 mM glucose) compared cellular respiration between diabetic and non-diabetic tissues. (**B**) Summary of all results, including retinal cellular ATP measurements (coupled processes). It is evident that the values of k_c between the diabetic and non-diabetic tissues are overlapping.

4. Discussion

In our study, the acute (early) onslaught of diabetes-induced hyperglycemia induced structural degeneration in the diabetic rat retina. This phenomenon is shown here to be associated with incretin and CAT overexpression, and with increased mitochondrial oxidative phosphorylation. Ultrastructural degeneration of the photoreceptor cell may damage the mitochondria of the diabetic rat retina thereby enhancing the overall process of cellular bioenergetics, possibly by attracting other compensatory mechanisms.

A profound degeneration of the photoreceptor discs was observed after 5 days of the induction of diabetes. These findings are in agreement with the observation of Énzsöly et al., who stated that the outer segment of the photoreceptor layer is most susceptible to diabetes-induced changes in the retina [1]. This implies that some structures within the retina are especially sensitive to the adverse effects of diabetes. Retinal degeneration in the photoreceptor cell layer of streptozotocin-induced diabetic rats has been previously reported by Aizu et al. [3] and Park et al. [2] and has been partially linked to the downregulation of GLP-1 receptors [12]. Studies conducted in our laboratory showed that structural and biochemical changes can be observed just hours after the induction of diabetes [23]. In other studies, the heart rate, temperature, and physical activity of rats declined three days after the induction of diabetes [29]. It is probably not surprising therefore to see biochemical and morphological alterations in the retina 5 days after the induction of diabetes.

Incretins and catalase are cytoprotective molecules, which intercept reactive oxygen species (ROS) that are generated from damaged mitochondria and increased cellular mitochondrial oxygen consumption (e.g., uncoupling oxidative phosphorylation). Their expression is known to inhibit apoptosis [12]; therefore, they block the 'mitochondrial pathway of apoptosis'. Consequently, incretins have been used as an adjunct therapy for diabetic retinopathy. While mild mitochondrial uncoupling has been shown to be beneficial for cell survival and ageing by reducing ROS production through reduced mitochondrial transmembrane potential [30–32], a severe mitochondrial uncoupling may produce a different effect. However, agents such as 2,4-dinitrophenol that are able to uncouple oxidative phosphorylation have been shown to improve cell metabolism and prolong lifespan [32]. In the present study, the extent of cell and tissue damage appeared to be sudden (acute) and severe, triggering a large variety of compensatory mechanisms before the body settles for the long-term effect of diabetes. In the in vivo and in vitro studies conducted by Ola [33], it was observed that high blood glucose causes a reduction in ROS release in the diabetic rat retina when compared to normal controls. This phenomenon was associated with a reduced glucose oxidation rate. The same study also reported an increase in ROS production, in vivo, in the retina of diabetic rats. The study concluded that non-mitochondrial structures may contribute to the ROS-induced diabetic retinopathy [33]. All of these observations showed that more studies are needed to completely elucidate the mechanisms by which diabetic retinopathy develops.

GLP-1 has been shown to be present in the pancreas [24] and brain [34], and we showed here that it is abundant in the retina along with its natural agonist exendin-4 [6,7]. The endogenous antioxidant catalase has been observed in rabbit eyes [35], but it has not been previously described in diabetic rat retinas. In this study, both EXE-4 and GLP-1 co-localized with catalase, mainly in the photoreceptor layer. Their levels were significantly higher in diabetic rat retinas. The increased production is likely a compensatory mechanism for neutralizing ROS. Thus, the overexpression of incretins and catalase in diabetic rat retina is expected to ameliorate apoptosis and increase mitochondrial oxidative phosphorylation (improving cellular function, including neuronal and non-neuronal elements). Enhanced oxidation in the mitochondrial respiratory chain provides the excess energy required for cellular repair mechanisms. The colocalization of incretins with catalase suggests that these cytoprotective proteins may be working together to combat diabetes-induced oxidative stress. Indeed, it been shown that antioxidants work together to mitigate the adverse effect of molecules released during oxidative stress [36].

A cellular respiration study has not been previously conducted in diabetic retinopathy in rats. We demonstrated here that retinal cellular mitochondrial oxygen consumption was significantly increased as a result of diabetes, and was associated with a concomitant elevation in retinal cellular ATP. The increase in both cellular respiration and ATP content could be considered as a compensatory mechanism against diabetes-induced oxidative stress. Since most of the mitochondria degenerated in the diabetic rat retina, the most likely explanation for the increase in the ATP level could probably be a shift in ADP/ATP balance due to a lower ATP. It could also be due to a higher rate of phosphorylation as a compensatory mechanism for the degenerated mitochondria. We propose that the overexpression of the studied endogenous cytoprotective molecules ameliorates apoptosis and enhances mitochondrial oxidative phosphorylation.

It is worth noting that the in vitro addition of GLP-1 at 1.0 μM did not directly affect the rate of retinal cellular respiration (data not shown). This result corroborates that of

Góralska et al., who reported that GLP-1 agonists did not affect mitochondrial electron chain transport in white adipose tissue [8]. In contrast, an increase in non-ATP associated O_2 consumption in human fat cells was noted [8]. Another study showed that short bouts of hyperglycemia (5.19 ± 0.83 h) did not influence O_2 consumption in feline photoreceptor cells [37]. This finding may have been due to the shorter periods of hyperglycemia. In our study, the serum of diabetic rats showed significantly elevated levels of glucose posing a significant challenge to the structure and the function of the retina. Indeed, glucotoxicity has been shown to cause deleterious effects and lesions to the retina [37]. The reason why GLP-1 did not improve retinal cellular respiration in this study is unknown. It is possible that a high concentration of GLP-1 or a longer stimulation period would have produced a beneficial effect. More studies are needed to make a definitive conclusion.

Biochemical studies showed that the creatinine and blood urea nitrogen levels increased when compared to non-diabetic controls. In contrast, low-density lipoprotein cholesterol, high-density lipoprotein cholesterol, and liver enzymes (aspartate aminotransferase and alanine aminotransferase) did not change significantly after 5 days of diabetes. These findings reveal the proneness of the liver and kidneys to early damage in diabetes, and suggest that the signs of diabetic nephropathy, such as microalbuminuria, may be seen relatively early in the course of the disease [38]. All of these indicate that a short duration of diabetes-induced hyperglycemia is enough to cause significant lesions in key organs such as the liver, kidney, and the retina. Indeed, it is worth noting that biochemical changes have been reported as early as one hour after the induction of diabetes [23].

The ATP content measured in the retinas could be due to either the increased activity of viable mitochondria, or to a larger contribution of ATP-generating mitochondria from other layers of the retina. More studies are needed to correlate the changes in the mitochondrial structure with the function in different layers of the retina. We agree with the conclusion that the cellular ATP content between diabetic and non-diabetic tissues did not reach the defined statistical significance of $p < 0.05$. This finding may suggest that the higher rate of respiration in the diabetic retina was only partly coupled to ATP synthesis, an effect that is usually referred to as "uncoupled cellular bioenergetics". This point is relevant to the above comment on retinal degeneration in diabetes. As this study, to the best of our knowledge, is the first to report on retinal bioenergetics in a diabetic rat model, future research is needed to investigate the size of mitochondrial uncoupling in the retina, and compare it to that of the diabetic retina.

It seems that diabetic retinas are slower at depleting O_2 compared to their non-diabetic counterparts (20 min difference). This appears to contradict the notion of higher respiration associated with the diabetic state. It is worth noting that the analysis of the rate of respiration was based on the 'best fit curve', a regression model that estimated the slope using a zero-order kinetic process. Time-dependent deviations from the expected zero-order kinetics require different models. This has to be completed in future research, as its mathematical approach needs to be validated in several tissues before the modeling of 'time-course of retinal degeneration in vitro'.

The observed retinal degeneration could reflect a marked excess of ROS that exceeded the protective capacities of incretins, catalase, and other endogenous antioxidants. This suggests that the early and optimal provision of cytoprotective agents, such as *N*-acetylcysteine, sodium methanethiolate (mesna), and 2-(3-aminopropylamino) ethylsulfanylphosphonic acid (amifostine) could complement endogenous mechanisms that are known to ameliorate the adverse events of diabetes in the retina.

In conclusion, we have demonstrated that several layers of the retina contain incretins and catalase, and that the retinal levels of these molecules increase significantly in early diabetes. These endogenous antioxidants may provide a protective mechanism against hyperglycemia-induced oxidative stress in the retina. In addition, these cytoprotective agents may ameliorate apoptosis and improve mitochondrial oxidative phosphorylation. Additional studies are needed to understand the dynamics of cellular respiration in relation to incretin and antioxidant levels, which could aid in the development of novel therapies for diabetic retinopathy. Our results can be summarized as follows: early (acute) diabetes induces the degeneration of the retina, which results in increased incretins and catalase production and increased retinal cellular bioenergetics.

Supplementary Materials: The following are available online at https://www.mdpi.com/article/10.3390/cells10081981/s1, Figure S1: Representative electron micrographs of photoreceptors cells.

Author Contributions: J.O.A.: study design, conduction of experiments, data analysis, literature review, manuscript writing and editing.; C.D.: conduction of experiments; O.K.: literature review, manuscript writing and editing; S.T.: conduction of experiments, provision of reagents; A.-K.S.: study design, provision of reagents, conduction of experiments, literature review, data analysis, manuscript writing and editing; E.A.: study design, provision of reagents, conduction of experiments, literature review, data analysis, manuscript writing and editing. All authors have read and agreed to the published version of the manuscript.

Funding: The present project was funded by the UAEU research grants from the College of Medicine and Health Sciences (G00003388), UAEU (G00003451) and the Zayed Bin Sultan Center for Health Sciences: G00003417. The ideas and views given in this manuscript are solely those of the contributing authors and do not reflect the ideas and wishes of the funding institutions.

Institutional Review Board Statement: The study was approved by the Animal Research Ethics Committee of the College of Medicine and Health Sciences at United Arab Emirates University (#A5-14). It is confirmed that all methods and procedures were performed in accordance with the National Institute of Health (NIH) guidelines and regulations for the care and use of laboratory animals.

Informed Consent Statement: Not applicable.

Data Availability Statement: All data generated or analyzed during this study are included in this published article.

Acknowledgments: The authors would like to thank Dhanya Saraswathiamma for her technical assistance.

Conflicts of Interest: The authors have declared that no conflict of interest exists.

References

1. Énzsöly, A.; Szabó, A.; Kántor, O.; Dávid, C.; Szalay, P.; Szabó, K.; Szél, Á.; Németh, J.; Lukáts, Á. Pathologic Alterations of the Outer Retina in Streptozotocin-Induced Diabetes. *Investig. Opthalmol. Vis. Sci.* **2014**, *55*, 3686–3699. [CrossRef] [PubMed]
2. Park, S.-J.; Park, J.-W.; Kim, K.-Y.; Chung, J.-W.; Chun, M.-H.; Oh, S.-J. Apoptotic death of photoreceptors in the streptozotocin-induced diabetic rat retina. *Diabetologia* **2003**, *46*, 1260–1268. [CrossRef]
3. Aizu, Y.; Oyanagi, K.; Hu, J.; Nakagawa, H. Degeneration of retinal neuronal processes and pigment epithelium in the early stage of the streptozotocin-diabetic rats. *Neuropathology* **2002**, *22*, 161–170. [CrossRef]
4. Padnick-Silver, L.; Linsenmeier, R.A. Effect of acute hyperglycemia on oxygen and oxidative metabolism in the intact cat retina. *Investig. Opthalmol. Vis. Sci.* **2003**, *44*, 745–750. [CrossRef]
5. Doliba, N. Bioenergetics and Type 2 Diabetes. *Biochem. Pharmacol.* **2017**, *5*, 222. [CrossRef]
6. Hernández, C.; Bogdanov, P.; Corraliza, L.; Garcia-Ramírez, M.; Solà-Adell, C.; Arranz, J.A.; Arroba, A.I.; Valverde, A.M.; Simó, R. Topical administration of GLP-1 receptor agonists prevents retinal neurodegeneration in experimental diabetes. *Diabetes* **2015**, *65*, 172–187. [CrossRef]
7. Abbate, M.; Cravedi, P.; Iliev, I.; Remuzzi, G.; Ruggenenti, P. Prevention and Treatment of Diabetic Retinopathy: Evidence from Clinical Trials and Perspectives. *Curr. Diabetes Rev.* **2011**, *7*, 190–200. [CrossRef] [PubMed]
8. Góralska, J.; Śliwa, A.; Gruca, A.; Razny, U.; Chojnacka, M.; Polus, A.; Solnica, B.; Malczewska-Malec, M. Glucagon-like peptide-1 receptor agonist stimulates mitochondrial bioenergetics in human adipocytes. *Acta Biochim. Pol.* **2017**, *64*, 423–429. [CrossRef] [PubMed]
9. Drucker, D.J.; Nauck, M.A. The incretin system: Glucagon-like peptide-1 receptor agonists and dipeptidyl peptidase-4 inhibitors in type 2 diabetes. *Lancet* **2006**, *368*, 1696–1705. [CrossRef]
10. Campbell, J.E.; Drucker, D.J. Pharmacology, Physiology, and Mechanisms of Incretin Hormone Action. *Cell Metab.* **2013**, *17*, 819–837. [CrossRef] [PubMed]
11. Puddu, A.; Sanguineti, R.; Montecucco, F.; Viviani, G.L. Retinal pigment epithelial cells express a functional receptor for gluca-gon-like peptide-1 (GLP-1). *Mediat. Inflamm.* **2013**, *2013*, 975032. [CrossRef]
12. Kim, D.-I.; Park, M.-J.; Choi, J.-H.; Lim, S.-K.; Choi, H.-J.; Park, S.-H. Hyperglycemia-induced GLP-1R downregulation causes RPE cell apoptosis. *Int. J. Biochem. Cell Biol.* **2015**, *59*, 41–51. [CrossRef] [PubMed]

13. Lotfy, M.; Singh, J.; Kalász, H.; Tekes, K.; Adeghate, E. Medicinal Chemistry and Applications of Incretins and DPP-4 Inhibitors in the Treatment of Type 2 Diabetes Mellitus. *Open Med. Chem. J.* **2011**, *5*, 82–92. [CrossRef]

14. Gonçalves, A.; Lin, C.-M.; Muthusamy, A.; Fontes-Ribeiro, C.; Ambrósio, A.F.; Abcouwer, S.F.; Fernandes, R.; Antonetti, D. Protective Effect of a GLP-1 Analog on Ischemia-Reperfusion Induced Blood–Retinal Barrier Breakdown and Inflammation. *Investig. Opthalmol. Vis. Sci.* **2016**, *57*, 2584–2592. [CrossRef] [PubMed]

15. Fan, Y.; Liu, K.; Wang, Q.; Ruan, Y.; Ye, W.; Zhang, Y. Exendin-4 alleviates retinal vascular leakage by protecting the blood–retinal barrier and reducing retinal vascular permeability in diabetic Goto-Kakizaki rats. *Exp. Eye Res.* **2014**, *127*, 104–116. [CrossRef]

16. Dorecka, M.; Siemianowicz, K.; Francuz, T.; Garczorz, W.; Chyra, A.; Klych, A.; Romaniuk, W. Exendin-4 and GLP-1 decreases induced expression of ICAM-1, VCAM-1 and RAGE in human retinal pigment epithelial cells. *Pharmacol. Rep.* **2013**, *65*, 884–890. [CrossRef]

17. Zhang, Y.; Zhang, J.; Wang, Q.; Lei, X.; Chu, Q.; Xu, G.-T.; Ye, W. Intravitreal Injection of Exendin-4 Analogue Protects Retinal Cells in Early Diabetic Rats. *Investig. Opthalmol. Vis. Sci.* **2011**, *52*, 278–285. [CrossRef]

18. Zhang, Y.; Wang, Q.; Zhang, J.; Lei, X.; Xu, G.-T.; Ye, W. Protection of exendin-4 analogue in early experimental diabetic retinopathy. *Graefe's Arch. Clin. Exp. Ophthalmol.* **2008**, *247*, 699–706. [CrossRef] [PubMed]

19. Giordano, C.R.; Roberts, R.; Krentz, K.A.; Bissig, D.; Talreja, D.; Kumar, A.; Terlecky, S.R.; Berkowitz, B.A. Catalase Therapy Corrects Oxidative Stress-Induced Pathophysiology in Incipient Diabetic Retinopathy. *Investig. Opthalmol. Vis. Sci.* **2015**, *56*, 3095–3102. [CrossRef] [PubMed]

20. Chen, B.H.; Jiang, D.Y.; Tang, L.S. Advanced glycation end-products induce apoptosis involving the signaling pathways of oxida-tive stress in bovine retinal pericytes. *Life Sci.* **2006**, *79*, 1040–1048. [CrossRef]

21. Ceriello, A. Postprandial Hyperglycemia and Diabetes Complications: Is It Time to Treat? *Diabetes* **2004**, *54*, 1–7. [CrossRef]

22. Adeghate, E.A.; Kalász, H.; Al Jaberi, S.; Adeghate, J.; Tekes, K. Tackling type 2 diabetes-associated cardiovascular and renal comorbidities: A key challenge for drug development. *Expert Opin. Investig. Drugs* **2021**, *30*, 85–93. [CrossRef]

23. Adeghate, E.; Hameed, R.S.; Ponery, A.S.; Tariq, S.; Sheen, R.S.; Shaffiullah, M.; Donáth, T. Streptozotocin Causes Pancreatic Beta Cell Failure via Early and Sustained Biochemical and Cellular Alterations. *Exp. Clin. Endocrinol. Diabetes* **2010**, *118*, 699–707. [CrossRef]

24. Lotfy, M.; Singh, J.; Rashed, H.; Tariq, S.; Zilahi, E.; Adeghate, E. Mechanism of the beneficial and protective effects of exenatide in diabetic rats. *J. Endocrinol.* **2013**, *220*, 291–304. [CrossRef]

25. Elabadlah, H.; Hameed, R.; D'Souza, C.; Mohsin, S.; Adeghate, E.A. Exogenous Ghrelin Increases Plasma Insulin Level in Diabetic Rats. *Biomolecules* **2020**, *10*, 633. [CrossRef] [PubMed]

26. Adeghate, E.; D'Souza, C.; Saeed, Z.; Al Jaberi, S.; Tariq, S.; Kalász, H.; Tekes, K.; Adeghate, E. Nociceptin Increases Antioxidant Expression in the Kidney, Liver and Brain of Diabetic Rats. *Biology* **2021**, *10*, 621. [CrossRef]

27. Adeghate, E. Host-graft circulation and vascular morphology in pancreatic tissue transplants in rats. *Anat. Rec. Adv. Integr. Anat. Evol. Biol.* **1998**, *251*, 448–459. [CrossRef]

28. Tao, Z.; Goodisman, J.; Souid, A.K. Oxygen measurement via phosphorescence: Reaction of sodium dithionite with dissolved ox-ygen. *J. Phys. Chem. A* **2008**, *112*, 1511–1518. [CrossRef] [PubMed]

29. Howarth, F.C.; Jacobson, M.; Naseer, O.; Adeghate, E. Short-term effects of streptozotocin-induced diabetes on the electrocardio-gram, physical activity and body temperature in rats. *Exp. Physiol.* **2005**, *90*, 237–245. [CrossRef]

30. Brand, M. Uncoupling to survive? The role of mitochondrial inefficiency in ageing. *Exp. Gerontol.* **2000**, *35*, 811–820. [CrossRef]

31. Miwa, S.; Brand, M. Mitochondrial matrix reactive oxygen species production is very sensitive to mild uncoupling. *Biochem. Soc. Trans.* **2003**, *31*, 1300–1301. [CrossRef] [PubMed]

32. Da Silva, C.C.C.; Cerqueira, F.M.; Barbosa, L.F.; de Medeiros, M.H.G.; Kowaltowski, A.J. Mild mitochondrial uncoupling in mice affects energy metabolism, redox balance and longevity. *Aging Cell* **2008**, *7*, 552–560. [CrossRef] [PubMed]

33. Ola, M. Does Hyperglycemia Cause Oxidative Stress in the Diabetic Rat Retina? *Cells* **2021**, *10*, 794. [CrossRef]

34. McKay, N.J.; Galante, D.L.; Daniels, D. Endogenous glucagon-like peptide-1 reduces drinking behavior and is differentially engaged by water and food intakes in rats. *J. Neurosci.* **2014**, *34*, 16417–16423. [CrossRef] [PubMed]

35. Atalla, L.R.; Sevanian, A.; A Rao, N. Immunohistochemical localization of peroxidative enzymes in ocular tissue. *CLAO J.* **1990**, *16*, S30–S33. [PubMed]

36. Kurutas, E.B. The importance of antioxidants which play the role in cellular response against oxidative/nitrosative stress: Current state. *Nutr. J.* **2015**, *15*, 71. [CrossRef]

37. Al-Hussaini, H.; Kilarkaje, N. Effects of trans-resveratrol on type 1 diabetes-induced inhibition of retinoic acid metabolism pathway in retinal pigment epithelium of Dark Agouti rats. *Eur. J. Pharmacol.* **2018**, *834*, 142–151. [CrossRef]

38. Jerums, G.; MacIsaac, R.J.; MacIsaac, R.J. Treatment of Microalbuminuria in Patients with Type 2 Diabetes Mellitus. *Treat. Endocrinol.* **2002**, *1*, 163–173. [CrossRef]

Article

Reduced Levels of Drp1 Protect against Development of Retinal Vascular Lesions in Diabetic Retinopathy

Dongjoon Kim [1,2], **Hiromi Sesaki** [3] **and Sayon Roy** [1,2,*]

1 Department of Medicine, Boston University School of Medicine, Boston, MA 02118, USA; djkim@bu.edu
2 Department of Ophthalmology, Boston University School of Medicine, Boston, MA 02118, USA
3 Department of Cell Biology, Johns Hopkins University School of Medicine, Baltimore, MD 21205, USA; hsesaki@jhmi.edu
* Correspondence: sayon@bu.edu; Tel.: +1-617-358-6801

Abstract: High glucose (HG)-induced Drp1 overexpression contributes to mitochondrial dysfunction and promotes apoptosis in retinal endothelial cells. However, it is unknown whether inhibiting Drp1 overexpression protects against the development of retinal vascular cell loss in diabetes. To investigate whether reduced Drp1 level is protective against diabetes-induced retinal vascular lesions, four groups of mice: wild type (WT) control mice, streptozotocin (STZ)-induced diabetic mice, Drp1$^{+/-}$ mice, and STZ-induced diabetic Drp1$^{+/-}$ mice were examined after 16 weeks of diabetes. Western Blot analysis indicated a significant increase in Drp1 expression in the diabetic retinas compared to those of WT mice; retinas of diabetic Drp1$^{+/-}$ mice showed reduced Drp1 level compared to those of diabetic mice. A significant increase in the number of acellular capillaries (AC) and pericyte loss (PL) was observed in the retinas of diabetic mice compared to those of the WT control mice. Importantly, a significant decrease in the number of AC and PL was observed in retinas of diabetic Drp1$^{+/-}$ mice compared to those of diabetic mice concomitant with increased expression of pro-apoptotic genes, Bax, cleaved PARP, and increased cleaved caspase-3 activity. Preventing diabetes-induced Drp1 overexpression may have protective effects against the development of vascular lesions, characteristic of diabetic retinopathy.

Keywords: Drp1; apoptosis; mitochondria; diabetic retinopathy

Citation: Kim, D.; Sesaki, H.; Roy, S. Reduced Levels of Drp1 Protect against Development of Retinal Vascular Lesions in Diabetic Retinopathy. *Cells* **2021**, *10*, 1379. https://doi.org/10.3390/cells10061379

Academic Editor: Maurice Ptito

Received: 17 May 2021
Accepted: 27 May 2021
Published: 3 June 2021

Publisher's Note: MDPI stays neutral with regard to jurisdictional claims in published maps and institutional affiliations.

1. Introduction

Diabetic retinopathy is the leading cause of blindness in the working-age population and, unfortunately, there is no cure for this ocular complication [1]. Diabetic retinopathy is characterized by retinal vascular cell loss, a characteristic early stage lesion [2] which manifests as acellular capillaries and pericyte ghosts [3,4]. Increasing evidence indicates that changes in mitochondrial morphology can promote mitochondrial dysfunction and contribute to apoptotic cell death associated with diabetic retinopathy [5–15]. Our recent study has identified mitochondrial fragmentation in vascular cells of retinal capillaries in diabetes [8]. Maintenance of mitochondrial morphology is regulated by fission and fusion events and is integral to mitochondrial functionality. Specifically, imbalance in mitochondrial dynamics through increased mitochondrial fission by dynamin-related protein 1 (Drp1) is known to compromise mitochondrial morphology and lead to mitochondrial dysfunction [16,17]. However, it is unclear whether abnormal changes in Drp1 contribute to the pathophysiology of diabetic retinopathy.

Drp1 is a GTPase that is considered to be a principal regulator of mitochondrial fission [18,19]. It is primarily localized in the cytosol, and upon activation through GTP hydrolysis, it oligomerizes around the mitochondrial outer membrane to initiate fission, mitochondrial fragmentation [20], and ultimately induce apoptosis [21]. Specifically, findings from previous studies overwhelmingly suggest excess fission leads to deleterious effects, including mitochondrial fragmentation and apoptotic cell death [22–24]. We have recently reported that retinal endothelial cells grown in 30 mM HG exhibit Drp1 overexpression, and that reducing Drp1 expression protects against HG-induced mitochondrial fragmentation and apoptosis in vitro [10]. However, it is unclear whether decreasing Drp1 upregulation provides beneficial effects against apoptotic cell death.

Drp1 overexpression has been widely reported in HG and diabetic conditions. Podocytes and glomerular mesangial cells grown in HG medium exhibit significant Drp1 upregulation and increased mitochondrial fission, which promote podocyte loss and compromise glomerular function, suggesting that elevated Drp1 plays a role in the pathogenesis of diabetic nephropathy [25–27]. In addition, Drp1 overexpression was observed in pancreatic β-islet cells grown in HG medium concomitant with increased mitochondrial fission, cytochrome c release, reduced mitochondrial membrane potential, caspase-3 activation, and reactive oxygen species (ROS) production [28]. However, these reported changes were not evident in pancreatic β-islet cells carrying a dominant negative mutant of Drp1 [28], suggesting that Drp1 plays a critical role in promoting HG-induced apoptosis of pancreatic β-islet cells. Moreover, increased Drp1 levels have been implicated in promoting mitochondrial fragmentation, ROS accumulation, and contributing to apoptotic cell death of endothelial cells in models of atherosclerosis and diabetic cardiomyopathy [29,30]. Elevated Drp1 expression and increased mitochondrial fission were also observed in the dorsal root ganglion and hippocampus of diabetic animals, suggesting that Drp1 overexpression may contribute to the pathogenesis of diabetic neuropathy [31,32]. Taken together, these findings indicate a critical role for Drp1 in promoting mitochondrial fragmentation and apoptosis under HG and diabetic conditions.

To determine whether increased levels of Drp1 contribute to the development of apoptotic death of vascular cells in the diabetic retina, in the present study, we induced diabetes in the Drp1$^{+/-}$ mouse and investigated whether reduced levels of Drp1 in these mice were protective against the development of acellular capillaries and pericyte loss. Specifically, proapoptotic genes Bax, cleaved PARP, and caspase-3 activity were examined in addition to TUNEL assays, which were performed to identify vascular cells undergoing apoptosis in retinal capillaries.

2. Materials and Methods

2.1. Animals

Studies involving animals were carried out following the guidelines of the ARVO Statement for the Use of Animals in Ophthalmic and Vision Research and approved by the IACUC Committee of Boston University (PROTO201800411; approved on March 9th, 2021). Male and female mice were used in the present study to address any sex-related differences. A total of 12 male and 12 female WT C57BL/6J mice (The Jackson laboratory, Bar Harbor, Maine), as well as 12 male and 12 female Drp1$^{+/-}$ mice bred into the C57/BL6J strain background provided by Dr. Hiromi Sesaki [33] were used to conduct experiments. A detailed methodology on how Drp1$^{+/-}$ mouse model was generated can be referred to in a previous study [33]. Polymerase chain reaction (PCR) was performed using tail tip DNA of animals to verify their genotype. PCR was carried out with primer sequences as follows: Primer 1, 5′-ACCAAAGTAAGGAATAGCTGTTG-3′; Primer 2, 5′-GAGTACCTAAAGTGGACAAGAGGTCC-3′; Primer 3, 5′-CACTGAGAGCTCTATATGTA GGC-3′). Drp1$^{+/-}$ allele is represented as a 539-bp fragment amplified by primers 1 and 2. Drp1$^{+/+}$ allele is represented as a 315-bp fragment amplified by primers 2 and 3. In the present study, Drp1$^{-/-}$ mice were not used as this genotype is embryonically lethal [33–35].

12 WT mice and 12 Drp1$^{+/-}$ mice were randomly assigned to receive 5 consecutive STZ injections intraperitoneally to induce diabetes at a concentration of 40 mg/kg body weight. Additionally, 12 WT mice and 12 Drp1$^{+/-}$ mice were randomly assigned to receive 5 consecutive citrate buffer injections intraperitoneally as vehicle, representing non-diabetic control groups. To verify the diabetes status in the animals, blood and urine glucose levels were measured 3 days post-STZ injection. Routine blood glucose assessment was performed 3 times per week. Depending on the hyperglycemic status, NPH insulin

injections were administered to achieve a level of ~350 mg/dL. A total of 16 weeks after the onset of diabetes, animals were sacrificed, blood was collected from each animal and blood glucose and HbA1c levels were measured. Following sacrifice, retinas from each animal were isolated, and total protein extracted from all samples.

2.2. Immunostaining of Drp1 in Retinal Capillary Networks

To study the expression and distribution of Drp1 in the retinal capillary networks, retinal trypsin digestion (RTD) preparations [36] were subjected to immunostaining with Drp1 antibody. The RTD preparations were washed several times with $1\times$ PBS and subjected briefly to ice-cold methanol, followed by additional PBS washes. Then, the RTDs were exposed to a 2% BSA solution diluted in $1\times$ PBS for 15 min at room temperature to block non-specific antibody binding. Following blocking, the RTDs were subjected to a primary antibody solution containing mouse monoclonal Drp1 antibody (1:200 in 2% BSA-PBS solution, Catalog #sc-271583, Santa Cruz Biotechnology, Dallas, TX, USA) and incubated overnight at 4 °C in a moist chamber. After the overnight incubation, the RTDs were washed in PBS and incubated at room temperature with FITC-conjugated rabbit anti–mouse IgG secondary antibody (1:100 in 2% BSA-PBS Solution, Jackson for 1 h). After three PBS washes, RTDs were counterstained with DAPI, and mounted in SlowFade Diamond Antifade Mountant reagent (SlowFade Diamond; Molecular Probes, Eugene, OR, USA). Digital images were captured, and relative Drp1 immunofluorescence was quantified using the NIH Image J software from 10 random representative fields from each RTD.

2.3. Western Blot Analysis

WB analysis was carried out as described [10]. Briefly, total protein was isolated from retinas of experimental animals using a lysis buffer solution containing 10 mmol/L Tris, pH 7.5 (Sigma, Temecula, CA, USA), 1 mmol/L EDTA, and 0.1% Triton X-100 (Sigma). Bicinchoninic acid assay (Pierce Chemical, Rockford, IL, USA) was used to obtain protein concentrations of the retinal lysates, which were then subjected to WB analysis for Drp1, Bax, and PARP activation. Equal amount of protein (20 µg) of retinal lysates was loaded into each lane in a 10% SDS-polyacrylamide gel and electrophoresed, followed by semi-dry transfer [37] using a PVDF membrane (Millipore, Billerica, MA, USA). Following transfer, the membrane was exposed to a blocking solution containing 5% non-fat dry milk for 1 h and subsequently incubated overnight at 4 °C with mouse monoclonal Drp1 antibody (1:1000, Catalog #sc-271583, Santa Cruz Biotechnology), rabbit Bax antibody (1:1000, Catalog #2772, Cell Signaling, Danvers, MA, USA), or PARP antibody (1:500, Catalog #9542, Cell Signaling) in a solution comprised of 5% bovine serum albumin dissolved in 0.1% Tween-20 (TTBS). The following day, the membrane was subjected to incubation with a secondary antibody solution (anti-rabbit IgG, AP-conjugated antibody (1:3000, Catalog #7054, Cell Signaling) or anti-mouse IgG, AP-conjugated antibody (1:3000, Catalog #7056, Cell Signaling)) for 1 h in room temperature. The membrane was then exposed to a chemiluminescent substrate (Bio-Rad, Hercules, CA, USA) and chemiluminescence signals were captured using a digital imager (Fujifilm LAS-4000). The membrane underwent Ponceau-S staining after transfer or was re-probed with β-actin antibody (1:1000, Catalog #4967, Cell Signaling) to confirm equal loading. NIH Image J software was used to conduct densitometric analysis of the chemiluminescent signal non-saturating exposures.

2.4. Assessment of Caspase-3 Activity

To evaluate caspase-3 activity in retinas of diabetic animals and Drp1$^{+/-}$ animals, fluorometric analysis was carried out using a commercially available caspase-3 assay kit (Abcam, Cambridge, UK; Catalog #ab39383). Lysates from retinal tissues were isolated using the kit's proprietary lysis buffer, incubated on ice for 10 min, and homogenized. Following BCA assay of the retinal tissue lysates, 20 µg of protein from each sample was used to perform the fluorometric evaluation of caspase-3 activity. The retinal lysates were mixed with reaction buffer containing DTT (10 mM final concentration) and Acetyl-Asp-

Glu-Val-Asp-7-amino-4 trifluoromethylcoumarin (DEVD-AFC) (50 µM final concentration), a fluorogenic substrate specific to caspase-3. The reaction mixture representing each sample was transferred to corresponding wells in a 96-well plate, incubated at 37 °C for 2 h, and subjected to fluorescent excitation and emission at 400 nm and 505 nm, respectively. Specifically, cleavage of the DEVD-AFC substrate is carried out by activated caspase-3 resulting in the formation of free AFC molecules, which can be detected at 400 nm excitation and 505 nm emission [38]. Therefore, relative difference in DEVD-AFC cleavage between experimental groups was used to analyze caspase-3 activity.

2.5. Retinal Trypsin Digestion and Assessment of Acellular Capillaries and Pericyte Loss

After animals were sacrificed, eyes were enucleated and placed in 10% formalin, and retinas isolated and exposed to 0.5 M glycine for 24 h. To isolate retinal capillaries, RTD was performed as described [36]. Briefly, retinas were subjected to 3% trypsin, glia removed through tapping with a single hair brush, and mounted on a silane-coated slide. RTDs were stained with periodic acid-Schiff and hematoxylin as described [39]. Using a digital camera attached to a microscope (Nikon Eclipse; TE2000-S, Nikon, Tokyo, Japan), ten representative fields were imaged assessed for AC and PL. Vessels without endothelial cells and pericytes represented ACs. PL was determined by counting pericyte ghosts, which appear as "empty shells" representing dead pericytes.

2.6. Terminal dUTP Nick-End Labeling Assay

To detect cells undergoing apoptosis in retinal capillaries, terminal deoxynucleotidyl transferase-mediated uridine 5′-triphosphate-biotin nick end labeling (TUNEL) assay was performed using a kit (ApopTag Fluorescein In Situ Apoptosis Detection; Millipore Sigma) as described previously [39]. Briefly, RTDs were fixed in paraformaldehyde, permeabilized in a pre-cooled mixture of a 2:1 ratio of ethanol/acetic acid, washed in PBS, exposed to equilibration buffer, and incubated for 1 h with deoxyribonucleotidyl transferase (TdT) enzyme in a moist chamber at 37 °C. Following incubation, RTDs were exposed to anti-digoxigenin peroxidase, washed in PBS, counterstained with DAPI, and mounted using anti-fade reagent (SlowFade Diamond Antifade, Cat#S36963; Invitrogen, Carlsbad, CA, USA). At least five images representing random fields of the RTD slide were captured using a digital microscope (Nikon Eclipse; TE2000-S) and TUNEL-positive cells per total number of cells per field were analyzed.

2.7. Statistical Analysis

Data are shown as mean $\pm$ standard deviation. Values representing experimental groups are shown as percentages of the control. The normalized values were subjected to Student's *t*-test for comparisons between two groups, or one-way ANOVA followed by Bonferroni's post-hoc test for comparisons between multiple groups. Statistical significance was considered at $p < 0.05$.

3. Results

3.1. Drp1$^{+/-}$ Animal Model

Genotypes of animals used in the present study were confirmed through PCR analysis using DNA derived from the animals' tail tips. PCR data was used to confirm that wild-type (WT) mice (Drp1$^{+/+}$) exhibited a band as expected at 0.54 kb whereas Drp1 heterozygous knockout (Drp1$^{+/-}$) mice exhibited a band as expected at 0.32 kb (Figure 1).

3.2. Effect of Diabetes on Drp1 Expression and Distribution in Retinal Capillary Networks

To determine whether the distribution of Drp1 is altered in retinal capillary networks of diabetic animals and Drp1$^{+/-}$ animals, Drp1 immunostaining was performed in RTDs from each experimental group. Interestingly, Drp1 immunostaining was significantly increased in RTDs of diabetic mice compared to that of non-diabetic WT mice (151 $\pm$ 14% of WT vs. 100 $\pm$ 15% of WT, $p < 0.01$; $n = 6$; Figure 2A,B). As expected, Drp1 immunostaining was significantly decreased in RTDs of Drp1$^{+/-}$ mice compared to that of non-diabetic WT mice ($73 \pm 7\%$ of WT vs. $100 \pm 15\%$ of WT, $p < 0.05$; $n = 6$; Figure 2A,B). Drp1 immunostaining was significantly decreased in RTDs of diabetic Drp1$^{+/-}$ mice compared to that of diabetic mice ($111 \pm 16\%$ of WT vs. $151 \pm 14\%$ of WT, $p < 0.01$; $n = 6$; Figure 2A,B), and significantly increased compared to non-diabetic Drp1$^{+/-}$ mice ($111 \pm 16\%$ of WT vs. $73 \pm 7\%$ of WT, $p < 0.01$; $n = 6$; Figure 2A,B).

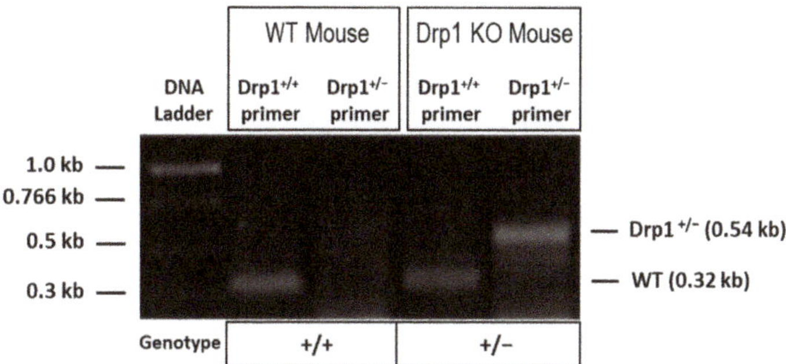

Figure 1. PCR analysis using mice tail tip DNA indicating genotypes of Drp1 heterozygous knockout ($^{+/-}$) and wild-type (WT) mice. The wild-type Drp1 allele (Drp1$^{+/+}$) is represented by a band at 0.32 kb, whereas the disrupted allele (Drp1$^{+/-}$) shows a band at 0.54 kb.

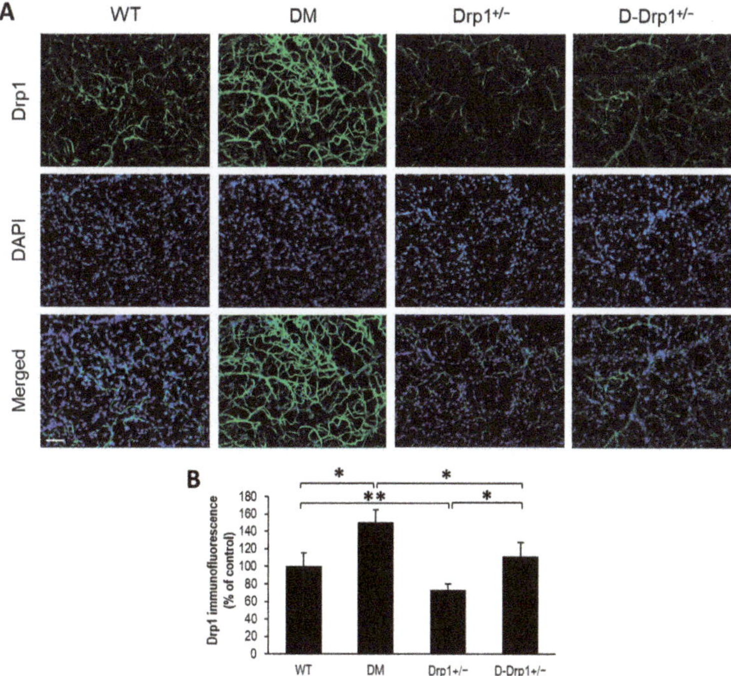

Figure 2. Drp1$^{+/-}$ mice exhibit reduced Drp1 immunofluorescence in the retinal capillary network. (**A**) Representative images of Drp1 immunofluorescence (green) and DAPI (blue) in retinal capillaries of wild-type (WT), diabetic (DM), Drp1$^{+/-}$, and diabetic Drp1$^{+/-}$ (D-Drp1$^{+/-}$) mice. Scale bar = 100 μm. (**B**) Graphical illustration of cumulative data shows decreased Drp1 immunofluorescence in retinal capillaries of Drp1$^{+/-}$ mice compared to that of WT mice. * $p < 0.01$, $n = 6$; ** $p < 0.05$, $n = 6$.

3.3. Normalization of Drp1 Expression in Retinas of Diabetic Drp1$^{+/-}$ Mice

Data from Western blot analysis showed that Drp1 expression level is significantly upregulated in diabetic mouse retinas compared to that of WT mouse retinas ($133 \pm 13\%$ of WT, $p < 0.01$; $n = 12$; Figure 3A,B). As expected, retinas of Drp1$^{+/-}$ exhibited a significant decrease in retinal Drp1 expression compared to that of WT mice ($63 \pm 10\%$ of WT, $p < 0.01$; $n = 6$; Figure 3A,B). In parallel, Drp1 expression was brought to near normal levels in retinas of diabetic Drp1$^{+/-}$ mice ($80 \pm 16\%$ of WT, $p < 0.01$; $n = 12$; Figure 3A,B) compared to diabetic mice.

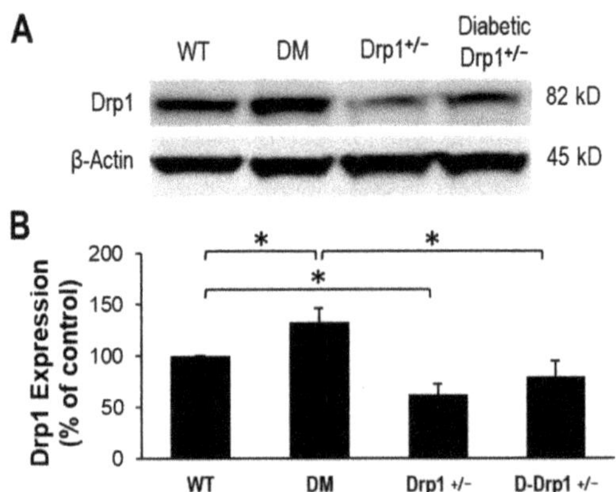

Figure 3. Drp1 expression is normalized in diabetic Drp1 $^{+/-}$ mouse retinas. (**A**) Representative WB image shows Drp1 protein levels in the retinas of WT, diabetic (DM), Drp1$^{+/-}$, and diabetic Drp1$^{+/-}$ (D-Drp1$^{+/-}$) mice. (**B**) Graphical illustration of cumulative data shows diabetes significantly upregulates Drp1 expression and that Drp1 expression is normalized in retinas of D-Drp1$^{+/-}$ mice. Data are expressed as mean $\pm$ SD. * $p < 0.01$, $n = 12$.

3.4. Diabetes-Induced Drp1 Upregulation Promotes Apoptosis

To assess whether reduced Drp1 level is protective against diabetes-induced pro-apoptotic genes, expression levels of Bax and cleaved PARP, as well as caspase-3 activity were monitored in retinal tissues. In retinas of diabetic mice, gene expression levels of pro-apoptotic Bax and cleaved PARP were significantly increased (Bax: $152 \pm 23\%$ of WT, $p < 0.01$; $n = 12$; Figure 4A,B; Cleaved PARP: $156 \pm 27\%$ of WT, $p < 0.01$; $n = 12$; Figure 4A,C) concomitant with increased caspase-3 activity ($120 \pm 16\%$ of WT, $p < 0.01$; $n = 12$; Figure 4D) compared to those of non-diabetic WT mice. Interestingly, reduced Drp1 level in retinas of diabetic Drp1$^{+/-}$ mice showed a decrease in diabetes-induced Bax expression ($113 \pm 19\%$ of WT, $p < 0.01$; $n = 12$; Figure 4A,B), PARP cleavage ($112 \pm 28\%$ of WT, $p < 0.01$; $n = 12$; Figure 4A,C), and caspase-3 activation ($102 \pm 8\%$ of WT, $p < 0.01$; $n = 12$; Figure 4D).

3.5. Drp1 Downregulation Inhibits Vascular Cell Apoptosis in the Diabetic Retina

To assess whether downregulating Drp1 expression protects apoptotic cell death in retinal vascular cells, TUNEL assay was performed in RTDs from each experimental group. Data indicate that there is a significant increase in the number of TUNEL-positive cells in RTDs of diabetic mice compared to that of non-diabetic WT mice ($284 \pm 28\%$ of WT vs. $100 \pm 33\%$ of WT, $p < 0.01$; $n = 6$; Figure 5A–M). Importantly, when diabetes-induced Drp1 overexpression was brought to near normal levels, the number of TUNEL-positive cells was significantly reduced in RTDs of diabetic Drp1$^{+/-}$ mice compared to that of diabetic mice ($180 \pm 29\%$ of WT vs. $284 \pm 28\%$ of WT, $p < 0.05$; $n = 6$; Figure 5A–M).

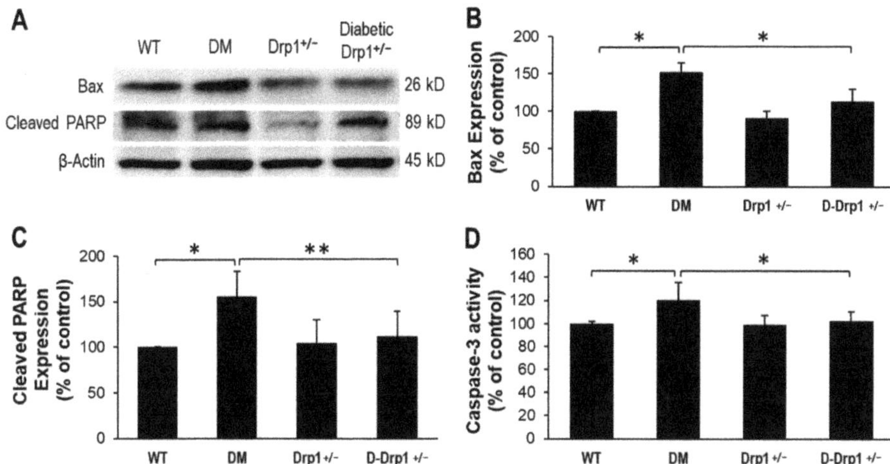

Figure 4. Reduced Drp1 expression lowers Bax activity in diabetic Drp1$^{+/-}$ mouse retinas. (**A**) Representative WB image shows Bax and cleaved PARP expression in the retinas of WT, diabetic (DM), Drp1$^{+/-}$, and diabetic Drp1$^{+/-}$ (D-Drp1$^{+/-}$) mice. Graphical illustrations of cumulative data suggest reduced Drp1 levels in D-Drp1$^{+/-}$ mice is protective against diabetes-induced increase in (**B**) Bax levels, (**C**) PARP cleavage and (**D**) caspase-3 activity. Data are expressed as mean $\pm$ SD. * $p < 0.01$, $n = 12$; ** $p < 0.05$, $n = 12$.

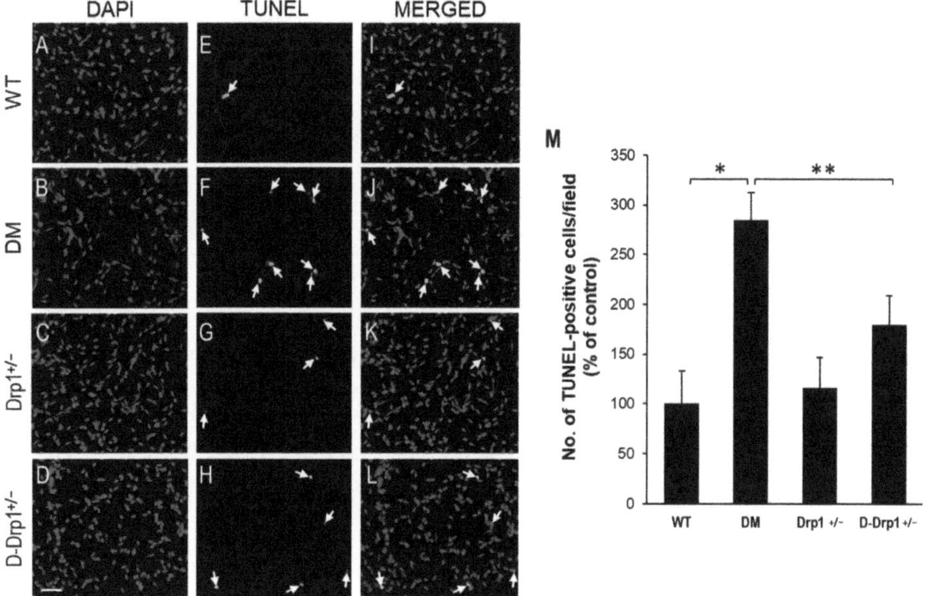

Figure 5. Reduced Drp1 level protects against diabetes-induced apoptosis of vascular cells in retinal capillary networks. Representative images of capillary networks showing DAPI-stained cells in the (**A**) WT, (**B**) diabetic (DM), (**C**) Drp1$^{+/-}$, and (**D**) diabetic Drp1$^{+/-}$ (D-Drp1$^{+/-}$) mice. (**E–H**) Corresponding images of TUNEL-positive cells (arrows) in the retinal capillary networks, respectively. (**I–L**) Merged images showing DAPI-stained cells superimposed with TUNEL-positive cells. Scale bar = 100 μm. (**M**) Graph of cumulative data showing that retinal capillary networks of diabetic mice exhibited an increase in number of TUNEL-positive cells compared to that of WT mice, while retinal capillary networks of D-Drp1$^{+/-}$ mice showed reduced number of TUNEL-positive cells compared to that of diabetic mice. Data are presented as mean $\pm$ SD. * $p < 0.01$, $n = 6$; ** $p < 0.05$, $n = 6$.

3.6. Reduced Drp1 Level Is Protective against Diabetes-Induced Development of AC and PL

To evaluate the effects of diabetes and reduced Drp1 levels in the development of AC and PL, retinal trypsin digestion was carried out and the numbers of AC and PL between the experimental groups were analyzed. RTD data indicate a significant increase in the numbers of AC and PL in retinal capillary networks of diabetic mice compared to those of non-diabetic WT mice (AC: 181 ± 23% of WT vs. 100 ± 17% of WT, $p < 0.01$; $n = 12$; Figure 6A–F, PL: 225 ± 23% of WT vs. 100 ± 31% of WT, $p < 0.01$; $n = 12$; Figure 6A–F). Importantly, when Drp1 levels were reduced in diabetic Drp1$^{+/-}$ mice, the numbers of AC and PL significantly decreased compared to those of diabetic mice (AC: 111 ± 7% of WT vs. 181 ± 23% of WT, $p < 0.01$; $n = 12$; Figure 6A–F, PL: 128 ± 11% of WT vs. 225 ± 23% of WT, $p < 0.01$; $n = 12$; Figure 6A–F). RTDs of Drp1$^{+/-}$ mice exhibited no significant difference in the numbers of AC and PL compared to those in non-diabetic WT mice (AC: 93 ± 13% of WT vs. 100 ± 17% of WT, $p > 0.05$; $n = 12$; Figure 6A–F, PL: 85 ± 17% of WT vs. 100 ± 31% of WT, $p > 0.05$; $n = 12$; Figure 6A–F).

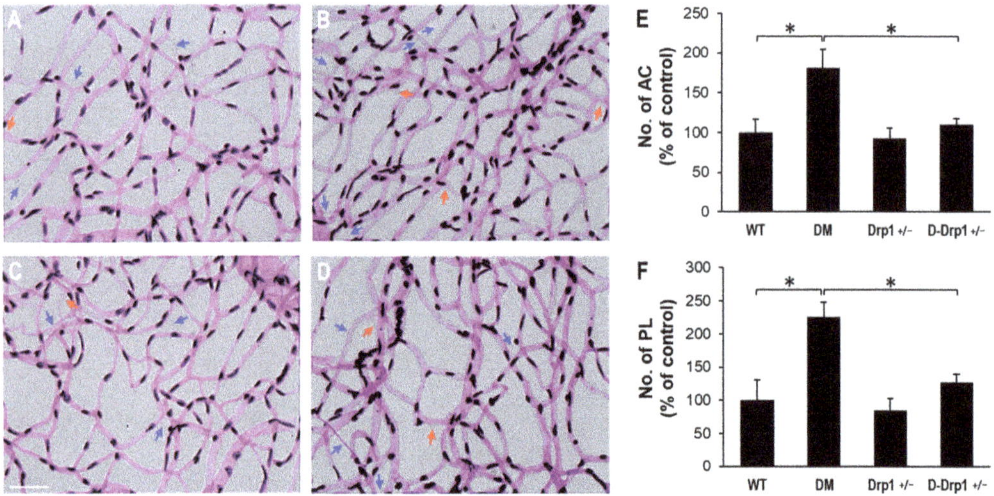

Figure 6. Effects of diabetes or decreased Drp1 level on AC and PL development in mouse retinas. (**A–D**) Representative images of retinal trypsin digestion of (**A**) WT, (**B**) DM, (**C**) Drp1$^{+/-}$, and D-Drp1$^{+/-}$ mice show the number of AC (blue arrows) and PL (red arrows) is increased in the diabetic mouse retina compared to that of control mouse retina. Importantly, the number of AC and PL is decreased in the D-Drp1$^{+/-}$ mouse retina compared to that of diabetic mouse retina. Scale bar = 100 μm. Graphical illustrations of cumulative data show that reduced DRP1 levels in the D-Drp1$^{+/-}$ mouse retina exhibited a protective effect against the development of (**E**) AC and (**F**) PL. Data are expressed as mean ± SD. * $p < 0.01$, $n = 12$.

4. Discussion

Present study provides novel evidence that Drp1 expression is significantly increased in retinas of diabetic mice, and that reduced levels of Drp1 provide beneficial effects in preventing apoptotic death of retinal vascular cells and subsequent development of acellular capillaries and pericyte loss characteristic of diabetic retinopathy. Of note, no significant adverse effects were observed in the retina and other tissues of Drp1$^{+/-}$ animals exhibiting ~40% reduction in Drp1 levels. Similarly, data from the present study showed no significant sex-related differences in neither WT or Drp1$^{+/-}$ animals. These findings indicate that increased Drp1 level is closely associated with the development of retinal vascular lesions, and that reducing it could prevent apoptotic vascular cell death associated with diabetic retinopathy.

Mitochondrial morphology and functionality are intrinsically linked [7]. Breakdown in mitochondrial morphology is known to compromise mitochondrial function [6,9,10].

Excess mitochondrial fission resulting from increased Drp1 levels can disturb the delicate balance between mitochondrial fission and fusion events, promote mitochondrial fragmentation and ultimately compromise mitochondrial functionality [16,17,40–45]. Additionally, Drp1-driven mitochondrial fragmentation can undermine mitochondrial respiration, alter calcium storage, and lead to increased ROS production [40,43]. Another study reported that changes due to increased Drp1-mediated mitochondrial fragmentation impair metabolic functions, which compromises mitochondrial homeostasis in neuroinflammation [44]. Importantly, inhibition of Drp1-induced mitochondrial fragmentation led to improvements in mitochondrial functionality as evidenced by restoration of mitochondrial membrane potential and mitochondrial respiration [10,42]. Taken together, these reports provide further evidence that excess Drp1-driven mitochondrial fragmentation contributes to impaired mitochondrial functionality in diabetic condition.

While studies have established that participation of Drp1 is essential in regulating mitochondrial fission, mechanisms underlying Drp1 upregulation and its role in inducing apoptosis remain unclear. Increased Drp1 activation triggers apoptosis by promoting translocation of Bax to mitochondria and ultimately activating caspase-3 signaling [46]. In addition, Drp1-driven mitochondrial fission contributes to mitochondrial fragmentation by promoting outer mitochondrial membrane permeabilization, hindering ATP production, and triggering release of pro-apoptotic factors [41] as well as facilitating mitochondrial division leading to oligomerization of Bax and cytochrome c release [47,48], ultimately inducing apoptosis. Interestingly, Drp1$^{-/-}$ cells are protected against apoptosis [49] and Drp1 inhibition reduced cleavage of caspase-3 and PARP in hepatocytes [50], suggesting that targeting Drp1 may be protective against apoptosis. Further studies are necessary to better understand the diverse mechanisms implicated in excess Drp1-mediated apoptosis.

Growing evidence shows that blocking Drp1 overexpression could be effective in preventing mitochondrial fission and protecting against apoptotic cell death [24,51–53]. Inhibition of Drp1-mediated mitochondrial fission reduced ER stress response in fibroblasts, which in turn reduced cellular stress and improved cell survival [54]. Importantly, selective inhibition of Drp1 using Mdivi-1 hindered Drp1 self-assembly, which effectively blocked Bax-dependent cytochrome c release and mitochondrial outer membrane permeabilization, ultimately preventing apoptosis [51]. Moreover, Drp1 downregulation inhibited mitochondrial fragmentation and prevented cytochrome c release and caspase activation [24,53]. Maintenance of Drp1 levels preserves mitochondrial cristae ultrastructure and prevents cytochrome c release and the downstream apoptotic signaling cascade [52]. Our previous report [10] as well as studies from other investigators [29,53,54] support our current finding that reducing Drp1 overexpression could be beneficial against diabetes-induced apoptosis in retinal vascular cells, and that targeting Drp1 overexpression could be a useful strategy against the development of retinal vascular lesions associated with diabetic retinopathy.

Author Contributions: Conceptualization, S.R.; methodology, D.K., H.S.; formal analysis, D.K. and S.R.; investigation, D.K. and S.R.; resources, H.S. and S.R.; writing—original draft preparation, D.K. and S.R; writing—review and editing, D.K, H.S. and S.R.; funding acquisition, S.R. All authors have read and agreed to the published version of the manuscript.

Funding: This research was supported by NEI, NIH grant EY027082 (SR).

Institutional Review Board Statement: The study was conducted according to the guidelines of the ARVO Statement for the Use of Animals in Ophthalmic and Vision Research and approved by the IACUC Committee of Boston University (PROTO201800411; approved on 9 March 2021).

Informed Consent Statement: Not applicable.

Data Availability Statement: Data presented in the article are available by request to corresponding author, Sayon Roy (sayon@bu.edu).

Acknowledgments: We acknowledge the technical support by Maeve Evans in the study.

Conflicts of Interest: The authors declare no conflict of interest.

References

1. Wong, T.Y.; Sabanayagam, C. Strategies to Tackle the Global Burden of Diabetic Retinopathy: From Epidemiology to Artificial Intelligence. *Ophthalmologica* **2020**, *243*, 9–20. [CrossRef]
2. Mizutani, M.; Kern, T.S.; Lorenzi, M. Accelerated death of retinal microvascular cells in human and experimental diabetic retinopathy. *J. Clin. Investig.* **1996**, *97*, 2883–2890. [CrossRef] [PubMed]
3. Shin, E.S.; Sorenson, C.M.; Sheibani, N. Diabetes and retinal vascular dysfunction. *J. Ophthalmic. Vis. Res.* **2014**, *9*, 362–373. [CrossRef] [PubMed]
4. Song, B.; Kim, D.; Nguyen, N.H.; Roy, S. Inhibition of Diabetes-Induced Lysyl Oxidase Overexpression Prevents Retinal Vascular Lesions Associated with Diabetic Retinopathy. *Investig. Ophthalmol. Vis. Sci.* **2018**, *59*, 5965–5972. [CrossRef] [PubMed]
5. Kowluru, R.A. Mitochondria damage in the pathogenesis of diabetic retinopathy and in the metabolic memory associated with its continued progression. *Curr. Med. Chem.* **2013**, *20*, 3226–3233. [CrossRef] [PubMed]
6. Mishra, M.; Kowluru, R.A. DNA Methylation-a Potential Source of Mitochondria DNA Base Mismatch in the Development of Diabetic Retinopathy. *Mol. Neurobiol.* **2019**, *56*, 88–101. [CrossRef] [PubMed]
7. Roy, S.; Kim, D.; Sankaramoorthy, A. Mitochondrial Structural Changes in the Pathogenesis of Diabetic Retinopathy. *J. Clin. Med.* **2019**, *8*, 1363. [CrossRef]
8. Kim, D.; Roy, S. Effects of Diabetes on Mitochondrial Morphology and Its Implications in Diabetic Retinopathy. *Investig. Ophthalmol. Vis. Sci.* **2020**, *61*, 10. [CrossRef]
9. Miller, D.J.; Cascio, M.A.; Rosca, M.G. Diabetic Retinopathy: The Role of Mitochondria in the Neural Retina and Microvascular Disease. *Antioxidants* **2020**, *9*, 905. [CrossRef]
10. Kim, D.; Sankaramoorthy, A.; Roy, S. Downregulation of Drp1 and Fis1 Inhibits Mitochondrial Fission and Prevents High Glucose-Induced Apoptosis in Retinal Endothelial Cells. *Cells* **2020**, *9*, 1662. [CrossRef]
11. Roy, S.; Trudeau, K.; Roy, S.; Tien, T.; Barrette, K.F. Mitochondrial dysfunction and endoplasmic reticulum stress in diabetic retinopathy: Mechanistic insights into high glucose-induced retinal cell death. *Curr. Clin. Pharm.* **2013**, *8*, 278–284. [CrossRef] [PubMed]
12. Tien, T.; Zhang, J.; Muto, T.; Kim, D.; Sarthy, V.P.; Roy, S. High Glucose Induces Mitochondrial Dysfunction in Retinal Muller Cells: Implications for Diabetic Retinopathy. *Investig. Ophthalmol. Vis. Sci.* **2017**, *58*, 2915–2921. [CrossRef]
13. Trudeau, K.; Molina, A.J.; Guo, W.; Roy, S. High glucose disrupts mitochondrial morphology in retinal endothelial cells: Implications for diabetic retinopathy. *Am. J. Pathol.* **2010**, *177*, 447–455. [CrossRef]
14. Trudeau, K.; Molina, A.J.; Roy, S. High glucose induces mitochondrial morphology and metabolic changes in retinal pericytes. *Investig. Ophthalmol. Vis. Sci.* **2011**, *52*, 8657–8664. [CrossRef]
15. Trudeau, K.; Muto, T.; Roy, S. Downregulation of mitochondrial connexin 43 by high glucose triggers mitochondrial shape change and cytochrome C release in retinal endothelial cells. *Investig. Ophthalmol. Vis. Sci.* **2012**, *53*, 6675–6681. [CrossRef]
16. Shen, Y.L.; Shi, Y.Z.; Chen, G.G.; Wang, L.L.; Zheng, M.Z.; Jin, H.F.; Chen, Y.Y. TNF-alpha induces Drp1-mediated mitochondrial fragmentation during inflammatory cardiomyocyte injury. *Int. J. Mol. Med.* **2018**, *41*, 2317–2327. [CrossRef] [PubMed]
17. Liu, Z.; Li, H.; Su, J.; Xu, S.; Zhu, F.; Ai, J.; Hu, Z.; Zhou, M.; Tian, J.; Su, Z.; et al. Numb Depletion Promotes Drp1-Mediated Mitochondrial Fission and Exacerbates Mitochondrial Fragmentation and Dysfunction in Acute Kidney Injury. *Antioxid. Redox Signal* **2019**, *30*, 1797–1816. [CrossRef] [PubMed]
18. Rovira-Llopis, S.; Banuls, C.; Diaz-Morales, N.; Hernandez-Mijares, A.; Rocha, M.; Victor, V.M. Mitochondrial dynamics in type 2 diabetes: Pathophysiological implications. *Redox Biol.* **2017**, *11*, 637–645. [CrossRef]
19. Purnell, P.R.; Fox, H.S. Autophagy-mediated turnover of dynamin-related protein 1. *BMC Neurosci.* **2013**, *14*, 86. [CrossRef]
20. Fonseca, T.B.; Sanchez-Guerrero, A.; Milosevic, I.; Raimundo, N. Mitochondrial fission requires DRP1 but not dynamins. *Nature* **2019**, *570*, E34–E42. [CrossRef]
21. Frank, S.; Gaume, B.; Bergmann-Leitner, E.S.; Leitner, W.W.; Robert, E.G.; Catez, F.; Smith, C.L.; Youle, R.J. The role of dynamin-related protein 1, a mediator of mitochondrial fission, in apoptosis. *Dev. Cell* **2001**, *1*, 515–525. [CrossRef]
22. Otera, H.; Mihara, K. Mitochondrial dynamics: Functional link with apoptosis. *Int. J. Cell Biol.* **2012**, *2012*, 821676. [CrossRef]
23. Serasinghe, M.N.; Chipuk, J.E. Mitochondrial Fission in Human Diseases. *Handb. Exp. Pharm.* **2017**, *240*, 159–188. [CrossRef]
24. Suen, D.F.; Norris, K.L.; Youle, R.J. Mitochondrial dynamics and apoptosis. *Genes Dev.* **2008**, *22*, 1577–1590. [CrossRef] [PubMed]
25. Ma, L.; Han, C.; Peng, T.; Li, N.; Zhang, B.; Zhen, X.; Yang, X. Ang-(1-7) inhibited mitochondrial fission in high-glucose-induced podocytes by upregulation of miR-30a and downregulation of Drp1 and p53. *J. Chin. Med. Assoc.* **2016**, *79*, 597–604. [CrossRef] [PubMed]
26. Zhang, L.; Ji, L.; Tang, X.; Chen, X.; Li, Z.; Mi, X.; Yang, L. Inhibition to DRP1 translocation can mitigate p38 MAPK-signaling pathway activation in GMC induced by hyperglycemia. *Ren. Fail.* **2015**, *37*, 903–910. [CrossRef] [PubMed]
27. Chen, Z.; Ma, Y.; Yang, Q.; Hu, J.; Feng, J.; Liang, W.; Ding, G. AKAP1 mediates high glucose-induced mitochondrial fission through the phosphorylation of Drp1 in podocytes. *J. Cell. Physiol.* **2020**, *235*, 7433–7448. [CrossRef] [PubMed]
28. Men, X.; Wang, H.; Li, M.; Cai, H.; Xu, S.; Zhang, W.; Xu, Y.; Ye, L.; Yang, W.; Wollheim, C.B.; et al. Dynamin-related protein 1 mediates high glucose induced pancreatic beta cell apoptosis. *Int. J. Biochem. Cell Biol.* **2009**, *41*, 879–890. [CrossRef] [PubMed]
29. Wang, Q.; Zhang, M.; Torres, G.; Wu, S.; Ouyang, C.; Xie, Z.; Zou, M.H. Metformin Suppresses Diabetes-Accelerated Atherosclerosis via the Inhibition of Drp1-Mediated Mitochondrial Fission. *Diabetes* **2017**, *66*, 193–205. [CrossRef]

30. Tao, A.; Xu, X.; Kvietys, P.; Kao, R.; Martin, C.; Rui, T. Experimental diabetes mellitus exacerbates ischemia/reperfusion-induced myocardial injury by promoting mitochondrial fission: Role of down-regulation of myocardial Sirt1 and subsequent Akt/Drp1 interaction. *Int. J. Biochem. Cell Biol.* **2018**, *105*, 94–103. [CrossRef]
31. Leininger, G.M.; Backus, C.; Sastry, A.M.; Yi, Y.B.; Wang, C.W.; Feldman, E.L. Mitochondria in DRG neurons undergo hyperglycemic mediated injury through Bim, Bax and the fission protein Drp1. *Neurobiol. Dis.* **2006**, *23*, 11–22. [CrossRef] [PubMed]
32. Huang, S.; Wang, Y.; Gan, X.; Fang, D.; Zhong, C.; Wu, L.; Hu, G.; Sosunov, A.A.; McKhann, G.M.; Yu, H.; et al. Drp1-mediated mitochondrial abnormalities link to synaptic injury in diabetes model. *Diabetes* **2015**, *64*, 1728–1742. [CrossRef] [PubMed]
33. Wakabayashi, J.; Zhang, Z.; Wakabayashi, N.; Tamura, Y.; Fukaya, M.; Kensler, T.W.; Iijima, M.; Sesaki, H. The dynamin-related GTPase Drp1 is required for embryonic and brain development in mice. *J. Cell Biol.* **2009**, *186*, 805–816. [CrossRef] [PubMed]
34. Manczak, M.; Sesaki, H.; Kageyama, Y.; Reddy, P.H. Dynamin-related protein 1 heterozygote knockout mice do not have synaptic and mitochondrial deficiencies. *Biochim. Biophys. Acta* **2012**, *1822*, 862–874. [CrossRef]
35. Rogers, M.A.; Maldonado, N.; Hutcheson, J.D.; Goettsch, C.; Goto, S.; Yamada, I.; Faits, T.; Sesaki, H.; Aikawa, M.; Aikawa, E. Dynamin-Related Protein 1 Inhibition Attenuates Cardiovascular Calcification in the Presence of Oxidative Stress. *Circ. Res.* **2017**, *121*, 220–233. [CrossRef]
36. Kuwabara, T.; Cogan, D.G. Studies of retinal vascular patterns. I. Normal architecture. *Arch. Ophthalmol.* **1960**, *64*, 904–911. [CrossRef]
37. Towbin, H.; Staehelin, T.; Gordon, J. Electrophoretic transfer of proteins from polyacrylamide gels to nitrocellulose sheets: Procedure and some applications. *Proc. Natl. Acad. Sci. USA* **1979**, *76*, 4350–4354. [CrossRef]
38. Xiang, J.; Chao, D.T.; Korsmeyer, S.J. BAX-induced cell death may not require interleukin 1 beta-converting enzyme-like proteases. *Proc. Natl. Acad. Sci. USA* **1996**, *93*, 14559–14563. [CrossRef]
39. Tien, T.; Muto, T.; Barrette, K.; Challyandra, L.; Roy, S. Downregulation of Connexin 43 promotes vascular cell loss and excess permeability associated with the development of vascular lesions in the diabetic retina. *Mol. Vis.* **2014**, *20*, 732–741. [PubMed]
40. Kim, B.; Park, J.; Chang, K.T.; Lee, D.S. Peroxiredoxin 5 prevents amyloid-beta oligomer-induced neuronal cell death by inhibiting ERK-Drp1-mediated mitochondrial fragmentation. *Free Radic. Biol. Med.* **2016**, *90*, 184–194. [CrossRef]
41. Huang, C.Y.; Lai, C.H.; Kuo, C.H.; Chiang, S.F.; Pai, P.Y.; Lin, J.Y.; Chang, C.F.; Viswanadha, V.P.; Kuo, W.W.; Huang, C.Y. Inhibition of ERK-Drp1 signaling and mitochondria fragmentation alleviates IGF-IIR-induced mitochondria dysfunction during heart failure. *J. Mol. Cell Cardiol.* **2018**, *122*, 58–68. [CrossRef]
42. Joshi, A.U.; Ebert, A.E.; Haileselassie, B.; Mochly-Rosen, D. Drp1/Fis1-mediated mitochondrial fragmentation leads to lysosomal dysfunction in cardiac models of Huntington's disease. *J. Mol. Cell Cardiol.* **2019**, *127*, 125–133. [CrossRef]
43. Park, J.; Won, J.; Seo, J.; Yeo, H.G.; Kim, K.; Kim, Y.G.; Jeon, C.Y.; Kam, M.K.; Kim, Y.H.; Huh, J.W.; et al. Streptozotocin Induces Alzheimer's Disease-Like Pathology in Hippocampal Neuronal Cells via CDK5/Drp1-Mediated Mitochondrial Fragmentation. *Front. Cell Neurosci.* **2020**, *14*, 235. [CrossRef]
44. Tian, H.; Wang, K.; Jin, M.; Li, J.; Yu, Y. Proinflammation effect of Mst1 promotes BV-2 cell death via augmenting Drp1-mediated mitochondrial fragmentation and activating the JNK pathway. *J. Cell Physiol.* **2020**, *235*, 1504–1514. [CrossRef]
45. Lutz, A.K.; Exner, N.; Fett, M.E.; Schlehe, J.S.; Kloos, K.; Lammermann, K.; Brunner, B.; Kurz-Drexler, A.; Vogel, F.; Reichert, A.S.; et al. Loss of parkin or PINK1 function increases Drp1-dependent mitochondrial fragmentation. *J. Biol. Chem.* **2009**, *284*, 22938–22951. [CrossRef] [PubMed]
46. Duan, C.; Kuang, L.; Xiang, X.; Zhang, J.; Zhu, Y.; Wu, Y.; Yan, Q.; Liu, L.; Li, T. Drp1 regulates mitochondrial dysfunction and dysregulated metabolism in ischemic injury via Clec16a-, BAX-, and GSH- pathways. *Cell Death Dis.* **2020**, *11*, 251. [CrossRef] [PubMed]
47. Friedman, J.R.; Lackner, L.L.; West, M.; DiBenedetto, J.R.; Nunnari, J.; Voeltz, G.K. ER tubules mark sites of mitochondrial division. *Science* **2011**, *334*, 358–362. [CrossRef] [PubMed]
48. Montessuit, S.; Somasekharan, S.P.; Terrones, O.; Lucken-Ardjomande, S.; Herzig, S.; Schwarzenbacher, R.; Manstein, D.J.; Bossy-Wetzel, E.; Basanez, G.; Meda, P.; et al. Membrane remodeling induced by the dynamin-related protein Drp1 stimulates Bax oligomerization. *Cell* **2010**, *142*, 889–901. [CrossRef]
49. Oettinghaus, B.; D'Alonzo, D.; Barbieri, E.; Restelli, L.M.; Savoia, C.; Licci, M.; Tolnay, M.; Frank, S.; Scorrano, L. DRP1-dependent apoptotic mitochondrial fission occurs independently of BAX, BAK and APAF1 to amplify cell death by BID and oxidative stress. *Biochim. Biophys. Acta* **2016**, *1857*, 1267–1276. [CrossRef] [PubMed]
50. Deng, X.; Liu, J.; Liu, L.; Sun, X.; Huang, J.; Dong, J. Drp1-mediated mitochondrial fission contributes to baicalein-induced apoptosis and autophagy in lung cancer via activation of AMPK signaling pathway. *Int. J. Biol. Sci.* **2020**, *16*, 1403–1416. [CrossRef]
51. Cassidy-Stone, A.; Chipuk, J.E.; Ingerman, E.; Song, C.; Yoo, C.; Kuwana, T.; Kurth, M.J.; Shaw, J.T.; Hinshaw, J.E.; Green, D.R.; et al. Chemical inhibition of the mitochondrial division dynamin reveals its role in Bax/Bak-dependent mitochondrial outer membrane permeabilization. *Dev. Cell* **2008**, *14*, 193–204. [CrossRef] [PubMed]
52. Cereghetti, G.M.; Costa, V.; Scorrano, L. Inhibition of Drp1-dependent mitochondrial fragmentation and apoptosis by a polypeptide antagonist of calcineurin. *Cell Death Differ.* **2010**, *17*, 1785–1794. [CrossRef] [PubMed]
53. Estaquier, J.; Arnoult, D. Inhibiting Drp1-mediated mitochondrial fission selectively prevents the release of cytochrome c during apoptosis. *Cell Death Differ.* **2007**, *14*, 1086–1094. [CrossRef] [PubMed]
54. Joshi, A.U.; Saw, N.L.; Vogel, H.; Cunnigham, A.D.; Shamloo, M.; Mochly-Rosen, D. Inhibition of Drp1/Fis1 interaction slows progression of amyotrophic lateral sclerosis. *Embo Mol. Med.* **2018**, *10*. [CrossRef] [PubMed]

cells

MDPI

Article

Betulinic Acid Protects from Ischemia-Reperfusion Injury in the Mouse Retina

Aytan Musayeva [1,2], Johanna C. Unkrig [1], Mayagozel B. Zhutdieva [1], Caroline Manicam [1], Yue Ruan [1], Panagiotis Laspas [1], Panagiotis Chronopoulos [1], Marie L. Göbel [1], Norbert Pfeiffer [1], Christoph Brochhausen [3,4], Andreas Daiber [5], Matthias Oelze [5], Huige Li [6], Ning Xia [6] and Adrian Gericke [1,*]

[1] Department of Ophthalmology, University Medical Center, Johannes Gutenberg University Mainz, Langenbeckstrasse 1, 55131 Mainz, Germany; aytan_musayeva@meei.harvard.edu (A.M.); charlotte.unkrig@gmail.com (J.C.U.); dr.zhutdieva@mail.ru (M.B.Z.); caroline.manicam@unimedizin-mainz.de (C.M.); yruan@uni-mainz.de (Y.R.); panagiotis.laspas@unimedizin-mainz.de (P.L.); panagiotis.chronopoulos@hotmail.com (P.C.); malu.goebel@yahoo.de (M.L.G.); norbert.pfeiffer@unimedizin-mainz.de (N.P.)

[2] Laboratory of Corneal Immunology, Transplantation and Regeneration, Schepens Eye Research Institute, Massachusetts Eye and Ear, Department of Ophthalmology, Harvard Medical School, Boston, MA 02114, USA

[3] Institute of Pathology, University Medical Center, Johannes Gutenberg University Mainz, Langenbeckstrasse 1, 55131 Mainz, Germany; christoph.brochhausen@ukr.de

[4] Institute of Pathology, University of Regensburg, Franz-Josef-Strauß-Allee 11, 93053 Regensburg, Germany

[5] Department of Cardiology 1, Laboratory of Molecular Cardiology, University Medical Center, Johannes Gutenberg University Mainz, Building 605, Langenbeckstrasse 1, 55131 Mainz, Germany; daiber@uni-mainz.de (A.D.); Matthias.oelze@unimedizin-mainz.de (M.O.)

[6] Department of Pharmacology, University Medical Center, Johannes Gutenberg University Mainz, Langenbeckstrasse 1, 55131 Mainz, Germany; huigeli@uni-mainz.de (H.L.); xianing@uni-mainz.de (N.X.)

* Correspondence: adrian.gericke@unimedizin-mainz.de; Tel.: +49-613-117-8276

check for updates

Citation: Musayeva, A.; Unkrig, J.C.; Zhutdieva, M.B.; Manicam, C.; Ruan, Y.; Laspas, P.; Chronopoulos, P.; Göbel, M.L.; Pfeiffer, N.; Brochhausen, C.; et al. Betulinic Acid Protects from Ischemia-Reperfusion Injury in the Mouse Retina. *Cells* **2021**, *10*, 2440. https://doi.org/10.3390/cells10092440

Academic Editors: Maurice Ptito and Joseph Bouskila

Received: 30 August 2021
Accepted: 14 September 2021
Published: 16 September 2021

Publisher's Note: MDPI stays neutral with regard to jurisdictional claims in published maps and institutional affiliations.

Abstract: Ischemia/reperfusion (I/R) events are involved in the pathophysiology of numerous ocular diseases. The purpose of this study was to test the hypothesis that betulinic acid protects from I/R injury in the mouse retina. Ocular ischemia was induced in mice by increasing intraocular pressure (IOP) to 110 mm Hg for 45 min, while the fellow eye served as a control. One group of mice received betulinic acid (50 mg/kg/day p.o. once daily) and the other group received the vehicle solution only. Eight days after the I/R event, the animals were killed and the retinal wholemounts and optic nerve cross-sections were prepared and stained with cresyl blue or toluidine blue, respectively, to count cells in the ganglion cell layer (GCL) of the retina and axons in the optic nerve. Retinal arteriole responses were measured in isolated retinas by video microscopy. The levels of reactive oxygen species (ROS) were assessed in retinal cryosections and redox gene expression was determined in isolated retinas by quantitative PCR. I/R markedly reduced cell number in the GCL and axon number in the optic nerve of the vehicle-treated mice. In contrast, only a negligible reduction in cell and axon number was observed following I/R in the betulinic acid-treated mice. Endothelial function was markedly reduced and ROS levels were increased in retinal arterioles of vehicle-exposed eyes following I/R, whereas betulinic acid partially prevented vascular endothelial dysfunction and ROS formation. Moreover, betulinic acid boosted mRNA expression for the antioxidant enzymes SOD3 and HO-1 following I/R. Our data provide evidence that betulinic acid protects from I/R injury in the mouse retina. Improvement of vascular endothelial function and the reduction in ROS levels appear to contribute to the neuroprotective effect.

Keywords: arterioles; betulinic acid; ischemia-reperfusion injury; reactive oxygen species; retina

1. Introduction

Ischemia/reperfusion (I/R) events have been implicated in the pathophysiology of various retinal diseases, such as retinal vascular occlusion, diabetic retinopathy and glaucoma [1]. Especially acute forms of vascular occlusion, such as central retinal artery occlusion (CRAO) are known to have a deleterious impact on visual acuity after an already short period of time [2]. The lack of oxygen supply to the retina often results in visual impairment and additional sequelae, such as retinal or vitreous hemorrhage, retinal neovascularization or neovascular glaucoma [3]. Arterial fibrinolysis failed to improve the clinical outcome of CRAO compared to conservative treatment, such as the application of acetylsalicylic acid and ocular massage, suggesting that deleterious, yet poorly understood, molecular processes are already activated in the early phase of retinal ischemia [4,5]. Hence, therapeutic approaches aimed at improving the resistance of retinal cells to I/R events are needed. We and others have previously demonstrated that oxidative stress plays a crucial role in mediating retinal tissue damage under hypoxic conditions and following I/R events [6–9].

The pentacyclic triterpenoid, betulinic acid, can be found in the peel of fruits, in leaves and in the stem bark of various plants, such as the white birch [10]. Initially, betulinic acid was found to exhibit biological activity against lymphocytic leukemia but was later found to exert effects against a variety of other tumors [11,12]. Moreover, the substance was reported to have anti-inflammatory, antiviral, antibacterial, antimalarial and antioxidant properties [13,14].

Recent studies have shown that betulinic acid protects against myocardial, renal and cerebral I/R injury [15–19]. However, the effects of betulinic acid on I/R are unknown in the retina; thus, the purpose of the present study was to test the hypothesis that betulinic acid protects from I/R injury in the mouse retina. Another goal of the study was to examine the involvement of oxidative stress in this process.

For our studies, we used a model in which intraocular pressure (IOP) was elevated by cannulation of the anterior chamber and administration of normal saline under high pressure, which leads to complete occlusion of blood vessels by compression [20,21].

2. Materials and Methods

2.1. Animals

All animal experiments were performed in accordance with the EU Directive 2010/63/EU for animal experiments and were approved by the Animal Care Committee of Rhineland-Palatinate, Germany (approval number: 23 177-07/G 13-1-064). Experiments were performed in 6-month-old, male C57Bl/6J mice. Mice were housed under standardized conditions with a 12 h light/dark cycle, a temperature of $22 \pm 2\,°C$, humidity of $55 \pm 10\%$ and with free access to food and tap water.

2.2. Application of Betulinic Acid and Induction of Ischemia-Reperfusion Injury

One day before induction of I/R, mice received either betulinic acid (BioSolutions Halle GmbH, Halle, Germany) at 50 mg/kg body weight diluted in dimethyl sulfoxide (DMSO, Carl Roth GmbH, Karlsruhe, Germany) or DMSO (vehicle solution) via gavage. Twenty-four hours later, mice received a second dose of betulinic acid or vehicle solution and were subsequently anesthetized with xylocaine (1 mg/mL, i.p.) and ketamine (10 mg/mL, i.p.). Body temperature was kept constant at 37 °C using a heating pad. Retinal ischemia was induced in a randomly chosen eye by introducing the tip of a glass micropipette (100 μm diameter) into the anterior chamber. The micropipette was attached via a silicon tube to a saline-filled (0.9% NaCl) reservoir that was raised above the mouse to increase intraocular pressure (IOP) to 110 mm Hg for 45 min. The fellow eye, which served as a control, was also cannulated in the same manner and maintained at an IOP of 15 mm Hg for 45 min. Retinal ischemia was considered complete when whitening of the anterior segment of the eye was observed by microscopic examination. Ofloxacin ophthalmic ointment (3 mg/g, Bausch + Lomb, Berlin, Germany) was applied on the ocular surface after needle removal. For the following seven days, mice received either betulinic acid or the vehicle solution once daily. Eight days after the I/R event, mice were sacrificed for further studies.

2.3. Retinal Wholemounts and Cell Counting

After mice had been sacrificed by CO_2 inhalation, the eye globes were removed using fine-point tweezers and Vannas scissors and fixed in 4% phosphate-buffered paraformaldehyde (Sigma-Aldrich, Munich, Germany) for one hour. Then, retinas were isolated from the eye globes in phosphate-buffered solution (PBS, Invitrogen, Karlsruhe, Germany) by using fine-point tweezers and Vannas scissors. After isolation, wholemounts were prepared and stained with cresyl blue using a standard protocol [22]. After de- and rehydration using increasing and decreasing concentrations of ethanol (70–100%), wholemounts were placed in distilled water and stained with 2% cresyl blue (Merck, Darmstadt, Germany). Next, wholemounts were dehydrated in ethanol, incubated in xylene and embedded in a quick-hardening mounting medium (Eukitt, Sigma-Aldrich). Subsequently, wholemounts were viewed under a light microscope (Vanox-T, Olympus, Hamburg, Germany) connected to a Hitachi CCD camera (Hitachi, Düsseldorf, Germany) and equipped with Diskus software (Carl H. Hilgers, Königswinter, Germany). Per wholemount, 16 pre-defined areas, eight central and eight peripheral, of 150 μm × 200 μm were photographed by a blinded investigator as previously described [22]. The proximal border of a central area was localized 0.75 mm from the center of the papilla. This distance corresponded to 5 heights of a photographed area. Each proximal border of a peripheral area was localized 0.75 mm from the distal border of a central photographed area. Thus, the distance from the center of the papilla and the proximal border of a peripheral area was 1.65 mm. In each photograph, cells were counted manually using the cell counter plug-in for ImageJ software (NIH, http://rsb.info.nih.gov/ij/) accessed on 11 March 2019. The mean cell density was calculated and the total number of cells per retina was assessed by multiplying the mean density by the area of the wholemount.

2.4. Optic Nerve Cross-Sections and Axon Counting

Optic nerves were dissected and placed in a fixative solution (2.5% glutaraldehyde and 2.0% paraformaldehyde in 0.15 M cacodylate buffer). Later, nerve segments were postfixed in 1% osmium tetroxide, dehydrated in ethanol and acetone, stained in 2% uranyl acetate, embedded in agar 100 resin (PLANO, Wetzlar, Germany) and submitted to polymerization at 60 °C for at least 48 h, according to standard protocols. Next, semithin cross-sections were cut with an ultramicrotome (Ultracut E, Leica, Bensheim, Germany), placed on conventional glass slides and stained with 1% toluidine blue in 1% sodium borate. Microscopical analysis and photomicroscopy of the cross-sections were performed with a light microscope (Vanox-T, Olympus) by a blinded investigator. The whole surface of each cross-section was assessed microscopically. Five non-overlapping fields of 80 μm × 60 μm (one central and four in the periphery) were photographed (Hitachi CCD camera) on every cross-section as previously described [22]. The axons were counted manually on these photographs using ImageJ software. The mean axon density was calculated and the total number of axons per optic nerve was assessed by multiplying the mean density by the cross-sectional area.

2.5. Measurements of Retinal Arteriole Reactivity

Retinal arteriole reactivity was measured in isolated retinas using video microscopy as previously described [23,24]. First, mice were sacrificed by CO_2 exposure and one eye per mouse was isolated and transferred into cold Krebs-Henseleit buffer. After preparation of the ophthalmic artery and isolation of the retina, the ophthalmic artery was canulated and the retina placed onto a transparent plastic platform. Next, retinal arterioles were pressurized by raising a reservoir connected to the micropipette to a level corresponding to 50 mmHg. Then, first-order retinal arterioles were imaged under brightfield conditions. After an equilibration period of 30 min, concentration-response curves for the thromboxane mimetic, U46619 (10^{-11} to 10^{-6} M; Cayman Chemical, Ann Arbor, MI, USA), were conducted. The arterioles were then preconstricted to 50–70% of the initial luminal diameter by titration of U46619, and responses to the endothelium-dependent vasodilator acetylcholine

(10^{-9} to 10^{-4} M; Sigma-Aldrich, Taufkirchen, Germany) and the endothelium-independent nitric oxide (NO) donor, sodium nitroprusside (SNP, 10^{-9} to 10^{-4} M; Sigma-Aldrich), were measured.

2.6. Assessment of ROS Levels

The fluorescent dye, dihydroethidium (DHE), was used to determine ROS levels in situ as described previously [25,26]. After mice had been sacrificed and their eyes harvested, frozen cross-sections of 10 μm thickness were prepared. After thawing, the tissue sections were immediately incubated with 1 μM of dihydroethidium (DHE, Thermo Fischer Scientific, Waltham, MA, USA). DHE is cell-permeable and reacts with superoxide to form ethidium, which in turn intercalates in deoxyribonucleic acid, thereby exhibiting red fluorescence. Using an Eclipse TS 100 microscope (Nikon, Tokyo, Japan) equipped with a DS–Fi1-U2 digital microscope camera (Nikon, Tokyo, Japan) and the imaging software NIS Elements (Version 1.10, Nikon, Tokyo, Japan) the fluorescence (518 nm/605 nm excitation/emission) was recorded and measured in retinal cross-sections by using ImageJ.

2.7. Quantitative PCR Analysis

Messenger RNA for the hypoxic markers, hypoxia-inducible factor 1α (HIF-1α) and vascular endothelial growth factor-A (VEGF-A), the prooxidant isoforms of the nicotinamide adenine dinucleotide phosphate oxidase, NOX1, NOX2 and NOX4, the antioxidant redox enzymes, catalase (CAT), glutathione peroxidase 1 (GPX1), heme oxygenase 1 (HO-1), the three isoforms of superoxide dismutase (SOD), SOD1, SOD2 and SOD3 and for the three nitric oxide synthase (NOS) isoforms, eNOS, iNOS and nNOS, was quantified in the retina by quantitative PCR (qPCR). After mice had been killed by CO_2 inhalation, one eye per mouse was immediately excised and transferred into cooled PBS (Invitrogen, Karlsruhe, Germany). Next, the retina was isolated by Vannas scissors and fine-point tweezers, transferred into a 1.5 mL plastic tube, rapidly frozen in liquid nitrogen and stored at $-80\ ^\circ$C. Within 3 months, tissue samples were homogenized (FastPrep, MP Biomedicals, Illkirch, France) and the expression of genes was measured by SYBR Green-based quantitative real-time PCR, as previously described [27]. RNA was isolated using peqGOLD TriFast™ (PEQLAB) and cDNA was generated with the High Capacity cDNA Reverse Transcription Kit (Applied Biosystems, Darmstadt, Germany). Real-time PCR reactions were performed on a StepOnePlus™ Real-Time PCR System (Applied Biosystems) using SYBR® Green JumpStart™ Taq ReadyMix™ (Sigma-Aldrich) and 20 ng cDNA. The relative mRNA levels of the target genes were quantified using comparative threshold (CT) normalized to the TATA-binding protein (TBP) housekeeping gene. Messenger RNA expression is presented as the fold-change to vehicle-treated eyes. The PCR primer sequences are listed in Table 1.

2.8. Statistical Analysis

Data are presented as the mean $\pm$ SE and n represents the number of mice per group. For the comparison of cell numbers, axon numbers, DHE staining intensity and mRNA expression levels, a one-way ANOVA and the Tukey's multiple comparisons test were used. Vasoconstrictor responses to U46619 are presented as percent change in luminal diameter from resting diameter, whereas responses to acetylcholine and SNP are presented as percent change in luminal diameter from preconstricted diameter. The comparison between concentration-responses was made using a two-way ANOVA for repeated measurements and the Tukey's multiple comparisons test. The level of significance was set at 0.05.

Table 1. Primer sequences used for quantitative PCR analysis.

Gene	Forward	Reverse
NOX1	GGAGGAATTAGGCAAAATGGATT	GCTGCATGACCAGCAATGTT
NOX2	CCAACTGGGATAACGAGTTCA	GAGAGTTTCAGCCAAGGCTTC
NOX4	TGTAACAGAGGGAAAACAGTTGGA	GTTCCGGTTACTCAAACTATGAAGAGT
eNOS	CCTTCCGCTACCAGCCAGA	CAGAGATCTTCACTGCATTGGCTA
iNOS	CAGCTGGGCTGTACAAACCTT	CATTGGAAGTGAAGCGTTTCG
nNOS	TCCACCTGCCTCGAAACC	TTGTCGCTGTTGCCAAAAAC
CAT	CAAGTACAACGCTGAGAAGCCTAAG	CCCTTCGCAGCCATGTG
GPX1	CCCGTGCGCAGGTACAG	GGGACAGCAGGGTTTCTATGTC
HO-1	GGTGATGCTGACAGAGGAACAC	TAGCAGGCCTCTGACGAAGTG
SOD1	CCAGTGCAGGACCTCATTTTAAT	TCTCCAACATGCCTCTCTTCATC
SOD2	CCTGCTCTAATCAGGACCCATT	CGTGCTCCCACACGTCAAT
SOD3	TTCTTGTTCTACGGCTTGCTACTG	AGCTGGACTCCCCTGGATTT
TBP	CTTCGTGCAAGAAATGCTGAAT	CAGTTGTCCGTGGCTCTCTTATT

3. Results

3.1. Number of Cells in the Retinal Ganglion Cell Layer and of Axons in the Optic Nerve

Ischemia-reperfusion markedly reduced the cell number in the retinal ganglion cell layer of vehicle-treated mice. Total cell number in the retinal ganglion cell layer was $129{,}378 \pm 6103$ cells and $92{,}053 \pm 6580$ cells in retinas from vehicle-treated and I/R + vehicle-treated eyes, respectively (*** $p < 0.001$), which constitutes a reduction of $\approx$29% following I/R. In contrast, only a negligible reduction of $\approx$10% in cell number was observed in the betulinic acid-treated eyes ($130{,}468 \pm 5791$ versus $117{,}836 \pm 5504$, betulinic acid versus I/R + betulinic acid-treated eyes, $p > 0.05$) (Figure 1A–E). Cells in the mouse retinal ganglion cell layer are mainly comprised of neurons, but also vascular endothelial cells and glial cells [22,28]. The neurons are composed primarily of retinal ganglion cells and displaced amacrine cells. Notably, retinal ganglion cells account for only about half of the neurons in the retinal ganglion cell layer of the mouse eye [22,29]. Since the cresyl blue staining method does not clearly distinguish between ganglion cells and other neurons because of some overlap in nuclear size and shape, we also calculated the axons of retinal ganglion cells in optic nerve cross-sections. Of note, I/R also reduced the number of optic nerve axons in the vehicle-treated mice. Axon number was $52{,}994 \pm 3411$ and $36{,}796 \pm 4079$ in vehicle-treated versus I/R + vehicle-treated eyes (* $p < 0.05$), which is a reduction of $\approx$31% following I/R. In contrast, I/R had only a negligible effect (reduction of $\approx$9%) on optic nerve axon number in the betulinic acid-treated mice ($58{,}019 \pm 3674$ and $52{,}603 \pm 3111$, betulinic acid versus I/R + betulinic acid, $p > 0.05$) (Figure 1F–J).

3.2. Retinal Arteriole Responses

U46619 (10^{-11}–10^{-6} M) elicited concentration-dependent vasoconstriction of retinal arterioles that was similar in all groups (Figure 2A). Likewise, endothelium-independent vasodilation induced by SNP (10^{-9}–10^{-4} M) was similar in all four groups (Figure 2B). In contrast, acetylcholine-induced (10^{-9}–10^{-4} M) vasodilation was greatly reduced in the arteries of mice exposed to I/R and the vehicle (Figure 2C). Of note, betulinic acid partially prevented endothelial dysfunction following I/R (Figure 2C).

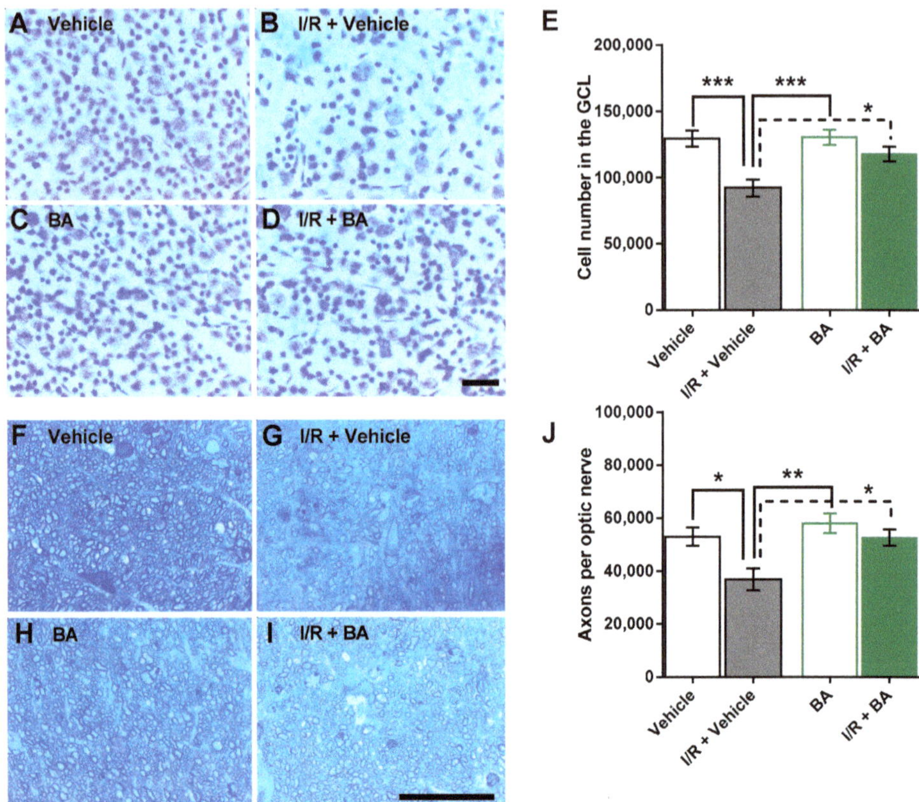

Figure 1. Total cell number in the ganglion cell layer (GCL) of the retina and axon number in the optic nerve. (**A–D**) Representative pictures of cells in the GCL stained with cresyl blue. Scale bar = 30 μm. (**E**) I/R markedly reduced the total cell number in the GCL in vehicle-treated mice but not in betulinic acid (BA)-treated mice (*** $p < 0.001$; * $p < 0.05$; $n = 8$ per group). (**F–I**) Representative pictures of optic nerve axons stained with toluidine blue. Scale bar = 30 μm. (**J**) I/R reduced axon number in the optic nerve in vehicle-treated mice but not in BA-treated mice (** $p < 0.01$; * $p < 0.05$; $n = 8$ per group).

3.3. ROS Levels in the Retina

The staining of retinal cross-sections with DHE revealed markedly increased staining intensity in retinal blood vessels from eyes exposed to I/R and the vehicle (Figure 3A–E), indicative of increased vascular ROS concentration. In contrast, DHE staining intensity did not differ between the four groups in any of the retinal layers (Figure 3F–J).

3.4. Messenger RNA Expression in the Retina

Notably, mRNA for the hypoxic genes, HIF-1α and VEGF-A, was not elevated following I/R (Figure 4A). In contrast, mRNA for NOX2 was elevated to a similar extent following I/R in the vehicle-exposed and betulinic acid-exposed mice (Figure 4B), suggesting that betulinic acid had no effect on I/R-induced NOX2 mRNA expression. Remarkably, betulinic acid boosted retinal mRNA expression for the antioxidant enzymes, SOD3 and HO-1 (Figure 4C) but had no effect on NOS mRNA expression (Figure 4D).

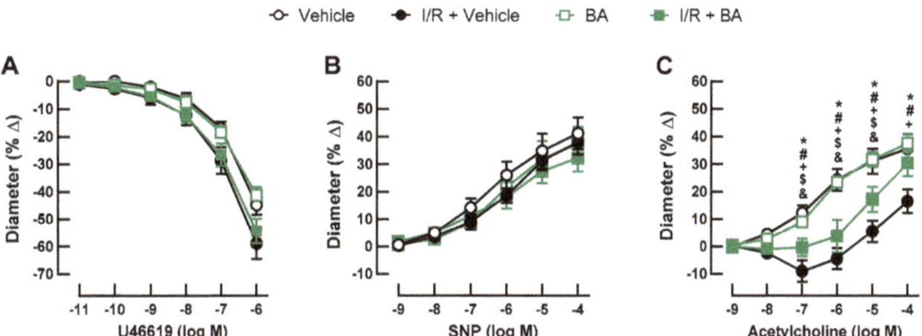

Figure 2. Responses of retinal arterioles to vasoactive substances. (**A**) The thromboxane mimetic, U46619, elicited concentration-dependent vasoconstriction in retinal arterioles that was similar in all groups. (**B**) Likewise, responses to the endothelium-independent vasodilator, sodium nitroprusside (SNP), did not differ between the four groups. (**C**) In contrast, retinal arterioles from mice subjected to I/R displayed blunted endothelium-dependent vasodilator responses to acetylcholine, which were partially improved by treatment with BA. Values are expressed as the mean $\pm$ SE (* $p < 0.05$, I/R + vehicle versus vehicle; # $p < 0.05$, I/R + vehicle versus BA; + $p < 0.05$, I/R + vehicle versus I/R + BA; $ $p < 0.05$, I/R + BA versus vehicle; & $p < 0.05$, I/R + BA versus BA; n = 8 per group).

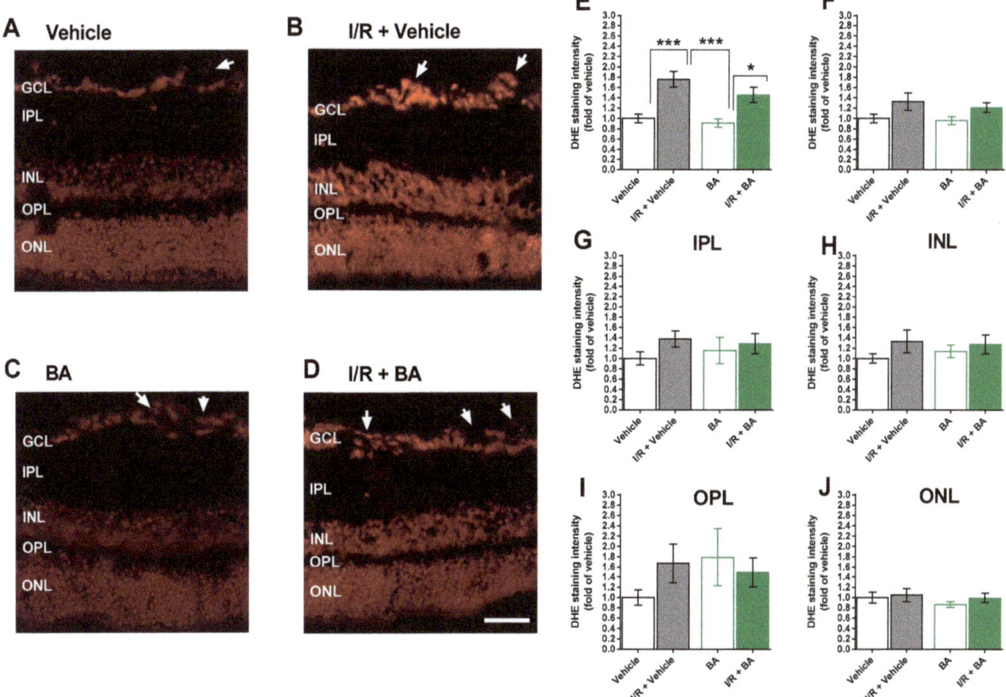

Figure 3. Dihydroethidium (DHE) staining in retinal cross-sections. (**A–D**) Representative pictures of retinal cross-sections from each group. Scale bar = 50 μm. (**E–J**) DHE staining intensity was markedly increased in retinal blood vessels from I/R- and vehicle-treated eyes (**E**). In none of the retinal layers, marked differences in DHE staining intensity were observed among groups (**F–J**) (GCL, ganglion cell layer; IPL, inner plexiform layer; INL, inner nuclear layer; OPL, outer plexiform layer; ONL, outer nuclear layer; *** $p < 0.001$; * $p < 0.05$; $n = 8$ per group).

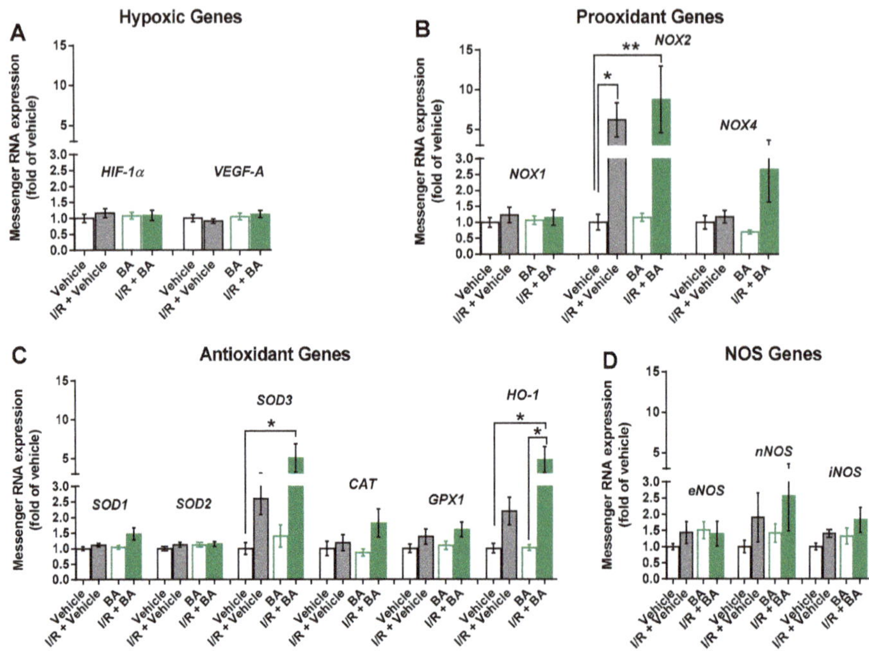

Figure 4. Messenger RNA expression of the hypoxic genes, *HIF-1α* and *VEGF-A* (**A**), the prooxidant genes (*NOX1*, *NOX2*, *NOX4*) (**B**), the antioxidant genes (*SOD1, SOD2, SOD3, CAT, GPX1, HO-1*) (**C**) and of the NOS genes (*eNOS, nNOS, iNOS*) (**D**) in the eyes treated with vehicle only, I/R + vehicle, betulinic acid (BA) only and I/R + BA. Notably, BA did not prevent upregulation of *NOX2* expression induced by I/R. However, mRNA expression of the antioxidant redox genes, *SOD3* and *HO-1*, was markedly increased in mice exposed to I/R and BA. Data are presented as the mean ± SE (** $p < 0.01$; * $p < 0.05$; $n = 8$ per group).

4. Discussion

There are several major new findings in the present study. First, following I/R, betulinic acid prevented cell loss in the retinal GCL and axon loss in the optic nerve, indicative of a protective effect on retinal ganglion cells. Second, I/R induced endothelial dysfunction in retinal arterioles, which was partially prevented by betulinic acid. Third, betulinic acid reduced the generation of ROS in retinal vessels following I/R. Fourth, treatment with betulinic acid-enhanced mRNA expression for the antioxidant enzymes, SOD3 and HO-1, while it did not prevent an increase in mRNA levels for the prooxidant NADPH oxidase subunit, NOX2, following I/R.

This is the first study to report on a protective effect of betulinic acid on I/R injury in the retina. Several previous studies reported on the protective effects of betulinic acid in I/R models of other organs. For example, in an ischemic heart model in which rats were pretreated for 7 days with betulinic acid (50, 100 and 200 mg/kg, i.g.) before cardiac ischemia was induced by 30 min of left anterior descending artery occlusion followed by 2 h of reperfusion, betulinic acid improved left ventricular function, suppressed myocardial apoptosis and reduced the release of lactate dehydrogenase and creatine kinase [16]. In a rat renal I/R model, the renal pedicle was occluded for 45 min to induce ischemia followed by reperfusion for 6 h. Rats that were treated with betulinic acid (250 mg/kg, i.p.) on two occasions, 30 min prior to ischemia and immediately before the reperfusion period, had attenuated I/R-induced oxidant responses, reduced microscopic damage and better renal function [17]. Likewise, a study in the rat brain reported that pretreatment with betulinic acid for seven days at 50 mg/kg i.g. reduced cerebral injury and oxidative stress after one hour of middle cerebral artery occlusion followed by 24 h of reperfusion by activation of the SIRT1/FoxO1 pathway and the suppression of autophagy [19]. Similarly, mouse brain pretreatment with betulinic acid for seven days at 50 mg/kg/day p.o. reduced I/R-induced infarct volume and ameliorated the neurological deficit after two hours of middle cerebral artery occlusion followed by 22 h of reperfusion in hypercholesterolemic apolipoprotein E knockout mice. This was accompanied by the prevention of NOX2, nNOS and iNOS upregulation and attenuation of oxidative stress [15]. Another study in the mouse brain reported that betulinic acid reduced ROS production together with mRNA levels for NOX4 following I/R [18].

Our study is in line with the previously reported observations in other organs by demonstrating that betulinic acid protects retinal cells from I/R-induced damage. Since only around half of the cells in the retinal GCL are actually retinal ganglion cells [29], a cell subgroup that transmits visual information from the retina to the brain, we calculated the axons of retinal ganglion cells in optic nerve cross-sections in order to specify its number. Of note, betulinic acid protected retinal ganglion cells from I/R injury.

The present study extends the previously reported observations by demonstrating that vascular endothelial function was impaired one week after the I/R event and that betulinic acid reduced the extent of endothelial dysfunction. We have previously shown in a pig model of ocular ischemia that retinal endothelial function was impaired after only 12 min of ischemia followed by 20 h of reperfusion [6]. The present study suggests that the vascular endothelial recovery is not finished one week after the I/R event. This finding may be important for at least a subgroup of patients with glaucoma, who seem to be predisposed to repeated I/R events of the retina and optic nerve [30–32]. Based on these findings, an acute IOP increase may cause sustained vascular dysfunction, which in turn may itself predispose to further I/R events resulting in damage of retinal ganglion cells.

Remarkably, betulinic acid ameliorated I/R-induced vascular endothelial dysfunction in retinal arterioles. Vasoprotective effects have already been reported for betulinic acid in larger blood vessels. In these studies, exposure to betulinic acid improved endothelial function and reduced vascular ROS levels [33–35].

Although we found positive effects of betulinic acid on neuron survival and retinal endothelial function, we did not find direct effects of betulinic acid on oxidative stress. However, our study protocol differed from protocols in the previously reported studies in several aspects. First, we started administration of betulinic acid one day prior to the I/R event and continued application until the seventh day after the event. We did so because the full extent of retinal neuronal damage is not visible directly after the I/R event. Second, we measured vascular reactivity, ROS levels and mRNA expression at one time point eight days after the I/R event. This may be the reason why we found ROS levels to be elevated significantly only in retinal blood vessels, but not in individual retinal layers. The oxidative stress, which was observed in the retina in the acute phase after I/R events in various studies, including our own, may return to normal after several days [6,8]. In support of this hypothesis, a study in mice that utilized the chemiluminescent probe, L-012, as a noninvasive in vivo ROS detection agent demonstrated that ROS levels were tremendously increased one day after the I/R event while they were already markedly lower after three and seven days [36].

However, we found indirect hints that an oxidative burst occurred following I/R, because mRNA for the prooxidant NADPH oxidase subunit, NOX2, was elevated eight days following I/R. We and others have previously reported that NOX2 mRNA and protein levels were elevated following I/R in the retina of various species, including mice and pigs [6,8]. Since in the present study, NOX2 mRNA levels were similarly elevated in the I/R + vehicle group and in the I/R + betulinic acid group, betulinic acid apparently had no major effects on NOX2 mRNA expression. We also did not find evidence for the downregulation of NOX4 mRNA expression by betulinic acid as previously suggested in the mouse brain [18].

However, we found that betulinic acid-enhanced mRNA expression for the antioxidant enzymes SOD3 and HO-1, which have both previously been demonstrated to exert antioxidant effects in the retina. For example, SOD3 was shown to reduce oxidative stress in the inner retina and at the vitreoretinal interface in mice [37]. Similarly, HO-1 was shown to exert potent antioxidant, antiapoptotic, anti-inflammatory and cytoprotective activities against I/R injury in various organs, including the retina [38].

One potential limitation of this study is that the samples for quantification of redox gene mRNA and oxidative stress were taken eight days after the I/R event, which may be too long to detect acute changes in oxidative stress and redox gene expression in response to I/R. Hence, the choice of this time point may underestimate the contribution of ROS and some redox genes to ischemic injury or to neuro- and vasoprotection. On the other hand, the choice of this time point gives us a picture of prolonged molecular changes following I/R. Moreover, in the present study, mice received betulinic acid one day before I/R and continued receiving the substance for seven days after the event because the aim of the study was to determine whether betulinic acid exerted neuroprotective properties at all. It remains to be established whether betulinic acid can protect from I/R when its administration is started after the I/R event, a situation typically seen in a clinical setting when patients come to the ophthalmologist after they experience visual problems due to an I/R event.

5. Conclusions

In conclusion, this is the first study demonstrating protective effects of betulinic acid following I/R in the retina, which is in line with previous studies in other organs, such as the brain, heart and kidney. Another new finding is that vascular endothelial function was markedly impaired eight days after the retinal I/R event, which suggests that even short periods of I/R, as observed in acute IOP increases, may lead to sustained functional deficits of the retinal vasculature. Remarkably, betulinic acid partially prevented endothelial dysfunction following I/R. From a clinical point of view, betulinic acid may become useful in treating ischemic diseases of the retina and optic nerve.

Author Contributions: Conceptualization, A.G., H.L. and N.X.; methodology, A.G., A.D., A.M., C.B., J.C.U., M.B.Z., C.M., M.O., N.X., P.L. and Y.R.; software, A.G., J.C.U., H.L., M.B.Z. and N.X.; validation, A.G., A.M. and J.C.U.; formal analysis, A.G., A.M., J.C.U., M.B.Z. and M.L.G.; investigation, A.G., A.M., J.C.U. and M.B.Z.; resources, A.G., N.P., C.B. and H.L.; data curation, A.G., A.M., M.B.Z. and N.X.; writing—original draft preparation, A.M.; writing—review and editing, A.G.; visualization, P.C. and Y.R.; supervision, A.G. and C.M.; project administration, A.G.; funding acquisition, A.G. All authors have read and agreed to the published version of the manuscript.

Funding: A.G. received financial support for the work from the intramural MAIFOR program of the University Medical Center Mainz.

Institutional Review Board Statement: All animals were treated in accordance with the guidelines of EU Directive 2010/63/EU for animal experiments and were approved by the Animal Care Committee of Rhineland-Palatinate, Germany (approval number: 23 177-07/G 13-1-064).

Informed Consent Statement: Not applicable.

Data Availability Statement: The data presented in this study are available on request from the corresponding author.

Acknowledgments: We thank Gisela Reifenberg, Department of Pharmacology, University Medical Center, Johannes Gutenberg University Mainz, for her technical assistance with PCR and Karin Molter, Institute of Pathology, University Medical Center, Johannes Gutenberg University Mainz, for her expert assistance with optic nerve embedding procedures. A part of the work described herein was carried out by J.C.U. as part of her doctoral thesis.

Conflicts of Interest: The authors declare no conflict of interest. The funders had no role in the design of the study; in the collection, analyses, or interpretation of data; in the writing of the manuscript, or in the decision to publish the results.

References

1. Ruan, Y.; Jiang, S.; Musayeva, A.; Gericke, A. Oxidative Stress and Vascular Dysfunction in the Retina: Therapeutic Strategies. *Antioxidants* **2020**, *9*, 761. [CrossRef] [PubMed]
2. Rumelt, S.; Dorenboim, Y.; Rehany, U. Aggressive systematic treatment for central retinal artery occlusion. *Am. J. Ophthalmol.* **1999**, *128*, 733–738. [CrossRef]
3. Terelak-Borys, B.; Skonieczna, K.; Grabska-Liberek, I. Ocular ischemic syndrome—A systematic review. *Med. Sci. Monit.* **2012**, *18*, RA138–RA144. [CrossRef] [PubMed]
4. Feltgen, N.; Neubauer, A.; Jurklies, B.; Schmoor, C.; Schmidt, D.; Wanke, J.; Maier-Lenz, H.; Schumacher, M. Multicenter study of the European Assessment Group for Lysis in the Eye (EAGLE) for the treatment of central retinal artery occlusion: Design issues and implications. EAGLE Study report no. 1: EAGLE Study report no. 1. *Graefes Arch. Clin. Exp. Ophthalmol.* **2006**, *244*, 950–956. [CrossRef]
5. Schumacher, M.; Schmidt, D.; Jurklies, B.; Gall, C.; Wanke, I.; Schmoor, C.; Maier-Lenz, H.; Solymosi, L.; Brueckmann, H.; Neubauer, A.S.; et al. Central Retinal Artery Occlusion: Local Intra-arterial Fibrinolysis versus Conservative Treatment, a Multicenter Randomized Trial. *Ophthalmology* **2010**, *117*, 1367–1375.e1. [CrossRef] [PubMed]
6. Zadeh, J.K.; Garcia-Bardon, A.; Hartmann, E.K.; Pfeiffer, N.; Omran, W.; Ludwig, M.; Patzak, A.; Xia, N.; Li, H.; Gericke, A. Short-Time Ocular Ischemia Induces Vascular Endothelial Dysfunction and Ganglion Cell Loss in the Pig Retina. *Int. J. Mol. Sci.* **2019**, *20*, 4685. [CrossRef]
7. Zadeh, J.K.; Ruemmler, R.; Hartmann, E.K.; Ziebart, A.; Ludwig, M.; Patzak, A.; Xia, N.; Li, H.; Pfeiffer, N.; Gericke, A. Responses of retinal arterioles and ciliary arteries in pigs with acute respiratory distress syndrome (ARDS). *Exp. Eye Res.* **2019**, *184*, 152–161. [CrossRef]
8. Yokota, H.; Narayanan, P.; Zhang, W.; Liu, H.; Rojas, M.; Xu, Z.; Lemtalsi, T.; Nagaoka, T.; Yoshida, A.; Brooks, S.E.; et al. Neuroprotection from Retinal Ischemia/Reperfusion Injury by NOX2 NADPH Oxidase Deletion. *Investig. Opthalmol. Vis. Sci.* **2011**, *52*, 8123–8131. [CrossRef]
9. Qin, X.; Li, N.; Zhang, M.; Lin, S.; Zhu, J.; Xiao, D.; Cui, W.; Zhang, T.; Lin, Y.; Cai, X. Tetrahedral framework nucleic acids prevent retina ischemia-reperfusion injury from oxidative stress via activating the Akt/Nrf2 pathway. *Nanoscale* **2019**, *11*, 20667–20675. [CrossRef]
10. Jäger, S.; Trojan, H.; Kopp, T.; Laszczyk, M.N.; Scheffler, A. Pentacyclic Triterpene Distribution in Various Plants—Rich Sources for a New Group of Multi-Potent Plant Extracts. *Molecules* **2009**, *14*, 2016–2031. [CrossRef]
11. Trumbull, E.R.; Bianchi, E.; Eckert, D.J.; Wiedhopf, R.M.; Cole, J.R. Tumor Inhibitory Agents from Vauquelinia corymbosa (Rosaceae). *J. Pharm. Sci.* **1976**, *65*, 1407–1408. [CrossRef]
12. Jiang, W.; Li, X.; Dong, S.; Zhou, W. Betulinic acid in the treatment of tumour diseases: Application and research progress. *Biomed. Pharmacother.* **2021**, *142*, 111990. [CrossRef]
13. Yogeeswari, P. Betulinic Acid and Its Derivatives: A Review on their Biological Properties. *Curr. Med. Chem.* **2005**, *12*, 657–666. [CrossRef]
14. Żwawiak, J.; Pawełczyk, A.; Olender, D.; Zaprutko, L. Structure and Activity of Pentacyclic Triterpenes Codrugs. A Review. *Mini-Rev. Med. Chem.* **2021**, *21*, 1–21. [CrossRef]
15. Lu, Q.; Xia, N.; Xu, H.; Guo, L.; Wenzel, P.; Daiber, A.; Münzel, T.; Förstermann, U.; Li, H. Betulinic acid protects against cerebral ischemia–reperfusion injury in mice by reducing oxidative and nitrosative stress. *Nitric Oxide* **2011**, *24*, 132–138. [CrossRef]
16. Xia, A.; Xue, Z.; Li, Y.; Wang, W.; Xia, J.; Wei, T.; Cao, J.; Zhou, W. Cardioprotective Effect of Betulinic Acid on Myocardial Ischemia Reperfusion Injury in Rats. *Evid.-Based Complement. Altern. Med.* **2014**, *2014*, 1–10. [CrossRef] [PubMed]
17. Ekşioğlu-Demiralp, E.; Kardaş, E.R.; Özgül, S.; Yağcı, T.; Bilgin, H.; Şehirli, Ö.; Ercan, F.; Şener, G. Betulinic acid protects against ischemia/reperfusion-induced renal damage and inhibits leukocyte apoptosis. *Phytother. Res.* **2010**, *24*, 325–332. [CrossRef] [PubMed]
18. Lu, P.; Zhang, C.-C.; Zhang, X.-M.; Li, H.-G.; Luo, A.-L.; Tian, Y.-K.; Xu, H. Down-regulation of NOX4 by betulinic acid protects against cerebral ischemia-reperfusion in mice. *Curr. Med. Sci.* **2017**, *37*, 744–749. [CrossRef] [PubMed]
19. Zhao, Y.; Shi, X.; Wang, J.; Mang, J.; Xu, Z. Betulinic Acid Ameliorates Cerebral Injury in Middle Cerebral Artery Occlusion Rats through Regulating Autophagy. *ACS Chem. Neurosci.* **2021**, *12*, 2829–2837. [CrossRef]
20. Hartsock, M.J.; Cho, H.; Wu, L.; Chen, W.-J.; Gong, J.; Duh, E.J. A Mouse Model of Retinal Ischemia-Reperfusion Injury Through Elevation of Intraocular Pressure. *J. Vis. Exp.* **2016**, e54065. [CrossRef] [PubMed]
21. Hein, T.W.; Ren, Y.; Potts, L.B.; Yuan, Z.; Kuo, E.; Rosa, R.H.; Kuo, L. Acute Retinal Ischemia Inhibits Endothelium-Dependent Nitric Oxide–Mediated Dilation of Retinal Arterioles via Enhanced Superoxide Production. *Investig. Opthalmol. Vis. Sci.* **2012**, *53*, 30–36. [CrossRef] [PubMed]
22. Laspas, P.; Zhutdieva, M.B.; Brochhausen, C.; Musayeva, A.; Zadeh, J.K.; Pfeiffer, N.; Xia, N.; Li, H.; Wess, J.; Gericke, A. The M1 muscarinic acetylcholine receptor subtype is important for retinal neuron survival in aging mice. *Sci. Rep.* **2019**, *9*, 5222. [CrossRef] [PubMed]
23. Gericke, A.; Sniatecki, J.J.; Goloborodko, E.; Steege, A.; Zavaritskaya, O.; Vetter, J.M.; Grus, F.H.; Patzak, A.; Wess, J.; Pfeiffer, N. Identification of the Muscarinic Acetylcholine Receptor Subtype Mediating Cholinergic Vasodilation in Murine Retinal Arterioles. *Investig. Opthalmol. Vis. Sci.* **2011**, *52*, 7479–7484. [CrossRef]

24. Gericke, A.; Goloborodko, E.; Pfeiffer, N.; Manicam, C. Preparation Steps for Measurement of Reactivity in Mouse Retinal Arterioles *Ex Vivo*. *J. Vis. Exp.* **2018**, e56199. [CrossRef] [PubMed]

25. Gericke, A.; Mann, C.; Zadeh, J.K.; Musayeva, A.; Wolff, I.; Wang, M.; Pfeiffer, N.; Daiber, A.; Li, H.; Xia, N.; et al. Elevated Intraocular Pressure Causes Abnormal Reactivity of Mouse Retinal Arterioles. *Oxidative Med. Cell. Longev.* **2019**, *2019*, 9736047. [CrossRef]

26. Zadeh, J.K.; Zhutdieva, M.B.; Laspas, P.; Yüksel, C.; Musayeva, A.; Pfeiffer, N.; Brochhausen, C.; Oelze, M.; Daiber, A.; Xia, N.; et al. Apolipoprotein E Deficiency Causes Endothelial Dysfunction in the Mouse Retina. *Oxidative Med. Cell. Longev.* **2019**, *2019*, 5181429. [CrossRef]

27. Musayeva, A.; Jiang, S.; Ruan, Y.; Zadeh, J.; Chronopoulos, P.; Pfeiffer, N.; Müller, W.; Ackermann, M.; Xia, N.; Li, H.; et al. Aged Mice Devoid of the M_3 Muscarinic Acetylcholine Receptor Develop Mild Dry Eye Disease. *Int. J. Mol. Sci.* **2021**, *22*, 6133. [CrossRef] [PubMed]

28. Dräger, U.C.; Olsen, J.F. Ganglion cell distribution in the retina of the mouse. *Investig. Ophthalmol. Vis. Sci.* **1981**, *20*, 285–293.

29. Schlamp, C.L.; Montgomery, A.D.; Mac Nair, C.E.; Schuart, C.; Willmer, D.J.; Nickells, R.W. Evaluation of the percentage of ganglion cells in the ganglion cell layer of the rodent retina. *Mol. Vis.* **2013**, *19*, 1387–1396.

30. Choi, J.; Kook, M.S. Systemic and Ocular Hemodynamic Risk Factors in Glaucoma. *BioMed. Res. Int.* **2015**, *2015*, 1–9. [CrossRef]

31. Flammer, J.; Mozaffarieh, M. What is the present pathogenetic concept of glaucomatous optic neuropathy? *Surv. Ophthalmol.* **2007**, *52* (Suppl. 2), S162–S173. [CrossRef]

32. Resch, H.; Garhofer, G.; Fuchsjäger-Mayrl, G.; Hommer, A.; Schmetterer, L. Endothelial dysfunction in glaucoma. *Acta Ophthalmol.* **2009**, *87*, 4–12. [CrossRef]

33. Bai, Y.-Y.; Yan, D.; Zhou, H.-Y.; Li, W.-X.; Lou, Y.-Y.; Zhou, X.-R.; Qian, L.-B.; Xiao, C. Betulinic acid attenuates lipopolysaccharide-induced vascular hyporeactivity in the rat aorta by modulating Nrf2 antioxidative function. *Inflammopharmacology* **2020**, *28*, 165–174. [CrossRef]

34. Xia, M.-L.; Qian, L.-B.; Fu, J.-Y.; Cai, X. Betulinic acid inhibits superoxide anion-mediated impairment of endothelium-dependent relaxation in rat aortas. *Indian J. Pharmacol.* **2012**, *44*, 588–592. [CrossRef] [PubMed]

35. Fu, J.-Y.; Qian, L.-B.; Zhu, L.-G.; Liang, H.-T.; Tan, Y.-N.; Lu, H.-T.; Lu, J.-F.; Wang, H.-P.; Xia, Q. Betulinic acid ameliorates endothelium-dependent relaxation in l-NAME-induced hypertensive rats by reducing oxidative stress. *Eur. J. Pharm. Sci.* **2011**, *44*, 385–391. [CrossRef] [PubMed]

36. Fan, N.; Silverman, S.M.; Liu, Y.; Wang, X.; Kim, B.-J.; Tang, L.; Clark, A.F.; Liu, X.; Pang, I.-H. Rapid repeatable in vivo detection of retinal reactive oxygen species. *Exp. Eye Res.* **2017**, *161*, 71–81. [CrossRef] [PubMed]

37. Wert, K.; Velez, G.; Cross, M.R.; Wagner, B.A.; Teoh-Fitzgerald, M.L.; Buettner, G.R.; McAnany, J.J.; Olivier, A.; Tsang, S.H.; Harper, M.; et al. Extracellular superoxide dismutase (SOD3) regulates oxidative stress at the vitreoretinal interface. *Free. Radic. Biol. Med.* **2018**, *124*, 408–419. [CrossRef]

38. Cheng, Y. Therapeutic Potential of Heme Oxygenase-1/carbon Monoxide System Against Ischemia-Reperfusion Injury. *Curr. Pharm. Des.* **2017**, *23*, 3884–3898. [CrossRef]

cells

MDPI

Review

Advancing Diabetic Retinopathy Research: Analysis of the Neurovascular Unit in Zebrafish

Chiara Simone Middel [1,2], Hans-Peter Hammes [2] and Jens Kroll [1,*]

[1] Department of Vascular Biology and Tumor Angiogenesis, European Center for Angioscience, Medical Faculty Mannheim, Heidelberg University, 68167 Mannheim, Germany; chiara.middel@medma.uni-heidelberg.de

[2] Fifth Medical Department and European Center for Angioscience, Medical Faculty Mannheim, Heidelberg University, 68167 Mannheim, Germany; hans-peter.hammes@medma.uni-heidelberg.de

* Correspondence: jens.kroll@medma.uni-heidelberg.de; Tel.: +49-(0)621-383-71455

Abstract: Diabetic retinopathy is one of the most important microvascular complications associated with diabetes mellitus, and a leading cause of vision loss or blindness worldwide. Hyperglycaemic conditions disrupt microvascular integrity at the level of the neurovascular unit. In recent years, zebrafish (*Danio rerio*) have come into focus as a model organism for various metabolic diseases such as diabetes. In both mammals and vertebrates, the anatomy and the function of the retina and the neurovascular unit have been highly conserved. In this review, we focus on the advances that have been made through studying pathologies associated with retinopathy in zebrafish models of diabetes. We discuss the different cell types that form the neurovascular unit, their role in diabetic retinopathy and how to study them in zebrafish. We then present new insights gained through zebrafish studies. The advantages of using zebrafish for diabetic retinopathy are summarised, including the fact that the zebrafish has, so far, provided the only animal model in which hyperglycaemia-induced retinal angiogenesis can be observed. Based on currently available data, we propose potential investigations that could advance the field further.

Keywords: diabetic retinopathy; zebrafish; neurovascular unit; microvascular complications and dysfunction; metabolism

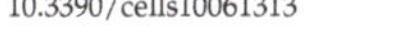

Citation: Middel, C.S.; Hammes, H.-P.; Kroll, J. Advancing Diabetic Retinopathy Research: Analysis of the Neurovascular Unit in Zebrafish. *Cells* **2021**, *10*, 1313. https://doi.org/10.3390/cells10061313

Academic Editor: Maurice Ptito

Received: 30 April 2021
Accepted: 18 May 2021
Published: 25 May 2021

Publisher's Note: MDPI stays neutral with regard to jurisdictional claims in published maps and institutional affiliations.

1. Introduction

Diabetes mellitus is one of the most prevalent metabolic conditions worldwide. The International Diabetes Foundation (IDF) estimated in 2015 that there were 415 million adults aged 20–79 living with diabetes. Due to increasing populations and the high prevalence of obesity in developed countries, this number is expected to rise to 642 million people by 2040 [1].

Diabetic retinopathy (DR) is a frequent microvascular complication occurring in patients with both type 1 or type 2 diabetes, and remains a leading cause of vision loss and blindness globally [2]. The probability of developing DR is highly dependent on the duration of diabetes and the level of glycaemic control. Furthermore, the management of other risk factors such as hypertension can also have a significant effect on the development of DR [3].

Due to earlier detection and improved treatment options, the prevalence of both retinopathy and sight-threatening stages has declined in recent years [3,4]. However, since the number of patients with diabetes and the average lifespan will increase globally in the coming decades, DR will continue to be a highly relevant condition in the foreseeable future [5,6].

The clinical aspects of DR have been thoroughly characterized [2,7,8]. There are several changes in the retinal vasculature that can be attributed to hyperglycaemia such as pericyte and endothelial cell loss. These are accompanied by altered blood flow and altered vascular permeability. The first ophthalmologically visible signs of DR are microaneurysms and haemorrhage, followed by hard exudates—the cardinal signs of non-proliferative DR (NPDR). In moderate stages, additional vascular abnormalities follow, such as the important intraretinal neovascularization. In later stages, due to increasing ischemia, retinal neovascularization extends through the inner limiting membrane (ILM) and along the surface of the retina or into the vitreous cavity. This stage is referred to as proliferative diabetic retinopathy (PDR). The complications associated with PDR, such as vitreous haemorrhage, retinal detachment or macular nonperfusion, may lead to vision loss. A common denominator in these complications is the associated photoreceptor dysfunction.

The UK Prospective Diabetes Study (UKPDS) showed that, at the turn of the century, 38% of newly diagnosed patients with type 2 diabetes already showed some stage of retinopathy [9]. In a large European population-based study published in 2016, 21% of patients with screening-detected type 2 diabetes already showed signs of DR [10]. This indicates that, while diabetes development has been overlooked in patients in the past, there has been a decline in this failure. A deeper understanding of the complex pathophysiology underlying the early stages of diabetic retinopathy is needed to develop new concepts for personalized medicine.

Zebrafish (*Danio rerio*) have been used as a model for human disease for decades. Their ease of maintenance and relatively short reproduction time make them a very attractive model organism. Their small size, fast development and their ability to produce up to 200 offspring per week underscore these advantages. Adult zebrafish measure 3 to 5 cm in length and can be housed in tanks with up to 30 other fish, depending on the tank size. Embryogenesis is almost complete, and most organs are developed at 72 h post fertilisation (hpf). Zebrafish are considered adults at 3 months of age. Since the larval development happens outside of the mother and the larvae are transparent, development can be closely monitored in vivo. Additionally, zebrafish show a high degree of genetic, anatomical and physiological similarities to humans [11–13].

In recent years, zebrafish have increasingly been used to investigate diabetes and other metabolic disorders. [14] Inducing diabetes in zebrafish can be performed in various ways. Diabetes can be induced through classical external approaches, such as injection of streptozotocin (STZ) [15]. Another approach in zebrafish is immersion in high-glucose solutions [16]. However, since the genome-wide association studies (GWAS) systematically identified various genetic loci that are associated with different kinds of disorders, including type 2 diabetes mellitus and obesity [17,18], zebrafish, as a well-established animal model for forward and reverse genetic methods, have gained attention in the field [19,20].

In this review, we discuss numerous pathologies associated with DR that have so far been identified in zebrafish models of diabetes and offer an overview of experimental techniques and perspectives for future investigators in the field.

2. Zebrafish in Diabetic Retinopathy Research

To investigate pathologies associated with DR in zebrafish, it is important to examine the various cell types that are implicated in the development of DR and to recognize differences between mammalian and zebrafish models of DR.

2.1. The Neurovascular Unit in Mammals and in Zebrafish—Similarities and Dissimilarities

The term neurovascular unit was first applied to the blood–brain barrier and, later, to the inner blood-retinal barrier. It defines the functional and structural coupling of vascular cells, i.e., endothelial and vascular mural cells (especially pericytes), neural cells, which encompass ganglion cells, amacrine cells, horizontal cells and bipolar cells as well as macro- and microglia [21]. These cell types work closely together to regulate the nutrient and oxygen levels in the functional retina through regulation of blood flow and trans- and paracellular transport. Several reviews have discussed this complex cellular crosstalk and its disruption by diabetes on various levels [22–25].

Zebrafish have long been accepted as a valuable model to study eye disease [26,27]. Both the anatomy and the function of the retina and the NVU have been highly conserved in vertebrates. The mammalian and the zebrafish retina both consist of three nuclear layers and two synaptic (plexiform) layers, and contain the same cell types (Figure 1). In zebrafish as well as in mammals, the phototransduction cascade is activated when light reaches the photoreceptors. They transport the information to the bipolar cells in the outer nuclear layer (ONL), with their synapses interacting in the outer plexiform layer (OPL). In the OPL, horizontal cells, which are local interneurons, regulate the photoreceptor output. The bipolar cells, which have their cell bodies in the inner nuclear layer (INL), activate retinal ganglion cells via synapses in the inner plexiform layer (IPL). This interaction is modulated by amacrine cells. The axons of the ganglion cells merge on the vitreous surface of the retina and form the optic nerve, through which they leave the retina and reach the visual cortex of the brain [28,29]. One key difference between the zebrafish retina and the mammalian retina is that the zebrafish retina can regenerate, a phenomenon discussed further in the chapter on immune cells.

Data describing the nature of the blood–retinal barrier in zebrafish are scarce. There are numerous studies confirming that the blood–brain barrier in zebrafish is highly conserved [30,31]. The main components of the NVU can be found in zebrafish, including a single, continuous endothelial cell layer with tight junctions to control paracellular transport, pericytes that cover the abluminal vessel wall and are covered by a basal lamina and radial glia processes that are in permanent contact with both the endothelial cells and neural cells to ensure proper vascular function.

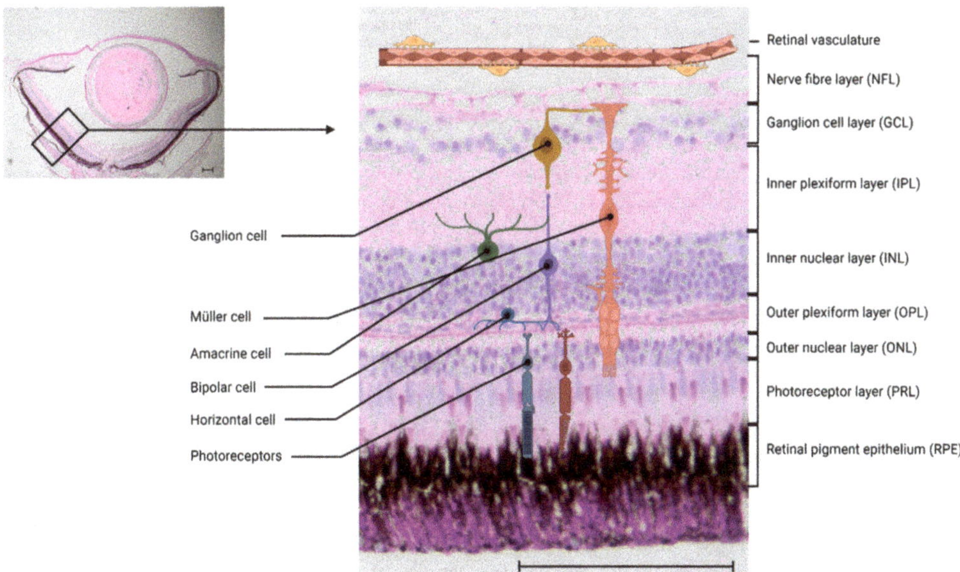

Figure 1. Anatomy of the zebrafish retina: Left: 4× magnification of a paraffin cut of the zebrafish retina, periodic-acid Schiff's (PAS) stain. Right: 20× magnification of the zebrafish retina with a schematic overview of the different cell types. Scale bar = 100 μm.

2.2. Endothelial Cells

Under physiological conditions, a single layer of non-fenestrated endothelial cells forms the luminal wall of vessels and ensures vessel integrity through communication with the surrounding cells. Especially in the brain, and by extension in the retina, the endothelium is very restrictive in order to protect the neurons from toxins and metabolites.

It ensures this selective permeability through inter-endothelial cell junctions such as tight junctions, adherence junctions and gap junctions, as well as tightly controlled transcellular pathways. Many molecules have been identified that play an important role in retinal endothelial junctions [32].

Critical features of DR include vascular dysfunction, which is associated with increased vascular permeability because of the loss of tight junctions, and loss of endothelial integrity [24]. This leads to hypoxia in the poorly perfused retina that induces an increase in levels of Angiopoietin-2 (Ang-2) and vascular endothelial growth factor (VEGF). Increased levels of VEGF, in turn, lead to the formation of new and more fragile blood vessels in the retina. This pathological neovascularization is a key component of irreversible causes of blindness in various retinopathies, as it leads to bleeding into the vitreous and retinal detachment due to macular oedema. Upregulation of VEGF is an important example of how the knowledge of pathological pathways can lead to the development of new treatments. An intravitreal injection of anti-VEGF antibodies is the only working treatment for advanced stages of retinal neovascularization. The discovery of the molecular pathways associated with the upregulation of VEGF, and the subsequent development of a new treatment option, show how crucial adequate animal models are for the development of new treatment methods [33,34].

Zebrafish have distinct advantages regarding their retinal vasculature because there is an extensive availability of reporter lines, e.g., the *Tg(fli1a:EGFP)* line, in which the whole endothelium is visualized due to the expression of enhanced green fluorescence protein (EGFP) [35]. Therefore, most of the research that has been performed on zebrafish in the context of diabetes has been focused on vascular pathologies. Quantification of endothelial cells and the whole retinal vasculature under different conditions is facilitated by the use of the appropriate reporter line. Reporter lines for other cell types are discussed below.

The development of zebrafish retinal vasculature has been well studied [36] and reviewed [37–39]. Therefore, we will only give a short overview of the development of the vessels and the key differences in comparison to mammals. The blood supply in the human retina and in some rodents is provided by two vascular plexuses, the choroid and the intraretinal vasculature. During development, the retinal vascular network undergoes intense remodelling. Blood supply during early development is provided by the hyaloid vasculature. Once the primary plexus of the retinal vasculature starts to grow into the retina, the hyaloid vasculature regresses. In humans, the growth of the primary plexus is controlled by astrocytes through the secretion of VEGF. The intraretinal vasculature grows from the primary plexus through angiogenesis, as the development of the retinal vasculature continues [40,41]. In humans, the hyaloid vasculature forms during late embryogenesis and the switch to retinal vasculature starts mid-gestation while, in mice, the switch happens at birth.

One major difference between mammalian and zebrafish vasculature is that zebrafish vasculature does not exhibit this switch in a vessel origin. The first endothelial cells are present by 48 hpf and are localised between the lens and the retina. They give rise to the hyaloid vasculature, which at first adheres tightly to the lens. By 5 days post fertilisation (dpf), the hyaloid vasculature enwraps the lens entirely and forms the peripheral circumferential vein, the inner optic circle (IOC). From 15 dpf on, they lose contact with the lens and become more and more attached to the retinal surface. There is no further invasion of vessels. Zebrafish only have vessels on the surface of the retina and on top of the inner limiting membrane (ILM). This suggests that, comparable to some mammalians such as the guinea pig, the relatively thin retina does not need an intraretinal plexus because oxygen supply can be ensured solely through diffusion [36]. This is another major difference between mammalian and zebrafish retinas, one which has consequences when analysing the vasculature of diabetic phenotypes. To visualize retinal vessels in mammals properly, the ILM needs to be removed during dissection. If the ILM is removed in zebrafish, this would also remove the retinal vasculature. This difference is essential in the context of DR research: the new blood vessels that develop in PDR break through the ILM. This is not possible in zebrafish.

2.2.1. Non-Genetic Zebrafish Models of Diabetic Retinopathy

Zebrafish have been used to model pathologies associated with DR for well over a decade now. The first approach was to induce hyperglycaemia in zebrafish through immersion in glucose solutions (Table 1). This was achieved by placing them in alternating solutions of 2% and 0% glucose every 24 h for up to 30 days, showing a decrease in the thickness of the IPL and the INL in the retina [16], which has also been observed in other animal models and diabetic patients through spectral domain optical coherence tomography [42]. This strengthens the position of zebrafish as a model to study the effects of high glucose levels on the retina, since the short duration of high glucose exposure, in combination with the ease of vascular visualization and the large breeding size, allow for convenient experimentation.

A re-evaluation of the model found various signs of retinopathy, apart from neurodegeneration. After a 30-day incubation period in alternating 2% and 0% glucose solutions, treated fish exhibited an increase in vessel diameter and a thickening of the vessel basement membrane, as well as prominent defects in cone photoreceptors with signs of photoreceptor degeneration, including impaired electroretinography (ERG) results. Other vascular pathologies were visible as well, such as wider tight and adherent junctions that suggest increased vessel permeability and upregulation of VEGF; however, those were visible in mannitol-treated control fish as well, which led the authors to reconcile these changes with hyperosmolarity rather than high glucose [43].

Since then, diabetes-like metabolic conditions have been induced through different methods such as a zebrafish model of experimental hypoxia. Hypoxia was achieved through a device that perfused N_2 gas directly into the water inside a sealed aquarium, preventing air from leaking out. The system automatically maintained a constant level of O_2 in the water and thus placed the zebrafish in 10% air saturation. The authors could show that 12 days of hypoxia treatment increased the density of capillary networks significantly, and that hypoxia, therefore, had an angiogenic effect in the zebrafish retina. They also established a dose-dependent relation of hypoxia to angiogenesis through exposure of zebrafish to different concentrations of air-saturated water and analyses of the different angiogenic responses. Most importantly, this neovascularization could be blocked by oral anti-VEGF agents (sunitinib and ZN323881), confirming that neovascularization in zebrafish is as VEGF-dependent as it is in mammals [44].

Experimental hypoxia is a strategy that has been used to model PDR in rodents, since diabetic mammalian models do not develop spontaneous neovascularization [45]. The best known rodent model is the oxygen-induced retinopathy (OIR) in mice, which has recently been reviewed, with all its advantages and disadvantages [46]. In rodent models, it is common to induce diabetes through STZ injection. STZ is an antibiotic that leads to sustained hyperglycaemia through the disruption of pancreatic islets of Langerhans and the destruction of beta-cells. Intraperitoneal or direct caudal fin injection of 300–350 mg/kg STZ in zebrafish leads to an increase in fasting blood glucose and, in addition, a marked decrease in retinal photoreceptor layer (PRL) and IPL thickness [47]. The administration of i.p. STZ-injections on days 1, 3 and 5, and subsequent booster injections once a week for two more weeks (day 12 and 19), can induce sustained hyperglycaemia for up to at least three weeks [47]. A later study, however, found that, when following the proposed protocol, the zebrafish showed a high mortality and increased levels of hypoglycaemia—indicators that this is an imperfect method of inducing diabetes in zebrafish [48].

In short, the most common ways of inducing diabetic metabolic states or retinal neovascularization in mammalian models of retinopathy (glucose exposure, hypoxia and STZ injections) can be used in adult zebrafish, and lead to pathologies that show similarities with DR in humans. Variants of the methods presented in this chapter have been used frequently, especially the high-glucose model, to study the effects of new potential anti-angiogenic drugs.

2.2.2. Genetic Zebrafish Models of Diabetic Retinopathy

The main advantage of zebrafish in diabetes research is the ease of genetic manipulation. The possibility of introducing targeted mutations using sequence-specific transcription activators such as effector nucleases (TALENs), or the clustered regularly interspaced short palindromic repeats (CRISPR) system, have made the zebrafish highly attractive for studying the consequences of loss-of-function alleles in an effective way (Table 1) [49,50].

After the successful establishment of the Zebrafish Mutation Project [51], the effects of the null mutation of *pdx1* (pancreatic and duodenal homeobox 1) in zebrafish were analysed. The analysis indicates that homozygous mutation leads to impaired pancreatic islet development and disrupted glucose homeostasis [52]. Subsequent analyses by the same group showed that these mutants exhibit distinct signs of retinal vasculature dysfunction, including vessel constriction, points of stenosis, a reduction of average vessel diameter, tortuous vessels with increased vessel density and increased sprouting and branching, as well as a reduced expression of ZO-1 (zonula occludens protein 1). ZO-1 is a molecule that is integral to tight junctions which are responsible for connecting endothelial cells and regulating permeability in vessels. Furthermore, GLUT1 expression was largely absent in mutants compared to wildtype controls [53]. Changes in GLUT1 expression have also been reported in DR patients and mouse models [54]. Parallel research on a CRISPR/Cas9-induced homozygous *pdx1* mutant by our group independently observed the same findings of hypersprouting and hyperbranching in the adult retina. This study additionally described that a pharmacological modulation of VEGF and Nitric Oxide signalling rescues the hyperglycaemia-induced changes in the vasculature [55]. Recently, work on a homozygous *aldh3a1* knockout zebrafish line showed a moderate retinal vasodilatory phenotype, which could be aggravated through experimental diabetic conditions achieved through *pdx1* expression silencing. This study provides evidence that 4-Hydroxynonenal (4-HNE), which has been implicated as a clinical feature in patients with diabetes and diabetic complications before, induces impaired glucose homeostasis and causes retinal vascular alterations [56]. Another interesting genetic model of hyperglycaemia associated retinal pathologies is the combination of a CRISPR/Cas9-induced mutation in the *glo1* (Glyoxalase 1) gene with an overfeeding protocol, which includes an overfeeding period of 8 weeks with artemia [57]. Glyoxalase 1 is an enzyme which catabolizes methylglyoxal (MG), a reactive metabolite that is a main precursor to advanced glycation end products (AGEs), and is elevated in the plasma and tissue of diabetic patients. Loss of glyoxalase 1 can therefore lead to increased levels of MG and a diabetic phenotype [58]. This protocol also leads to an increased angiogenic sprout formation of the retinal vasculature [57], and was recently scored as the best type 2 diabetes model in zebrafish that is directly followed by both *pdx1* mutants [59].

These findings are exciting from multiple perspectives. The confirmation that two different *pdx1* mutants by two independent groups produce a phenotype resembling DR in zebrafish establishes this zebrafish mutant line as a reliable and reproducible model for further research into the mechanisms associated with DR, particularly retinal angiogenesis in adults [53,55]. Furthermore, the homozygous *pdx1* mutant is the first animal model to show retinal angiogenesis under hyperglycaemia. Additionally, these results show that genes potentially involved in the pathogenesis of diabetes can be tested in zebrafish to find out whether and in which dimension they are involved in the development of DR. This could catalyse research into potential pathways and treatments, such as high-throughput screenings for potential targets of personalized treatment methods.

2.2.3. Small Molecule Testing on Zebrafish Larvae

The main principle of the ethical use of animal studies in clinical research is that of the 3 R's: researchers should always work to replace, reduce and refine animal studies. Zebrafish have increasingly come into focus to achieve these goals.

In the context of DR, this especially applies to small molecules that are supposed to stop or delay neovascularization. Currently, the gold standard for treating sight-threatening PDR is pan-retinal laser coagulation, supplemented by intravitreal injections of anti-VEGF agents when clinically significant macular oedema (CSME) is present. Clinicians need alternative therapies targeting other pathways and potentially allowing for non-invasive application methods. Using zebrafish larvae to study the effects of antiangiogenic drugs can replace some animal toxicity studies, since zebrafish larvae start being considered independent organisms at 5 dpf. Studying the effect of new agents on the developing zebrafish hyaloid vasculature can screen them for their function and thus leave a smaller number of potential pharmaceutical agents which have already proven their efficiency in zebrafish larvae to be tested on mammals [60,61].

A retinal phenotype in the *pdx1* zebrafish line has been observed, even in the larval hyaloid vasculature. The phenotype could be rescued through incubation in metformin and PK11195, providing evidence that the phenotype is caused by hyperglycaemia [55]. This highlights the suitability of the model: shortly after the induction of the mutation, there is already in vivo imaging available to confirm whether the changes in the retinal vasculature can be rescued through the application of a specific drug.

Wildtype zebrafish larvae that are incubated in glucose for 3 days, starting at 3 dpf, show an increased diameter in hyaloid vessels and upregulated expression of VEGF RNA [62]. This can easily be used to evaluate the efficacy of an antiangiogenic drug. However, screening for antiangiogenic agents does not necessarily require a model associated with hyperglycaemia [63]. The physiological development of the hyaloid vessels can also be disrupted by anti-angiogenic agents, and this disruption can be observed in vivo in a matter of days. As already proposed in 2003, zebrafish, with their rapid development and optical transparency, are exceptionally convenient for high-throughput in vivo screening of anti-angiogenic agents [64]. Their use in such experiments has been on the rise for years [65–67] because researchers want to develop treatments to inhibit neovascularization as the main reason for vision loss in various ocular diseases including, but not limited to, DR. Most recently, a protocol for drug pooling has been established. In this study, the authors evaluated the usefulness of pooling various agents and incubating zebrafish larvae with multiple agents at a time [68]. This enhances and facilitates the screening of ocular anti-angiogenic drugs in zebrafish larvae, making it possible to use even fewer animals in a first line screening of novel agents.

2.3. Pericytes

Pericytes are specialised mural cells (MCs) which occupy the abluminal side of the vessel wall and are in constant communication with endothelial cells, microglia and neurons. Their location and morphology are very distinct. They sit within the basal membrane with long processes covering the vessel walls [69]. Pericytes contribute to the regulation of blood flow through the retinal vessels and thereby to the oxygen supply for the retina, as well as the anatomical stabilization of the BRB [70]. Additionally, they are important for the formation of new vessels. As such, they are recruited for developing vessels through chemotactic factors such as platelet-derived growth factor B (PDGF-B), which is secreted by endothelial cells, and the interaction of Angiopoietin 1 (Ang1), which pericytes express, and the endothelial cell derived tyrosine-kinase receptor Tie-2. A lack of pericytes leads to severe endothelial dysfunction and even perinatal death in PDGF-B-deficient mice through the absence of functional blood vessels [71]. Angiopoietin 2 (Ang2), on the other hand, which is upregulated in patients with DR, has been found to disrupt the PDGF-B stimulated pathway and subsequently impair communication and the recruiting of pericytes to vessel walls.

In the retina, the pericyte-to-endothelial-cell ratio is 1:1, which emphasizes the importance of their role in ensuring the structural and functional integrity of the vascular architecture and the BRB [69,72]. An early feature of DR is the depletion of capillary pericytes. Several mechanisms have been suggested to play a role in pericyte loss. These include a reduction in PDGF-B signalling due to hyperglycaemia and increased secretion of Ang2 by endothelial cells. Further mechanisms implicated in pericyte depletion are microglial pro-inflammatory-mediated activation of pro-apoptotic molecules, reactive oxygen species (ROS) damage to pericyte mitochondria by activating apoptotic cascades, induction of apoptosis through generation of tumour necrosis factor-alpha (TNF-alpha) and advanced glycation end-product (AGE) and glutamate excitotoxicity [73]. Pericyte loss contributes to the eventual formation of acellular capillaries and microaneurysms, as well as haemorrhage, and consequently hypoperfusion in the retina.

Pericytes are present in the retinal vasculature of zebrafish. It was confirmed through electron microscopic analysis of the ultrastructure of retinal vessels that zebrafish vitreo-retinal vessels carry mature pericytes that are located on top of the vascular endothelium within the basal lamina, which is the same location that they inhabit in mammalian vessels. This is observed both in young and senescent specimen but not in larvae [36]. Neither in mammals nor in zebrafish do pericytes express one specific cell marker [74], and not all markers can be used across different species. Of the most prominent and commonly used pericyte markers, pericytes in zebrafish express PDGF receptor beta (PDGFRβ) and Notch3 [75]. More recently, transgenic reporter lines have been developed for live imaging of MCs. In one study, EGFP, mCherry or the Gal4FF drivers are expressed under the control of the *pdgfrβ* promotor and indicate that the first *pdgfrβ* positive cells, which are most likely pericytes, can be observed at the 8-somite stage in the cranial neural crests of zebrafish [76]. Two zebrafish smooth muscle actin (*sma*) homologues have been reported: *acta2* and *transgelin*. SMA is used to stain vascular smooth muscle cells (vSMCs) [77]. These can include pericytes, *acta2* and *transgelin*; however, they do not seem to be specific for pericytes. Specifically, one study found that the retinal vessels in the centre of the optic disc were arteries and densely covered by *transgelin1* and *acta2* positive cells. This suggests that those cells were vSMCs, while capillaries and venous vessels were covered with PDGFRβ positive cells. This confirmed the finding previously established in the mammalian retina, as mentioned above, that there is no one pan-pericyte marker [78]. These findings were in accordance with the observation that vSMCs are typically found in large blood vessels and are separated from the endothelium by the basement membrane, controlling vessel contractility and regulating blood flow. Pericytes as specialised MCs are rather found in microvessels, especially in the brain and the eye, and are embedded in the basement membrane. Even though both vSMCs and pericytes may come from the same cell line and express similar molecular markers at various time points, they are defined as two different cell types. [69]

To study the development of vSMCs in vivo, reporter lines were established that express GFP or mCherry under the mural cell promotor *acta 2*. In these reporter lines, it was established that vascular mural cells turned on *acta2:EGFP* several days after the initiation of circulation, and were morphologically similar to pericytes in early development. The larger head vessels were associated with *acta2:EGFP* positive cells at 7 dpf. In this study, the authors suggested that zebrafish do not need as many mural cells as mammals, which thus explains the difference in expression of acta2 and the late association of mural cells with the vessels, as mammalian blood pressure is much higher than that in zebrafish. Accordingly, this could mean that zebrafish vessels do not need the stabilizing effect of mural cells as much as mammals do. [79] We have, however, observed that, at least in the retina, the ratio of one pericyte to one endothelial cell is most likely conserved in zebrafish.

Recently, to research the function of the *frizzled4* gene, which is implicated in the development of familial exudative vitreoretinopathy (FEVR), one study found that pericytes in the zebrafish retina contain *frizzled4* mRNA and have a very unique position on the retinal vasculature [80]. These observations could finally make the quantification of pericyte numbers in the zebrafish retina possible. To quantify pericyte numbers in rodent models of DR, or in diabetic donor eyes, the retinal digest method is used [81]. In Figure 2, we provide an image of a retinal trypsin digest of the zebrafish retina and indicate the different cell types (endothelial cells and pericytes) that can be visualized through this method.

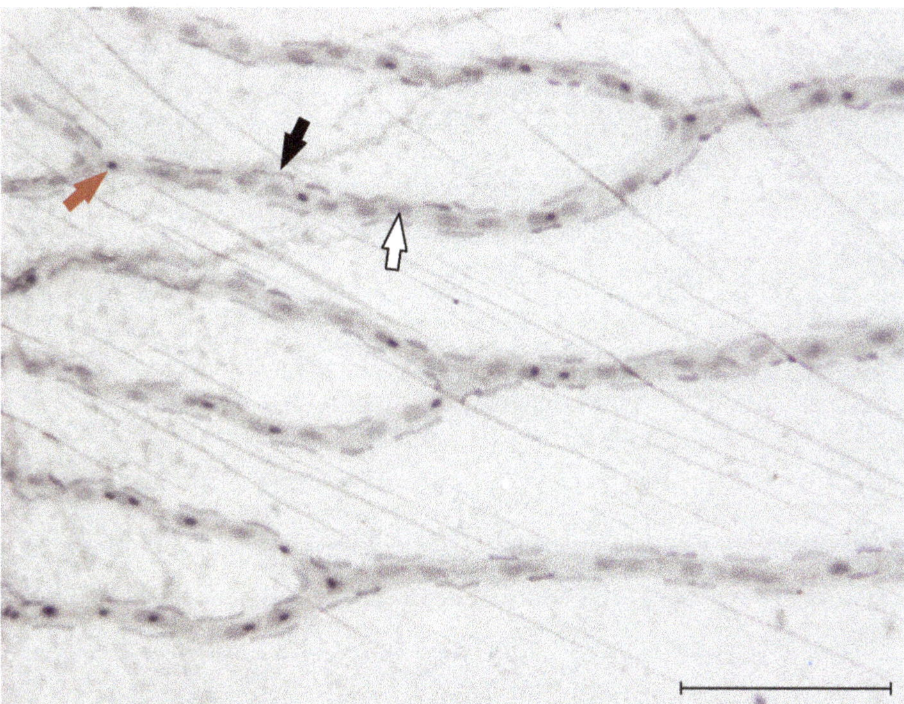

Figure 2. Visualization of endothelial cells, pericytes and erythrocytes in the zebrafish retina: 20× magnification of the zebrafish retina after digestion in 3% trypsin [81] and haemalum stain. Red arrow: erythrocyte, black arrow: pericyte, white arrow: endothelial cell. Scale bar = 100 μm.

2.4. Microglia

Microglia are the resident macrophages of the central nervous system (CNS) and an important part of the neurovascular unit. Their morphology is unique, with small cell bodies in the plexiform layers and long cell processes that may span all the nuclear layers. They monitor and control the surrounding microenvironment in the CNS, and they are able to synthesize and release multiple cytokines, chemokines, neurotrophic factors and neurotransmitters. Under physiological conditions, microglia receive inhibitory signals from the surrounding microenvironment, such as secretion of transforming growth factor beta (TGFβ) by the retinal pigment epithelium (RPE) and expression of CD200 on several retinal cells and the release of CX3CL1 by healthy neurons or endothelial cells. TGFβ even induces an anti-inflammatory effect [82].

The activation of microglia, noticeable through morphological transformation and migration of residential microglia, is a common hallmark sign of retinal disease. Microglia are involved in all stages of DR. In early stages, there is a moderate increase in perivascular microglia and they appear to be slightly hypertrophic. During non-proliferative stages of DR, microglia tend to cluster around lesions such as microaneurysms [83]. In proliferative stages, the new vessels, which are highly fragile and dilated, are surrounded by microglia [82,84]. If chronically activated, microglia play a part in neuroinflammation by constantly releasing cytokines, and thus lead to an increased invasion of immune cells into, and further damage to, the retina through inflammatory processes [84,85].

At present, there are no studies on microglia or inflammatory processes in zebrafish models of DR in general. There are various reporter lines in which microglia/macrophages express fluorescence proteins, as well as the possibility of staining microglia with antibodies against lymphocyte cytosolic *plastin 1* (*lcp1*, a pan-leukocyte maker) and 4C4 (which stains microglia exclusively in zebrafish) [86,87].

2.5. Müller Glia

Microglia are involved in a constant active crosstalk with Müller cells. Müller cells and astrocytes are the resident macroglia in the retina, with Müller cells being the major glial cell type in the mammalian retina. In all, 90% of retinal glia are Müller cells [84]. They are in intimate contact with all other cell types, with their end-feet in close contact with ganglion cells and endothelial cells on the vitreal side of the retina, and with photoreceptors on the outer side of the retina. Their cell bodies are found in the inner nuclear layer, but their processes span the entire retina.

There are three main functions that are being attributed to Müller glia: the uptake and recycling of neurotransmitters and retinoic acid compounds, the control over the metabolism and the supply of nutrients to the retina, as well as the regulation of blood flow and maintenance of the BRB [88].

In diabetic conditions, Müller cells are a potential source of growth factors, e.g., VEGF, that are effectors of neovascularization, and cytokines, which lead to the activation and migration of microglia. After prolonged periods of overstimulation, Müller glia begin to die, leading to photoreceptor degeneration, vascular leakage and intraretinal neovascularization [84,89,90]. The first marker of activation of Müller glia is an increase in expression of glial fibrillary acidic protein (GFAP), which is the most common marker of gliosis.

Müller cells become activated in hyperglycaemic conditions in adult zebrafish, as shown by antibody staining against GFAP or glutamine synthetase (GS) [43,53,78]. However, while in rodent models and in human eyes, healthy Müller cells do not express GFAP, Müller glia in zebrafish always express GFAP, thus rendering the demonstration of Müller cell activation difficult. This may be caused by the zebrafish retina's regeneration capability after injury, which has been reviewed elsewhere [91,92]. The source of regenerated neurons in zebrafish is the Müller cells themselves; upon detecting an injury, they re-enter the cell cycle and undergo asymmetric division, ultimately generating multipotent progenitors that replace the lost neurons [93]. This may pose an interesting addition to DR research in mammalian models, since mammalian Müller glia cells seem to have retained some of those regenerating abilities. Defining the mechanisms underlying the regenerative process in zebrafish may offer opportunities and new directions in the future of DR treatments [94].

There are established transgenic reporter lines for Müller glia as well, e.g., the *Tg(gfap:GFP)* line, in which Müller cells express the green fluorescence protein under the control of the *gfap* promotor. It was recently shown that larvae that are exposed to 4% glucose from 24 to 48 hpf show a significantly reduced number of Müller glial cells in the retina, which cannot be rescued post glucose exposure [95]. This reporter line has yet to be used in adult zebrafish models of diabetic retinopathy to evaluate its use in researching Müller glia activation or loss in diabetic conditions.

2.6. Photoreceptors/Neurodegeneration

Visual information is encoded in photoreceptors. In contrast to rodents, which are nocturnal animals and therefore have a rod dominated retina, zebrafish have a cone-dominated retina. Rods are highly sensitive and are most useful in dim-light conditions; therefore, they are mainly found in the peripheral parts of the human retina. The point of best visual acuity in the human retina, the fovea, which is the central part of the macula lutea, is mainly populated by cones, which are most active during daytime. Macular degeneration is one of the most important reasons for vision loss in humans. Even though the increased cone ratio in the zebrafish retina in comparison with the rodent retina makes the zebrafish an interesting model to study photoreceptor degeneration, it should be remembered that the fovea does not exist in the zebrafish retina (nor in other animal models). The effect of photoreceptor degeneration on the development of DR has only recently come into focus [96].

The hallmarks of diabetes-induced neuroglial degeneration include reactive gliosis (as discussed above), diminished neuronal function and neural-cell apoptosis. All of those have been observed to occur well before the first signs of microangiopathy in experimental models of DR or before the retina of diabetic human donors become visible [97,98]. Retinal ganglion cells and amacrine cells may be the first neuronal cells in which apoptosis becomes detectable, but photoreceptors also have an increased apoptotic rate [99]. The visible consequences of apoptotic cell death are reduced thickness of retinal layers, specifically the inner plexiform layer and the nerve fibre layer, as well as impaired ERG results due to photoreceptor degeneration and consequent dysfunction. Both of these effects can be observed in zebrafish models of DR. The data suggest that there are some discrepancies when analysing the effect of neurodegeneration in the zebrafish retina. Some studies found that the thickness of the inner plexiform layer significantly decreased, for example, in models of immersion-induced hyperglycaemia or STZ injection [16,47]. A decrease of nuclei in the outer nuclear layer in the genetic *pdx1* model was observed compared to age matched controls [53]. However, another study described an increase in retinal layer thickness in a model of immersion-induced hyperglycaemia [100], and it has even been shown that, in the *pdx1* model, there is an increase in nuclei in the inner nuclear layer [53]. These conflicting results clearly indicate that neurodegeneration in diabetic zebrafish models has not been sufficiently studied. A potential explanation could be that the regenerative capabilities of the zebrafish retina make it impossible to reliably quantify neurodegeneration, since the retinal layer thickness and the number of nuclei changes dynamically throughout the time of the experiment.

Photoreceptor degeneration has been widely and consistently observed in zebrafish, both morphologically [43,47,53] and functionally through abnormal ERG results [43,53,100]. This may be an interesting topic for further studies, with the new shift in research to consider the role that photoreceptors play in the pathogenesis of DR.

3. Perspectives and Conclusions

Zebrafish have gained popularity as models for complications associated with diabetes and other metabolic diseases in recent years [101], with a wide array of different methods for inducing diabetic conditions published to this date [59]. Table 1 provides an overview of the various zebrafish models of DR published to date.

The focus of DR research in zebrafish rests on endothelial cell dysfunction and angiogenesis, since the ease of analysis of vasculature in transgenic lines is one of the main advantages of using zebrafish in research. Thus far, not much has been achieved in regard to the role of pericytes in zebrafish models of DR. One study found that, in the *pdx1* mutant, transgelin1 expression is reduced in mutants compared with age matched wildtype controls [53]. However, as mentioned above, this marker cannot reliably stain pericytes on zebrafish retinal vessels. With the discussed transgenic reporter lines for vascular mural cells in zebrafish, and the recent confirmation of the morphology and localisation of pericytes in zebrafish [80], this should change in the future. Future research should elucidate how pericytes behave in diabetic or other pathological conditions in zebrafish, revealing more about their function and potentially making them a target for further research. Apart from that, the role of inflammation in zebrafish models with impaired glucose metabolism has not yet been uncovered. This should also be taken into account for future investigations in the field, as there are multiple reporter lines for immune cells, including microglia [87].

In conclusion, zebrafish are no longer an exotic alternative model in diabetes research. Instead, research has focused on studying mechanisms and pathologies associated with hyperglycaemic conditions in zebrafish. Zebrafish demonstrate many advantages in such research. For example, hyperglycaemic conditions can be induced through easy and fast protocols, and the first effects on the hyaloid vasculature can be studied at larval stages. This is especially interesting when studying the effect of antiangiogenic drugs on the formation of new blood vessels. Adult zebrafish also show a reaction to hyperglycaemic conditions, with the *pdx1* mutants being the only known model organism in which retinal angiogenesis due to hyperglycaemia can be studied. Other important retinal phenotypes such as photoreceptor degeneration, increased vascular permeability, and the activation of Müller glial cells have been shown in various zebrafish models of DR, highlighting the similarities between mammalian and zebrafish models. Photoreceptor degeneration can be reliably modelled in zebrafish, including photoreceptor dysfunction, which can be quantified in an ERG. Furthermore, the zebrafish genome shows a high amount of shared genetic identity with humans. Through the possibility of inducing targeted mutations, this leads to an extensive number of possibilities for researchers in the field.

Important differences must also be considered. The ability of zebrafish to regenerate injured parts of the retina through the Müller glial cells is a relevant factor that differentiates zebrafish from mammals. Thus far, what role this regenerative capability plays in the development of DR-like pathologies is still unknown. Even though most of the anatomy of the retina and the individual cell types involved in the neurovascular unit are highly conserved, in contrast to the mammalian retina, zebrafish retinal vessels lay on top of the inner limiting membrane and do not form intraretinal plexuses. When choosing a zebrafish model of DR or analysing results of studies, researchers must be aware of these factors. In conclusion, zebrafish are a reliable model for various pathologies associated with DR, and they can be used to extend and improve the toolbox that mammalian models have provided for DR research so far.

Table 1. Zebrafish models of diabetic retinopathy. Abbreviations: 4-HNE = 4-hydroxynonenal, dpf = days post fertilization, GCL = ganglion cell layer, GS = glutamine synthetase, hpf = hours post fertilization, i.p. = intraperitoneal, INL = inner nuclear layer, IPL = inner plexiform layer, MG = methylglyoxal, n.e. = not evaluated, ONL = outer nuclear layer, OPL = outer plexiforme layer, RGC = retinal ganglion cell, STZ = streptozotocin, "+" indicates positive findings.

Model	Induction	Angiogenesis	Endothelial Cell Dysfunction	Pericyte Loss	Müller Glia Activation	Photoreceptor Degeneration	Neurodegeneration
Gleeson, Connaughton et al. 2007 [16]	Exposure to alternating glucose/water solutions for 28 days (adult zf)	n.e.	n.e.	n.e.	n.e.	n.e.	+(decreased IPL thickness)
Cao, Jensen et al. 2008 [44]	Experimental hypoxia for up to 15 days (adult zf)	+	n.e.	n.e.	n.e.	n.e.	n.e.
Alvarez, Chen et al. 2010 [43]	Exposure to alternating glucose/water solutions for 30 days (adult zf)	n.e.	+ (thickening of vessel basement membrane, wider tight and adherens junctions)	n.e.	+	+(abnormal retinal histology, impaired cone ERGs)	-
Olsen, Sarras et al. 2010 [47]	i.p. or direct caudal fin injection of STZ (adult zf)	n.e.	n.e.	n.e.	n.e.	+(decreased PRL thickness)	+(decreased IPL thickness)

Table 1. *Cont.*

Model	Induction	Angiogenesis	Endothelial Cell Dysfunction	Pericyte Loss	Müller Glia Activation	Photoreceptor Degeneration	Neurodegeneration
Carnovali, Luzi et al. 2016 [102]	Exposure to 4% glucose solution for 28 days (adult zf)	n.e.	+(increased vessel diameter, aneurysm-like structure, marked fragility of the anatomical structure)	n.e.	n.e.	n.e.	n.e.
Jung, Kim et al. 2016 [62]	Treatment with 130 mM glucose from 3–6 days post fertilisation (zf larvae)	-	+(increased vessel diameter, irregular and discontinuous staining of ZO-1)	n.e.	n.e.	n.e.	n.e.
Tanvir, Nelson et al. 2018 [100]	Exposure to alternating glucose/water solutions for 28 days (adult zf)	n.e.	n.e.	n.e.	n.e.	+(impaired ERG)	+(increased IPL and OPL thickness)
Ali, Mukawaya et al. 2019 [78]	Experimental hypoxia for up to 15 days (adult zf)	-(however: remodelling by intussusception)	+(decrease in ZO-1 abundance, increased vessel permeability)	n.e.	n.e.	n.e.	n.e.
Li, Zhao et al. 2019 [103]	Incubation with 500µM methylglyoxal with or without 30 mM glucose starting at 10 hpf to 4 dpf (zf larvae)	+(MG induces an increase in vascular area and branch points)	n.e.	n.e.	n.e.	n.e.	n.e.
Lodd, Wiggenhauser et al. 2019 [57]	CRISPR/Cas9 generated knockout zebrafish line for *glo1* + overfeeding of artemia (adult zf)	+	n.e.	n.e.	n.e.	n.e.	n.e.
Singh, Castillo et al. 2019 [95]	Exposure to 4 and 5% D-Glucose in a pulsatile manner from 3 hpf to 5 dpf (zf larvae, adult zf)	+(adult zf show an increased number of hyaloid blood vessel sprouts at 100 dpf after glucose treatment from 3 hpf to 5 dpf)	+(increased vessel permeability)	n.e.	(+) (reduced number of Müller glia cells)	n.e.	+(decreased IPL thickness, increased INL thickness, increased GCL thickness; decreased number of RGC)
Ali, Zang et al. 2020 [53]	*Pdx1* mutant fish generated through the Zebrafish Mutation Project (as described by Kimmel, Dobler et al. 2015) (adult zf)	+	+(vessel constriction and stenosis, reduction of average vessel diameter, reduced ZO-1 expression, reduced GLUT1 expression, increased vessel permeability)	(+) (reduced expression of transgelin1)	+(enhance expression of GS, hypertrophic changes)	+(reduced numbers of rods and cones, impaired ERG)	+(increased nuclei in the INL, decreased nuclei in the ONL)

Table 1. *Cont.*

Model	Induction	Angiogenesis	Endothelial Cell Dysfunction	Pericyte Loss	Müller Glia Activation	Photoreceptor Degeneration	Neurodegeneration
Wiggenhauser, Qi et al. 2020 [55]	CRISPR/Cas9 generated knockout line for *pdx1* (zf larvae, adult zf)	+(at 6 dpf and in the adult retina)	+(increased number of endothelial cell nuclei, increased vessel permeability)	n.e.	n.e.	n.e.	n.e.
Lou, Boger et al. 2020 [56]	Incubation with 4-HNE (zebrafish larvae)	+(elevated vascular sprout formation)	+(increased branch diameters)	n.e.	n.e.	n.e.	n.e.

Author Contributions: Conceptualization, C.S.M., H.-P.H. and J.K.; writing—original draft preparation, C.S.M.; writing—review and editing, H.-P.H. and J.K.; visualization, C.S.M.; supervision, J.K.; project administration, H.-P.H. and J.K.; funding acquisition, H.-P.H. and J.K. All authors have read and agreed to the published version of the manuscript.

Funding: The work was supported by Deutsche Forschungsgemeinschaft (CRC1118 and IRTG 1874/2 DIAMICOM) and by the Rolf M. Schwiete Stiftung.

Institutional Review Board Statement: Not applicable.

Informed Consent Statement: Not applicable.

Data Availability Statement: Not applicable.

Acknowledgments: The authors acknowledge the support of the Zebrafish Core Facility Mannheim and the Hammes lab (Fifth Medical Department, Medical Faculty Mannheim, Heidelberg University, Mannheim, Germany). The critical reading of the manuscript by Luke Kurowski is acknowledged. The figures were created with BioRender.com.

Conflicts of Interest: The authors declare no conflict of interest.

References

1. Ogurtsova, K.; da Rocha Fernandes, J.D.; Huang, Y.; Linnenkamp, U.; Guariguata, L.; Cho, N.H.; Cavan, D.; Shaw, J.E.; Makaroff, L.E. IDF Diabetes Atlas: Global estimates for the prevalence of diabetes for 2015 and 2040. *Diabetes Res. Clin. Pract.* **2017**, *128*, 40–50. [CrossRef] [PubMed]
2. Antonetti, D.A.; Silva, P.S.; Stitt, A.W. Current understanding of the molecular and cellular pathology of diabetic retinopathy. *Nat. Rev. Endocrinol.* **2021**, *17*, 195–206. [CrossRef] [PubMed]
3. Antonetti, D.A.; Klein, R.; Gardner, T.W. Diabetic retinopathy. *N. Engl. J. Med.* **2012**, *366*, 1227–1239. [CrossRef] [PubMed]
4. Gangaputra, S.; Lovato, J.F.; Hubbard, L.; Davis, M.D.; Esser, B.A.; Ambrosius, W.T.; Chew, E.Y.; Greven, C.; Perdue, L.H.; Wong, W.T.; et al. Comparison of standardized clinical classification with fundus photograph grading for the assessment of diabetic retinopathy and diabetic macular edema severity. *Retina* **2013**, *33*, 1393–1399. [CrossRef] [PubMed]
5. Leasher, J.L.; Bourne, R.R.; Flaxman, S.R.; Jonas, J.B.; Keeffe, J.; Naidoo, K.; Pesudovs, K.; Price, H.; White, R.A.; Wong, T.Y.; et al. Global Estimates on the Number of People Blind or Visually Impaired by Diabetic Retinopathy: A Meta-analysis from 1990 to 2010. *Diabetes Care* **2016**, *39*, 1643–1649. [CrossRef] [PubMed]
6. Bourne, R.; Steinmetz, J.D.; Flaxman, S.; Briant, P.S.; Taylor, H.R.; Resnikoff, S.; Casson, R.J.; Abdoli, A.; Abu-Gharbieh, E.; Afshin, A.; et al. Trends in prevalence of blindness and distance and near vision impairment over 30 years: An analysis for the Global Burden of Disease Study. *Lancet Glob. Health* **2021**, *9*, e130–e143. [CrossRef]
7. Early Treatment Diabetic Retinopathy Study Research Group. Fundus photographic risk factors for progression of diabetic retinopathy. ETDRS report number 12. *Ophthalmology* **1991**, *98* (Suppl. S5), 823–833. [CrossRef]
8. Stewart, J.M.; Coassin, M.; Schwartz, D.M. Diabetic Retinopathy. In *Endotext*; Feingold, K.R., Anawalt, B., Boyce, A., Chrousos, G., de Herder, W.W., Dungan, K., Grossman, A., et al., Eds.; South Dartmouth: Dartmouth, MA, USA, 2000.
9. UK Prospective Diabetes Study (ukpds) Group; Kohner, E.M.; Stratton, I.M.; Aldington, S.J.; Holman, R.R.; Matthews, D.R. Relationship between the severity of retinopathy and progression to photocoagulation in patients with Type 2 diabetes mellitus in the UKPDS (UKPDS 52). *Diabet. Med.* **2001**, *18*, 178–184. [CrossRef]
10. Ponto, K.A.; Koenig, J.; Peto, T.; Lamparter, J.; Raum, P.; Wild, P.S.; Lackner, K.J.; Pfeiffer, N.; Mirshahi, A. Prevalence of diabetic retinopathy in screening-detected diabetes mellitus: Results from the Gutenberg Health Study (GHS). *Diabetologia* **2016**, *59*, 1913–1919. [CrossRef]

11. Dooley, K.; Zon, L.I. Zebrafish: A model system for the study of human disease. *Curr. Opin. Genet. Dev.* **2000**, *10*, 252–256. [CrossRef]
12. Parng, C.; Seng, W.L.; Semino, C.; McGrath, P. Zebrafish: A Preclinical Model for Drug Screening. *ASSAY Drug Dev. Technol.* **2002**, *1*, 41–48. [CrossRef] [PubMed]
13. Menke, A.L.; Spitsbergen, J.M.; Wolterbeek, A.P.M.; Woutersen, R.A. Normal Anatomy and Histology of the Adult Zebrafish. *Toxicol. Pathol.* **2011**, *39*, 759–775. [CrossRef] [PubMed]
14. Zang, L.; Maddison, L.A.; Chen, W. Zebrafish as a Model for Obesity and Diabetes. *Front. Cell Dev. Biol.* **2018**, *6*, 91. [CrossRef] [PubMed]
15. Intine, R.V.; Olsen, A.S.; Sarras, M.P., Jr. A Zebrafish Model of Diabetes Mellitus and Metabolic Memory. *J. Vis. Exp.* **2013**, *2013*, e50232. [CrossRef]
16. Gleeson, M.; Connaughton, V.; Arneson, L.S. Induction of hyperglycaemia in zebrafish (Danio rerio) leads to morphological changes in the retina. *Acta Diabetol.* **2007**, *44*, 157–163. [CrossRef]
17. Lawlor, N.; Khetan, S.; Ucar, D.; Stitzel, M.L. Genomics of Islet (Dys)function and Type 2 Diabetes. *Trends Genet.* **2017**, *33*, 244–255. [CrossRef] [PubMed]
18. Loos, R.J. The genetics of adiposity. *Curr. Opin. Genet. Dev.* **2018**, *50*, 86–95. [CrossRef]
19. Lidster, K.; Readman, G.D.; Prescott, M.J.; Owen, S. International survey on the use and welfare of zebrafish Danio rerio in research. *J. Fish Biol.* **2017**, *90*, 1891–1905. [CrossRef]
20. Wiggenhauser, L.M.; Kroll, J. Vascular Damage in Obesity and Diabetes: Highlighting Links Between Endothelial Dysfunction and Metabolic Disease in Zebrafish and Man. *Curr. Vasc. Pharmacol.* **2019**, *17*, 476–490. [CrossRef]
21. Metea, M.R.; Newman, E.A. Signalling within the neurovascular unit in the mammalian retina. *Exp. Physiol.* **2007**, *92*, 635–640. [CrossRef]
22. Zlokovic, B.V. Neurovascular pathways to neurodegeneration in Alzheimer's disease and other disorders. *Nat. Rev. Neurosci.* **2011**, *12*, 723–738. [CrossRef]
23. Feng, Y.; Busch, S.; Gretz, N.; Hoffmann, S.; Hammes, H.-P. Crosstalk in the Retinal Neurovascular Unit—Lessons for the Diabetic Retina. *Exp. Clin. Endocrinol. Diabetes* **2012**, *120*, 199–201. [CrossRef]
24. Duh, E.J.; Sun, J.K.; Stitt, A.W. Diabetic retinopathy: Current understanding, mechanisms, and treatment strategies. *JCI Insight* **2017**, *2*, 1–13. [CrossRef] [PubMed]
25. Gardner, T.W.; Davila, J.R. The neurovascular unit and the pathophysiologic basis of diabetic retinopathy. *Graefe's Arch. Clin. Exp. Ophthalmol.* **2017**, *255*, 1–6. [CrossRef] [PubMed]
26. Fadool, J.M.; Dowling, J.E. Zebrafish: A model system for the study of eye genetics. *Prog. Retin. Eye Res.* **2008**, *27*, 89–110. [CrossRef] [PubMed]
27. Bilotta, J.; Saszik, S. The zebrafish as a model visual system. *Int. J. Dev. Neurosci.* **2001**, *19*, 621–629. [CrossRef]
28. Angueyra, J.M.; Kindt, K.S. Leveraging Zebrafish to Study Retinal Degenerations. *Front. Cell Dev. Biol.* **2018**, *6*, 110. [CrossRef]
29. Bibliowicz, J.; Tittle, R.K.; Gross, J.M. Toward a better understanding of human eye disease insights from the zebrafish, Danio rerio. *Prog. Mol. Biol. Transl. Sci.* **2011**, *100*, 287–330.
30. Eliceiri, B.P.; Gonzalez, A.M.; Baird, A. Zebrafish Model of the Blood-Brain Barrier: Morphological and Permeability Studies. *Methods Mol. Biol.* **2011**, *686*, 371–378. [CrossRef]
31. O'Brown, N.M.; Pfau, S.J.; Gu, C. Bridging barriers: A comparative look at the blood–brain barrier across organisms. *Genes Dev.* **2018**, *32*, 466–478. [CrossRef]
32. Klaassen, I.; Van Noorden, C.J.; Schlingemann, R.O. Molecular basis of the inner blood-retinal barrier and its breakdown in diabetic macular edema and other pathological conditions. *Prog. Retin. Eye Res.* **2013**, *34*, 19–48. [CrossRef]
33. Carmeliet, P.; Jain, R.K. Molecular mechanisms and clinical applications of angiogenesis. *Nat. Cell Biol.* **2011**, *473*, 298–307. [CrossRef] [PubMed]
34. Campochiaro, P.A. Molecular pathogenesis of retinal and choroidal vascular diseases. *Prog. Retin. Eye Res.* **2015**, *49*, 67–81. [CrossRef] [PubMed]
35. Lawson, N.D.; Weinstein, B.M. In Vivo Imaging of Embryonic Vascular Development Using Transgenic Zebrafish. *Dev. Biol.* **2002**, *248*, 307–318. [CrossRef] [PubMed]
36. Alvarez, Y.; Cederlund, M.L.; Cottell, D.C.; Bill, B.R.; Ekker, S.C.; Torres-Vazquez, J.; Weinstein, B.M.; Hyde, D.R.; Vihtelic, T.S.; Kennedy, B.N. Genetic determinants of hyaloid and retinal vasculature in zebrafish. *BMC Dev. Biol.* **2007**, *7*, 114. [CrossRef]
37. Gestri, G.; Link, B.A.; Neuhauss, S.C. The visual system of zebrafish and its use to model human ocular Diseases. *Dev. Neurobiol.* **2011**, *72*, 302–327. [CrossRef]
38. Gore, A.V.; Monzo, K.; Cha, Y.R.; Pan, W.; Weinstein, B.M. Vascular Development in the Zebrafish. *Cold Spring Harb. Perspect. Med.* **2012**, *2*, a006684. [CrossRef]
39. Richardson, R.; Tracey-White, D.; Webster, A.; Moosajee, M. The zebrafish eye—A paradigm for investigating human ocular genetics. *Eye* **2017**, *31*, 68–86. [CrossRef]
40. Fruttiger, M. Development of the retinal vasculature. *Angiogenesis* **2007**, *10*, 77–88. [CrossRef]
41. Kur, J.; Newman, E.A.; Chan-Ling, T. Cellular and physiological mechanisms underlying blood flow regulation in the retina and choroid in health and disease. *Prog. Retin. Eye Res.* **2012**, *31*, 377–406. [CrossRef]

42. Van Dijk, H.W.; Kok, P.H.; Garvin, M.; Sonka, M.; DeVries, J.H.; Michels, R.P.; van Velthoven, M.E.; Schlingemann, R.O.; Verbraak, F.D.; Abramoff, M.D. Selective loss of inner retinal layer thickness in type 1 diabetic patients with minimal diabetic retinopathy. *Investig. Ophthalmol. Vis. Sci.* **2009**, *50*, 3404–3409. [CrossRef]

43. Alvarez, Y.; Chen, K.; Reynolds, A.; Waghorne, N.; O'Connor, J.J.; Kennedy, B.N. Predominant cone photoreceptor dysfunction in a hyperglycaemic model of non-proliferative diabetic retinopathy. *Dis. Model. Mech.* **2010**, *3*, 236–245. [CrossRef]

44. Cao, R.; Jensen, L.D.E.; Söll, I.; Hauptmann, G.; Cao, Y. Hypoxia-Induced Retinal Angiogenesis in Zebrafish as a Model to Study Retinopathy. *PLoS ONE* **2008**, *3*, e2748. [CrossRef] [PubMed]

45. Olivares, A.M.; Althoff, K.; Chen, G.F.; Wu, S.; Morrisson, M.A.; DeAngelis, M.M.; Haider, N. Animal Models of Diabetic Retinopathy. *Curr. Diabetes Rep.* **2017**, *17*, 1–17. [CrossRef] [PubMed]

46. Kim, C.B.; D'Amore, P.; Connor, K.M. Revisiting the mouse model of oxygen-induced retinopathy. *Eye Brain* **2016**, *8*, 67–79. [CrossRef] [PubMed]

47. Olsen, A.S.; Sarras, M.P., Jr.; Intine, R.V. Limb regeneration is impaired in an adult zebrafish model of diabetes mellitus. *Wound Repair Regen.* **2010**, *18*, 532–542. [CrossRef]

48. Benchoula, K.; Khatib, A.; Quzwain, F.M.C.; Mohamad, C.A.C.; Sulaiman, W.M.A.W.; Wahab, R.A.; Ahmed, Q.U.; Ghaffar, M.A.; Saiman, M.Z.; Alajmi, M.F.; et al. Optimization of Hyperglycemic Induction in Zebrafish and Evaluation of Its Blood Glucose Level and Metabolite Fingerprint Treated with Psychotria malayana Jack Leaf Extract. *Molecules* **2019**, *24*, 1506. [CrossRef] [PubMed]

49. Hwang, W.Y.; Peterson, R.T.; Yeh, J.-R.J. Methods for targeted mutagenesis in zebrafish using TALENs. *Methods* **2014**, *69*, 76–84. [CrossRef]

50. Shah, A.N.; Davey, C.F.; Whitebirch, A.C.; Miller, A.C.; Moens, C.B. Rapid reverse genetic screening using CRISPR in zebrafish. *Nat. Methods* **2015**, *12*, 535–540. [CrossRef]

51. Kettleborough, R.N.W.; Busch-Nentwich, E.M.; Harvey, S.A.; Dooley, C.M.; De Bruijn, E.; Van Eeden, F.; Sealy, I.; White, R.J.; Herd, C.; Nijman, I.J.; et al. A systematic genome-wide analysis of zebrafish protein-coding gene function. *Nat. Cell Biol.* **2013**, *496*, 494–497. [CrossRef]

52. Kimmel, R.A.; Dobler, S.; Schmitner, N.; Walsen, T.; Freudenblum, J.; Meyer, D. Diabetic pdx1-mutant zebrafish show conserved responses to nutrient overload and anti-glycemic treatment. *Sci. Rep.* **2015**, *5*, 14241. [CrossRef] [PubMed]

53. Ali, Z.; Zang, J.; Lagali, N.; Schmitner, N.; Salvenmoser, W.; Mukwaya, A.; Neuhauss, S.C.; Jensen, L.D.; Kimmel, R.A. Photoreceptor Degeneration Accompanies Vascular Changes in a Zebrafish Model of Diabetic Retinopathy. *Investig. Opthalmol. Vis. Sci.* **2020**, *61*, 43. [CrossRef]

54. Koepsell, H. Glucose transporters in brain in health and disease. *Pflügers Arch. Eur. J. Physiol.* **2020**, *472*, 1299–1343. [CrossRef]

55. Wiggenhauser, L.M.; Qi, H.; Stoll, S.J.; Metzger, L.; Bennewitz, K.; Poschet, G.; Krenning, G.; Hillebrands, J.L.; Hammes, H.P.; Kroll, J. Activation of Retinal Angiogenesis in Hyperglycemic pdx1 (-/-) Zebrafish Mutants. *Diabetes* **2020**, *69*, 1020–1031. [CrossRef] [PubMed]

56. Lou, B.; Boger, M.; Bennewitz, K.; Sticht, C.; Kopf, S.; Morgenstern, J.; Fleming, T.; Hell, R.; Yuan, Z.; Nawroth, P.P.; et al. Elevated 4-hydroxynonenal induces hyperglycaemia via Aldh3a1 loss in zebrafish and associates with diabetes progression in humans. *Redox Biol.* **2020**, *37*, 101723. [CrossRef] [PubMed]

57. Lodd, E.; Wiggenhauser, L.M.; Morgenstern, J.; Fleming, T.H.; Poschet, G.; Büttner, M.; Tabler, C.T.; Wohlfart, D.P.; Nawroth, P.P.; Kroll, J. The combination of loss of glyoxalase1 and obesity results in hyperglycemia. *JCI Insight* **2019**, *4*, 1–17. [CrossRef] [PubMed]

58. Rabbani, N.; Thornalley, P.J. Glyoxalase in diabetes, obesity and related disorders. *Semin. Cell Dev. Biol.* **2011**, *22*, 309–317. [CrossRef]

59. Salehpour, A.; Rezaei, M.; Khoradmehr, A.; Tahamtani, Y.; Tamadon, A. Which Hyperglycemic Model of Zebrafish (Danio rerio) Suites My Type 2 Diabetes Mellitus Research? A Scoring System for Available Methods. *Front. Cell Dev. Biol.* **2021**, *9*, 652061. [CrossRef] [PubMed]

60. Geisler, R.; Köhler, A.; Dickmeis, T.; Strähle, U. Archiving of zebrafish lines can reduce animal experiments in biomedical research. *EMBO Rep.* **2017**, *18*, 1–2. [CrossRef]

61. Cassar, S.; Adatto, I.; Freeman, J.; Gamse, J.T.; Iturria, I.; Lawrence, C.; Muriana, A.; Peterson, R.T.; Van Cruchten, S.; Zon, L.I. Use of Zebrafish in Drug Discovery Toxicology. *Chem. Res. Toxicol.* **2020**, *33*, 95–118. [CrossRef]

62. Jung, S.-H.; Kim, Y.S.; Lee, Y.-R.; Kim, J.S. High glucose-induced changes in hyaloid-retinal vessels during early ocular development of zebrafish: A short-term animal model of diabetic retinopathy. *Br. J. Pharmacol.* **2015**, *173*, 15–26. [CrossRef] [PubMed]

63. Van de Venter, M.; Didloff, J.; Reddy, S.; Swanepoel, B.; Govender, S.; Dambuza, N.S.; Williams, S.; Koekemoer, T.C.; Venables, L. Wild-Type Zebrafish (Danio rerio) Larvae as a Vertebrate Model for Diabetes and Comorbidities: A Review. *Animals* **2020**, *11*, 54. [CrossRef] [PubMed]

64. Kidd, K.R.; Weinstein, B.M. Fishing for novel angiogenic therapies. *Br. J. Pharmacol.* **2003**, *140*, 585–594. [CrossRef]

65. Sasore, T.; Kennedy, B. Deciphering Combinations of PI3K/AKT/mTOR Pathway Drugs Augmenting Anti-Angiogenic Efficacy In Vivo. *PLoS ONE* **2014**, *9*, e105280. [CrossRef] [PubMed]

66. Galvin, O.; Srivastava, A.; Carroll, O.; Kulkarni, R.; Dykes, S.; Vickers, S.; Dickinson, K.; Reynolds, A.; Kilty, C.; Redmond, G.; et al. A sustained release formulation of novel quininib-hyaluronan microneedles inhibits angiogenesis and retinal vascular permeability in vivo. *J. Control. Release* **2016**, *233*, 198–207. [CrossRef] [PubMed]
67. Sulaiman, R.S.; Merrigan, S.; Quigley, J.; Qi, X.; Lee, B.; Boulton, M.E.; Kennedy, B.; Seo, S.-Y.; Corson, T.W. A novel small molecule ameliorates ocular neovascularisation and synergises with anti-VEGF therapy. *Sci. Rep.* **2016**, *6*, 25509. [CrossRef] [PubMed]
68. Ohnesorge, N.; Sasore, T.; Hillary, D.; Alvarez, Y.; Carey, M.; Kennedy, B.N. Orthogonal Drug Pooling Enhances Phenotype-Based Discovery of Ocular Antiangiogenic Drugs in Zebrafish Larvae. *Front. Pharmacol.* **2019**, *10*, 508. [CrossRef] [PubMed]
69. Armulik, A.; Genové, G.; Betsholtz, C. Pericytes: Developmental, Physiological, and Pathological Perspectives, Problems, and Promises. *Dev. Cell* **2011**, *21*, 193–215. [CrossRef]
70. Caporarello, N.; D'Angeli, F.; Cambria, M.T.; Candido, S.; Giallongo, C.; Salmeri, M.; Lombardo, C.; Longo, A.; Giurdanella, G.; Anfuso, C.D.; et al. Pericytes in Microvessels: From "Mural" Function to Brain and Retina Regeneration. *Int. J. Mol. Sci.* **2019**, *20*, 6351. [CrossRef]
71. Leveen, P.; Pekny, M.; Gebre-Medhin, S.; Swolin, B.; Larsson, E.; Betsholtz, C. Mice deficient for PDGF B show renal, cardiovascular, and hematological abnormalities. *Genes Dev.* **1994**, *8*, 1875–1887. [CrossRef]
72. Alarcon-Martinez, L.; Villafranca-Baughman, D.; Quintero, H.; Kacerovsky, J.B.; Dotigny, F.; Murai, K.K.; Prat, A.; Drapeau, P.; Di Polo, A. Interpericyte tunnelling nanotubes regulate neurovascular coupling. *Nat. Cell Biol.* **2020**, *585*, 1–5. [CrossRef] [PubMed]
73. Spencer, B.G.; Estevez, J.J.; Liu, E.; Craig, J.E.; Finnie, J.W. Pericytes, inflammation, and diabetic retinopathy. *Inflammopharmacology* **2020**, *28*, 697–709. [CrossRef] [PubMed]
74. Pfister, F.; Przybyt, E.; Harmsen, M.C.; Hammes, H.-P. Pericytes in the eye. *Pflügers Arch. Eur. J. Physiol.* **2013**, *465*, 789–796. [CrossRef] [PubMed]
75. Wang, Y.; Pan, L.; Moens, C.B.; Appel, B. Notch3 establishes brain vascular integrity by regulating pericyte number. *Development* **2014**, *141*, 307–317. [CrossRef] [PubMed]
76. Ando, K.; Fukuhara, S.; Izumi, N.; Nakajima, H.; Fukui, H.; Kelsh, R.; Mochizuki, N. Clarification of mural cell coverage of vascular endothelial cells by live imaging of zebrafish. *Development* **2016**, *143*, 1328–1339. [CrossRef] [PubMed]
77. Santoro, M.M.; Pesce, G.; Stainier, D.Y. Characterization of vascular mural cells during zebrafish development. *Mech. Dev.* **2009**, *126*, 638–649. [CrossRef]
78. Ali, Z.; Mukwaya, A.; Biesemeier, A.; Ntzouni, M.; Ramsköld, D.; Giatrellis, S.; Mammadzada, P.; Cao, R.; Lennikov, A.; Marass, M.; et al. Intussusceptive Vascular Remodeling Precedes Pathological Neovascularization. *Arter. Thromb. Vasc. Biol.* **2019**, *39*, 1402–1418. [CrossRef] [PubMed]
79. Whitesell, T.R.; Kennedy, R.M.; Carter, A.D.; Rollins, E.L.; Georgijevic, S.; Santoro, M.M. An alpha-smooth muscle actin (acta2/alphasma) zebrafish transgenic line marking vascular mural cells and visceral smooth muscle cells. *PLoS ONE* **2014**, *9*, e90590. [CrossRef]
80. Caceres, L.; Prykhozhij, S.V.; Cairns, E.; Gjerde, H.; Duff, N.M.; Collett, K.; Ngo, M.; Nasrallah, G.K.; McMaster, C.R.; Litvak, M.; et al. Frizzled 4 regulates ventral blood vessel remodeling in the zebrafish retina. *Dev. Dyn.* **2019**, *248*, 1243–1256. [CrossRef]
81. Dietrich, N.; Hammes, H.P. Retinal Digest Preparation: A Method to Study Diabetic Retinopathy. *Methods Mol. Biol.* **2012**, *933*, 291–302.
82. Karlstetter, M.; Scholz, R.; Rutar, M.; Wong, W.T.; Provis, J.M.; Langmann, T. Retinal microglia: Just bystander or target for therapy? *Prog. Retin. Eye Res.* **2015**, *45*, 30–57. [CrossRef] [PubMed]
83. Zeng, H.-Y.; Green, W.R.; Tso, M.O.M. Microglial Activation in Human Diabetic Retinopathy. *Arch. Ophthalmol.* **2008**, *126*, 227–232. [CrossRef]
84. Rübsam, A.; Parikh, S.; Fort, P.E. Role of Inflammation in Diabetic Retinopathy. *Int. J. Mol. Sci.* **2018**, *19*, 942. [CrossRef] [PubMed]
85. Arroba, A.I.; Valverde, A.M. Modulation of microglia in the retina: New insights into diabetic retinopathy. *Acta Diabetol.* **2017**, *54*, 527–533. [CrossRef] [PubMed]
86. Var, S.R.; Byrd-Jacobs, C.A. Role of Macrophages and Microglia in Zebrafish Regeneration. *Int. J. Mol. Sci.* **2020**, *21*, 4768. [CrossRef]
87. Van Dyck, A.; Bollaerts, I.; Beckers, A.; Vanhunsel, S.; Glorian, N.; van Houcke, J.; van Ham, T.J.; De Groef, L.; Andries, L.; Moons, L. Müller glia–myeloid cell crosstalk accelerates optic nerve regeneration in the adult zebrafish. *Glia* **2021**, *69*, 1444–1463. [CrossRef]
88. Reichenbach, A.; Bringmann, A. Glia of the human retina. *Glia* **2019**, *68*, 768–796. [CrossRef]
89. Lundkvist, A.; Reichenbach, A.; Betsholtz, C.; Carmeliet, P.; Wolburg, H.; Pekny, M. Under stress, the absence of intermediate filaments from Muller cells in the retina has structural and functional consequences. *J. Cell Sci.* **2004**, *117*, 3481–3488. [CrossRef]
90. Coughlin, B.A.; Feenstra, D.J.; Mohr, S. Müller cells and diabetic retinopathy. *Vis. Res.* **2017**, *139*, 93–100. [CrossRef]
91. Goldman, D. Müller glial cell reprogramming and retina regeneration. *Nat. Rev. Neurosci.* **2014**, *15*, 431–442. [CrossRef]
92. Gorsuch, R.A.; Hyde, D.R. Regulation of Müller glial dependent neuronal regeneration in the damaged adult zebrafish retina. *Exp. Eye Res.* **2014**, *123*, 131–140. [CrossRef] [PubMed]
93. Mitchell, D.M.; Lovel, A.G.; Stenkamp, D.L. Dynamic changes in microglial and macrophage characteristics during degeneration and regeneration of the zebrafish retina. *J. Neuroinflamm.* **2018**, *15*, 1–20. [CrossRef] [PubMed]

94. White, D.T.; Sengupta, S.; Saxena, M.T.; Xu, Q.; Hanes, J.; Ding, D.; Ji, H.; Mumm, J.S. Immunomodulation-accelerated neuronal regeneration following selective rod photoreceptor cell ablation in the zebrafish retina. *Proc. Natl. Acad. Sci. USA* **2017**, *114*, E3719–E3728. [CrossRef] [PubMed]
95. Singh, A.; Castillo, H.A.; Brown, J.; Kaslin, J.; Dwyer, K.M.; Gibert, Y. High glucose levels affect retinal patterning during zebrafish embryogenesis. *Sci. Rep.* **2019**, *9*, 4121. [CrossRef]
96. Kern, T.S.; Berkowitz, B.A. Photoreceptors in diabetic retinopathy. *J. Diabetes Investig.* **2015**, *6*, 371–380. [CrossRef]
97. Hammes, H.P.; Federoff, H.J.; Brownlee, M. Nerve growth factor prevents both neuroretinal programmed cell death and capillary pathology in experimental diabetes. *Mol. Med.* **1995**, *1*, 527–534. [CrossRef]
98. Barber, A.J.; Lieth, E.; Khin, S.; Antonetti, D.; Buchanan, A.G.; Gardner, T.W. Neural apoptosis in the retina during experimental and human diabetes. Early onset and effect of insulin. *J. Clin. Investig.* **1998**, *102*, 783–791. [CrossRef]
99. Simó, R.; Stitt, A.W.; Gardner, T.W. Neurodegeneration in diabetic retinopathy: Does it really matter? *Diabetologia* **2018**, *61*, 1902–1912. [CrossRef]
100. Tanvir, Z.; Nelson, R.F.; DeCicco-Skinner, K.; Connaughton, V.P. One month of hyperglycemia alters spectral responses of the zebrafish photopic electroretinogram. *Dis. Model. Mech.* **2018**, *11*, dmm035220. [CrossRef]
101. Seth, A.; Stemple, D.L.; Barroso, I. The emerging use of zebrafish to model metabolic disease. *Dis. Model. Mech.* **2013**, *6*, 1080–1088. [CrossRef]
102. Carnovali, M.; Luzi, L.; Banfi, G.; Mariotti, M. Chronic hyperglycemia affects bone metabolism in adult zebrafish scale model. *Endocrine* **2016**, *54*, 808–817. [CrossRef] [PubMed]
103. Li, Y.; Zhao, Y.; Sang, S.; Leung, T. Methylglyoxal-Induced Retinal Angiogenesis in Zebrafish Embryo: A Potential Animal Model of Neovascular Retinopathy. *J. Ophthalmol.* **2019**, *2019*, 2746735. [CrossRef] [PubMed]

cells

Article

Mesenchymal Stem Cell Induced Foxp3(+) Tregs Suppress Effector T Cells and Protect against Retinal Ischemic Injury

Mona Agrawal [1], Pratheepa Kumari Rasiah [1], Amandeep Bajwa [2,3,4], Johnson Rajasingh [4,5] and Rajashekhar Gangaraju [1,6,*]

1   Department of Ophthalmology, University of Tennessee Health Science Center, Memphis, TN 38163, USA; magrawa1@uthsc.edu (M.A.); prasiah@uthsc.edu (P.K.R.)
2   James D. Eason Transplant Institute, Department of Surgery, University of Tennessee Health Science Center, Memphis, TN 38163, USA; abajwa@uthsc.edu
3   Department of Genetics, Genomics, and Informatics, University of Tennessee Health Science Center, Memphis, TN 38163, USA
4   Department of Microbiology, Immunology, and Biochemistry, University of Tennessee Health Science Center, Memphis, TN 38163, USA; rjohn186@uthsc.edu
5   Department of Bioscience Research, University of Tennessee Health Science Center, Memphis, TN 38163, USA
6   Department of Anatomy & Neurobiology, University of Tennessee Health Science Center, Memphis, TN 38163, USA
*   Correspondence: sgangara@uthsc.edu; Tel.: +1-901-448-2721

Citation: Agrawal, M.; Rasiah, P.K.; Bajwa, A.; Rajasingh, J.; Gangaraju, R. Mesenchymal Stem Cell Induced Foxp3(+) Tregs Suppress Effector T Cells and Protect against Retinal Ischemic Injury. *Cells* **2021**, *10*, 3006. https://doi.org/10.3390/cells10113006

Academic Editors: Maurice Ptito, Joseph Bouskila and Karl-Wilhelm Koch

Received: 14 September 2021
Accepted: 29 October 2021
Published: 4 November 2021

Publisher's Note: MDPI stays neutral with regard to jurisdictional claims in published maps and institutional affiliations.

Abstract: Mesenchymal stem/stromal cells (MSC) are well known for immunomodulation; however, the mechanisms involved in their benefits in the ischemic retina are unknown. This study tested the hypothesis that MSC induces upregulation of transcription factor forkhead box protein P3 (Foxp3) in T cells to elicit immune modulation, and thus, protect against retinal damage. Induced MSCs (iMSCs) were generated by differentiating the induced pluripotent stem cells (iPSC) derived from urinary epithelial cells through a noninsertional reprogramming approach. In in-vitro cultures, iMSC transferred mitochondria to immune cells via F-actin nanotubes significantly increased oxygen consumption rate (OCR) for basal respiration and ATP production, suppressed effector T cells, and promoted differentiation of CD4+CD25+ T regulatory cells (Tregs) in coculture with mouse splenocytes. In in-vivo studies, iMSCs transplanted in ischemia-reperfusion (I/R) injured eye significantly increased Foxp3+ Tregs in the retina compared to that of saline-injected I/R eyes. Furthermore, iMSC injected I/R eyes significantly decreased retinal inflammation as evidenced by reduced gene expression of *IL1β*, *VCAM1*, *LAMA5*, and *CCL2* and improved b-wave amplitudes compared to that of saline-injected I/R eyes. Our study demonstrates that iMSCs can transfer mitochondria to immune cells to suppress the effector T cell population. Additionally, our current data indicate that iMSC can enhance differentiation of T cells into Foxp3 Tregs in vitro and therapeutically improve the retina's immune function by upregulation of Tregs to decrease inflammation and reduce I/R injury-induced retinal degeneration in vivo.

Keywords: CD4+CD25+; retinopathy; inflammation; iPSC; mitochondria

1. Introduction

Ischemic retinopathies, including diabetic retinopathy (DR), retinopathy of prematurity (ROP), and retinal vascular occlusion (RVO), are increasing in prevalence, represent a significant economic burden, and are major causes of vision loss and blindness worldwide [1–3]. A wide variety of traditional treatment therapies, including photocoagulation and anti-VEGF therapies during the neovascularization phase, showed benefits, with no treatments currently approved that address underlying proinflammatory pathways that are known to trigger neurovascular degeneration [2]. Stem cell therapies, mainly multipotent mesenchymal stem cells (MSCs), recently gained significant attention as a potential therapy for the treatment of ischemic retinopathies [3,4]. Our previous studies utilizing MSCs derived from the stromal vascular fraction of adult human adipose tissue (adipose-derived stem cells, ASC) [5,6], and independently corroborated with bone marrow (BM-MSC) [7] and umbilical cord (UC-MSC) [8], demonstrated substantial regeneration and recovery of the damaged retina after treatment, although the exact mechanisms by which MSCs may protect against vision loss, remains unclear.

MSCs are well known for being involved in immunomodulation, in part by donating mitochondria to damaged tissue or cells [9–11]. Recent evidence also suggests that MSCs can modulate T regulatory cells (Tregs) [12,13], in particular, Tregs expressing the Fork-head box P3 (Foxp3) transcription factor, which is part of the adaptive immune system and are principal regulators of inflammation and immune homeostasis [14]. The Tregs can migrate to the diseased tissue and dampen inflammation by increasing the milieu of anti-inflammatory cytokines and activating macrophages to clear the debris and restore the damaged tissue [15–17]. Since CD4+ T cells may mediate retinal ganglion cell (RGC) degeneration and loss of retinal function after injury [18], it may be possible to reprogram the CD4+ T cells at the site of injury to acquire a Treg phenotype. This might aid in the rescue of retinal damage and, as a result, decrease cell loss and enhance immunotoler-ance [19]. Therefore, in the present study, we hypothesized that intravitreal injection of MSCs in ischemia-reperfusion (I/R) injured retina reprogram CD4+ T cells to Tregs, dampen inflammation, and improve visual function. Interestingly, our data visualized that MSC actively transferred mitochondria to immune cells to suppress effector T cells and promoted differentiation of CD4+CD25+ Tregs in coculture with mouse splenocytes. Intravitreal injection of MSCs in the I/R eye significantly increased Tregs in the retina, decreased retinal inflammation, and improved visual function compared to saline-injected I/R eyes. These findings indicate that harnessing the immunosuppressive capacity of MSCs is a potential therapy for the treatment of ischemic retinal diseases.

2. Materials and Methods

2.1. Cell Isolation and Culture

iMSCs were prepared from human urinary tubular epithelial cells (UEs) through the generation of iPSCs via reprogramming with a cocktail of Oct-4, Sox-2, Klf-4, c-Myc, and Lin-28 mRNAs and subsequently differentiated into MSC as described by us pre-viously [20,21]. Briefly, iMSCs from healthy human urine tubular epithelial cells (UEs) expressing epithelial markers CK19 and ZO1 were used to generate iPSCs via reprogram-ming with a cocktail of Oct-4, Sox-2, Klf-4, c-Myc and Lin-28 mRNAs [20]. To induce MSC differentiation, the UE-iPSCs were cultured under conditions conducive for mesenchymal differentiation in Mesencult ACF plus medium for 18–21 days and were characterized for MSC markers. iMSC were positive for CD105, CD90, and CD73 while negative for CD31 and CD45 (Supplemental Figure S1). To validate the mechanism of action of iMSC, ASC was used as a well-known control in in-vitro studies. ASCs used in the current study were obtained from Lonza (Cat#PT-5006), cultured in EGM-2MV media, and used between p2 and p7 in all experiments as previously described [22]. All studies involving human ASC and iMSC were approved for research as per the University of Tennessee Institutional Biosafety and as an exempt study by the Institutional Review Board. Human monocytes, THP-1 cells were purchased from ATCC (Cat#TIB-202) and cultured in RPMI 1640 complete medium as a suspension culture. In addition, primary mouse splenocytes were prepared as described previously [23]. Briefly, 6–8 weeks old wild type (C57BL/6) mice were euthanized, the spleen was collected and washed in PBS. The tissue was crushed on a 70-micron cell strainer with 5 mL syringe plunger, rinsed with RPMI 1640 media; cells were pelleted by centrifugation. Subsequently, RBCs were lysed using ACK lysis buffer, inactivated the reaction using 10% FBS. Finally, splenocytes were pelleted and resuspended in RPMI 1640 complete media and filtered through 40-micron cell strainer. Trypan blue negative splenocytes were counted as live cells and used for experiments. All cell cultures were maintained at 37 °C and 5% CO_2 in a humidified atmosphere.

2.2. Coculture and Microscopy

iMSC or ASC were plated at 1×10^6 cells/cm^2 in a 60 mm dish and we let them adhere. Following this, 1×10^6 cells were stained with fluorescent MitoTracker Red CMXRos (100 nm, Life Technologies, Grand Island, NE, USA) for 45 min at 37 °C. Labeled cells were washed 2× with PBS, trypsinized, and seeded into a 6-well plate containing 10 mm coverslips at 1×10^5 cells per well. On the second day, 1×10^6 THP-1 cells labeled with CellTracker fluorescent probe green (0.5 μM, Life Technologies) were cocultured with iMSC or ASC in a 1:10 ratio for 24 h. After 24 h of coculture, cells were fixed with 4% paraformaldehyde (PFA) and stained with DAPI nuclear stain, mounted using ProLongTM diamond antifade mountant (Life Technologies). Similarly, 1×10^6 iMSC or ASC were stained with fluorescent MitoTracker Green (100 nm, Life Technologies) to assess the mitochondrial transfer through F-actin nanotubes. Following this, 1×10^5 cells were seeded into a 6 well plate containing 10 mm coverslips. On the second day, 1×10^6 THP-1 cells labeled with CellTracker CMAC blue (0.5 μM, Life Technologies) were cocultured with iMSC or ASC in 1:10 ratio for 24 h and treated with 350 nm of cytochalasin B (CytoB, Sigma-Aldrich, Inc., St. Louis, MO, USA). Tubular microstructure tunneling nanotubes were assessed by costaining for F-actin using Phalloidin- Tetramethylrhodamine B isothiocyanate (Sigma) in the presence or absence of CytoB. The mitochondrial transfer was assessed from images captured with an EVOS fluorescence microscope (Life Technologies) or captured with a laser scanning confocal microscope (Zeiss LSM 710, Carl Zeiss Microscopy, LLC, White Plains, NY, USA). For quantification of mitochondrial transfer, THP-1 positive for CMAC blue in the coculture experiments were identified with a region of interest (ROI) and the pixel intensities of MitoTracker Green were computed for each ROI using ImageJ. At least 20 cells/image were considered, and the data were expressed as Mean fluorescent intensity (MFI) values/cell.

2.3. Seahorse Flux BioanalyzerV

THP-1 cells were pretreated for 2 h with 500 nM Rotenone (acts as a potent inhibitor of complex I of the mitochondrial respiratory chain). Following this, cells were washed (2×) and cocultured with iMSC or cultured as a monoculture. Cocultured THP-1 cells were transferred to a Seahorse 24-well tissue culture plates and oxygen consumption rate (OCR) was measured, and parameters were calculated as previously described [24]. Briefly, prior to the assay, the media was changed to unbuffered DMEM (Gibco #12800-017, pH 7.4, 37 °C), and cells were equilibrated for 1 h at 37 °C. After measuring basal respiratory rate, Oligomycin (Sigma; 1 μM; uncouples ATP-coupled respiration by inhibiting ATP synthase), FCCP (Sigma; 1 μM; carbonyl cyanide 4-(trifluoromethoxy)-phenylhydrazone (FCCP), mitochondrial uncoupling agent; uncouples mitochondrial respiration from ATP to determine maximal respiratory rate), and electron transport chain (complex I and III) inhibitors, rotenone (Sigma; 0.5 μM) and antimycin A (Sigma; 0.5 μM; to eliminate all mitochondrial respiration) were injected sequentially during the assay. Basal mitochondrial respiration and ATP-linked respiration were determined in whole cells.

2.4. Coculture and Flow Cytometry

About 1×10^5 MitoTracker green-labeled iMSC or ASC seeded in 12-well cell culture plate for 24 h. The next day, 1×10^6 freshly prepared splenocytes (1:10 ratio as described above) were added to the top of the stem cell monolayer in complete RPMI 1640 media. After 24 h of coculture, non-adherent and loosely bound splenocytes were collected, washed 2× with PBS. Single-cell suspension of splenocytes was stained in FACS buffer (2% FBS in PBS), incubated with the Fc-block anti-CD16/CD32 (2.4G2) antibody followed by a panel of fluorochrome-coupled antibodies (Table 1). While viability dye eFluor 506 (eBioscience, San Diego, CA, USA) excluded dead cells, a MitoTracker dye uptake assessed mitochondrial transfer by flow cytometry. Stem cells were pretreated for 2 h with 500 nM Rotenone to study the functional relevance of mitochondria transfer and cocultured with primary mouse splenocytes.

Table 1. List of fluorophores-coupled FC antibody.

Antibody	Catalog Number	Manufacturer
CD4 (RM4-5)-PE	12-0042-82	eBioscience, San Diego, USA
CD8a (53-6.7)-PerCp/Cyanine 5.5	100733	Bio-legend, San Diego, USA
B220 (RA3-6B2)-AF700	103231	Bio-legend, San Diego, USA
CD69 (H1.2F3)-FITC	11-0691-82	eBioscience, San Diego, USA
CD25 (PC61)-V450	561257	BD Biosciences, Franklin Lakes, USA
Foxp3 (FJK-16s)-APC	17-5773-82	eBioscience, San Diego, USA

To assess Treg differentiation, freshly prepared splenocytes were activated in the presence of coated anti-CD3 (BD, 10 µg/mL) and soluble anti-CD28 antibody (BD, 1 µg/mL) and recombinant IL-2 (PeproTech, Cranbury, NJ, USA, 1 ng/mL) for 24 h. Activated splenocytes were cocultured on the monolayer of iMSC or ASC in differentiation medium [RPMI 1640 medium, 10% FBS, 1 ng/mL IL-2 (PeproTech) and 5 ng/mL TGF-β (R&D)] for five days with media change at day 3. After 5 days of coculture, nonadherent and loosely bound splenocytes were collected, washed 2× with PBS. Single-cell suspension of splenocytes was then analyzed by flow cytometry as described above using specific surface markers (Table 1). For Foxp3 intracellular staining, cells were fixed and permeabilized using Foxp3/Transcription Factor Staining Buffer (eBioscience). Following this, cells were incubated with anti-Foxp3 (eBioscience) antibody for 50 min, washed and resuspended in the FACS buffer, kept at 4 °C, protected from light till acquisition. Data were acquired using a Bio-Rad ZE5 cell analyzer and analyzed by FlowJo software v10.8 (Flowo, Ashland, Wilmington, DE, USA).

2.5. RNA Isolation and Quantitative RT–PCR

Total RNA was isolated from cells using Nucleospin miRNA isolation kit (Macherey-Nagel Inc, Allentown, PA, USA), following the manufacturer's protocol. RNA quality and integrity were measured by absorbance at 260/280 and 260/230 nm using NanoDrop 1000 spectrophotometer (Thermo Fisher Scientific, Waltham, MA, USA). cDNA was prepared in a reverse transcription reaction using 500 ng of RNA and a high-capacity cDNA reverse transcription kit (Thermo Fisher Scientific) following the manufacturer's instructions. About 100 nanograms of template RNA and 5 nM of each forward and reverse human-specific primers related to nanotube formation (*ACTN1*, *NEXEN*, *CAP2*, and *18S*) (Table 2) were used in SYBR green-based qPCR. Total RNA was isolated from Sham, I/R injured, and I/R injured with iMSC treated mice retina at day 7 postinjury, and gene expression was quantified using TaqMan probe-based gene-specific mouse primers (Table 3) and accompanying Master Mix (Applied Biosystems, Foster City, USA) using QuantStudio 3 (Applied Biosystems) Real-Time PCR system. Data were expressed as relative gene expression or fold mRNA expression using the $2^{-\Delta\Delta Ct}$ method and normalized to *18S* housekeeping gene.

Table 2. List of Primer Sequences Used for SYBR green-based qPCR.

Gene	Forward	Reverse
ACTN1	5′-ACATGCAGCCAGAAGAGGAC-3′	5′-ACACCATGCCGTGAATGTCT-3′
NEXN	5′-ACGGAGGAGGAACGAAAACG-3′	5′-TGTCCTCAATCTGTTCAGCCC-3′
CAP2	5′-AGCTGTGTCTCCCAAACCTG-3′	5′-ACCCAATCCACATGACGCAA-3′
18S	5′-GCAATTATTCCCCATGAACG-3′	5′-GGCCTCACTAAACCATCCAA-3′

Table 3. List of Taqman assay IDs for qPCR.

Gene	Assay ID	Reference
18S ribosomal RNA (*18S*)	Mm04277571	NR_003278.3
Laminin, alpha 5 (*LAMA5*)	Mm01222029	NM_001081171.2
Chemokine (C–C motif) ligand 2 (*CCL2*)	Mm00441242	NM_011333.3
Vascular cell adhesion molecule 1 (*VCAM-1*)	Mm01320973_m1	NM_011693.3
Interleukin 1 β (*IL1β*)	Mm00434228_m1	NM_008361.3

2.6. Mice, Retinal IR Injury, Electroretinography, and Immunohistology

Animal studies were approved by the Institutional Animal Care and Use Committee, University of Tennessee Health Sciences Center (UTHSC), Memphis (IACUC ID: 20-0152, Approved 16 June 2020) following the guidelines as per the Association for Research in Vision and Ophthalmology (ARVO) Statement for the Use of Animals in Ophthalmic and Vision Research. C57BL/6 wild-type (B6) mice, between 12- and 16-weeks-old, were used for the study. Animals were housed under a 12-h light/dark cycle and kept under pathogen-free conditions with unlimited food and water supply. Mice were anesthetized with a mixture of 70–90 mg/kg of ketamine and 0.04–0.08 mg/kg of dexdomitor (Orion Pharma Animal Health, FI-02101 Espoo, Finland). Retinal I/R injury was induced unilaterally in the right eye. The pupil was dilated with 1% tropicamide (Akorn, Inc., Lake Forest, CA, USA), and 0.5% proparacaine hydrochloride (Alcon Laboratories, Inc., Fort Worth, TX, USA) was applied topically onto the cornea. The eye's anterior chamber was cannulated under microscopic guidance with a 32 1/2 inch-gauge needle connected to a silicone infusion line providing a balanced salt solution (Baxter, Deerfield, IL, USA), avoiding injury to the corneal endothelium, iris, and lens. Retinal ischemia was induced by raising intraocular pressure of cannulated eyes to 70 mm Hg (as measured by iCare Tonovet) for 60 min by elevating the saline reservoir. Whitening of the fundus was observed to ensure the induction of retinal ischemia. After 24 h of injury, about 1000 iMSCs/2 μL saline were intravitreally injected into the IR injured eye. Following this, on day 7, ERG (Celeris Rodent Electrophysiology system, Diagnosys LLC, Lowell, MA, USA) was recorded as described in previous publications [25]. Briefly, animals were dark-adapted overnight and anesthetized with ketamine (50 mg/kg) and dexmedetomidine (0.25 mg/kg) cocktail. Pupil dilation was achieved with 1% tropicamide. The electrodes were positioned on the surface of both the corneas. Light pulses were delivered at a frequency at 0.01, 0.1, and 1 cd-s/m2, and the responses were recorded simultaneously from both eyes. All the offline analyses were done with Diagnosys software to calculate b-wave amplitudes. At least three-five responses to light stimuli were averaged to determine the b-wave amplitude. Following ERG, mice were euthanized, enucleated the eye, fixed in paraformaldehyde; the retinal cup was isolated. Retinas were permeabilized and blocked before incubation with the primary antibody. Retinas were immuno-stained with anti-Foxp3 antibody (1:200, Cell signaling) for 48 h followed by incubation with goat antimouse IgG Alexa Fluor 546 secondary antibody. To distinguish vasculature, retinas were incubated with Alexa Fluor 488-Isolectin B (Invitrogen, Carlsbad, USA) and flat-mounted on a glass slide using ProLongTM diamond antifade mountant (Life Technologies). Imaging of retinal flat mounts was examined under a laser scanning confocal microscope (Zeiss LSM 710). The number of Foxp3 positive cells were counted in a blinded fashion from each image from all groups, and results were expressed as Foxp3 positive cells per square millimeter of the retina.

2.7. Statistical Analysis

Results are expressed as mean $\pm$ SEM for all experiments. One-way ANOVA followed by post hoc *t*-tests with the Bonferroni correction was used for multiple group comparisons using GraphPad Prism software 6.0. Comparisons of ERG data between the groups were performed using the Student's *t*-test. Values of $p < 0.05$ were considered statistically significant.

3. Results

3.1. Mitochondrial Transfer from MSCs to Immune Cells

To explore whether MSC can transfer their mitochondria to the immune cells, donor cells (iMSC or ASC) labeled with mitochondria-specific fluorescent probes (CMXRos red) were cocultured with CellTracker (green) labeled THP-1 cells in 1:10 ratio for 24 h (Figure 1A). Fluorescence imaging of cocultures revealed both iMSC and ASC can successfully transfer their mitochondria to recipient (THP-1) cells, as shown by colocalization of red fluorescent mitochondria in green labeled THP-1 cells (yellow, Figure 1B). To study the functional relevance of mitochondria transfer, rotenone challenged MSC were cocultured with primary mouse splenocytes at a 1:10 ratio. Mitochondria-specific fluorescent probe (MitoTracker green) labeled iMSC or ASC were treated with rotenone (500 nm, 2 h) and without, cocultured with primary mouse splenocytes, and analyzed by flow cytometry after 24 h of coculture (Figure 1C). Flow cytometry analysis demonstrated mitochondria transfer from stem cells to splenocytes (Figure 1D). While splenocytes cocultured with native iMSC had 43.4 ± 0.38 percent of MitoTracker green positive splenocytes, rotenone treated cocultures showed only 30 ± 0.81 percent positive cells (Figure 1D; $p < 0.001$). Similarly, native and rotenone treated ASC when cocultured with splenocytes demonstrated 30.4 ± 0.3 and 18.67 ± 1.3 (Figure 1E; $p < 0.001$), percent of MitoTracker green positive splenocytes, respectively. Incubation of primary mouse splenocytes with culture medium obtained by MitoTracker-labeled ASCs, but without ASCs, failed to demonstrate any fluorescence signal in splenocytes (Supplemental Figure S2), ruling out the possibility of passive transfer of the mitochondrial stain to splenocytes due to MitoTracker probe leak. To determine if the changes in mitochondrial content also altered mitochondrial function, bioenergetic analysis was undertaken. THP-1 cells pre-incubated with rotenone significantly blunted oxygen consumption compared to untreated cells as expected (Figure 1F, blue to red line). Upon coculture with iMSC, those cells exposed to rotenone demonstrated near-normal basal respiration. When ATP production was quantified, rotenone treated THP-1 cells showed a reduced ATP production though the data did not reach statistical significance while those cells cocultured with iMSC demonstrated significantly greater ATP production, which is not significantly different from untreated THP-1 cells (Figure 1F; $p < 0.05$).

3.2. MSC Efficiently Transfer Mitochondria in a Dose-Dependent Manner to CD4+ and CD8+ T Cells

To further explore the mitochondrial transfer to specific cell populations in splenocytes, mitochondria-specific fluorescent probes (MitoTracker green) labeled iMSC or ASC were cocultured with primary mouse splenocytes at different ratios and analyzed by flow cytometry. After 24 h of coculture, nonadherent splenocytes analyzed demonstrated mitochondria transfer from stem cells into all studied lymphocyte subsets, mainly directed to B220+ B lymphocytes (86.8%), T helper CD4+ (50.68%) rather than T cytotoxic CD8+ (10.29%) lymphocytes (Figure 2A and Supplemental Figure S3; $p < 0.001$). Interestingly, donor iMSCs increased mitochondria transfer to T cells (CD4+ and CD8+) with splenocytes at increasing ratios (1:100, 1:25, and 1:10), with an average number of mitochondria of 31.2, 37, and 49.4% respectively in CD4+ T cells and with an average number of mitochondria of 4.6, 5.1 and 9.9% respectively for in CD8+ T cells (Figure 2B; $p < 0.001$). Similarly, when donor ASCs were cocultured with recipient splenocytes at increasing ratios (1:100, 1:25, and 1:10), an average of 31.7, 59.2, and 79.5% cells demonstrated mitochondria in CD4+ T cells and with an average of 5.7, 16.9, and 30.3% for mitochondria in CD8+ T cells, respectively (Figure 2B; $p < 0.001$). The level of mitochondria transferred to B220+ B cells from iMSCs or ASCs cocultured with primary mouse splenocytes ranged from 78–85% with minimal change noted with increasing coculture ratio (Figure 2B; $p < 0.001$). Altogether, these results indicate that both iMSCs and ASCs can transfer their mitochondria efficiently into primary mouse splenocytes.

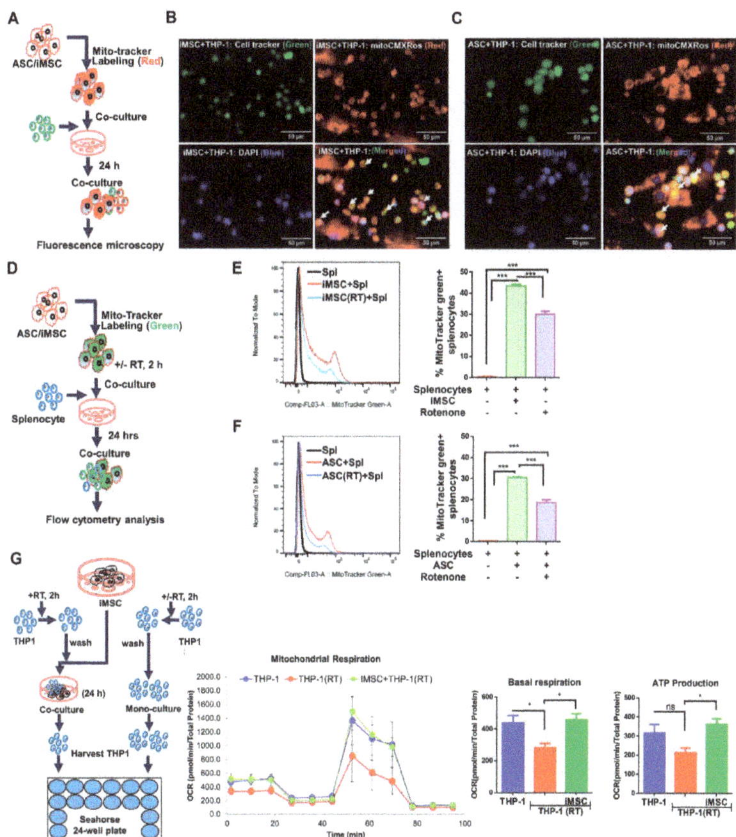

Figure 1. Both iMSC and ASC effectively transfer mitochondria to immune cells. (**A**) Schematic representation of stem cells and THP-1 coculture setup and fluorescence microscopy analysis. (**B**) MitoTracker CMXRos labeled iMSC (Red) or (**C**) ASC (Red) independently cocultured with cell tracker green-labeled THP-1 cells (Green) for 24 h, stained with nuclear dye DAPI (Blue), and images captured under fluorescence microscope shows increased donor-derived mitochondria in THP-1 cells (yellow; arrows). (**D**) Schematic representation of stem cells and THP-1 coculture setup and flow cytometry analysis. (**E**) Overlays and histograms showing flow cytometry analysis of mitochondria transfer from iMSC and (**F**) ASC to mouse splenocytes cocultured at 1:10 ratio. MitoTracker green-labeled iMSC and ASC pretreated with rotenone (RT) (500 nm, 2 h) and without rotenone treatment cocultured with splenocytes for 24 h. Mitochondria transfer to splenocytes is expressed as percent MitoTracker positive green splenocytes. Flow cytometry data were analyzed using Flowjo (v10.8) software and represented as mean $\pm$ SEM from triplicates of same experiment and statistical analysis by one-way ANOVA with Bonferroni correction (* $p < 0.01$, *** $p < 0.0001$, ns-not significant). (**G**) Schematic representation of stem cells and THP-1 coculture setup and OCR analysis. iMSC and RT treated THP-1 coculture was set up at 1:1 ratio for 24 h. From coculture, THP-1 were harvested and seeded onto a Cell-Tak coated Seahorse XF-24e V7 PS cell culture microplate, and oxygen consumption rate (OCR) was determined, followed by quantification of basal OCR and ATP production. Data shown as mean $\pm$ SEM from a single experiment repeated independently with similar results (* $p < 0.01$, *** $p < 0.0001$ one-way ANOVA).

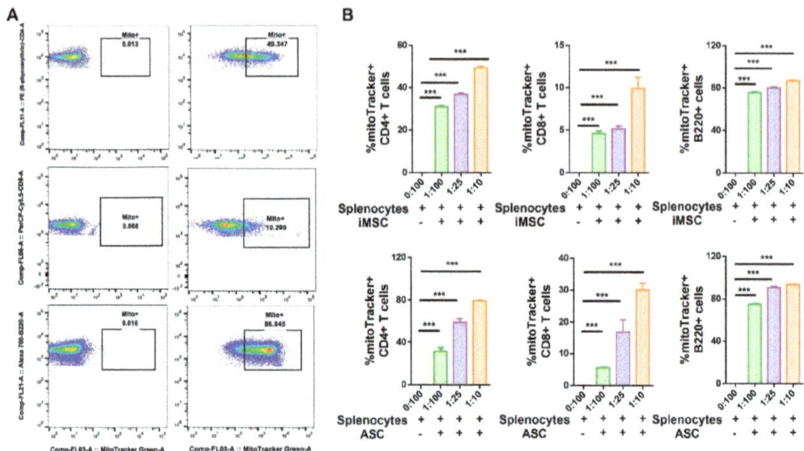

Figure 2. Dose-dependent transfer of mitochondria from iMSC or ASC to mouse T and B cells *in vitro*. (**A**) Representative flow cytometry dot plots showing increased MitoTracker positive CD4+ T, CD8+ T, and B220+ cells in splenocyte-iMSC coculture compared to that of monoculture. (**B**) Quantification of percent mitochondria transfer to CD4+ T, CD8+ T, and B220+ B cells increased with decreased ratio with both iMSC (upper panel) and ASC (lower panel). Data shown as mean ± SEM from a single experiment repeated independently with similar results (*** $p < 0.0001$ one-way ANOVA).

3.3. Tunneling Nanotubes Mediate Mitochondrial Transfer from MSC to THP-1 Cells

After confirming the mitochondrial transfer from iMSCs or ASCs to immune cells next, we investigated whether F-actin-positive tubular microstructure known as tunneling nanotubes (TNTs) are involved in intercellular mitochondrial transfer from donor to recipient cells. To this end, MitoTracker green-labeled iMSCs or ASCs were cocultured with CMAC blue labeled THP-1 cells with and without CytoB (Figure 3A–D). As expected, fluorescence imaging revealed the transfer of mitochondria (green) from iMSC or ASC into THP-1 cells (blue) via F-actin positive TNT's (red; Figure 3E,G,J). Interestingly, those cells that were pre-incubated with CytoB demonstrated a substantial reduction in mitochondrial transfer to THP-1 cells (Figure 3F,H,I). To further confirm TNT-mediated mitochondrial transfer, cocultures in the presence or absence of CytoB were assessed for the gene expression of CAP2, NEXN, and ACTN1 that are known to associate with F-actin synthesis. Whereas the expression of CAP2, NEXN, and ACTN1 increased by 1.5–45-fold ($p < 0.001$) in cocultures as compared to monocultures, CytoB treatment significantly decreased gene expression of all three markers (Figure 3K; $p < 0.01$). Taken together, the data suggest that mitochondrial transfer from iMSCs or ASCs to immune cells occurs via F-actin-positive tubular tunneling nanotubes.

3.4. MSC Suppresses T Cell Population

Different T cell populations of splenocytes interact closely; their ratio at a given time results from a balance between their mutual effects. Mitochondrial transfer to recipient cells can increase cell metabolism, resulting in cell division or cell differentiation. To better understand the impact of mitochondrial transfer to immune cells, we first assessed cell viability in mouse splenocyte coculture with either iMSC or ASCs. While the viability of splenocytes in monoculture is 48.7%, the viability in cocultures with iMSC and ASC increased to 65.7 and 64.3%, respectively (Figure 4B; $p < 0.001$). Similarly, the frequency of B220+ B lymphocyte cells in splenocytes cocultures with iMSC and ASC also significantly increased to 64.3 ($p < 0.001$) and 58.5%, respectively (Figure 4B; $p < 0.01$). Next, to better understand the effects of mitochondria transfer from stem cells to immune cells on their subpopulation, we assessed non-adherent splenocytes from cocultures with monoclonal antibodies for CD4, CD8, and B220 and compared them to monoculture in the presence or absence of rotenone. While the helper CD4+ T cells of splenocytes in monoculture is 21%, the coculture levels with iMSC and ASC significantly reduced to 15.4 and 15.9%, respectively (Figure 4B; $p < 0.001$). Similarly, 17% cytotoxic CD8+ T cells in splenocytes monoculture reduced to 12.9 and 12.8% in iMSC and ASC coculture with splenocytes, respectively (Figure 4B; $p < 0.001$). Interestingly, rotenone challenged cocultures demonstrated a small but significant increase in percent immune cells in both iMSC and ASC as compared to cells without rotenone. Altogether, these results indicate that both iMSCs and ASCs suppress the effector T cell population upon transferring their mitochondria to immune cells.

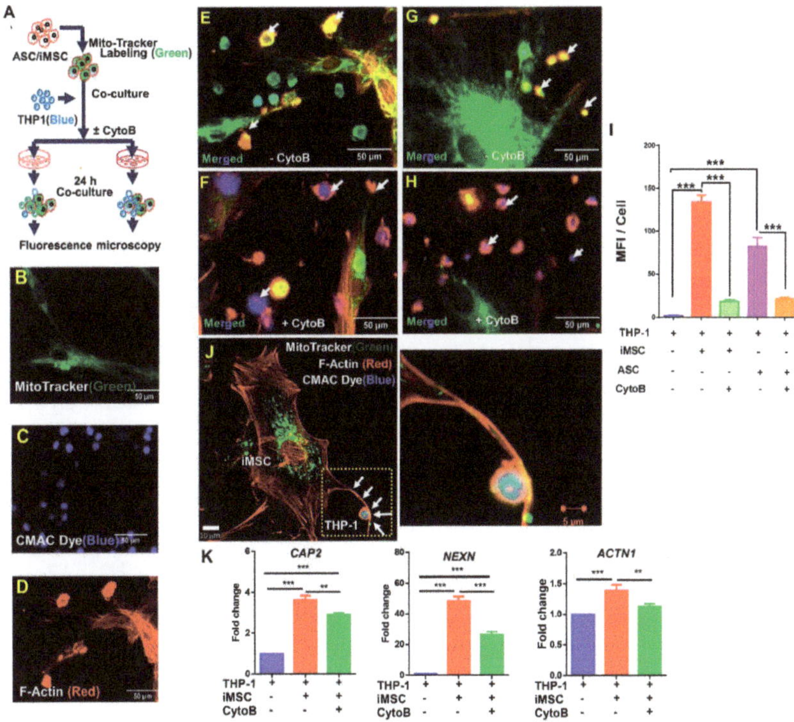

Figure 3. Tunneling nanotubes mediate mitochondrial transfer from iMSC to THP-1 cells. (**A**) Schematic representation of experimental design of iMSC or ASC coculture with THP-1 cells in presence and absence of cytochalasin-B. (**B–H**) Representative fluorescence microscopy images showing MitoTracker green positive iMSC (**B**), CMAC blue positive THP-1 cells (**C**), F-Actin red positive ASC (**D**), ASC and THP-1 coculture in absence (**E**) and in presence (**F**) of cytochalasin-B and, iMSC and THP-1 coculture in absence (**G**) and presence (**H**) of cytochalasin-B. (**I**) Mitochondria transfer from iMSC or ASC to THP-1 cells reduced with cytochalasin-B. Mean fluorescent intensity (MFI) values calculated using ImageJ. Data shown as mean $\pm$ SEM from a single experiment (*** $p < 0.001$ one-way ANOVA). (**J**) Representative Confocal images of intercellular mitochondrial transfer between iMSC and THP-1 via F-Actin positive nanotubes. Marked area magnified to show MitoTracker positive mitochondria in F-actin positive nanotube. (**K**) qRT-PCR analysis of genes related to nanotubes formation (CAP2, NEXN, and ACTN1) increased significantly during mitochondria transfer from donor (iMSC) to recipient (THP-1) cells. On other hand, cells exposed to cytochalasin-B significantly reduced CAP2, NEXN, and ACTN1 expression. Data shown as mean $\pm$ SEM from a single experiment repeated independently with similar results (** $p < 0.001$, *** $p < 0.0001$ one-way ANOVA).

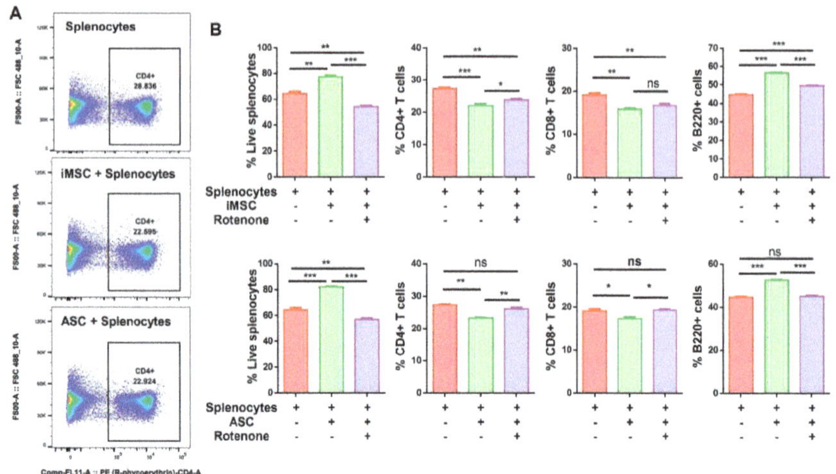

Figure 4. Both iMSC and ASC suppress effector T cell population. (**A**) Representative flow cytometry dot plots showing decreased CD4+ T cells in iMSC-splenocytes and ASC-splenocytes coculture compared to monoculture. (**B**) Quantification data represented as bar graphs after flow cytometry analysis of live cells, CD4+ T cells, CD8+ T cells, and B220+ cells in mouse splenocytes monocultures and cocultures with iMSC, ASC, as well as iMSC, and ASC pretreated with rotenone (500 nm, 2 h) at 1:10 (iMSC/ASC and splenocytes) ratio. Data shown as mean ± SEM from a single experiment repeated independently with similar results (* $p < 0.01$, ** $p < 0.001$, *** $p < 0.0001$ one-way ANOVA).

3.5. MSC Differentiates T Cells into Tregs and Suppresses CD69 Expression

The suppression of immune cell activation is one of the manifestations of MSC-mediated immunomodulation [12,26]. To assess whether mitochondrial transfer from stem cells to primary mouse splenocytes impacts T-cell differentiation, activated splenocytes were cocultured with iMSCs or ASCs and subsequently assessed for the expression of T regulatory cells (Figure 5A and Supplemental Figure S4). The dot plots from flow cytometry analysis clearly showed double-positive CD25+Foxp3+ cells in the upper right quadrant that were gated on CD4+ T cells (Figure 5B). While the level of CD25+Foxp3+ cells in monoculture is 2.7%, the levels in cocultures with iMSC and ASC increased to 7.1 and 7.4%, respectively (Figure 5B,C; $p < 0.001$). To further confirm the immunosuppression capability of iMSCs and ASCs, the expression of CD69, a potent immune activation marker, was evaluated. While the level of CD69+ cells in monoculture is 18.43%, the levels in cocultures with iMSC and ASC significantly decreased to 7.2 and 6.5%, respectively (Figure 5D; $p < 0.001$). Taken together, our data suggests that both iMSC and ASC manifest their immunomodulation via increased Treg population with a significant reduction in CD69+ cells in splenocyte coculture.

3.6. iMSC Significantly Increases Regulatory T Cells in the Retina of I/R Injured Mice

MSC are well known to protect against retinal I/R damage [27–30]. To better understand if iMSC also can protect against retinal damage, we tested the intravitreal injection of iMSCs in the retinal I/R injury model in-vivo (Figure 6A). After 7 days, the retinal function, assessed by Electroretinogram (ERG), demonstrated improved b-wave amplitudes in I/R mice receiving iMSC as compared to saline-injected I/R eyes (at 1cd.s.m2 139 ± 48 v/s 48 ± 10 µvolt, $p = 0.05$) (Figure 6B). To further correlate the improved visual function observed with iMSC in I/R mice, gene expression analysis of proinflammatory markers was performed. Figure 6D shows normalized data of individual genes in all 3 groups of mice. I/R mice receiving saline had a significantly increased abundance of gene transcripts involved in microglial activation (IL1β) [31], endothelial activation (VCAM1, CCL2) [31,32], and T-cell regulation (LAMA5) [33] compared to sham mice. Interestingly, I/R mice receiving iMSC significantly ameliorated the increased gene expression (Figure 6C). The normalized fold change in expression of IL1β (I/R, 27.08 ± 8.69 vs. I/R+iMSC, 8.69 ± 1.54, $p < 0.01$), CCL2 (I/R, 13.63 ± 2.7 vs. I/R+iMSC, 3.52 ± 0.96, $p < 0.0001$), LAMA5 (I/R, 2.4 ± 0.37 vs. I/R+iMSC, 1.09 ± 0.09, $p < 0.001$) and VCAM1 (I/R, 7.4 ± 1.48 vs. I/R+iMSC, 2.2 ± 0.96, $p < 0.01$).

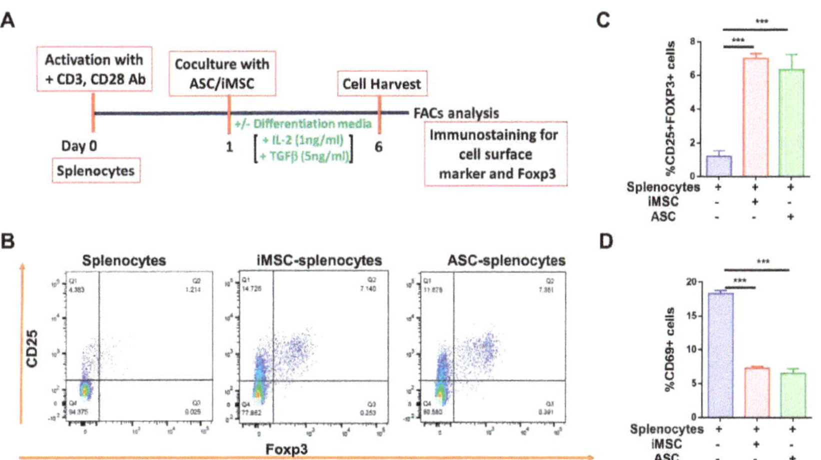

Figure 5. Both iMSC and ASC increase differentiation of T cells into Tregs and suppress CD69 expression. (**A**) Timeline and experimental details of iMSC/ASC coculture with mouse splenocytes (**B**) Representative flow cytometry dot plots showing increased CD25+Foxp3+ Tregs in iMSC/ASC cocultures with activated splenocytes. (**C**) Quantification of a percent increase in CD25+Foxp3+ cells with (**D**) a reduction in CD69+ cells in cocultures. Data shown as mean ± SEM from a single experiment repeated independently with similar results (*** $p < 0.0001$ one-way ANOVA).

Subsequently, to correlate the iMSC ability to provide immunomodulation through increased Tregs, retinal flat mounts analyzed for Tregs in the retina by confocal microscopy revealed positive immunostaining (red) only in I/R and I/R+iMSC groups (Figure 6D). Neither the uninjured contralateral eye nor the iMSC injected into the uninjured eye was positive for Foxp3 expression (Supplemental Figure S5), suggesting the specificity of immunostaining and the correlation of Foxp3 upregulation to I/R injury. While the number of positive Foxp3 cells in the Sham group was 3.6 ± 1.2, the I/R injury retina demonstrated a significant upregulation with 57 ± 8.7 Tregs per mm2 area ($p < 0.001$). Interestingly, those I/R injury animals that received iMSC demonstrated a further significant increase in Tregs to 112 ± 11 cells per mm2 area ($p < 0.0001$; Figure 6E).

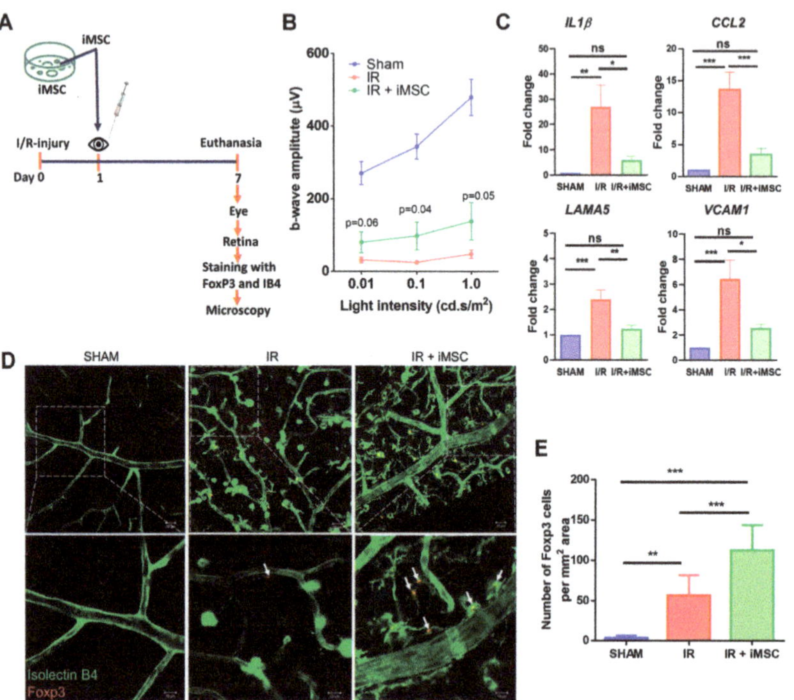

Figure 6. iMSC improves b-wave amplitudes, reduces inflammation correlated with increased regulatory T cells in retina of I/R injured mice. (**A**) Schematic representation of in-vivo experimental timeline and analyses. (**B**) B-wave amplitudes as measured by ERG show an expected decrease in I/R with a significant improvement in iMSC group. Data are shown as mean $\pm$ SEM, $n = 7$–13/group. *t*-test. (**C**) An increase in proinflammatory markers in I/R decreased with iMSC. Data are shown as mean $\pm$ SEM, one-way ANOVA. (**D**) Immunofluorescence images of Sham, I/R injured, and I/R injured with iMSC stained with isolectin B4 (green) and anti-Foxp3 (red) at day 7 post-injury. White arrows indicate Foxp3 cells. (**E**) Quantification of Foxp3 cells significantly increased in I/R injured animals with iMSC compared to both Sham and I/R. Data shown as mean $\pm$ SEM, one-way ANOVA with Bonferroni correction (* $p < 0.01$, ** $p < 0.001$, *** $p < 0.0001$, ns = not significant).

4. Discussion

Our study is the first to demonstrate that iMSC reprograms mouse CD4-T cells into Foxp3 regulatory Tregs to the best of our knowledge. Additionally, we show that intravitreal delivery of iMSCs in a retinal ischemia-reperfusion (I/R) injury model display therapeutic potential and suggests the mechanism of action may involve recruitment of Foxp3Tregs. Our in-vitro culture data show that iMSCs transferred mitochondria to immune T cells via F-actin nanotubes, suppressed effector T cells, and promoted differentiation CD4+CD25+Foxp3+ Tregs in coculture with mouse splenocytes on par with ASC. Importantly, we also demonstrated that the increased recruitment of Foxp3+Tregs in the retina correlated with dampened retinal inflammation and improved b-wave amplitudes in the I/R injury model. Our findings are in keeping with published data by others that have demonstrated Tregs act as part of the adaptive immune system, and thereby (A) serve as important regulators of inflammation and play a critical role in immune homeostasis [14]; (B) are inversely correlated with retinal ischemia [34]; and (C) are well known for immunomodulation, with recent evidence suggesting that MSCs can enhance Foxp3+Treg differentiation and stability from activated T cells in part through mitochondrial transfer [12,13,35].

Recent evidence suggests that MSCs can modulate Tregs [12], particularly Tregs expressing the Foxp3 transcription factor, part of the adaptive immune system and principal regulators of inflammation and immune homeostasis [14]. Although Tregs primarily originate in the thymus (tTregs), during tissue damage, the CD4+ (T_H) cells at the site of injury can also be reprogrammed and acquire the Treg phenotype, also known as peripheral Tregs (pTregs) and regulate immunotolerance [19]. Vice-versa tTregs acquire CD4+ (T_H) cell phenotype and become defective to perform its immunomodulatory function [35]. Thus, the ability of Tregs to migrate to the damaged tissue and dampen inflammation is an attractive strategy to curtail the ongoing inflammation [15–17]. Following this, we show increased Foxp3 positive cells in the retina after retinal damage that further increased with intravitreal injection of iMSC in the ischemic retina. This initial increase in Tregs under acute ischemia compared to sham injury aligns with a previous observation in the oxygen-induced retinopathy mouse model [28]; however, Treg numbers are likely insufficient to repair the retinal damage. On the other hand, those animals that received iMSC probably reached the optimal levels of Foxp3 cells to dampen the retinal damage. One limitation of our initial study is that it is unclear if the increased Foxp3 Tregs are reprogrammed from the local CD4+ T cells that were shown to be upregulated and causally linked to retinal damage [19] or were recruited from elsewhere. Another limitation is that how the increased Foxp3 Tregs regulate retinal tissue inflammation is not explored. Future studies beyond the scope of this study need further analysis on how iMSC induces Tregs and, thus, regulates retinal tissue damage.

Understanding the molecular basis of Treg generation and its stability is an active area of research, with several agents proposed to upregulate and stabilize Foxp3 expression [36,37]. Notably, some agents have even been shown to be effective under hypoxic conditions [38]. MSCs grown as "feeder cells" with Tregs significantly increased Treg cell number, suppressive function, and ex vivo expansion, primarily through mitochondrial transfer from adjacent MSCs, coupled with the promotion of Foxp3 cotranscriptional proteins [13]. Thus, a primary mechanism of action of MSCs may involve an increase in Treg cell number and/or function via Foxp3 stability. To this end, we show increased differentiation of Foxp3 cells from a mixed population of mouse splenocytes cocultured with iMSC, a feature also observed with ASC cocultured with whole human peripheral blood mononuclear cells [12]. One limitation of our current study is that it is unclear if the reprogramming of Foxp3 Tregs primarily occurs through mitochondrial transfer from adjacent iMSCs or other unknown mechanism(s). To this end, genome-wide patterns of DNA methylation of Foxp3 locus were implicated with FOXP3 gene expression, which determines the tTreg generation, pTreg generation, Treg stability, and tTreg and pTreg proliferation [39]. One possibility is that iMSC affects the epigenetic stability of the FOXP3 gene in Tregs, thus might increase its stability. Future studies need to explore these hypotheses.

MSCs help in the intercellular exchange of mitochondria to restore and regenerate the damaged tissue [9], specifically in various ocular cells [10]. Therefore, iMSC by mitochondrial transfer to Tregs or immune cells may modulate them towards more stable and functional pTregs phenotype in-vivo and aid in the suppressive ability to protect from I/R-injury. To this end, we show that iMSCs can transfer mitochondria to CD4+ T-cells and suppress the CD4+ cell population. Interestingly, both these activities are dependent on mitochondria transfer as evidenced by either blocking mitochondrial function via rotenone or the use of Cytochalasin-B, a known agent that blocks actin polymerization that caused a significant reduction in tunneling nanotubes and a significant reduction in mitochondria transfer to recipient cells. In support of our study, a previous study showed such tunnel nanotube-mediated transfer of mitochondria occurs in MSC [13]. The predominant mechanism of cytochalasin B is the inhibition of actin filament polymerization by binding to the end of growing filaments. Since F-actin filaments are dynamic in nature and their formation between donor and recipient cells is influenced by the activity of many actin-binding genes/proteins, we studied the transcriptional regulation of F-actin genes as also shown previously by Jiang et al. [10]. Based on our gene expression data, it is conceivable that cytochalasin B might indirectly affect transcriptional regulation of the F-actin-related genes and thus influences the mitochondrial transfer. Future studies are required to decipher the causal link of these actin-related genes to mitochondria transfer in our studies as mitochondria from MSCs are transferred to other cells via their exosomes [40] or gap junctions [41,42] or simply by non-mitochondrial paracrine factors [43]. Finally, one alternate hypothesis could be that a direct transfer of mitochondria, a phenomenon known as mitoception, shown for bone-marrow MSC, [44] should be explored for iMSC.

iMSC obtained in the current study were obtained by reprogramming human urine-derived epithelial cells with mRNA reprogramming, the fastest and most reliable reprogramming method to date [45]. Furthermore, our study demonstrates an iPSC line that has unlimited proliferation potential [21] and has the ease of obtaining without any surgical interventions will likely address the current challenges of cell therapies in the ischemic retina [46]. Furthermore, since iPSC lines can also be obtained from diseased or aged individuals [47], our studies will likely benefit future personalized medicine studies. In conclusion, our study demonstrates that iPSC-derived MSCs can transfer mitochondria to T cells to enhance differentiation into Foxp3 Tregs. Additionally, our current data indicate that MSC can improve the retina's immune function by upregulation of Tregs to decrease inflammation and reduce I/R injury-induced retinal degeneration.

Supplementary Materials: The following are available online at https://www.mdpi.com/article/10.3390/cells10113006/s1, Figure S1: Gating strategy for iMSC characterization and representative plots. Overlay plots showing MSC characteristics positive and negative surface markers percent expression on iMSC.; Figure S2: Flow cytometry representative plots showing splenocytes incubated with media ob-tained from mitoTrackergreen stained ASC were negative for MitoTrackerGreen. ASC were stained with mitoTrackerGreen and cultured for 2 h. The cell supernatant was then incubated with splenocytes for 24 h, washed and assessed for mitoTrackerGreen.; Figure S3: Gatingstrategy for flow cytometry analysis and representative plots for transfer of mitochondria from iMSC to mouse T and B cells in vitro.; Figure S4: Representative gating strategy and flow cytometry dot plots showing CD25+Foxp3+ Tregs in iMSC/ASC co-cultures with activated splenocytes.; Figure S5: Retinal flat mounts analyzed for Tregs in the retin the iMSC injected contralateral eye (left) and un-injured cina by confocal microscopy revealed no posi-tive immunostaining ontralateral eye (right) for Foxp3 expression. IsolectinB4 staining was used to identify blood vessels. The data shown is representative of n = 3–4 eyes/group.

Author Contributions: Conceived and designed the experiments: M.A., P.K.R., A.B. and R.G.; performed the experiments: M.A., P.K.R. and A.B; analyzed the data: M.A., P.K.R., A.B. and R.G.; contributed reagents/materials/analysis tools: A.B., J.R. and R.G; wrote the paper: M.A., P.K.R., A.B., J.R. and R.G.; reviewed and final approval: all authors. All authors have read and agreed to the published version of the manuscript.

Funding: This study was funded by grants from the Department of Defense (W81XWH-16-1-0761), National Eye Institute (EY023427), and unrestricted funds from Research to Prevent Blindness to R.G., NIH(R01DK117183) to A.Band NIH (HL141345) to J.R. The funders played no role in the conduct of the study, collection of data, management of the study, analysis of data, interpretation of data, or preparation of the manuscript.

Institutional Review Board Statement: UTHSC Institutional Animal Care Committee, approval # 20-0152.

Informed Consent Statement: Not applicable.

Data Availability Statement: Data are available from the authors upon request.

Acknowledgments: Authors wish to acknowledge Vinoth Sigamani, MTech, for technical assistance, Sheeja Rajasingh, MSc, Mphil for helpful discussions. Authors wish to acknowledge Vaishnavi Sowmya Narayn, BA, MSc. LIS, for editorial assistance.

Conflicts of Interest: R.G. is cofounder and holds equity in Cell Care Therapeutics Inc., whose interest is in using adipose-derived stromal cells in visual disorders. None of the other authors declare any financial conflicts.

References

1. Park, S.S. Cell Therapy Applications for Retinal Vascular Diseases: Diabetic Retinopathy and Retinal Vein Occlusion. *Investig. Ophthalmol. Vis. Sci.* **2016**, *57*, ORSFj1–ORSFj10. [CrossRef]
2. Rivera, J.C.; Dabouz, R.; Noueihed, B.; Omri, S.; Tahiri, H.; Chemtob, S. Ischemic Retinopathies: Oxidative Stress and Inflammation. *Oxid. Med. Cell. Longev.* **2017**, *2017*, 3940241. [CrossRef] [PubMed]
3. Bertelli, P.M.; Pedrini, E.; Guduric-Fuchs, J.; Peixoto, E.; Pathak, V.; Stitt, A.W.; Medina, R.J. Vascular Regeneration for Ischemic Retinopathies: Hope from Cell Therapies. *Curr. Eye Res.* **2020**, *45*, 372–384. [CrossRef] [PubMed]
4. Gaddam, S.; Periasamy, R.; Gangaraju, R. Adult Stem Cell Therapeutics in Diabetic Retinopathy. *Int. J. Mol. Sci.* **2019**, *20*, 4876. [CrossRef]
5. Rajashekhar, G.; Ramadan, A.; Abburi, C.; Callaghan, B.; Traktuev, D.O.; Evans-Molina, C.; Maturi, R.; Harris, A.; Kern, T.S.; March, K.L. Regenerative therapeutic potential of adipose stromal cells in early stage diabetic retinopathy. *PLoS ONE* **2014**, *9*, e84671. [CrossRef]
6. Elshaer, S.L.; Evans, W.; Pentecost, M.; Lenin, R.; Periasamy, R.; Jha, K.A.; Alli, S.; Gentry, J.; Thomas, S.M.; Sohl, N.; et al. Adipose stem cells and their paracrine factors are therapeutic for early retinal complications of diabetes in the Ins2(Akita) mouse. *Stem Cell Res. Ther.* **2018**, *9*, 322. [CrossRef] [PubMed]
7. Cerman, E.; Akkoc, T.; Eraslan, M.; Sahin, O.; Ozkara, S.; Vardar Aker, F.; Subasi, C.; Karaoz, E.; Akkoc, T. Correction: Retinal Electrophysiological Effects of Intravitreal Bone Marrow Derived Mesenchymal Stem Cells in Streptozotocin Induced Diabetic Rats. *PLoS ONE* **2016**, *11*, e0165219. [CrossRef]
8. Zhang, W.; Wang, Y.; Kong, J.; Dong, M.; Duan, H.; Chen, S. Therapeutic efficacy of neural stem cells originating from umbilical cord-derived mesenchymal stem cells in diabetic retinopathy. *Sci. Rep.* **2017**, *7*, 408. [CrossRef]
9. Ahmad, T.; Mukherjee, S.; Pattnaik, B.; Kumar, M.; Singh, S.; Kumar, M.; Rehman, R.; Tiwari, B.K.; Jha, K.A.; Barhanpurkar, A.P.; et al. Miro1 regulates intercellular mitochondrial transport & enhances mesenchymal stem cell rescue efficacy. *EMBO J.* **2014**, *33*, 994–1010. [CrossRef] [PubMed]
10. Jiang, D.; Chen, F.X.; Zhou, H.; Lu, Y.Y.; Tan, H.; Yu, S.J.; Yuan, J.; Liu, H.; Meng, W.; Jin, Z.B. Bioenergetic Crosstalk between Mesenchymal Stem Cells and various Ocular Cells through the intercellular trafficking of Mitochondria. *Theranostics* **2020**, *10*, 7260–7272. [CrossRef]
11. Babenko, V.A.; Silachev, D.N.; Popkov, V.A.; Zorova, L.D.; Pevzner, I.B.; Plotnikov, E.Y.; Sukhikh, G.T.; Zorov, D.B. Miro1 Enhances Mitochondria Transfer from Multipotent Mesenchymal Stem Cells (MMSC) to Neural Cells and Improves the Efficacy of Cell Recovery. *Molecules* **2018**, *23*, 687. [CrossRef] [PubMed]
12. Fiori, A.; Uhlig, S.; Klüter, H.; Bieback, K. Human Adipose Tissue-Derived Mesenchymal Stromal Cells Inhibit CD4+ T Cell Proliferation and Induce Regulatory T Cells as Well as CD127 Expression on CD4+CD25+ T Cells. *Cells* **2021**, *10*, 58. [CrossRef] [PubMed]
13. Do, J.S.; Zwick, D.; Kenyon, J.D.; Zhong, F.; Askew, D.; Huang, A.Y.; Van't Hof, W.; Finney, M.; Laughlin, M.J. Mesenchymal stromal cell mitochondrial transfer to human induced T-regulatory cells mediates FOXP3 stability. *Sci. Rep.* **2021**, *11*, 10676. [CrossRef] [PubMed]
14. Walker, L.S. Treg and CTLA-4: Two intertwining pathways to immune tolerance. *J. Autoimmun.* **2013**, *45*, 49–57. [CrossRef] [PubMed]
15. Taams, L.S.; van Amelsfort, J.M.; Tiemessen, M.M.; Jacobs, K.M.; de Jong, E.C.; Akbar, A.N.; Bijlsma, J.W.; Lafeber, F.P. Modulation of monocyte/macrophage function by human CD4+CD25+ regulatory T cells. *Hum. Immunol.* **2005**, *66*, 222–230. [CrossRef]
16. Maloy, K.J.; Salaun, L.; Cahill, R.; Dougan, G.; Saunders, N.J.; Powrie, F. CD4+CD25+ T(R) cells suppress innate immune pathology through cytokine-dependent mechanisms. *J. Exp. Med.* **2003**, *197*, 111–119. [CrossRef] [PubMed]
17. Andre, S.; Tough, D.F.; Lacroix-Desmazes, S.; Kaveri, S.V.; Bayry, J. Surveillance of antigen-presenting cells by CD4+ CD25+ regulatory T cells in autoimmunity: Immunopathogenesis and therapeutic implications. *Am. J. Pathol.* **2009**, *174*, 1575–1587. [CrossRef] [PubMed]
18. Khanh Vu, T.H.; Chen, H.; Pan, L.; Cho, K.S.; Doesburg, D.; Thee, E.F.; Wu, N.; Arlotti, E.; Jager, M.J.; Chen, D.F. CD4(+) T-Cell Responses Mediate Progressive Neurodegeneration in Experimental Ischemic Retinopathy. *Am. J. Pathol.* **2020**, *190*, 1723–1734. [CrossRef] [PubMed]
19. Josefowicz, S.Z.; Lu, L.F.; Rudensky, A.Y. Regulatory T cells: Mechanisms of differentiation and function. *Annu. Rev. Immunol.* **2012**, *30*, 531–564. [CrossRef] [PubMed]
20. Rajasingh, S.; Thangavel, J.; Czirok, A.; Samanta, S.; Roby, K.F.; Dawn, B.; Rajasingh, J. Generation of Functional Cardiomyocytes from Efficiently Generated Human iPSCs and a Novel Method of Measuring Contractility. *PLoS ONE* **2015**, *10*, e0134093. [CrossRef] [PubMed]
21. Rajasingh, S.; Sigamani, V.; Selvam, V.; Gurusamy, N.; Kirankumar, S.; Vasanthan, J.; Rajasingh, J. Comparative analysis of human induced pluripotent stem cell-derived mesenchymal stem cells and umbilical cord mesenchymal stem cells. *J. Cell. Mol. Med.* **2021**, *25*, 8904–8919. [CrossRef] [PubMed]
22. Periasamy, R.; Elshaer, S.L.; Gangaraju, R. CD140b (PDGFRbeta) signaling in adipose-derived stem cells mediates angiogenic behavior of retinal endothelial cells. *Regen. Eng. Transl. Med.* **2019**, *5*, 1–9. [CrossRef] [PubMed]
23. Lim, J.F.; Berger, H.; Su, I.H. Isolation and Activation of Murine Lymphocytes. *J. Vis. Exp.* **2016**, *116*. [CrossRef] [PubMed]

24. Namwanje, M.; Bisunke, B.; Rousselle, T.V.; Lamanilao, G.G.; Sunder, V.S.; Patterson, E.C.; Kuscu, C.; Kuscu, C.; Maluf, D.; Kiran, M.; et al. Rapamycin Alternatively Modifies Mitochondrial Dynamics in Dendritic Cells to Reduce Kidney Ischemic Reperfusion Injury. *Int. J. Mol. Sci.* **2021**, *22*, 5386. [CrossRef] [PubMed]

25. Jha, K.A.; Pentecost, M.; Lenin, R.; Gentry, J.; Klaic, L.; Del Mar, N.; Reiner, A.; Yang, C.H.; Pfeffer, L.M.; Sohl, N.; et al. TSG-6 in conditioned media from adipose mesenchymal stem cells protects against visual deficits in mild traumatic brain injury model through neurovascular modulation. *Stem Cell Res. Ther.* **2019**, *10*, 318. [CrossRef] [PubMed]

26. Kronsteiner, B.; Wolbank, S.; Peterbauer, A.; Hackl, C.; Redl, H.; van Griensven, M.; Gabriel, C. Human mesenchymal stem cells from adipose tissue and amnion influence T-cells depending on stimulation method and presence of other immune cells. *Stem Cells Dev.* **2011**, *20*, 2115–2126. [CrossRef] [PubMed]

27. Roth, S.; Dreixler, J.C.; Mathew, B.; Balyasnikova, I.; Mann, J.R.; Boddapati, V.; Xue, L.; Lesniak, M.S. Hypoxic-Preconditioned Bone Marrow Stem Cell Medium Significantly Improves Outcome After Retinal Ischemia in Rats. *Investig. Ophthalmol. Vis. Sci.* **2016**, *57*, 3522–3532. [CrossRef] [PubMed]

28. Noueihed, B.; Rivera, J.C.; Dabouz, R.; Abram, P.; Omri, S.; Lahaie, I.; Chemtob, S. Mesenchymal Stromal Cells Promote Retinal Vascular Repair by Modulating Sema3E and IL-17A in a Model of Ischemic Retinopathy. *Front. Cell Dev. Biol.* **2021**, *9*, 630645. [CrossRef] [PubMed]

29. Mathew, B.; Ravindran, S.; Liu, X.; Torres, L.; Chennakesavalu, M.; Huang, C.C.; Feng, L.; Zelka, R.; Lopez, J.; Sharma, M.; et al. Mesenchymal stem cell-derived extracellular vesicles and retinal ischemia-reperfusion. *Biomaterials* **2019**, *197*, 146–160. [CrossRef] [PubMed]

30. Moisseiev, E.; Anderson, J.D.; Oltjen, S.; Goswami, M.; Zawadzki, R.J.; Nolta, J.A.; Park, S.S. Protective Effect of Intravitreal Administration of Exosomes Derived from Mesenchymal Stem Cells on Retinal Ischemia. *Curr. Eye Res.* **2017**, *42*, 1358–1367. [CrossRef] [PubMed]

31. Gustavsson, C.; Agardh, C.D.; Hagert, P.; Agardh, E. Inflammatory markers in nondiabetic and diabetic rat retinas exposed to ischemia followed by reperfusion. *Retina* **2008**, *28*, 645–652. [CrossRef]

32. Wang, L.; Li, C.; Guo, H.; Kern, T.S.; Huang, K.; Zheng, L. Curcumin inhibits neuronal and vascular degeneration in retina after ischemia and reperfusion injury. *PLoS ONE* **2011**, *6*, e23194. [CrossRef] [PubMed]

33. Zhang, X.; Wang, Y.; Song, J.; Gerwien, H.; Chuquisana, O.; Chashchina, A.; Denz, C.; Sorokin, L. The endothelial basement membrane acts as a checkpoint for entry of pathogenic T cells into the brain. *J. Exp. Med.* **2020**, *217*. [CrossRef] [PubMed]

34. Deliyanti, D.; Talia, D.M.; Zhu, T.; Maxwell, M.J.; Agrotis, A.; Jerome, J.R.; Hargreaves, E.M.; Gerondakis, S.; Hibbs, M.L.; Mackay, F.; et al. Foxp3(+) Tregs are recruited to the retina to repair pathological angiogenesis. *Nat. Commun.* **2017**, *8*, 748. [CrossRef] [PubMed]

35. Zhou, X.; Bailey-Bucktrout, S.L.; Jeker, L.T.; Penaranda, C.; Martinez-Llordella, M.; Ashby, M.; Nakayama, M.; Rosenthal, W.; Bluestone, J.A. Instability of the transcription factor Foxp3 leads to the generation of pathogenic memory T cells in vivo. *Nat. Immunol.* **2009**, *10*, 1000–1007. [CrossRef]

36. Colamatteo, A.; Carbone, F.; Bruzzaniti, S.; Galgani, M.; Fusco, C.; Maniscalco, G.T.; Di Rella, F.; de Candia, P.; De Rosa, V. Molecular Mechanisms Controlling Foxp3 Expression in Health and Autoimmunity: From Epigenetic to Post-translational Regulation. *Front. Immunol.* **2019**, *10*, 3136. [CrossRef]

37. Kanamori, M.; Nakatsukasa, H.; Okada, M.; Lu, Q.; Yoshimura, A. Induced Regulatory T Cells: Their Development, Stability, and Applications. *Trends Immunol.* **2016**, *37*, 803–811. [CrossRef] [PubMed]

38. Someya, K.; Nakatsukasa, H.; Ito, M.; Kondo, T.; Tateda, K.I.; Akanuma, T.; Koya, I.; Sanosaka, T.; Kohyama, J.; Tsukada, Y.I.; et al. Improvement of Foxp3 stability through CNS2 demethylation by TET enzyme induction and activation. *Int. Immunol.* **2017**, *29*, 365–375. [CrossRef]

39. Zheng, Y.; Josefowicz, S.; Chaudhry, A.; Peng, X.P.; Forbush, K.; Rudensky, A.Y. Role of conserved non-coding DNA elements in the Foxp3 gene in regulatory T-cell fate. *Nature* **2010**, *463*, 808–812. [CrossRef]

40. Monsel, A.; Zhu, Y.G.; Gennai, S.; Hao, Q.; Hu, S.; Rouby, J.J.; Rosenzwajg, M.; Matthay, M.A.; Lee, J.W. Therapeutic Effects of Human Mesenchymal Stem Cell-derived Microvesicles in Severe Pneumonia in Mice. *Am. J. Respir. Crit. Care Med.* **2015**, *192*, 324–336. [CrossRef] [PubMed]

41. Islam, M.N.; Das, S.R.; Emin, M.T.; Wei, M.; Sun, L.; Westphalen, K.; Rowlands, D.J.; Quadri, S.K.; Bhattacharya, S.; Bhattacharya, J. Mitochondrial transfer from bone-marrow-derived stromal cells to pulmonary alveoli protects against acute lung injury. *Nat. Med.* **2012**, *18*, 759–765. [CrossRef]

42. Li, C.; Cheung, M.K.H.; Han, S.; Zhang, Z.; Chen, L.; Chen, J.; Zeng, H.; Qiu, J. Mesenchymal stem cells and their mitochondrial transfer: A double-edged sword. *Biosci. Rep.* **2019**, *39*, BSR20182417. [CrossRef] [PubMed]

43. Tasso, R.; Ilengo, C.; Quarto, R.; Cancedda, R.; Caspi, R.R.; Pennesi, G. Mesenchymal Stem Cells Induce Functionally Active T-Regulatory Lymphocytes in a Paracrine Fashion and Ameliorate Experimental Autoimmune Uveitis. *Investig. Ophthalmol. Vis. Sci.* **2012**, *53*, 786–793. [CrossRef] [PubMed]

44. Court, A.C.; Le-Gatt, A.; Luz-Crawford, P.; Parra, E.; Aliaga-Tobar, V.; Bátiz, L.F.; Contreras, R.A.; Ortúzar, M.I.; Kurte, M.; Elizondo-Vega, R.; et al. Mitochondrial transfer from MSCs to T cells induces Treg differentiation and restricts inflammatory response. *EMBO Rep.* **2020**, *21*, e48052. [CrossRef] [PubMed]

45. Gaignerie, A.; Lefort, N.; Rousselle, M.; Forest-Choquet, V.; Flippe, L.; Francois-Campion, V.; Girardeau, A.; Caillaud, A.; Chariau, C.; Francheteau, Q.; et al. Urine-derived cells provide a readily accessible cell type for feeder-free mRNA reprogramming. *Sci. Rep.* **2018**, *8*, 14363. [CrossRef] [PubMed]

46. Bhattacharya, S.; Gangaraju, R.; Chaum, E. Recent Advances in Retinal Stem Cell Therapy. *Curr. Mol. Biol. Rep.* **2017**, *3*, 172–182. [CrossRef] [PubMed]

47. Spitzhorn, L.-S.; Megges, M.; Wruck, W.; Rahman, M.S.; Otte, J.; Degistirici, Ö.; Meisel, R.; Sorg, R.V.; Oreffo, R.O.C.; Adjaye, J. Human iPSC-derived MSCs (iMSCs) from aged individuals acquire a rejuvenation signature. *Stem Cell Res. Ther.* **2019**, *10*, 100. [CrossRef]

Article

Anti-VEGF Drugs Influence Epigenetic Regulation and AMD-Specific Molecular Markers in ARPE-19 Cells

Mohamed A. Hamid [1,2,†], M. Tarek Moustafa [1,2,†], Sonali Nashine [1], Rodrigo Donato Costa [1,3], Kevin Schneider [1], Shari R. Atilano [1], Baruch D. Kuppermann [1,4] and M. Cristina Kenney [1,5,*]

[1]   Gavin Herbert Eye Institute, University of California Irvine, Irvine, CA 92697, USA; drmohamedhamid83@mu.edu.eg (M.A.H.); mohamedtarek@mu.edu.eg (M.T.M.); snashine@uci.edu (S.N.); rodrigodonato@yahoo.com.br (R.D.C.); kschneid@hs.uci.edu (K.S.); satilano@hs.uci.edu (S.R.A.); bdkupper@uci.edu (B.D.K.)

[2]   Ophthalmology Department, Faculty of Medicine, Minia University, Minia 61111, Egypt

[3]   Instituto Donato Oftalmologia, Poços de Caldas, MG 37701-528, Brazil

[4]   Department of Biomedical Engineering, University of California Irvine, Irvine, CA 92697, USA

[5]   Department of Pathology and Laboratory Medicine, University of California Irvine, Irvine, CA 92697, USA

*   Correspondence: mkenney@hs.uci.edu; Tel.: +1-949-824-7603

†   Both authors contributed equally.

Citation: Hamid, M.A.; Moustafa, M.T.; Nashine, S.; Costa, R.D.; Schneider, K.; Atilano, S.R.; Kuppermann, B.D.; Kenney, M.C. Anti-VEGF Drugs Influence Epigenetic Regulation and AMD-Specific Molecular Markers in ARPE-19 Cells. *Cells* **2021**, *10*, 878. https://doi.org/10.3390/cells10040878

Academic Editor: Maurice Ptito

Received: 2 February 2021
Accepted: 6 April 2021
Published: 12 April 2021

Publisher's Note: MDPI stays neutral with regard to jurisdictional claims in published maps and institutional affiliations.

Abstract: Our study assesses the effects of anti-VEGF (Vascular Endothelial Growth Factor) drugs and Trichostatin A (TSA), an inhibitor of histone deacetylase (HDAC) activity, on cultured ARPE-19 (Adult Retinal Pigment Epithelial-19) cells that are immortalized human retinal pigment epithelial cells. ARPE-19 cells were treated with the following anti-VEGF drugs: aflibercept, ranibizumab, or bevacizumab at $1\times$ and $2\times$ concentrations of the clinical intravitreal dose (12.5 µL/mL and 25 µL/mL, respectively) and analyzed for transcription profiles of genes associated with the pathogenesis age-related macular degeneration (AMD). HDAC activity was measured using the Fluorometric Histone Deacetylase assay. TSA downregulated *HIF-1α* and *IL-1β* genes, and upregulated *BCL2L13*, *CASPASE-9*, and *IL-18* genes. TSA alone or bevacizumab plus TSA showed a significant reduction of HDAC activity compared to untreated ARPE-19 cells. Bevacizumab alone did not significantly alter HDAC activity, but increased gene expression of *SOD2*, *BCL2L13*, *CASPASE-3*, and *IL-18* and caused downregulation of *HIF-1α* and *IL-18*. Combination of bevacizumab plus TSA increased gene expression of *SOD2*, *HIF-1α*, *GPX3A*, *BCL2L13*, and *CASPASE-3*, and reduced *CASPASE-9* and *IL-β*. In conclusion, we demonstrated that anti-VEGF drugs can: (1) alter expression of genes involved in oxidative stress (*GPX3A* and *SOD2*), inflammation (*IL-18* and *IL-1β*) and apoptosis (*BCL2L13*, *CASPASE-3*, and *CASPASE-9*), and (2) TSA-induced deacetylation altered transcription for angiogenesis (*HIF-1α*), apoptosis, and inflammation genes.

Keywords: AMD; age-related macular degeneration; trichostatin A (TSA); HDAC; histone deacetylase; vascular endothelial growth factor (VEGF)

1. Introduction

Pathological angiogenesis, which subsequently leads to choroidal neovascularization, subretinal fibrosis, and exudative hemorrhage, is an underlying cause of the severe, late-stage, wet form of AMD (Age-related Macular Degeneration) [1].

Wet or neovascular AMD, which accounts for 10–20% of cases, is the less common of the two types of AMD. However, 90% of AMD-associated irreversible vision loss is attributed to wet AMD [2]. Dry AMD is characterized by degeneration of Retinal Pigment Epithelial (RPE) cells and accounts for 80% of AMD cases.

VEGF (Vascular Endothelial Growth Factor) is a signaling growth factor for vascular endothelial cells and a critical angiogenic factor that stimulates ocular neovascularization. Therefore, the most widely used wet AMD treatment targets the pro-angiogenic activity of

VEGF to inhibit ocular neovascularization [3]. Administration of intravitreal injections of anti-VEGF drugs, such as ranibizumab (Lucentis), bevacizumab (Avastin), and aflibercept (Eylea), to wet AMD patients is successfully and routinely being used as a wet AMD therapy worldwide [4,5]. Despite the widespread use of anti-VEGF drugs, 10–15% of patients fail to respond to rigorous treatment protocols in clinical trial settings [6–10]. This variability in AMD patients' response to therapy has been attributed to several clinical, behavioral, and genetic factors [11]. Pharmacogenetic studies have identified VEGFA, VEGFR2 (VEGF Receptor 2), CFH (Complement Factor H), and ARMS2 (Age-Related Maculopathy Susceptibility 2) as potential biomarkers for response to anti-VEGF drugs [12]. Many investigators, including the CATT (Comparison of Age-related Macular Degeneration Treatments) and IVAN (Inhibition of VEGF in Age-related choroidal Neovascularisation) research groups, did not find any significant association between genes polymorphism and visual or anatomic responses to treatment [13–17]. This inconsistency in findings by pharmacogenetic studies could be explained in part by possible gene–gene or gene–environmental interactions [18,19].

Genetic and environmental factors contribute to the development and progression of AMD. Genome-wide association studies (GWAS) have identified 52 genetic variants distributed across 34 loci associated with AMD [20]. Furthermore, epigenetic modifications, which include DNA methylation, histone acetylation/deacetylation, non-coding RNA-mediated gene silencing, and chromatin remodeling [21] have been implicated in the pathogenesis of AMD by selective transcription of genes involved in angiogenesis, inflammation, and oxidative stress pathways [22,23]. Epigenetic mechanisms result in covalent modifications in the DNA and regulate gene transcription either by activation or repression, in response to environmental stimuli and are often heritable [24]. Epigenetics can elucidate gene–environment interactions and explain why a certain genotype frequently results in different phenotypes [25]. Histone acetylation is catalyzed by histone transferases (HATs) and acts to destabilize nucleosomes and unwrap DNA to make it accessible to transcription factors. Conversely, histone deacetylation, carried out by histone deacetylases (HDACs), stabilizes nucleosomes and represses DNA transcription [26]. Histone acetylation is known to regulate the expression of 2–10% of genes. Other non-histone proteins, particularly transcription factors, are also regulated by acetylation/deacetylation. This could explain the fact that gene expression is not always silenced by deacetylation [27].

AMD being a leading cause of blindness in the United States and the third major cause of visual impairment worldwide, [28] poses a major health risk to the elderly population, and AMD risk is projected to increase by 54% in the United States in the next five year [29]. Therefore, we speculate that delving into the mechanisms of action of the currently used anti-VEGF drugs might contribute to the design of more effective therapeutic strategies for wet AMD. To this end, the current in vitro study was designed to examine the effects of anti-VEGF drugs on epigenetic regulation in immortalized human ARPE-19 cell lines. The ARPE-19 cell line used in this study was originally developed from the retinal pigment epithelium (RPE) of a human donor eye and resembles the phenotype and properties characteristic of aged native human RPE cells, lack of pigmentation, weak tight junctions, reduced expression of all-trans retinol, Pigment-Epithelium-Derived Factor (PEDF), and RPE markers, and hypersensitivity to VEGF activity, thereby making the ARPE-19 cell line an ideal in vitro AMD model [30]. However, it should be emphasized that the ARPE-19 cell line loses some of the aging RPE characteristics, especially with increasing passages, such as morphology, retinoid metabolism, and VEGF secretion. Furthermore, it should be mentioned that depigmentation in the RPE of AMD eyes is different from that of ARPE-19 cells. Melanosome density in the RPE decreases significantly with normal aging and more evidently in AMD, but melanin is not completely lost. More importantly, RPE melanin in AMD loses its antioxidant properties [31].

RPE cells in vivo form the outer blood-brain barrier and support photoreceptor cells and enable phototransduction. The outer blood-brain barrier is formed of a continuous layer of tight junctions that enable transepithelial transport and phototransduction.

Pigment-Epithelium-Derived Factor (PEDF), which maintains barrier integrity and is endogenously secreted by RPE in vivo in large amounts, shows only basal levels in aged eyes. The in vitro ARPE-19 cell line is a rapidly growing immortalized human cell line derived from primary RPE cells from the globes of a 19-year-old male donor. ARPE-cell lines express RPE cell-specific markers CRALBP and RPE65 and form a viable cuboidal to columnar epithelium monolayer in culture media. Our recent study characterized and confirmed the expression of the following RPE-specific markers in the ARPE-19 cell line used in our lab: Cellular retinaldehyde binding protein-1 (CRALBP), Bestrophin1 (BEST1), and Keratin-18 (KRT18) [32]. Although ARPE-19 cells retain the characteristic features of RPE, including defined cell borders and pigmentation, they require considerable time for differentiation and are unable to completely differentiate into RPE-like layers found in vivo. ARPE-19 cells show partial polarization as some of the cells in the monolayer resemble the morphology of differentiated RPE cells such as apical microvilli, polarized distribution of organelles, basolateral infoldings, and junctional complexes on the apical plasma membrane. ARPE-19 cells exhibit low transepithelial resistance (TER) that reaches a maximum value of 50–100 Ω cm^2 after 28 days of culture in low-serum media in laminin-coated Transwell-COL filters. It is speculated that the low TER might be due to heterogeneity of the cell line since some of the cells show polarization [33].

The present study demonstrated that treatment of ARPE-19 cells with anti-VEGF drugs altered the total HDAC protein activity and the gene expression levels of apoptotic, inflammatory, and oxidative stress-related genes. Moreover, addition of Trichostatin-A (TSA), an HDAC inhibitor, along with an anti-VEGF drug modulated the gene expression of *VEGF*, apoptotic and inflammatory markers. These results suggest that epigenetics modulation in ARPE-19 cells is strongly influenced by anti-VEGF drug treatment.

2. Materials and Methods

2.1. ARPE-19 Cell Culture

Human RPE cells (ARPE-19 cells, ATCC, Manassas, VA, USA) were cultured till confluent in 175 cm^2-flasks containing DMEM/F-12 culture medium (Dulbecco's Modification of Eagle's Medium, Mediatech, Inc., Manassas, VA, USA), 10% dialyzed fetal bovine serum, antibiotics (penicillin G 100 U/mL, streptomycin sulfate 0.1 mg/mL, gentamicin 10 µg/mL, amphotericin B 2.5 µg/mL) and 17.5 mM glucose. All ARPE-19 cells were at passage 10 and cultured side-by-side under identical standard conditions of 37 °C in 5% CO$_2$ and 95% relative humidity, in order to avoid any potential technical variability.

ARPE-19 cells are a spontaneously arising RPE cell line derived by Amy Aotaki-Keen from the normal eyes of a 19-year-old male who died from head trauma in a motor vehicle accident in 1986. The ARPE-19 cell line, established using the cuboidal basal cell layer cultured in specific culture media, expresses the RPE-specific markers cellular retinaldehyde binding protein and RPE-65.

In our study we utilized ARPE-19 cells at passage 10 in all our experiments to ensure the cells retained acceptable fidelity.

2.2. Drug Treatment of ARPE-19 Cell Cultures

ARPE-19 cells were plated in triplicate for 24 h in 6-well plates at a density of 500,000 cells per well, culture media were removed and replaced with the same media containing anti-VEGF drugs: aflibercept, ranibizumab or bevacizumab at $1\times$ and $2\times$ concentrations of the clinical intravitreal dose (12.5 µL/mL and 25 µL/mL, respectively). The clinical dose was calculated by assuming that the amount of each drug clinically used in intravitreal injections distributes equally throughout the 4 mL human vitreous. Untreated cells were used as control.

In order to further explore the potential relationship between anti-angiogenic treatment and the acetylation status of the target genes expression, a subset of ARPE-19 cells was treated with trichostatin A (TSA), an inhibitor of histone deacetylase (HDAC) activity, at 0.3 µM concentration, or a combination of 1X bevacizumab plus 0.3 µM TSA.

The control and anti-VEGF treated cultures were incubated for an additional 24 h, then RNA was isolated to be used for Quantitative Real-Time PCR (qRT-PCR) analyses.

Proteins extracted from cultures of untreated ARPE-19 cells, as well as cells treated with 1X bevacizumab, 0.3 μM TSA, and a combination of both drugs were analyzed for HDAC activity as described below.

2.3. RNA Extraction, Amplification of cDNA, and Quantitative Real-Time PCR (qRT-PCR) Analysis

ARPE-19 cells were pelleted for RNA isolation using a PureLink RNA Mini Kit (Ambion, Thermo Fisher Scientific, Waltham, MA, USA). For qRT-PCR analyses, 100 ng of individual RNA samples were reverse transcribed into complementary DNA (cDNA) using SuperScript VILO Master Mix (Thermo Fisher Scientific, Waltham, MA, USA).

We investigated transcription profiles of downstream genes known to play a role in AMD pathogenesis. RNA samples were isolated from cells that were (a) untreated; (b) treated with $1\times$ or $2\times$ concentrations of the three anti-VEGF drugs; (c) treated with 0.3μM TSA alone; and (d) treated with $1\times$ bevacizumab plus TSA (bevacizumab/TSA). qRT-PCR was performed using primers for downstream pathway genes, including angiogenesis (*VEGF-A* and *HIF-1α*), apoptosis (*BCL2L13*, *CASPASE-3*, and *CASPASE-9*), inflammation (*IL-18* and *IL-1β*) and oxidative stress (*GPX3A* and *SOD2*) (Table 1). The qRT-PCR was performed on individual samples using QuantiFast SYBR Green PCR Kit (Qiagen, Inc., Germantown, MD, USA) on StepOne Plus Real-Time PCR system (Thermo Fisher Scientific, Waltham, MA, USA). For the various target genes, housekeeping genes that had comparable amplification efficiencies were chosen in order to maximize the accuracy of our ΔΔCT values. The housekeeper genes were either hypoxanthine phosphorbosyltransferase 1 (HPRT1) or hydroxymethylbilane synthase (HMBS). Untreated samples were used as control. ΔΔCts (differences in cycle thresholds) were obtained and folds calculated using the formula $2^{ΔΔCt}$.

2.4. Protein Extraction and HDAC Activity Assay

ARPE-19 cell samples were lysed using RIPA buffer (Cat. #89900, Life Technologies Corp., Calsbad, CA, USA), supernatants were transferred to a new microfuge tube and protein concentrations were measured using the Bio-Rad DC Protein Assay system (Bio-Rad Laboratories, Richmond, CA, USA) according to the manufacturer's instructions. Protein samples were kept in a $-80\ ^{\circ}\text{C}$ freezer until time of use in the HDAC activity assay.

HDAC activity in the protein samples was measured using the Fluorometric Histone Deacetylase Assay Kit (Sigma–Aldrich, St. Louis, MO, USA) according to the manufacturer's protocol. Briefly, protein samples, assay buffer and HDAC substrate solution were added to the wells of a 96-well plate. Each well contained 20 μL of the protein sample, 30 μL of assay buffer and 50 μL of HDAC substrate solution. The plate was incubated at $30\ ^{\circ}\text{C}$ for 30 min, then 10 μL of Developer solution added to each well. The plate was incubated 10 min at room temperature, and fluorescence measured with a microplate spectrofluorometer (Gemini XPS, Molecular Devices, Sunnyvale, CA, USA) using an excitation wavelength of 350 nm and an emission wavelength of 440 nm. Samples were plated in triplicate and Hela cell lysate used as a positive control for HDAC activity. Experiments were performed on three replicates i.e., in triplicate. The plate in which the cells were plated was read three times and the fluorescence intensity was averaged. The entire experiment, i.e., treatment and reading the fluorescence was repeated three times to ensure reproducibility. Protein samples from untreated ARPE-19 cells were used as control and were normalized to 100%.

2.5. Statistical Analyses

Statistical analyses of the data were performed by unpaired t-test using GraphPad Prism, Version 5 (GraphPad Software, Inc., La Jolla, CA, USA). $p < 0.05$ was considered statistically significant. Untreated samples (controls) were normalized to a value of 100% for comparison to treated samples.

Table 1. Gene symbols, names, gene bank accession numbers, and functions.

Gene Symbol [a]	Gene Name [b]	Gene Bank Accession Num. [c]	Function [d]
VEGF-A	Vascular endothelial growth factor A	NM_001025366, NM_001025367, NM_001025368, NM_001033756, NM_001171623, NM_001171624, NM_001171625, NM_001171626, NM_001171629, NM_003376, NM_001287044	Member of PDGF/VEGF growth factor family and encodes a protein that specifically acts on endothelial cells, mediating increased vascular permeability, inducing angiogenesis, vasculogenesis, and endothelial cell growth, promoting cell migration, and inhibiting apoptosis.
HIF-1α	Hypoxia inducible factor 1 alpha	NM_001243084, NM_001530, NM_181054	Master regulator of cellular and systemic homeostatic response to hypoxia by activating transcription of many genes, including those involved in energy metabolism, angiogenesis, apoptosis, and other genes whose protein products increase oxygen delivery or facilitate metabolic adaptation to hypoxia.
SOD2	Superoxide dismutase 2, mitochondria	NM_000636 NM_001024465	Binds to the superoxide byproducts of oxidative phosphorylation and converts them to H_2O_2 and O_2.
GPX3	Glutathione peroxidase 3	NM_002084	Catalyzes the reduction of hydrogen peroxide.
BCL2L13	BCL2-like 13 (apoptosis facilitator)	NM_015367	Encodes a mitochondrially-localized protein with conserved B-cell lymphoma 2 homology motifs. Overexpression of the encoded protein results in apoptosis.
CASPASE-3	Caspase 3	NM_004346, NM_032991	The protein encoded by this gene is a cysteine-aspartic acid protease that plays a central role in the execution phase of cell apoptosis. It cleaves and inactivates poly (ADP-ribose) polymerase while it cleaves and activates sterol regulatory element binding proteins, as well as caspases 6, 7, and 9. This protein itself is processed by caspases 8, 9, and 10. It is the predominant caspase involved in the cleavage of amyloid-beta 4A precursor protein, which is associated with neuronal death in Alzheimer's disease
CASPASE-9	Caspase 9	NM_001229, NM_032996	This gene encodes a member of the cysteine-aspartic acid protease (caspase) family that plays a central role in the execution phase of cell apoptosis. This protein can undergo autoproteolytic processing and activation by the apoptosome, a protein complex of cytochrome c and the apoptotic peptidase activating factor 1; this step is thought to be one of the earliest in the caspase activation cascade. This protein is thought to play a central role in apoptosis and to be a tumor suppressor.
IL-18	Interleukin 18	NM_001243211 NM_001562	Proinflammatory cytokine that augments natural killer cell activity in spleen cells and stimulates interferon gamma production in T-helper type I cells.
IL-1β	Interleukin 1, beta (also known as IL-1β)	NM_000576, XM_006712496	Produced by activated macrophages as a proprotein and processed to its active form by caspase 1 (CASP1/ICE). It is an important mediator of the inflammatory response; and is involved in cell proliferation, differentiation, and apoptosis

[a] Official gene symbol by HUGO (Human Genome Organization) Gene Nomenclature Committee (HGNC). [b] Official gene name by HUGO Gene Nomenclature Committee (HGNC). [c] Gene Accession Bank Number from the primers used (Qiagen, Valencia, CA). [d] Gene function modified from PubMed gene.

3. Results

3.1. Measurement of HDAC Activity in ARPE-19 Cells before and after Treatment with Anti-VEGF and TSA

Treatment of ARPE-19 cells with 0.3 µM TSA resulted in significant reduction of HDAC activity (p = 0.0003), as did the combination of bevacizumab 1× plus TSA (p < 0.0001) (Figure 1). Although treatment with bevacizumab 1X alone did not significantly alter HDAC activity (p = 0.15), its addition to TSA significantly potentiated its inhibition of HDAC activity (p = 0.0001).

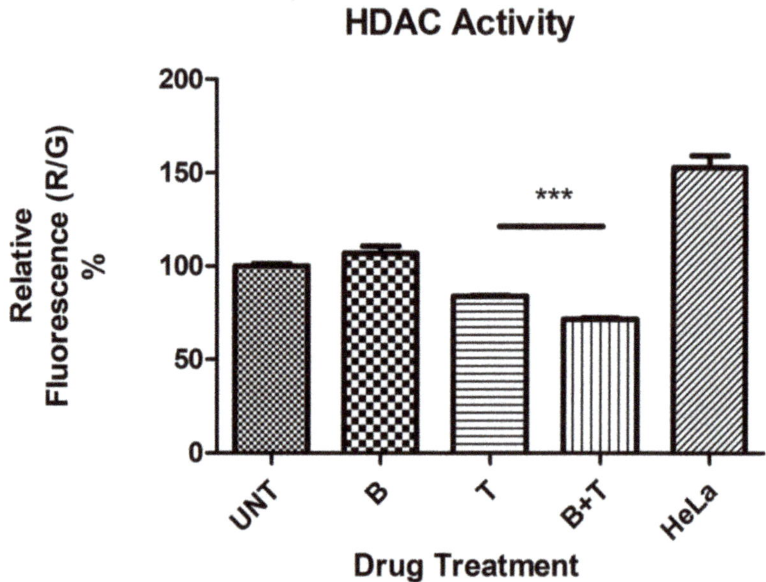

Figure 1. Histone deacetylase (HDAC) activity as determined by relative fluorescence (%) in untreated and treated ARPE-19 cultures. *** p < 0.001; Bars with no asterisk represent nonsignificant difference. UNT: Untreated; B: Bevacizumab 1× conc.; T: Trichostatin A 0.3 µM; B+T: Bevacizumab 1× conc. + Trichostatin A 0.3 µM. HeLa: Hela cell lysate positive control for HDAC activity. Error bars represent 'Mean ± SEM'.

3.2. Effect of Anti-VEGF Treatment and HDAC Inhibition on the Expression Profiles of Downstream Genes

Aflibercept 1×-treated ARPE-19 cells showed upregulation of *VEGF-A* (1.12-fold, p = 0.67), *HIF-1α* (1.03-fold, p = 0.63), *GPX3A* (1.13-fold, p = 0.49), *BCL2L13* (1.26-fold, p = 0.05), *CASPASE-3* (1.21-fold, p = 0.04), *CASPASE-9* (1.32-fold, p = 0.003) and *IL-1β* (1.38-fold, p = 0.22); and downregulation of *SOD2* (0.76-fold, p = 0.14) and *IL-18* (0.95-fold, p = 0.051), compared to untreated cells (Table 2, Figure 2).

Treatment with aflibercept 2× downregulated the expression of *VEGF-A* (0.98-fold, p = 0.94), *HIF-1α* (0.67-fold, p = 0.002), *SOD2* (0.52-fold, p = 0.005), *GPX3A* (0.96-fold, p = 0.87), and *IL-18* (0.78-fold, p = 0.002), and led to upregulation of *BCL2L13* (1.21-fold, p < 0.0001), *CASPASE-3* (1.36-fold, p = 0.006), *CASPASE-9* (1.19-fold, p = 0.004) and *IL-1β* (1.14-fold, p = 0.55) (Table 2, Figure 2).

ARPE-19 cells treated with ranibizumab 1× showed downregulation of *VEGF-A* (0.75-fold, p = 0.40), *HIF-1α* (0.85-fold, p = 0.05), *GPX3A* (0.96-fold, p = 0.89), and *IL-1β* (0.84-fold, p = 0.43) and upregulation of *SOD2* (1.22-fold, p = 0.16), *BCL2L13* (1.21-fold, p > 0.0001), *CASPASE-3* (1.73-fold, p = 0.004), *CASPASE-9* (1.61-fold, p < 0.0001) and *IL-18* (1.67-fold, p < 0.0001) (Table 2, Figure 2).

Table 2. Expression folds for downstream genes in untreated and anti-vascular endothelial growth factor (anti-VEGF)-treated ARPE-19 cultures *.

| | Aflibercept | | | | Ranibizumab | | | | Bevacizumab | | | |
| | 1× | | 2× | | 1× | | 2× | | 1× | | 2× | |
Gene	Fold	*p*-value	Fold	*p*-value	Fold	*p*-value	Fold	*p*-value	Fold	*p*-value	Fold	*p*-value
Angiogenesis												
VEGF-A	1.12	0.67	0.98	0.94	0.75	0.40	0.72	0.20	0.95	0.80	0.82	0.41
HIF-1α	1.03	0.63	0.67	*0.002*	0.85	0.05	0.76	*0.008*	0.85	0.26	0.63	*0.001*
Antioxidant												
SOD2	0.76	0.14	0.52	*0.005*	1.22	0.16	1.28	0.09	1.80	*0.009*	1.54	0.06
GPX3A	1.13	0.49	0.96	0.87	0.96	0.89	1.09	0.60	1.28	0.23	0.91	0.63
Apoptosis												
BCL2L13	1.26	*0.05*	1.21	*<0.0001*	1.21	*<0.0001*	1.40	*<0.0001*	1.80	*<0.0001*	1.07	0.16
CASPASE-3	1.21	*0.04*	1.36	*0.006*	1.73	*0.004*	1.26	*0.04*	1.50	*0.003*	1.04	0.39
CASPASE-9	1.32	*0.003*	1.19	*0.004*	1.61	*<0.0001*	2.82	*<0.0001*	1.04	0.52	2.16	*<0.0001*
Inflammation												
IL-18	0.95	0.051	0.78	*0.002*	1.67	*<0.0001*	1.32	*0.002*	1.33	*0.0003*	0.86	*0.02*
IL-1β	1.38	0.22	1.14	0.55	0.84	0.43	0.54	*0.04*	0.81	0.42	0.68	0.12

* Fold change was calculated using the formula: $2^{\Delta\Delta CT}$. Untreated samples had a value of 1.

Ranibizumab 2×-treated cells showed downregulation of *VEGF-A* (0.72-fold, $p = 0.20$), *HIF-1α* (0.76-fold, $p = 0.008$) and *IL-1β* (0.54-fold, $p = 0.04$), and upregulation of *SOD2* (1.28-fold, $p = 0.09$), *GPX3A* (1.09-fold, $p = 0.60$), *BCL2L13* (1.4-fold, $p < 0.0001$), *CASPASE-3* (1.26-fold, $p = 0.04$), *CASPASE-9* (2.82-fold, $p < 0.0001$) and *IL-18* (1.32-fold, $p = 0.002$) (Table 2, Figure 2).

Treatment with bevacizumab 1× decreased the expression of *VEGF-A* (0.95-fold, $p = 0.80$), *HIF-1α* (0.85-fold, $p = 0.26$) and *IL-1β* (0.81-fold, $p = 0.42$), and increased the expression of *SOD2* (1.8-fold, $p = 0.009$), *GPX3A* (1.28-fold, $p = 0.23$), *BCL2L13* (1.8-fold, $p < 0.0001$), *CASPASE-3* (1.5-fold, $p = 0.003$), *CASPASE-9* (1.04-fold, $p = 0.52$) and *IL-18* (1.33-fold, $p = 0.0003$) (Table 2, Figure 2).

Bevacizumab 2× resulted in downregulation of *VEGF-A* (0.82-fold, $p = 0.41$), *HIF-1α* (0.63-fold, $p = 0.001$), *GPX3A* (0.91-fold, $p = 0.63$), *IL-18* (0.86-fold, $p = 0.02$), and *IL-1β* (0.68-fold, $p = 0.12$); and upregulation of *SOD2* (1.54-fold, $p = 0.06$), *BCL2L13* (1.07-fold, $p = 0.16$), *CASPASE-3* (1.04-fold, $p = 0.39$), and *CASPASE-9* (2.16-fold, $p < 0.0001$) (Table 2, Figure 2).

TSA treated (0.3 μM) ARPE-19 cultures resulted in downregulation of *VEGF-A* (0.75-fold, $p = 0.24$), *HIF-1α* (0.66-fold, $p = 0.001$) and *IL-1β* (0.2-fold, $p = 0.001$), and caused upregulation of *SOD2* (1.69-fold, $p = 0.11$), *GPX3A* (2.11-fold, $p = 0.06$), *BCL2L13* (1.13-fold, $p = 0.0003$), *CASPASE-3* (1.14-fold, $p = 0.07$), *CASPASE-9* (1.49-fold, $p = 0.003$) and *IL-18* (1.69-fold, $p < 0.0001$) compared to untreated cells (Table 3, Figure 3).

The combination of bevacizumab 1× plus TSA resulted in upregulation of *VEGF-A* (1.42-fold, $p = 0.25$), *HIF-1α* (1.60-fold, $p = 0.006$) *SOD2* (1.97-fold, $p = 0.005$), *GPX3A* (4.03-fold, $p = 0.03$), *BCL2L13* (1.28-fold, $p = 0.0007$) and *CASPASE-3* (1.2-fold, $p = 0.05$), and decreased expression of *CASPASE-9* (0.59-fold, $p = 0.09$), *IL-18* (0.89-fold, $p = 0.09$) and *IL-1β* (0.37-fold, $p = 0.008$) compared to untreated cells (Table 3, Figure 3).

When comparing cells treated with both drugs to cells treated with TSA alone, the addition of bevacizumab 1× significantly reversed the effect of TSA on the expression of *VEGF-A* ($p = 0.02$), *HIF-1α* ($p = 0.0003$), *CASPASE-9* ($p < 0.0001$), and *IL-18* ($p = 0.0003$), reduced its effect on an *IL-1β* ($p = 0.0006$), and potentiated its effect on *BCL2L13* ($p = 0.008$) (Figure 3).

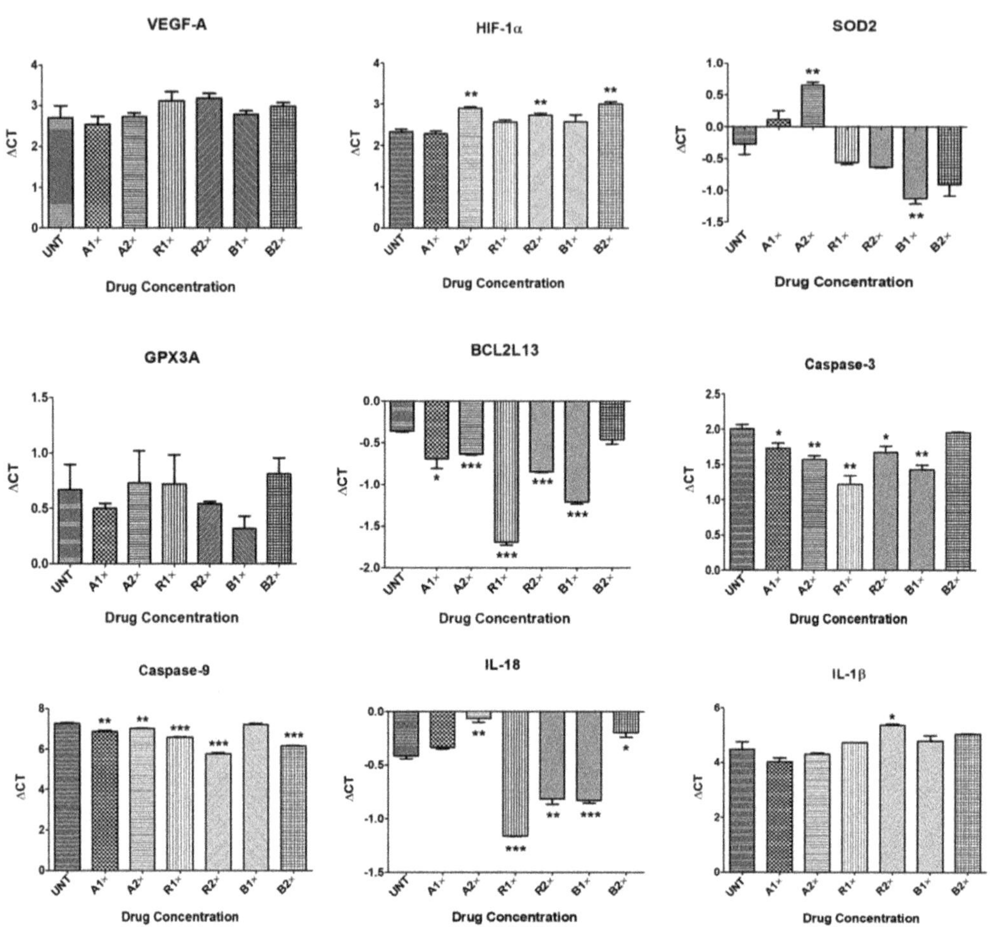

Figure 2. Quantitative Real-Time PCR (qPCR) data showing Delta Cts for downstream genes in untreated and anti-VEGF treated ARPE-19 cultures. * $p < 0.05$; ** $p < 0.01$; *** $p < 0.001$; Bars with no asterisk represent nonsignificant difference. UNT: Untreated; A1×: Aflibercept 1× conc.; A2×: Aflibercept 2× conc.; R1×: Ranibizumab 1× conc.; R2×: Ranibizumab 2× conc.; B1×: Bevacizumab 1× conc.; B2×: Bevacizumab 2× conc. Error bars represent 'Mean ± SEM'.

Table 3. Expression folds of downstream genes after treatment with Trichostatin A alone and in combination with Bevacizumab 1× *.

Gene	TSA Compared to Untreated		Bevacizumab 1×+TSA Compared to Untreated	
	Fold	*p*-value	Fold	*p*-value
Angiogenesis				
VEGF-A	0.75	0.24	1.42	0.25
HIF-1α	0.66	*0.001*	1.60	*0.006*
Antioxidant				
SOD2	1.69	0.11	1.97	*0.005*
GPX3A	2.11	0.06	4.03	*0.03*
Apoptosis				
BCL2L13	1.13	*0.0003*	1.28	*0.0007*
CASPASE-3	1.14	0.07	1.20	*0.05*
CASPASE-9	1.49	*0.003*	0.59	*<0.0001*

Table 3. *Cont.*

	TSA Compared to Untreated		Bevacizumab 1×+TSA Compared to Untreated	
Gene	Fold	*p*-value	Fold	*p*-value
Inflammation				
IL18	1.69	*<0.0001*	0.89	0.09
IL-1β	0.20	*0.001*	0.37	*0.008*

* Fold change was calculated using the formula: $2^{\Delta\Delta CT}$. Untreated samples had a value of 1.

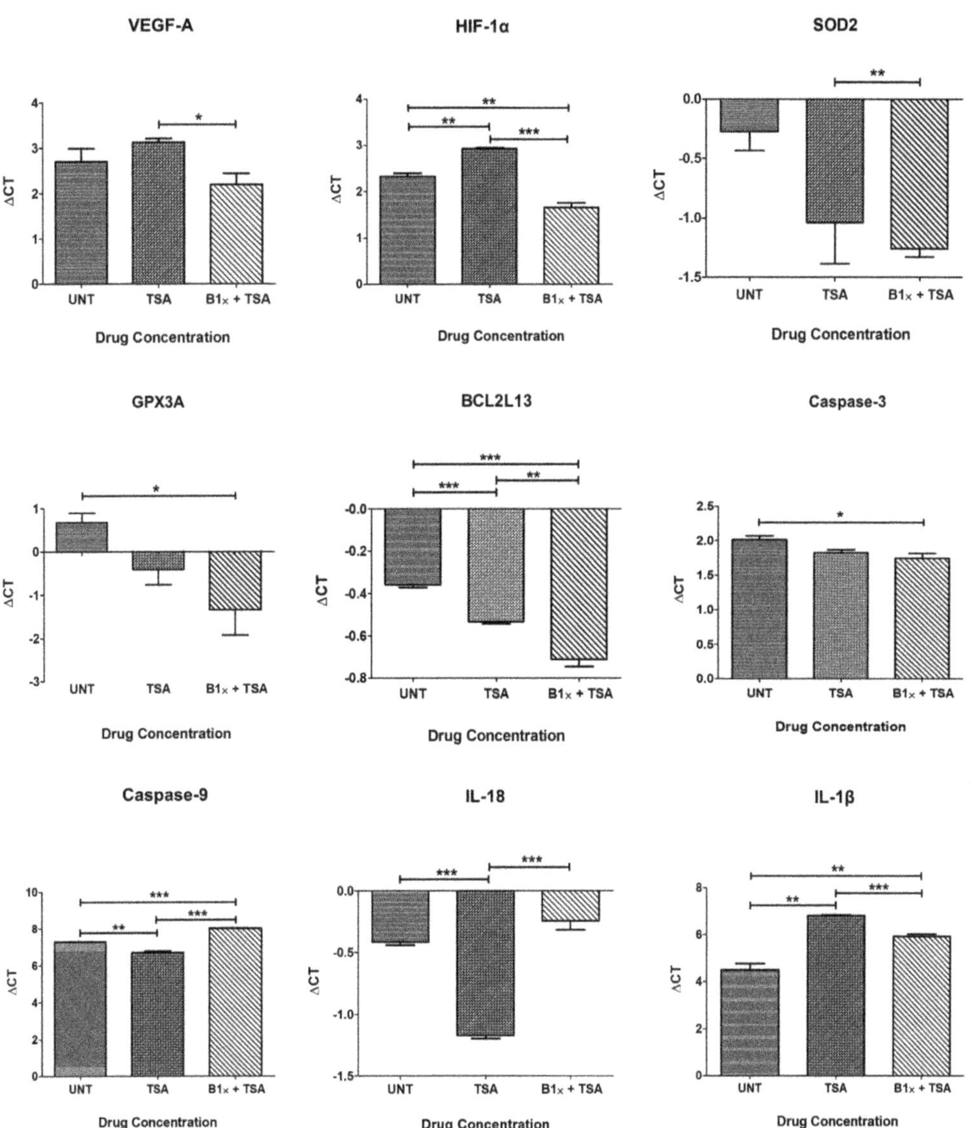

Figure 3. Quantitative Real-Time PCR (qPCR) data showing Delta Cts for downstream genes after treatment with Trichostatin A and Bevacizumab. * $p < 0.05$; ** $p < 0.01$; *** $p < 0.001$; Bars with no asterisk represent nonsignificant difference. UNT: Untreated; TSA: Trichostatin A 0.3 μM; B1×+TSA: Bevacizumab 1× conc. + Trichostatin A 0.3 μM. Error bars represent 'Mean ± SEM'.

4. Discussion

In this study, we demonstrated differential HDAC total protein activity in response to anti-VEGF drug treatment in ARPE-19 cells. Treatment of ARPE-19 cells with anti-VEGF drugs alone in varying concentrations as well as additive effects of TSA and anti-VEGF drug significantly modulated the gene expression profiles of apoptotic, inflammatory, and oxidative stress markers. Therefore, our results suggest that addition of anti-VEGF drugs to cultured ARPE-19 cells strongly influence the regulation of epigenetic markers and downstream molecular markers.

Environmental stimuli are thought to induce epigenetic changes that accumulate in the cell with increasing age, as evidenced by age being the major risk factor for AMD, as well as the discordance of disease phenotype in identical twins that possess similar risk profiles for AMD [34]. Hypermethylation of genes encoding for reactive oxygen species (ROS) scavengers, namely *GSTM1* and *GSTM5*, was demonstrated in AMD RPE and choroid, downregulating their expression and rendering the cells more susceptible to oxidative damage. Additionally, several micro RNAs (miRNAs) have been implicated in AMD pathogenesis through their contribution to aberrant angiogenesis, inflammation, and apoptosis in response to oxidative stress both in vitro and in vivo [22]. Hypomethylation of the promoter region of interleukin 17 receptor C (*IL17RC*), a pro-inflammatory gene, promotes its expression and increases inflammation in AMD patients [35]. *Oliver* et al. conducted the first genome-wide epigenetic study in AMD and found hypomethylation at *HTRA/ARMS2* locus, which is a major susceptibility locus for AMD. They also observed hypermethylation of the *PRSS50* locus which had not been previously associated with AMD [23].

Over the last decade, intravitreal injection of anti-angiogenic drugs has been the mainstay of therapy for neovascular AMD as well as for macular edema associated with diabetic retinopathy and retinal vein occlusion [36]. The three drugs commonly used in clinical practice are bevacizumab, ranibizumab and aflibercept. Bevacizumab is a full-length monoclonal antibody that has a molecular weight of 149 kilodalton (kDa) and binds all isoforms of VEGF-A rendering them inactive. Ranibizumab is an antigen-binding Fab fragment of the same parent antibody as bevacizumab. It lacks the Fc domain and has a molecular weight of 48 kDa. Aflibercept, a 96.9 kDa recombinant fusion protein, contains immunoglobulin fragments from both VEGF receptors: VEGFR1 and 2, combined with an Fc antibody fragment. It acts as a decoy receptor that not only binds VEGF-A, but also, unlike the former two drugs, can bind VEGF-B and placental-like growth factor (PlGF) [37–39]. The different molecular weights and structures of the three drugs may have an effect on their ocular and systemic pharmacokinetics and pharmacodynamics [40]. All three drugs have been shown to accumulate intracellularly. Uptake of bevacizumab and aflibercept is mediated through their Fc portion, but ranibizumab is likely internalized through a VEGF/VEGFR2-mediated mechanism [30].

RPE dysfunction is central to the pathogenesis of AMD, and RPE cells are the main source of VEGF in the retina [41]. Many investigators have tested the safety of anti-VEGF agents on RPE cells both in vitro and in animal models. Our group previously demonstrated a good safety profile of the three anti-angiogenic drugs on ARPE-19 cell cultures at the clinical dose, but mild cytotoxic effects were found at higher doses [42,43]. Other studies have shown similar results [44,45].

A study by Dinc et al. in which APRE-19 cells were subjected to H_2O_2-induced oxidative stress and levels of miRNA expression were evaluated, suggested an epigenetic role for anti-VEGF drugs [46]. In that study, several miRNAs were dysregulated in response to oxidative stress compared to untreated samples. Preincubation of cells with any of the three anti-VEGF drugs before H_2O_2 treatment significantly altered the miRNA dysregulation induced by H_2O_2. The authors suggested that anti-VEGF drugs could protect RPE cells from oxidative stress through their effect on miRNAs.

We tried to determine whether the changes in *HDAC* expression would affect the activity of HDAC enzymes in ARPE-19 cells using an assay kit designed to measure the collective activity of all HDACs. Bevacizumab treatment alone failed to alter HDAC activity to a significant extent compared to untreated cultures. Our inability to observe a significant net effect on HDAC enzyme activity may be because bevacizumab had opposite effects on the expression of both *HDAC1* and *HDAC6* genes, so the overall HDAC activity after treatment was neutral. As expected, Trichostatin A (TSA) significantly inhibited HDAC activity. TSA is known to inhibit both class I and II HDACs. Interestingly, the combination of TSA plus bevacizumab increased significantly the frank inhibitory effect on HDAC activity, but the mechanisms for additive effects are not clear and need to be further investigated.

Next, the expression levels for genes known to regulate important pathogenetic pathways for AMD after treatment with anti-VEGF drugs and TSA were measured.

4.1. Angiogenesis Genes

HIF-1α is the oxygen-sensitive subunit of HIF-1, a transcription factor upregulated in cells in response to hypoxia. Activation of HIF-1α results in overexpression of downstream genes, including pro-angiogenic genes, mainly VEGF-A [47]. Both class I and II HDACs are upregulated by hypoxia and induce angiogenesis, particularly HDAC1 [48,49]. HIF-1α is a non-histone protein target for acetylation. It is acetylated under normoxic conditions, which decreases its stability by facilitating its proteasomal degradation. Under hypoxic conditions, HIF-1α becomes deacetylated by HDACs, which makes it more stable and prolongs its half-life [50].

We found that all three anti-VEGF drugs downregulated the expression of *HIF1-α* gene at 2$\times$, but not 1$\times$, concentration. This suggests that these drugs might protect cells against oxidative damage. Treatment with TSA had the same inhibitory effect. This observation comes in agreement with previous studies demonstrating that TSA could downregulate both gene and protein expression of HIF-1α in vitro and in animal models [51,52]. Other HDAC inhibitors demonstrated a protective effect against oxidative damage in vitro and animal models. These include valproic acid (VPA), which is another broad-spectrum HDACi, and tubastatin A (TST); a specific inhibitor of HDAC6 [53,54]. The inhibitory effect displayed by anti-VEGF drugs and TSA in our study was reversed by the combination of bevacizumab and TSA, which significantly upregulated *HIF1-α* expression.

VEGF-A is the main pro-angiogenic factor implicated in the pathogenesis of wet AMD [55]. The expression levels of VEGF, as well as tissue response to its secretion, are modulated by both genetic and epigenetic factors. Epigenetic regulation of angiogenesis has been extensively studied in cancer cells. Both HDAC7 and SIRT1 promote angiogenesis during development and disease by inducing pro-angiogenic factors and suppressing anti-angiogenic ones [56–58]. HDAC6 has also been associated with angiogenesis due to its ability to bind several cytoskeletal proteins in the cytoplasm and stimulate vascular proliferation and sprouting. The anti-angiogenic properties of HDAC inhibitors are the basis for their use as anti-cancer agents and extend beyond gene silencing to directly acetylating angiogenic factors in the cytoplasm. Chan and colleagues demonstrated that TSA could downregulate *VEGF-A* by epigenetically silencing its expression in the presence of CoCl$_2$, a hypoxia-inducing agent, in human RPE cells in vitro. Furthermore, they showed that TSA could attenuate a laser induced CNV in a mouse model. However, we found that TSA did not alter *VEGF-A* gene expression in ARPE-19 cells. The ability of TSA to silence gene expression might be more pronounced under conditions of hypoxic stress that induce aberrant upregulation of VEGF-A, which was not the case in our experiment.

Two studies explored the effect of either ranibizumab or bevacizumab on *VEGF-A* gene expression in primary RPE cells. In one experiment, bevacizumab had no effect on the baseline expression levels of *VEGF-A* [59]. Similarly, in another study ranibizumab treatment did not significantly alter *VEGF-A* mRNA overexpression induced by white light illumination of RPE cells [60]. In ARPE-19 cells, both aflibercept and ranibizumab induced *VEGF-A* mRNA expression after 24 h of treatment. The authors postulated that the cells upregulated the transcription of *VEGF-A* to compensate for blocking the VEGF-A protein in the culture media [43]. Similarly, another study demonstrated that ranibizumab and bevacizumab induced overexpression of *VEGF-A* inARPE-19 cells that were subjected to oxidative stress by preincubation in a hypoxic chamber [61]. Another study showed a similar compensatory upregulation in response to bevacizumab treatment in murine RPE cells in vivo, but not in ARPE-19 cells in vitro. These investigators suggested that the compensatory upregulation of *VEGF-A* expression in complex in vivo systems might not be captured by the simpler APRE-19 cell model. It could also be that the RPE of healthy young animals was able to compensate for VEGF-A neutralization, unlike ARPE-19 cells, which carry more similarities to aged RPE [62]. Other studies showed that anti-VEGF drugs could bring down *VEGF-A* gene expression to control levels in human retinal pericytes [63] and in ARPE-19 cells [19].

VEGF-A expression levels were not affected by any of the used drugs in our experiment, although its expression was significantly lower in cells treated with a combination of bevacizumab plus TSA compared to treatment with TSA alone. It seems that the effect of anti-angiogenic drugs on *VEGF-A* gene expression varies according to cell type, the nature of biologic stress that the cells are exposed to as well as the in vivo versus in vitro environment.

4.2. Oxidative Stress Genes

Oxidative stress occurs when ROS accumulation overwhelms the capacity of the cell to detoxify them. Neural and RPE cells in the retina have a high metabolic demand and are most prone to oxidative damage with aging. Aging is associated with differential gene expression and chromatin reorganization mediated by epigenetic mechanisms, ultimately leading to impaired ability of the cells to adapt to environmental stress [64].

The main antioxidant in the retina is the superoxide dismutase (SOD) family [65]. *SOD2* gene encodes for a mitochondrial enzyme, which deactivates superoxide free radicals. Polymorphisms of *SOD2* have been associated with exudative AMD [66]. SOD2 expression depends on acetylation. Being a mitochondrial enzyme, SOD2 is directly deacetylated by SIRT3, a mitochondrial Class III HDAC, resulting in its activation [67]. Forkhead box O3a (FoxO3a) activates the promoter of *SOD2* gene inducing its expression. TSA was shown to increase the acetylation of *FoxO3a* promoter region, and upregulating its expression, as well as its target protein, SOD2, expression in vitro [68]. Treatment with bevacizumab $1\times$ in our study significantly upregulated *SOD2* expression, as did combined treatment with bevacizumab and TSA. Conversely, aflibercept significantly reduced *SOD2* expression.

Glutathione peroxidase (GPX) is another antioxidant enzyme found in the RPE and photoreceptors that protects the retina from oxidative damage. GPX expression is upregulated in AMD patients, most likely due to oxidative stress [69]. VPA induced the expression of *SOD2* and *GPX* genes in ARPE-19 cells in normal conditions and maintained their expression hypoxic conditions. A similar effect was demonstrated in rat retina [70]. In our study, only the combination of bevacizumab plus TSA was able to significantly increase *GPX3A* expression.

4.3. Inflammation Genes

Inflammation has been recently recognized as a key player in the pathogenesis of AMD. Both drusen components and intracellular lipofuscin can incite inflammasome activation and the release of the pro-inflammatory cytokines IL-18 and IL-1β in retinal tissues [71,72]. It is thought that epigenetic mechanisms are involved in initiating the immune response by altering the gene expression of immune cells to allow for cytokine production and chemotaxis [24]. HDAC6 inhibition by TST suppressed mRNA expression of *IL-1β* in an inflammation model of mammary epithelial cells in vitro [73]. We found that TSA treatment had the same effect on *IL-1β* gene expression in ARPE-19 cells. Cultures treated with ranibizumab $2\times$ and the TSA+bevacizumab combination also displayed downregulation of *IL-1β*.

Although inflammation significantly contributes to tissue damage in AMD, a certain degree of inflammation might be needed to protect against neovascularization. IL-18 has exhibited anti-angiogenic properties in tumors and post-ischemic injury and is being investigated as a potential anti-angiogenic therapy in wet AMD [74]. Shen et al. demonstrated reciprocal suppression between VEGF and IL-18. They were able to detect increased levels of IL-18 in the aqueous of patients receiving intravitreal ranibizumab injections for macular edema secondary to retinal vein occlusion. Furthermore, they found that intravitreal injection of anti-VEGF antibody in a mouse model of ischemic retinopathy upregulated mRNA expression of *IL-18*. IL-18 was able to block VEGF-induced vascular leakage and neovascularization in mice. Thus, each agent can suppress both the production and function of the other [75]. This observation could explain our findings that *IL-18* was significantly upregulated by both concentrations of ranibizumab and bevacizumab. TSA also upregulated *IL-18* expression, possibly owing to its anti-angiogenic properties. Aflibercept 2×, however, suppressed its expression. Aflibercept has a different molecular structure than both ranibizumab and bevacizumab and could have triggered another signaling mechanism that reduced *IL-18* transcription.

4.4. Apoptosis Genes

RPE cell loss characteristic of advanced AMD is thought to represent cell death by apoptosis. The BCL-2 family regulates the intrinsic mitochondrial pathway of apoptosis. Accumulation of oxidized low-density lipoproteins (LDL) in drusen and basal linear deposits, both hallmark lesions of AMD, results in upregulation of BAX, a pro-apoptotic BCL-2 family member, and downregulation of BCL-2, an anti-apoptotic member. The increased BAX/BCL-2 ratio tips the balance in favor of apoptosis [76]. Activation of pro-apoptotic BCL-2 members results in opening of mitochondrial membrane pores with subsequent release of pro-apoptotic factors, such as cytochrome c, into the cytoplasm. Cytochrome c can recruit and activate caspase-9, an initiator caspase, which activates effector caspases, such as caspase-3, that eventually cause degradation of genomic DNA and cell death [77].

Anti-VEGF drug treatment of ARPE-19 cells resulted in upregulation of the 4 pro-apoptotic genes in this study. This effect was seen with the three drugs tested at both 1× and 2× concentrations. We previously demonstrated some degree of reduced mitochondrial membrane potential (MMP), which is an early sign of apoptosis, in ARPE-19 cells after 24 h of treatment with higher-than-clinical concentrations of ranibizumab and aflibercept. Only bevacizumab decreased MMP at 1× concentration [42]. Another study found that bevacizumab significantly increased apoptosis in an ARPE-19 cell model of oxidative stress and as stress levels increased, the dose of bevacizumab capable of inducing apoptosis decreased. The authors postulated that bevacizumab blocked the protective effects of VEGF under high oxidative stress conditions and downregulated *BCL-2* gene expression [78]. Another study showed that ranibizumab could enhance the anti-proliferation effects of oxidative stress on ARPE-19 cells [42].

These results warrant further in vivo investigations since the net effect on retinal cells in vivo is subject to a complex interplay of many protective and detrimental factors. Anti-angiogenic therapy has demonstrated protective effects in our in vitro study, as evidenced by suppressing oxidative stress and inflammatory cytokine gene transcription. However, further research is warranted as concerns have been raised about the development of geographic atrophy in 98% of wet AMD patients receiving chronic anti-VEGF injections over prolonged periods of time [79].

5. Conclusions

In conclusion, we demonstrated that anti-VEGF drugs can (1) alter expression profiles for genes involved in oxidative stress, inflammation, and apoptosis pathways and (2) modulate intracellular signal transduction in addition to blocking VEGF-A. This could have implications in management of resistance or nonresponse to anti-VEGF therapy in some AMD patients. The phenomenon of individual variation in response to anti-VEGF treatment has also been observed in different cancers treated with bevacizumab [80–82]. Genetic variations may render vascular tissue more responsive or resistant to drug effects. Epigenetic mechanisms may render tissues less sensitive to the anti-VEGF treatment and influence pharmacogenetic interactions, as evidenced by miRNA regulation of enzymes involves in drug uptake and metabolism [83,84]. The fact that these drugs could influence the epigenome might guide precision medicine in the future by obtaining an "epigenetic" profile for wet AMD patients to predict resistance and direct the choice of therapy.

Another avenue we believe is worth exploring is the possibility of adding HDAC inhibitors to the therapeutic armamentarium of AMD. Epigenetic drugs have shown a great promise in immunomodulation, neuroprotection, and angiogenesis suppression [85]. To date, 6 epigenetic drugs have been approved by the FDA for cancer therapy [51]. A variety of epigenetic drugs, including DNMT and HDAC inhibitors are currently under investigation as potential therapeutic agents in AMD, owing to their ability to reverse inflammation and angiogenesis [55,86]. Specific HDAC inhibitors might be preferable to pan-inhibitors, such as TSA, as the latter can cause undesirable alterations in gene expression by inducing histone hyperacetylation [53].

Further studies including in vivo tests are required. Other retinal cell types are involved in the evolution of AMD pathogenesis, and the impact of the studied drugs on these cells needs to be explored as well. Human pluripotent stem cell (hPSC)-derived retinal organoids could provide an alternative platform to study drug interactions and intracellular signaling mechanisms that more closely approximates the retinal environment in vivo [87].

Author Contributions: Conceptualization: M.C.K. and B.D.K.; methodology: M.A.H., M.T.M., S.N., R.D.C., K.S. and S.R.A.; investigation: M.A.H., M.T.M., S.N., R.D.C., K.S. and S.R.A.; validation: M.A.H., M.T.M. and S.N.; data curation: M.A.H., M.T.M., S.N. and R.D.C.; formal analysis: M.A.H. and M.T.M.; resources: M.C.K. and B.D.K.; writing—original draft preparation: M.A.H., M.T.M. and S.N.; writing—review and editing: S.N., M.C.K. and B.D.K.; supervision: B.D.K. and M.C.K.; project administration: B.D.K. and M.C.K.; funding acquisition: B.D.K. and M.C.K. All authors have read and agreed to the published version of the manuscript.

Funding: This research was funded by Discovery Eye Foundation, Polly and Michael Smith Foundation, Iris and B. Gerald Cantor Foundation, Beckman Initiative for Macular Research, Max Factor Family Foundation, and Guenther Foundation. This study was also supported in part by an unrestricted grant from Research to Prevent Blindness, Inc. The authors acknowledge the support of the Institute for Clinical and Translational Science grant number: ICTS (ULI TR001414/TR/NCATS) at University of California, Irvine.

Institutional Review Board Statement: Not applicable.

Informed Consent Statement: Not applicable.

Data Availability Statement: All data are presented within the manuscript.

Acknowledgments: NEI grant: NEI R01 EY027363 (MCK).

Conflicts of Interest: B.D.K. is consultant to Alcon, Alimera, Allegro, Allergan, Genentech, Glaukos, GSK, Neurotech, Novagali, Novartis, Ophthotech, Pfizer, Regeneron, Santen, SecondSight, Teva, ThromboGenics. The rest of the authors declare that they have no competing interests. The funders had no role in the design of the study; in the collection, analyses, or interpretation of data; in the writing of the manuscript, or in the decision to publish the results.

References

1. Gehrs, K.M.; Anderson, D.H.; Johnson, L.V.; Hageman, G.S. Age-related macular degeneration–emerging pathogenetic and therapeutic concepts. *Ann. Med.* **2006**, *38*, 450–471. [CrossRef]
2. Wong, T.Y.; Chakravarthy, U.; Klein, R.; Mitchell, P.; Zlateva, G.; Buggage, R.; Fahrbach, K.; Probst, C.; Sledge, I. The natural history and prognosis of neovascular age-related macular degeneration: A systematic review of the literature and meta-analysis. *Ophthalmology* **2008**, *115*, 116–126. [CrossRef]

3. Amadio, M.; Govoni, S.; Pascale, A. Targeting VEGF in eye neovascularization: What's new? A comprehensive review on current therapies and oligonucleotide-based interventions under development. *Pharmacol. Res.* **2016**, *103*, 253–269. [CrossRef]
4. Fogli, S.; Del Re, M.; Rofi, E.; Posarelli, C.; Figus, M.; Danesi, R. Clinical pharmacology of intravitreal anti-VEGF drugs. *Eye* **2018**, *32*, 1010–1020. [CrossRef]
5. Grisanti, S.; Tatar, O. The role of vascular endothelial growth factor and other endogenous interplayers in age-related macular degeneration. *Prog. Retin. Eye Res.* **2008**, *27*, 372–390. [CrossRef]
6. Brown, D.M.; Kaiser, P.K.; Michels, M.; Soubrane, G.; Heier, J.S.; Kim, R.Y.; Sy, J.P.; Schneider, S. Ranibizumab versus Verteporfin for Neovascular Age-Related Macular Degeneration. *N. Engl. J. Med.* **2006**, *355*, 1432–1444. [CrossRef]
7. Chakravarthy, U.; Harding, S.P.; Rogers, C.A.; Downes, S.M.; Lotery, A.J.; Wordsworth, S.; Reeves, B.C. Ranibizumab versus Bevacizumab to Treat Neovascular Age-related Macular Degeneration: One-Year Findings from the IVAN Randomized Trial. *Ophthalmology* **2012**, *119*, 1399–1411. [CrossRef]
8. Group, T.C.R. Ranibizumab and Bevacizumab for Neovascular Age-Related Macular Degeneration. *N. Engl. J. Med.* **2011**, *364*, 1897–1908.
9. Heier, J.S.; Brown, D.M.; Chong, V.; Korobelnik, J.-F.; Kaiser, P.K.; Nguyen, Q.D.; Kirchhof, B.; Ho, A.; Ogura, Y.; Yancopoulos, G.D.; et al. Intravitreal Aflibercept (VEGF Trap-Eye) in Wet Age-related Macular Degeneration. *Ophthalmology* **2012**, *119*, 2537–2548. [CrossRef]
10. Rosenfeld, P.J.; Brown, D.M.; Heier, J.S.; Boyer, D.S.; Kaiser, P.K.; Chung, C.Y.; Kim, R.Y. Ranibizumab for Neovascular Age-Related Macular Degeneration. *N. Engl. J. Med.* **2006**, *355*, 1419–1431. [CrossRef]
11. Shah, A.R.; Williams, S.; Baumal, C.R.; Rosner, B.; Duker, J.S.; Seddon, J.M. Predictors of Response to Intravitreal Anti-Vascular Endothelial Growth Factor Treatment of Age-Related Macular Degeneration. *Am. J. Ophthalmol.* **2016**, *163*, 154–166.e158. [CrossRef] [PubMed]
12. Cascella, R.; Strafella, C.; Caputo, V.; Errichiello, V.; Zampatti, S.; Milano, F.; Potenza, S.; Mauriello, S.; Novelli, G.; Ricci, F.; et al. Towards the application of precision medicine in Age-Related Macular Degeneration. *Prog. Retin. Eye Res.* **2018**, *63*, 132–146. [CrossRef]
13. Hagstrom, S.A.; Ying, G.-S.; Pauer, G.J.T.; Sturgill-Short, G.M.; Huang, J.; Callanan, D.G.; Kim, I.K.; Klein, M.L.; Maguire, M.G.; Martin, D.F. Pharmacogenetics for Genes Associated with Age-related Macular Degeneration in the Comparison of AMD Treatments Trials (CATT). *Ophthalmology* **2013**, *120*, 593–599. [CrossRef]
14. Hagstrom, S.A.; Ying, G.; Pauer, G.T.; Sturgill-Short, G.M.; Huang, J.; Maguire, M.G.; Martin, D.F. Vegfa and vegfr2 gene polymorphisms and response to anti–vascular endothelial growth factor therapy: Comparison of age-related macular degeneration treatments trials (catt). *JAMA Ophthalmol.* **2014**, *132*, 521–527. [CrossRef] [PubMed]
15. Hagstrom, S.A.; Ying, G.S.; Maguire, M.G.; Martin, D.F.; Gibson, J.; Lotery, A.; Chakravarthy, U. VEGFR2 Gene Polymorphisms and Response to Anti-Vascular Endothelial Growth Factor Therapy in Age-Related Macular Degeneration. *Ophthalmology* **2015**, *122*, 1563–1568. [CrossRef]
16. Hagstrom, S.A.; Ying, G.S.; Pauer, G.J.; Huang, J.; Maguire, M.G.; Martin, D.F. Endothelial PAS domain-containing protein 1 (EPAS1) gene polymorphisms and response to anti-VEGF therapy in the comparison of AMD treatments trials (CATT). *Ophthalmology* **2014**, *121*, 1663–1664.e1661. [CrossRef]
17. Lotery, A.J.; Gibson, J.; Cree, A.J.; Downes, S.M.; Harding, S.P.; Rogers, C.A.; Reeves, B.C.; Ennis, S.; Chakravarthy, U. Pharmacogenetic Associations with Vascular Endothelial Growth Factor Inhibition in Participants with Neovascular Age-related Macular Degeneration in the IVAN Study. *Ophthalmology* **2013**, *120*, 2637–2643. [CrossRef] [PubMed]
18. Chen, G.; Tzekov, R.; Li, W.; Jiang, F.; Mao, S.; Tong, Y. Pharmacogenetics of Complement Factor H Y402H Polymorphism and Treatment of Neovascular AMD with Anti-VEGF Agents: A Meta-Analysis. *Sci. Rep.* **2015**, *5*, 1–9. [CrossRef]
19. Park, U.C.; Shin, J.Y.; McCarthy, L.C.; Kim, S.J.; Park, J.H.; Chung, H.; Yu, H.G. Pharmacogenetic associations with long-term response to anti-vascular endothelial growth factor treatment in neovascular AMD patients. *Mol. Vis.* **2014**, *20*, 1680–1694.
20. Fritsche, L.G.; Igl, W.; Bailey, J.N.; Grassmann, F.; Sengupta, S.; Bragg-Gresham, J.L.; Burdon, K.P.; Hebbring, S.J.; Wen, C.; Gorski, M.; et al. A large genome-wide association study of age-related macular degeneration highlights contributions of rare and common variants. *Nat. Genet.* **2016**, *48*, 134–143. [CrossRef]
21. Hjelmeland, L.M. Dark matters in AMD genetics: Epigenetics and stochasticity. *Investig. Ophthalmol. Vis. Sci.* **2011**, *52*, 1622–1631. [CrossRef]
22. Gemenetzi, M.; Lotery, A.J. The role of epigenetics in age-related macular degeneration. *Eye* **2014**, *28*, 1407–1417. [CrossRef]
23. Oliver, V.F.; Jaffe, A.E.; Song, J.; Wang, G.; Zhang, P.; Branham, K.E.; Swaroop, A.; Eberhart, C.G.; Zack, D.J.; Qian, J.; et al. Differential DNA methylation identified in the blood and retina of AMD patients. *Epigenetics* **2015**, *10*, 698–707. [CrossRef]
24. Wen, X.; Hu, X.; Miao, L.; Ge, X.; Deng, Y.; Bible, P.W.; Wei, L. Epigenetics, microbiota, and intraocular inflammation: New paradigms of immune regulation in the eye. *Prog. Retin. Eye Res.* **2018**, *64*, 84–95. [CrossRef]
25. Wu, C.; Morris, J.R. Genes, Genetics, and Epigenetics: A Correspondence. *Science* **2001**, *293*, 1103–1105. [CrossRef] [PubMed]
26. Struhl, K. Histone acetylation and transcriptional regulatory mechanisms. *Genes Dev.* **1998**, *12*, 599–606. [CrossRef] [PubMed]
27. Eckschlager, T.; Plch, J.; Stiborova, M.; Hrabeta, J. Histone Deacetylase Inhibitors as Anticancer Drugs. *Int. J. Mol. Sci.* **2017**, *18*, 1414. [CrossRef] [PubMed]
28. Resnikoff, S.; Pascolini, D.; Etya'ale, D.; Kocur, I.; Pararajasegaram, R.; Pokharel, G.P.; Mariotti, S.P. Global data on visual impairment in the year 2002. *Bull. World Health Organ.* **2004**, *82*, 844–851.

29. Klein, R.; Klein, B.E.K.; Knudtson, M.D.; Meuer, S.M.; Swift, M.; Gangnon, R.E. Fifteen-Year Cumulative Incidence of Age-Related Macular Degeneration: The Beaver Dam Eye Study. *Ophthalmology* **2007**, *114*, 253–262. [CrossRef]
30. Ranjbar, M.; Brinkmann, M.P.; Tura, A.; Rudolf, M.; Miura, Y.; Grisanti, S. Ranibizumab interacts with the VEGF-A/VEGFR-2 signaling pathway in human RPE cells at different levels. *Cytokine* **2016**, *83*, 210–216. [CrossRef]
31. Mahendra, C.K.; Tan, L.T.H.; Pusparajah, P.; Htar, T.T.; Chuah, L.H.; Lee, V.S.; Low, L.E.; Tang, S.Y.; Chan, K.G.; Goh, B.H. Detrimental Effects of UVB on Retinal Pigment Epithelial Cells and Its Role in Age-Related Macular Degeneration. *Oxid Med. Cell Longev.* **2020**, *2020*, 1904178. [CrossRef]
32. Moustafa, M.T.; Ramirez, C.; Schneider, K.; Atilano, S.R.; Limb, G.A.; Kuppermann, B.D.; Kenney, M.C. Protective Effects of Memantine on Hydroquinone-Treated Human Retinal Pigment Epithelium Cells and Human Retinal Müller Cells. *J. Ocul. Pharmacol. Ther.* **2017**, *33*, 610–619. [CrossRef] [PubMed]
33. Dunn, K.; Aotaki-Keen, A.; Putkey, F.; Hjelmeland, L. ARPE-19, A Human Retinal Pigment Epithelial Cell Line with Differentiated Properties. *Exp. Eye Res.* **1996**, *62*, 155–170. [CrossRef] [PubMed]
34. Pennington, K.L.; DeAngelis, M.M. Epigenetic Mechanisms of the Aging Human Retina. *J. Exp. Neurosci.* **2015**, *9*, 51–79. [CrossRef] [PubMed]
35. Sobrin, L.; Seddon, J.M. Nature and nurture- genes and environment- predict onset and progression of macular degeneration. *Prog. Retin. Eye Res.* **2014**, *40*, 1–15. [CrossRef]
36. Stewart, M.W. Extended Duration Vascular Endothelial Growth Factor Inhibition in the Eye: Failures, Successes, and Future Possibilities. *Pharmaceutics* **2018**, *10*, 21. [CrossRef] [PubMed]
37. Ferrara, N.; Damico, L.; Shams, N.; Lowman, H.; Kim, R. Development of ranibizumab, an anti-vascular endothelial growth factor antigen binding fragment, as therapy for neovascular age-related macular degeneration. *Retina* **2006**, *26*, 859–870. [CrossRef] [PubMed]
38. Kramer, I.; Lipp, H.P. Bevacizumab, a humanized anti-angiogenic monoclonal antibody for the treatment of colorectal cancer. *J. Clin. Pharm. Ther.* **2007**, *32*, 1–14. [CrossRef]
39. Ohr, M.; Kaiser, P.K. Aflibercept in wet age-related macular degeneration: A perspective review. *Ther. Adv. Chronic Dis.* **2012**, *3*, 153–161. [CrossRef]
40. Avery, R.L.; Castellarin, A.A.; Steinle, N.C.; Dhoot, D.S.; Pieramici, D.J.; See, R.; Couvillion, S.; Nasir, M.A.; Rabena, M.D.; Le, K.; et al. Systemic pharmacokinetics following intravitreal injections of ranibizumab, bevacizumab or aflibercept in patients with neovascular AMD. *Br. J. Ophthalmol.* **2014**, *98*, 1636–1641. [CrossRef]
41. Puddu, A.; Sanguineti, R.; Traverso, C.E.; Viviani, G.L.; Nicolo, M. Response to anti-VEGF-A treatment of retinal pigment epithelial cells in vitro. *Eur. J. Ophthalmol.* **2016**, *26*, 425–430. [CrossRef]
42. Luthra, S.; Sharma, A.; Dong, J.; Neekhra, A.; Gramajo, A.L.; Seigel, G.M.; Kenney, M.C.; Kuppermann, B.D. Effect of bevacizumab (Avastin (TM)) on mitochondrial function of in vitro retinal pigment epithelial, neurosensory retinal and microvascular endothelial cells. *Indian J. Ophthalmol.* **2013**, *61*, 705–710.
43. Malik, D.; Tarek, M.; Caceres del Carpio, J.; Ramirez, C.; Boyer, D.; Kenney, M.C.; Kuppermann, B.D. Safety profiles of anti-VEGF drugs: Bevacizumab, ranibizumab, aflibercept and ziv-aflibercept on human retinal pigment epithelium cells in culture. *Br. J. Ophthalmol.* **2014**, *98* (Suppl. S1), i11–i16. [CrossRef]
44. Ammar, D.A.; Mandava, N.; Kahook, M.Y. The effects of aflibercept on the viability and metabolism of ocular cells in vitro. *Retina* **2013**, *33*, 1056–1061. [CrossRef]
45. Schnichels, S.; Hagemann, U.; Januschowski, K.; Hofmann, J.; Bartz-Schmidt, K.U.; Szurman, P.; Spitzer, M.S.; Aisenbrey, S. Comparative toxicity and proliferation testing of aflibercept, bevacizumab and ranibizumab on different ocular cells. *Br. J. Ophthalmol.* **2013**, *97*, 917–923. [CrossRef] [PubMed]
46. Dinc, E.; Ayaz, L.; Kurt, A.H. Effects of Bevacizumab, Ranibizumab, and Aflibercept on MicroRNA Expression in a Retinal Pigment Epithelium Cell Culture Model of Oxidative Stress. *J. Ocul. Pharmacol. Ther.* **2018**, *34*, 346–353. [CrossRef]
47. Masoud, G.N.; Li, W. HIF-1alpha pathway: Role, regulation and intervention for cancer therapy. *Acta Pharm. Sin. B* **2015**, *5*, 378–389. [CrossRef]
48. Dai, Y.; Dong, F.; Zhang, X.; Yang, Z. Internal limiting membrane transplantation for unclosed and large macular holes. *Graefes Arch. Clin. Ex. P Ophthalmol.* **2016**, *254*, 2095–2099. [CrossRef] [PubMed]
49. Elmasry, K.; Mohamed, R.; Sharma, I.; Elsherbiny, N.M.; Liu, Y.; Al-Shabrawey, M.; Tawfik, A. Epigenetic modifications in hyperhomocysteinemia: Potential role in diabetic retinopathy and age-related macular degeneration. *Oncotarget* **2018**, *9*, 12562–12590. [CrossRef]
50. Glozak, M.A.; Sengupta, N.; Zhang, X.; Seto, E. Acetylation and deacetylation of non-histone proteins. *Gene* **2005**, *363*, 15–23. [CrossRef]
51. Berndsen, R.H.; Abdul, U.K.; Weiss, A.; Zoetemelk, M.; Te Winkel, M.T.; Dyson, P.J.; Griffioen, A.W.; Nowak-Sliwinska, P. Epigenetic approach for angiostatic therapy: Promising combinations for cancer treatment. *Angiogenesis* **2017**, *20*, 245–267. [CrossRef] [PubMed]
52. Turtoi, A.; Peixoto, P.; Castronovo, V.; Bellahcene, A. Histone deacetylases and cancer-associated angiogenesis: Current understanding of the biology and clinical perspectives. *Crit. Rev. Oncog.* **2015**, *20*, 119–137.
53. Kothary, P.C.; Rossi, B.; Del Monte, M.A. Valproic Acid Induced Human Retinal Pigment Epithelial Cell Death as Well as its Survival after Hydrogen Peroxide Damage is Mediated by P38 Kinase. *Adv. Exp. Med. Biol.* **2016**, *854*, 765–772. [PubMed]

54. Leyk, J.; Daly, C.; Janssen-Bienhold, U.; Kennedy, B.N.; Richter-Landsberg, C. HDAC6 inhibition by tubastatin A is protective against oxidative stress in a photoreceptor cell line and restores visual function in a zebrafish model of inherited blindness. *Cell Death Dis.* **2017**, *8*, e3028. [CrossRef]

55. Chan, N.; He, S.; Spee, C.K.; Ishikawa, K.; Hinton, D.R. Attenuation of choroidal neovascularization by histone deacetylase inhibitor. *PLoS ONE* **2015**, *10*, e0120587. [CrossRef]

56. Kwa, F.A.; Thrimawithana, T.R. Epigenetic modifications as potential therapeutic targets in age-related macular degeneration and diabetic retinopathy. *Drug. Discov. Today* **2014**, *19*, 1387–1393. [CrossRef]

57. Mottet, D.; Bellahcene, A.; Pirotte, S.; Waltregny, D.; Deroanne, C.; Lamour, V.; Lidereau, R.; Castronovo, V. Histone deacetylase 7 silencing alters endothelial cell migration, a key step in angiogenesis. *Circ. Res.* **2007**, *101*, 1237–1246. [CrossRef]

58. Potente, M.; Ghaeni, L.; Baldessari, D.; Mostoslavsky, R.; Rossig, L.; Dequiedt, F.; Haendeler, J.; Mione, M.; Dejana, E.; Alt, F.W.; et al. SIRT1 controls endothelial angiogenic functions during vascular growth. *Genes Dev.* **2007**, *21*, 2644–2658. [CrossRef]

59. Subramani, M.; Murugeswari, P.; Dhamodaran, K.; Chevour, P.; Gunasekaran, S.; Kumar, R.S.; Jayadev, C.; Shetty, R.; Begum, N.; Das, D. Short Pulse of Clinical Concentration of Bevacizumab Modulates Human Retinal Pigment Epithelial Functionality. *Investig. Ophthalmol. Vis. Sci.* **2016**, *57*, 1140–1152. [CrossRef]

60. Kernt, M.; Thiele, S.; Neubauer, A.S.; Koenig, S.; Hirneiss, C.; Haritoglou, C.; Ulbig, M.W.; Kampik, A. Inhibitory activity of ranibizumab, sorafenib, and pazopanib on light-induced overexpression of platelet-derived growth factor and vascular endothelial growth factor A and the vascular endothelial growth factor A receptors 1 and 2 and neuropilin 1 and 2. *Retina* **2012**, *32*, 1652–1663. [CrossRef]

61. Golan, S.; Entin-Meer, M.; Semo, Y.; Maysel-Auslender, S.; Mezad-Koursh, D.; Keren, G.; Loewenstein, A.; Barak, A. Gene profiling of human VEGF signaling pathways in human endothelial and retinal pigment epithelial cells after anti VEGF treatment. *BMC Res. Notes* **2014**, *7*, 617. [CrossRef]

62. Ford, K.M.; Saint-Geniez, M.; Walshe, T.; Zahr, A.; D'Amore, P.A. Expression and role of VEGF in the adult retinal pigment epithelium. *Investig. Ophthalmol. Vis. Sci.* **2011**, *52*, 9478–9487. [CrossRef]

63. Giurdanella, G.; Anfuso, C.D.; Olivieri, M.; Lupo, G.; Caporarello, N.; Eandi, C.M.; Drago, F.; Bucolo, C.; Salomone, S. Aflibercept, bevacizumab and ranibizumab prevent glucose-induced damage in human retinal pericytes in vitro, through a PLA2/COX-2/VEGF-A pathway. *Biochem. Pharmacol.* **2015**, *96*, 278–287. [CrossRef]

64. Corso-Diaz, X.; Jaeger, C.; Chaitankar, V.; Swaroop, A. Epigenetic control of gene regulation during development and disease: A view from the retina. *Prog. Retin Eye Res.* **2018**, *65*, 1–27. [CrossRef] [PubMed]

65. Pujol-Lereis, L.M.; Schäfer, N.; Kuhn, L.B.; Rohrer, B.; Pauly, D. Interrelation Between Oxidative Stress and Complement Activation in Models of Age-Related Macular Degeneration. *Adv. Exp. Med. Biol.* **2016**, *854*, 87–93. [PubMed]

66. Kan, M.; Liu, F.; Weng, X.; Ye, J.; Wang, T.; Xu, M.; He, L.; Liu, Y. Association study of newly identified age-related macular degeneration susceptible loci SOD2, MBP, and C8orf42 in Han Chinese population. *Diagn. Pathol.* **2014**, *9*, 1–4. [CrossRef]

67. Gao, J.; Zheng, Z.; Gu, Q.; Chen, X.; Liu, X.; Xu, X. Deacetylation of MnSOD by PARP-regulated SIRT3 protects retinal capillary endothelial cells from hyperglycemia-induced damage. *Biochem. Biophys. Res. Commun.* **2016**, *472*, 425–431. [CrossRef] [PubMed]

68. Guo, Y.; Li, Z.; Shi, C.; Li, J.; Yao, M.; Chen, X. Trichostatin A attenuates oxidative stress-mediated myocardial injury through the FoxO3a signaling pathway. *Int. J. Mol. Med.* **2017**, *40*, 999–1008. [CrossRef]

69. Zhao, B.; Wang, M.; Xu, J.; Li, M.; Yu, Y. Identification of pathogenic genes and upstream regulators in age-related macular degeneration. *BMC Ophthalmol.* **2017**, *17*, 102. [CrossRef]

70. Tokarz, P.; Kaarniranta, K.; Blasiak, J. Inhibition of DNA methyltransferase or histone deacetylase protects retinal pigment epithelial cells from DNA damage induced by oxidative stress by the stimulation of antioxidant enzymes. *Eur. J. Pharm.* **2016**, *776*, 167–175. [CrossRef]

71. Brandstetter, C.; Mohr, L.K.M.; Latz, E.; Holz, F.G.; Krohne, T.U. Light induces NLRP3 inflammasome activation in retinal pigment epithelial cells via lipofuscin-mediated photooxidative damage. *J. Mol. Med.* **2015**, *93*, 905–916. [CrossRef]

72. Doyle, S.L.; Campbell, M.; Ozaki, E.; Salomon, R.G.; Mori, A.; Kenna, P.F.; Farrar, G.J.; Kiang, A.-S.; Humphries, M.M.; Lavelle, E.C.; et al. NLRP3 has a protective role in age-related macular degeneration through the induction of IL-18 by drusen components. *Nat. Med.* **2012**, *18*, 791–798. [CrossRef]

73. Wang, J.; Zhao, L.; Wei, Z.; Zhang, X.; Wang, Y.; Li, F.; Fu, Y.; Liu, B. Inhibition of histone deacetylase reduces lipopolysaccharide-induced-inflammation in primary mammary epithelial cells by regulating ROS-NF- B signaling pathways. *Int. Immunopharmacol.* **2018**, *56*, 230–234. [CrossRef] [PubMed]

74. Campbell, M.; Doyle, S.; Humphries, P. IL-18: A new player in immunotherapy for age-related macular degeneration? *Expert Rev. Clin. Immunol.* **2014**, *10*, 1273–1275. [CrossRef]

75. Shen, J.; Choy, D.F.; Yoshida, T.; Iwase, T.; Hafiz, G.; Xie, B.; Hackett, S.F.; Arron, J.R.; Campochiaro, P.A. Interleukin-18 has antipermeablity and antiangiogenic activities in the eye: Reciprocal suppression with VEGF. *J. Cell Physiol.* **2014**, *229*, 974–983. [CrossRef]

76. Yating, Q.; Yuan, Y.; Wei, Z.; Qing, G.; Xingwei, W.; Qiu, Q.; Lili, Y. Oxidized LDL induces apoptosis of human retinal pigment epithelium through activation of ERK-Bax/Bcl-2 signaling pathways. *Curr. Eye Res.* **2015**, *40*, 415–422. [CrossRef]

77. Hanus, J.; Anderson, C.; Wang, S. RPE necroptosis in response to oxidative stress and in AMD. *Ageing Res. Rev.* **2015**, *24Pt. B*, 286–298. [CrossRef]

78. Kim, S.; Kim, Y.J.; Kim, N.R.; Chin, H.S. Effects of Bevacizumab on Bcl-2 Expression and Apoptosis in Retinal Pigment Epithelial Cells under Oxidative Stress. *Korean J. Ophthalmol.* **2015**, *29*, 424–432. [CrossRef]

79. Rofagha, S.; Bhisitkul, R.B.; Boyer, D.S.; Sadda, S.R.; Zhang, K. SEVEN-UP Study Group. Seven-year outcomes in ranibizumab-treated patients in ANCHOR, MARINA, and HORIZON: A multicenter cohort study (SEVEN-UP). *Ophthalmology* **2013**, *120*, 2292–2299. [CrossRef] [PubMed]

80. Hurwitz, H.; Fehrenbacher, L.; Novotny, W.; Cartwright, T.; Hainsworth, J.; Heim, W.; Berlin, J.; Baron, A.; Griffing, S.; Holmgren, E.; et al. Bevacizumab plus irinotecan, fluorouracil, and leucovorin for metastatic colorectal cancer. *N. Engl. J. Med.* **2004**, *350*, 2335–2342. [CrossRef]

81. Miller, K.D.; Chap, L.I.; Holmes, F.A.; Cobleigh, M.A.; Marcom, P.K.; Fehrenbacher, L.; Dickler, M.; Overmoyer, B.A.; Reimann, J.D.; Sing, A.P.; et al. Randomized phase III trial of capecitabine compared with bevacizumab plus capecitabine in patients with previously treated metastatic breast cancer. *J. Clin. Oncol.* **2005**, *23*, 792–799. [CrossRef]

82. Sandler, A.; Gray, R.; Perry, M.C.; Brahmer, J.; Schiller, J.H.; Dowlati, A.; Lilenbaum, R.; Johnson, D.H. Paclitaxel-carboplatin alone or with bevacizumab for non-small-cell lung cancer. *N. Engl. J. Med.* **2006**, *355*, 2542–2550, Erratum in: *N. Engl. J. Med.* **2007**, *356*, 318. [CrossRef]

83. Buysschaert, I.; Schmidt, T.; Roncal, C.; Carmeliet, P.; Lambrechts, D. Genetics, epigenetics and pharmaco-(epi)genomics in angiogenesis. *J. Cell Mol. Med.* **2008**, *12*, 2533–2551. [CrossRef]

84. Yu, A.-M.; Tian, Y.; Tu, M.-J.; Ho, P.Y.; Jilek, J.L. MicroRNA Pharmacoepigenetics: Posttranscriptional Regulation Mechanisms behind Variable Drug Disposition and Strategy to Develop More Effective Therapy. *Drug. Metab. Dispos.* **2016**, *44*, 308–319. [CrossRef]

85. Ahmadi, M.; Gharibi, T.; Dolati, S.; Rostamzadeh, D.; Aslani, S.; Baradaran, B.; Younesi, V.; Yousefi, M. Epigenetic modifications and epigenetic based medication implementations of autoimmune diseases. *Biomed. Pharmacother.* **2017**, *87*, 596–608. [CrossRef]

86. Christen, W.G.; Glynn, R.J.; Chew, E.Y.; Albert, C.M.; Manson, J.E. Folic Acid, Vitamin B6, and Vitamin B12 in Combination and Age-Related Cataract in a Randomized Trial of Women. *Ophthalmic. Epidemiol.* **2016**, *23*, 32–39. [CrossRef] [PubMed]

87. Browne, A.W.; Arnesano, C.; Harutyunyan, N.; Khuu, T.; Martinez, J.C.; Pollack, H.A.; Koos, D.S.; Lee, T.C.; Fraser, S.E.; Moats, R.A.; et al. Structural and Functional Characterization of Human Stem-Cell-Derived Retinal Organoids by Live Imaging. *Investig. Ophthalmol. Vis. Sci.* **2017**, *58*, 3311–3318. [PubMed]

Review

Mesenchymal Stem Cell-Based Therapy for Retinal Degenerative Diseases: Experimental Models and Clinical Trials

Vladimir Holan [1,2,*], Katerina Palacka [1,2] and Barbora Hermankova [1]

[1] Department of Nanotoxicology and Molecular Epidemiology, Institute of Experimental Medicine of the Czech Academy of Sciences, 14220 Prague, Czech Republic; katerina.palacka@iem.cas.cz (K.P.); barbora.hermankova@iem.cas.cz (B.H.)

[2] Department of Cell Biology, Faculty of Science, Charles University, 12843 Prague, Czech Republic

* Correspondence: vladimir.holan@iem.cas.cz

Abstract: Retinal degenerative diseases, such as age-related macular degeneration, retinitis pigmentosa, diabetic retinopathy or glaucoma, represent the main causes of a decreased quality of vision or even blindness worldwide. However, despite considerable efforts, the treatment possibilities for these disorders remain very limited. A perspective is offered by cell therapy using mesenchymal stem cells (MSCs). These cells can be obtained from the bone marrow or adipose tissue of a particular patient, expanded in vitro and used as the autologous cells. MSCs possess potent immunoregulatory properties and can inhibit a harmful inflammatory reaction in the diseased retina. By the production of numerous growth and neurotrophic factors, they support the survival and growth of retinal cells. In addition, MSCs can protect retinal cells by antiapoptotic properties and could contribute to the regeneration of the diseased retina by their ability to differentiate into various cell types, including the cells of the retina. All of these properties indicate the potential of MSCs for the therapy of diseased retinas. This view is supported by the recent results of numerous experimental studies in different preclinical models. Here we provide an overview of the therapeutic properties of MSCs, and their use in experimental models of retinal diseases and in clinical trials.

Keywords: retinal degenerative diseases; mesenchymal stem cells; stem cell therapy; experimental models; clinical trials

Citation: Holan, V.; Palacka, K.; Hermankova, B. Mesenchymal Stem Cell-Based Therapy for Retinal Degenerative Diseases: Experimental Models and Clinical Trials. *Cells* **2021**, *10*, 588. https://doi.org/10.3390/cells10030588

Academic Editor: Maurice Ptito

Received: 3 February 2021
Accepted: 2 March 2021
Published: 7 March 2021

Publisher's Note: MDPI stays neutral with regard to jurisdictional claims in published maps and institutional affiliations.

1. Introduction

The retina is a highly specialized structure composed of several layers of morphologically and functionally different cell types. The individual layers are mutually interconnected and their primary function is to capture a light signal via the photoreceptors and to convert it into electrical impulses. These impulses are relayed to ganglion cells and then pass through the optic nerve into the visual cortex of the brain.

Individual retinal layers have an irreplaceable role in the capture and transduction of light signals. A disease or damage to any particular cell layer has a negative impact on the surrounding cell types and is reflected by the impairment of the vision. The progression of the retinal damage results in the development of retinal degenerative disorders. Although the exact etiology, causes and starting mechanisms of these diseases are mostly unknown, many factors, such as oxidative stress, light-induced damage, chemical insults, vascular defects, cytokine imbalance, damage of blood–retinal barrier and infiltration with immune cells or aging, have been suggested to contribute to the development of retinal degeneration [1–3]. Irrespectively of the different etiologies and various causes of retinal disorders, cumulative damage and loss of retinal cells, chronic inflammation, immune cell infiltration and enhanced cytokine secretion by immune and retinal cells represent the main pathological signs of retinal degenerative diseases, which represent the leading cause of blindness worldwide.

2. Retinal Degenerative Diseases

Retinal diseases are a heterogenous and multifactorial group of light-threatening disorders, which include age-related macular degeneration (AMD), retinitis pigmentosa (RP), diabetic retinopathy (DR), pediatric Stargardt's macular dystrophy, glaucoma and many other similar forms. Although retinal diseases have various causes and different etiologies, a common characteristic is the death or dying of the specialized retinal cells and the loss of integrity of the retina or the degeneration of the photoreceptors; and this process then results in a visual impairment and ultimately in blindness.

In several of these disorders, including AMD, the earliest changes observed are caused by a loss of the cells of the retinal pigment epithelium (RPE), which play a major role in photoreceptor nutrition and in the maintenance of homeostasis. The degeneration of RPE and photoreceptors is the main cause of AMD which may have its onset in choroidal neovascularization or in accumulation of amorphous deposits. Both these causes lead to alterations in the retina and to the impairment of its functions. Early AMD is characterized by the appearance of soft drusen and pigmentary changes in the RPE, which can progress into two forms of advanced AMD—dry and wet AMD. Both of these forms result in the loss of central vision [1]. AMD is a leading cause of vision loss, affecting tens of millions of elderly people worldwide. Similarly, in DR, which is primarily caused by hyperglycemia in diabetes mellitus, the reduction in the number of pericytes at the vascular level, and the decreased number of retinal neurons and glial cells result in the interruption of retinal integrity and a progressive loss of vision. Nearly all patients with type I diabetes and over 60% of patients with type II diabetes have some degree of retinopathy after 20 years duration. DR is a leading cause of blindness in developed countries. RP is a genetic disorder of the eye which is caused by a progressive loss of the rod photoreceptor cells on the back of the eye [4]. Similarly, Stargardt's macular dystrophy is also an inherited retinal disease that begins in childhood or adolescence and that affects the macula. On the contrary, the main risk for a group of eye diseases called glaucoma is increasing age and high pressure in the eye. The mechanism of glaucoma is believed to be a slow exit of the aqueous humor through the trabecular meshwork, which results in damage to the optic nerve and causes vision loss. In addition to these degenerative processes associated with the loss of the specialized retinal cells, local inflammation significantly contributes to the triggering and development of retinal diseases. Among the main contributors to inflammation belong the various types of infiltrating immune cells and activated microglia. Increased numbers of glial cells have been observed in the retina with the degeneration of the photoreceptors [5]. Activated microglia can contribute to the production of proinflammatory factors and to the damage of the hematoretinal barrier [6]. Furthermore, different populations of glial cells expressing genes associated with AMD (such as *VEGFA* and *HTRA1*), have been identified in the retina with this type of disease [7]. Alternatively, it has been shown that the inhibition of the microglia delays retinal degeneration in the experimental retinal vein occlusion in mice [8].

To date, the treatment options for retinal diseases have been very limited. In the advanced stages, laser photocoagulation remains the main method of treatment for DR. Other therapeutic approaches are represented by a vitrectomy or different microsurgery interventions, which involve complicated surgery and are highly invasive procedures. Recently, less invasive treatment of some forms of retinal degenerative diseases has been based on the administration of inhibitors of vascular endothelial growth factor (VEGF) or other drugs. However, these inhibitors only induce short-term effects and just slow down the progression of the disease. Therefore, the need for a safe and less invasive approach to prevent development and to treat these sight-threatening manifestations of retinal diseases is vital.

3. Perspectives of Cell Therapy for Retinal Diseases

Since the loss of specialized retinal cells and local inflammatory reactions are the main causes contributing to the progression of retinal degenerative diseases, the inhibition of inflammation and a support for the surviving retinal cells appear to be prospective approaches to manage these diseases. Recent studies have indicated that various types of stem cells could contribute, by paracrine effects, to the support of the survival of the residual retinal cells, and to the inhibition of inflammation [9]. A therapeutic possibility is offered by embryonic stem cells (ESCs), which can be isolated from blastocysts and which have a high differentiation potential. Another possibility is represented by the induced pluripotent stem cells (iPSCs), prepared by the reprogramming of normal adult fibroblasts or other cells. Both ESCs and iPSCs have the potential for differentiation into various retinal cell types [10,11]. However, the use of ESCs or iPSCs is limited by the possibility of immune rejection, teratogenicity and ethical restrictions in the case of ESCs. For these reasons, mesenchymal stem cells (MSCs) show great potential and could be a prospective tool for the treatment of retinal diseases. MSCs can be obtained from the bone marrow or adipose tissue of a particular patient and after separation and culturing in vitro could be used as autologous cells without the danger of immune rejection. It has been shown that after an injection of MSCs into the vitreous body, the cells can survive for a long period of time and can protect retinal ganglion cell survival or stimulate axon regeneration after optic nerve crush [12,13].

MSCs are multipotent stem cells which can be obtained relatively easily in a sufficient amount from various types of tissues and expanded in vitro for autologous application. It has been shown that MSCs retain their differentiation potential during their in vitro expansion, and that they can be differentiated into different cell types including cells expressing RPE or photoreceptor cell markers [14–16]. Similarly, the anti-inflammatory properties of MSCs [17,18] and their ability to support ocular surface healing [19–23] have been well documented. An advantage of MSCs is also their safety in use. The experiments in animal models confirmed that the subcutaneous administration of MSCs did not induce tumor growth during several months of observation [24]. Similarly, an extensive meta-analysis of studies using MSCs in over 1000 patients did not reveal a significant association between MSC treatment and the toxicity of infusions, internal organ infection, cancer or death [25].

4. Mesenchymal Stem Cells

4.1. Characteristics of MSCs

MSCs currently represent the most frequently studied type of adult stem cells. Originally, these cells were described as a population of bone marrow-derived cells that adhere to plastic and form fibrocyte-like colonies [26]. They have differentiation potential, which they retain during their in vitro expansion, as was demonstrated by their differentiation into other cell types of the mesenchymal cell line [27,28]. For therapeutic purposes, MSCs are mainly isolated from the bone marrow or adipose tissue. However, no specific marker that could characterize these cells has been identified. According to the International Society of Cellular Therapy, human MSCs are characterized by the ability to adhere to plastic surfaces in standard culture conditions, by being positive for the surface markers CD105, CD73 and CD90 and negative for hematopoietic markers CD45, CD34, CD14, CD19 and CD11b, and by their ability to differentiate into adipocytes, chondroblasts and osteoblasts [29]. It has been shown that MSCs possess potent immunmodulatory and anti-inflammatory properties, produce a number of cytokines and growth factors, and contribute to tissue healing and regeneration. The great advantage of these cells is their relatively easy isolation from the bone marrow or adipose tissue, good growth properties during their propagation in vitro and the possibility to use them as autologous (patient´s own) cells. It has also been demonstrated that MSCs from different sources (bone marrow, adipose tissue, umbilical cord blood, etc.) have similar function properties [30–32]. All these characteristics make them a promising candidate for the cell therapy of inflammatory and degenerative diseases.

4.2. Immunoregulatory and Anti-Inflammatory Properties of MSCs

The immunomodulatory properties of MSCs are mediated by multiple mechanisms including regulation by direct cell-to-cell contact, the production of various immunomodulatory molecules, the negative effects on antigen-presenting cells or the activation of regulatory T cells (Tregs). The complexity of the immunoregulatory effects of MSCs is also evident from the observation that MSCs inhibited lymphocyte proliferation induced by mitogens and alloantigens by different mechanisms [33]. In general, MSCs have potent immunosuppressive properties. It has been shown that MSCs inhibit T and B cell proliferation, the production of cytokines and activity of cytotoxic T and NK cells [17,34]. In in vivo experimental models, the administration of MSCs prolonged the survival of skin allografts in baboons [35] and mice [36], prevented the rejection of corneal allografts [37,38], decreased the incidence of graft-versus-host disease in mice and humans [39,40] attenuated septic complications [41] and suppressed the incidence and severity of autoimmune diseases [42,43]. These suppressive effects of MSCs can be mediated by multiple mechanisms. It has been shown that MSCs express numerous molecules contributing to the immunosuppression, such as indoleamine 2,3-deoxygenase (IDO), cyclooxygenase-2 (Cox-2), TNF-α stimulated gene 6 protein (TSG-6), programmed death-ligand 1 (PDL-1) or Fas-L molecule [38,44–46]. Furthermore, MSCs produce a number of cytokines which can negatively influence in immune reaction. It has been shown that MSCs produce transforming growth factor-β (TGF-β) and interleukin-6 (IL-6) which are the principal cytokines regulating the development of anti-inflammatory Tregs and proinflammatory Th17 cells [47,48]. The spectrum of cytokines produced by MSCs depends on the state of their activation. We have demonstrated that a cytokine environment, where MSCs reside, considerably influences their secretory and immunoregulatory potential [49]. The beneficial effects of MSCs after their systemic application in vivo are supported by the demonstration of their ability to migrate to the site of injury or inflammation and to contribute to tissue healing and regeneration [50–52]. In this respect we showed that mouse bone marrow-derived MSCs (BM-MSCs) administered intravenously migrated in a significantly higher number to the injured eye than into the contralateral healthy eye [53], and that adipose tissue-derived MSCs (A-MSCs) delivered intraperitoneally into transplanted mice were detected in a significantly higher amount in skin allografts than in healthy skin [54]. It has been suggested that the cytokines and chemokines produced by immune cells in the site of an injury attract MSCs to migrate to the damage site, where they participate in the attenuation of inflammation [55,56].

4.3. Antiapoptotic Properties of MSCs

Degenerative and inflammatory reactions in the diseased retina are regularly associated with a locally enhanced production of a variety of cytokines. These molecules can be produced either by inflammatory immune cells or by the activated cells of the retina [57]. It has been shown in vitro and in vivo that increased levels of proinflammatory cytokines can induce apoptosis of the surrounding cells [58,59]. Moreover, chronic inflammation is associated with endoplasmic reticulum stress, which also promotes the induction of apoptosis [60]. Furthermore, proinflammatory cytokines induce changes in the expression of various genes (such as *Bcl-2*, *Bax*, *p53*) associated with apoptosis. Any damage in the retina attracts the cells of the immune system which produce chemokines and cytokines, and thus potentiate inflammatory and apoptotic reactions. Therefore, the inhibition of a local inflammatory reaction and attenuation of apoptosis might be promising approaches to alleviate and inhibit the development of retinal injury. In this respect, MSCs by their immunoregulatory, anti-inflammatory and antiapoptotic properties could also be a promising therapeutic tool for developing retinal disorders [18,61]. We have recently shown that MSCs inhibit the expression of proapoptotic genes and decrease the number of apoptotic cells in the corneal explants cultured in the presence of apoptosis-inducing proinflammatory cytokines [62].

4.4. The Production of Growth Factors by MSCs

MSCs are potent producers of various growth and trophic factors. Some of these factors are produced by MSCs constitutively, while others are only secreted after activation with proinflammatory cytokines, mitogens or other signals. It has been suggested that the production of growth factors and their paracrine action are the main mechanisms of the therapeutic action of MSCs. Among the growth factors which are produced by

MSCs and that could contribute to retinal regeneration are hepatocyte growth factor (HGF), nerve growth factor (NGF), glial cell-derived neurotrophic factor (GDNF), insulin-like growth factor-1 (IGF-1), pigment epithelium growth factor (PEGF), fibrocyte growth factor (FGF), platelet-derived growth factor (PDGF), epidermal growth factor (EGF), angiopoietin-1, erythropoietin, VEGF and TGF-β [16,57,63–66]. Some of these factors are secreted spontaneously by untreated MSCs, and their production is enhanced after stimulation with proinflammatory cytokines [16]. On the contrary, the production of some other cytokines which are produced spontaneously (such as TGF-β, HGF) is significantly decreased in the presence of proinflammatory cytokines [16]. We also showed that higher levels of some growth factors are produced by MSCs after their differentiation into cells expressing retinal cell markers [16]. The expression of genes *Ngf*, *Gdnf* and *Il-6* was enhanced in differentiated MSCs, which suggests a higher potential of differentiated MSCs for the regeneration of diseased retinal tissue. It was demonstrated that the supernatants from light-injured retina significantly promote the secretion of neurotrophic factors by MSCs and slow down the process of apoptosis in damaged retinal cells [67]. Another study showed that the secretion of neurotrophic factors by MSCs promoted the viability of photoreceptors in vitro, and supported their survival after the subretinal transplantation of MSCs in a retinal degeneration model [68]. All these observations indicate that MSCs differentiated into cells with characteristics of retinal cells have a higher secretory activity than untreated MSCs, and could have a better regenerative potential than primary MSCs.

4.5. The Ability of MSCs to Differentiate into Cells with Retinal Cell Characteristics

One of the characteristics of stem cells is the ability to differentiate or even transdifferentiate into different cell types. With regards to differentiation into ocular cells, relatively extensive data exist about the differentiation of MSCs and other stem cells into cornea-like cells [69–72], but the data are less abundant on the differentiation of MSCs into neurons [73] or various types of retinal cells [74,75].

The ability of different types of stem cells and MSCs to differentiate into retinal cells has been reviewed by Salehi et al. [76]. For example, MSCs isolated from rat conjunctiva and cultured in the presence of taurine expressed markers characteristic of photoreceptors and bipolar cells [75]. Taurine, together with activin A and EGF, has been used in other studies to differentiate MSCs to photoreceptors. The cells cultured in differentiation conditions for 8–10 days expressed the *Rho* and *Rlbp* genes [74]. The same authors also showed that MSCs injected into the subretinal space are able to integrate into the retina and express markers specific for photoreceptors. Other studies demonstrated that the transplantation of MSCs into the damaged retina induced the expression of markers typical for photoreceptors, bipolar and amacrine cells in grafted MSCs [77–79]. Several other studies also showed the differentiation of MSCs into RPE cells [79,80], which play an important role in the nourishment of photoreceptors.

In our study, to simulate the environment of the damaged retina, we cultured mouse BM-MSCs with the retinal cell extract and with supernatant from Concanavalin A-stimulated mouse spleen cells. MSCs cultured for 7 days in such conditions differentiated to cells expressing retinal cell markers such as rhodopsin, S antigen, retinaldehyde binding protein, calbindin 2, recoverin and retinal pigment epithelium 65 [16]. Interferon-γ, present in the supernatant from activated spleen cells was identified as the main factor supporting the retinal differentiation of MSCs. In addition, the differentiated MSCs produced a number of neurotrophic factors which are important for retinal regeneration. This study, and the results of other authors [78–80], indicate that the signals from the damaged retina induce the differentiation of MSCs into cells expressing retinal cell markers, and that the MSC differentiation is supported by cytokines produced by activated immune cells [81,82].

4.6. Additional Mechanisms Contributing to the Therapeutic Action of MSCs

In addition to the ability of MSCs to produce several growth, immunoregulatory or neurotrophic factors, MSCs release various types of extracellular vesicles (EVs). These particles encapsulate different functional molecules which could support the survival of cells [83,84]. For example, it has been shown that intravitreally injected EVs were as effective as MSCs in improving vision in experimental model of retinal laser injury [85]. Similarly, Mead and Tomarev [86] showed that MSC-derived exosomes protected retinal ganglion cell function in a rat optic nerve crush model.

Furthermore, mitochondrial transfer has been described as additional mechanism which MSCs can use to support anti-inflammatory conditions and cell survival [87,88]. Since mitochondrial disfunction has been proved in many retinal diseases, the mitochondrial transfer therapy might have an impact on the treatment of retinal diseases [89].

Finally, the ability of MSCs to fuse with other cell types has been documented in various models [90,91]. Therefore, the possibility of the fusion of MSCs and the cells of diseased retina cannot be excluded, and should be considered as another mechanism contributing to the therapeutic action of intraocularly administered MSCs.

5. The Potential of MSCs for the Treatment of Retinal Diseases

Abundant experimental data demonstrate the beneficial therapeutic effects of MSCs on retinal diseases [92–94]. It has been shown that MSC transplantation significantly delays retinal degeneration, supports the regeneration of RPE, cone cells and axons, and improves the survival of retinal ganglion cells. On the basis of these encouraging results, the potential to use MSCs for the treatment of retinal diseases has been proposed and tested [95–99]. The main mechanisms of the therapeutic action of MSCs for the treatment of retinal diseases are shown in Figure 1.

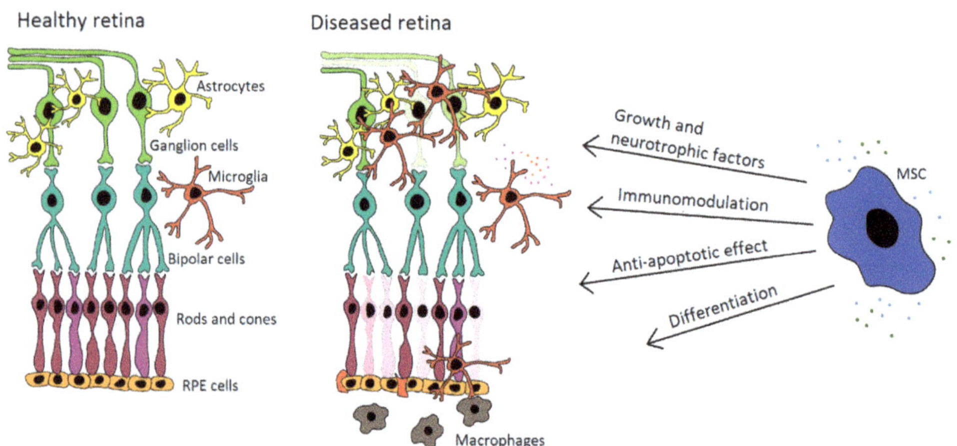

Figure 1. The main mechanisms of the therapeutic effect of mesenchymal stem cells (MSCs) for retinal diseases. MSCs contribute to treatment of retinal disorders by multiple mechanisms involving the production of growth and neurotrophic factors, immunomodulatory actions, by antiapoptotic effect and by direct cell differentiation.

Although several questions about the clinical use of MSCs still remain unanswered, for the purpose of great interest to use stem cells for the treatment of currently incurable retinal diseases, the first clinical trials using MSCs have been initiated [100,101]. However, before the introduction of the stem cell-based therapies into clinical practice, extensive research is needed to optimize the therapeutic procedures. For this reason, experimental animal models using pharmacologically induced degeneration of the individual retinal cell types or using animals with genetically induced retinal diseases, have been introduced. The pharmacological models use the application of sodium iodate ($NaIO_3$) for the destruction of RPE cells that mimic progression of macular degeneration [102,103], or the application of *N*-methyl D-aspartic acid (NMDA) or *N*-methyl-*N*-nitrosourea (MNU) that induces apoptosis and the selective degeneration of ganglion cells and photoreceptors that are processes resembling hereditary RP or glaucoma [104,105]. The studies using these preclinical experimental models support the idea that the treatment of retinal diseases with stem cells could represent the most modern and prospective approach for the treatment of currently incurable severe retinal disorders, and to improve the patient´s quality of life [106–108].

6. Possible Problems and Limitations Associated with MSC-Based Therapy

Although stem cell therapy is safe, as shown in both animal studies and clinical trials [24,25], there are still several issues that have to be taken into account before final translation of MSC-therapy from preclinical models into clinical therapy.

First at all, there is a heterogeneity of individual MSC samples, based on differences in the cell source, isolation and culture procedure. MSCs are used at different time intervals after their isolation and different doses of cells are used. It has been documented that a longer cell culture duration has an impact on MSC morphology, secretory potential and migratory properties [109,110]. Therefore, there should be an agreement about the preparation of MSCs for individual types of application.

There is still controversy about in vivo survival of in vitro cultivated MSCs. Although MSCs are considered immune privileged cells which do not express costimulatory molecules and MHC Class II molecules, in the presence of some cytokines they can express these molecules and become a target for immune cells. Without respect to these observations, numerous studies suggested a long-term survival of allogeneic or even xenogeneic MSCs in immunocompetent recipients. In contrast to these studies, Eggenhofer et al. [111] claimed that in vitro cultured MSCs are extremely short-lived and do not survive in vivo. However, there is a possibility that in immunologically privileged sites, such as those in the eye, MSCs could survive.

Another unresolved issue is the fate and immunological functions of MSCs after their transfer into the inflammatory environment of the diseased retina. There is a possibility that immunosuppressive MSCs, transferred into an environment where there are proinflammatory cytokines present, can turn out into a cell population supporting the development of aggressive proinflammatory Th17 cells.

The route of MSC administration is also very important. After the intravenous injection of MSCs, only a small proportion of the delivered cells can be found in the eye (our preliminary observations). Therefore, for the treatment of the retina, the intraocular delivery of cells appears more effective. Using experimental models, especially based on small animals, the intravitreal injection of MSCs is the most common way. However, in the healthy eye, only a few cells can be detected in the vitreous cavity, and the occurrence of side effects after such delivery of MSCs was reported [112]. Therefore, other routes of MSC application have been tested. These approaches include subretinal application [113], suprachoroidal delivery [114] and subtenon injection [115].

7. The Use of MSCs for the Treatment of Retinal Diseases in Experimental Models

To study the mechanisms of retinal diseases and to validate new therapeutic approaches for these diseases, various experimental models resembling different types of retinal damage have been established and tested. These models are based on pharmacological interventions which induce the degeneration of specialized retinal cells, or utilize mutant animal strains, genetically modified recipients or various mechanical damages or injuries [116]. We review here the selected experimental models that have been used to test the therapeutic potential of the various types of MSCs.

7.1. Experimental Models of AMD

AMD is characterized by a progressive degeneration of the RPE and photoreceptors, and this process represents the major cause of visual impairment and irreversible blindness in the elderly population. Numerous experimental models have been established to study the individual steps of AMD progression. Transgenic experimental animal models provide systems to explore the cellular and molecular mechanisms of this disease. Some advantages are offered by laser-induced models. Other approaches are based on the application of pharmacological agents inducing pathological changes in the retina. One of the well-established pharmacological models is based on the systemic or a local administration of $NaIO_3$. $NaIO_3$ is a chemical which selectively induces the degeneration and death of RPE cells. It was shown in vitro that the exposure of human RPE cell line ARPE-19 to $NaIO_3$ induces the activation of inflammasome, changes the expression of molecules involved in the apoptosis, induces cell dysfunctions resembling conditions in AMD and finally causes RPE cell death [117,118]. In this model, human A-MSCs decreased the levels of mRNA for proapoptotic molecules and provided a rescue effect for ARPE-19 cells cultivated in the presence of $NaIO_3$ [118]. In our recent study we have observed that $NaIO_3$ increases the expression of genes for proinflammatory cytokines IL-1α and IL-6 or for proapoptotic Bax and p53 molecules in cultured mouse retinal explants. This increase was inhibited in the presence of mouse BM-MSCs, or by using a supernatant obtained after the cultivation of MSCs (Palacka et al., preliminary observations).

The intravenous or intraperitoneal application of $NaIO_3$ in vivo causes a rapid degeneration of the RPE cells and consequent damage to the outer nuclear layer. Increased levels of mRNA for Htra-1 a C3, the genes associated with the development of AMD, were detected in the retina of the $NaIO_3$-treated recipients [103,119,120]. The intravitreal or subretinal application of $NaIO_3$ thus provides a suitable experimental model for study of the late phase of nonvascular AMD called geographic atrophy. The subretinal delivery of $NaIO_3$ in rats causes the formation of an atrophic area characterized by the degeneration of RPE cells and photoreceptors [121]. The intravitreal administration of $NaIO_3$ in rabbits after vitrectomy induced retinal atrophy and diffused outer retinal degeneration [122]. In these in vivo models, MSCs provided protection of the retinal cells from degeneration. The intravitreal injection of human A-MSCs in mice treated with $NaIO_3$ protected the RPE layer, photoreceptors and other nuclear cells from the damage [123]. Gong et al. [124] showed that rat BM-MSCs transplanted into the subretinal space can differentiate into cells expressing retinal markers, and can protect the retina in the experimental models of $NaIO_3$-induced retinal damage. Since the RPE cells are the first damaged cell type in the progression of AMD, the protective effect of MSCs may be a promising option for the treatment of this condition.

7.2. Experimental Models of DR

DR represents a common complication of diabetes which is caused by hyperglycemia and by injury in retinal microvasculature and neurons. This disorder represents one of the leading causes of blindness globally. Despite the high prevalence of DR and extensive research, the treatment options for this disease are still strongly limited. Various experimental animal models have been established for the study of treatment possibilities. These models have been generated by a selective inbreeding or genetic modifications, the feeding of a galactose diet, or by a pharmacological induction using streptozotocin. This chemical selectively damages the β cells of the pancreas, increases blood glucose level and decreases the number of ganglion cells. To date, various animal models have been used to test the possibilities of treating DR by the application of stem cells [125]. In the majority of these models, the beneficial effects of a systemic or local application of MSCs were observed.

In models of streptozotocin-induced diabetes and DR, the application of MSCs had a positive effect on the retinal architecture. For example, Ezquer et al. [126] showed that the local application of mouse A-MSCs prevented the loss of retinal ganglion cells in diabetic mice. Levels of neurotrophic factors, such as NGF, GDNF and bFGF were increased in the eyes treated with A-MSCs. Although donor A-MSCs were found integrated into the host retina, these authors did not observe the differentiation of MSCs into retinal cells. In other studies, the intravitreal administration of MSCs obtained from the human umbilical cord attenuated capillary damage in streptozotocin-induced DR and increased levels of BDNF and NGF in the treated eyes. Donor MSCs also restored the visual function measured by ERG [94,127,128]. Yang et al. [92] showed that the administration of human A-MSCs improved the integrity of the blood–retinal barrier and ameliorated DR in streptozotocin diabetic rats. Slightly enhanced levels of BDNF in the retina were also obtained after the transplantation of neural stem cells differentiated from umbilical cord MSCs, thus suggesting that this type of cells originated from MSCs may represent another suitable option for neuroprotection in DR [127]. Since MSCs isolated from mice with DR have lower proliferative abilities and higher levels of apoptosis compared to cells from healthy individuals [129,130], attempts were made to improve their therapeutic properties with the aim of using these cells for autologous transplantation in patients with DR. It has been shown that the treatment of BM-MSCs from mice with streptozotocin-induced diabetes with Wharton's jelly extract (containing a number of growth factors and other cytokines) significantly improved their proliferative abilities and therapeutic potential [130]. It suggests that the preconditioning of diabetic MSCs could improve their therapeutic properties.

In addition to models of pharmacologically induced diabetes, several studies used models of spontaneously or genetically induced diabetes. These models were described in detail by Robinson et al. [131] and Lai and Lo [132]. For example, the Akita ($Ins2^{Akita}$) mice were created by a point mutation in the *insulin-2* gene and represent a spontaneous type-1 diabetes model. It was shown that hyperglycemia in these mice causes neurodegenerative effects in the retina resulting in retinal thickness [133]. In addition, elevated levels of VEGF, PEGF and placental growth factor (PlGF) and an increased expression of Iba-1 (activated glial marker) and monocyte chemoattractant protein-1 (MCP-1) were observed in the neural retina and RPE layer in $Ins2^{Akita}$ mice during the progression of the diseases [133]. The therapeutic administration of human A-MSCs into the vitreal cavity of $Ins2^{Akita}$ mice improved vascular permeability and vision in this model of nonproliferative DR. Similar results were also obtained after the application of the conditioned medium from human A-MSCs which were pretreated with TNF-α and IFN-γ. The conditioned medium from A-MSCs also reduced the retinal expression of GFAP, the gene associated with neuroinflammation [134]. The experimental mouse model of proliferative DR was created by the mutation causing an overexpression of VEGFa (Akimba mice). In the retina of the Akimba mice, hemorrhage and neovascularization, the degeneration of photoreceptors, the activation of microglial cells and infiltration with monocytes and macrophages were detected. The inflammatory environment is manifested by a local increase in the expression of genes for IL-1β and IL-6 and by the upregulated activation of the NLRP3 inflammasome in the retina [135,136]. Locally administered A-MSCs obtained from mice without mutation in the Insulin 2 gene were mainly found in the perivascular space and improved the vascular density in the retina [129].

7.3. Experimental Models for RP

RP is a group of inherited neurodegenerative diseases characterized by a loss of photoreceptor cells, leading to visual impairment and eventually to blindness. The experimental models (natural and transgenic) of this disease are based on the use of spontaneous or genetically induced degeneration of the photoreceptors, and on the administration of chemicals inducing degeneration of the retinal cells [137–139]. A frequently used model for RP is the rd mouse with a mutation causing the early loss of the photoreceptors. For example, the rd1 mouse is characterized by mutation in the *PDE6b* gene which is, under physiological conditions, important for the signal transmission. In addition, it was shown that activated microglia with proinflammatory polarization occur in the rd1 retina [140]. It is also possible to use an rd10 mouse which has a spontaneous mutation in the *PDE* gene for rod-phosphodiesterase. This mutation causes the degeneration of photoreceptors and other retinal cell types [141]. Moreover, it has been shown that (as with rd1 mice) activated microglia can play a role in the development of a pathological condition [142]. Another example is an rd6 mouse carrying a mutation in the *Mrfp* gene, which is expressed in the RPE layer of cells [143]. In addition to mouse experimental models, the Royal College of Surgeons (RCS) rat is often used to study RP. The RCS rat carries a mutation in the *Merkt* gene, causing photoreceptor damage and an increase in microglial activation in the retina, resulting in inherited retinal degeneration [144,145]. Some of these models have been used to study the therapeutic effects of MSCs. Treatment with MSCs has supported the survival of photoreceptors and showed therapeutic benefits. For example, the application of MSCs to the eyes of rd1 and rd10 mice provided a rescue effect for retinal cells [146]. The administration of genetically modified MSCs with an overexpression of BDNF resulted in increased antiapoptotic signaling in the retina, and in a reduction in cell damage in the rd6 mouse [147]. Moreover, the donor cells preferentially integrated into the outer retinal layers. In addition, the combined transplantation of the human retinal progenitor cells and BM-MSCs into the subretinal space provided an effective immunomodulation in the eye of RCS rats and prevented pathological changes more effectively than with a single therapy [148]. Decreased levels of TNF-α and IL-1β and an increased expression of growth factors, such as BDNF and NGF, were observed in the treated eye.

Another approach to imitate the retinal degeneration observed in RP is based on the administration of *N*-methyl-*N*-nitrosourea (MNU). A single systemic administration of MNU causes retinal degeneration in various species [139]. Deng et al. [149] showed that the treatment of mice with MNU induces retinal degeneration that can be attenuated by the administration of MSCs, and that the therapeutic effect was decreased if MSCs were prepared from the aging mice with bone progeria.

Thus, as in the case of other experimental models of retinal degeneration, the positive therapeutic effects of MSC therapy were also demonstrated in RP models.

7.4. Experimental Models for Glaucoma

Glaucoma is a heterogenous group of eye diseases mainly caused by increased intraocular pressure and characterized by the progressive loss of retinal ganglion cells. So far, numerous experimental models have been established to study this disease. They include the intracameral injection of microbeads, laser photocoagulation, episcleral vein cauterization, the injection of hyaluronic acid and various models based on genetically modified rodents [150–153]. These models lead to increased intraocular pressure, the degeneration of retinal ganglion cells, the activation of glial cells in the retina and increased levels of inflammatory factors in the retina [154,155]. All these models, having their advantages and limitations, were used to study new therapeutic approaches involving MSC-based therapy. Various types of MSCs have been tested with a positive impact on the decrease of intraocular pressure and on the protection of the retina. For example, Mead et al. [156] showed that an intravitreal administration of A-MSCs, BM-MSCs or dental pulp stem cells decreased ocular pressure and offered a neuroprotective effect.

A trabecular meshwork regeneration observed after intraocular administration of BM-MSCs was described in a model of a laser-induced retinal damage model [157,158]. The enhanced neuroprotective effects were observed in these models using MSCs with an increased secretion of BDNF. The neuroprotective effect of A-MSCs was also described in a model of hyaluronic acid-induced glaucoma in rats [159]. Another therapeutic approach is provided by the application of BM-MSC-derived exosomes. It has been shown that the administration of the exosomes secreted by BM-MSCs promoted the survival of retinal ganglion cells and improved the retinal structure in the eye of rats after optic nerve crush injury, and in glaucoma models with ocular pressure induced by an intracameral injection of microbeads [86]. Similarly, the application of umbilical cord MSC-derived exosomes in a glaucoma model induced by an optic nerve crush injury in rats promoted retinal ganglion cell survival and glial cell activation [160]. Thus, MSC-derived exosomes injected into the vitreous provide a significant therapeutic benefit for glaucomatous eyes and for other types of retinal degenerative diseases.

Altogether, the animal models indicated positive therapeutic effects of MSC-based therapies in various types of retinal diseases. Selected experimental models where MSCs were used are summarized in Table 1.

Table 1. Selected experimental models of MSC-based therapy for retinal degenerative disorders.

Induction of Retinal Diseases	Species	Treatment	Result	Reference
NaIO$_3$	Mouse	Human A-MSCs	Protection of RPE cells, photoreceptors and outer nuclear layer	[123]
	Rat	Rat BM-MSCs	Differentiation of transplanted MSCs into cells with retinal markers	[124]
Streptozotocin	Mouse	Mouse A-MSCs	Enhanced levels of neurotrohpic factors, protection of retinal ganglion cells	[126]
	Rat	Neural stem cells (derived from humal umbilical MSCs)	Enhanced levels of neurotrohpic factors, protection of retinal ganglion cells	[127]
		Human A-MSCs	Decreased apoptosis, decrease in expression of genes related to DR	[94]
		Human umbilical MSCs	Increased expression of NGF	[128]
		Rat BM-MSCs	Improvement in visual function	[161]
Insulin 2 gene mutation	Mouse	Human A-MSCs	Decreased vascular permeability	[134]
		Conditioned medium from human A-MSCs	Decreased vascular permeability, improvement in visual function	[134]
Insulin 2/VEGFa gene mutation	Mouse	Mouse A-MSCs	Increased vascular density, incorporation of host MSCs into the retina	[129]
Cauterization of 3 episcleral veins	Rat	Rat BM-MSCs	Regulation of intraocular pressure, protection of retinal ganglion cells	[162]
Laser damage	Rat	Rat BM-MSCs	Protection of retinal ganglion cells	[157]
		Rat BM-MSCs (engineered to express BDNF)	Improvement in ERG function, protection of retinal ganglion cells	[163]
Optic nerve crush injury	Rat	Exosomes from human BM-MSCs	Protection of retinal ganglion cells	[96]
PDE gene mutation (rd 10 mouse)	Mouse	Mouse BM-MSCs	Protection of photoreceptors	[146]
Mfrp mutation (rd 6 mouse)	Mouse	Mouse BM-MSCs (engineered to express BDNF)	Induction of antiapoptotic signaling, improvement in ERG	[147]
Mertk gene mutation (RCS rats)	Rat	Human BM-MSCs with human progenitor retinal cells	Inflammatory modulation, promoting differentiation of donors cells into photoreceptor	[148]

8. Clinical Trials Using MSCs for Retinal Diseases

Currently, several clinical trials are in progress to test the potential of MSCs for the treatment of retinal degenerative diseases. Most of these studies are in the phase 1 or 2 focused on the safety of MSC application. The first finished studies showed that the administration of MSCs is not associated with serious complications [164,165]. Moreover, some studies have also showed an improvement in visual function based on the examination of in visual acuity, visual field and electroretinography. The effects of the treatment have been studied in patients with variety types of retinal diseases, such as AMD, DR, RP, glaucoma, inherited retinal dystrophy, optic nerve diseases or macular holes. MSCs obtained from various sources have been used for the treatments and BM-MSC or A-MSC were tested as an option for autologous transplantation. In addition, the effects of the administration of the conditioned medium obtained from cultured MSCs or the application of exosomes prepared by the ultracentrifugation of a conditioned medium are also examined.

The application of MSCs appears safe and no serious treatment-related problems were observed in the eyes of patients with AMD, RP or retinal vascular occlusion six months after the administration of autologous BM-MSCs in the phase 1 testing. An improvement in visual function was also noted, but as the studies were designed to assess the safety of the treatment, it was not possible to definitely confirm whether this improvement was caused by the MSC application [164]. Similar results were observed in clinical trials after the application of autologous BM-MSCs in patients with RP [165]. There were no severe complications associated with cell transplantation in the treated eye. This conclusion is also supported by the study of Gu et al. [166] who showed that autologous BM-MSCs were beneficial in DR subjects with correction in macular thickness, and the improvement in visual acuity was also observed. In addition, the administration of autologous BM-MSCs improved visual acuity in patients with RP [167]. A similar improvement in visual function after the injection of autologous BM-MSC was observed in patients with optic nerve diseases and nonarteritic ischemic optic neuropathy [168,169]. In addition to the application of autologous BM-MSC, Wharton's jelly-derived MSC transplantation improved outer retinal thickness and visual acuity in patients with RP in phase 3 clinical study [115].

However, further clinical trials with a higher number of patients and a longer follow-up are still needed to evaluate the efficacy of MSC therapy. It will also be necessary to evaluate the benefits and advantages of autologous MSCs and the transplantation of stem cells obtained from another source, such as Wharton's jelly or the umbilical cord, or the use of a conditioned medium or exosomes. Selected clinical trials with MSCs for retinal degenerative disorders and their results are shown in Table 2.

Table 2. Selected examples of clinical trials using MSCs for retinal degenerative diseases.

Retinal Disease	Cells For Treatment	Result	Reference
AMD, RP, retinal vascular occlusion	Autologous BM-MSC (intravitreal)	Phase 1, no severe safety issues associated with treatment	[164]
RP, cone-rod dystrophy	Autologous BM-MSC(intravitreal)	Phase 1, no severe safety issues associated with treatment	[165]
RP	Autologous BM-MSC (retrobulbar, subtenons, intravitreal, intravenous)	Improvement in visual function	[167]
Optic nerve diseases	Autologous BM-MSC (retrobulbar, subtenons, intravitreal, intravenous)	Improvement in visual function	[168]
Ischemic optic neuropathy	Autologous BM-MSC (retrobulbar, subtenons, intravitreal, intravenous)	Improvement in visual function	[169]
RP, inherited retinal dystrophy	Wharton's jelly-derived MSC (subtenons)	Improvement in visual acuity and in outer retinal thickness	[115]
DR	Autologous MSCs (intravenous)	Improvements in macular thickness and in visual acuity	[166]
RP	Umbilical cord- derived MSC (suprachorodial)	Improvements in best corrected visual acuity, electroretinography and visual field	[114]
RP	A-MSC (subretinal)	Minor ocular complicaions, no severe safety issues associated with the treatment	[113]

9. Conclusions

Sight-threatening retinal degenerative diseases represent the main cause of visual impairment or even blindness worldwide. Despite great endeavors, there is still a lack of effective therapeutic approaches to stop or even cure these disorders. A prospective option has recently been offered by stem cell therapy. Experimental data from numerous animal models and using different types of stem cells offer promising results. Although the data from the experimental models are encouraging, numerous questions about the use of stem cell therapy have to be resolved to make this therapy more effective and safe [170]. Nevertheless, abundant clinical trials on the use of stem cells for retinal diseases have already been initiated [101,171,172]. These trials are focused on the study of the safety of the therapy, the selection of the optimal stem cells and their activation or modification prior to application, the optimalization of the dose of cells, the routes of application and the possibility of replacing the cells with their paracrine products.

One of the most important issues associated with MSC-based therapy, the safety of MSC administration, has been tested in numerous preclinical studies and clinical trials. Various experimental models and clinical studies using an intravitreal administration of MSCs demonstrated the safety of this therapy without any undesirable side effects [100,101]. Another issue that deserves attention is the mechanism of MSC therapy. Although some authors showed a long-time survival of therapeutically applied MSCs and demonstrated their presence in the eye even a few months after application, other studies suggested that MSCs are short-lived, do not survive in the recipients and can be only detected for a few days [111]. Moreover, in some experimental animal studies human MSCs were administered intravitreally and their biocompatibility, long survival and positive therapeutic effects were observed [173–175]. There arises a question about the mechanisms of this therapeutic effect across the interspecies barrier. Namely, Lohan et al. [176] demonstrated that human MSCs injected into rats do not have the same therapeutic effect as rat MSCs have, and that the immunoregulatory action of human MSCs is strongly limited by the interspecies barrier. These differences in the therapeutic effect between autologous/syngeneic and xenogeneic MSCs have to be taken into consideration, when human MSCs are applied therapeutically in rodent recipients and the knowledge from such experimental studies is translated to the clinical situation.

Finally, the immunoregulatory action of MSCs could strongly depend on the cytokine environment [49]. The inflammatory conditions in the diseased retina can significantly change the immunoregulatory properties of MSCs. We showed that unstimulated MSCs are immunosuppressive and spontaneously produce high levels of TGF-β, but not IL-6 [48]. TGF-β is a negative regulator of immunity and is also a factor determining the development of suppressive Tregs. However, in the presence of proinflammatory cytokines MSCs secrete, in addition to TGF-β, high levels of IL-6 [48,177]. The combination of TGF-β and IL-6 determines the development of proinflammatory Th17 cells [178]. Therefore, there is a danger that the application of MSCs into the inflammatory environment of the diseased retina could result in the inhibition of the immunosuppressive action of MSCs and in the preferential activation of proinflammatory Th17 cell population.

Although there are still many issues to be addressed before the final approval of MSC therapy for retinal degenerative diseases, the results obtained so far in preclinical animal models and in clinical trials are promising and encouraging. Therefore, the stem cell-based therapy offers a prospective option, especially for the patients without alternative therapeutic options. However, before the definitive expansion of the clinical use of MSC-based therapies, several questions, such as the sources of MSCs, the conditions of the in vitro propagation of MSCs, the routes of the cell applications or the possibility of the use of MSC products have to be answered and carefully verified.

Author Contributions: All authors participated equally in the preparation and writhing of the manuscript. All authors have also read and agreed to the published version of the manuscript.

Funding: This work was supported the grant No. 19-02290S from the Grant Agency of the Czech Republic and by the Charles University programs SVV 260435 and 20604315 PROGRES Q43.

Institutional Review Board Statement: Not applicable.

Informed Consent Statement: Not applicable.

Data Availability Statement: Not Applicable.

Conflicts of Interest: The authors declare no conflict of interest.

References

1. Shaw, P.X.; Stiles, T.; Douglas, C.; Ho, D.; Fan, W.; Du, H.; Xiao, X. Oxidative stress, innate immunity, and age-related macular degeneration. *AIMS Mol. Sci.* **2016**, *3*, 196–221. [CrossRef]
2. Semeraro, F.; Cancarini, A.; Dell'Omo, R.; Rezzola, S.; Romano, M.R.; Costagliola, C. Diabetic Retinopathy: Vascular and Inflammatory Disease. *J. Diabetes Res.* **2015**, *2015*, 1–16. [CrossRef]
3. Van Norren, D.; Vos, J.J. Light damage to the retina: An historical approach. *Eye* **2016**, *30*, 169–172. [CrossRef] [PubMed]
4. Daiger, S.P.; Sullivan, L.S.; Bowne, S.J. Genes and mutations causing retinitis pigmentosa. *Clin. Genet.* **2013**, *84*, 132–141. [CrossRef] [PubMed]
5. Voigt, A.P.; Binkley, E.; Flamme-Wiese, M.J.; Zeng, S.; DeLuca, A.P.; Scheetz, T.E.; Tucker, B.A.; Mullins, R.F.; Stone, E.M. Single-Cell RNA Sequencing in Human Retinal Degeneration Reveals Distinct Glial Cell Populations. *Cells* **2020**, *9*, 438. [CrossRef] [PubMed]
6. Jo, D.H.; Yun, J.-H.; Cho, C.S.; Kim, J.H.; Kim, J.H.; Cho, C.-H. Interaction between microglia and retinal pigment epithelial cells determines the integrity of outer blood-retinal barrier in diabetic retinopathy. *Glia* **2019**, *67*, 321–331. [CrossRef]
7. Menon, M.; Mohammadi, S.; Davila-Velderrain, J.; Goods, B.A.; Cadwell, T.D.; Xing, Y.; Stemmer-Rachamimov, A.; Shalek, A.K.; Love, J.C.; Kellis, M.; et al. Single-cell transcriptomic atlas of the human retina identifies cell types associated with age-related macular degeneration. *Nat. Commun.* **2019**, *10*, 1–9. [CrossRef]
8. Jovanovic, J.; Liu, X.; Kokona, D.; Zinkernagel, M.S.; Ebneter, A. Inhibition of inflammatory cells delays retinal degeneration in experimental retinal vein occlusion in mice. *Glia* **2019**, *68*, 574–588. [CrossRef] [PubMed]
9. Jones, M.K.; Lu, B.; Girman, S.; Wang, S. Cell-based therapeutic strategies for replacement and preservation in retinal degenerative diseases. *Prog. Retin. Eye Res.* **2017**, *58*, 1–27. [CrossRef]
10. Garcia, J.M.; Mendonça, L.; Brant, R.; Abud, M.; Regatieri, C.; Diniz, B. Stem cell therapy for retinal diseases. *World J. Stem Cells* **2015**, *7*, 160–164. [CrossRef] [PubMed]
11. Alonso-Alonso, M.L.; Srivastava, G.K. Current focus of stem cell application in retinal repair. *World J. Stem Cells* **2015**, *7*, 641–648. [CrossRef]
12. Chen, M.; Xiang, Z.; Cai, J. The anti-apoptotic and neuro-protective effects of human umbilical cord blood mesenchymal stem cells (hUCB-MSCs) on acute optic nerve injury is transient. *Brain Res.* **2013**, *1532*, 63–75. [CrossRef] [PubMed]
13. Mesentier-Louro, L.A.; Zaverucha-Do-Valle, C.; Da Silva-Junior, A.J.; Nascimento-Dos-Santos, G.; Gubert, F.; De Figueirêdo, A.B.P.; Torres, A.L.; Paredes, B.D.; Teixeira, C.; Tovar-Moll, F.; et al. Distribution of Mesenchymal Stem Cells and Effects on Neuronal Survival and Axon Regeneration after Optic Nerve Crush and Cell Therapy. *PLoS ONE* **2014**, *9*, e110722. [CrossRef] [PubMed]
14. Duan, P.; Xu, H.; Zeng, Y.; Wang, Y.; Yin, Z.Q. Human Bone Marrow Stromal Cells can Differentiate to a Retinal Pigment Epithelial Phenotype when Co-Cultured with Pig Retinal Pigment Epithelium using a Transwell System. *Cell. Physiol. Biochem.* **2013**, *31*, 601–613. [CrossRef] [PubMed]
15. Rezanejad, H.; Soheili, Z.-S.; Haddad, F.; Matin, M.M.; Samiei, S.; Manafi, A.; Ahmadieh, H. In vitro differentiation of adipose-tissue-derived mesenchymal stem cells into neural retinal cells through expression of human PAX6 (5a) gene. *Cell Tissue Res.* **2014**, *356*, 65–75. [CrossRef] [PubMed]
16. Hermankova, B.; Kossl, J.; Javorkova, E.; Bohacova, P.; Hajkova, M.; Zajicova, A.; Krulova, M.; Holan, V. The Identification of Interferon-γ as a Key Supportive Factor for Retinal Differentiation of Murine Mesenchymal Stem Cells. *Stem Cells Dev.* **2017**, *26*, 1399–1408. [CrossRef]
17. Le Blanc, K.; Ringdén, O. Immunomodulation by mesenchymal stem cells and clinical experience. *J. Intern. Med.* **2007**, *262*, 509–525. [CrossRef]
18. Abumaree, M.; Al Jumah, M.; Pace, R.A.; Kalionis, B. Immunosuppressive Properties of Mesenchymal Stem Cells. *Stem Cell Rev. Rep.* **2011**, *8*, 375–392. [CrossRef]
19. Reinshagen, H.; Sorg, R.V.; Boehringer, D.; Eberwein, P.; Sundmacher, R.; Reinhard, T.; Auw-Haedrich, C.; Schwartzkopff, J. Corneal surface reconstruction using adult mesenchymal stem cells in experimental limbal stem cell deficiency in rabbits. *Acta Ophthalmol.* **2009**, *89*, 741–748. [CrossRef]
20. Holan, V.; Javorkova, E. Mesenchymal Stem Cells, Nanofiber Scaffolds and Ocular Surface Reconstruction. *Stem Cell Rev. Rep.* **2013**, *9*, 609–619. [CrossRef] [PubMed]

21. Čejková, J.; Trosan, P.; Čejka, Č.; Lencova, A.; Zajicova, A.; Javorkova, E.; Kubinová, Š.; Syková, E.; Holan, V. Suppression of alkali-induced oxidative injury in the cornea by mesenchymal stem cells growing on nanofiber scaffolds and transferred onto the damaged corneal surface. *Exp. Eye Res.* **2013**, *116*, 312–323. [CrossRef] [PubMed]

22. Holan, V.; Trosan, P.; Cejka, C.; Javorkova, E.; Zajicova, A.; Hermankova, B.; Chudickova, M.; Cejkova, J. A Comparative Study of the Therapeutic Potential of Mesenchymal Stem Cells and Limbal Epithelial Stem Cells for Ocular Surface Reconstruction. *Stem Cells Transl. Med.* **2015**, *4*, 1052–1063. [CrossRef] [PubMed]

23. Sahu, A.; Foulsham, W.; Amouzegar, A.; Mittal, S.K.; Chauhan, S.K. The therapeutic application of mesenchymal stem cells at the ocular surface. *Ocul. Surf.* **2019**, *17*, 198–207. [CrossRef]

24. Wang, Y.; Han, Z.-B.; Ma, J.; Zuo, C.; Geng, J.; Gong, W.; Sun, Y.; Li, H.; Wang, B.; Zhang, L.; et al. A Toxicity Study of Multiple-Administration Human Umbilical Cord Mesenchymal Stem Cells in Cynomolgus Monkeys. *Stem Cells Dev.* **2012**, *21*, 1401–1408. [CrossRef] [PubMed]

25. Lalu, M.M.; McIntyre, L.; Pugliese, C.; Fergusson, D.; Winston, B.W.; Marshall, J.C.; Granton, J.; Stewart, D.J. Canadian Critical Care Trials Group. Safety of Cell Therapy with Mesenchymal Stromal Cells (SafeCell): A Systematic Review and Meta-Analysis of Clinical Trials. *PLoS ONE* **2012**, *7*, e47559. [CrossRef]

26. Friedenstein, A.J.; Chailakhjan, R.K.; Lalykina, K.S. The development of fibroblast colonies in monolayer cultures of guinea-pig bone marrow and spleen cells. *Cell Prolif.* **1970**, *3*, 393–403. [CrossRef] [PubMed]

27. Pittenger, M.F.; Mackay, A.M.; Beck, S.C.; Jaiswal, R.K.; Douglas, R.; Mosca, J.D.; Moorman, M.A.; Simonetti, D.W.; Craig, S.; Marshak, D.R. Multilineage Potential of Adult Human Mesenchymal Stem Cells. *Science* **1999**, *284*, 143–147. [CrossRef]

28. Phinney, D.G.; Prockop, D.J. Concise Review: Mesenchymal Stem/Multipotent Stromal Cells: The State of Transdifferentiation and Modes of Tissue Repair-Current Views. *Stem Cells* **2007**, *25*, 2896–2902. [CrossRef]

29. Dominici, M.; Le Blanc, K.; Mueller, I.; Slaper-Cortenbach, I.; Marini, F.; Krause, D.; Deans, R.; Keating, A.; Prockop, D.; Horwitz, E. Minimal criteria for defining multipotent mesenchymal stromal cells. The International Society for Cellular Therapy position statement. *Cytotherapy* **2006**, *8*, 315–317. [CrossRef] [PubMed]

30. Musina, R.A.; Bekchanova, E.S.; Sukhikh, G.T. Comparison of Mesenchymal Stem Cells Obtained from Different Human Tissues. *Bull. Exp. Biol. Med.* **2005**, *139*, 504–509. [CrossRef] [PubMed]

31. Strioga, M.; Viswanathan, S.; Darinskas, A.; Slaby, O.; Michalek, J. Same or Not the Same? Comparison of Adipose Tissue-Derived Versus Bone Marrow-Derived Mesenchymal Stem and Stromal Cells. *Stem Cells Dev.* **2012**, *21*, 2724–2752. [CrossRef]

32. Isobe, Y.; Koyama, N.; Nakao, K.; Osawa, K.; Ikeno, M.; Yamanaka, S.; Okubo, Y.; Fujimura, K.; Bessho, K. Comparison of human mesenchymal stem cells derived from bone marrow, synovial fluid, adult dental pulp, and exfoliated deciduous tooth pulp. *Int. J. Oral Maxillofac. Surg.* **2016**, *45*, 124–131. [CrossRef] [PubMed]

33. Rasmusson, I.; Ringdén, O.; Sundberg, B.; Le Blanc, K. Mesenchymal stem cells inhibit lymphocyte proliferation by mitogens and alloantigens by different mechanisms. *Exp. Cell Res.* **2005**, *305*, 33–41. [CrossRef]

34. Di Nicola, M.; Carlo-Stella, C.; Magni, M.; Milanesi, M.; Longoni, P.D.; Matteucci, P.; Grisanti, S.; Gianni, A.M. Human bone marrow stromal cells suppress T-lymphocyte proliferation induced by cellular or nonspecific mitogenic stimuli. *Blood* **2002**, *99*, 3838–3843. [CrossRef] [PubMed]

35. Bartholomew, A.; Sturgeon, C.; Siatskas, M.; Ferrer, K.; McIntosh, K.; Patil, S.; Hardy, W.; Devine, S.; Ucker, D.; Deans, R.; et al. Mesenchymal stem cells suppress lymphocyte proliferation in vitro and prolong skin graft survival in vivo. *Exp. Hematol.* **2002**, *30*, 42–48. [CrossRef]

36. Xu, G.; Zhang, L.; Ren, G.; Yuan, Z.; Zhang, Y.; Zhao, R.C.; Shi, Y. Immunosuppressive properties of cloned bone marrow mesenchymal stem cells. *Cell Res.* **2007**, *17*, 240–248. [CrossRef] [PubMed]

37. Jia, Z.; Jiao, C.; Zhao, S.; Li, X.; Ren, X.; Zhang, L.; Han, Z.C.; Zhang, X. Immunomodulatory effects of mesenchymal stem cells in a rat corneal allograft rejection model. *Exp. Eye Res.* **2012**, *102*, 44–49. [CrossRef] [PubMed]

38. Oh, J.Y.; Lee, R.H.; Yu, J.M.; Ko, J.H.; Lee, H.J.; Ko, A.Y.; Roddy, G.W.; Prockop, D.J. Intravenous Mesenchymal Stem Cells Prevented Rejection of Allogeneic Corneal Transplants by Aborting the Early Inflammatory Response. *Mol. Ther.* **2012**, *20*, 2143–2152. [CrossRef] [PubMed]

39. Le Blanc, K.; Rasmusson, I.; Sundberg, B.; Götherström, C.; Hassan, M.; Uzunel, M.; Ringdén, O. Treatment of severe acute graft-versus-host disease with third party haploidentical mesenchymal stem cells. *Lancet* **2004**, *363*, 1439–1441. [CrossRef]

40. Lazarus, H.M.; Koc, O.N.; Devine, S.M.; Curtin, P.; Maziarz, R.T.; Holland, H.K.; Shpall, E.J.; McCarthy, P.; Atkinson, K.; Cooper, B.W.; et al. Cotransplantation of HLA-Identical Sibling Culture-Expanded Mesenchymal Stem Cells and Hematopoietic Stem Cells in Hematologic Malignancy Patients. *Biol. Blood Marrow Transplant.* **2005**, *11*, 389–398. [CrossRef]

41. Luo, C.-J.; Zhang, F.-J.; Zhang, L.; Geng, Y.-Q.; Li, Q.-G.; Hong, Q.; Fu, B.; Zhu, F.; Cui, S.-Y.; Feng, Z.; et al. Mesenchymal Stem Cells Ameliorate Sepsis-associated Acute Kidney Injury in Mice. *Shock* **2014**, *41*, 123–129. [CrossRef] [PubMed]

42. Zappia, E.; Casazza, S.; Pedemonte, E.; Benvenuto, F.; Bonanni, I.; Gerdoni, E.; Giunti, D.; Ceravolo, A.; Cazzanti, F.; Frassoni, F.; et al. Mesenchymal stem cells ameliorate experimental autoimmune encephalomyelitis inducing T-cell anergy. *Blood* **2005**, *106*, 1755–1761. [CrossRef] [PubMed]

43. Augello, A.; Tasso, R.; Negrini, S.M.; Cancedda, R.; Pennesi, G. Cell therapy using allogeneic bone marrow mesenchymal stem cells prevents tissue damage in collagen-induced arthritis. *Arthritis Rheum.* **2007**, *56*, 1175–1186. [CrossRef]

44. English, K.; Ryan, J.M.; Tobin, L.M.; Murphy, M.J.; Barry, F.P.; Mahon, B.P. Cell contact, prostaglandin E2and transforming growth factor beta 1 play non-redundant roles in human mesenchymal stem cell induction of CD4+CD25Highforkhead box P3+regulatory T cells. *Clin. Exp. Immunol.* **2009**, *156*, 149–160. [CrossRef] [PubMed]

45. Meisel, R.; Zibert, A.; Laryea, M.; Göbel, U.; Däubener, W.; Dilloo, D. Human bone marrow stromal cells inhibit allogeneic T-cell responses by indoleamine 2,3-dioxygenase–mediated tryptophan degradation. *Blood* **2004**, *103*, 4619–4621. [CrossRef]

46. Akiyama, K.; Chen, C.; Wang, D.; Xu, X.; Qu, C.; Yamaza, T.; Cai, T.; Chen, W.; Sun, L.; Shi, S. Mesenchymal-Stem-Cell-Induced Immunoregulation Involves FAS-Ligand-/FAS-Mediated T Cell Apoptosis. *Cell Stem Cell* **2012**, *10*, 544–555. [CrossRef]

47. Ghannam, S.; Pène, J.; Torcy-Moquet, G.; Jorgensen, C.; Yssel, H. Mesenchymal Stem Cells Inhibit Human Th17 Cell Differentiation and Function and Induce a T Regulatory Cell Phenotype. *J. Immunol.* **2010**, *185*, 302–312. [CrossRef]

48. Svobodova, E.; Krulova, M.; Zajicova, A.; Pokorna, K.; Prochazkova, J.; Trosan, P.; Holan, V. The Role of Mouse Mesenchymal Stem Cells in Differentiation of Naive T-Cells into Anti-Inflammatory Regulatory T-Cell or Proinflammatory Helper T-Cell 17 Population. *Stem Cells Dev.* **2012**, *21*, 901–910. [CrossRef] [PubMed]

49. Holan, V.; Hermankova, B.; Bohacova, P.; Kossl, J.; Chudickova, M.; Hajkova, M.; Krulova, M.; Zajicova, A.; Javorkova, E. Distinct Immunoregulatory Mechanisms in Mesenchymal Stem Cells: Role of the Cytokine Environment. *Stem Cell Rev. Rep.* **2016**, *12*, 654–663. [CrossRef]

50. Sasaki, M.; Abe, R.; Fujita, Y.; Ando, S.; Inokuma, D.; Shimizu, H. Mesenchymal Stem Cells Are Recruited into Wounded Skin and Contribute to Wound Repair by Transdifferentiation into Multiple Skin Cell Type. *J. Immunol.* **2008**, *180*, 2581–2587. [CrossRef] [PubMed]

51. Lan, Y.; Kodati, S.; Lee, H.S.; Omoto, M.; Jin, Y.; Chauhan, S.K. Kinetics and Function of Mesenchymal Stem Cells in Corneal Injury. *Investig. Opthalmol. Vis. Sci.* **2012**, *53*, 3638–3644. [CrossRef] [PubMed]

52. Assis, A.C.M.; Carvalho, J.L.; Jacoby, B.A.; Ferreira, R.L.B.; Castanheira, P.; Diniz, S.O.F.; Cardoso, V.N.; Goes, A.M.; Ferreira, A.J. Time-Dependent Migration of Systemically Delivered Bone Marrow Mesenchymal Stem Cells to the Infarcted Heart. *Cell Transplant.* **2010**, *19*, 219–230. [CrossRef]

53. Javorkova, E.; Trosan, P.; Zajicova, A.; Krulová, M.; Hajkova, M.; Holan, V. Modulation of the Early Inflammatory Microenvironment in the Alkali-Burned Eye by Systemically Administered Interferon-γ-Treated Mesenchymal Stromal Cells. *Stem Cells Dev.* **2014**, *23*, 2490–2500. [CrossRef] [PubMed]

54. Holan, V.; Echalar, B.; Palacka, K.; Kossl, J.; Bohacova, P.; Krulova, M.; Brejchova, J.; Svoboda, P.; Zajicova, A. The Altered Migration and Distribution of Systemically Administered Mesenchymal Stem Cells in Morphine-Treated Recipients. *Stem Cell Rev. Rep.* **2021**. [CrossRef]

55. Ponte, A.L.; Marais, E.; Gallay, N.; Langonné, A.; Delorme, B.; Hérault, O.; Charbord, P.; Domenech, J. The In Vitro Migration Capacity of Human Bone Marrow Mesenchymal Stem Cells: Comparison of Chemokine and Growth Factor Chemotactic Activities. *Stem Cells* **2007**, *25*, 1737–1745. [CrossRef]

56. Li, L.; Jiang, J. Regulatory factors of mesenchymal stem cell migration into injured tissues and their signal transduction mechanisms. *Front. Med.* **2011**, *5*, 33–39. [CrossRef] [PubMed]

57. Hermankova, B.; Kossl, J.; Bohacova, P.; Javorkova, E.; Hajkova, M.; Krulova, M.; Zajicova, A.; Holan, V. The Immunomodulatory Potential of Mesenchymal Stem Cells in a Retinal Inflammatory Environment. *Stem Cell Rev. Rep.* **2019**, *15*, 880–891. [CrossRef]

58. Grunnet, L.G.; Aikin, R.; Tonnesen, M.F.; Paraskevas, S.; Blaabjerg, L.; Storling, J.; Rosenberg, L.; Billestrup, N.; Maysinger, D.; Mandrup-Poulsen, T. Proinflammatory Cytokines Activate the Intrinsic Apoptotic Pathway in -Cells. *Diabetes* **2009**, *58*, 1807–1815. [CrossRef]

59. Yang, L.; Zhang, S.; Duan, H.; Dong, M.; Hu, X.; Zhang, Z.; Wang, Y.; Zhang, X.; Shi, W.; Zhou, Q. Different Effects of Pro-Inflammatory Factors and Hyperosmotic Stress on Corneal Epithelial Stem/Progenitor Cells and Wound Healing in Mice. *Stem Cells Transl. Med.* **2019**, *8*, 46–57. [CrossRef] [PubMed]

60. Woodward, A.M.; Di Zazzo, A.; Bonini, S.; Argüeso, P. Endoplasmic reticulum stress promotes inflammation-mediated proteolytic activity at the ocular surface. *Sci. Rep.* **2020**, *10*, 1–9. [CrossRef] [PubMed]

61. Khubutiya, M.S.; Vagabov, A.V.; Temnov, A.A.; Sklifas, A.N. Paracrine mechanisms of proliferative, anti-apoptotic and anti-inflammatory effects of mesenchymal stromal cells in models of acute organ injury. *Cytotherapy* **2014**, *16*, 579–585. [CrossRef]

62. Kossl, J.; Bohacova, P.; Hermankova, B.; Javorkova, E.; Zajicova, A.; Holan, V. Anti-Apoptotic Properties of Mesenchymal Stem Cells in a Mouse Model of Corneal Inflammation. *Stem Cells Dev.* **2021**. [CrossRef] [PubMed]

63. García, R.; Aguiar, J.; Alberti, E.; De La Cuétara, K.; Pavón, N. Bone marrow stromal cells produce nerve growth factor and glial cell line-derived neurotrophic factors. *Biochem. Biophys. Res. Commun.* **2004**, *316*, 753–754. [CrossRef]

64. Zhang, Y.; Wang, W. Effects of Bone Marrow Mesenchymal Stem Cell Transplantation on Light-Damaged Retina. *Investig. Opthalmol. Vis. Sci.* **2010**, *51*, 3742–3748. [CrossRef] [PubMed]

65. Zwart, I.; Hill, A.J.; Al-Allaf, F.; Shah, M.; Girdlestone, J.; Sanusi, A.B.; Mehmet, H.; Navarrete, R.; Navarrete, C.; Jen, L.-S. Umbilical cord blood mesenchymal stromal cells are neuroprotective and promote regeneration in a rat optic tract model. *Exp. Neurol.* **2009**, *216*, 439–448. [CrossRef] [PubMed]

66. Meirelles, L.D.S.; Fontes, A.M.; Covas, D.T.; Caplan, A.I. Mechanisms involved in the therapeutic properties of mesenchymal stem cells. *Cytokine Growth Factor Rev.* **2009**, *20*, 419–427. [CrossRef]

67. Xu, W.; Wang, X.; Xu, G.; Guo, J. Light-induced retinal injury enhanced neurotrophins secretion and neurotrophic effect of mesenchymal stem cells in vitro. *Arq. Bras. Oftalmol.* **2013**, *76*, 105–110. [CrossRef] [PubMed]

68. Inoue, Y.; Iriyama, A.; Ueno, S.; Takahashi, H.; Kondo, M.; Tamaki, Y.; Araie, M.; Yanagi, Y. Subretinal transplantation of bone marrow mesenchymal stem cells delays retinal degeneration in the RCS rat model of retinal degeneration. *Exp. Eye Res.* **2007**, *85*, 234–241. [CrossRef]

69. Notara, M.; Hernandez, D.; Mason, C.; Daniels, J.T. Characterization of the phenotype and functionality of corneal epithelial cells derived from mouse embryonic stem cells. *Regen. Med.* **2012**, *7*, 167–178. [CrossRef]

70. Gu, S.; Xing, C.; Han, J.; Tso, M.O.; Hong, J. Differentiation of rabbit bone marrow mesenchymal stem cells into corneal epithelial cells in vivo and ex vivo. *Mol. Vis.* **2009**, *15*, 99–107.

71. Jiang, T.-S.; Cai, L.; Ji, W.-Y.; Hui, Y.-N.; Wang, Y.-S.; Hu, D.; Zhu, J. Reconstruction of the corneal epithelium with induced marrow mesenchymal stem cells in rats. *Mol. Vis.* **2010**, *16*, 1304–1316. [PubMed]

72. Trosan, P.; Svobodova, E.; Chudickova, M.; Krulova, M.; Zajicova, A.; Holan, V. The Key Role of Insulin-Like Growth Factor I in Limbal Stem Cell Differentiation and the Corneal Wound-Healing Process. *Stem Cells Dev.* **2012**, *21*, 3341–3350. [CrossRef]

73. Tropel, P.; Platet, N.; Platel, J.-C.; Noël, D.; Albrieux, M.; Benabid, A.-L.; Berger, F. Functional Neuronal Differentiation of Bone Marrow-Derived Mesenchymal Stem Cells. *Stem Cells* **2006**, *24*, 2868–2876. [CrossRef] [PubMed]

74. Kicic, A.; Shen, W.-Y.; Wilson, A.S.; Constable, I.J.; Robertson, T.; Rakoczy, P.E. Differentiation of Marrow Stromal Cells into Photoreceptors in the Rat Eye. *J. Neurosci.* **2003**, *23*, 7742–7749. [CrossRef] [PubMed]

75. Nadri, S.; Kazemi, B.; Eeslaminejad, M.B.; Yazdani, S.; Soleimani, M.; Eslaminejad, M.B. High yield of cells committed to the photoreceptor-like cells from conjunctiva mesenchymal stem cells on nanofibrous scaffolds. *Mol. Biol. Rep.* **2013**, *40*, 3883–3890. [CrossRef] [PubMed]

76. Salehi, H.; Amirpour, N.; Razavi, S.; Esfandiari, E.; Zavar, R. Overview of retinal differentiation potential of mesenchymal stem cells: A promising approach for retinal cell therapy. *Ann. Anat. Anat. Anz.* **2017**, *210*, 52–63. [CrossRef]

77. Castanheira, P.; Torquetti, L.; Nehemy, M.B.; Goes, A.M. Retinal incorporation and differentiation of mesenchymal stem cells intravitreally injected in the injured retina of rats. *Arq. Bras. Oftalmol.* **2008**, *71*, 644–650. [CrossRef] [PubMed]

78. Huo, D.-M.; Dong, F.-T.; Yu, W.-H.; Gao, F. Differentiation of mesenchymal stem cell in the microenviroment of retinitis pigmentosa. *Int. J. Ophthalmol.* **2010**, *3*, 216–219. [PubMed]

79. Huang, C.; Zhang, J.; Ao, M.; Li, Y.; Zhang, C.; Xu, Y.; Li, X.; Wang, W. Combination of retinal pigment epithelium cell-conditioned medium and photoreceptor outer segments stimulate mesenchymal stem cell differentiation toward a functional retinal pigment epithelium cell phenotype. *J. Cell. Biochem.* **2011**, *113*, 590–598. [CrossRef]

80. Mathivanan, I.; Trepp, C.M.; Brunold, C.; Baerlocher, G.M.; Enzmann, V. Retinal differentiation of human bone marrow-derived stem cells by co-culture with retinal pigment epithelium in vitro. *Exp. Cell Res.* **2015**, *333*, 11–20. [CrossRef] [PubMed]

81. Croitoru-Lamoury, J.; Lamoury, F.M.J.; Caristo, M.; Suzuki, K.; Walker, D.; Takikawa, O.; Taylor, R.; Brew, B.J. Interferon-γ Regulates the Proliferation and Differentiation of Mesenchymal Stem Cells via Activation of Indoleamine 2,3 Dioxygenase (IDO). *PLoS ONE* **2011**, *6*, e14698. [CrossRef]

82. Wong, G.; Goldshmit, Y.; Turnley, A.M. Interferon-γ but not TNFα promotes neuronal differentiation and neurite outgrowth of murine adult neural stem cells. *Exp. Neurol.* **2004**, *187*, 171–177. [CrossRef]

83. Liang, X.; Ding, Y.; Zhang, Y.; Tse, H.F.; Lian, Q. Paracrine Mechanisms of Mesenchymal Stem Cell-Based Therapy: Current Status and Perspectives. *Cell Transplant.* **2014**, *23*, 1045–1059. [CrossRef] [PubMed]

84. Adak, S.; Magdalene, D.; Deshmukh, S.; Das, D.; Jaganathan, B.G. A Review on Mesenchymal Stem Cells for Treatment of Retinal Diseases. *Stem Cell Rev. Rep.* **2021**, 1–20. [CrossRef]

85. Yu, B.; Shao, H.; Su, C.; Jiang, Y.; Chen, X.; Bai, L.; Zhang, Y.; Li, Q.; Zhang, X.; Li, X. Exosomes derived from MSCs ameliorate retinal laser injury partially by inhibition of MCP-1. *Sci. Rep.* **2016**, *6*, srep34562. [CrossRef] [PubMed]

86. Mead, B.; Tomarev, S. Bone Marrow-Derived Mesenchymal Stem Cells-Derived Exosomes Promote Survival of Retinal Ganglion Cells Through miRNA-Dependent Mechanisms. *Stem Cells Transl. Med.* **2017**, *6*, 1273–1285. [CrossRef]

87. Islam, M.N.; Das, S.R.; Emin, M.T.; Wei, M.; Sun, L.; Westphalen, K.; Rowlands, D.J.; Quadri, S.K.; Bhattacharya, S.; Bhattacharya, J. Mitochondrial transfer from bone-marrow–derived stromal cells to pulmonary alveoli protects against acute lung injury. *Nat. Med.* **2012**, *18*, 759–765. [CrossRef]

88. Paliwal, S.; Chaudhuri, R.; Agrawal, A.; Mohanty, S. Regenerative abilities of mesenchymal stem cells through mitochondrial transfer. *J. Biomed. Sci.* **2018**, *25*, 1–12. [CrossRef] [PubMed]

89. Eells, J.T. Mitochondrial Dysfunction in the Aging Retina. *Biology* **2019**, *8*, 31. [CrossRef]

90. Freeman, B.T.; Kouris, N.A.; Ogle, B.M. Tracking Fusion of Human Mesenchymal Stem Cells After Transplantation to the Heart. *Stem Cells Transl. Med.* **2015**, *4*, 685–694. [CrossRef]

91. Azizi, Z.; Lange, C.; Paroni, F.; Ardestani, A.; Meyer, A.; Wu, Y.; Zander, A.R.; Westenfelder, C.; Maedler, K. β-MSCs: Successful fusion of MSCs with β-cells results in a β-cell like phenotype. *Oncotarget* **2016**, *7*, 48963–48977. [CrossRef]

92. Yang, Z.; Li, K.; Yan, X.; Dong, F.; Zhao, C. Amelioration of diabetic retinopathy by engrafted human adipose-derived mesenchymal stem cells in streptozotocin diabetic rats. *Graefe's Arch. Clin. Exp. Ophthalmol.* **2010**, *248*, 1415–1422. [CrossRef]

93. Mendel, T.A.; Clabough, E.B.D.; Kao, D.S.; Demidova-Rice, T.N.; Durham, J.T.; Zotter, B.C.; Seaman, S.A.; Cronk, S.M.; Rakoczy, E.P.; Katz, A.J.; et al. Pericytes Derived from Adipose-Derived Stem Cells Protect against Retinal Vasculopathy. *PLoS ONE* **2013**, *8*, e65691. [CrossRef]

94. Rajashekhar, G.; Ramadan, A.; Abburi, C.; Callaghan, B.; Traktuev, D.O.; Evans-Molina, C.; Maturi, R.; Harris, A.; Kern, T.S.; March, K.L. Regenerative Therapeutic Potential of Adipose Stromal Cells in Early Stage Diabetic Retinopathy. *PLoS ONE* **2014**, *9*, e84671. [CrossRef]

95. Ezquer, F.; Ezquer, M.; Conget, P.; Arango-Rodriguez, M. Could donor multipotent mesenchymal stromal cells prevent or delay the onset of diabetic retinopathy? *Acta Ophthalmol.* **2013**, *92*, e86–e95. [CrossRef]

96. Mead, B.; Berry, M.; Logan, A.; Scott, R.A.; Leadbeater, W.; Scheven, B.A. Stem cell treatment of degenerative eye disease. *Stem Cell Res.* **2015**, *14*, 243–257. [CrossRef] [PubMed]

97. Holan, V.; Hermankova, B.; Kossl, J. Perspectives of Stem Cell–Based Therapy for Age-Related Retinal Degenerative Diseases. *Cell Transplant.* **2017**, *26*, 1538–1541. [CrossRef]

98. Holan, V.; Hermankova, B.; Krulova, M.; Zajicova, A. Cytokine interplay among the diseased retina, inflammatory cells and mesenchymal stem cell-a clue to stem cell-based therapy. *World J. Stem Cells* **2019**, *11*, 957–967. [CrossRef]

99. Park, S.S.; Moisseiev, E.; Bauer, G.; Anderson, J.D.; Grant, M.B.; Zam, A.; Zawadzki, R.J.; Werner, J.S.; Nolta, J.A. Advances in bone marrow stem cell therapy for retinal dysfunction. *Prog. Retin. Eye Res.* **2017**, *56*, 148–165. [CrossRef] [PubMed]

100. Ng, T.K.; Fortino, V.R.; Pelaez, D.; Cheung, H.S. Progress of mesenchymal stem cell therapy for neural and retinal diseases. *World J. Stem Cells* **2014**, *6*, 111–119. [CrossRef] [PubMed]

101. Labrador-Velandia, S.; Alonso-Alonso, M.L.; Alvarez-Sanchez, S.; González-Zamora, J.; Carretero-Barrio, I.; Pastor, J.C.; Fernandez-Bueno, I.; Srivastava, G.K. Mesenchymal stem cell therapy in retinal and optic nerve diseases: An update of clinical trials. *World J. Stem Cells* **2016**, *8*, 376–383. [CrossRef] [PubMed]

102. Zhang, X.-Y.; Ng, T.K.; Brelén, M.E.; Wu, D.; Wang, J.X.; Chan, K.P.; Yung, J.S.Y.; Cao, D.; Wang, Y.; Zhang, S.; et al. Continuous exposure to non-lethal doses of sodium iodate induces retinal pigment epithelial cell dysfunction. *Sci. Rep.* **2016**, *6*. [CrossRef] [PubMed]

103. Chowers, G.; Cohen, M.; Marks-Ohana, D.; Stika, S.; Eijzenberg, A.; Banin, E.; Obolensky, A. Course of Sodium Iodate–Induced Retinal Degeneration in Albino and Pigmented Mice. *Investig. Opthalmol. Vis. Sci.* **2017**, *58*, 2239–2249. [CrossRef]

104. Tao, Y.; Chen, T.; Fang, W.; Peng, G.; Wang, L.; Qin, L.; Liu, B.; Huang, Y.F. The temporal topography of the N-Methyl-N-nitrosourea induced photoreceptor degeneration in mouse retina. *Sci. Rep.* **2015**, *5*. [CrossRef] [PubMed]

105. Pirmardan, E.R.; Soheili, Z.-S.; Samiei, S.; Ahmadieh, H.; Mowla, S.J.; Naseri, M.; Daftarian, N. In Vivo Evaluation of PAX6 Overexpression and NMDA Cytotoxicity to Stimulate Proliferation in the Mouse Retina. *Sci. Rep.* **2018**, *8*, 17700. [CrossRef]

106. Jin, Z.-B.; Gao, M.-L.; Deng, W.-L.; Wu, K.-C.; Sugita, S.; Mandai, M.; Takahashi, M. Stemming retinal regeneration with pluripotent stem cells. *Prog. Retin. Eye Res.* **2019**, *69*, 38–56. [CrossRef]

107. Ludwig, P.E.; Freeman, S.C.; Janot, A.C. Novel stem cell and gene therapy in diabetic retinopathy, age related macular degeneration, and retinitis pigmentosa. *Int. J. Retin. Vitr.* **2019**, *5*, 1–14. [CrossRef]

108. Nuzzi, R.; Tridico, F. Perspectives of Autologous Mesenchymal Stem-Cell Transplantation in Macular Hole Surgery: A Review of Current Findings. *J. Ophthalmol.* **2019**, *2019*, 1–8. [CrossRef]

109. Wagner, W.; Ho, A.D.; Zenke, M. Different Facets of Aging in Human Mesenchymal Stem Cells. *Tissue Eng. Part B Rev.* **2010**, *16*, 445–453. [CrossRef] [PubMed]

110. Rombouts, W.J.C.; Ploemacher, E.R. Primary murine MSC show highly efficient homing to the bone marrow but lose homing ability following culture. *Leukemia* **2003**, *17*, 160–170. [CrossRef] [PubMed]

111. Eggenhofer, E.; Benseler, V.; Kroemer, H.; Popp, F.; Geissler, E.; Schlitt, H.; Baan, C.; Dahlke, M.; Hoogduijn, M.J. Mesenchymal stem cells are short-lived and do not migrate beyond the lungs after intravenous infusion. *Front. Immunol.* **2012**, *3*, 297. [CrossRef]

112. Huang, H.; Kolibabka, M.; Eshwaran, R.; Chatterjee, A.; Schlotterer, A.; Willer, H.; Bieback, K.; Hammes, H.-P.; Feng, Y. Intravitreal injection of mesenchymal stem cells evokes retinal vascular damage in rats. *FASEB J.* **2019**, *33*, 14668–14679. [CrossRef] [PubMed]

113. Oner, A.; Gonen, Z.B.; Sinim, N.; Cetin, M.; Ozkul, Y. Subretinal adipose tissue-derived mesenchymal stem cell implantation in advanced stage retinitis pigmentosa: A phase I clinical safety study. *Stem Cell Res. Ther.* **2016**, *7*. [CrossRef] [PubMed]

114. Kahraman, N.S. Umbilical cord derived mesenchymal stem cell implantation in retinitis pigmentosa: A 6-month follow-up results of a phase 3 trial. *Int. J. Ophthalmol.* **2020**, *13*, 1423–1429. [CrossRef] [PubMed]

115. Oumlzmert, E.; Arslan, U. Management of retinitis pigmentosa by Wharton's jelly-derived mesenchymal stem cells: Prospective analysis of 1-year results. *Stem Cell Res. Ther.* **2020**, *11*, 353. [CrossRef]

116. Niwa, M.; Aoki, H.; Hirata, A.; Tomita, H.; Green, P.G.; Hara, A. Retinal Cell Degeneration in Animal Models. *Int. J. Mol. Sci.* **2016**, *17*, 110. [CrossRef]

117. Hanus, J.; Anderson, C.; Sarraf, D.; Ma, J.; Wang, S. Retinal pigment epithelial cell necroptosis in response to sodium iodate. *Cell Death Discov.* **2016**, *2*, 16054. [CrossRef]

118. Mao, X.; Pan, T.; Shen, H.; Xi, H.; Yuan, S.; Liu, Q. The rescue effect of mesenchymal stem cell on sodium iodate-induced retinal pigment epithelial cell death through deactivation of NF-κB-mediated NLRP3 inflammasome. *Biomed. Pharmacother.* **2018**, *103*, 517–523. [CrossRef]

119. Moriguchi, M.; Nakamura, S.; Inoue, Y.; Nishinaka, A.; Nakamura, M.; Shimazawa, M.; Hara, H. Irreversible Photoreceptors and RPE Cells Damage by Intravenous Sodium Iodate in Mice Is Related to Macrophage Accumulation. *Investig. Opthalmol. Vis. Sci.* **2018**, *59*, 3476–3487. [CrossRef] [PubMed]

120. Liu, Y.; Li, Y.; Wang, C.; Zhang, Y.; Su, G. Morphologic and histopathologic change of sodium iodate-induced retinal degeneration in adult rats. *Int. J. Clin. Exp. Pathol.* **2019**, *12*, 443–454.

121. Bhutto, I.A.; Ogura, S.; Baldeosingh, R.; McLeod, D.S.; Lutty, G.A.; Edwards, M.M. An Acute Injury Model for the Phenotypic Characteristics of Geographic Atrophy. *Investig. Opthalmol. Vis. Sci.* **2018**, *59*, AMD143–AMD151. [CrossRef] [PubMed]
122. Ahn, S.M.; Ahn, J.; Cha, S.; Yun, C.; Park, T.K.; Kim, Y.-J.; Goo, Y.S.; Kim, S.-W. The effects of intravitreal sodium iodate injection on retinal degeneration following vitrectomy in rabbits. *Sci. Rep.* **2019**, *9*, 15696–15710. [CrossRef] [PubMed]
123. Barzelay, A.; Algor, S.W.; Niztan, A.; Katz, S.; Benhamou, M.; Nakdimon, I.; Azmon, N.; Gozlan, S.; Mezad-Koursh, D.; Neudorfer, M.; et al. Adipose-Derived Mesenchymal Stem Cells Migrate and Rescue RPE in the Setting of Oxidative Stress. *Stem Cells Int.* **2018**, *2018*, 1–11. [CrossRef]
124. Gong, L.; Wu, Q.; Song, B.; Lu, B.; Zhang, Y. Differentiation of rat mesenchymal stem cells transplanted into the subretinal space of sodium iodate-injected rats. *Clin. Exp. Ophthalmol.* **2008**, *36*, 666–671. [CrossRef]
125. Fiori, A.; Terlizzi, V.; Kremer, H.; Gebauer, J.; Hammes, H.-P.; Harmsen, M.C.; Bieback, K. Mesenchymal stromal/stem cells as potential therapy in diabetic retinopathy. *Immunobiology* **2018**, *223*, 729–743. [CrossRef]
126. Ezquer, M.; Urzua, C.A.; Montecino, S.; Leal, K.; Conget, P.; Ezquer, F. Intravitreal administration of multipotent mesenchymal stromal cells triggers a cytoprotective microenvironment in the retina of diabetic mice. *Stem Cell Res. Ther.* **2016**, *7*, 42. [CrossRef] [PubMed]
127. Zhang, W.; Wang, Y.; Kong, J.; Dong, M.; Duan, H.; Chen, S. Therapeutic efficacy of neural stem cells originating from umbilical cord-derived mesenchymal stem cells in diabetic retinopathy. *Sci. Rep.* **2017**, *7*, 1–8. [CrossRef]
128. Kong, J.-H.; Zheng, D.; Chen, S.; Duan, H.-T.; Wang, Y.-X.; Dong, M.; Song, J. A comparative study on the transplantation of different concentrations of human umbilical mesenchymal cells into diabetic rats. *Int. J. Ophthalmol.* **2015**, *8*, 257–262. [PubMed]
129. Cronk, S.M.; Kelly-Goss, M.R.; Ray, H.C.; Mendel, T.A.; Hoehn, K.L.; Bruce, A.C.; Dey, B.K.; Guendel, A.M.; Tavakol, D.N.; Herman, I.M.; et al. Adipose-Derived Stem Cells From Diabetic Mice Show Impaired Vascular Stabilization in a Murine Model of Diabetic Retinopathy. *Stem Cells Transl. Med.* **2015**, *4*, 459–467. [CrossRef]
130. Nagaishi, K.; Mizue, Y.; Chikenji, T.; Otani, M.; Nakano, M.; Saijo, Y.; Tsuchida, H.; Ishioka, S.; Nishikawa, A.; Saito, T.; et al. Umbilical cord extracts improve diabetic abnormalities in bone marrow-derived mesenchymal stem cells and increase their therapeutic effects on diabetic nephropathy. *Sci. Rep.* **2017**, *7*, 1–17. [CrossRef]
131. Robinson, R.; Barathi, V.A.; Chaurasia, S.S.; Wong, T.Y.; Kern, T.S. Update on animal models of diabetic retinopathy: From molecular approaches to mice and higher mammals. *Dis. Model. Mech.* **2012**, *5*, 444–456. [CrossRef] [PubMed]
132. Lai, A.K.W.; Lo, A.C.Y. Animal Models of Diabetic Retinopathy: Summary and Comparison. *J. Diabetes Res.* **2013**, *2013*, 1–29. [CrossRef] [PubMed]
133. Araújo, R.S.; Silva, M.S.; Santos, D.F.; Silva, G.A. Dysregulation of trophic factors contributes to diabetic retinopathy in the Ins2Akita mouse. *Exp. Eye Res.* **2020**, *194*. [CrossRef] [PubMed]
134. Elshaer, S.L.; Evans, W.; Pentecost, M.; Lenin, R.; Periasamy, R.; Jha, K.A.; Alli, S.; Gentry, J.; Thomas, S.M.; Sohl, N.; et al. Adipose stem cells and their paracrine factors are therapeutic for early retinal complications of diabetes in the Ins2Akita mouse. *Stem Cell Res. Ther.* **2018**, *9*, 1–18. [CrossRef]
135. Van Hove, I.; De Groef, L.; Boeckx, B.; Modave, E.; Hu, T.-T.; Beets, K.; Etienne, I.; Van Bergen, T.; Lambrechts, D.; Moons, L.; et al. Single-cell transcriptome analysis of the Akimba mouse retina reveals cell-type-specific insights into the pathobiology of diabetic retinopathy. *Diabetologia* **2020**, *63*, 2235–2248. [CrossRef]
136. Chaurasia, S.S.; Lim, R.R.; Parikh, B.H.; Wey, Y.S.; Tun, B.B.; Wong, T.Y.; Luu, C.D.; Agrawal, R.; Ghosh, A.; Mortellaro, A.; et al. The NLRP3 Inflammasome May Contribute to Pathologic Neovascularization in the Advanced Stages of Diabetic Retinopathy. *Sci. Rep.* **2018**, *8*, 1–15. [CrossRef]
137. Rivas, M.A.; Vecino, E. Animal models and different therapies for treatment of retinitis pigmentosa. *Histol. Histopathol.* **2009**, *24*, 1295–1322. [CrossRef]
138. He, Y.; Zhang, Y.; Liu, X.; Ghazaryan, E.; Li, Y.; Xie, J.; Su, G. Recent Advances of Stem Cell Therapy for Retinitis Pigmentosa. *Int. J. Mol. Sci.* **2014**, *15*, 14456–14474. [CrossRef]
139. Tsubura, A.; Yoshizawa, K.; Kuwata, M.; Uehara, N. Animal models for retinitis pigmentosa induced by MNU, disease progression, mechanisms and therapeutic trials. *Histol. Histopathol.* **2010**, *25*, 933–944. [CrossRef]
140. Zhou, T.; Huang, Z.; Sun, X.; Zhu, X.; Zhou, L.; Li, M.; Cheng, B.; Liu, X.; He, C. Microglia Polarization with M1/M2 Phenotype Changes in rd1 Mouse Model of Retinal Degeneration. *Front. Neuroanat.* **2017**, *11*, 77. [CrossRef]
141. Gargini, C.; Terzibasi, E.; Mazzoni, F.; Strettoi, E. Retinal organization in the retinal degeneration 10 (rd10) mutant mouse: A morphological and ERG study. *J. Comp. Neurol.* **2006**, *500*, 222–238. [CrossRef] [PubMed]
142. Zhao, L.; Zabel, M.K.; Wang, X.; Ma, W.; Shah, P.; Fariss, R.N.; Qian, H.; Parkhurst, C.N.; Gan, W.; Wong, W.T. Microglial phagocytosis of living photoreceptors contributes to inherited retinal degeneration. *EMBO Mol. Med.* **2015**, *7*, 1179–1197. [CrossRef]
143. Kameya, S.; Hawes, N.L.; Chang, B.; Heckenlively, J.R.; Naggert, J.K.; Nishina, P.M. Mfrp, a gene encoding a frizzled related protein, is mutated in the mouse retinal degeneration 6. *Hum. Mol. Genet.* **2002**, *11*, 1879–1886. [CrossRef]
144. D'Cruz, P.M.; Yasumura, D.; Weir, J.; Matthes, M.T.; Abderrahim, H.; Lavail, M.M.; Vollrath, D. Mutation of the receptor tyrosine kinase gene Mertk in the retinal dystrophic RCS rat. *Hum. Mol. Genet.* **2000**, *9*, 645–651. [CrossRef] [PubMed]
145. Di Pierdomenico, J.; García-Ayuso, D.; Pinilla, I.; Cuenca, N.; Vidal-Sanz, M.; Agudo-Barriuso, M.; Villegas-Pérez, M.P. Early Events in Retinal Degeneration Caused by Rhodopsin Mutation or Pigment Epithelium Malfunction: Differences and Similarities. *Front. Neuroanat.* **2017**, *11*, 14. [CrossRef] [PubMed]

146. Otani, A.; Dorrell, M.I.; Kinder, K.; Moreno, S.K.; Nusinowitz, S.; Banin, E.; Heckenlively, J.; Friedlander, M. Rescue of retinal degen-eration by intravitreally injected adult bone marrow-derived lineage-negative hematopoietic stem cells. *J. Clin. Investig.* **2004**, *114*, 765–774. [CrossRef] [PubMed]

147. Lejkowska, R.; Kawa, M.P.; Pius-Sadowska, E.; Rogińska, D.; Łuczkowska, K.; Machaliński, B.; Machalińska, A. Preclinical Evaluation of Long-Term Neuroprotective Effects of BDNF-Engineered Mesenchymal Stromal Cells as Intravitreal Therapy for Chronic Retinal Degeneration in Rd6 Mutant Mice. *Int. J. Mol. Sci.* **2019**, *20*, 777. [CrossRef] [PubMed]

148. Qu, L.; Gao, L.; Xu, H.; Duan, P.; Zeng, Y.; Liu, Y.; Yin, Z.Q. Combined transplantation of human mesenchymal stem cells and human retinal progenitor cells into the subretinal space of RCS rats. *Sci. Rep.* **2017**, *7*, 1–14. [CrossRef]

149. Deng, C.-L.; Hu, C.-B.; Wang, B.-Y.; Xiong, Y.-C.; Chen, T.; Zhao, N.; Bao, L.-H.; Quan, R.; Du, F.-Y.; Sui, B.-D.; et al. Bone progeria diminished the therapeutic effects of bone marrow mesenchymal stem cells on retinal degeneration. *Biochem. Biophys. Res. Commun.* **2020**, *531*, 180–186. [CrossRef]

150. Johnson, T.V.; Tomarev, S.I. Rodent models of glaucoma. *Brain Res. Bull.* **2010**, *81*, 349–358. [CrossRef] [PubMed]

151. Harada, C.; Kimura, A.; Guo, X.; Namekata, K.; Harada, T. Recent advances in genetically modified animal models of glaucoma and their roles in drug repositioning. *Br. J. Ophthalmol.* **2018**, *103*, 161–166. [CrossRef]

152. Overby, D.R.; Clark, A.F. Animal models of glucocorticoid-induced glaucoma. *Exp. Eye Res.* **2015**, *141*, 15–22. [CrossRef]

153. Biswas, S.; Wan, K.H. Review of rodent hypertensive glaucoma models. *Acta Ophthalmol.* **2019**, *97*, e331–e340. [CrossRef] [PubMed]

154. Bai, Y.; Zhu, Y.; Chen, Q.; Xu, J.; Sarunic, M.V.; Saragovi, U.H.; Zhuo, Y. Validation of glaucoma-like features in the rat episcleral vein cauterization model. *Chin. Med. J.* **2014**, *127*, 359–364. [PubMed]

155. Huang, W.; Hu, F.; Wang, M.; Gao, F.; Xu, P.; Xing, C.; Sun, X.; Zhang, S.; Wu, J. Comparative analysis of retinal ganglion cell damage in three glaucomatous rat models. *Exp. Eye Res.* **2018**, *172*, 112–122. [CrossRef]

156. Mead, B.; Hill, L.J.; Blanch, R.J.; Ward, K.; Logan, A.; Berry, M.; Leadbeater, W.; Scheven, B.A. Mesenchymal stromal cell–mediated neuroprotection and functional preservation of retinal ganglion cells in a rodent model of glaucoma. *Cytotherapy* **2016**, *18*, 487–496. [CrossRef]

157. Johnson, T.V.; Bull, N.D.; Hunt, D.P.; Marina, N.; Tomarev, S.I.; Martin, K.R. Neuroprotective Effects of Intravitreal Mesenchymal Stem Cell Transplantation in Experimental Glaucoma. *Investig. Opthalmol. Vis. Sci.* **2010**, *51*, 2051–2059. [CrossRef]

158. Manuguerra-Gagné, R.; Boulos, P.R.; Ammar, A.; Leblond, F.A.; Krosl, G.; Pichette, V.; Lesk, M.R.; Roy, D.-C. Transplantation of Mesenchymal Stem Cells Promotes Tissue Regeneration in a Glaucoma Model Through Laser-Induced Paracrine Factor Secretion and Progenitor Cell Recruitment. *Stem Cells* **2013**, *31*, 1136–1148. [CrossRef] [PubMed]

159. Emre, E.; Yüksel, N.; Duruksu, G.; Pirhan, D.; Subaşi, C.; Erman, G.; Karaöz, E. Neuroprotective effects of intravitreally transplanted adipose tissue and bone marrow–derived mesenchymal stem cells in an experimental ocular hypertension model. *Cytotherapy* **2015**, *17*, 543–559. [CrossRef] [PubMed]

160. Pan, D.; Chang, X.; Xu, M.; Zhang, M.; Zhang, S.; Wang, Y.; Luo, X.; Xu, J.; Yang, X.; Sun, X. UMSC-derived exosomes promote retinal ganglion cels survival in a rat model of optic nerve crush. *J. Chem. Neuroanat.* **2019**, *96*, 134–139. [CrossRef]

161. Çerman, E.; Akkoc, T.; Eraslan, M.; Şahin, O.; Ozkara, S.; Aker, F.V.; Subaşı, C.; Karaoz, E.; Akkoç, T. Retinal Electrophysiological Effects of Intravitreal Bone Marrow Derived Mesenchymal Stem Cells in Streptozotocin Induced Diabetic Rats. *PLoS ONE* **2016**, *11*, e0156495. [CrossRef]

162. Roubeix, C.; Godefroy, D.; Mias, C.; Sapienza, A.; Riancho, L.; Degardin, J.; Fradot, V.; Ivkovic, I.; Picaud, S.; Sennlaub, F.; et al. Intraocular pressure reduction and neuroprotection conferred by bone marrow-derived mesenchymal stem cells in an animal model of glaucoma. *Stem Cell Res. Ther.* **2015**, *6*, 1–13. [CrossRef]

163. Harper, M.M.; Grozdanic, S.D.; Blits, B.; Kuehn, M.H.; Zamzow, D.; Buss, J.E.; Kardon, R.H.; Sakaguchi, D.S. Transplantation of BDNF-Secreting Mesenchymal Stem Cells Provides Neuroprotection in Chronically Hypertensive Rat Eyes. *Investig. Opthalmol. Vis. Sci.* **2011**, *52*, 4506–4515. [CrossRef] [PubMed]

164. Park, S.S.; Bauer, G.; Abedi, M.; Pontow, S.; Panorgias, A.; Jonnal, R.S.; Zawadzki, R.J.; Werner, J.S.; Nolta, A.J. Intravitreal Autologous Bone Marrow CD34+ Cell Therapy for Ischemic and Degenerative Retinal Disorders: Preliminary Phase 1 Clinical Trial Findings. *Investig. Opthalmol. Vis. Sci.* **2014**, *56*, 81–89. [CrossRef]

165. Siqueira, R.C.; Messias, A.; Voltarelli, J.C.; Scott, I.U.; Jorge, R. Intravitreal injection of autologous bone marrow–derived mononuclear cells for hereditary retinal dystrophy. *Retina* **2011**, *31*, 1207–1214. [CrossRef]

166. Gu, X.; Yu, X.; Zhao, C.; Duan, P.; Zhao, T.; Liu, Y.; Li, S.; Yang, Z.; Li, Y.; Qian, C.; et al. Efficacy and Safety of Autologous Bone Marrow Mesenchymal Stem Cell Transplantation in Patients with Diabetic Retinopathy. *Cell. Physiol. Biochem.* **2018**, *49*, 40–52. [CrossRef] [PubMed]

167. Weiss, J.N.; Levy, S. Stem Cell Ophthalmology Treatment Study: Bone marrow derived stem cells in the treatment of Retinitis Pigmentosa. *Stem Cell Investig.* **2018**, *5*, 18. [CrossRef]

168. Levy, S.; Weiss, J.N.; Malkin, A. Stem Cell Ophthalmology Treatment Study (SCOTS) for retinal and optic nerve diseases: A preliminary report. *Neural Regen. Res.* **2015**, *10*, 982–988. [CrossRef]

169. Weiss, J.N.; Levy, S.; Benes, S.C. Stem Cell Ophthalmology Treatment Study: Bone marrow derived stem cells in the treatment of non-arteritic ischemic optic neuropathy (NAION). *Stem Cell Investig.* **2017**, *4*, 94. [CrossRef]

170. Hoogduijn, M.J.; Dor, F.J.M.F. Mesenchymal Stem Cells: Are We Ready for Clinical Application in Transplantation and Tissue Regeneration? *Front. Immunol.* **2013**, *4*, 144. [CrossRef] [PubMed]

171. Bhattacharya, S.; Gangaraju, R.; Chaum, E. Recent Advances in Retinal Stem Cell Therapy. *Curr. Mol. Biol. Rep.* **2017**, *3*, 172–182. [CrossRef] [PubMed]
172. Wang, Y.; Tang, Z.; Gu, P. Stem/progenitor cell-based transplantation for retinal degeneration: A review of clinical trials. *Cell Death Dis.* **2020**, *11*, 1–14. [CrossRef]
173. Tzameret, A.; Sher, I.; Belkin, M.; Treves, A.J.; Meir, A.; Nagler, A.; Levkovitch-Verbin, H.; Rotenstreich, Y.; Solomon, A.S. Epiretinal transplantation of human bone marrow mesenchymal stem cells rescues retinal and vision function in a rat model of retinal degeneration. *Stem Cell Res.* **2015**, *15*, 387–394. [CrossRef]
174. Ji, S.; Lin, S.; Chen, J.; Huang, X.; Wei, C.-C.; Li, Z.; Tang, S. Neuroprotection of Transplanting Human Umbilical Cord Mesenchymal Stem Cells in a Microbead Induced Ocular Hypertension Rat Model. *Curr. Eye Res.* **2018**, *43*, 810–820. [CrossRef]
175. Velandia, S.L.; Di Lauro, S.; Alonso-Alonso, M.L.; Bartolomé, S.T.; Srivastava, G.K.; Pastor, J.C.; Fernandez-Bueno, I. Biocompatibility of intravitreal injection of human mesenchymal stem cells in immunocompetent rabbits. *Graefe's Arch. Clin. Exp. Ophthalmol.* **2017**, *256*, 125–134. [CrossRef]
176. Lohan, P.; Treacy, O.; Morcos, M.; Donohoe, E.; O'Donoghue, Y.; Ryan, A.E.; Elliman, S.J.; Ritter, T.; Griffin, M.D. Interspecies Incompatibilities Limit the Immunomodulatory Effect of Human Mesenchymal Stromal Cells in the Rat. *Stem Cells* **2018**, *36*, 1210–1215. [CrossRef]
177. Oh, J.Y.; Kim, M.K.; Shin, M.S.; Wee, W.R.; Lee, J.H. Cytokine secretion by human mesenchymal stem cells cocultured with damaged corneal epithelial cells. *Cytokine* **2009**, *46*, 100–103. [CrossRef]
178. Zhou, L.; Lopes, J.E.; Chong, M.M.W.; Ivanov, I.I.; Min, R.; Victora, G.D.; Shen, Y.; Du, J.; Rubtsov, Y.P.; Rudensky, A.Y.; et al. TGF-β-induced Foxp3 inhibits TH17 cell differentiation by antagonizing RORγt function. *Nat. Cell Biol.* **2008**, *453*, 236–240. [CrossRef] [PubMed]

cells

Review

Kinins and Their Receptors as Potential Therapeutic Targets in Retinal Pathologies

Rahmeh Othman [1,2,*], Gael Cagnone [3], Jean-Sébastien Joyal [3], Elvire Vaucher [1,*] and Réjean Couture [2,*]

1 School of Optometry, Université de Montréal, Montreal, QC H3T 1P1, Canada
2 Department of Pharmacology and Physiology, Faculty of Medicine, Université de Montréal, Montreal, QC H3T 1J4, Canada
3 Department of Pediatry, Faculty of Medicine, CHU St Justine, Université de Montréal, Montreal, QC H3T 1J4, Canada; gael.cagnone@umontreal.ca (G.C.); js.joyal@umontreal.ca (J.-S.J.)
* Correspondence: rahmeh.othman@umontreal.ca (R.O.); elvire.vaucher@umontreal.ca (E.V.); rejean.couture@umontreal.ca (R.C.)

Abstract: The kallikrein-kinin system (KKS) contributes to retinal inflammation and neovascularization, notably in diabetic retinopathy (DR) and neovascular age-related macular degeneration (AMD). Bradykinin type 1 (B1R) and type 2 (B2R) receptors are G-protein-coupled receptors that sense and mediate the effects of kinins. While B2R is constitutively expressed and regulates a plethora of physiological processes, B1R is almost undetectable under physiological conditions and contributes to pathological inflammation. Several KKS components (kininogens, tissue and plasma kallikreins, and kinin receptors) are overexpressed in human and animal models of retinal diseases, and their inhibition, particularly B1R, reduces inflammation and pathological neovascularization. In this review, we provide an overview of the KKS with emphasis on kinin receptors in the healthy retina and their detrimental roles in DR and AMD. We highlight the crosstalk between the KKS and the renin–angiotensin system (RAS), which is known to be detrimental in ocular pathologies. Targeting the KKS, particularly the B1R, is a promising therapy in retinal diseases, and B1R may represent an effector of the detrimental effects of RAS (Ang II-AT1R).

Keywords: kallikrein-kinin system; kinin receptors; diabetic retinopathy; age-related macular degeneration

check for updates

Citation: Othman, R.; Cagnone, G.; Joyal, J.-S.; Vaucher, E.; Couture, R. Kinins and Their Receptors as Potential Therapeutic Targets in Retinal Pathologies. *Cells* **2021**, *10*, 1913. https://doi.org/10.3390/cells10081913

Academic Editors: Maurice Ptito and Joseph Bouskila

Received: 14 June 2021
Accepted: 24 July 2021
Published: 28 July 2021

Publisher's Note: MDPI stays neutral with regard to jurisdictional claims in published maps and institutional affiliations.

1. Preface

Ocular pathologies involving chronic inflammation of the retina are particularly devastating in terms of visual acuity. Among these, age-related macular degeneration (AMD) and diabetic retinopathy (DR) are the leading cause of severe vision loss in the elderly and active population of industrialized countries, respectively. In addition to the chronic inflammation, vascular dysfunction and neovascularization, which correspond to the formation of new pathological branches from pre-existing retinal or choroidal vessels, occur. The inflammatory process includes a breakdown of the blood–retinal barrier, leukocyte adhesion on the blood vessel wall, macrophage and microglial activation, and cytokine and chemokine production. Current treatments of these diseases are only compensatory and consist commonly of invasive treatments such as quarterly intravitreal (ITV) injections of anti-angiogenesis agents (anti-VEGF antibodies) or laser coagulation to prevent loss of sight due to aberrant neovascularization. Moreover, a large population of patients does not respond to anti-VEGF therapy. To offer alternative and comfortable treatment to nonresponders, such as a topical approach, our team's ongoing research effort has shown that the kallikrein-kinin system (KKS)—involved in inflammation—is overexpressed in the human AMD and DR retina and contributes to the development of pathological events in animal models of these diseases. Moreover, we were able to specifically target the KKS via topical ocular kinin B1 receptor (B1R) antagonist administration, which decreased neovascularization and retinal inflammatory responses. The purpose of this review is thus to better describe the possible involvement of the KKS in retinal diseases and therapeutical approaches that can prevent deleterious events that lead to blindness.

2. The kallikrein-kinin system

The kallikrein-kinin system (KKS) is a complex multi-enzymatic and peptidergic system known to play a critical role in human physiology, but also in pain and inflammation [1–3]. Its physiological functions encompass nociception, cardiovascular and renal functions, vasomotricity, and host defense to infectious diseases [2,4]. The KKS is constituted by a panel of vasoactive peptides (kinins), synthesized and metabolized by different enzymes (kallikreins and kininases), and two G-protein-coupled receptors (GPCR) (Figure 1).

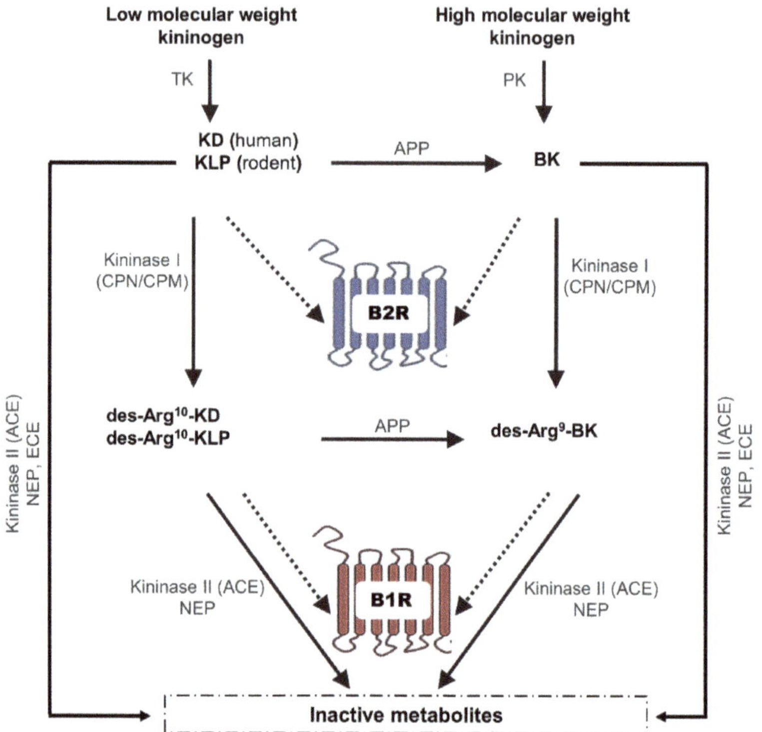

Figure 1. Biosynthesis and metabolism of kinins. Low- and high-molecular-weight kininogens are cleaved by tissue kallikrein and plasma kallikrein, respectively, into kallidin (in humans) and kallidin-like peptide (in rodents) [5,6], and bradykinin. Bradykinin, kallidin, and kallidin-like peptide are then either converted by the action of kininase I to des-Arg^9-bradykinin, des-Arg^{10}-kallidin, and des-Arg^{10}-kallidin-like peptide, respectively, or inactivated by kininase II, neutral endopeptidase 24.11 (neprilysin, NEP), and the endothelin-converting enzyme [4,7–11]. ACE, angiotensin-1- converting enzyme (also known as kininase II); APP, aminopeptidase P; B1R, bradykinin type 1 receptor; B2R, bradykinin type 2 receptor; BK, bradykinin; ECE, endothelin-converting enzyme; KD, kallidin; KLP, kallidin-like peptide, which is Arg(1)-kallidin (Arg(0)-bradykinin); NEP, neutral endopeptidase; PK, plasma kallikrein; TK, tissue kallikrein.

2.1. Kinins Generation

Kinins are small peptides of 9–11 amino acids, including bradykinin (BK), kallidin (KD or Lys-BK), kallidin-like peptide (Arg(1)-kallidin (Arg(0)-bradykinin)), and T-kinin

(Ile-Ser-BK; expressed exclusively in rats), which are generated from high-molecular-weight kininogen (88 to 120 kDa) (HK) and low-molecular-weight kininogen (50 to 68 kDa) (LK) under the action of plasma kallikrein (PK) and tissue kallikrein (TK) [4–6]. BK generation in plasma takes part in the intrinsic coagulation pathway activation, involving the interaction of Factor XII (Hageman factor), prekallikrein (PPK), and Factor XI with HK on negatively charged surfaces, such as components of the extracellular matrix or other negatively charged particles (cholesterol sulfate, urate, or phospholipid acid), leading to prothrombotic and inflammatory effects [2,12]. Aminopeptidase P transforms KD and KLP into BK, while kininase I that includes carboxypeptidases N (CPN) and M (CPM) transforms BK, KD, and KLP into des-Arg^9-BK, lys-des-Arg^9-BK (or des-Arg^{10}-KD), and des-Arg^{10}-KLP, respectively. Alternatively, kininase II (also called angiotensin-1-converting enzyme (ACE)), neutral endopeptidase 24.11 (neprilysin, NEP), and the endothelin-converting enzyme (ECE) degrade BK, KD, and KLP into inactive fragments on the canonical B1 and B2 receptors [4,7–11]. Moreover, ACE and NEP can metabolize des-Arg^9-BK, des-Arg^{10}-KD, and des-Arg^{10}-KLP into inactive metabolites. It is worth noting that the enzymes involved in the catabolism of kinins are also involved in the metabolism of other peptides belonging to other systems such as angiotensin, endothelin, anaphylatoxins C3a and C5a, substance P, neurotensin, enkephalins, atrial natriuretic peptides, and chemotactic peptide [5,8,13–20].

2.2. Kinin Receptors

The KKS operates through the activation of two GPCR, bradykinin type 1 (B1R), and type 2 (B2R) receptors. While BK, KD, and KLP are the endogenous agonists of B2R, their kininase I metabolites (deprived of C-terminal Arginine) are the preferred agonists of B1R [3,21]. The agonist selectivity of mouse B1R differs from human and rabbit B1R; des-Arg^9-BK is the preferred B1R agonist in mice, while des-Arg^{10}-KD displays much higher selectivity for human and rabbit B1R [22]. B2R can activate a plethora of signaling pathways either indirectly by interacting with guanine nucleotide-binding proteins, mainly with Gαq and less commonly with Gαs, Gαi, and G$\alpha_{12/13}$, as reviewed in [23]; or directly by interacting with endothelial nitric oxide synthase (eNOS), phospholipase A 2 (PLA2), and tyrosine phosphatase (SHP2) [4]. B1R interacts with the same guanine nucleotide-binding proteins as those of B2R, but preferentially with Gαq to activate the phosphatidyl inositol-mitogen-activated protein kinase (MAPK) pathway, and with Gαi to activate the extracellular signal-regulated kinase (ERK)-nducible nitric oxide synthase (iNOS) pathway [4,23,24].

Most physiological effects of kinins are mediated by the constitutive B2R, since B1R is virtually absent in healthy tissues. BK is a potent endothelium-dependent vasodilator that has important cardiovascular and renal functions via the B2R [25]. Moreover, B2R contributes to the therapeutic effects of angiotensin-1-converting enzyme inhibitors (ACEI) and angiotensin AT1 receptor blockers [26]. These benefits derive primarily from its vasodilatory, antiproliferative, antihypertrophic, antifibrotic, antithrombic, and antioxidant properties [4,26–36]. However, it is worth noting that B2R can also contribute to inflammation. Indeed, uncontrolled production of kinins and excessive activation of B2R may lead to unwanted pro-inflammatory side effects as observed in angioedema, septic shock, stroke, hypertension, and Chagas vasculopathy, in which B2R antagonism is salutary [4,26,37–41].

B1R, however, is induced and upregulated during tissue injury involving the cytokine pathway, oxidative stress, and the transcriptional nuclear factor NF-κB [2,38,39,42–44]. The highly inducible character of B1R is often symptomatic of the occurrence of autoimmune, infectious, cardiac, kidney, and bowel inflammatory diseases [2,45–49]. However, B1R may play a compensatory role for the lack of B2R, and its upregulation during tissue damage may be a useful mechanism of host defense [25,50–52].

2.3. Kinin Receptors in Inflammation and Neovascularization

B1R antagonism or deletion plays a protective role in inflammation, organ damage, and lethal thrombosis in septic shock in diabetes [53]; lipopolysaccharide (LPS) mediated acute renal inflammation [54]; renal ischemia-reperfusion injury [55]; and in cardiovascular [56] and retinal [57–60] inflammatory diseases. B1R inhibition reversed vascular [61] and retinal [58,60] inflammation induced by diabetes mellitus. Moreover, genetic deletion of B1R or administration of B1R antagonist in mice reduced pro-inflammatory mediators' expression and increased anti-inflammatory mediators [55]. Besides the well-described pro-inflammatory roles of both kinin receptors, an anti-inflammatory effect has been attributed to B2R. For instance, intramyocardial injections of tissue kallikrein reduced the expression of many inflammatory mediators through B2R activation [62]. Moreover, BK can counteract the inflammation in the brain [63]. Indeed, BK reduced LPS-induced TNF-α release from microglia activated by B1R [63]. Recently, a neuroprotective role for B2R was highlighted, and the use of B2R agonists was proposed as a possible therapeutic option for patients diagnosed with Alzheimer's disease [64]. Altogether, these findings support a dual role of B2R in inflammation, whereas B1R is mainly involved in the inflammatory responses, especially those triggered by cytokines or pathogens [65–67]. Because B1R is a potent activator of iNOS and NADPH oxidase, it is associated with vascular inflammation, increased vascular permeability, insulin resistance, endothelial dysfunction, and diabetic complications [24,43,44,68–70].

The contribution of kinin receptors to neovascularization has been widely studied in various models and diseases. In some vascular diseases, drugs are used to inhibit neovascularization (i.e., cancer, neovascular retinal pathologies, etc.), while in others such as ischemia, treatments aim to stimulate neovascularization. Therefore, both activation and inhibition of kinin receptors are important drug targets of vascular diseases. For instance, the activation of B1 and/or B2 receptors may be beneficial, notably in neovascularization and angiogenesis in diabetic mice, renal ischemia/reperfusion injury, diabetic nephropathy, and cerebral and heart ischemia [38,71–78]. B1R deletion or antagonism was shown to impair neovascularization, while B1R agonist had a positive outcome in a model of hindlimb ischemia in diabetic mice [77]. In the same model, B1R or B2R agonists administration induced revascularization by stimulating the mobilization of monocytes and proangiogenic CD34/VEGFR-2 mononuclear cells, and the infiltration of macrophages [76]. Moreover, B1R inhibition prevented the revascularization, as well as VEGF, eNOS, and basic fibroblast growth factor (FGF2) upregulation, induced by ACE inhibitor [79]. While the proangiogenic effect of ACE inhibitor was attributed to an increase in BK generation (Figure 1) and the activation of B2R in diabetic ischemia [80], B1R was more implicated than the B2R in ACE inhibitor mediated angiogenesis in Ang II type 1a receptor knockout (AT1aKO) mice after hindlimb ischemia [81]. Indeed, the B1R antagonist reversed the neovascularization and reduced VEGF-A and VEGFR-2 expression, while the B2R antagonist had less impact [81].

Cancer is among the diseases for which inhibiting kinin receptors would be beneficial. Indeed, the role of kinin receptors in promoting angiogenesis was supported by many experimental studies using cancer cells/tissues. For instance, B1R activation was shown to increase IL-4 and VEGF generation from human keratinocytes and to stimulate endothelial cell migration, thus promoting neovascularization [82]. Furthermore, when human endothelial cells were co-cultured with neuroblastoma cells, B1R and B2R expression was observed at the sites of interaction between these two cell types, regulating angiogenesis and tumorigenesis [83]. Interestingly, blockade of either B1R or B2R reduced tumor vascularization in vivo and significantly inhibited proliferation and migration of colorectal cancer cells in vitro [84]. In studies of mice bearing sarcoma 180 cells, it was suggested that BK promotes angiogenesis in the early phase of tumor development by increasing vascular permeability via B2R, expressed in the endothelial cells and not via B1R, and in the late phase by stimulating the upregulation of VEGF via B2R in the stromal fibroblasts [85–87]. BK was also found to increase VEGF expression in human prostate cancer cells and further promote tumor angiogenesis. Interestingly, B2R blockade using antagonists or genetic deletion reduced VEGF expression and abolished prostate cancer cell conditional medium-mediated angiogenesis [88]. Altogether, these studies suggest that kinins play a pivotal role in angiogenesis through B1R, B2R, or both.

The dual beneficial and deleterious effects of kinin receptors raise questions about the therapeutic value of B1R/B2R agonists or antagonists in various diseases. Hence, the Janus face of kinin receptors needs to be seriously addressed in each pathological setting. The discovery of the expression of kinin receptors and other KKS components in the eye led many investigators to address their physiological and pathological roles, particularly in the retina.

3. kallikrein-kinin system in the Eye

Similarly to other organs, the KKS in the eye is a double-edged sword, as it contributes to many physiological processes including blood-flow regulation and vascular tone control, but also partakes in the complex processes of inflammation [4,57,89]. It was reported that the KKS underlies a number of ocular pathologies (DR, AMD, choroidal neovascularization, macular edema) associated with inflammation and pathological neovascularization, particularly in the human and rat retina [57–60,69,90–93]. For instance, PK and HK, by binding to the vascular endothelium, release BK and subsequently activate B2R, which plays a key role in the control of vascular tone [4]. However, in diabetic rats, an increase in PK mediates retinal vascular dysfunction and induces retinal thickening [91]. Moreover, tissue kallikrein (TK) was expressed in the human retina, cornea, and ciliary body [94]. TK does not seem to be implicated in retinal pathologies, particularly in diabetic retinopathy, as it was slightly detectable in vitreous fluids of patients with severe proliferative DR [95]. An expression of TK, B1R, and B2R was also reported at multiple tissue sites in the anterior portion of the human eye [96]. Nevertheless, B2R but not B1R was expressed in the control human retinae [93]. BK produces B2R-mediated vasodilatation of retinal vessels in control rats [97]. This response involves the COX-2 pathway, including prostacyclin [97]. Hence, B2R contributes to retinal blood flow control. On the other hand, the vasodilatation mediated by kinins is associated with B2R and B1R in streptozotocin (STZ)-diabetic rats and involves both NO and prostacyclin [97]. A protective compensatory role on retinal microcirculation was attributed to B1R at day 4 but not at 6 weeks following diabetes induction [98]. Likewise, both B1R and B2R contribute to the increased retinal vascular permeability in STZ-diabetic rats [58,60,99]. Collectively, these studies support the presence of the KKS throughout the eye and its ability to influence ocular function in health and disease.

It is still unclear whether the KKS expression is generated locally in the eye, or if it is a result of a systemic infiltration of KKS components. While the observation of some KKS components in the healthy eye [94] suggests a local production of these components, Phipps and Feener have suggested an infiltration of these components from the systemic circulation that could happen in DR [100]. This was explained by the increase of KKS components expression in the plasma of diabetic patients, and their infiltration in the retinal interstitium and vitreous that may occur following the increase in vascular permeability and hemorrhages in the retinal vessels [100]. Nonetheless, whether the origin of the KKS expression is local or a result of its infiltration from the systemic circulation in the eye, all these studies support an implication of the KKS in the pathogenesis and development of retinal diseases, such as DR and age-related macular degeneration (AMD).

4. kallikrein-kinin system in Diabetic Retinopathy

DR is one of the most common microvascular complications of diabetes, observed in up to 90% of patients with type 1 diabetes and 50 to 60% of patients with type 2 diabetes, despite a tight glycemic control [101–103]. If left untreated, DR can cause severe vision loss. Current therapeutic strategies target the advanced stages of the disease and aim to slow its progression without really reversing its outcome [104]. Among the current treatments for proliferative DR and macular edema are laser photocoagulation, vitrectomy, and intravitreal injections of corticosteroids or anti-VEGF that could prevent further vision loss [105]. However, the curative activity of these treatments is limited by side effects. For instance, pan-retinal photocoagulation can cause a loss of peripheral vision, color vision, and night vision [106]; intravitreal injection of anti-VEGF has a short effect duration and can cause a tractional retinal detachment and endophthalmitis [107], and many patients are refractory to it [108–110]. Importantly, there is no effective treatment for the highly widespread early stages of the disease [111]. Thus, there is an urgent need for less-invasive and more-effective therapeutic strategies.

Kallikreins and Kinin Receptors in Diabetic Retinopathy

A decrease in the concentration of kallikrein-binding protein (KBP), a serine protease that binds to tissue kallikrein and inhibits kallikrein activity, was reported in the vitreous humor of patients with proliferative DR [112]. Parallel to this study, the levels of KBP were reported to be decreased by 60% for at least 4 months in the retina of STZ-diabetic rats [113]. Moreover, tissue kallikrein was significantly elevated in vitreous fluid in proliferative DR patients when compared with control patients [114]. Interestingly, intravitreal injection of kallistatin, a tissue kallikrein inhibitor, in STZ-diabetic rats reduced retinal neovascularization; however, these effects have been attributed to the tissue kallikrein effects on the VEGF system [115]. Other components of the plasma KKS, including PK, FXII, and HK were also found in the vitreous fluid of patients with advanced DR [116,117]. Increased levels of PK and PK activity were observed in the retina of diabetic rats compared with nondiabetic controls [90,91]. PK injection increased vascular permeability in the healthy retina, and further in the diabetic retina, yet these effects were reversed by the inhibition of PK [91]. Furthermore, the retinal thickening, as well as the increase in vascular permeability caused by intravitreal injections of VEGF, were reduced (by 47% and 68%, respectively) in plasma prekallikrein knockout mice [118]. In phase I.B of a recent clinical trial, PK inhibition by one-time intravitreal injection of KVD001 was shown to be effective in treating macular edema without creating a safety concern. The injection improved visual acuity and central retinal thickness, and no exacerbation of the severity of DR was observed [119]. PK contribution to DR pathogenesis was, however, attributed to B2R activation. Indeed, C1 inhibitor-deficient mice caused vasogenic edema due to increases in PK expression, BK synthesis, and activation of B2R [120]. Given the fact that PK is a constitutive enzyme involved in other systems, including thrombosis and blood hemostasis, its inhibition may risk interfering with its physiological role [57].

Alternatively, B1R expression was shown to be significantly increased in retinae of rats and humans affected by type 1 and type 2 diabetes [57,58,60,69,92]. B1R expression was enhanced on the 4th day of STZ-diabetic retina [97], and it remained upregulated even 6 months after the induction of type 1 diabetes [58]. B1R upregulation in STZ-diabetic rats leads to retinal microvessel vasodilation [97], vascular hyperpermeability, and inflammation [60]. Importantly, these responses were reversed by eye-drop application of B1R antagonists (LF22-0542 and R-954) [58,60]. B1R was strongly expressed in vascular endothelial cells and in the retinal pigment epithelium of human and rats' retinae, suggesting its implication in altering the integrity of the internal and external blood–retinal barrier (BRB) in DR and AMD [58,59,69,92,93]. B1R might disrupt the BRB [59,60,69] either by the suppression of tight junction components (occludin, claudin, and zonula occludens-1), or by a rearrangement of the cytoskeleton filamentous actin [121]. In human cerebral microvascular endothelial cells, B1R agonist (des-Arg^9-BK) was shown to decrease the expression of zonula occludens-1 and occludin in vitro [122]. Altogether, these data support a strong implication of B1R in DR.

One additional mechanism by which B1R contributes to the pathogenesis of DR has been recently suggested that involves the activation of the iNOS pathway [69]. In HEK293 cells, it was shown that B1R associated with Gαi can activate iNOS through ERK [24,123], thereby producing sustained amounts of NO. Interestingly, iNOS inhibition in the retina of diabetic mice caused a decrease in occludin and zonula occludens-1 expression, thus protecting the dissociation of BRB [124]. The elevated concentrations of NO, nitration of proteins, prostaglandin E2, superoxide, leukostasis, and retinal thickness induced by diabetes were significantly inhibited in diabetic iNOS $(-/-)$ mice [125]. In addition, diabetes-induced acellular capillaries and pericyte ghosts were significantly inhibited in diabetic iNos $(-/-)$ mice [125]. Given that B1R can also enhance the production of superoxide anion through PLC and the activation of NADPH oxidase [44], NO produced by iNOS upon B1R activation can react with superoxide anion to yield peroxynitrite, a highly toxic molecule [126–128], which causes endothelial and neuronal cell apoptosis, neuronal degeneration, and BRB breakdown in DR [125,126,129–133]. Peroxynitrite can also activate NF-κB, and thereby can increase the expression of several pro-inflammatory mediators, including B1R [4,43]. Hence, B1R activation can further amplify and perpetuate the inflammatory response, as well as the oxidative stress, through a positive feedback loop [43,69] In resonance with this hypothesis, pharmacological iNOS inhibition in the retina of STZ-diabetic rats reversed peroxynitrite formation, the upregulation of inflammatory mediators (notably B1R), and the enhanced vascular hyperpermeability induced by B1R agonist [69]. Collectively, these data support a robust implication of the B1R in DR, mainly by increasing and perpetuating inflammation and oxidative stress. Hence, targeting B1R represents a promising therapeutic approach in DR, and deserves further investigation.

In the retina of 2-week STZ-diabetic rats, B2R mRNA and protein expression did not change when compared to the control retina [69], yet a significant increase in B2R mRNA was observed at 24 weeks in the retina of diabetic rats [58]. B2R contributes to the increased retinal vascular permeability in STZ-diabetic rats [99]. Indeed, BK induces vascular endothelial cadherin phosphorylation and a subsequent rapid internalization and ubiquitination, leading to an opening of endothelial cell junctions and plasma leakage [134] (Figure 2). However, more studies are needed in DR using recently developed selective and stable B2R antagonists or biostable kinin analogs [41,46,135].

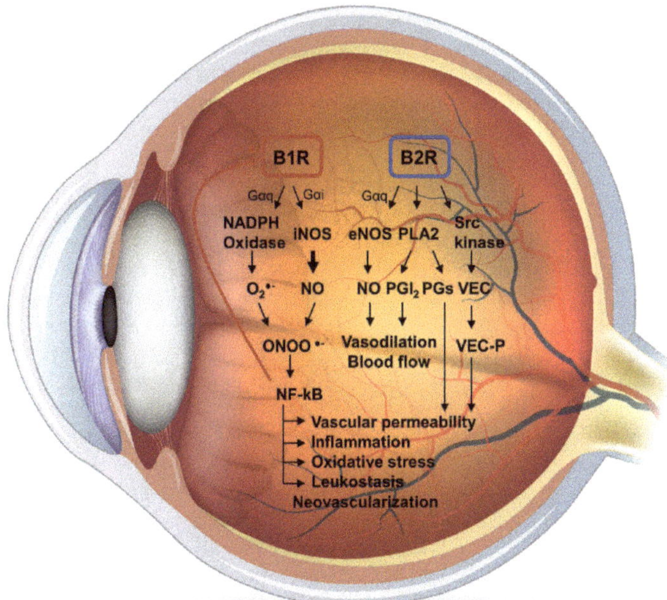

Figure 2. kallikrein-kinin system in diabetic retinopathy. Schematic proposal of the signaling pathways activated by B1R and B2R in diabetic retinopathy. PGs, prostaglandins; PGI2, prostacyclin; PLA2, phospholipase A2; Src, kinase proto-oncogene tyrosine-protein kinase; VEC, vascular endothelial cadherin; VEC-P, phosphorylated vascular endothelial cadherin. The human eye anatomy diagram was acquired from Shutterstock (http://www.shutterstock.com, accessed on 16 July 2021).

5. kallikrein-kinin system in Age-Related Macular Degeneration

AMD is a multifactorial disorder, highly heritable, and caused by an interplay of many factors including age, genetic, and environmental risk factors. The prevalence of AMD is rising worldwide, and it is expected to increase from 196 million in 2020 to 288 million by 2040 [136,137]. In its early stages, AMD is characterized by pigmentary abnormalities and deposits of lipoproteinaceous debris (soft drusen) between the basal lamina of the retinal pigment epithelium (RPE) and the inner collagenous layer of Bruch's membrane (BM) of the central retina [138]. Early and late forms of AMD include wet (exudative) AMD and dry (nonexudative) AMD. The late form of dry AMD is also called geographic atrophy AMD. Exudative/wet AMD is mainly characterized by neovascularization that arises from the choroid, but in about 10–15% of the cases, originates from the retinal vasculature in the subretinal space [139,140]. Dry AMD is more prevalent, affecting 85–90% of patients suffering from AMD [136,137], and is characterized by an extending lesion of the RPE and photoreceptors [141]. Current treatments target only the neovascular AMD mainly by anti-angiogenic therapy (anti-VEGF), which aims to decrease the vascular permeability and to inhibit the formation of new vessels without treating the degenerative processes and the vision loss in 30% of patients that occur in the long term [142]. On the other hand, no effective treatment options are available for dry AMD, besides lifestyle modification and nutrient supplementation [143].

Similarly to DR, the pathogenesis of AMD is driven by both inflammation and microvascular alterations leading to BRB dysfunction and pathological neovascularization. Indeed, an increase of diverse transcriptional factors (NF-kB, HIF-1α) and pro-inflammatory mediators (cyclooxygenase-2 (COX-2) products, IL-1β, TNF-α, iNOS, NO) has been reported in different models of DR and AMD [58–60,69,144]. Consistently with the roles of kinin receptors in both inflammation and neovascularization, we showed that most upregulated inflammatory mediators were blocked by B1R inhibition in DR and AMD [58–60]. B1R was shown to be expressed on Müller cells and astrocytes in these retinal pathologies in rat and post-mortem human retina [58,59,93], and on microglia in post-mortem human wet AMD retina [93]. Macroglia play a primary role in vascular function and neuronal integrity of the retina [145]. These results deserve closer scrutiny and encourage further investigations to assess the impact of an ocular treatment with a B1R antagonist on macro- and microglial reactivity in DR and AMD.

B1R expression was upregulated in a rat model of choroidal neovascularization (CNV), and B1R blockade reduced the size of the neovascularization [59]. B1R contribution to retinal neovascularization in humans was also suggested in a recent study in post-mortem human wet AMD retinae. In these retinae, B1R was strongly expressed in endothelial/vascular smooth muscle cells, and co-localized with iNOS and fibrosis markers. Its presence on vascular smooth muscle cells can induce prolonged vessel constriction and consequently contribute to retinal ischemia, a main trigger of neovascularization, mainly by activating the VEGF-A pathway [146]. Altogether, these data highlight a contribution of B1R to retinal pathologies associated with neovascularization. By analogy with another ocular pathology, B1R agonist administration in the rabbit eye induced corneal neovascularization, an effect that was reversed by B1R inhibition with the same efficacy as VEGF-A inhibition [147]. The implication of B2R in ocular neovascular pathologies has also been suggested. For instance, in an ischemic retinopathy model, B2R antagonist (Fasitibant) significantly decreased the expression of VEGF and FGF2, as well as pathological retinal neovascularization [148]. In a mice model of CNV, B2R blockade with Icatibant had a limited effect, yet concomitant inhibition of B2R and kininase II had additive suppression of the CNV size [149]. We reported no significant modification of B2R mRNA and protein expression in human neovascular AMD retinae [93].

In addition to KKS gene expression in the ocular pathologies reviewed above, we also mined a recent public single-cell transcriptomics database of post-mortem choroid tissues from neovascular AMD human patients [150], using previously described analyses [151,152]. KKS genes were detected in fibroblasts and immune, RPE, and endothelial cells (Figure 3, unpublished original results). Choroidal endothelial cell specifically expressed *KLKB1*, *BDKRB1*, and *BDKRB2* (genes for prekallikrein, B1R, and B2R, respectively), albeit at low expression levels. Subclustering of the heterogenous choroidal endothelial cell population identified four subtypes (see legend of Figure 3), including vein clusters 1 and 2, discriminated by the higher expression of *SELECTIN E* and *VCAM1* (Figure 3f), a pattern reminiscent of post-capillary venous identity [153]. Interestingly, vein cluster 2 showed greater expression of KKS genes, notably *BDKRB1*, *BDKRB2*, and *MME* (genes for B1R, B2R, and neprilysin, respectively) in choroid endothelial cells from AMD patients (Figure 3g). Although the relatively low detection levels for these three genes (less than 10%) requires cautious interpretation, their specific expression in post-capillary venous endothelial cells of neovascular AMD patients is intriguing and warrants further investigation of kinin receptors in AMD.

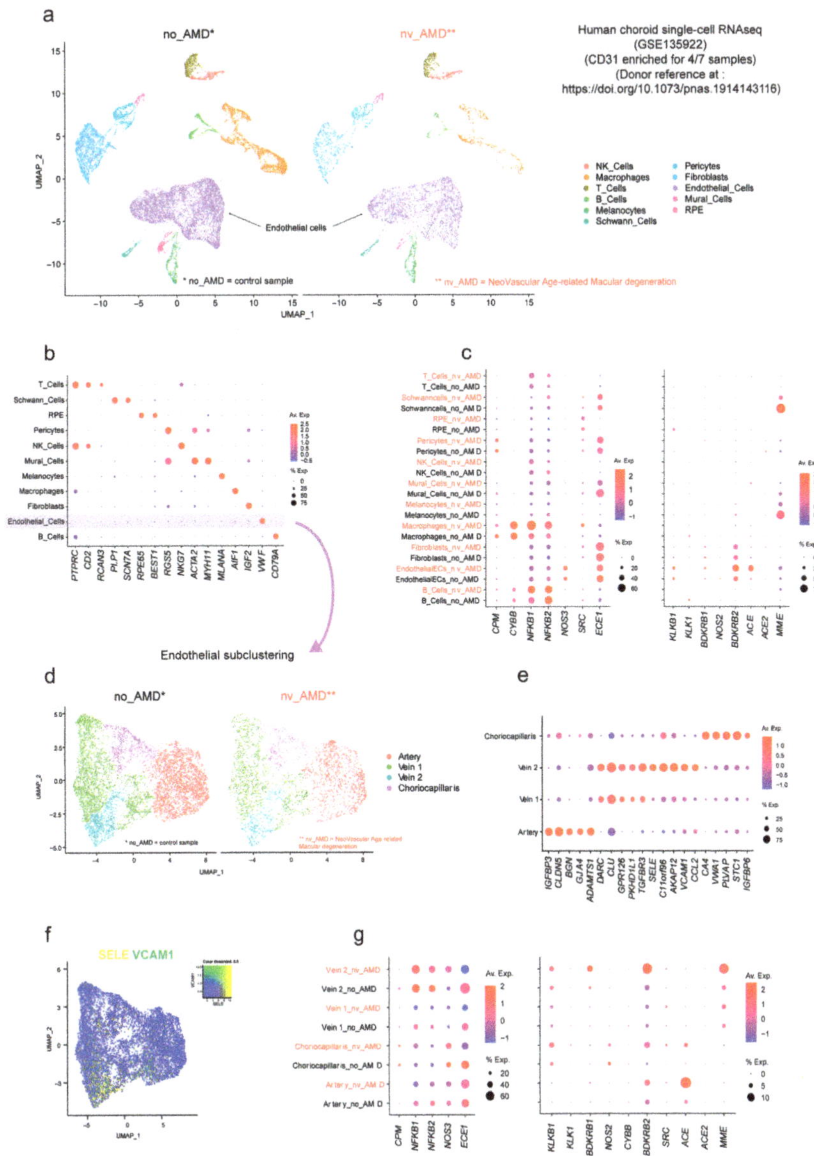

Figure 3. Transcriptomic impact of age-related macular degeneration on the kallikrein-kinin system

in human choroid tissues by single-cell RNA seq. (**a**) Dimensionality reduction and cluster visualization with UMAP plot. Color-coded clusters represent the different choroid cell types (see legend in bottom right corner) identified by single-cell RNAseq analysis of post-mortem control (left panel) and neovascular AMD (right panel) choroids (public data deposited on GEO with reference number GSE135922). (**b**) Dotplot of the expression of the gene markers used to identify choroid cell types. (**c**) Dotplot of the expression of the genes involved in the KSS pathway across choroidal cell types from control (black legend) and neovascular (nv)-AMD (red legend) samples. As shown in (**a–c**), KLKB1, BDKRB1, and BDKRB2 are mainly expressed in choroidal endothelial cells, albeit at low expression levels; some expression was also detected in fibroblasts. (**d**) Dimensionality reduction and cluster visualization with UMAP plot of the subpopulations of choroid endothelial cells of post-mortem control (left panel) and neovascular AMD (right panel) choroids. Choroid cells clustered into four distinct endothelial cell subtypes: two venous subtypes, one choriocapillaris subtype, and one arterial subtype (see legend on right side). (**e**) Dotplot of the expression of the specific gene markers in these endothelial subcluster, as annotated by Voigt et al., [150]. (**f**) Visualization with UMAP plot of E SELECTIN (SELE) and VCAM1 expression co-localizing to vein 2 subcluster, a signature reminiscent of post-capillary venous identity. (**g**) Dotplot of the expression of genes involved in the KSS pathway across choroidal endothelial cell subtypes from control (black legend) and neovascular (nv)-AMD (red legend) samples. Vein cluster 2 showed greater expression of KKS genes, notably BDKRB1, BDKRB2, and MME, across all choroidal endothelial cells of control and nv-AMD choroid samples. In all the dotplots, the size of the dots encodes the percentage of cells within a class, and the color scale encodes the average expression level across all cells within a class (red being the strongest value). Av. Exp., average gene expression across all cells within each cluster; % Exp., percentage of cells with detectable gene expression within each cluster.

Although recent studies support the implication of the KKS in wet neovascular AMD, it is still not clear if the KKS is implicated in the dry form. In the retina of aged rats, an increase of KKS components was demonstrated, where 4-month-old rats showed a significant decrease in KBP, and consequently an increase in tissue kallikrein compared to 2-week-old rats [113]. Recent data using post-mortem human retinal sections showed only a weak expression of B1R and no changes of B2R in dry AMD [93].

6. kallikrein-kinin system in Other Retinal Damage

This review highlights the implication of the KKS in retinal pathologies associated with inflammation and neovascularization. However, KKS can also be implicated in ocular pathologies such as glaucoma and ocular ischemia. For instance, BK alters the shape of cells in both bovine and human trabecular meshwork [154–156]. Moreover, FR-190997, a B2R agonist, was shown to lower the intraocular pressure by promoting uveoslceral outflow in monkeys [157]. Taken together, these results suggest that the KKS can also be implicated in ocular diseases with elevated intraocular pressure. Intravenous administration of TK protected against retinal ischemic damage in a retinal ischemia/reperfusion model in mice [158]. In this model, TK administration inhibited retinal ganglion cell death, counteracted the retinal permeability induced by ischemia, and improved the visual function [158]. However, these protective effects seem to be independent of blood flow and might be mediated by eNOS activation and subsequent NF-κB silencing.

7. Crosstalk between the kallikrein-kinin system and the Renin–Angiotensin System (RAS) in Ocular Pathologies

There is compelling evidence for a local renin–angiotensin system (RAS) within the human eye that is activated in ocular disorders and DR [159–163]. Multiple interactions (crosstalk) exist between the RAS and the KKS [4,25,164–166] (Figure 4). In addition to the implication of ACE (kininase II) in the degradation of kinins (acting on B1R and B2R) and the formation of angiotensin II (Ang II) from angiotensin I (Ang I) [165], the activation of the angiotensin II type 2 receptor (AT2R) leads to BK generation, which promotes vasodilation through the NO/cGMP system [167]. Under the action of angiotensin-converting enzyme

2 (ACE 2), Ang I is cleaved into angiotensin-(1-9) this is corrected (Ang-(1-9)), a peptide that elicits vasodilation and anti-inflammatory effects through activation of AT2R [168,169]. ACE 2 can also cleave Ang II to Ang-(1-7), an agonist of AT2R and Mas-receptor (MasR) that elicits the release of BK, vasodilatory, antiproliferative, anticoagulation, anti-inflammatory, and antifibrotic activity, thus counterbalancing the adverse effects of Ang II mediated by AT1R [170–172]. Importantly, ACE 2 hydrolyses B1R agonists (des-Arg^9-BK and Lys-des-Arg^9-BK) into inactive metabolites and therefore impairment of ACE 2 (as under COVID-19 infection) is expected to enhance the pro-inflammatory effects of the des-Arg^9-BK/B1R axis [164,173] (Figure 4). Moreover, the pro-inflammatory effects of Ang II was attributed to AT1R and B1R activation [174]. Following AT1R activation, Ang II enhances B1R expression in vitro [175,176] and in vivo [174,177] by activating NADPH oxidase, IL-1β, IL-6, TNFα, and NF-κB [174,176]. Besides ACE2, neutral endopeptidase 24.11 (NEP) was described to be biochemically capable of producing Ang-(1-7) from Ang I and Ang-(1-9) [178]. NEP can also hydrolyze Ang-(1-7) to form angiotensin-(1-4) (Ang-(1-4)), an inactive metabolite [179]. Hence the reciprocal interaction between the RAS and the KKS must be considered in the development of novel therapeutic approaches in the treatment of retinal diseases.

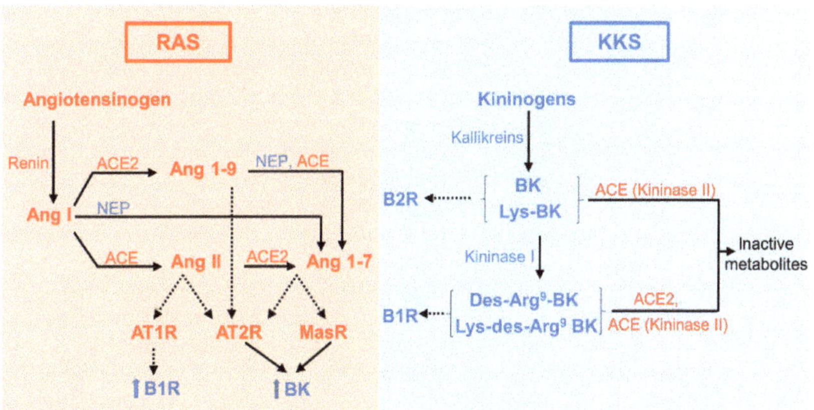

Figure 4. Crosstalk between the kallikrein-kinin system (KKS) and the renin–angiotensin system (RAS). ACE, angiotensin -1 -converting enzyme (kininase II); ACE2, angiotensin-converting enzyme 2; Ang I, angiotensin I; Ang II, angiotensin II; AT1R, angiotensin type 1 receptor; AT2R, angiotensin type 2 receptor; BK, bradykinin; MasR, Mas receptor; NEP, neutral endopeptidase 24.11 (neprilysin, NEP).

7.1. Renin–Angiotensin System in Diabetic Retinopathy

The RAS is implicated in inflammation, vascular alterations, neovascularization, and edema in retinal pathologies, notably in DR and retinopathy of prematurity [163]. An increase in prorenin level was reported in the vitreous fluid of patients with proliferative DR [161]. Ang II induces pericyte apoptosis in the retina in vivo and in vitro in hypertensive rats by increasing the expression of RAGE receptor for advanced glycation end products (AGEs); these effects were reversed by an Ang II-AT1R blocker [180]. An AngII-AT1R blocker (Candesartan) inhibits the development of DR by reducing the accumulation of AGEs and the expression of VEGF in the retina in a rat model of type 2 diabetes [181]. This AT1R blocker reduces retinal vascular permeability induced by diabetes and Ang II in rats [182]. Importantly, the DIRECT study based on more than 1400 patients found that Candesartan reduces the progression of microaneurysms in both type 1 and type 2 diabetic patients, yet no effects were observed on the DR regression and progression, or on the prevention of diabetic macular edema risk [183]. Another multicenter study of 285 patients with type 1 diabetes reported that Losartan, another AT1R blocker, slows the progression of DR [184]. Together, these studies suggest that angiotensin AT1R blockers may be effective against DR independently of their anti-hypertensive action.

ACE inhibition was also shown to lower the risk and prevent the development and the evolution of DR in humans [185,186]. ACE inhibition reduces retinal VEGF overexpression and hyperpermeability in experimental diabetes [187] and vitreous VEGF concentrations in patients with proliferative DR [188]. Interestingly, changes in circulating VEGF do not account for the beneficial effect of ACE inhibition on retinopathy in patients with type 1 diabetes [189]. Previous clinical trials have associated the decrease in DR progression in type 1 or type 2 diabetic patients with a reduction of hypertension [190,191]. The United Kingdom Prospective Diabetes Study (UKPDS) with more than 1000 patients reported a reduction in the progression of DR with ACE inhibitor and β1-adrenergic receptor blocker, suggesting that the beneficial effect may be related to the anti-hypertensive and not to the ACE-inhibition-specific effect [190]. Nevertheless, other studies have reported a slowdown in DR progression in normotensive diabetic patients taking an ACE inhibitor, suggesting a possible therapeutic effect of ACE inhibitors not related to the anti-hypertensive effect [184,191]. In resonance with this, a meta-analysis of 21 clinical trials with more than 13,000 patients disclosed no effects of RAS inhibitors on DR progression in hypertensive patients, but a reduced risk of DR, and increased possibility of DR regression in normotensive patients [186]. In rank order of anti-hypertensive drug classes, the association with risk of DR progression was lowest with ACE inhibitors, followed by Ang II-AT1R blockers, β-blockers, and finally with calcium-channel blockers [186].

While ACE inhibitors show promising results against DR and diabetic macular edema, several safety questions related to increased kinin levels can be raised, such as hypotension, angioedema, and pain associated with inflammation, which are B2R-mediated [192–195]. Increased kinin levels are also associated with retinal vascular permeability, inflammation, and neovascularization (Figure 2). A decrease in the degradation of endogenous B1R agonist (des-Arg^9-BK) was also observed in the plasma of patients treated with an ACE inhibitor [196]. In theory, the use of kinin receptor antagonists can overcome the side effects of ACE inhibitors in the retina.

Furthermore, Ang II-AT1R is a potent enhancer of the pro-inflammatory B1R [174–177] and ACE inhibition ablated B1R expression in diabetic vessels [197], suggesting that B1R acts as an effector of the RAS (Figure 4). Therefore, targeting the RAS (AT1R and ACE) in DR may be a promising approach to prevent the induction and deleterious effects of B1R. Nonetheless, further studies are needed to unveil the exact mechanism(s) and crosstalk with other components of the RAS/KKS (ACE2, AT2R, and MasR) to address the beneficial versus the detrimental effects of the dual pro- and anti-inflammatory role of B2R in retinal disorders. Until these questions are fully answered, targeting B1R in retinal pathologies associated with inflammation and/or vascular alterations remains by far the best asset, with less possible interaction with other axes involved in physiological signaling pathways.

7.2. Renin–Angiotensin System in Age-Related Macular Degeneration

The implication of RAS was also reported to contribute to CNV pathogenesis. Indeed, our single-cell RNA seq showed a high expression of ACE in the neovascular AMD arteries and choriocapillaries (Figure 3). Furthermore, prorenin receptor blockade in a murine model of laser-induced CNV exhibited a significant reduction of CNV, macrophage infiltration, and the upregulation of ICAM-1, monocyte chemotactic protein-1, (MCP-1), VEGF, VEGFR1, and VEGFR2 [198]. Moreover, AT1R inhibition pharmacologically or genetically inhibited CNV and macrophage infiltration [198]. VEGF, ICAM-1, and MCP-1 levels, elevated by CNV induction, were significantly suppressed by ACE inhibition, which led to significant suppression of CNV development to the level seen in AT1R-deficient mice [149]. Despite these significant beneficial effects in rodents, antihypertensive drugs (ACE inhibitors and angiotensin receptor blockers) failed to show any positive effects on AMD in humans [199–201].

8. Conclusions

Inflammatory and neovascular retinal diseases, including DR and AMD, can lead to severe vision loss if left untreated. Current treatments for these pathologies are invasive and can sometimes worsen the pathology. Besides these side effects, many patients do not respond well or become refractory to these treatments, thus there is an urgent need to identify new therapeutic targets and new treatment strategies. Interestingly, the pro-angiogenic, pro-inflammatory, and vasoactive effects of the KKS make it a promising therapeutic target for treating retinal pathologies associated with inflammation and neovascularization. However, KKS targeting needs to be carefully documented before clinical application, as this system is also involved in physiological functions (such as organ blood-flow perfusion and blood coagulation) [4]. To minimize as much as possible the side effects of a complete shutdown of this system that may lead to ischemia and thrombotic events, it is advisable to use a more selective approach by targeting directly kinin receptors in retinal pathologies. Conflicting data are available regarding the implication of B2R in retinal pathologies. This may be related to its important physiological role on the vasculature and the regulation of blood flow. Thus, the inhibition of this receptor may cause unwanted side effects, notably ischemia, and its role in retinal pathology warrants further investigation. In contrast, currently available data strongly support the contribution of B1R in inflammatory and neovascular retinal diseases. Inhibiting the inducible B1R, by topical eye-drop treatment represents a promising noninvasive therapeutic approach in retinal diseases. This is keeping with the finding that B1R acts as an effector of the RAS (Ang II-AT1R) and may subserve its deleterious effects in ocular diseases.

Author Contributions: Writing—original draft and designing figures, R.O.; analysis of scRNAseq (Figure 3), G.C. and J.-S.J.; editing the manuscript, R.O., G.C., J.-S.J., E.V., and R.C. All authors have read and agreed to the published version of the manuscript.

Funding: This research was funded by the Canadian Institutes of Health Research (PJT-175061) and the FRQS Vision Health Research Network, in partnership with the Antoine-Turmel Foundation, to E.V. and R.C. R.O. received PhD Studentship Awards from the Graduate Program of Physiology, the Faculty of Graduate and Postdoctoral Studies, and the Faculty of Medicine, Université de Montréal.

References

1. Couture, R.; Harrisson, M.; Vianna, R.M.; Cloutier, F. Kinin receptors in pain and inflammation. *Eur. J. Pharmacol.* **2001**, *429*, 161–176. [CrossRef]
2. Dagnino, A.P.A.; Campos, M.M.; Silva, R.B.M. Kinins and Their Receptors in Infectious Diseases. *Pharmaceuticals* **2020**, *13*, 215. [CrossRef]
3. Marceau, F.; Regoli, D. Bradykinin receptor ligands: Therapeutic perspectives. *Nat. Rev. Drug Discov.* **2004**, *3*, 845–852. [CrossRef] [PubMed]
4. Couture, R.; Blaes, N.; Girolami, J.P. Kinin receptors in vascular biology and pathology. *Curr. Vasc. Pharmacol.* **2014**, *12*, 223–248. [CrossRef]
5. Campbell, D.J. The renin-angiotensin and the kallikrein-kinin systems. *Int. J. Biochem. Cell Biol.* **2003**, *35*, 784–791. [CrossRef]
6. Hilgenfeldt, U.; Stannek, C.; Lukasova, M.; Schnolzer, M.; Lewicka, S. Rat tissue kallikrein releases a kallidin-like peptide from rat low-molecular-weight kininogen. *Br. J. Pharmacol.* **2005**, *146*, 958–963. [CrossRef] [PubMed]
7. Decarie, A.; Raymond, P.; Gervais, N.; Couture, R.; Adam, A. Serum interspecies differences in metabolic pathways of bradykinin and [des-Arg9]BK: Influence of enalaprilat. *Am. J. Physiol.* **1996**, *271*, H1340–H1347. [CrossRef] [PubMed]
8. Erdös, E. Kininases. In *Bradykinin, Kallidin and Kallikrein*; Springer: Berlin, Germany, 1979; pp. 427–487.
9. Blais, C., Jr.; Marc-Aurele, J.; Simmons, W.H.; Loute, G.; Thibault, P.; Skidgel, R.A.; Adam, A. Des-Arg9-bradykinin metabolism in patients who presented hypersensitivity reactions during hemodialysis: Role of serum ACE and aminopeptidase P. *Peptides* **1999**, *20*, 421–430. [CrossRef]
10. Sheikh, I.A.; Kaplan, A.P. Studies of the digestion of bradykinin, Lys-bradykinin, and des-Arg9-bradykinin by angiotensin converting enzyme. *Biochem. Pharmacol.* **1986**, *35*, 1951–1956. [CrossRef]
11. Hoang, M.V.; Turner, A.J. Novel activity of endothelin-converting enzyme: Hydrolysis of bradykinin. *Biochem. J.* **1997**, 23–26. [CrossRef]

12. Kaplan, A.P.; Joseph, K.; Silverberg, M. Pathways for bradykinin formation and inflammatory disease. *J. Allergy Clin. Immunol.* **2002**, *109*, 195–209. [CrossRef]
13. Campbell, W.D.; Lazoura, E.; Okada, N.; Okada, H. Inactivation of C3a and C5a octapeptides by carboxypeptidase R and carboxypeptidase N. *Microbiol. Immunol.* **2002**, *46*, 131–134. [CrossRef]
14. Bhoola, K.D.; Figueroa, C.D.; Worthy, K. Bioregulation of kinins: Kallikreins, kininogens, and kininases. *Pharmacol. Rev.* **1992**, *44*, 1–80.
15. Briggs, J.P.; Marin-Grez, M.; Steipe, B.; Schubert, G.; Schnermann, J. Inactivation of atrial natriuretic substance by kallikrein. *Am. J. Physiol.* **1984**, *247*, F480–F484. [CrossRef] [PubMed]
16. Vijayaraghavan, J.; Scicli, A.G.; Carretero, O.A.; Slaughter, C.; Moomaw, C.; Hersh, L.B. The hydrolysis of endothelins by neutral endopeptidase 24.11 (enkephalinase). *J. Biol. Chem.* **1990**, *265*, 14150–14155. [CrossRef]
17. Skidgel, R.A.; Schulz, W.W.; Tam, L.T.; Erdos, E.G. Human renal angiotensin I converting enzyme and neutral endopeptidase. *Kidney Int. Suppl.* **1987**, *20*, S45–S48.
18. Skidgel, R.A.; Engelbrecht, S.; Johnson, A.R.; Erdos, E.G. Hydrolysis of substance p and neurotensin by converting enzyme and neutral endopeptidase. *Peptides* **1984**, *5*, 769–776. [CrossRef]
19. Couture, R.; Regoli, D. Inactivation of substance P and its C-terminal fragments in rat plasma and its inhibition by Captopril. *Can. J. Physiol. Pharmacol.* **1981**, *59*, 621–625. [CrossRef]
20. Schwartz, J.C.; Gros, C.; Lecomte, J.M.; Bralet, J. Enkephalinase (EC 3.4.24.11) inhibitors: Protection of endogenous ANF against inactivation and potential therapeutic applications. *Life Sci.* **1990**, *47*, 1279–1297. [CrossRef]
21. Menke, J.G.; Borkowski, J.A.; Bierilo, K.K.; MacNeil, T.; Derrick, A.W.; Schneck, K.A.; Ransom, R.W.; Strader, C.D.; Linemeyer, D.L.; Hess, J.F. Expression cloning of a human B1 bradykinin receptor. *J. Biol. Chem.* **1994**, *269*, 21583–21586. [CrossRef]
22. Hess, J.F.; Derrick, A.W.; MacNeil, T.; Borkowski, J.A. The agonist selectivity of a mouse B1 bradykinin receptor differs from human and rabbit B1 receptors. *Immunopharmacology* **1996**, *33*, 1–8. [CrossRef]
23. Leeb-Lundberg, L.M.; Marceau, F.; Muller-Esterl, W.; Pettibone, D.J.; Zuraw, B.L. International union of pharmacology. XLV. Classification of the kinin receptor family: From molecular mechanisms to pathophysiological consequences. *Pharmacol. Rev.* **2005**, *57*, 27–77. [CrossRef]
24. Brovkovych, V.; Zhang, Y.; Brovkovych, S.; Minshall, R.D.; Skidgel, R.A. A novel pathway for receptor-mediated post-translational activation of inducible nitric oxide synthase. *J. Cell. Mol. Med.* **2011**, *15*, 258–269. [CrossRef]
25. Girolami, J.-P.; Bouby, N.; Richer-Giudicelli, C.; Alhenc-Gelas, F. Kinins and Kinin Receptors in Cardiovascular and Renal Diseases. *Pharmaceuticals* **2021**, *14*, 240. [CrossRef]
26. Hamid, S.; Rhaleb, I.A.; Kassem, K.M.; Rhaleb, N.E. Role of Kinins in Hypertension and Heart Failure. *Pharmaceuticals* **2020**, *13*, 347. [CrossRef]
27. Erdos, E.G.; Marcic, B.M. Kinins, receptors, kininases and inhibitors–where did they lead us? *Biol. Chem.* **2001**, *382*, 43–47. [CrossRef]
28. Freeman, E.J.; Chisolm, G.M.; Ferrario, C.M.; Tallant, E.A. Angiotensin-(1-7) inhibits vascular smooth muscle cell growth. *Hypertension* **1996**, *28*, 104–108. [CrossRef]
29. Jalowy, A.; Schulz, R.; Heusch, G. AT1 receptor blockade in experimental myocardial ischemia/reperfusion. *J. Am. Soc. Nephrol.* **1999**, *10* (Suppl. 11), S129–S136.
30. Linz, W.; Wiemer, G.; Gohlke, P.; Unger, T.; Scholkens, B.A. Contribution of kinins to the cardiovascular actions of angiotensin-converting enzyme inhibitors. *Pharmacol. Rev.* **1995**, *47*, 25–49.
31. McDonald, K.M.; Mock, J.; D'Aloia, A.; Parrish, T.; Hauer, K.; Francis, G.; Stillman, A.; Cohn, J.N. Bradykinin antagonism inhibits the antigrowth effect of converting enzyme inhibition in the dog myocardium after discrete transmural myocardial necrosis. *Circulation* **1995**, *91*, 2043–2048. [CrossRef] [PubMed]
32. Sato, M.; Engelman, R.M.; Otani, H.; Maulik, N.; Rousou, J.A.; Flack, J.E., 3rd; Deaton, D.W.; Das, D.K. Myocardial protection by preconditioning of heart with losartan, an angiotensin II type 1-receptor blocker: Implication of bradykinin-dependent and bradykinin-independent mechanisms. *Circulation* **2000**, *102*. [CrossRef]
33. Shen, B.; Harrison-Bernard, L.M.; Fuller, A.J.; Vanderpool, V.; Saifudeen, Z.; El-Dahr, S.S. The Bradykinin B2 receptor gene is a target of angiotensin II type 1 receptor signaling. *J. Am. Soc. Nephrol.* **2007**, *18*, 1140–1149. [CrossRef]
34. Stauss, H.M.; Zhu, Y.C.; Redlich, T.; Adamiak, D.; Mott, A.; Kregel, K.C.; Unger, T. Angiotensin-converting enzyme inhibition in infarct-induced heart failure in rats: Bradykinin versus angiotensin II. *J. Cardiovasc. Risk* **1994**, *1*, 255–262. [CrossRef]
35. Tan, Y.; Hutchison, F.N.; Jaffa, A.A. Mechanisms of angiotensin II-induced expression of B2 kinin receptors. *Am. J. Physiol. Heart Circ. Physiol.* **2004**, *286*, H926–H932. [CrossRef]
36. Tschope, C.; Gohlke, P.; Zhu, Y.Z.; Linz, W.; Scholkens, B.; Unger, T. Antihypertensive and cardioprotective effects after angiotensin-converting enzyme inhibition: Role of kinins. *J. Card. Fail.* **1997**, *3*, 133–148. [CrossRef]
37. Blaes, N.; Girolami, J.P. Targeting the 'Janus face' of the B2-bradykinin receptor. *Expert Opin. Ther. Targets* **2013**, *17*, 1145–1166. [CrossRef]
38. Desposito, D.; Zadigue, G.; Taveau, C.; Adam, C.; Alhenc-Gelas, F.; Bouby, N.; Roussel, R. Neuroprotective effect of kinin B1 receptor activation in acute cerebral ischemia in diabetic mice. *Sci. Rep.* **2017**, *7*, 9410. [CrossRef]
39. Albert-Weissenberger, C.; Siren, A.L.; Kleinschnitz, C. Ischemic stroke and traumatic brain injury: The role of the kallikrein-kinin system. *Prog. Neurobiol.* **2013**, *101–102*, 65–82. [CrossRef]

40. Mukherjee, S.; Huang, H.; Weiss, L.M.; Costa, S.; Scharfstein, J.; Tanowitz, H.B. Role of vasoactive mediators in the pathogenesis of Chagas' disease. *Front. Biosci.* **2003**, *8*, e410–e419. [CrossRef]

41. Marceau, F.; Bachelard, H.; Charest-Morin, X.; Hebert, J.; Rivard, G.E. In Vitro Modeling of Bradykinin-Mediated Angioedema States. *Pharmaceuticals* **2020**, *13*, 201. [CrossRef]

42. Campos, M.M.; Souza, G.E.; Calixto, J.B. In vivo B1 kinin-receptor upregulation. Evidence for involvement of protein kinases and nuclear factor kappaB pathways. *Br. J. Pharmacol.* **1999**, *127*, 1851–1859. [CrossRef]

43. Haddad, Y.; Couture, R. Interplay between the kinin B1 receptor and inducible nitric oxide synthase in insulin resistance. *Br. J. Pharmacol.* **2016**, *173*, 1988–2000. [CrossRef]

44. Haddad, Y.; Couture, R. Localization and Interaction between Kinin B1 Receptor and NADPH Oxidase in the Vascular System of Diabetic Rats. *Front. Physiol.* **2017**, *8*, 861. [CrossRef]

45. Gobel, K.; Pankratz, S.; Schneider-Hohendorf, T.; Bittner, S.; Schuhmann, M.K.; Langer, H.F.; Stoll, G.; Wiendl, H.; Kleinschnitz, C.; Meuth, S.G. Blockade of the kinin receptor B1 protects from autoimmune CNS disease by reducing leukocyte trafficking. *J. Autoimmun.* **2011**, *36*, 106–114. [CrossRef]

46. Lau, J.; Rousseau, J.; Kwon, D.; Benard, F.; Lin, K.S. A Systematic Review of Molecular Imaging Agents Targeting Bradykinin B1 and B2 Receptors. *Pharmaceuticals* **2020**, *13*, 199. [CrossRef]

47. Qin, L.; Du, Y.; Ding, H.; Haque, A.; Hicks, J.; Pedroza, C.; Mohan, C. Bradykinin 1 receptor blockade subdues systemic autoimmunity, renal inflammation, and blood pressure in murine lupus nephritis. *Arthritis Res. Ther.* **2019**, *21*, 12. [CrossRef]

48. Stadnicki, A.; Pastucha, E.; Nowaczyk, G.; Mazurek, U.; Plewka, D.; Machnik, G.; Wilczok, T.; Colman, R.W. Immunolocalization and expression of kinin B1R and B2R receptors in human inflammatory bowel disease. *Am. J. Physiol. Gastrointest. Liver Physiol.* **2005**, *289*, G361–G366. [CrossRef]

49. Westermann, D.; Walther, T.; Savvatis, K.; Escher, F.; Sobirey, M.; Riad, A.; Bader, M.; Schultheiss, H.P.; Tschope, C. Gene deletion of the kinin receptor B1 attenuates cardiac inflammation and fibrosis during the development of experimental diabetic cardiomyopathy. *Diabetes* **2009**, *58*, 1373–1381. [CrossRef]

50. Duka, I.; Kintsurashvili, E.; Gavras, I.; Johns, C.; Bresnahan, M.; Gavras, H. Vasoactive potential of the b(1) bradykinin receptor in normotension and hypertension. *Circ. Res.* **2001**, *88*, 275–281. [CrossRef]

51. Kakoki, M.; Takahashi, N.; Jennette, J.C.; Smithies, O. Diabetic nephropathy is markedly enhanced in mice lacking the bradykinin B2 receptor. *Proc. Natl. Acad. Sci. USA.* **2004**, *101*, 13302–13305. [CrossRef]

52. Tan, Y.; Keum, J.S.; Wang, B.; McHenry, M.B.; Lipsitz, S.R.; Jaffa, A.A. Targeted deletion of B2-kinin receptors protects against the development of diabetic nephropathy. *Am. J. Physiol. Renal Physiol.* **2007**, *293*, F1026–F1035. [CrossRef] [PubMed]

53. Tidjane, N.; Hachem, A.; Zaid, Y.; Merhi, Y.; Gaboury, L.; Girolami, J.P.; Couture, R. A primary role for kinin B1 receptor in inflammation, organ damage, and lethal thrombosis in a rat model of septic shock in diabetes. *Eur J Inflamm* **2015**, *13*, 40–52. [CrossRef] [PubMed]

54. Bascands, J.L.; Bachvarova, M.; Neau, E.; Schanstra, J.P.; Bachvarov, D. Molecular determinants of LPS-induced acute renal inflammation: Implication of the kinin B1 receptor. *Biochem. Biophys. Res. Commun.* **2009**, *386*, 407–412. [CrossRef] [PubMed]

55. Wang, P.H.; Campanholle, G.; Cenedeze, M.A.; Feitoza, C.Q.; Goncalves, G.M.; Landgraf, R.G.; Jancar, S.; Pesquero, J.B.; Pacheco-Silva, A.; Camara, N.O. Bradykinin [corrected] B1 receptor antagonism is beneficial in renal ischemia-reperfusion injury. *PLoS ONE* **2008**, *3*, e3050. [CrossRef]

56. Duchene, J.; Ahluwalia, A. The kinin B(1) receptor and inflammation: New therapeutic target for cardiovascular disease. *Curr. Opin. Pharmacol.* **2009**, *9*, 125–131. [CrossRef]

57. Bhat, M.; Pouliot, M.; Couture, R.; Vaucher, E. The kallikrein-kinin system in diabetic retinopathy. *Prog. Drug Res.* **2014**, *69*, 111–143. [PubMed]

58. Hachana, S.; Bhat, M.; Senecal, J.; Huppe-Gourgues, F.; Couture, R.; Vaucher, E. Expression, distribution and function of kinin B1 receptor in the rat diabetic retina. *Br. J. Pharmacol.* **2018**, *175*, 968–983. [CrossRef]

59. Hachana, S.; Fontaine, O.; Sapieha, P.; Lesk, M.; Couture, R.; Vaucher, E. The effects of anti-VEGF and kinin B1 receptor blockade on retinal inflammation in laser-induced choroidal neovascularization. *Br. J. Pharmacol.* **2020**, *177*, 1949–1966. [CrossRef]

60. Pouliot, M.; Talbot, S.; Senecal, J.; Dotigny, F.; Vaucher, E.; Couture, R. Ocular application of the kinin B1 receptor antagonist LF22-0542 inhibits retinal inflammation and oxidative stress in streptozotocin-diabetic rats. *PLoS ONE* **2012**, *7*, e33864. [CrossRef]

61. Dias, J.P.; Couture, R. Suppression of vascular inflammation by kinin B1 receptor antagonism in a rat model of insulin resistance. *J. Cardiovasc. Pharmacol.* **2012**, *60*, 61–69. [CrossRef]

62. Yao, Y.Y.; Yin, H.; Shen, B.; Chao, L.; Chao, J. Tissue kallikrein infusion prevents cardiomyocyte apoptosis, inflammation and ventricular remodeling after myocardial infarction. *Regul. Pept.* **2007**, *140*, 12–20. [CrossRef] [PubMed]

63. Noda, M.; Sasaki, K.; Ifuku, M.; Wada, K. Multifunctional effects of bradykinin on glial cells in relation to potential anti-inflammatory effects. *Neurochem. Int.* **2007**, *51*, 185–191. [CrossRef] [PubMed]

64. Nunes, M.A.; Toricelli, M.; Schowe, N.M.; Malerba, H.N.; Dong-Creste, K.E.; Farah, D.; De Angelis, K.; Irigoyen, M.C.; Gobeil, F.; Araujo Viel, T.; et al. Kinin B2 Receptor Activation Prevents the Evolution of Alzheimer's Disease Pathological Characteristics in a Transgenic Mouse Model. *Pharmaceuticals* **2020**, *13*, 288. [CrossRef]

65. Marceau, F.; Hess, J.F.; Bachvarov, D.R. The B1 receptors for kinins. *Pharmacol. Rev.* **1998**, *50*, 357–386.

66. Tonussi, C.R.; Ferreira, S.H. Bradykinin-induced knee joint incapacitation involves bradykinin B2 receptor mediated hyperalgesia and bradykinin B1 receptor-mediated nociception. *Eur. J. Pharmacol.* **1997**, *326*, 61–65. [CrossRef]

67. Walker, K.; Dray, A.; Perkins, M. Hyperalgesia in rats following intracerebroventricular administration of endotoxin: Effect of bradykinin B1 and B2 receptor antagonist treatment. *Pain* **1996**, *65*, 211–219. [CrossRef]
68. Haddad, Y.; Couture, R. Kininase 1 as a Preclinical Therapeutic Target for Kinin B1 Receptor in Insulin Resistance. *Front. Pharmacol.* **2017**, *8*, 509. [CrossRef] [PubMed]
69. Othman, R.; Vaucher, E.; Couture, R. Bradykinin Type 1 Receptor - Inducible Nitric Oxide Synthase: A New Axis Implicated in Diabetic Retinopathy. *Front. Pharmacol.* **2019**, *10*, 300. [CrossRef]
70. Tidjane, N.; Gaboury, L.; Couture, R. Cellular localisation of the kinin B1R in the pancreas of streptozotocin-treated rat and the anti-diabetic effect of the antagonist SSR240612. *Biol. Chem.* **2016**, *397*, 323–336. [CrossRef]
71. Emanueli, C.; Madeddu, P. Targeting kinin receptors for the treatment of tissue ischaemia. *Trends Pharmacol. Sci.* **2001**, *22*, 478–484. [CrossRef]
72. Kakoki, M.; McGarrah, R.W.; Kim, H.S.; Smithies, O. Bradykinin B1 and B2 receptors both have protective roles in renal ischemia/reperfusion injury. *Proc. Natl. Acad. Sci. USA* **2007**, *104*, 7576–7581. [CrossRef]
73. Kakoki, M.; Smithies, O. The kallikrein-kinin system in health and in diseases of the kidney. *Kidney Int.* **2009**, *75*, 1019–1030. [CrossRef]
74. Sanchez de Miguel, L.; Neysari, S.; Jakob, S.; Petrimpol, M.; Butz, N.; Banfi, A.; Zaugg, C.E.; Humar, R.; Battegay, E.J. B2-kinin receptor plays a key role in B1-, angiotensin converting enzyme inhibitor-, and vascular endothelial growth factor-stimulated in vitro angiogenesis in the hypoxic mouse heart. *Cardiovasc. Res.* **2008**, *80*, 106–113. [CrossRef] [PubMed]
75. Tomita, H.; Sanford, R.B.; Smithies, O.; Kakoki, M. The kallikrein-kinin system in diabetic nephropathy. *Kidney Int.* **2012**, *81*, 733–744. [CrossRef]
76. Desposito, D.; Potier, L.; Chollet, C.; Gobeil, F., Jr.; Roussel, R.; Alhenc-Gelas, F.; Bouby, N.; Waeckel, L. Kinin receptor agonism restores hindlimb postischemic neovascularization capacity in diabetic mice. *J. Pharmacol. Exp. Ther.* **2015**, *352*, 218–226. [CrossRef]
77. Emanueli, C.; Bonaria Salis, M.; Stacca, T.; Pintus, G.; Kirchmair, R.; Isner, J.M.; Pinna, A.; Gaspa, L.; Regoli, D.; Cayla, C.; et al. Targeting kinin B(1) receptor for therapeutic neovascularization. *Circulation* **2002**, *105*, 360–366. [CrossRef] [PubMed]
78. Ji, B.; Cheng, B.; Pan, Y.; Wang, C.; Chen, J.; Bai, B. Neuroprotection of bradykinin/bradykinin B2 receptor system in cerebral ischemia. *Biomed. Pharmacother.* **2017**, *94*, 1057–1063. [CrossRef]
79. Gao, L.; Yu, D.M. Molecular mechanism of limbs' postischemic revascularization improved by perindopril in diabetic rats. *Chin. Med. J. (Engl.)* **2008**, *121*, 2129–2133. [CrossRef]
80. Ebrahimian, T.G.; Tamarat, R.; Clergue, M.; Duriez, M.; Levy, B.I.; Silvestre, J.S. Dual effect of angiotensin-converting enzyme inhibition on angiogenesis in type 1 diabetic mice. *Arterioscler. Thromb. Vasc. Biol.* **2005**, *25*, 65–70. [CrossRef] [PubMed]
81. Li, P.; Kondo, T.; Numaguchi, Y.; Kobayashi, K.; Aoki, M.; Inoue, N.; Okumura, K.; Murohara, T. Role of bradykinin, nitric oxide, and angiotensin II type 2 receptor in imidapril-induced angiogenesis. *Hypertension* **2008**, *51*, 252–258. [CrossRef] [PubMed]
82. Mejia, A.J.; Matus, C.E.; Pavicic, F.; Concha, M.; Ehrenfeld, P.; Figueroa, C.D. Intracellular signaling pathways involved in the release of IL-4 and VEGF from human keratinocytes by activation of kinin B1 receptor: Functional relevance to angiogenesis. *Arch. Dermatol. Res.* **2015**, *307*, 803–817. [CrossRef]
83. Naidoo, S.; Raidoo, D.M. Tissue kallikrein and kinin receptor expression in an angiogenic co-culture neuroblastoma model. *Metab. Brain Dis.* **2006**, *21*, 253–265. [CrossRef]
84. da Costa, P.L.N.; Wynne, D.; Fifis, T.; Nguyen, L.; Perini, M.; Christophi, C. The kallikrein-Kinin system modulates the progression of colorectal liver metastases in a mouse model. *BMC Cancer* **2018**, *18*, 382. [CrossRef] [PubMed]
85. Ikeda, Y.; Hayashi, I.; Kamoshita, E.; Yamazaki, A.; Endo, H.; Ishihara, K.; Yamashina, S.; Tsutsumi, Y.; Matsubara, H.; Majima, M. Host stromal bradykinin B2 receptor signaling facilitates tumor-associated angiogenesis and tumor growth. *Cancer Res.* **2004**, *64*, 5178–5185. [CrossRef]
86. Ishihara, K.; Hayash, I.; Yamashina, S.; Majima, M. A potential role of bradykinin in angiogenesis and growth of S-180 mouse tumors. *Jpn. J. Pharmacol.* **2001**, *87*, 318–326. [CrossRef] [PubMed]
87. Ishihara, K.; Kamata, M.; Hayashi, I.; Yamashina, S.; Majima, M. Roles of bradykinin in vascular permeability and angiogenesis in solid tumor. *Int. Immunopharmacol.* **2002**, *2*, 499–509. [CrossRef]
88. Yu, H.S.; Wang, S.W.; Chang, A.C.; Tai, H.C.; Yeh, H.I.; Lin, Y.M.; Tang, C.H. Bradykinin promotes vascular endothelial growth factor expression and increases angiogenesis in human prostate cancer cells. *Biochem. Pharmacol.* **2014**, *87*, 243–253. [CrossRef] [PubMed]
89. Jeppesen, P.; Aalkjaer, C.; Bek, T. Bradykinin relaxation in small porcine retinal arterioles. *Invest. Ophthalmol. Vis. Sci.* **2002**, *43*, 1891–1896. [PubMed]
90. Catanzaro, O.; Labal, E.; Andornino, A.; Capponi, J.A.; Di Martino, I.; Sirois, P. Blockade of early and late retinal biochemical alterations associated with diabetes development by the selective bradykinin B1 receptor antagonist R-954. *Peptides* **2012**, *34*, 349–352. [CrossRef]
91. Clermont, A.; Chilcote, T.J.; Kita, T.; Liu, J.; Riva, P.; Sinha, S.; Feener, E.P. Plasma kallikrein mediates retinal vascular dysfunction and induces retinal thickening in diabetic rats. *Diabetes* **2011**, *60*, 1590–1598. [CrossRef]
92. Hachana, S.; Pouliot, M.; Couture, R.; Vaucher, E. Diabetes-Induced Inflammation and Vascular Alterations in the Goto-Kakizaki Rat Retina. *Curr. Eye Res.* **2020**, 1–10. [CrossRef] [PubMed]
93. Othman, R.; Berbari, S.; Vaucher, E.; Couture, R. Differential Expression of Kinin Receptors in Human Wet and Dry Age-Related Macular Degeneration Retinae. *Pharmaceuticals* **2020**, *13*, 130. [CrossRef]

94. Ma, J.X.; Song, Q.; Hatcher, H.C.; Crouch, R.K.; Chao, L.; Chao, J. Expression and cellular localization of the kallikrein-kinin system in human ocular tissues. *Exp. Eye Res.* **1996**, *63*, 19–26. [CrossRef]

95. Pinna, A.; Emanueli, C.; Dore, S.; Salvo, M.; Madeddu, P.; Carta, F. Levels of human tissue kallikrein in the vitreous fluid of patients with severe proliferative diabetic retinopathy. *Ophthalmologica* **2004**, *218*, 260–263. [CrossRef]

96. Webb, J.G.; Yang, X.; Crosson, C.E. Expression of the kallikrein/kinin system in human anterior segment. *Exp. Eye Res.* **2009**, *89*, 126–132. [CrossRef]

97. Abdouh, M.; Khanjari, A.; Abdelazziz, N.; Ongali, B.; Couture, R.; Hassessian, H.M. Early upregulation of kinin B1 receptors in retinal microvessels of the streptozotocin-diabetic rat. *Br. J. Pharmacol.* **2003**, *140*, 33–40. [CrossRef]

98. Pouliot, M.; Hetu, S.; Lahjouji, K.; Couture, R.; Vaucher, E. Modulation of retinal blood flow by kinin B(1) receptor in Streptozotocin-diabetic rats. *Exp. Eye Res.* **2011**, *92*, 482–489. [CrossRef] [PubMed]

99. Abdouh, M.; Talbot, S.; Couture, R.; Hassessian, H.M. Retinal plasma extravasation in streptozotocin-diabetic rats mediated by kinin B(1) and B(2) receptors. *Br. J. Pharmacol.* **2008**, *154*, 136–143. [CrossRef] [PubMed]

100. Phipps, J.A.; Feener, E.P. The kallikrein-kinin system in diabetic retinopathy: Lessons for the kidney. *Kidney Int.* **2008**, *73*, 1114–1119. [CrossRef]

101. Klein, R.; Klein, B.E.; Moss, S.E.; Davis, M.D.; DeMets, D.L. The Wisconsin epidemiologic study of diabetic retinopathy. III. Prevalence and risk of diabetic retinopathy when age at diagnosis is 30 or more years. *Arch. Ophthalmol.* **1984**, *102*, 527–532. [CrossRef]

102. Klein, R.; Klein, B.E.; Moss, S.E.; Davis, M.D.; DeMets, D.L. The Wisconsin epidemiologic study of diabetic retinopathy. II. Prevalence and risk of diabetic retinopathy when age at diagnosis is less than 30 years. *Arch. Ophthalmol.* **1984**, *102*, 520–526. [CrossRef]

103. Klein, R.; Klein, B.E.; Moss, S.E.; Davis, M.D.; DeMets, D.L. The Wisconsin epidemiologic study of diabetic retinopathy. IV. Diabetic macular edema. *Ophthalmology* **1984**, *91*, 1464–1474. [CrossRef]

104. Ciulla, T.A.; Amador, A.G.; Zinman, B. Diabetic retinopathy and diabetic macular edema: Pathophysiology, screening, and novel therapies. *Diabetes Care* **2003**, *26*, 2653–2664. [CrossRef] [PubMed]

105. ValdezGuerrero, A.S.; Quintana-Perez, J.C.; Arellano-Mendoza, M.G.; Castaneda-Ibarra, F.J.; Tamay-Cach, F.; Aleman-Gonzalez-Duhart, D. Diabetic Retinopathy: Important Biochemical Alterations and the Main Treatment Strategies. *Can. J. Diabetes* **2020**, in press. [CrossRef]

106. Al-Shabrawey, M.; Zhang, W.; McDonald, D. Diabetic retinopathy: Mechanism, diagnosis, prevention, and treatment. *Biomed Res Int* **2015**, *2015*, 854593. [CrossRef] [PubMed]

107. Cheung, N.; Wong, I.Y.; Wong, T.Y. Ocular anti-VEGF therapy for diabetic retinopathy: Overview of clinical efficacy and evolving applications. *Diabetes Care* **2014**, *37*, 900–905. [CrossRef]

108. Khan, Z.; Kuriakose, R.K.; Khan, M.; Chin, E.K.; Almeida, D.R. Efficacy of the Intravitreal Sustained-Release Dexamethasone Implant for Diabetic Macular Edema Refractory to Anti-Vascular Endothelial Growth Factor Therapy: Meta-Analysis and Clinical Implications. *Ophthalmic Surg Lasers Imaging Retina* **2017**, *48*, 160–166. [CrossRef] [PubMed]

109. Pacella, F.; Romano, M.R.; Turchetti, P.; Tarquini, G.; Carnovale, A.; Mollicone, A.; Mastromatteo, A.; Pacella, E. An eighteen-month follow-up study on the effects of Intravitreal Dexamethasone Implant in diabetic macular edema refractory to anti-VEGF therapy. *Int. J. Ophthalmol.* **2016**, *9*, 1427–1432. [CrossRef]

110. Yilmaz, T.; Weaver, C.D.; Gallagher, M.J.; Cordero-Coma, M.; Cervantes-Castaneda, R.A.; Klisovic, D.; Lavaque, A.J.; Larson, R.J. Intravitreal triamcinolone acetonide injection for treatment of refractory diabetic macular edema: A systematic review. *Ophthalmology* **2009**, *116*, 902–911. [CrossRef] [PubMed]

111. Simo, R.; Stitt, A.W.; Gardner, T.W. Neurodegeneration in diabetic retinopathy: Does it really matter? *Diabetologia* **2018**, *61*, 1902–1912. [CrossRef]

112. Ma, J.X.; King, L.P.; Yang, Z.; Crouch, R.K.; Chao, L.; Chao, J. Kallistatin in human ocular tissues: Reduced levels in vitreous fluids from patients with diabetic retinopathy. *Curr. Eye Res.* **1996**, *15*, 1117–1123. [CrossRef] [PubMed]

113. Hatcher, H.C.; Ma, J.X.; Chao, J.; Chao, L.; Ottlecz, A. Kallikrein-binding protein levels are reduced in the retinas of streptozotocin-induced diabetic rats. *Invest. Ophthalmol. Vis. Sci.* **1997**, *38*, 658–664. [PubMed]

114. Nakamura, S.; Morimoto, N.; Tsuruma, K.; Izuta, H.; Yasuda, Y.; Kato, N.; Ikeda, T.; Shimazawa, M.; Hara, H. Tissue kallikrein inhibits retinal neovascularization via the cleavage of vascular endothelial growth factor-165. *Arterioscler. Thromb. Vasc. Biol.* **2011**, *31*, 1041–1048. [CrossRef]

115. Gao, G.; Shao, C.; Zhang, S.X.; Dudley, A.; Fant, J.; Ma, J.X. Kallikrein-binding protein inhibits retinal neovascularization and decreases vascular leakage. *Diabetologia* **2003**, *46*, 689–698. [CrossRef] [PubMed]

116. Gao, B.B.; Chen, X.; Timothy, N.; Aiello, L.P.; Feener, E.P. Characterization of the vitreous proteome in diabetes without diabetic retinopathy and diabetes with proliferative diabetic retinopathy. *J. Proteome Res.* **2008**, *7*, 2516–2525. [CrossRef]

117. Kim, T.; Kim, S.J.; Kim, K.; Kang, U.B.; Lee, C.; Park, K.S.; Yu, H.G.; Kim, Y. Profiling of vitreous proteomes from proliferative diabetic retinopathy and nondiabetic patients. *Proteomics* **2007**, *7*, 4203–4215. [CrossRef] [PubMed]

118. Clermont, A.; Murugesan, N.; Zhou, Q.; Kita, T.; Robson, P.A.; Rushbrooke, L.J.; Evans, D.M.; Aiello, L.P.; Feener, E.P. Plasma Kallikrein Mediates Vascular Endothelial Growth Factor-Induced Retinal Dysfunction and Thickening. *Invest. Ophthalmol. Vis. Sci.* **2016**, *57*, 2390–2399. [CrossRef]

119. Sun, J.K.; Maturi, R.K.; Boyer, D.S.; Wells, J.A.; Gonzalez, V.H.; Tansley, R.; Hernandez, H.; Maetzel, A.; Feener, E.P.; Aiello, L.P. One-Time Intravitreal Injection of KVD001, a Plasma Kallikrein Inhibitor, in Patients with Central-Involved Diabetic Macular Edema and Reduced Vision: An Open-Label Phase 1B Study. *Ophthalmol Retina* **2019**, *3*, 1107–1109. [CrossRef]

120. Han, E.D.; MacFarlane, R.C.; Mulligan, A.N.; Scafidi, J.; Davis, A.E., 3rd. Increased vascular permeability in C1 inhibitor-deficient mice mediated by the bradykinin type 2 receptor. *J. Clin. Invest.* **2002**, *109*, 1057–1063. [CrossRef] [PubMed]

121. Sriramula, S. Kinin B1 receptor: A target for neuroinflammation in hypertension. *Pharmacol. Res.* **2020**, *155*, 104715. [CrossRef]

122. Mugisho, O.O.; Robilliard, L.D.; Nicholson, L.F.B.; Graham, E.S.; O'Carroll, S.J. Bradykinin receptor-1 activation induces inflammation and increases the permeability of human brain microvascular endothelial cells. *Cell Biol. Int.* **2019**. [CrossRef] [PubMed]

123. Kuhr, F.; Lowry, J.; Zhang, Y.; Brovkovych, V.; Skidgel, R.A. Differential regulation of inducible and endothelial nitric oxide synthase by kinin B1 and B2 receptors. *Neuropeptides* **2010**, *44*, 145–154. [CrossRef] [PubMed]

124. Leal, E.C.; Manivannan, A.; Hosoya, K.; Terasaki, T.; Cunha-Vaz, J.; Ambrosio, A.F.; Forrester, J.V. Inducible nitric oxide synthase isoform is a key mediator of leukostasis and blood-retinal barrier breakdown in diabetic retinopathy. *Invest. Ophthalmol. Vis. Sci.* **2007**, *48*, 5257–5265. [CrossRef]

125. Zheng, L.; Du, Y.; Miller, C.; Gubitosi-Klug, R.A.; Kern, T.S.; Ball, S.; Berkowitz, B.A. Critical role of inducible nitric oxide synthase in degeneration of retinal capillaries in mice with streptozotocin-induced diabetes. *Diabetologia* **2007**, *50*, 1987–1996. [CrossRef] [PubMed]

126. Canto, A.; Olivar, T.; Romero, F.J.; Miranda, M. Nitrosative Stress in Retinal Pathologies: Review. *Antioxidants* **2019**, *8*, 543. [CrossRef]

127. Huie, R.E.; Padmaja, S. The reaction of no with superoxide. *Free Radic. Res. Commun.* **1993**, *18*, 195–199. [CrossRef]

128. Pacher, P.; Beckman, J.S.; Liaudet, L. Nitric oxide and peroxynitrite in health and disease. *Physiol. Rev.* **2007**, *87*, 315–424. [CrossRef]

129. Al-Shabrawey, M.; Bartoli, M.; El-Remessy, A.B.; Ma, G.; Matragoon, S.; Lemtalsi, T.; Caldwell, R.W.; Caldwell, R.B. Role of NADPH oxidase and Stat3 in statin-mediated protection against diabetic retinopathy. *Invest. Ophthalmol. Vis. Sci.* **2008**, *49*, 3231–3238. [CrossRef]

130. Al-Shabrawey, M.; Rojas, M.; Sanders, T.; Behzadian, A.; El-Remessy, A.; Bartoli, M.; Parpia, A.K.; Liou, G.; Caldwell, R.B. Role of NADPH oxidase in retinal vascular inflammation. *Invest. Ophthalmol. Vis. Sci.* **2008**, *49*, 3239–3244. [CrossRef] [PubMed]

131. El-Remessy, A.B.; Abou-Mohamed, G.; Caldwell, R.W.; Caldwell, R.B. High glucose-induced tyrosine nitration in endothelial cells: Role of eNOS uncoupling and aldose reductase activation. *Invest. Ophthalmol. Vis. Sci.* **2003**, *44*, 3135–3143. [CrossRef]

132. El-Remessy, A.B.; Al-Shabrawey, M.; Khalifa, Y.; Tsai, N.T.; Caldwell, R.B.; Liou, G.I. Neuroprotective and blood-retinal barrier-preserving effects of cannabidiol in experimental diabetes. *Am. J. Pathol.* **2006**, *168*, 235–244. [CrossRef]

133. el-Remessy, A.B.; Bartoli, M.; Platt, D.H.; Fulton, D.; Caldwell, R.B. Oxidative stress inactivates VEGF survival signaling in retinal endothelial cells via PI 3-kinase tyrosine nitration. *J. Cell Sci.* **2005**, *118*, 243–252. [CrossRef]

134. Orsenigo, F.; Giampietro, C.; Ferrari, A.; Corada, M.; Galaup, A.; Sigismund, S.; Ristagno, G.; Maddaluno, L.; Koh, G.Y.; Franco, D.; et al. Phosphorylation of VE-cadherin is modulated by haemodynamic forces and contributes to the regulation of vascular permeability in vivo. *Nat. Commun.* **2012**, *3*, 1208. [CrossRef]

135. Sikpa, D.; Whittingstall, L.; Savard, M.; Lebel, R.; Cote, J.; McManus, S.; Chemtob, S.; Fortin, D.; Lepage, M.; Gobeil, F. Pharmacological Modulation of Blood-Brain Barrier Permeability by Kinin Analogs in Normal and Pathologic Conditions. *Pharmaceuticals* **2020**, *13*, 279. [CrossRef]

136. Klein, R.; Klein, B.E.K.; Linton, K.L.P. Prevalence of Age-related Maculopathy: The Beaver Dam Eye Study. *Ophthalmology* **2020**, *127*, S122–S132. [CrossRef]

137. Wong, W.L.; Su, X.; Li, X.; Cheung, C.M.; Klein, R.; Cheng, C.Y.; Wong, T.Y. Global prevalence of age-related macular degeneration and disease burden projection for 2020 and 2040: A systematic review and meta-analysis. *Lancet Glob Health* **2014**, *2*, e106–e116. [CrossRef]

138. Guillonneau, X.; Eandi, C.M.; Paques, M.; Sahel, J.A.; Sapieha, P.; Sennlaub, F. On phagocytes and macular degeneration. *Prog. Retin. Eye Res.* **2017**, *61*, 98–128. [CrossRef]

139. Ghazi, N.G. Retinal angiomatous proliferation in age-related macular degeneration. *Retina* **2002**, *22*, 509–511. [CrossRef]

140. Yannuzzi, L.A.; Negrao, S.; Iida, T.; Carvalho, C.; Rodriguez-Coleman, H.; Slakter, J.; Freund, K.B.; Sorenson, J.; Orlock, D.; Borodoker, N. Retinal angiomatous proliferation in age-related macular degeneration. 2001. *Retina* **2012**, *32* (Suppl. 1), 416–434. [CrossRef]

141. Sarks, S.H. Ageing and degeneration in the macular region: A clinico-pathological study. *Br. J. Ophthalmol.* **1976**, *60*, 324–341. [CrossRef]

142. Rofagha, S.; Bhisitkul, R.B.; Boyer, D.S.; Sadda, S.R.; Zhang, K.; Group, S.-U.S. Seven-year outcomes in ranibizumab-treated patients in ANCHOR, MARINA, and HORIZON: A multicenter cohort study (SEVEN-UP). *Ophthalmology* **2013**, *120*, 2292–2299. [CrossRef]

143. Choudhary, M.; Malek, G. A Review of Pathogenic Drivers of Age-Related Macular Degeneration, Beyond Complement, with a Focus on Potential Endpoints for Testing Therapeutic Interventions in Preclinical Studies. *Adv. Exp. Med. Biol.* **2019**, *1185*, 9–13. [CrossRef]

144. Kern, T.S.; Miller, C.M.; Du, Y.; Zheng, L.; Mohr, S.; Ball, S.L.; Kim, M.; Jamison, J.A.; Bingaman, D.P. Topical administration of nepafenac inhibits diabetes-induced retinal microvascular disease and underlying abnormalities of retinal metabolism and physiology. *Diabetes* **2007**, *56*, 373–379. [CrossRef] [PubMed]

145. Le, Y.Z. VEGF production and signaling in Muller glia are critical to modulating vascular function and neuronal integrity in diabetic retinopathy and hypoxic retinal vascular diseases. *Vision Res.* **2017**, *139*, 108–114. [CrossRef]

146. Penn, J.S.; Madan, A.; Caldwell, R.B.; Bartoli, M.; Caldwell, R.W.; Hartnett, M.E. Vascular endothelial growth factor in eye disease. *Prog. Retin. Eye Res.* **2008**, *27*, 331–371. [CrossRef]

147. Parenti, A.; Morbidelli, L.; Ledda, F.; Granger, H.J.; Ziche, M. The bradykinin/B1 receptor promotes angiogenesis by up-regulation of endogenous FGF-2 in endothelium via the nitric oxide synthase pathway. *FASEB J.* **2001**, *15*, 1487–1489. [CrossRef] [PubMed]

148. Terzuoli, E.; Morbidelli, L.; Nannelli, G.; Giachetti, A.; Donnini, S.; Ziche, M. Involvement of Bradykinin B2 Receptor in Pathological Vascularization in Oxygen-Induced Retinopathy in Mice and Rabbit Cornea. *Int. J. Mol. Sci.* **2018**, *19*, 330. [CrossRef] [PubMed]

149. Nagai, N.; Oike, Y.; Izumi-Nagai, K.; Koto, T.; Satofuka, S.; Shinoda, H.; Noda, K.; Ozawa, Y.; Inoue, M.; Tsubota, K.; et al. Suppression of choroidal neovascularization by inhibiting angiotensin-converting enzyme: Minimal role of bradykinin. *Invest. Ophthalmol. Vis. Sci.* **2007**, *48*, 2321–2326. [CrossRef]

150. Voigt, A.P.; Mulfaul, K.; Mullin, N.K.; Flamme-Wiese, M.J.; Giacalone, J.C.; Stone, E.M.; Tucker, B.A.; Scheetz, T.E.; Mullins, R.F. Single-cell transcriptomics of the human retinal pigment epithelium and choroid in health and macular degeneration. *Proc. Natl. Acad. Sci. USA* **2019**, *116*, 24100–24107. [CrossRef]

151. Binet, F.; Cagnone, G.; Crespo-Garcia, S.; Hata, M.; Neault, M.; Dejda, A.; Wilson, A.M.; Buscarlet, M.; Mawambo, G.T.; Howard, J.P.; et al. Neutrophil extracellular traps target senescent vasculature for tissue remodeling in retinopathy. *Science* **2020**, *369*. [CrossRef]

152. Fu, Z.; Chen, C.T.; Cagnone, G.; Heckel, E.; Sun, Y.; Cakir, B.; Tomita, Y.; Huang, S.; Li, Q.; Britton, W.; et al. Dyslipidemia in retinal metabolic disorders. *EMBO Mol. Med.* **2019**, *11*, e10473. [CrossRef] [PubMed]

153. Rohlenova, K.; Goveia, J.; García-Caballero, M.; Subramanian, A.; Kalucka, J.; Treps, L.; Falkenberg, K.D.; de Rooij, L.; Zheng, Y.; Lin, L.; et al. Single-Cell RNA Sequencing Maps Endothelial Metabolic Plasticity in Pathological Angiogenesis. *Cell Metab.* **2020**, *31*, 862–877. [CrossRef]

154. Llobet, A.; Gual, A.; Pales, J.; Barraquer, R.; Tobias, E.; Nicolas, J.M. Bradykinin decreases outflow facility in perfused anterior segments and induces shape changes in passaged BTM cells in vitro. *Invest. Ophthalmol. Vis. Sci.* **1999**, *40*, 113–125.

155. Sharif, N.A.; Xu, S.X. Pharmacological characterization of bradykinin receptors coupled to phosphoinositide turnover in SV40-immortalized human trabecular meshwork cells. *Exp. Eye Res.* **1996**, *63*, 631–637. [CrossRef]

156. Webb, J.G.; Shearer, T.W.; Yates, P.W.; Mukhin, Y.V.; Crosson, C.E. Bradykinin enhancement of PGE2 signalling in bovine trabecular meshwork cells. *Exp. Eye Res.* **2003**, *76*, 283–289. [CrossRef]

157. Sharif, N.A.; Katoli, P.; Scott, D.; Li, L.; Kelly, C.; Xu, S.; Husain, S.; Toris, C.; Crosson, C. FR-190997, a nonpeptide bradykinin B2-receptor partial agonist, is a potent and efficacious intraocular pressure lowering agent in ocular hypertensive cynomolgus monkeys. *Drug Dev Res* **2014**, *75*, 211–223. [CrossRef]

158. Masuda, T.; Shimazawa, M.; Ishizuka, F.; Nakamura, S.; Tsuruma, K.; Hara, H. Tissue kallikrein (kallidinogenase) protects against retinal ischemic damage in mice. *Eur. J. Pharmacol.* **2014**, *738*, 74–82. [CrossRef] [PubMed]

159. Choudhary, R.; Kapoor, M.S.; Singh, A.; Bodakhe, S.H. Therapeutic targets of renin-angiotensin system in ocular disorders. *J Curr. Ophthalmol.* **2017**, *29*, 7–16. [CrossRef]

160. Danser, A.H.; Derkx, F.H.; Admiraal, P.J.; Deinum, J.; de Jong, P.T.; Schalekamp, M.A. Angiotensin levels in the eye. *Invest. Ophthalmol. Vis. Sci.* **1994**, *35*, 1008–1018. [PubMed]

161. Danser, A.H.; van den Dorpel, M.A.; Deinum, J.; Derkx, F.H.; Franken, A.A.; Peperkamp, E.; de Jong, P.T.; Schalekamp, M.A. Renin, prorenin, and immunoreactive renin in vitreous fluid from eyes with and without diabetic retinopathy. *J. Clin. Endocrinol. Metab.* **1989**, *68*, 160–167. [CrossRef] [PubMed]

162. White, A.J.; Cheruvu, S.C.; Sarris, M.; Liyanage, S.S.; Lumbers, E.; Chui, J.; Wakefield, D.; McCluskey, P.J. Expression of classical components of the renin-angiotensin system in the human eye. *J. Renin Angiotensin Aldosterone Syst.* **2015**, *16*, 59–66. [CrossRef]

163. Wilkinson-Berka, J.L.; Suphapimol, V.; Jerome, J.R.; Deliyanti, D.; Allingham, M.J. Angiotensin II and aldosterone in retinal vasculopathy and inflammation. *Exp. Eye Res.* **2019**, *187*, 107766. [CrossRef]

164. Abassi, Z.; Skorecki, K.; Hamo-Giladi, D.B.; Kruzel-Davila, E.; Heyman, S.N. Kinins and chymase: The forgotten components of the renin-angiotensin system and their implications in COVID-19 disease. *Am. J. Physiol. Lung Cell Mol. Physiol.* **2021**, *320*, L422–L429. [CrossRef]

165. Regoli, D.; Gobeil, F. Kallikrein-kinin system as the dominant mechanism to counteract hyperactive renin-angiotensin system. *Can. J. Physiol. Pharmacol.* **2017**, *95*, 1117–1124. [CrossRef]

166. Igic, R. Four decades of ocular renin-angiotensin and kallikrein-kinin systems (1977-2017). *Exp. Eye Res.* **2018**, *166*, 74–83. [CrossRef]

167. Tsutsumi, Y.; Matsubara, H.; Masaki, H.; Kurihara, H.; Murasawa, S.; Takai, S.; Miyazaki, M.; Nozawa, Y.; Ozono, R.; Nakagawa, K.; et al. Angiotensin II type 2 receptor overexpression activates the vascular kinin system and causes vasodilation. *J. Clin. Invest.* **1999**, *104*, 925–935. [CrossRef]

168. Ferrario, C.M. Role of angiotensin II in cardiovascular disease therapeutic implications of more than a century of research. *J. Renin Angiotensin Aldosterone Syst.* **2006**, *7*, 3–14. [CrossRef]

169. Paz Ocaranza, M.; Riquelme, J.A.; García, L.; Jalil, J.E.; Chiong, M.; Santos, R.A.S.; Lavandero, S. Counter-regulatory renin–angiotensin system in cardiovascular disease. *Nature Reviews Cardiology* **2020**, *17*, 116–129. [CrossRef] [PubMed]

170. Forrester, S.J.; Booz, G.W.; Sigmund, C.D.; Coffman, T.M.; Kawai, T.; Rizzo, V.; Scalia, R.; Eguchi, S. Angiotensin II Signal Transduction: An Update on Mechanisms of Physiology and Pathophysiology. *Physiol. Rev.* **2018**, *98*, 1627–1738. [CrossRef]

171. Santos, R.A.; Ferreira, A.J.; Verano-Braga, T.; Bader, M. Angiotensin-converting enzyme 2, angiotensin-(1-7) and Mas: New players of the renin-angiotensin system. *J. Endocrinol.* **2013**, *216*, R1–R17. [CrossRef] [PubMed]

172. Zimmerman, D.; Burns, K.D. Angiotensin-(1–7) in kidney disease: A review of the controversies. *Clin. Sci.* **2012**, *123*, 333–346. [CrossRef] [PubMed]

173. Sodhi, C.P.; Wohlford-Lenane, C.; Yamaguchi, Y.; Prindle, T.; Fulton, W.B.; Wang, S.; McCray, P.B., Jr.; Chappell, M.; Hackam, D.J.; Jia, H. Attenuation of pulmonary ACE2 activity impairs inactivation of des-Arg(9) bradykinin/BKB1R axis and facilitates LPS-induced neutrophil infiltration. *Am. J. Physiol. Lung Cell Mol. Physiol.* **2018**, *314*, L17–L31. [CrossRef]

174. Parekh, R.U.; Robidoux, J.; Sriramula, S. Kinin B1 Receptor Blockade Prevents Angiotensin II-induced Neuroinflammation and Oxidative Stress in Primary Hypothalamic Neurons. *Cell. Mol. Neurobiol.* **2019**. [CrossRef] [PubMed]

175. Kintsurashvili, E.; Duka, I.; Gavras, I.; Johns, C.; Farmakiotis, D.; Gavras, H. Effects of ANG II on bradykinin receptor gene expression in cardiomyocytes and vascular smooth muscle cells. *Am. J. Physiol. Heart Circ. Physiol.* **2001**, *281*, H1778–H1783. [CrossRef] [PubMed]

176. Morand-Contant, M.; Anand-Srivastava, M.B.; Couture, R. Kinin B1 receptor upregulation by angiotensin II and endothelin-1 in rat vascular smooth muscle cells: Receptors and mechanisms. *Am. J. Physiol. Heart Circ. Physiol.* **2010**, *299*, H1625–H1632. [CrossRef]

177. Fernandes, L.; Ceravolo, G.S.; Fortes, Z.B.; Tostes, R.; Santos, R.A.; Santos, J.A.; Mori, M.A.; Pesquero, J.B.; de Carvalho, M.H. Modulation of kinin B1 receptor expression by endogenous angiotensin II in hypertensive rats. *Regul. Pept.* **2006**, *136*, 92–97. [CrossRef] [PubMed]

178. Yamamoto, K.; Chappell, M.C.; Brosnihan, K.B.; Ferrario, C.M. In vivo metabolism of angiotensin I by neutral endopeptidase (EC 3.4.24.11) in spontaneously hypertensive rats. *Hypertension* **1992**, *19*, 692–696. [CrossRef]

179. Allred, A.J.; Diz, D.I.; Ferrario, C.M.; Chappell, M.C. Pathways for angiotensin-(1—7) metabolism in pulmonary and renal tissues. *Am. J. Physiol. Renal Physiol.* **2000**, *279*, F841–F850. [CrossRef]

180. Yamagishi, S.; Takeuchi, M.; Matsui, T.; Nakamura, K.; Imaizumi, T.; Inoue, H. Angiotensin II augments advanced glycation end product-induced pericyte apoptosis through RAGE overexpression. *FEBS Lett.* **2005**, *579*, 4265–4270. [CrossRef] [PubMed]

181. Sugiyama, T.; Okuno, T.; Fukuhara, M.; Oku, H.; Ikeda, T.; Obayashi, H.; Ohta, M.; Fukui, M.; Hasegawa, G.; Nakamura, N. Angiotensin II receptor blocker inhibits abnormal accumulation of advanced glycation end products and retinal damage in a rat model of type 2 diabetes. *Exp. Eye Res.* **2007**, *85*, 406–412. [CrossRef]

182. Phipps, J.A.; Clermont, A.C.; Sinha, S.; Chilcote, T.J.; Bursell, S.E.; Feener, E.P. Plasma kallikrein mediates angiotensin II type 1 receptor-stimulated retinal vascular permeability. *Hypertension* **2009**, *53*, 175–181. [CrossRef] [PubMed]

183. Sjolie, A.K.; Klein, R.; Porta, M.; Orchard, T.; Fuller, J.; Parving, H.H.; Bilous, R.; Aldington, S.; Chaturvedi, N. Retinal microaneurysm count predicts progression and regression of diabetic retinopathy. Post-hoc results from the DIRECT Programme. *Diabet. Med.* **2011**, *28*, 345–351. [CrossRef]

184. Mauer, M.; Zinman, B.; Gardiner, R.; Suissa, S.; Sinaiko, A.; Strand, T.; Drummond, K.; Donnelly, S.; Goodyer, P.; Gubler, M.C.; et al. Renal and retinal effects of enalapril and losartan in type 1 diabetes. *N. Engl. J. Med.* **2009**, *361*, 40–51. [CrossRef] [PubMed]

185. Kiire, C.A.; Porta, M.; Chong, V. Medical management for the prevention and treatment of diabetic macular edema. *Surv. Ophthalmol.* **2013**, *58*, 459–465. [CrossRef] [PubMed]

186. Wang, B.; Wang, F.; Zhang, Y.; Zhao, S.H.; Zhao, W.J.; Yan, S.L.; Wang, Y.G. Effects of RAS inhibitors on diabetic retinopathy: A systematic review and meta-analysis. *Lancet Diabetes Endocrinol* **2015**, *3*, 263–274. [CrossRef]

187. Gilbert, R.E.; Kelly, D.J.; Cox, A.J.; Wilkinson-Berka, J.L.; Rumble, J.R.; Osicka, T.; Panagiotopoulos, S.; Lee, V.; Hendrich, E.C.; Jerums, G.; et al. Angiotensin converting enzyme inhibition reduces retinal overexpression of vascular endothelial growth factor and hyperpermeability in experimental diabetes. *Diabetologia* **2000**, *43*, 1360–1367. [CrossRef]

188. Hogeboom van Buggenum, I.M.; Polak, B.C.; Reichert-Thoen, J.W.; de Vries-Knoppert, W.A.; van Hinsbergh, V.W.; Tangelder, G.J. Angiotensin converting enzyme inhibiting therapy is associated with lower vitreous vascular endothelial growth factor concentrations in patients with proliferative diabetic retinopathy. *Diabetologia* **2002**, *45*, 203–209. [CrossRef]

189. Chaturvedi, N.; Fuller, J.H.; Pokras, F.; Rottiers, R.; Papazoglou, N.; Aiello, L.P.; Group, E.S. Circulating plasma vascular endothelial growth factor and microvascular complications of type 1 diabetes mellitus: The influence of ACE inhibition. *Diabet. Med.* **2001**, *18*, 288–294. [CrossRef]

190. UK Prospective Diabetes Study Group. Tight blood pressure control and risk of macrovascular and microvascular complications in type 2 diabetes: UKPDS 38. *BMJ* **1998**, *317*, 703–713. [CrossRef]

191. Chaturvedi, N.; Sjolie, A.K.; Stephenson, J.M.; Abrahamian, H.; Keipes, M.; Castellarin, A.; Rogulja-Pepeonik, Z.; Fuller, J.H. Effect of lisinopril on progression of retinopathy in normotensive people with type 1 diabetes. The EUCLID Study Group. EURODIAB Controlled Trial of Lisinopril in Insulin-Dependent Diabetes Mellitus. *Lancet* **1998**, *351*, 28–31. [CrossRef]

192. Baş, M.; Greve, J.; Stelter, K.; Havel, M.; Strassen, U.; Rotter, N.; Veit, J.; Schossow, B.; Hapfelmeier, A.; Kehl, V.; et al. A randomized trial of icatibant in ACE-inhibitor-induced angioedema. *N. Engl. J. Med.* **2015**, *372*, 418–425. [CrossRef]
193. Beavers, C.J.; Dunn, S.P.; Macaulay, T.E. The role of angiotensin receptor blockers in patients with angiotensin-converting enzyme inhibitor-induced angioedema. *Ann. Pharmacother.* **2011**, *45*, 520–524. [CrossRef]
194. Bezalel, S.; Mahlab-Guri, K.; Asher, I.; Werner, B.; Sthoeger, Z.M. Angiotensin-converting enzyme inhibitor-induced angioedema. *Am. J. Med.* **2015**, *128*, 120–125. [CrossRef]
195. Borsook, D.; Sava, S. Do ACE inhibitors exacerbate complex regional pain syndrome? *Nat. Rev. Neurol.* **2009**, *5*, 306–308. [CrossRef] [PubMed]
196. Molinaro, G.; Cugno, M.; Perez, M.; Lepage, Y.; Gervais, N.; Agostoni, A.; Adam, A. Angiotensin-converting enzyme inhibitor-associated angioedema is characterized by a slower degradation of des-arginine(9)-bradykinin. *J. Pharmacol. Exp. Ther.* **2002**, *303*, 232–237. [CrossRef]
197. Ismael, M.A.; Talbot, S.; Carbonneau, C.L.; Beausejour, C.M.; Couture, R. Blockade of sensory abnormalities and kinin B(1) receptor expression by N-acetyl-L-cysteine and ramipril in a rat model of insulin resistance. *Eur. J. Pharmacol.* **2008**, *589*, 66–72. [CrossRef] [PubMed]
198. Satofuka, S.; Ichihara, A.; Nagai, N.; Noda, K.; Ozawa, Y.; Fukamizu, A.; Tsubota, K.; Itoh, H.; Oike, Y.; Ishida, S. (Pro)renin receptor promotes choroidal neovascularization by activating its signal transduction and tissue renin-angiotensin system. *Am. J. Pathol.* **2008**, *173*, 1911–1918. [CrossRef]
199. Kolomeyer, A.M.; Maguire, M.G.; Pan, W.; VanderBeek, B.L. Systemic Beta-Blockers and Risk of Progression to Neovascular Age-Related Macular Degeneration. *Retina* **2019**, *39*, 918–925. [CrossRef]
200. Lee, H.; Jeon, H.L.; Park, S.J.; Shin, J.Y. Effect of Statins, Metformin, Angiotensin-Converting Enzyme Inhibitors, and Angiotensin II Receptor Blockers on Age-Related Macular Degeneration. *Yonsei Med. J.* **2019**, *60*, 679–686. [CrossRef]
201. Thomas, A.S.; Redd, T.; Hwang, T. Effect of Systemic Beta-Blockers, Ace Inhibitors, and Angiotensin Receptor Blockers on Development of Choroidal Neovascularization in Patients with Age-Related Macular Degeneration. *Retina* **2015**, *35*, 1964–1968. [CrossRef] [PubMed]

Article

The Timecourses of Functional, Morphological, and Molecular Changes Triggered by Light Exposure in Sprague–Dawley Rat Retinas

Serena Riccitelli [1,†], Mattia Di Paolo [1,†], James Ashley [2], Silvia Bisti [1,3,4] and Stefano Di Marco [1,3,4,5,*]

1   Department of Biotechnological and Applied Clinical Sciences, University of L'Aquila, 67100 L'Aquila, Italy; serena.riccitelli@weizmann.ac.il (S.R.); m.dipaolo@bio-aurum.it (M.D.P.); Silvia.Bisti@iit.it (S.B.)
2   School of Biological Sciences, The University of Manchester, Manchester M13 9PL, UK; jamash22@gmail.com
3   Istituto Nazionale di Biostrutture e Biosistemi (INBB), 00136 Roma, Italy
4   Center for Synaptic Neuroscience and Technology, Istituto Italiano di Tecnologia, 16132 Genova, Italy
5   IRCCS, Ospedale Policlinico San Martino, 16132 Genova, Italy
*   Correspondence: stefano.dimarco@iit.it
†   These authors contributed equally to this work.

Abstract: Retinal neurodegeneration can impair visual perception at different levels, involving not only photoreceptors, which are the most metabolically active cells, but also the inner retina. Compensatory mechanisms may hide the first signs of these impairments and reduce the likelihood of receiving timely treatments. Therefore, it is essential to characterize the early critical steps in the neurodegenerative progression to design adequate therapies. This paper describes and correlates early morphological and biochemical changes in the degenerating retina with in vivo functional analysis of retinal activity and investigates the progression of neurodegenerative stages for up to 7 months. For these purposes, Sprague–Dawley rats were exposed to 1000 lux light either for different durations (12 h to 24 h) and examined seven days afterward (7d) or for a fixed duration (24 h) and monitored at various time points following the exposure (up to 210d). Flash electroretinogram (fERG) recordings were correlated with morphological and histological analyses to evaluate outer and inner retinal disruptions, gliosis, trophic factor release, and microglial activation. Twelve hours or fifteen hours of exposure to constant light led to a severe retinal dysfunction with only minor morphological changes. Therefore, early pathological signs might be hidden by compensatory mechanisms that silence retinal dysfunction, accounting for the discrepancy between photoreceptor loss and retinal functional output. The long-term analysis showed a transient functional recovery, maximum at 45 days, despite a progressive loss of photoreceptors and coincident increases in glial fibrillary acidic protein (GFAP) and basic fibroblast growth factor-2 (bFGF-2) expression. Interestingly, the progression of the disease presented different patterns in the dorsal and ventral retina. The information acquired gives us the potential to develop a specific diagnostic tool to monitor the disease's progression and treatment efficacy.

Keywords: light damage; neurodegeneration; functional analysis; early detection; remodeling

check for **updates**

Citation: Riccitelli, S.; Di Paolo, M.; Ashley, J.; Bisti, S.; Di Marco, S. The Timecourses of Functional, Morphological, and Molecular Changes Triggered by Light Exposure in Sprague–Dawley Rat Retinas. *Cells* **2021**, *10*, 1561. https://doi.org/10.3390/cells10061561

Academic Editors: Karl Matter, Maurice Ptito and Joseph Bouskila

Received: 27 March 2021
Accepted: 16 June 2021
Published: 21 June 2021

Publisher's Note: MDPI stays neutral with regard to jurisdictional claims in published maps and institutional affiliations.

1. Introduction

The quality of vision dramatically impacts the quality of human life [1]. Many ocular pathologies induced by several factors, such as gene mutations, environmental stresses, metabolic dysfunction, and aging, lead to reduced visual performance and eventually complete blindness caused by inflammation, mitochondrial dysfunction, synaptic remodeling, and neuronal death [2]. These pathologies include a heterogeneous group of photoreceptor degenerations whereby populations of rods and cones are distinctly affected, i.e., rods are primarily involved in retinitis pigmentosa (RP) [3], cones in age-related macular degenerations (AMD) [3], and both rods and cones simultaneously in Leber congenital amaurosis (LCA) [4] and Stargardt's disease (STGD1, autosomal recessive [5]). Although a variety of new therapeutic strategies have been suggested (including stem cell transplantation to replace photoreceptors [6,7], bionic implants [8,9], and optogenetics approaches [10–12]), currently there are no effective cures. Common pharmacological approaches studied on rodents [13,14] and tested on humans [15–18] have as their primary objective slowing down photoreceptor death to preserve surviving retinal function, increase tissue resilience, and prevent retinal reorganization. Indeed, it is well known that photoreceptors stress initiate an unavoidable chain of events, collectively termed retinal remodeling [19,20], in the remnant inner retina, independently of the molecular defects that initially trigger retinal degeneration [21–24]. Abnormal reprogramming occurs in all pathology phases and can begin even before photoreceptor death [22,25]. In their final states, these modifications could lead to considerable changes in the receptive field properties of retinal ganglion cells, the retina's output cells, impinging consequently on the transmission of visual information to the brain [26–28].

Retinal degeneration has at least three related and recognized phases [29]. The first is the irreversible death of photoreceptors—cells with high metabolic activity predisposed to a wide range of insults, including high-intensity light [3]. The second is the dysfunction of surviving photoreceptors and their subsequent loss [30–32]. The third is related to profound structural and functional abnormalities in the inner retina (i.e., neuronal cell death, microglia migration, rewiring of retinal circuits, and glial hypertrophy) [33].

With these factors in mind, there is a need to diagnose visual impairments at early stages to increase the chances of an effective pharmacological treatment. From a therapeutic perspective, it is also essential to define the critical steps during neurodegenerative progression to design adequate therapies that arrest remodeling events or target specific cell subtypes in the remnant tissue.

To address these clinical needs, we planned an experimental protocol with two primary aims:

1. To characterize and correlate early morphological and biochemical changes occurring in the degenerating retina with in vivo functional analysis of retinal activity.
2. To investigate the progression of neurodegenerative stages.

In the current work, we used a consolidated rat model of light-induced retinal degeneration, previously shown to mimic some aspects of AMD (the presence of oxidative stress [34], inflammatory processes [35], and photoreceptor death [36]). Indeed, apart from the relevance of this model in AMD research, we exploited its well-known characteristics to follow a neurodegenerative process that starts in the dorsal retina and spreads over time. Specifically, by using two different protocols (<24 h and 24 h of light exposure) and a combination of histological and electrophysiological investigations, we first determined the minimal duration of bright light exposure (1000 lux, white light) necessary to induce detectable retinal functional alterations. Interestingly, our findings revealed that functional changes precede anatomical ones. In the second stage, we investigated the temporal course of the neurodegeneration following 24 h of light exposure while extending our analyses to the inner retina by relying on antibodies against cell-type-specific markers and trophic factors.

Altogether, the knowledges acquired will allow us to extrapolate information (i.e., emergent ERG signals) that might help to detect the pathology through functional tests before a large portion of photoreceptors is already injured, the retina severely damaged, and neuronal circuitry remodeled. Secondly, following alterations over time will provide valuable insights for dissecting general retinal degeneration pathways that might help develop treatments based on different dystrophy stages.

2. Materials and Methods

2.1. Animals

According to the ARVO Statement for the "Use of Animals in Ophthalmic and Vision Research", all experiments were carried out with authorization numbers 448/2016-PR and 862/2018-PR, issued by the Ministry of Health, and approved by the local Ethical Committee of the University of L'Aquila.

2.2. Light Exposure Protocols

Albino Sprague–Dawley (SD) rats were born and housed at the University of L'Aquila animal facilities in dim cyclic light conditions (12 h light, 12 h dark), at an ambient light level of approximately 5 lux, with food and water available ad libitum.

Data reported in the present study were obtained from experiments conducted on healthy control (HC) animals and animals treated with intense white light, commonly referred to as "light damaged" (LD), which were divided into twelve groups, as depicted in Table 1. Light damage was generated by placing adult albino rats into singular plexiglass cages with lights positioned at the top and at the bottom to ensure an iso-luminance environment (1000 lux, monitored through a lux meter) inside the cage. The litter was removed from the cage to prevent rats from hiding their eyes from the light. Light exposure started at the beginning of the day, immediately after 12 h of darkness. We analyzed different combinations of exposure durations and periods of time following the end of light exposure in normal light conditions. Accordingly, animals were exposed to 1000 lux light for different durations (12 h, 15 h, 18 h, and 24 h) and allowed to recover for a post-exposure period of 7 days (7d) in dim cyclic illumination conditions in standard cages (5 lux, cyclic light). The second group of animals was exposed to the same light intensity (1000 lux) for 24 h and sacrificed at different recovery periods (after 0, 15, 30, 45, 60, 90, and 210 days (d)). The same healthy control (HC) group, raised at 5 lux in cyclic light, and the group exposed to light for 24 h and left to recover for 7 days (LD24h7d), were used in the two experiments. In the text, we refer to short-term exposure to indicate 12 h of light exposure.

Table 1. Experimental groups.

Group	Treatment	Light Exposure Duration (hours)	Days (d) after Light Exposure
HC	Healthy Control	-	-
LD12h7d		12 (short-term)	
LD15h7d		15	7
LD18h7d		18	
LD24h0d			0
LD24h7d			7
LD24h15d	LD 1000 lux		15
LD24h30d			30
LD24h45d		24 (long-term)	45
LD24h60d			60
LD24h90d			90
LD24h210d			210

2.3. Electrophysiological Recordings and Data Analysis

We performed flash electroretinogram (fERG) recordings in controls and evaluated retinal function in each experimental group following the light exposure treatment. To minimize variability among groups, electrophysiological recordings were performed before exposing animals to LD (a pull of these recordings were used as HC, assuming no age-related changes in our temporal window of interest [37,38]). Albino rats were previously dark-adapted overnight, and electroretinograms were recorded in a dark room [14]. Briefly, animals were anesthetized with an intraperitoneal injection of ketamine/xylazine (keta-vet 100 mg/mL, Intervet production Srl, and xylazine 1 g, Sigma Co., at 10 and 1.2 mg/100 g, respectively). Corneas were anesthetized with a drop of Novocaine, and pupils were dilated with 1.0% mydriacyl tropicamide. Animals were mounted on a stereotaxic apparatus and positioned inside the Ganzfeld dome's opening (Biomedica Mangoni, Pisa, Italy). Body temperature was maintained at 37.5 °C using a heating pad controlled by a rectal temperature probe. Recordings were carried out for both eyes simultaneously, with a platinum electrode loop being placed on each cornea, and individually considered. The reference electrodes were inserted subcutaneously in the proximity of the eyes, and the ground electrode was inserted in the anterior scalp, between the eyes. The standard ERG protocol advocated by the ISCEV (International Society for Clinical Electrophysiology of Vision) was used [39]. Responses were recorded at progressively brighter short flashes ($0.001–100$ cd*s/m^2 range, from scotopic to photopic) over 450 ms, plus 50 ms of pre-trial baseline. Responses were amplified differentially, bandpass filtered at $0.3–300$ Hz, and sampled at 16.3 kHz. To reduce variability and background noise, responses were averaged (n = 3 per each luminance), with an inter-stimulus interval (ISI) ranging from 60 s for lower intensities to 5 min for the three brightest flashes. We evaluated the amplitudes (in μV) of the a-wave (baseline to the first negative deflection), b-wave (from a-wave peak to positive b-wave peak), and oscillatory potentials (the high-frequency wavelengths residing on the leading edge of the b-wave separated from the ERG using a high-pass filter. The amplitudes of wavelets OP1–4, measured from the trough to the peak of each response component, were summed) (see Figure 5a,f). For photopic recordings (light-adapted ERG, cone response), rats were light-adapted to background illumination (30 cd*s/m^2, 10 min), and 20 replicate responses elicited by white light (100 cd*s/m^2) at a pulse frequency of 1 Hz were averaged. A 30 Hz photopic flicker response was also obtained using the same background illumination and stimulus intensity. One hundred replicate responses were averaged. The first and the second harmonic magnitudes were determined by performing FFT analysis with a compiled routine (see Figure 6a,c).

In some cases, measurements were obtained from the same animal at successive survival times following light exposure cessation (not more than three ERG sessions, including the initial one, were performed in total for a single animal). When the amplitude of either an a-wave or a b-wave was not higher than at least 1 SD (not measurable) across the entire recording session, the retina was discarded from the subsequent analyses (more common in the later stages of the degeneration, i.e., LD24h210d). Custom-written procedures in IGOR Pro 6.3 (Wavemetrics, Lake Oswego, OR, USA) software were used to analyze electrophysiological data. The distributions of ERG response across several experimental groups at different light intensities were described by means and standard errors. Two-way ANOVA followed by the Tukey post hoc test was used to compare the ERG parameters (a, b, and OP amplitudes) between groups of animals for each flash intensity used. One-way ANOVA followed by the Tukey post hoc test was used to analyze statistical significance in the photopic recording (photopic ERG and flicker ERG, a single flash intensity used).

2.4. Tissue Processing, Morphological Analyses, and Immunostaining Protocols

At the end of the last recording session, rats were sacrificed by CO_2 inhalation. Both eyes were enucleated and fixed in 4% paraformaldehyde for 6 h at 4°C and washed in 0.1 M phosphate-buffered saline (PBS, pH 7.4). The eyes were further processed to obtain an "eyecup" consisting of the sclera, choroid, and retina. Eyecups were cryoprotected by immersion in 10%, 20%, and 30% sucrose solution overnight, embedded in OCT compound (Tissue Tek; Qiagen, Valencia, CA, USA), and frozen in liquid nitrogen. With eyecups marked to maintain proper orientation, cross-sections 20 μm thick were obtained for each retina (CM1850 Cryostat; Leica, Wetzlar, Germany), collected on gelatine poly-L-lysine coated slides, and stored at -20 °C until processed. The analyses were performed on the sections crossing the optic nerve in the dorsal-ventral direction (~8–10 central sections per retina were collected on different slides) to minimize variations in the retinal length and position.

The morphological analysis was carried out by labeling the nuclei with the DNA-specific dye bisbenzimide and measuring the thickness of the outer nuclear layer (ONL) across the entire retinal extension from dorsal to ventral crossing the papilla, following a previously described procedure [40]. Shortly, histological reconstructions were obtained by joining consecutive acquired micrographs (each long ~380 µm, usually 10–12 per hemiretina). To facilitate comparison between retinas, each retina was divided into 10 dorsal and 10 ventral fields while taking the optic nerve as a reference. Analyses included: (1) the ratio of ONL to the total retinal thickness to measure ONL thickness (rather than the absolute thickness of the ONL (µm), to compensate for possible oblique sectioning); (2) the number of photoreceptor rows in the ONL; (3) damaged ONL length. ONL thickness and ONL rows were calculated as the averages of 4 equi-spaced measurements from each of the 20 fields (to account for slight differences within the same area). The averages of the ONL thickness and rows measured in the 10 dorsal or ventral fields were calculated for each section. We then evaluated the extent of damage in the photoreceptor layer by calculating the ratio between "damaged ONL length" (clearly detectable because of the presence of "rosette" structures [41]) and the total dorsal retinal length. Measurements were performed using ImageJ 2.0 software (Rasband, W.S., ImageJ, U. S. National Institutes of Health, Bethesda, MD, USA).

Together with nuclear labeling, the slides were immunolabeled with antibodies according to Table 2. Bovine serum albumin (BSA) (Sigma-Aldrich, St. Louis, MO, USA) or specific sera were used to block nonspecific binding sites, and Triton X−100 (Sigma-Aldrich) was used as a detergent, when appropriate. Primary antibodies were omitted to control for nonspecific binding of the secondary antibody. Only one of the two eyes was used for a single marker or histological analyses; therefore, the number of analyzed sections is equivalent to the number of animals (consequently, in figure legends, we reported N. as the number of analyzed retinal slices).

Table 2. Primary and secondary antibodies used on retinal cryosections in this study.

Primary Antibodies	Supplier	Host Organism	Dilution	Product#	Ref
Anti-ionized calcium-binding adaptor molecule 1 (Iba1)	Wako	Rabbit	1:1000	019–19741	[42]
Anti-glial fibrillary acidic protein (GFAP)	Dako	Rabbit	1:500	Z0334	[41]
Anti-choline acetyltransferase (ChAT)	Millipore (Chemicon)	Goat	1:100	AB144P	[43]
Anti-L/M Opsin	Millipore (Chemicon)	Rabbit	1:100	AB5405	[41]
Anti-fibroblast growth factor (FGF)−2/basic FGF	Millipore	Mouse	1:200	05–117	[14]
Anti-synaptophysin (SYN)	Osenses	Rabbit	1:300	OSS00021W	
Secondary Antibodies	**Supplier**	**Host Organism**	**Dilution**	**Product#**	
Anti-rabbit IgG (H + L) Alexa Fluor 488	Thermo Fisher Scientific	Goat	1:500	A−11008	
Anti-rabbit IgG (H + L) Alexa Fluor 594	Thermo Fisher Scientific	Goat	1:500	A−11012	
Anti-goat IgG (H + L) biotinylated	Thermo Fisher Scientific	Rabbit	1:300	31732	
Anti-mouse IgG (H + L) Alexa Fluor 488	Thermo Fisher Scientific	Goat	1:500	A−11001	

2.5. Microscopy and Image Analysis

Images of the retina were acquired with Leica TCS SP5 (Leica Microsystems, Wetzlar, Germany) or Nikon80i (Nikon, Tokyo), and acquisition parameters were kept constant throughout each imaging session. Measurements of fluorescence intensity and signal localization were made with ImageJ Fiji software (National Institutes of Health, Bethesda, MD, USA).

The L/M opsin signal was calculated following the method used by Rutar et al. [41]. Briefly, ten images (5 dorsal and 5 ventral) equally spaced were taken under 20× magnification from a total of 5 or 6 retinal sections from different rats for each experimental condition. We counted external outer segments (OSs) of cone photoreceptors. The number of positive cells is expressed as a mean of labeled cells per field or as their sum, respectively, for the dorsal and ventral retina.

Spatial distribution and the total number of Iba1$^+$ microglial cells were analyzed in the ganglion cell layer and inner plexiform layer (GCL + IPL), in the inner nuclear layer, and outer plexiform layer (INL + OPL), and in the outer nuclear layer (ONL). Each retinal layer was determined according to the nuclear staining. The number of Iba1$^+$ cells was manually counted throughout the entire section from dorsal to ventral (10 images per hemisphere acquired at 40× magnification, each micrograph covered ~380 μm of retinal length). Results are given as the total number of microglial cells (Iba1$^+$) per layer and per section.

Densitometric analysis of GFAP and bFGF-2 fluorescent signals (N = 3–6 and N = 5–7 retinal sections, quantified in Figures 3 and 7, respectively) was performed using custom-written algorithms in ImageJ Fiji software (NIH, Maryland, USA). The percentage of areas positive for bFGF-2 or GFAP over the total area of the ONL or the entire retinal area, respectively, was determined by applying a threshold and calculated as described in Antognazza et al. [44].

Cholinergic neurons in the retina, known as starburst amacrine cells (SBAC), were labeled with anti-ChAT antibody. The integrity of their processes, located in the inner plexiform layer (IPL), was quantified in two areas of the retina (dorsal and ventral, respectively) relatively to the total length (N = 5 retinal sections from different rats for each experimental condition; HC and 24 h light-exposed rats). Cell bodies density was also quantified (data not shown because of lack of statistical significance).

Synaptophysin, a protein expressed in the presynaptic vesicles, was qualitatively assessed (acquired under 60x magnification) in 5 retinal sections for each experimental condition, and representative immunofluorescence images are shown in the figure.

2.6. Statistical Analysis

One or two-way ANOVA was used to evaluate the effects of different LD durations and recovery periods. When the significance level was 0.05 or less, post hoc pairwise comparisons were performed using Tukey's test. Data are reported as mean ± standard error of the mean (SEM). Values of $p < 0.05$ were considered to be statistically significant. All statistical analyses were performed using Prism7 software (Graph Pad, San Diego, CA, USA). A detailed description of each test can be found in figure legends.

3. Results

3.1. Different Durations of Light Exposure: Discrepancies between Functional Output and the Extent of Structural Damage

This section examines the effects of different light exposure durations (i.e., 12 h, 15 h, 18 h, 24 h at 1000 lux) on retina functional, morphological, and molecular states, assessed 7 days after the treatment. For the sake of simplicity, we refer to 12 h as short-term exposure (compared to 24 h).

3.1.1. Exposure to Constant Light for 12 h Led to Functional but Not Severe Morphological Retinal Alterations

Figure 1 describes the effects of different exposure regimes on the retinal functional state, as assessed through in vivo flash electroretinography (fERG). Surprisingly, our analyses revealed no major functional differences between light damaged experimental groups, with a few exceptions. Indeed, even the short-term exposure induced a significant reduction of a-wave and b-wave amplitudes compared to the healthy control (Figure 1a,c). Interestingly, only the a-wave amplitude, obtained in response to the brightest stimulus (100 cd*s/m^2), was better preserved in LD12h and LD15h compared to LD24h (Figure 1a,b for magnification; a-wave amplitude was ~2.5 times higher in LD12h and LD15h vs. LD24h, $p = 0.0011$ and $p = 0.0177$ respectively). Instead, our evaluation did not show any significant difference between retinas exposed to light with regard to the b-wave amplitude; as expected, they were all different compared to the HC (Figure 1c,d; $p < 0.0001$ for all stimulus intensities). In particular, fERG b-wave amplitude decreased by 30% or 15% in rats exposed to 12 or 24 h, respectively, compared to HC (LD12h7d = 287 ± 26 µV and LD24h7d = 150 ± 13 µV vs. HC = 1061 ± 76 µV at 100 cd*s/m^2).

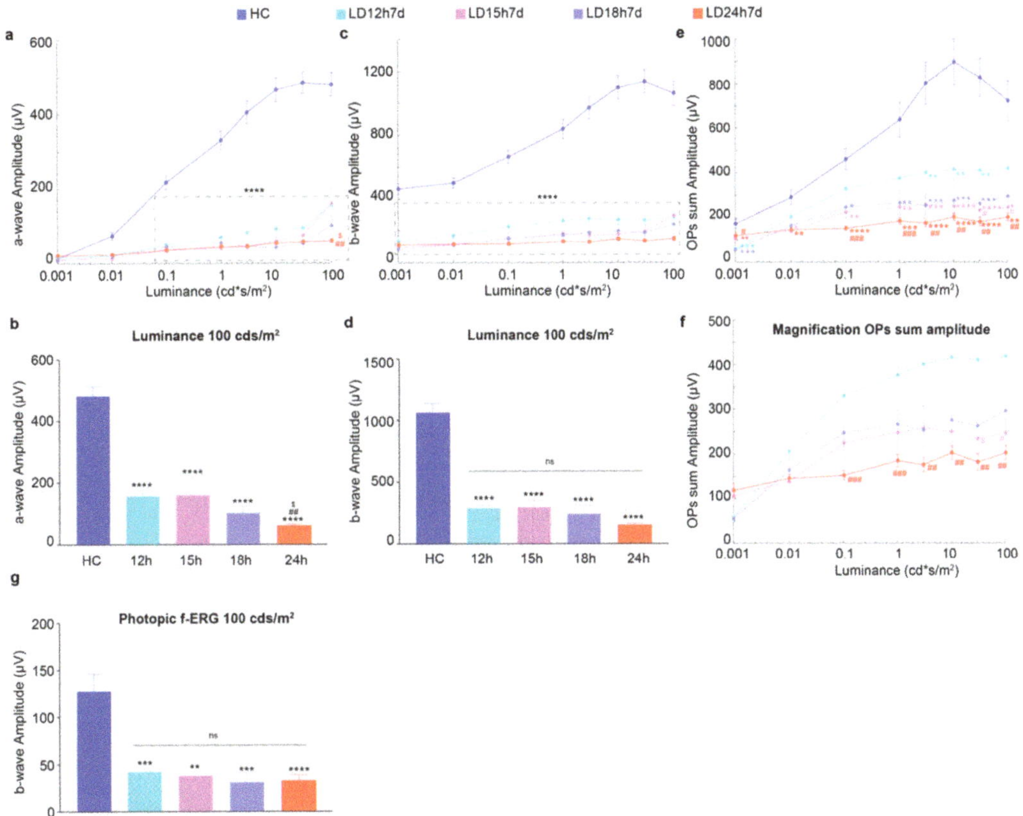

Figure 1. Retinal functional impairment following different durations of light exposure. (**a,c**) Amplitudes of the a-wave and b-wave in healthy control (HC) and different experimental groups (LD12h7d, LD15h7d, LD18h7d, and LD24h7d) are plotted as functions of stimulus intensity (0.001–100 cd*s/m^2). A-wave amplitude in the HC was significantly higher than all groups exposed to light for all luminances brighter than 0.1 cd*s/m^2. B-wave amplitude was significantly higher than all experimental groups at all luminances. (**b,d**) Panels b and d represent a-wave and b-wave amplitudes measured at 100 cd*s/m^2. LD24h7d a-wave amplitude at 100 cd*s/m^2 was significantly reduced compared to LD12h7d, [##] p = 0.0011, and LD15h7d, [$] p = 0.0177; two-way ANOVA with Tukey's post hoc test. (**e**) Amplitudes of the OPs sum in HC and different experimental groups are plotted as functions of stimulus intensity (0.001–100 cd*s/m^2). On average, OPs sum amplitudes were significantly affected by all durations of light exposure. (**f**) Panel f represents a large magnification of panel (**e**). The effect of light exposure was milder after 12 h; indeed, OP amplitude was significantly preserved in LD12h7d vs. LD24h7d at intensities brighter than 0.1 cd*s/m^2. (**g**) The values represent the mean response amplitudes of photopic b-waves. Statistical significance is represented as follows: [*], [#], [$] $p \leq 0.05$, [**], [##], [++] $p \leq 0.01$, [***], [###] $p \leq 0.001$, [****] $p \leq 0.0001$; [*], [#], [$], [+] are in comparison to HC, LD12h7d, LD15h7d, and LD18h7d, respectively; two-way and one-way ANOVA with Tukey's post hoc test were used to assess differences in the panels (**a–g**), respectively. Values represent means ± standard errors of the means (SEM). Between 6 and 12 retinas were recorded per group.

It is worth highlighting that by studying oscillatory potentials (OPs), an indicator of the inner retinal activity [45], some differences were unexpectedly found between the light-exposed experimental groups. As observed for the other two analyzed ERG parameters, the light damage paradigm caused a significant reduction of OP sum amplitude at all stimuli intensities (i.e., to ~35%, ~32%, and ~44% respectively in LD15 h, 18 h, and 24 h groups compared to HC at 10 cd*s/m^2; LD15h7d = 250 ± 32 µV, LD18h7d = 276 ± 48 µV and LD24h7d = 203 ± 15 µV vs. HC = 896 ± 100 µV, in Figure 1e), but with a few exceptions. Indeed 12 h led to a significant reduction of the OP sum compared to HC only at stimuli intensities between 3 and 30 cd*s/m^2 (reduction by ~22% at 10 cd*s/m^2; LD12h7d = 415 ± 45 µV and HC = 896 ± 100 µV). What was more interesting was the difference between the two groups LD12h7d and LD24h7d (Figure 1f for magnification; at light intensities higher than 0.1 cd*s/m^2), revealing evidence for a different time course and extent of light-induced lesions that interests the inner retina.

In contrast, no differences between groups (except for the reduction compared to HC) were observed when cone functionality was assessed by photopic-ERG. Indeed, 7 days after 12 h, 15 h, 18 h, and 24 h of light damage, the photopic b-wave amplitudes represented ~33%, ~30%, ~24%, and ~26% of the HC one, respectively (LD12h7d = 42 ± 5 µV, LD15h7d = 38 ± 5 µV, LD18h7d = 31 ± 10 µV, LD24h7d = 33 ± 7 µV and HC = 128 ± 18 µV; all significant vs. HC, Figure 1g).

3.1.2. Outer Nuclear Layer (ONL) Disruption Does Not Correlate with Detectable Functional Alterations

Figure 2a indicates that exposure to high intensity light for different durations of time (from at least 12 h to 24 h) caused a gradual alteration of the photoreceptor layer integrity in the dorsal retina, which has extensively been shown to be the more susceptible to induced neuronal death [41,46]. To analyze tissue structure, the cross-sectioned retinas, from the dorsal to ventral quadrant, were stained with Hoechst and the ONL thickness relative to the entire retina, and the number of nuclei rows in the ONL were analyzed. In the representative retinal sections in Figure 2a, we observed that the ONL underwent a reduction in thickness and structural changes correlated with the light exposure duration. Surprisingly, despite showing no apparent signs of retinal damage (Figure 2a, second micrograph) and a preserved ONL/retinal thickness ratio (Figure 2b,c; cyan, LD12h7d), the number of photoreceptor nuclei rows was slightly reduced (not significant) in the dorsal retina following 12 h of light exposure (Figure 2d). It is plausible to think that the loss of a limited number of photoreceptors does not change the ratio of ONL/total retinal thickness; indeed, the photoreceptor nuclei rows count may be a more convenient measure of mild damage, though does not account for possible oblique sectioning. As expected, the increase in light exposure duration produced a more significant thinning of the outer retinal layers, specifically in the hot spot area in the dorsal hemiretina (Figure 2c; LD24h7d significantly different compared to HC and LD12h7d, p = 0.004 and 0.0038, respectively).

In contrast, a significantly reduced number of nuclei rows compared to the HC was observed within the dorsal hemiretina also in the LD15h7d and LD18h7d groups (Figure 2e; LD15h7d, LD18h7d, and LD24h7d vs. HC, p = 0.0136, 0.0006 and <0.0001). This result is also significant when comparing LD24h7d to LD12h7d (respectively 4.6 vs. 8.4 rows, p = 0.0104). ONL disruption was then correlated with the a-wave amplitude (photoreceptors driven), and no correlation was found (Figure 2h, p = 0.20, and p = 0.6617 for the ONL thickness and N. rows vs. a-wave amplitude, respectively. Only 19 (LD12h7d = 5, LD15h7d = 4, LD18h7d = 4, and LD24h7d = 6) and 13 retinas (LD12h7d = 3, LD15h7d = 3, LD18h7d = 3, and LD24h7d = 4), for which both functional and morphological features were assessed, were included in this analysis. No HC was included to study the effect of light exposure). This result demonstrates that ERG prematurely diagnoses ONL matrix disorganization. Further analyses have been performed to measure the hot spot extension, evaluated as damaged ONL length/dorsal retinal length. Figure 2f presents a histological reconstruction of the dorsal hemiretina central section, with retinal damage extensions labeled with white arrows. We observed that the hot spot length increased with light exposure durations, covering up to 60% of the dorsal retina in the LD24h7d (Figure 2g).

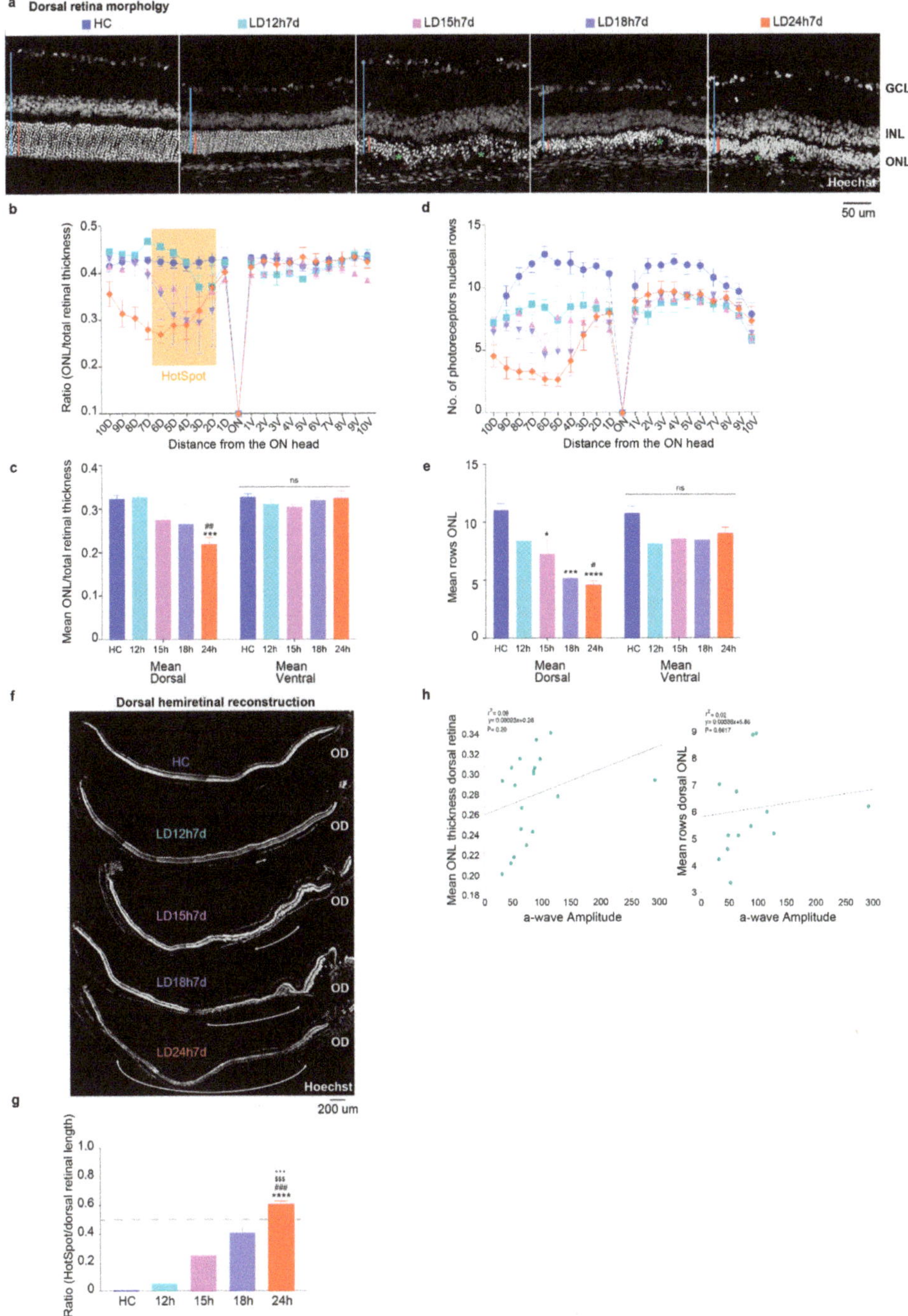

Figure 2. Morphometric evaluation of the light damage resulting from different exposure times. (**a**) Photomicrographs of the hot spot in the dorsal central retina of an undamaged control rat (HC) and rats exposed to 12 h, 15 h, 18 h, and 24 h of constant light; nuclei are stained with Hoechst. The red bars show the absolute thickness of the ONL, and the blue one the entire retinal thickness. Scale bar = 50 μm. Green asterisks signify disruptions in the ONL. Abbreviations: GCL: ganglion cell layer; INL: inner nuclear layer; ONL: outer nuclear layer. (**b**,**d**) ONL thickness and number of rows of photoreceptor

nuclei (±SEM) quantified in 20 equidistant retinal locations (10 dorsal, 10 ventral). (**c,e**) Average across all dorsal or ventral retinal locations of the ONL thickness (ratio ONL/total retinal thickness) and the number of photoreceptor nuclei rows. (**f**) Representative dorsal hemiretina reconstruction after nuclei staining. The white arrows depict the damage extensions defined as "hot spots" on light damaged retinas. Scale bar = 200 μm. OD: optic disc. (**g**) Damaged ONL length/dorsal retina length for different light damage durations. (**h**) Relationship between a-wave amplitudes and ONL thickness/rows (Pearson correlation). Data in all graphs are shown as mean ± standard error of the mean (SEM). Statistically significant differences are represented as follows: *, # $p \leq 0.05$, ## $p \leq 0.01$, ***, ###, \$\$\$, +++ $p \leq 0.001$, **** $p \leq 0.0001$; *, #, \$, + compared to HC, LD12h7d, LD15h7d, and LD18h7d, respectively; one-way ANOVA with Tukey's post hoc test. $N = 3/12$ retinas from different rats for each experimental condition.

3.1.3. Upregulation of GFAP and bFGF-2 Was Triggered after Short Periods of Constant Light Exposure

Figure 3 describes a gradual increase in the expression of stress (glial fibrillary acidic protein, GFAP) and trophic markers (basic fibroblast growth factor-2, bFGF-2) in the retina, linked to light exposure. Under normal conditions, astrocytes and Müller glia contact retinal neurons, providing stability to the neural tissue. Stress of any origin induces an astrocyte response with increased expression of GFAP in the radially oriented Müller cells (MC). GFAP expression was confined to the GCL layer in the HC retinal section to form a homogeneous plexus (Figure 3a, first micrograph). Different light damage protocols induced upregulation of GFAP: the protein was visible along the entire length of MCs. To evaluate the effect of different light exposure durations, corresponding retina areas (one in the dorsal and one in the ventral) were selected to determine the relative GFAP-labelled regions [47]. For this purpose, the images were processed with ImageJ software's threshold tool (see Materials and Methods). Areas of the retina marked with the threshold overlay (GFAP+ astrocytes plus GFAP+ end-foot of the MC; Figure 3a) were included in the measurement among study groups. Results obtained at 7d following 12 h, 15 h, 18 h, and 24 h light damage are shown quantitatively in Figure 3c. From the one-way ANOVA, performed separately for the dorsal and ventral retina, significant differences were found in the dorsal retina of all light-exposed groups compared to the HC ($p = 0.0155, 0.0078, 0.0023$, and 0.0018 for LD12,15,18, and 24 h vs. HC, respectively). Hence, even 12 h of light exposure was sufficient to produce retinal stress, in accordance with the ERG results. The upregulation of GFAP also extended in the ventral retina, which was significant only for the LD24h7d group compared to the HC ($p = 0.0048$). As expected, stressed retinas also increase the expression levels of trophic factors, such as bFGF-2. Notably, the release of bFGF-2 in the ONL negatively correlates with the b-wave amplitude [40]. In control retinas, bFGF-2 expression was confined to the Müller cell bodies in the INL (Figure 3b, first micrograph). Light damage induced its progressive release in the ONL (the protein was visible around photoreceptor cell bodies), and even though exclusively significant for the LD24h7d ($p = 0.0288$ vs. HC), it is noticeable that bFGF-2 levels increased in the dorsal retina after light exposure treatment. Indeed, a significant trend ($p = 0.0168$, one-way ANOVA with Trend) relates bFGF-2 release in the ONL to the duration of light exposure (no linear trend was revealed in the ventral side).

3.1.4. Microglia Increased in the Dorsal Retina after Exposure to Constant Light

The immune response is triggered along with the degeneration of the outer retina and macrogliosis (assessed by anti-GFAP staining) [48]. Thus, we characterized the effect of light exposure on microglia distribution in the retina using Iba1 as a marker. Representative images of the dorsal retina from HC and the four experimental groups (different light exposure durations) are shown in Figure 4a (Iba1+ cells in green). As expected, [49], under physiological conditions, microglia that showed typical resting characteristics were located in the inner retina in HC (Figure 4b,d, GCL + IPL, and INL + OPL), and no cells were found in the ONL in the dorsal and ventral retina (GCL + IPL > ONL + OPL > ONL; $p < 0.01$). Exposure to light promoted their morphological change (from ramified to ameboid, depicted in Figure 4a with the orange and white arrows, respectively) and migration to the ONL (green dashed arrow in Figure 4a, LD12h7d) [50]. Detailed analyses of Iba1$^+$ cells in the dorsal and ventral retina and across retinal layers are reported in Figure 4b–e. In the retinas of LD groups, we found that Iba1$^+$ cells accumulate significantly in the ONL in the dorsal retina after 7 days from 18 h or 24 h of light exposure (LD18h7d vs. HC, $p = 0.0013$; LD24h7d vs. HC, $p < 0.0001$; one-way ANOVA). Overall, when the total number of microglia across all layers was assessed, a significant increase in Iba1$^+$ cells was found only in the dorsal retina in LD24h7d compared to HC, LD12h7d, and LD15h7d ($p < 0.0001$, $p = 0.0052$, $p = 0.0287$, respectively). Only 24 h of light exposure induced a significant difference between the dorsal and ventral retina (t-test, LD24h7d dorsal vs. ventral, $p = 0.04$).

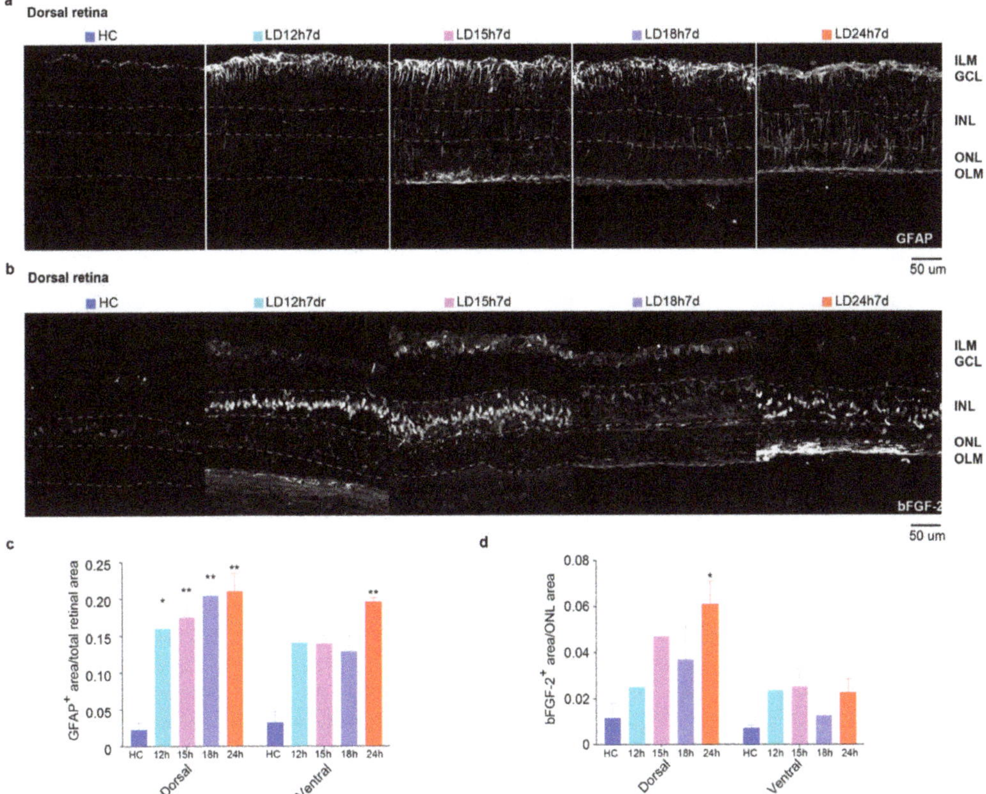

Figure 3. GFAP and bFGF-2 expression in retinas of rats exposed to light for different durations of time. (**a**) Representative photomicrographs illustrating GFAP$^+$ astrocytes in dorsal retinal sections of HC and 7 days after 12 h, 15 h, 18 h, and 24 h of light damage (in the area around the degenerating hot spot, defined "penumbra" [41]), respectively. Light exposure induced the upregulation of GFAP in the radially oriented Müller cells; the protein was visible along the full length of the Müller cells, from the ILM to the OLM. (**b**) Representative photomicrographs illustrating bFGF-2$^+$ signal in the dorsal retinal areas of control and light-exposed rats. Light exposure induced the upregulation of bFGF-2, released in the ONL by Müller cells; the protein is visible around photoreceptor bodies in the ONL and Müller cell bodies in the INL. Abbreviations: GCL: ganglion cell layer; INL: inner nuclear layer; ONL: outer nuclear layer; ILM: inner limiting membrane; OLM: outer limiting membrane. Scale bar = 50 μm. (**c**) Mean GFAP$^+$ area averaged across dorsal and ventral retinal selected areas. (**d**) Mean bFGF$^+$ area/ONL area averaged across the dorsal and ventral retina. Statistical analyses were performed with the analysis of variance (one-way ANOVA) followed by a Tukey's post hoc test; one-way ANOVA with Trend was also used to analyze data in panel d. Statistically significant differences are represented as follows: * $p \leq 0.05$, ** $p \leq 0.01$, respectively, compared to HC. Data are shown as the means ± SEM of at least $N = 3$ retinas from different rats for each experimental condition.

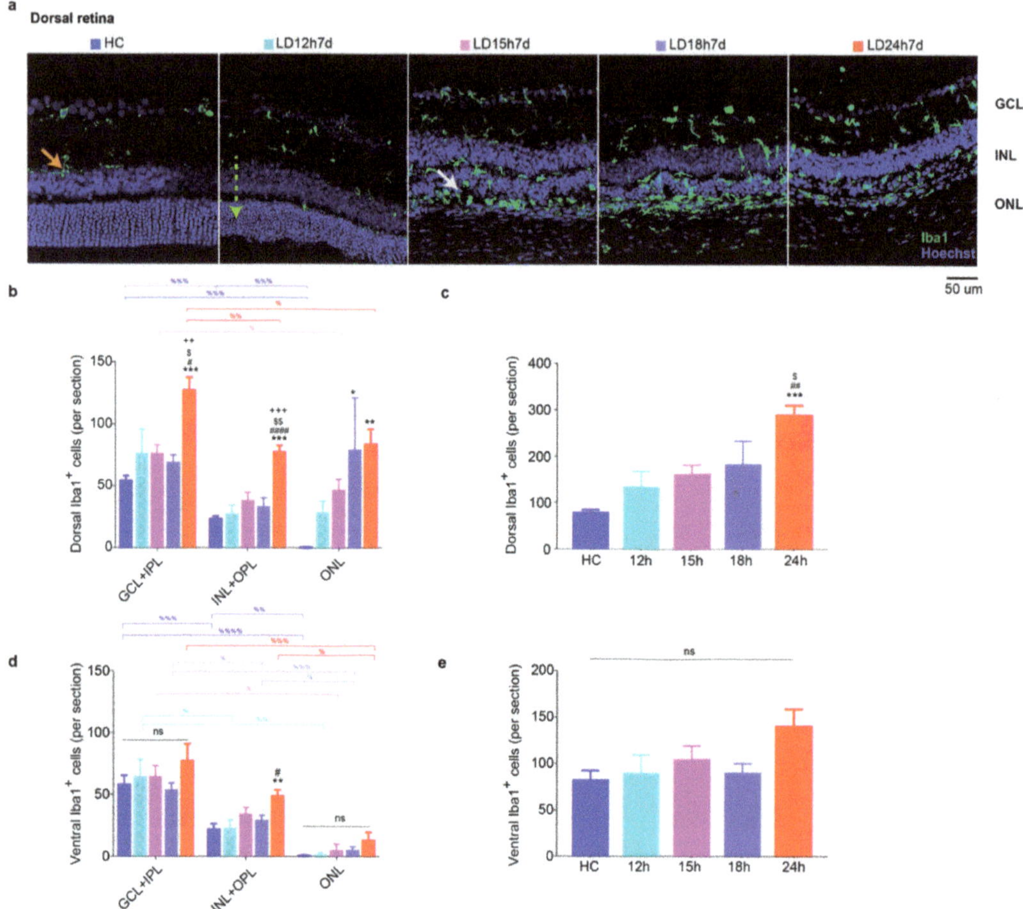

Figure 4. Microglia quantification in retinas of rats exposed to light for different lengths of time. (**a**) Representative photomicrographs illustrating Iba1$^+$ cells in dorsal retinal sections (hot spot region) of HC and light-exposed retinas tested after 7 days. Light damage induced migration of resident Iba1$^+$ cells to the ONL (significant after 18 or 24 h of light exposure). Orange arrows: ramified microglia; white arrows: ameboid shape; dashed green arrow: migration from the inner to the outer retina. Abbreviations: GCL: ganglion cell layer; INL: inner nuclear layer; ONL: outer nuclear layer. Scale bar = 50 μm. (**b**,**d**) Quantification of Iba1$^+$ cells in dorsal and ventral retinal sections in each retinal layer. (**c**,**e**) Total numbers of microglia along the dorsal and the ventral retina (summarized from b, d data). Subretinal Iba1$^+$ cells were not included in the count. Statistical analyses were performed with the analysis of variance (one-way ANOVA) followed by a Tukey's post hoc test. Statistically significant differences are represented as follows: *, #, $ $p \leq 0.05$, **, ##, $$, ++ $p \leq 0.01$, ***, +++ $p \leq 0.001$, #### $p \leq 0.0001$; *, #, $, + vs. HC, LD12h7d, LD15h7d, and LD18h7d, respectively; % was used to depict differences within layers. Data are shown as means ± SEM of at least $N = 3$ retinas from different rats for each experimental condition.

3.2. Long-Term Morpho-Functional Changes in the Neurodegenerative Progression

Given that 24 h of white light exposure resulted in well-established retinal structural and functional impairment after 7 days, we decided to investigate how the neurodegenerative process evolves using this experimental protocol. After inducing damage (24 h, 1000 lux), critical stages were determined within different retinal regions for up to 7 months.

3.2.1. Timecourse Effects of 24 h of Bright Light Exposure Highlighted a Transient Functional Recovery

We analyzed the time course of ERG response variations following 24 h of light-induced damage. Long-term monitoring of neurodegenerative processes underlined a transient increase in retinal function, followed by a further decline. Figure 5a depicts the response series of dark-adapted flash ERG recordings (at luminance 100 cd*s/m^2, representative waveforms) before light damage (HC, age 60 days), and 7, 15, 30, 45, 60, 90, and 210 days after the light damage (24 h, 1000 lux). Quantitative analysis indicated a significant decrease of ERG components (a-wave and b-wave amplitudes) recorded at 7 days after light exposure and described a partial recovery, maximum at 45 days, followed by an irreversible drop. In fact, at 7d after LD, the dark-adapted a-wave maximum amplitude (at 10 cd*s/m^2) decreased significantly compared to HC (LD24h7d = 55 $\pm$ 4 μV vs. HC = 466 $\pm$ 33 μV, $p < 0.001$) to ~12% of the pre-LD level (HC), and the b-wave maximum amplitude decreased to ~14% (HC = 1098 $\pm$ 74 μV and LD24h7d = 149 $\pm$ 10 μV, $p < 0.001$). Figure 5b–e indicates that by 45d, the dark-adapted a and b-waves maximum amplitudes (at 10 cd*s/m^2) recovered ~34% and ~45% of their initial values (LD24h45d a-wave = 159 $\pm$ 21 μV, and b-wave = 503 $\pm$ 49 μV vs. HC = 466 $\pm$ 33 μV and HC = 1098 $\pm$ 74 μV). This amelioration was found to be transient: significant decrements of a-wave and b- wave amplitudes were evident at longer recovery periods (LD24h45d vs. LD24h60d, $p = 0.001$ and vs. 90d, and 210d, $p < 0.001$ with respect to the b-wave amplitude; and LD24h45d vs. LD24h60d, $p = 0.001$, vs. 90d, $p < 0.001$, and 210d, $p = 0.0008$). Figure 5f illustrates the filtered oscillatory potentials (OPs) from control and light-exposed retinas at different recovery times following the same light exposure protocol. The four analyzed OPs are enumerated on the HC trace (OP 1–4). Fifteen days were enough to detect significant partial recovery (LD24h7d vs. LD24h15d, $p \leq 0.05$ at 0.1–10 cd*s/m^2). By 45d the OP sum amplitude recovered to over 50% of the initial HC value and was significantly higher than LD24h7d (HC = 897 $\pm$ 100 μV; LD24h7d = 203 $\pm$ 15 μV vs. LD24h45d = 427 $\pm$ 53 μV at 10 cd*s/m^2, $p = 0.006$). After 45d, the significant difference from LD24h7d was lost, meaning that OPs amplitudes diminished and deteriorated until 210d (Figure 5h). Light adapted responses (cone-driven activity) followed a similar pattern: at 7d after the light-damage, the LD24h7d average light-adapted b-wave amplitude reached only 26% of its baseline (HC = 128 $\pm$ 19 μV and LD24h7d = 33 $\pm$ 7 μV, $p < 0.001$), but recovered almost completely by 45 days (indeed LD24h45d = 92 $\pm$ 14 μV non-significant vs. HC, but significant vs. LD24h90d = 41 $\pm$ 7 μV, $p = 0.07$) (Figure 6b). Moreover, responses to brief flashes at 30 Hz were recorded (with flash intensities of 100 cd*s/m^2). We analyzed the flicker ERG responses in the frequency domain (Figure 6c), and the averaged results are illustrated in Figure 6d,e. The two graphs illustrate the mean amplitudes of the fundamental and the second harmonics of the flicker responses at 100 cd*s/m^2. Light exposure flattened the amplitude of the fundamental and the second harmonic in all experimental conditions. In addition, we observed a significant worsening of function at later stages of the neurodegeneration (indeed, the first fundamental harmonic amplitudes at LD24h90d and 210d were significantly reduced compared to LD24h45d; Figure 6d).

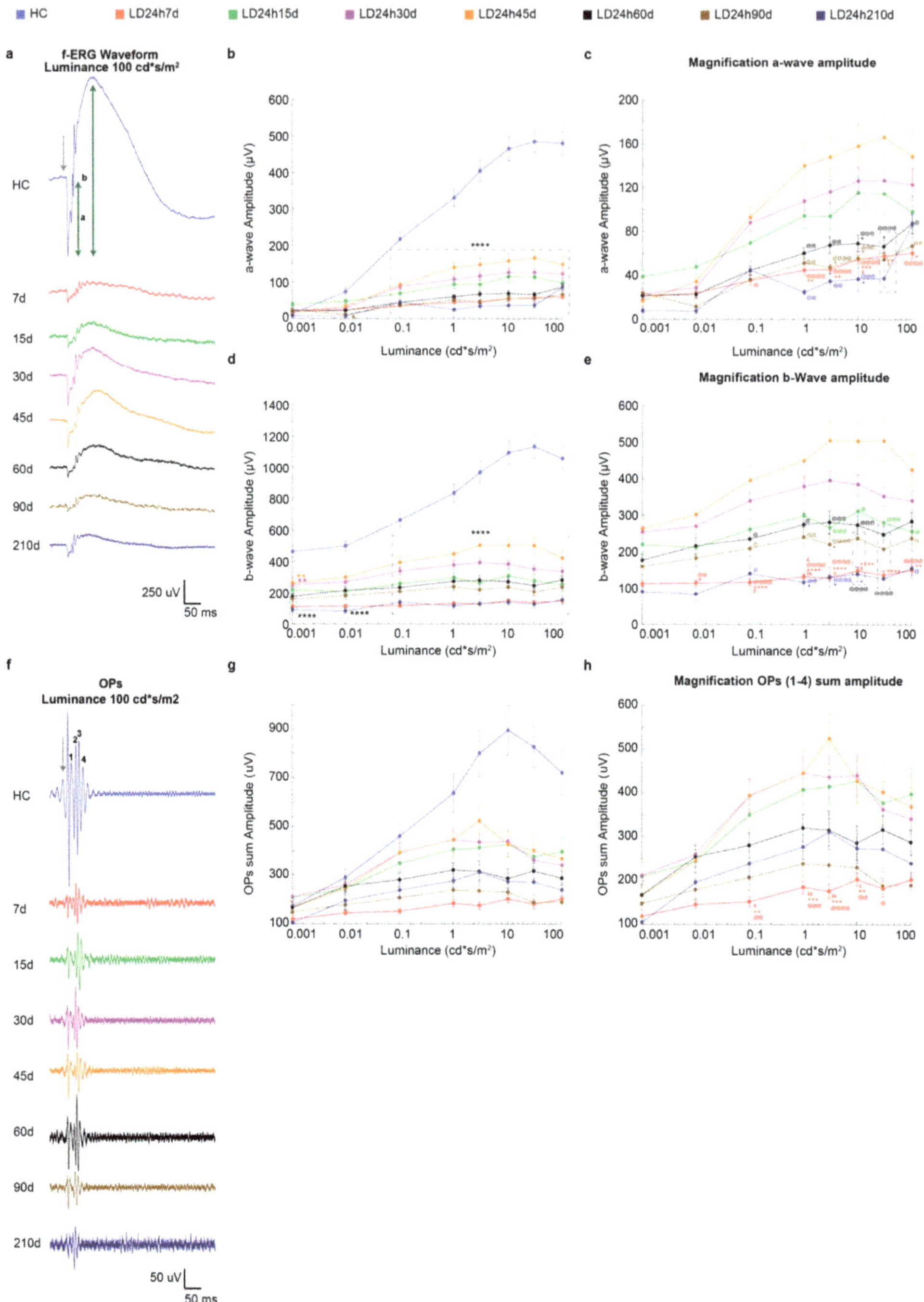

Figure 5. In vivo functional assessment at different recovery times following 24 h of light exposure. (**a**) Representative fERG waveforms in response to 100 cd*s/m² flash intensity obtained from control and all experimental groups (7, 15, 30, 45, 60, 90, and 210 days following the end of light exposure). Calibration: horizontal 50 ms, vertical 250 µV. The grey arrow indicates when the flash stimulus was sent. (**b**,**d**) Amplitudes of the a-wave and b-wave in HC (blue line) and different experimental

groups as a function of stimulus intensity (0.001–100 cd*s/m^2). (**c,e**) represent magnifications of the graphs in (**b,d**). (**f**) Representative OP waveforms, isolated by bandpass filtering the dark-adapted ERGs to bright light (100 cd*s/m^2), from HC and all experimental groups (7, 15, 30, 45, 60, 90, and 210 days following the end of light exposure). Calibration: horizontal 50 ms, vertical 50 μV. The grey arrow indicates when the flash stimulus was sent. 1–4 depict, respectively, OP1–4, whose amplitude was summed. (**g**) Amplitudes of the OP sum in HC (blue line) and different experimental groups as a function of stimulus intensity (0.001–100 cd*s/m^2). (**h**) represents a magnification of the graph in panel (**g**). A single value represents the mean ± standard error of the mean (SEM). Statistical significance is represented as follows: *, $, +, @, & $p \leq 0.05$, **, $$, ++, @@ $p \leq 0.01$, +++, @@@ $p \leq 0.001$, ****, ++++, @@@@ $p \leq 0.0001$; *, $, +, @, & represent comparisons with HC, LD24h15d, LD24h30d, LD24h45d, and LD24h60d, respectively; two-way ANOVA with Tukey's post hoc test. The control group was significantly higher than all light damaged groups at all intensities higher apart 0.001 cd*s/m^2 (only statistical results vs. HC are shown in (**b,d,g**); differences between light-exposed groups are shown in (**c,e,f**)). $N = 15/30$ recorded retinas per group.

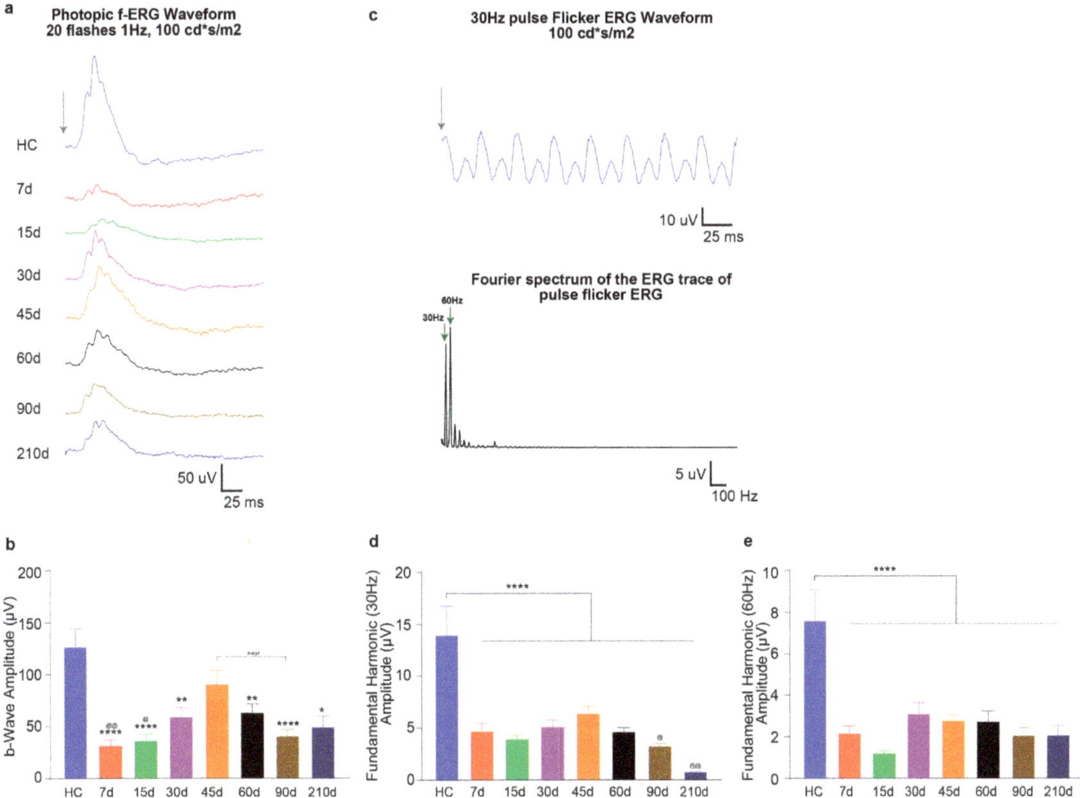

Figure 6. In vivo photopic functional assessment at different recovery times following 24 h of light exposure. (**a**) Representative averaged light-adapted ERGs in response to 20 bright flashes (100 cd*s/m^2) of light at 1 Hz presented on a rod-saturating background. Calibration: horizontal 25 ms, vertical 50 μV. No stimulus baseline is included in the traces. (**b**) The values are the mean response amplitudes of photopic b-waves recorded from control animals before light exposure (HC) and 7, 15, 30, 45, 60, 90, or 210 days after 24 h light exposure. The photopic response is affected by light exposure but recovers almost completely by 45 days. Values represent mean ± standard error of the mean (SEM). (**c**) Representative photopic flicker response in control rats elicited at 30 Hz. Each waveform was recorded for 200 ms, and responses are the averages of 100 tests. (**d,e**) Amplitudes of the photopic flicker responses in LD rats are lower than those of the HC. Values represent mean ± standard error of the mean (SEM). Statistical significance is represented as follows: *, @ $p \leq 0.05$, **, @@ $p \leq 0.01$, **** $p \leq 0.0001$; * and @ are in comparison to HC and LD24h45d, respectively; one-way ANOVA with Tukey's post hoc test. $N = 15/30$ recorded retinas per group.

3.2.2. The Ventral Retinal Structural Layer Organization Was Better Preserved Compared to the Dorsal One

The loss of photoreceptors induced by 24 h light exposure accompanied a progressive decrease in the ONL thickness over time. Figure 7 displays the histological appearance of the dorsal (a) and ventral (b) retina along the vertical meridian from control rats and those exposed to 24 h of high-intensity light at different time-points following the light exposure. The dorsal retina was more severely affected than the ventral one, whose retinal layered structure appeared better preserved (no rosette structures or matrix disorganization). Photoreceptor nuclei were few and dispersed in the dorsal side. The INL directly neighboring the RPE, and several holes were present in the retina 30 days after light exposure (see also Figure 9a). The average ratios and the mean number of photoreceptor rows for all twenty quadrants (Figure 7d,f) also decreased in the ventral side in LD animals.

Furthermore, despite photoreceptor rows' progressive reduction (Figure 7f; all LD groups are significantly different vs. HC), relative ratios in the ventral retina were conserved, underling a homogeneous thinning among retinal layers up to 60 days of recovery (mean ONL/total retinal thickness overall ns vs. HC till later stages of neurodegeneration). A further analysis was done to measure the extent of damage, evaluated as damaged ONL length/dorsal retinal length. In the central section, crossing the optic nerve, the amount of dorsal ONL damage increased over time, reaching 80% of the dorsal retina after 60 days (Figure 7g).

3.2.3. Light-Induced Cone-Photoreceptor Damage

Following light exposure, the first sign of damage is observed at the level of photoreceptor OSs. Retinal sections were stained with an antibody for L/M opsin, allowing differentiation of individual cones for measuring cone density. L/M cones (in general also referred to as L-cones, because of the opsin they express) are mainly localized in the medial and central retina, where light exposure effects are more pronounced, and they are more abundant than S cones (densest in the retinal rims and periphery [51]). Figure 8a shows a composite of photographs taken from the control with uniform cone photoreceptor staining and light-exposed retinas characterized by a diminished density of cones and almost an absence of their corresponding external segments (cone OSs) on the dorsal side. As previously shown [41], the L/M opsin immunostaining revealed a sparse and disorganized population of surviving cones in the hot spot. After 7days, the dorsal retina's remaining cones lost their normal elongated morphology and were immunoreactive to L/M opsin throughout the soma and axon terminals [52].

The remaining L and M-cones in retinal sections are plotted in Figure 8b,c. Twenty-four hours of light damage resulted in the loss of approximately 75% of the L/M-cone population by 45 days: detected cones were 134.3 ± 41.81 ($N = 3$) at 45 days compared to 535.40 ± 62.55 ($N = 5$) in the control ($p = 0.0014$) (Figure 8c). Sometimes, even though it was difficult to identify associated Hoechst-stained cell nuclei within the ONL, some residual L/M opsin immunoreactivity was present. Overall, in the ventral retina, no significant differences were found between HC and experimental groups up to the time point evaluated; a decrease was detectable only in the first two acquired fields (1- and 2-V) of ventral hemiretina at LD24h45d, in proximity to the damaged area (Figure 8b,c). Animals sacrificed after 45d were not included in cone density analysis. The number of lost cones progressed further from 7 to 45 days after light damage, without any significant rescue.

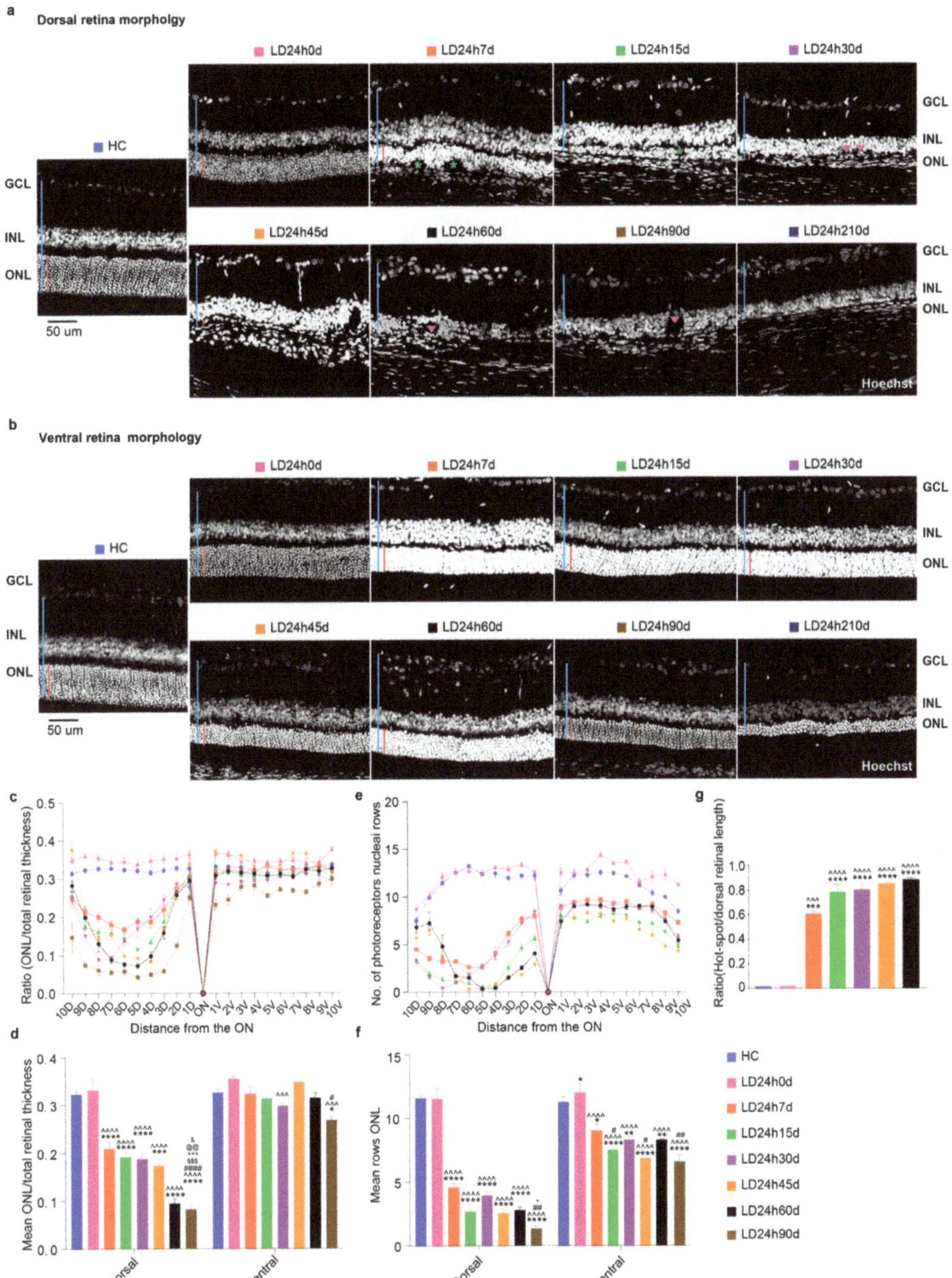

Figure 7. Morphological alterations were assessed at different recovery times following 24 h of light damage. (**a,b**) Representative Hoechst-stained retinal cross-sections from the dorsal and ventral central retina. Light damage was more severe in the dorsal retina, considering the matrix disorganization. Abbreviations: GCL: ganglion cell layer; INL: inner nuclear layer; ONL: outer nuclear layer. The red and blue vertical bars show the absolute thickness of the ONL and total retinal thickness,

respectively. Green asterisks mark the rosette structures in the ONL, and pink triangles the holes in the INL. Scale bar = 50 μm. (**c,e**) Graphs showing the ONL thickness and number of photoreceptor nuclei rows (±SEM) in the ONL in 20 equidistant retinal locations (10 dorsal, 10 ventral). (**d,f**) The average and standard error of ONL thickness (ratio ONL/total retinal length) and numbers of photoreceptor rows. Photoreceptor loss (cell rows) in the retina was significant after 7 days of recovery and progressed for up to 7 months (210 days are not shown on the graph because it was hard to quantify these two parameters because of the tissue's damage). (**g**) Damaged ONL length analyses for different recovery times. The extension of the damaged area significantly increased over time, never reaching the retina's ventral side. Histograms show means ± SEM. Statistical significance is represented as follows: *, #, &, + $p \leq 0.05$, **, ##, @@ $p \leq 0.01$, ***, ˆˆˆ, \$\$\$, +++ $p \leq 0.001$, ****, ˆˆˆˆ, #### $p \leq 0.0001$; *, ˆ, #, \$, +, @, & refer to HC, LD24h0d, LD24h7d, LD24h15d, LD24h30d, LD24h45d, and LD24h60d, respectively; one-way ANOVA with Tukey's post hoc test. Values represent the means ± standard error of the mean (SEM) for at least $N = 5$ retinas from different rats for each experimental condition.

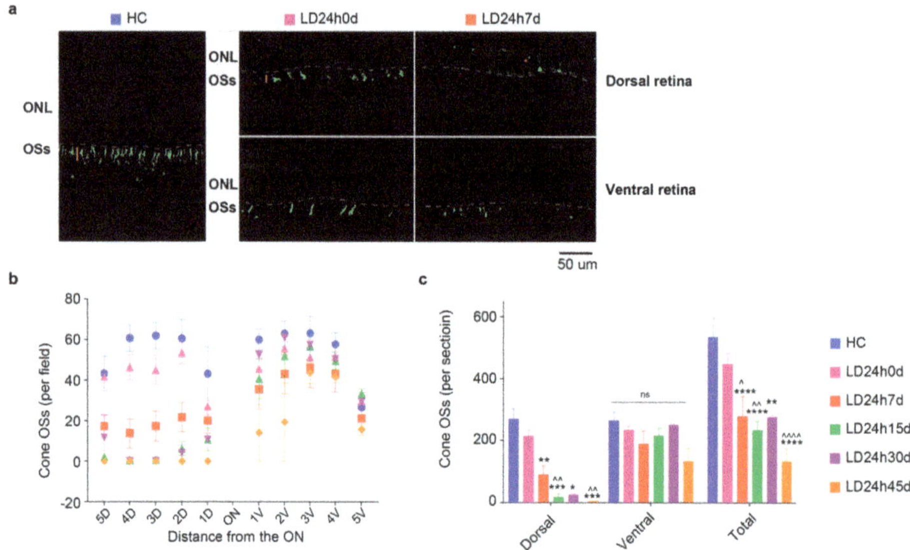

Figure 8. Light-induced damage to the L/M cone-photoreceptors population. (**a**) Retinal cross-sections from a representative HC, and 0 and 7 days after light-induced retinal damage. Sections are labeled with an antibody against L/M opsin. Outer segments (OSs) appear normally distributed throughout the retina in the control condition. Light-exposed retinas immediately showed a partial reduction of cone density (0 d) depicted by the length of the orange bar and redistribution of the signal into the soma (*, orange) in the dorsal retina, shortening the remaining OSs in the ventral at 7d. Abbreviations: ONL: outer nuclear layer; OSs: outer segments. Scale bar = 50 μm. (**b,c**) Numbers of cones distributed along the retinal sections passing through the optic nerve (ON), the totals on the dorsal and ventral sides, and their sum for HC, LD24h0d, LD24h7d, LD24h15d, LD24h30d, and LD24h45d. Statistical significance is represented as follows: *, ˆ $p \leq 0.05$, **, ˆˆ $p \leq 0.01$, *** $p \leq 0.001$, ****, ˆˆˆˆ $p \leq 0.0001$; * and ˆ refer to HC and LD24h0d, respectively; two-way ANOVA with Tukey's post hoc test. Data represent the means ± SEM of at least $N = 5$ retinas from different rats, for each experimental condition.

3.2.4. Light Exposure Induced Modifications in the Inner Nuclear Layer (INL), Detectable at Least after 15 Days Following Light Exposure

Morphological analyses mostly underlined a longitudinal extension of the hot spot area. However, as previously described from the ratio analysis (ONL thickness/total retinal thickness) and in other studies [28,29,34,53], the inner retina is also severely implicated. Figure 9a illustrates representative nuclei-stained retinal cross-sections after 7, 30, and 60 days post light exposure treatment. Gradually, with increasing time (after 7 days), the INL structure appeared highly disorganized, considerably thinned, and characterized by the presence of hole-like areas (Figure 9a; see also Figure 8a,b for a comparison between dorsal and ventral). Furthermore, in light of our functional data indicating a partial recovery of the a-wave and the b-wave, peaking at 45 days, we speculated that there could be compensatory retinal changes to account for the functional recovery. To assess this possibility, we studied the expression pattern of synaptophysin (SYN), an integral membrane protein expressed in presynaptic vesicles. In control retinas, SYN was expressed in the IPL and OPL, whereas after light exposure, the OPL signal was fainter compared to control (Figure 9b). Accordingly, we looked for other biochemical and immunohistochemical changes in the inner retina. Therefore, we aimed to establish whether neuronal modifications in the inner retina occurred in the light damaged rat model using an amacrine-specific marker. We stained retinal sections with an antibody against choline-acetyltransferase (ChAT), revealing the so-called ON and OFF starburst amacrine cells. These are critical elements in the encoding of motion, heavily contributing to direction-selective ganglion cells properties. We observed that although there was no reduction in the number of marked cell bodies either in the INL or GCL (data not shown), their dendritic arborization was disrupted 30 days after light damage (Figure 9c,d; significant at 210d vs. all experimental groups assessed at an earlier time point). As with other markers, ChAT expression in the ventral retina was mainly unaffected by neurodegenerative progression (data not shown).

3.2.5. Spatial Differences in Glial Cells' Reaction, bFGF-2, and Microglia Modulation: Non-Neuronal Cells' Involvement in Retinal Degeneration

Light-induced gliosis was assessed by studying the evolution of astrocyte alterations during retinal degeneration through immunocytochemical localization of GFAP. Figure 10a illustrates the representative distribution of the GFAP signal in the marginal hot spot area (penumbra [41]). Due to the highly disorganized structure typical of the chronic phase, GFAP expression measurements were performed in the penumbra area. Quantification of the signal across all retinal layers (Figure 10c) discloses a significant increase in GFAP immunoreactivity after 7 days both on the dorsal and ventral sides compared to the control. Anyway, in the ventral retina, whose structure was better preserved, the GFAP$^+$ area peaked at 30 days. It slightly decreased by 45 days, with significant differences between LD24h7d, 15d, and 30d vs. LD24h90d (one-way ANOVA, Tukey's test, $p = 0.0067, 0.0192$ 0.0265, respectively). Notice that at 90d GFAP$^+$ signal in the dorsal retina was significantly higher compared to the ventral retina ($p = 0.0032$).

On the contrary, in the penumbra (dorsal retina), glial cell reaction was less exacerbated at 15 d, 30 d, and 45 d than at 60 d and 90 d. Müller cells' support to photoreceptors was monitored by analyzing the distribution and the protein level of bFGF-2 in the ONL using immunohistochemistry, at successive survival times after 24 h LD (Figure 10b). No significant differences among groups were detected, although bFGF-2 expression was linearly augmenting in the dorsal retina up to 60d (Figure 10d; one-way ANOVA with Trend, $p = 0.0099$. Non-significant trend in the ventral side was found, $p = 0.2592$). On the ventral side, bFGF-2 expression increased immediately after LD, became prominent at about 15 d, and then decreased to control levels by 60d (indeed LD24h60d dorsal is significantly different compared to LD2460d ventral). The immune response also accompanies neurodegeneration for a considerable time. The onset of photoreceptor degeneration activated the recruitment of retinal microglia from the plexiform layer to the injured site (see Figure 3). The analysis of Iba1$^+$ cells distribution among layers described a consistent migration of resident microglia from the inner retina (GCL-IPL and INL) to the ONL, which required more than 24 h to occur (indeed LD24h0d not significant vs. HC). Iba1$^+$ cells (both microglia and macrophages) decreased in the inner retina during more extended recovery periods from the onset of neurodegeneration and increased in the ONL. Following the initial typical acute neuroinflammatory response, from 30 to 60 days, we noticed a significant decrease of microglial cells in the dorsal retina (Figure 11d). No differences were detected in the ventral retina when microglia cells were summed across layers (Figure 11f).

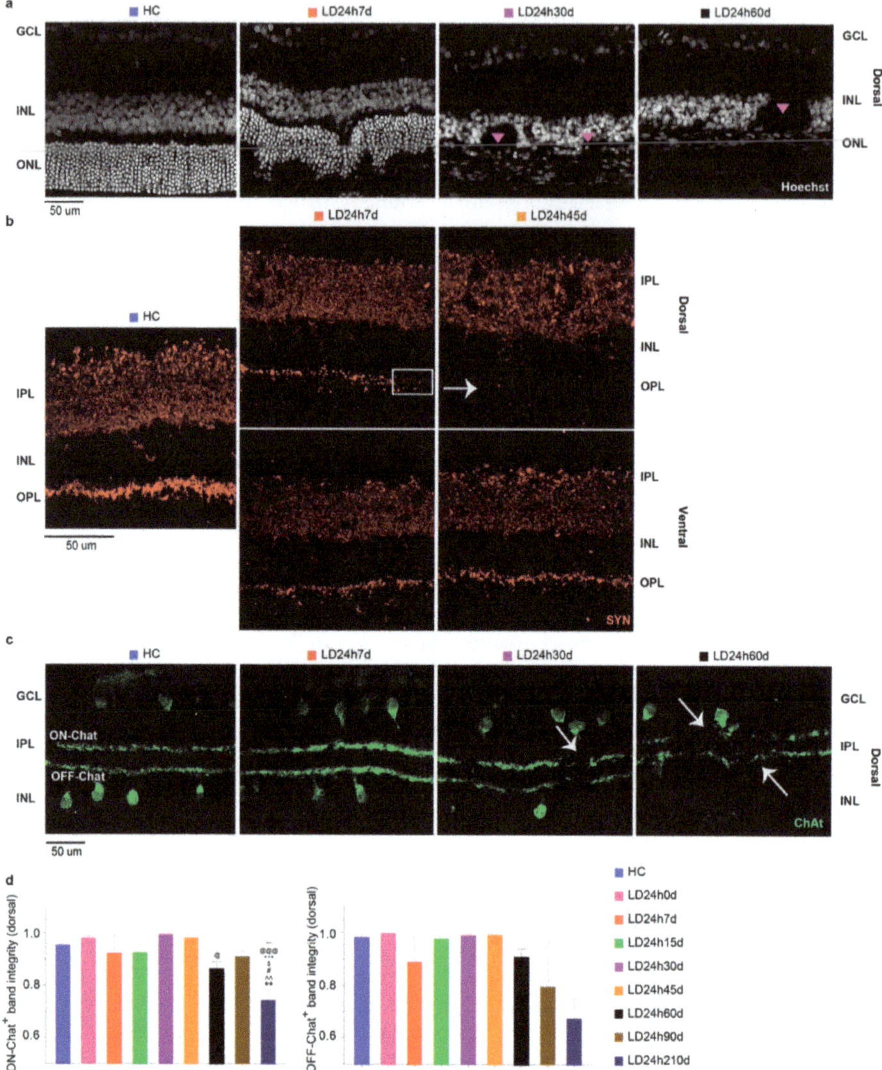

Figure 9. Inner retinal changes at different recovery times following 24 h of light exposure. (**a**) Hoechst-stained retinal cross-sections from the dorsal (hot spot) retina. A considerable disruption of the integrity of the INL is depicted. Structures like rosettes were formed (pink triangles) (see Figure 7b; INL holes were not present in the ventral retina). (**b**) Changes in synaptophysin (SYN) expression after 24 h light exposure were assessed on retinal cross-sections stained with an anti-SYN antibody. Compared to HC, there was a fainter expression in the dorsal OPL at 7 d (white rectangle) and almost no expression at 45 d (white arrowhead). (**c**) Retinal sections immunolabeled with anti-ChAT antibody revealed positive cell bodies in the IPL and GCL (OFF and ON Chat, respectively) and two narrowly stratified immunoreactive bands appearing in the IPL that presented interruptions from 30 days on after the light exposure (white arrowheads). (**d**) ON and OFF Chat bands were "integrity" normalized to the analyzed section length. Abbreviations: GCL: ganglion cell layer; IPL: inner plexiform layer; INL: inner nuclear layer; OPL: outer plexiform layer; IPL: inner plexiform layer; ONL: outer nuclear layer. Scale bar = 50 µm. Statistical significance is represented as follows: #, \$, @ $p \leq 0.05$, **, ^^, ~~ $p \leq 0.01$, +++, @@@ $p \leq 0.001$; *, ^, #, \$, +, @, ~ refer to HC, LD24h0d, LD24h7d, LD24h15d, LD24h30d, LD24h45d, and LD24h90d, respectively; one-way ANOVA with Tukey's post hoc test. Data are shown as means ± SEM of $N = 5$ retinas from different rats for each experimental condition in panel d.

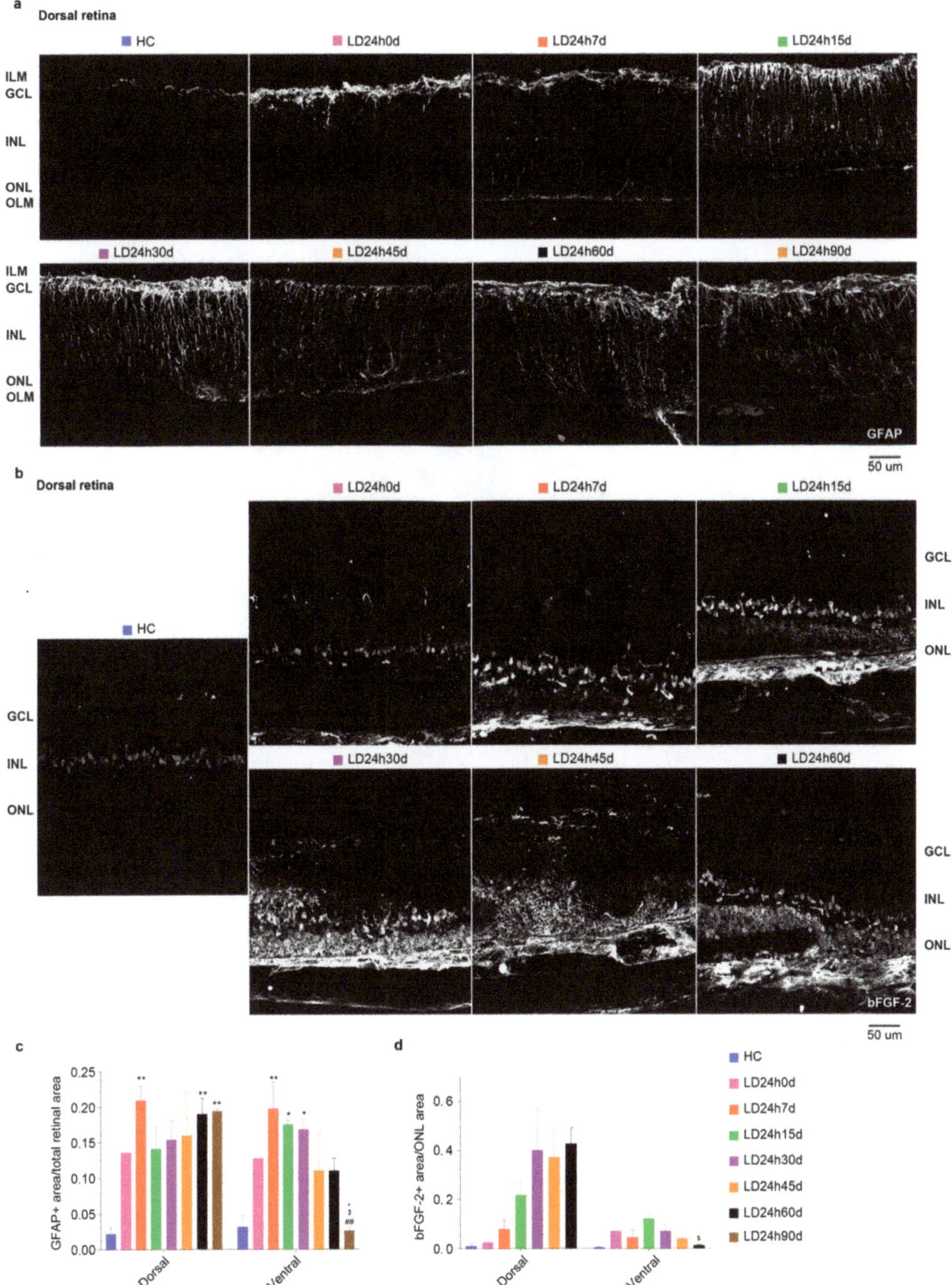

Figure 10. Distribution and density of GFAP and bFGF-2 in light-exposed retinas at different recovery times following 24 h of light damage. (**a**) Representative photomicrographs illustrating GFAP$^+$ astrocytes in dorsal retinal sections (penumbra, around the degenerating hot spot) of control retinas and retinas monitored several days after light exposure. Light exposure triggered an increased expression of GFAP in the radially oriented Müller cells; the protein was visible along the full length of the Müller cells, from the ILM to the OLM. (**b**) Representative photomicrographs illustrating bFGF-2$^+$ cells in dorsal retinal

sections (penumbra) of control and light-exposed retinas were assessed several days after. Light exposure increased the expression of bFGF-2 released in the ONL by Müller cells; the protein was visible around photoreceptor bodies in the ONL and Müller cell bodies in the INL. (**c**) Mean GFAP+ area (retinal area occupied by GFAP+ signal) averaged across the dorsal and ventral retina. (**d**) Mean bFGF+ area (ONL area occupied by bFGF+ signal) averaged across the dorsal and ventral retina. Abbreviations: GCL: ganglion cell layer; INL: inner nuclear layer; ONL: outer nuclear layer; ILM: inner limiting membrane; OLM: outer limiting membrane. Scale bar = 50 μm. Statistical significance is represented as follows: *, $, + $p \leq 0.05$, **, ## $p \leq 0.01$; *, #, $, + refer to HC, LD24h7d, LD24h15d, LD24h30d, respectively; one-way ANOVA with Tukey's post hoc test. One-way ANOVA with Trend was also used to analyze data in panel (**d**). Data are shown as means ± SEM of at least $N = 5$ retinas from different rats for each experimental condition in panel (**d**).

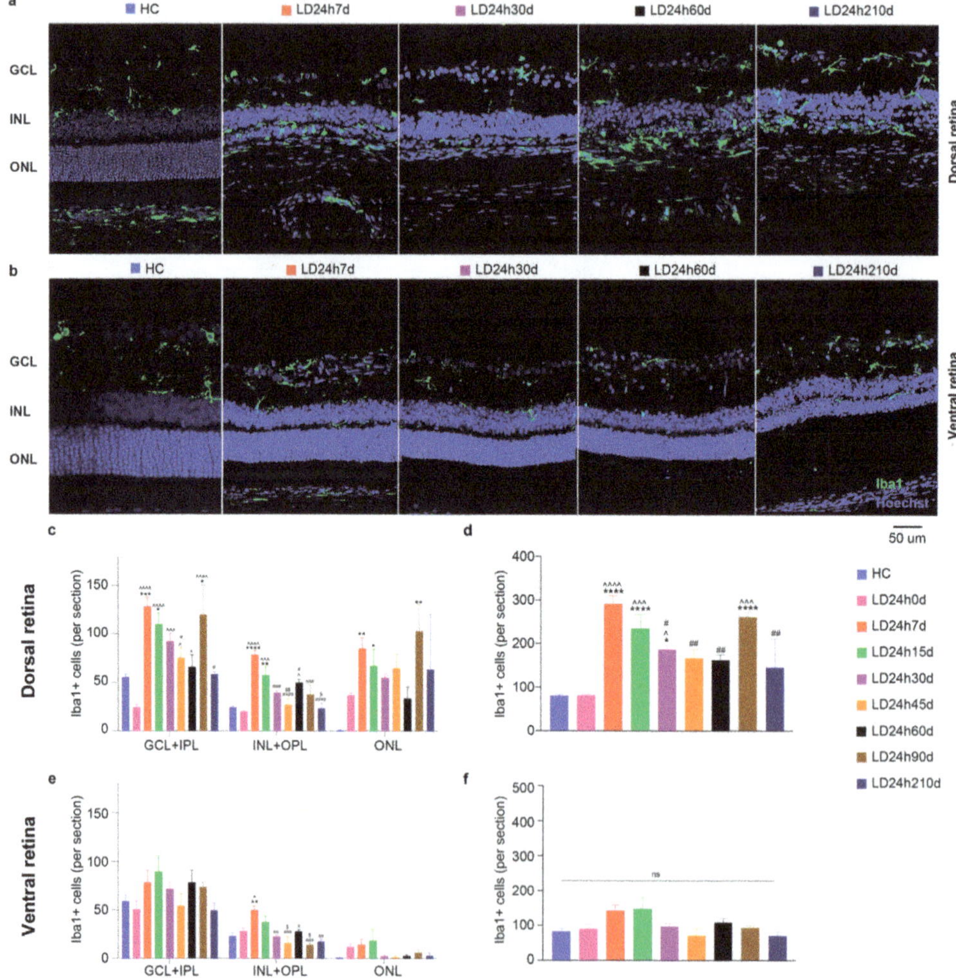

Figure 11. Quantification of microglia cells at different recovery times after 24 h of light exposure. (**a,b**) Representative photomicrographs illustrating Iba1+ cells in dorsal retinal sections (hot spot) of control and light-damaged retinas after light exposure. Light damage increased numbers of Iba1+ cells in the ONL by 7 d in the dorsal retina. Abbreviations: GCL: ganglion cell layer; INL: inner nuclear layer; ONL: outer nuclear layer. Scale bar = 50 μm. (**c–f**) Total numbers of microglia in the dorsal and ventral retina in each retinal layer (**c,e**) and summed across the entire retinal thickness (**d,f**). Statistical significance is represented as follows: *, ^, #, $ $p \leq 0.05$, **, ##, $$ $p \leq 0.01$, ***, ^^^, ### $p \leq 0.001$, ****, ^^^^, #### $p \leq 0.0001$; *, ^, #, $ refer to HC, LD24h0d, LD24h7d, LD24h15d, respectively; one-way ANOVA with Tukey's post hoc test. Data are shown as mean ± SEM of at least $N = 5$ retinas from different rats for each experimental condition.

4. Discussion and Conclusions

Most neurodegenerative retinal diseases are due to progressive degeneration of rods or cones. Current treatment strategies and paradigms manage to slow down photoreceptor loss or replace them using prosthetic devices or stem cells [6–9,11,12]. The therapeutic efficacy of these approaches depends on the diagnosis of the disease, the timing of which is usually related to the symptom's occurrence. At later neurodegenerative stages, however, therapeutical approaches are less effective. Moreover, early dysfunctions of retinal diseases might be overcome by compensatory mechanisms of the retina, which could silence pathological signs [54,55]. Therefore, it is paramount to achieve rapid diagnosis and effective, targeted treatment at the earliest possible stage.

Aiming to understand better the critical steps involved, we exposed adult albino rats to bright light to induce retinal neurodegeneration and followed the disease's morphological and functional progression. It was previously shown that this model mimics some aspects of human retinal degenerative disorders [56]. Our experiments first goal was to determine the minimal duration of bright light exposure necessary to induce retinal damage. Likewise, the second goal was to study the pathophysiological changes of the diseased retina over time, extending the focus to the inner retina.

4.1. Different Durations of Bright Light Exposure Provided Insights into Early Events in the Pathology of Light-Induced Retinal Damage

The first aim of our experiments was to characterize retinal pathophysiological processes following different durations of light exposure. We revealed that 12 h of homogeneous light exposure is sufficient to severely reduce the retinal response to light (assessed by in vivo analyses). However, this was not associated with critical morphological changes. Indeed, functional changes precede anatomical ones.

Specifically, the dark-adapted response is more severely impaired than the light-adapted response, in agreement with the higher survival rate of cones compared to rods, confirming previous studies about the etiology (rhodopsin-mediated [57,58]) of this light-induced retinopathy [59]. Interestingly, OP sum amplitude analysis revealed that the deterioration of OPs correlated better with the duration of light exposure than did the amplitudes of the a-wave and b-wave. The OPs are described as low-voltage, high-frequency components of the ERG [45] and are believed to reflect the synaptic activity of inhibitory feedback processes generated mainly by the amacrine cells. However, contributions of other inner retinal cells cannot be excluded [45]. Our findings indicated that 12 h of bright light exposure did not strongly affect the inner retinal circuitry, as suggested by the more preserved OP amplitude at LD12h7d (at least at specific stimuli intensities) despite continuing a-wave and b-wave amplitudes reduction.

Remarkably, by analyzing morphological features, we demonstrated that a minimum of 15 h is required to trigger the development of a hot spot (matrix disorganization and rosette structures formation) in the dorsal retina at 7 days after damage [41], which characterizes the model of light damage. Conversely, retinal morphology seemed better preserved after 12 h of light exposure, given the more subtle changes in photoreceptor rows and microglia numbers. Aside from the absence of a prominent hot spot in the LD12h7d group, the functional response (a-wave and b-wave amplitude) was similar to those of all other groups. Therefore, histological results at the ONL level did not correlate with those achieved at the functional level. Given this, we examined how different mechanisms might occur even before ONL disorganization and could influence the functional in vivo test. One such factor, the cytoskeletal type III intermediate filament protein GFAP of astrocytes, is considered an early marker for retinal injury and is commonly used as an index of gliosis-hypertrophy [60,61]. Twelve hours of exposure to light was enough to activate Müller cells, which are well-known to control several retinal trophic factors [62]. Simultaneously, activation of self-protective mechanisms might impinge on the retinal response to light, as supported by studies showing that an increase in bFGF-2 causes a reduction of the b-wave [40,63]. Moreover, it was interesting to notice that while the architecture of the ventral retina is affected more moderately than the dorsal retina (the layered structure of the retina was maintained), the inflammatory markers are upregulated compared to their baseline values. The upregulation of a range of protective factors might contribute to photoreceptor stability and protection; however, this does not extend to retinal performance, as ERG reduction shows.

This study disclosed that 12 h of light damage was sufficient to elicit clear signs of retinal dysfunction with minor morphological signs of damage in the dorsal hot spot area. For some parameters, such as photopic ERG, the functional impairment was irrespective of the exposure duration. Overall, this indicates that functional alterations precede morphological ones. These findings could be used to develop a diagnostic test that highlights an inflammatory condition in the retina that impinges on retinal functional output, as demonstrated. Indeed, ERG responses, recorded with an active extracellular electrode, according to Ohm's law, might be affected by a change in the resistance. We are currently investigating whether neuroinflammation can bias the ERG output by changing the resistance of the eye. Consequently, the substantial initial reduction in retinal function is likely due to both electro-mechanical problems and the neuroinflammatory state that causes a change in tissue resistances.

4.2. Analyses Performed after Different Recovery Periods Following 24 h of Bright Light Exposure Showed Discrepancies between the Morphological and Functional Timecourses of the Disease

To achieve our second goal, retinas exposed for 24 h to light at 1000 lux were analyzed at different time points after the cessation of light exposure (acute phase)—up to 7 months (chronic phase). This study allowed us to categorize the pathophysiological process at different stages. The immediate consequence of light exposure was characterized by a progressive and irreversible loss of photoreceptors and a significant attenuation of retinal function in response to light. The chronic phase was characterized by a significant partial recovery, maximal at 45 days, despite the hot spot region gradually enlarging. Our findings suggest that during this stage, the retina attempted to recover from the light insult. Functional rescue following a phototoxic insult is a well-known phenomenon in light-damaged rats [36,64]. It was shown by Rutar et al. [41], as well as others, that once the animal is returned to standard lighting conditions, regrowth of the outer segment occurs, although we found a reduction in the total number of cones expressing L/M-opsin up to 45 days after light exposure. Schremser and Williams [65] showed that growing outer segments to their maximal length could take 20 days, but our study observed the maximal ERG amplitude value at 45 days. While the a-wave's (linked to photoreceptor function) maximum amplitude recovered to ~34% (LD24h45d a-wave = 159 ± 21 μV and HC = 466 ± 33 μV, at 10 cd*s/m^2) [66], the b-wave's (second-order neurons and Müller cells) amplitude recovered to ~45% of its pre-LD value (LD24h45d b-wave = 503 ± 49 μV and HC = 1098 ± 74 μV, at 10 cd*s/m^2). Conversely, OP amplitudes recovered significantly (~50% at 10 cd*s/m^2), indicating a more pronounced rescue of the inner retina at 45 days.

We recorded ERG responses for up to 7 months and observed that retinal function deteriorates again after 45 days. As already described, astrocyte retinal response is triggered after bright light exposure and is followed by an upregulation of neurotrophic factors [67]. We demonstrated that after 24 h of light damage, the retinas of adult rats enhanced the expression of GFAP. This increase was maintained for up to 7 days in the dorsal side and up to 30 days in the ventral, and there was a subsequent ventral reduction at 45 days (ONL structure was more preserved) that became significant at 90 days, suggesting that GFAP may contribute, along with bFGF-2, to improving retinal function. bFGF-2 expression increased over time (significant trend) in the dorsal ONL for up to 60 days, but in the ventral side, its increase started immediately after LD and became prominent at approximately 15 days. Subsequently, it declined to control levels by 60 days (indeed LD24h60d significantly different only compared to LD24h15d). This slower downregulation of bFGF-2 in the ventral retina correlated with the increased b-wave amplitude at 45 days, as reported in Valter et al. [40]. More studies will be necessary to unveil the mechanisms by which worsening of the retinal function occurred after this time point.

Thus, the initial decrease in retinal function might have been due to a combination of OS shortening, the inhibitory effect of bFGF-2, and the change in retinal resistance due to gliosis. In other words, as the decreases in the quantified GFAP, bFGF-2, and Iba1[+] cells accompanied an increase in OS length, a gradual gain in function occurred. Afterward, in the last stage (at 210 days), severe photoreceptor degeneration seemed to overcome retinal functional restoration. Interestingly, the inner retina appeared disarranged at later stages; synaptophysin expression was reduced, and dendritic arborization of ChAT[+] cells became disorganized. These results suggest that the damage initially triggered by the bright light exposure was too severe such that the intrinsic protective mechanisms of the retina were insufficient to stop its deterioration. Interestingly, the evaluation of synaptophysin expression was discordant with the results obtained in a mouse model by Luis Montalbán-Soler et al. [64]. This difference could be explained by species-dependent differences or by different mechanisms activated through various light damaging protocols; it must be noted that mice (differences exist among strains) do not present a hot spot.

Nonetheless, it is worth noting that most of the retinal disorganization seen in our model was observed in the dorsal retina for up to 7 months, with a rapid loss of photoreceptors followed by substantial disorganization of the inner retina. Instead, the ventral side showed parallel reductions in the thickness of both the ONL and the inner retina. In this chronic phase (i.e., months after the cessation of light exposure), the degree of retinal damage was proportional to the length of the recovery period. The surviving photoreceptors were responsible for maintaining the a-wave. Future studies should look more profoundly at the synaptic and cellular reorganization in the animals inner retina to better understand the retinal network changes. If the outer retina is affected, the signals sent to the inner retina would also be influenced, and consequently, the response of second-order neurons would be impaired. These modifications must lead to substantial changes in the receptive field properties of retinal ganglion cells that are ultimately reflected in altered visual performance [26,68]. Are these changes occurring only at a later stage of the pathology? Are these permanent changes, or can they be restored after the substitution of defective photoreceptors?

It is clear that the ability to detect these pathologies at an early stage is essential for any attempt at restoring retinal function, reducing the impairment of the inner retina, or increasing the chance of successful response to future therapies. Additionally, our data may provide valuable information for developing retinal prosthetics more efficiently and improving their integration with the patient's retina. Furthermore, it will be fundamental to adopt a more realistic approach that considers the degeneration in the outer retina and the rewiring of the remaining circuit.

Author Contributions: Conceptualization, S.D.M. and S.B.; methodology, S.D.M., S.R., M.D.P. and S.B.; formal analysis, S.D.M., S.R., M.D.P. and S.B.; data curation, S.D.M., S.R., M.D.P., J.A.; writing—original draft preparation, S.R., M.D.P. and S.B.; writing—review and editing, all authors; supervision, S.D.M. and S.B.; project administration, S.D.M.; funding acquisition, S.D.M. and S.B. All authors have read and agreed to the published version of the manuscript.

Funding: This research was funded by ESSSE caffè Srl, Hortus Novus Srl and ENEA.

Institutional Review Board Statement: The study was conducted according to the guidelines of the ARVO Statement for the Use of Animals in Ophthalmic and Vision Research. All experiments were carried out with authorization numbers 448/2016-PR and 862/2018-PR, issued by the Ministry of Health, and approved by the local Ethical Committee of the University of L'Aquila.

Data Availability Statement: The data that support the findings of this study are available on request from the corresponding author, S.d.M.

Conflicts of Interest: The authors declare no conflict of interest.

References

1. Lamoureux, E.; Pesudovs, K. Vision-specific quality-of-life research: A need to improve the quality. *Am. J. Ophthalmol.* **2011**, *151*, 195–197.e2. [CrossRef]
2. Athanasiou, D.; Aguilà, M.; Bevilacqua, D.; Novoselov, S.S.; Parfitt, D.A.; Cheetham, M.E. The cell stress machinery and retinal degeneration. *FEBS Lett.* **2013**, *587*, 2008–2017. [CrossRef]
3. Wright, A.F.; Chakarova, C.F.; Abd El-Aziz, M.M.; Bhattacharya, S.S. Photoreceptor degeneration: Genetic and mechanistic dissection of a complex trait. *Nat. Rev. Genet.* **2010**, *11*, 273–284. [CrossRef]
4. Chacon-Camacho, O.F.; Zenteno, J.C. Review and update on the molecular basis of Leber congenital amaurosis. *World J. Clin. Cases* **2015**, *3*, 112–124. [CrossRef]
5. Song, H.; Rossi, E.A.; Latchney, L.; Bessette, A.; Stone, E.; Hunter, J.J.; Williams, D.R.; Chung, M. Cone and rod loss in Stargardt disease revealed by adaptive optics scanning light ophthalmoscopy. *JAMA Ophthalmol.* **2015**, *133*, 1198–1203. [CrossRef]
6. Young, M.J.; Ray, J.; Whiteley, S.J.; Klassen, H.; Gage, F.H. Neuronal differentiation and morphological integration of hippocampal progenitor cells transplanted to the retina of immature and mature dystrophic rats. *Mol. Cell. Neurosci.* **2000**, *16*, 197–205. [CrossRef] [PubMed]
7. Gagliardi, G.; Ben M'Barek, K.; Goureau, O. Photoreceptor cell replacement in macular degeneration and retinitis pigmentosa: A pluripotent stem cell-based approach. *Prog. Retin. Eye Res.* **2019**, *71*, 1–25. [CrossRef] [PubMed]
8. Ghezzi, D.; Antognazza, M.R.; Maccarone, R.; Bellani, S.; Lanzarini, E.; Martino, N.; Mete, M.; Pertile, G.; Bisti, S.; Lanzani, G.; et al. A polymer optoelectronic interface restores light sensitivity in blind rat retinas. *Nat. Photonics* **2013**, *7*, 400–406. [CrossRef]
9. Maya-Vetencourt, J.F.; Ghezzi, D.; Antognazza, M.R.; Colombo, E.; Mete, M.; Feyen, P.; Desii, A.; Buschiazzo, A.; Di Paolo, M.; Di Marco, S.; et al. A fully organic retinal prosthesis restores vision in a rat model of degenerative blindness. *Nat. Mater.* **2017**, *16*, 681–689. [CrossRef] [PubMed]
10. Henriksen, B.S.; Marc, R.E.; Bernstein, P.S. Optogenetics for retinal disorders. *J. Ophthalmic Vis. Res.* **2014**, *9*, 374–382. [PubMed]
11. Garita-Hernandez, M.; Lampič, M.; Chaffiol, A.; Guibbal, L.; Routet, F.; Santos-Ferreira, T.; Gasparini, S.; Borsch, O.; Gagliardi, G.; Reichman, S.; et al. Restoration of visual function by transplantation of optogenetically engineered photoreceptors. *Nat. Commun.* **2019**, *10*, 4524. [CrossRef] [PubMed]
12. Sahel, J.-A.; Boulanger-Scemama, E.; Pagot, C.; Arleo, A.; Galluppi, F.; Martel, J.N.; Esposti, S.D.; Delaux, A.; de Saint Aubert, J.-B.; de Montleau, C.; et al. Partial recovery of visual function in a blind patient after optogenetic therapy. *Nat. Med.* **2021**.
13. Maccarone, R.; Di Marco, S.; Bisti, S. Saffron supplement maintains morphology and function after exposure to damaging light in mammalian retina. *Investig. Ophthalmol. Vis. Sci.* **2008**, *49*, 1254–1261. [CrossRef]
14. Di Marco, S.; Riccitelli, S.; Di Paolo, M.; Campos, E.; Buzzi, M.; Bisti, S.; Versura, P. Cord Blood Serum (CBS)-Based Eye Drops Modulate Light-Induced Neurodegeneration in Albino Rat Retinas. *Biomolecules* **2020**, *10*, 678. [CrossRef] [PubMed]
15. Finger, R.P.; Daien, V.; Eldem, B.M.; Talks, J.S.; Korobelnik, J.-F.; Mitchell, P.; Sakamoto, T.; Wong, T.Y.; Pantiri, K.; Carrasco, J. Anti-vascular endothelial growth factor in neovascular age-related macular degeneration-a systematic review of the impact of anti-VEGF on patient outcomes and healthcare systems. *BMC Ophthalmol.* **2020**, *20*, 294. [CrossRef] [PubMed]
16. Kniggendorf, V.; Dreyfuss, J.L.; Regatieri, C.V. Age-related macular degeneration: A review of current therapies and new treatments. *Arq. Bras. Oftalmol.* **2020**, *83*, 552–561.
17. Piccardi, M.; Marangoni, D.; Minnella, A.M.; Savastano, M.C.; Valentini, P.; Ambrosio, L.; Capoluongo, E.; Maccarone, R.; Bisti, S.; Falsini, B. A longitudinal follow-up study of saffron supplementation in early age-related macular degeneration: Sustained benefits to central retinal function. *Evid. Based Complement. Alternat. Med.* **2012**, *2012*, 429124. [CrossRef]
18. Marangoni, D.; Falsini, B.; Piccardi, M.; Ambrosio, L.; Minnella, A.M.; Savastano, M.C.; Bisti, S.; Maccarone, R.; Fadda, A.; Mello, E.; et al. Functional effect of Saffron supplementation and risk genotypes in early age-related macular degeneration: A preliminary report. *J. Transl. Med.* **2013**, *11*, 228. [CrossRef]
19. Jones, B.W.; Kondo, M.; Terasaki, H.; Lin, Y.; McCall, M.; Marc, R.E. Retinal remodeling. *Jpn. J. Ophthalmol.* **2012**, *56*, 289–306. [CrossRef]
20. Krishnamoorthy, V.; Cherukuri, P.; Poria, D.; Goel, M.; Dagar, S.; Dhingra, N.K. Retinal remodeling: Concerns, emerging remedies and future prospects. *Front. Cell Neurosci.* **2016**, *10*, 38. [CrossRef]
21. Gargini, C.; Terzibasi, E.; Mazzoni, F.; Strettoi, E. Retinal organization in the retinal degeneration 10 (rd10) mutant mouse: A morphological and ERG study. *J. Comp. Neurol.* **2007**, *500*, 222–238. [CrossRef]
22. Marc, R.E.; Jones, B.W. Retinal remodeling in inherited photoreceptor degenerations. *Mol. Neurobiol.* **2003**, *28*, 139–147. [CrossRef]
23. Strettoi, E.; Porciatti, V.; Falsini, B.; Pignatelli, V.; Rossi, C. Morphological and functional abnormalities in the inner retina of the rd/rd mouse. *J. Neurosci.* **2002**, *22*, 5492–5504. [CrossRef]
24. Gupta, C.L.; Nag, T.C.; Jha, K.A.; Kathpalia, P.; Maurya, M.; Kumar, P.; Gupta, S.; Roy, T.S. Changes in the Inner Retinal Cells after Intense and Constant Light Exposure in Sprague-Dawley Rats. *Photochem. Photobiol.* **2020**, *96*, 1061–1073. [CrossRef]
25. Marc, R.E.; Jones, B.W.; Anderson, J.R.; Kinard, K.; Marshak, D.W.; Wilson, J.H.; Wensel, T.; Lucas, R.J. Neural reprogramming in retinal degeneration. *Investig. Ophthalmol. Vis. Sci.* **2007**, *48*, 3364–3371. [CrossRef] [PubMed]
26. García-Ayuso, D.; Di Pierdomenico, J.; Vidal-Sanz, M.; Villegas-Pérez, M.P. Retinal ganglion cell death as a late remodeling effect of photoreceptor degeneration. *Int. J. Mol. Sci.* **2019**, *20*, 4649. [CrossRef]

27. Marco-Gomariz, M.A.; Hurtado-Montalbán, N.; Vidal-Sanz, M.; Lund, R.D.; Villegas-Pérez, M.P. Phototoxic-induced photoreceptor degeneration causes retinal ganglion cell degeneration in pigmented rats. *J. Comp. Neurol.* **2006**, *498*, 163–179. [CrossRef] [PubMed]

28. García-Ayuso, D.; Salinas-Navarro, M.; Agudo-Barriuso, M.; Alarcón-Martínez, L.; Vidal-Sanz, M.; Villegas-Pérez, M.P. Retinal ganglion cell axonal compression by retinal vessels in light-induced retinal degeneration. *Mol. Vis.* **2011**, *17*, 1716–1733.

29. Marc, R.E.; Jones, B.W.; Watt, C.B.; Strettoi, E. Neural remodeling in retinal degeneration. *Prog. Retin. Eye Res.* **2003**, *22*, 607–655. [CrossRef]

30. Chrysostomou, V.; Stone, J.; Stowe, S.; Barnett, N.L.; Valter, K. The status of cones in the rhodopsin mutant P23H-3 retina: Light-regulated damage and repair in parallel with rods. *Investig. Ophthalmol. Vis. Sci.* **2008**, *49*, 1116–1125. [CrossRef]

31. Demontis, G.C.; Longoni, B.; Marchiafava, P.L. Molecular steps involved in light-induced oxidative damage to retinal rods. *Investig. Ophthalmol. Vis. Sci.* **2002**, *43*, 2421–2427.

32. Stone, J.; Maslim, J.; Valter-Kocsi, K.; Mervin, K.; Bowers, F.; Chu, Y.; Barnett, N.; Provis, J.; Lewis, G.; Fisher, S.K.; et al. Mechanisms of photoreceptor death and survival in mammalian retina. *Prog. Retin. Eye Res.* **1999**, *18*, 689–735. [CrossRef]

33. Pfeiffer, R.L.; Marc, R.E.; Jones, B.W. Persistent remodeling and neurodegeneration in late-stage retinal degeneration. *Prog. Retin. Eye Res.* **2020**, *74*, 100771. [CrossRef] [PubMed]

34. Marc, R.E.; Jones, B.W.; Watt, C.B.; Vazquez-Chona, F.; Vaughan, D.K.; Organisciak, D.T. Extreme retinal remodeling triggered by light damage: Implications for age related macular degeneration. *Mol. Vis.* **2008**, *14*, 782–806.

35. Johnson, L.V.; Leitner, W.P.; Staples, M.K.; Anderson, D.H. Complement activation and inflammatory processes in Drusen formation and age related macular degeneration. *Exp. Eye Res.* **2001**, *73*, 887–896. [CrossRef]

36. Noell, W.K.; Walker, V.S.; Kang, B.S.; Berman, S. Retinal damage by light in rats. *Investig. Ophthalmol.* **1966**, *5*, 450–473.

37. Chaychi, S.; Polosa, A.; Lachapelle, P. Differences in Retinal Structure and Function between Aging Male and Female Sprague-Dawley Rats are Strongly Influenced by the Estrus Cycle. *PLoS ONE* **2015**, *10*, e0136056. [CrossRef]

38. Nadal-Nicolás, F.M.; Vidal-Sanz, M.; Agudo-Barriuso, M. The aging rat retina: From function to anatomy. *Neurobiol. Aging* **2018**, *61*, 146–168. [CrossRef]

39. McCulloch, D.L.; Marmor, M.F.; Brigell, M.G.; Hamilton, R.; Holder, G.E.; Tzekov, R.; Bach, M. ISCEV Standard for full-field clinical electroretinography (2015 update). *Doc. Ophthalmol.* **2015**, *130*, 1–12. [CrossRef]

40. Valter, K.; Bisti, S.; Gargini, C.; Di Loreto, S.; Maccarone, R.; Cervetto, L.; Stone, J. Time course of neurotrophic factor upregulation and retinal protection against light-induced damage after optic nerve section. *Investig. Ophthalmol. Vis. Sci.* **2005**, *46*, 1748–1754. [CrossRef]

41. Rutar, M.; Provis, J.M.; Valter, K. Brief exposure to damaging light causes focal recruitment of macrophages, and long-term destabilization of photoreceptors in the albino rat retina. *Curr. Eye Res.* **2010**, *35*, 631–643. [CrossRef]

42. Fiorani, L.; Passacantando, M.; Santucci, S.; Di Marco, S.; Bisti, S.; Maccarone, R. Cerium oxide nanoparticles reduce microglial activation and neurodegenerative events in light damaged retina. *PLoS ONE.* **2015**, *10*, e0140387. [CrossRef]

43. Huang, P.-C.; Hsiao, Y.-T.; Kao, S.-Y.; Chen, C.-F.; Chen, Y.-C.; Chiang, C.-W.; Lee, C.-F.; Lu, J.-C.; Chern, Y.; Wang, C.-T. Adenosine A(2A) receptor up-regulates retinal wave frequency via starburst amacrine cells in the developing rat retina. *PLoS ONE* **2014**, *9*, e95090.

44. Antognazza, M.R.; Di Paolo, M.; Ghezzi, D.; Mete, M.; Di Marco, S.; Maya-Vetencourt, J.F.; Maccarone, R.; Desii, A.; Di Fonzo, F.; Bramini, M.; et al. Characterization of a Polymer-Based, Fully Organic Prosthesis for Implantation into the Subretinal Space of the Rat. *Adv. Healthc. Mater.* **2016**, *5*, 2271–2282. [CrossRef] [PubMed]

45. Wachtmeister, L. Oscillatory potentials in the retina: What do they reveal. *Prog. Retin. Eye Res.* **1998**, *17*, 485–521. [CrossRef]

46. Tanito, M.; Kaidzu, S.; Ohira, A.; Anderson, R.E. Topography of retinal damage in light-exposed albino rats. *Exp. Eye Res.* **2008**, *87*, 292–295. [CrossRef] [PubMed]

47. Gallego, B.I.; Salazar, J.J.; de Hoz, R.; Rojas, B.; Ramírez, A.I.; Salinas-Navarro, M.; Ortín-Martínez, A.; Valiente-Soriano, F.J.; Avilés-Trigueros, M.; Villegas-Perez, M.P.; et al. IOP induces upregulation of GFAP and MHC-II and microglia reactivity in mice retina contralateral to experimental glaucoma. *J. Neuroinflamm.* **2012**, *9*, 92. [CrossRef] [PubMed]

48. Li, L.; Eter, N.; Heiduschka, P. The microglia in healthy and diseased retina. *Exp. Eye Res.* **2015**, *136*, 116–130. [CrossRef] [PubMed]

49. Bruera, M.G.; Benedetto, M.M.; Guido, M.E.; Degano, A.L.; Contin, M.A. Glial Cell Responses to constant low Light exposure in Rat Retina. *BioRxiv* **2021**.

50. Noailles, A.; Fernández-Sánchez, L.; Lax, P.; Cuenca, N. Microglia activation in a model of retinal degeneration and TUDCA neuroprotective effects. *J. Neuroinflamm.* **2014**, *11*, 186. [CrossRef]

51. Ortín-Martínez, A.; Jiménez-López, M.; Nadal-Nicolás, F.M.; Salinas-Navarro, M.; Alarcón-Martínez, L.; Sauvé, Y.; Villegas-Pérez, M.P.; Vidal-Sanz, M.; Agudo-Barriuso, M. Automated quantification and topographical distribution of the whole population of S- and L-cones in adult albino and pigmented rats. *Investig. Ophthalmol. Vis. Sci.* **2020**, *51*, 3171–3183. [CrossRef]

52. Edward, D.P.; Lim, K.; Sawaguchi, S.; Tso, M.O. An immunohistochemical study of opsin in photoreceptor cells following light-induced retinal degeneration in the rat. *Graefes Arch. Clin. Exp. Ophthalmol.* **1993**, *231*, 289–294. [CrossRef] [PubMed]

53. D'Orazi, F.D.; Suzuki, S.C.; Wong, R.O. Neuronal remodeling in retinal circuit assembly, disassembly, and reassembly. *Trends Neurosci.* **2014**, *37*, 594–603. [CrossRef] [PubMed]

54. Gidday, J.M.; Zhang, L.; Chiang, C.-W.; Zhu, Y. Enhanced retinal ganglion cell survival in glaucoma by hypoxic postconditioning after disease onset. *Neurotherapeutics* **2015**, *12*, 502–514. [CrossRef]

55. Gidday, J.M. Adaptive plasticity in the retina: Protection against acute injury and neurodegenerative disease by conditioning stimuli. *Cond. Med.* **2018**, *1*, 85–97.
56. Pennesi, M.E.; Neuringer, M.; Courtney, R.J. Animal models of age related macular degeneration. *Mol. Asp. Med.* **2012**, *33*, 487–509. [CrossRef]
57. Noell, W.K. Possible mechanisms of photoreceptor damage by light in mammalian eyes. *Vision Res.* **1980**, *20*, 1163–1171. [CrossRef]
58. Tso, M.O.; Woodford, B.J. Effect of photic injury on the retinal tissues. *Ophthalmology* **1983**, *90*, 952–963. [CrossRef]
59. Cicerone, C.M. Cones survive rods in the light-damaged eye of the albino rat. *Science* **1976**, *194*, 1183–1185. [CrossRef]
60. Bringmann, A.; Pannicke, T.; Grosche, J.; Francke, M.; Wiedemann, P.; Skatchkov, S.N.; Osborne, N.N.; Reichenbach, A. Müller cells in the healthy and diseased retina. *Prog. Retin. Eye Res.* **2006**, *25*, 397–424. [CrossRef] [PubMed]
61. Ramírez, J.M.; Ramírez, A.I.; Salazar, J.J.; de Hoz, R.; Triviño, A. Changes of astrocytes in retinal ageing and age-related macular degeneration. *Exp. Eye Res.* **2001**, *73*, 601–615. [CrossRef] [PubMed]
62. Puro, D.G. Growth factors and Müller cells. *Prog. Retin. Eye Res.* **1995**, *15*, 89–101. [CrossRef]
63. Gargini, C.; Bisti, S.; Demontis, G.C.; Valter, K.; Stone, J.; Cervetto, L. Electroretinogram changes associated with retinal upregulation of trophic factors: Observations following optic nerve section. *Neuroscience* **2004**, *126*, 775–783. [CrossRef] [PubMed]
64. Montalbán-Soler, L.; Alarcón-Martínez, L.; Jiménez-López, M.; Salinas-Navarro, M.; Galindo-Romero, C.; Bezerra de Sá, F.; García-Ayuso, D.; Avilés-Trigueros, M.; Vidal-Sanz, M.; Agudo-Barriuso, M.; et al. Retinal compensatory changes after light damage in albino mice. *Mol. Vis.* **2012**, *18*, 675–693.
65. Schremser, J.L.; Williams, T.P. Rod outer segment (ROS) renewal as a mechanism for adaptation to a new intensity environment. I. Rhodopsin levels and ROS length. *Exp. Eye Res.* **1995**, *61*, 17–23. [CrossRef]
66. Sugawara, T.; Sieving, P.A.; Bush, R.A. Quantitative relationship of the scotopic and photopic ERG to photoreceptor cell loss in light damaged rats. *Exp. Eye Res.* **2000**, *70*, 693–705. [CrossRef]
67. Cao, W.; Wen, R.; Li, F.; Lavail, M.M.; Steinberg, R.H. Mechanical injury increases bFGF and CNTF mRNA expression in the mouse retina. *Exp. Eye Res.* **1997**, *65*, 241–248. [CrossRef]
68. Garcia-Ayuso, D.; Di Pierdomenico, J.; Agudo-Barriuso, M.; Vidal-Sanz, M.; Villegas-Pérez, M.P. Retinal remodeling following photoreceptor degeneration causes retinal ganglion cell death. *Neural Regen. Res.* **2018**, *13*, 1885–1886. [CrossRef]

Article

Expression of the Endoplasmic Reticulum Stress Marker GRP78 in the Normal Retina and Retinal Degeneration Induced by Blue LED Stimuli in Mice

Yong Soo Park [1,2], Hong-Lim Kim [3], Seung Hee Lee [1,2], Yan Zhang [1,4] and In-Beom Kim [1,2,3,4,5,*]

1   Department of Anatomy, College of Medicine, The Catholic University of Korea, 222 Banpo-daero, Seocho-gu, Seoul 06591, Korea; yongsoopark88@gmail.com (Y.S.P.); seunghui6310@daum.net (S.H.L.); 00zhangyan00@naver.com (Y.Z.)
2   Catholic Neuroscience Institute, College of Medicine, The Catholic University of Korea, 222 Banpo-daero, Seocho-gu, Seoul 06591, Korea
3   Integrative Research Support Center, College of Medicine, The Catholic University of Korea, 222 Banpo-daero, Seocho-gu, Seoul 06591, Korea; wgwkim@catholic.ac.kr
4   Department of Biomedicine and Health Sciences, Graduate School, The Catholic University of Korea, 222 Banpo-daero, Seocho-gu, Seoul 06591, Korea
5   Catholic Institute for Applied Anatomy, College of Medicine, The Catholic University of Korea, 222 Banpo-daero, Seocho-gu, Seoul 06591, Korea
*   Correspondence: ibkimmd@catholic.ac.kr; Tel.: +82-2-2258-7263

Abstract: Retinal degeneration is a leading cause of blindness. The unfolded protein response (UPR) is an endoplasmic reticulum (ER) stress response that affects cell survival and death and GRP78 forms a representative protective response. We aimed to determine the exact localization of GRP78 in an animal model of light-induced retinal degeneration. Dark-adapted mice were exposed to blue light-emitting diodes and retinas were obtained at 24 h and 72 h after exposure. In the normal retina, we found that GRP78 was rarely detected in the photoreceptor cells while it was expressed in the perinuclear space of the cell bodies in the inner nuclear and ganglion cell layers. After injury, the expression of GRP78 in the outer nuclear and inner plexiform layers increased in a time-dependent manner. However, an increased GRP78 expression was not observed in damaged photoreceptor cells in the outer nuclear layer. GRP78 was located in the perinuclear space and ER lumen of glial cells and the ER developed in glial cells during retinal degeneration. These findings suggest that GRP78 and the ER response are important for glial cell activation in the retina during photoreceptor degeneration.

Keywords: retinal degeneration; endoplasmic reticulum; stress response; unfolded protein response; GRP78; retinal glial cell

Citation: Park, Y.S.; Kim, H.-L.; Lee, S.H.; Zhang, Y.; Kim, I.-B. Expression of the Endoplasmic Reticulum Stress Marker GRP78 in the Normal Retina and Retinal Degeneration Induced by Blue LED Stimuli in Mice. *Cells* **2021**, *10*, 995. https://doi.org/10.3390/cells10050995

Academic Editor: Maurice Ptito

Received: 26 March 2021
Accepted: 22 April 2021
Published: 23 April 2021

Publisher's Note: MDPI stays neutral with regard to jurisdictional claims in published maps and institutional affiliations.

1. Introduction

Retinal degeneration (RD) is a leading cause of blindness and is characterized by the irreversible and progressive loss of photoreceptor cells. Age-related macular degeneration (AMD) is the most common RD with a multifactorial cause of progression. Various factors including a genetic contribution, light-induced stress, oxidative stress and inflammation affect photoreceptor cell death during AMD [1–3]. For the treatment of wet AMD, anti-vascular endothelial growth factor (VEGF) therapy is a unique, effective treatment option. However, long-term intravitreal injections of an anti-VEGF agent is needed, which can cause complications including ocular hypertension, inflammation, retinal detachment and hemorrhage [4,5]. Moreover, a therapeutic strategy for dry AMD has not yet been established [6]. Thus, studies on the pathogenesis of RD including AMD are needed.

The endoplasmic reticulum (ER) stress response is an important intracellular mechanism of neuronal cell death [7,8]. Stress conditions in the ER can trigger the unfolded protein response (UPR), which increases the number of endoplasmic chaperones to clear the accumulated unfolded proteins and to maintain homeostasis [7,9]. However, when ER stress exceeds the UPR, ER stress undergoes a cell death pathway via the activation of the C/EBP homologous protein (CHOP) [10]. There is increasing evidence that ER stress and ER responses are involved in the pathogenesis of AMD and its progression [11,12]. Previous animal studies revealed that ER stress markers were increased in the retina during photoreceptor degeneration [13,14]. In addition, a drug-induced ER stress model could promote photoreceptor loss in mice [15]. However, there is a contrary report that ER stress could protect photoreceptors in Drosophila [16]. Thus, the relationship between photoreceptor cell death and the ER stress of photoreceptors remains unclear.

Immunoglobulin heavy chain-binding protein (BiP), also known as 78-kDa glucose-regulated protein (GRP78), is a representative ER stress marker that belongs to the ER chaperone. GRP78 regulates protein folding to prevent the accumulation of misfolded proteins and maintains ER homeostasis and cell protection [17,18]. Previous studies have shown increased GRP78 expression in RD models induced by light injury [13,19] and retinal detachment [13,19] and in inner retinal degeneration induced by N-methyl-D-aspartate (NMDA) toxicity [20,21], suggesting the involvement of GRP78 in the pathogenesis of various types of retinal degeneration. However, its localization in the normal retina remains unclear. A few investigators have reported that GRP78 is expressed in the inner nuclear layer (INL) and ganglion cell layer (GCL) in the normal retina [20,21] while others have not detected GRP78 expression in cells in the outer nuclear layer (ONL), INL and GCL in a normal retina [19]. Therefore, detailed information on GRP78 expression at cellular and subcellular levels in normal and RD retinas needs to be elucidated to understand the role of ER stress in the pathogenesis of RD.

We aimed to determine the cellular and subcellular localization of GRP78 in normal retinas and to examine the changes in the GRP78 expression profile in RD induced by light-emitting diodes (LED) using immunohistochemistry and an advanced immuno-electron microscopic technique, correlative light and electron microscopy (CLEM).

2. Materials and Methods

2.1. Animals

A total of 30 male BALB/c mice at seven weeks of age were used in this study. The mice were kept in a climate-controlled condition with a 12 h light and dark cycle and divided into three groups: normal, 24 h and 72 h after LED exposure (n = 10 for each group). All procedures followed the regulations established by the Institutional Animal Care and Use Committee of the School of Medicine, The Catholic University of Korea (Approval number: CUMS-2017-0241-03), which acquired AAALAC International full accreditation in 2018.

2.2. Exposure to a Blue LED

The blue LED-induced RD model was described in detail in our previous studies [22–24]. Mice were kept in a dark room for 24 h before LED exposure and their pupils were dilated with Mydrin P (Santen Pharmaceutical Co., Osaka, Japan) under dim red-light conditions 30 min before LED exposure. Afterwards, mice were exposed to a 2000-lux blue LED for 2 h. After LED exposure, the mice were kept in a dark room for 1 h and then the mice were moved to a climate-controlled condition with a 12 h light and dark cycle.

2.3. Tissue Preparation

At 24 h and 72 h after LED exposure, the mice were anesthetized by an intraperitoneal injection of zolazepam (20 mg/kg) and xylazine (7.5 mg/kg) for tissue preparation. The anterior portion of the eye was dissected and eye cups were fixed in 4% paraformaldehyde for 2 h. After fixation, the eye cups were washed with a phosphate buffer (PB, 0.1 M) and then transferred to 30% sucrose in PB 0.1 M for one night. The tissues were embedded in an OCT compound to prepare the frozen tissue. Sections of the eye cups at the center of the retina of 8 μm thickness were obtained.

2.4. Terminal Deoxynucleotidyl Transferase dUTP Nick End Labeling (TUNEL) Assay

After washing, the retinal sections were incubated in a permeabilization solution containing 0.1% Triton X-100 and 0.1% sodium citrate in 0.01 M phosphate buffered saline (PBS) for 2 min. The TUNEL reaction mixture from the in situ cell death detection kit (Roche, Basel, Switzerland) was used to treat the permeabilized tissue sections for 1 h at 37 °C in a humidified atmosphere in the dark. After the end of the TUNEL reaction, the sections were counterstained with DAPI for 5 min.

2.5. Immunohistochemistry

We used 0.01 M PBS in every procedure. After washing with 0.01 M PBS, the retinal sections were blocked in 10% normal donkey serum for 1 h with 0.1% Triton X-100. Subsequently, they were incubated with primary antibodies for 18 h at 4 °C. The sections were washed in PBS and incubated with a secondary antibody for 2 h at room temperature. Cell nuclei were counterstained with DAPI for 5 min. Anti-GRP78 (1:1000; Abcam, Cambridge, UK), anti-ionized calcium binding adaptor molecule 1 (IBA1) (1:500; Wako Chemical, Osaka, Japan), anti-glutamine synthase (GS) (1:1000; Chemicon, Temecula, CA, USA), anti-glial fibrillary acidic protein (GFAP; 1:1000; Chemicon) and Cy3 (1:1500; Jackson Immunoresearch, West Grove, PA, USA) or Alexa 488-conjugated antibodies (1:1000; Molecular Probes, Eugene, OR, USA) were used as secondary antibodies. Images were obtained using a Zeiss LSM 800 confocal microscope (Carl Zeiss Co. Ltd., Oberkochen, Germany).

2.6. Immuno-Electron Microscopy (EM) and CLEM

For immuno-EM and CLEM, we followed the protocol described by Jin et al. [25] with modifications. The retinas were dissected from the eye cup and fixed with 4% PFA for 2 h. Fixed retinas were washed with 0.1 M PB and cryoprotected with 2.3 M sucrose in 0.1 M PB for 24 h. Retinas with 2.3 M sucrose were frozen in liquid nitrogen. Semi-thin cryosections (2 µm) of the frozen retina were obtained at −100 °C using a Leica EM UC7 ultramicrotome equipped with an FC7 cryochamber (Leica). For immunohistochemistry, sections were incubated with 10% normal donkey serum for 1 h without Triton X-100 at room temperature and then double-labeled with GRP78 and GS overnight at 4 °C. To detect GRP78 in the EM, FluoroNanogold-conjugated Alexa 488 goat anti-rabbit (1:100; Nanoprobes, Yaphank, NY, USA) was used as a secondary antibody for 2 h at room temperature. The sections were counterstained with DAPI for 3 min and confocal images were obtained. After confocal microscopy, the sections were washed with 0.1 M PB and post-fixed with 2.5% glutaraldehyde and 1% osmium tetroxide for 30 min. Following post-fixation, silver enhancement was performed using an HQ silver enhancement kit (Nanoprobes) and then sections were dehydrated in graded alcohols. After the completion of all procedures, the tissues were embedded in Epon 812. Interested areas were selected in the confocal image and selected areas were cut into ultrathin sections (80–90 nm) and observed under an electron microscope (JEM 1010, Tokyo, Japan).

2.7. Image Analysis

Confocal images were analyzed using ZEN 2.3 software (Carl Zeiss). The immunohistochemistry intensity was measured by a histogram function and co-localization ratio automatically calculated by ZEN's co-localization function.

2.8. Statistical Analysis

Quantified intensity values were statistically analyzed using Prism 8.0 software (GraphPad, San Diego, CA, USA). A one-way ANOVA was used to determine statistical significance with a *p*-value < 0.05.

3. Results

3.1. GRP78 Expression in a Normal Retina

First, we determined the GRP78 expression pattern and its localization in the normal retina at cellular and subcellular levels using immunohistochemistry and immuno-EM methods. In the normal retina, a strong immunoreactivity of GRP78 was observed in cell bodies in the INL and GCL and in the inner and outer segments (IS/OS) of photoreceptors and the outer plexiform layer (OPL) while weak punctate labeling was observed in the ONL and inner plexiform layer (IPL) (Figure 1A). To examine its localization at the subcellular level, we performed immuno-EM using a 1.4 nm gold-conjugated secondary antibody. In the ONL, gold particles were rarely detected in the cell bodies of the photoreceptor while they were mainly localized in thin cytoplasmic components within the intercellular spaces (Figure 1B), which might be the process of Müller glial cells, which are the main glial cells in the retina. In the INL and GCL, gold particles were distributed through the nuclear membrane (Figure 1C,D) and membranous structures near the nucleus (Figure 1D), which might be the ER.

3.2. Changes in GRP78 Expression in Blue LED-Induced RD

We subsequently examined the GRP78 expression in RD retinas at 24 h and 72 h after blue LED exposure (Figure 2) when photoreceptor degeneration started and peaked, respectively [22–24]. GRP78 immunoreactivity was increased in the ONL and IPL after LED exposure in a time-dependent manner (yellow rectangles in Figure 2B,F,J). In the RD retina 72 h after blue LED exposure, many GRP78-labeled lumps were observed in the ONL (yellow circles in Figure 2J). Considering that GRP78 is mainly localized in a cellular component of the intercellular spaces between photoreceptors in the normal retina (Figure 1B), which might be Müller cell processes, its cellular identity appeared to be Müller cells not photoreceptors. To confirm this point, we performed a double-label immunofluorescence experiment with anti-GRP78 and anti-GS, a representative Müller cell marker. In the normal retina, GS immunoreactivity was found in Müller cells whose bodies were situated in the middle of the INL and processes in the ONL, OPL, IPL and GCL from cell bodies to the outer and inner limiting membranes with a variety of lines in size (Figure 2C). In the RD retinas, GS immunoreactivity was remarkably increased in the ONL and thus it appeared as circles and outwardly extending thick irregular lines (Figure 2G,K,M), which suggested that Müller cell processes became hypertrophied and occupied the ONL where photoreceptors degenerated. In merged images (Figure 2D,H,L), the co-localization of GRP78 with GS in the ONL and INL was observed in approximately 23% and 41% of the normal retina, respectively, and significantly increased to 42% and 66%, respectively, in RD at 72 h after blue LED exposure ($p < 0.05$, $n = 7$ in Figure 2M,N).

However, GRP78-labeled lumps in the ONL in RD at 72 h after blue LED exposure did not show GS immunoreactivity (Figure 2L, yellow circles) suggesting that they belonged to different cellular profiles. As microglial cells occurred in similar locations in this RD model with the same morphology [22–24], we tested whether GRP78 was expressed in microglial cells in RD retinas by double-labeling with anti-GRP78 and anti-IBA1, a microglial cell marker. In the normal retina, IBA1-labeled microglial cells were mainly observed in the IPL and OPL (red in Figure 3A). However, in RD induced by blue LED exposure, IBA1-labeled microglial cells were mainly found in the ONL and inner and outer segment layers of the photoreceptors (red in Figure 3B,C). These microglial cells showed weak GRP78 immunoreactivity in their cell bodies and processes in normal retinas (Figure 3A) while it was increased in activated microglial cells, which migrated into the ONL where photoreceptors degenerated followed by the time course of RD (Figure 3B–C). In addition, GRP78-labeled lumps in the ONL observed in RD at 72 h after blue LED exposure were co-labeled with IBA1 (Figure 3C). These results indicated that resting and activated microglial cells expressed GRP78 in the normal and RD retinas and GRP78 expression was increased as microglial cells were activated in RD.

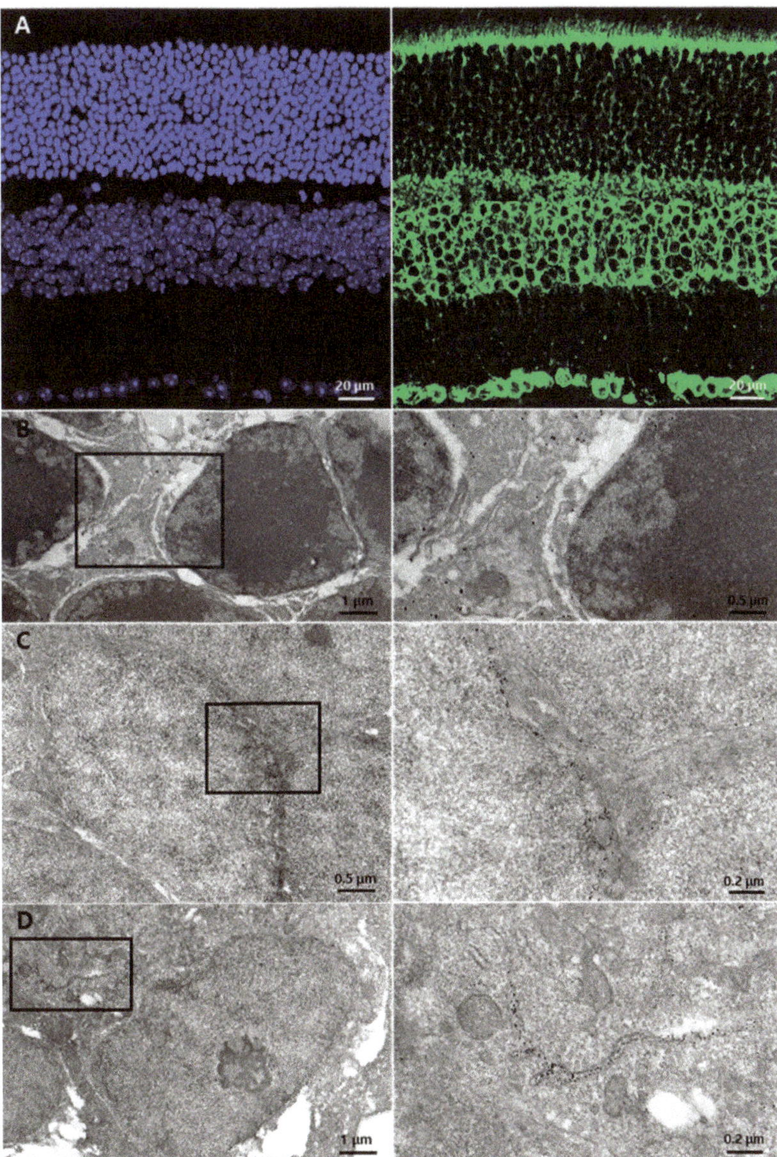

Figure 1. GRP78 expression in normal retinas: (**A**) Representative normal retina labeled with DAPI (blue) and GRP78 (green). GRP78 was mainly labeled in the rod and cone layers and the cell bodies of the INL and GCL. In the ONL, the photoreceptor cell bodies were weakly surrounded by GRP78. (**B–D**) Representative EM images of the normal retina with immuno-gold-labeled GRP78. (**B**) Normal photoreceptor cells in ONL. Immuno-gold labeled-GRP78 was detected in the interspace of the photoreceptor cell bodies but not in the photoreceptor cells. (**C**) Cell bodies in INL. Immuno-gold-labeled GRP78 was detected in the perinuclear spaces of INL cells. (**D**) Representative EM images of GCL. Immuno-gold-labeled GRP78 was detected in the perinuclear spaces and ER lumen of the GC. INL, inner nuclear layer; GCL, ganglion cell layer; ONL, outer nuclear layer; EM, electron microscopy; ER, endoplasmic reticulum; GC, ganglion cell.

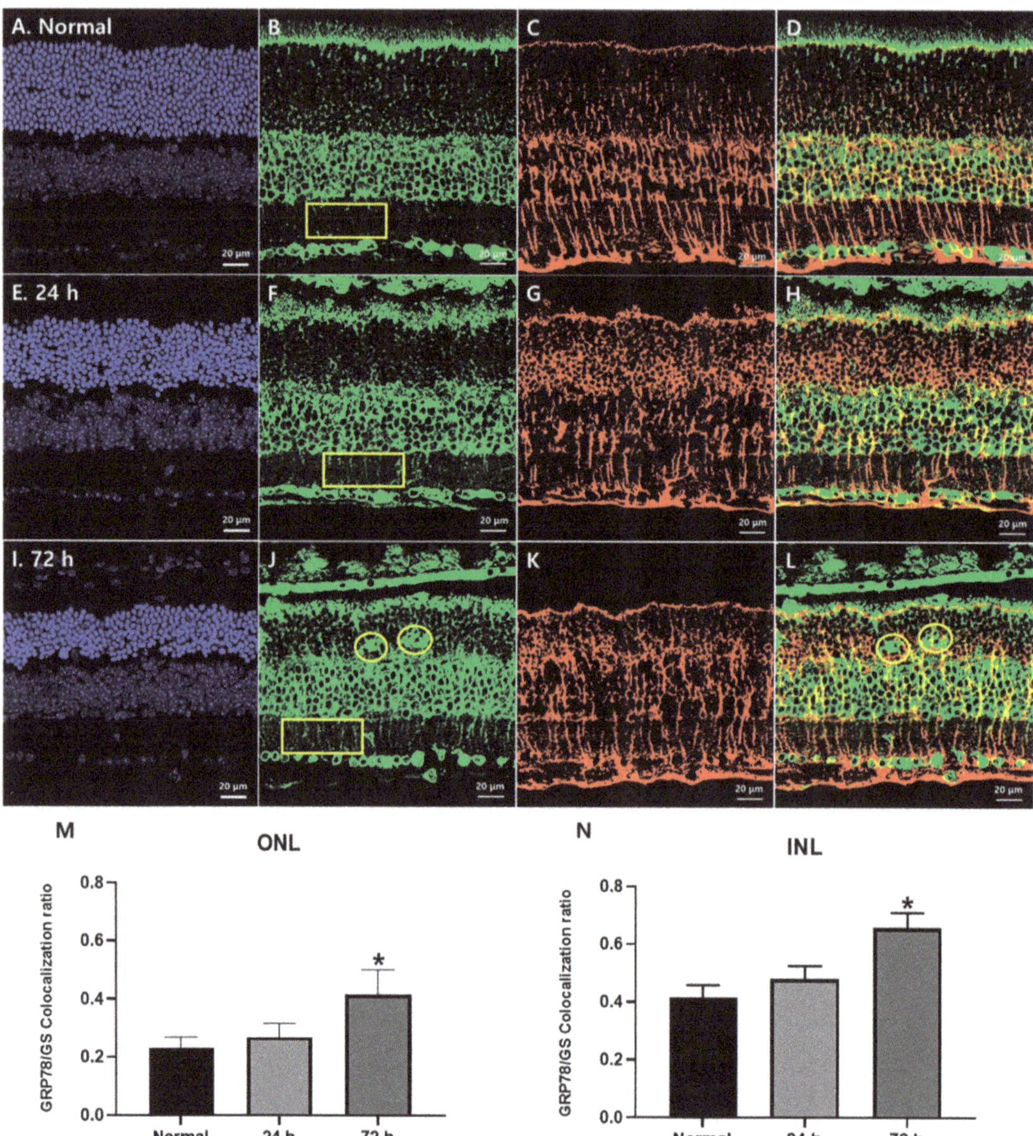

Figure 2. GRP78 expression in Müller glial cells in blue LED-induced RD retinas. (**A,E,I**) DAPI (blue) stained retina sections of the normal (**A**), 24 h (**E**) and 72 h (**I**) retinas after blue LED exposure. ONL thickness prominently decreased at 72 h. (**B,F,J**) GRP78 (green)-labeled sections of normal (**B**), 24 h (**F**) and 72 h (**J**) retinas after blue LED exposure. GRP78 was increased in the ONL and IPL in a time-dependent manner after LED exposure (yellow rectangles) and labeled in giant cells in the ONL (yellow circles). (**C,G,K**) GS (red)-labeled retina sections of the normal (**C**), 24 h (**G**) and 72 h (**K**) retinas after blue LED exposure. GS-labeled Müller cell processes enlarged and encircled photoreceptor cells in the ONL after injury. (**D,H,L**) Merged results of GRP78 and GS. GRP78 was increased in the Müller cell processes from IPL to ONL in a time-dependent manner while GRP78-labeled giant cells in the OPL showed no GS immunoreactivity (yellow circles). (**M,N**) Co-localization ratio of the GRP78/GS in the ONL (**M**) and IPL (**N**). In both the ONL and INL, the GRP78 co-localization ratio in GS-labeled Müller glial cells was significantly increased at 72 h after LED exposure ($n = 7$, $p < 0.05$ (*), ANOVA).

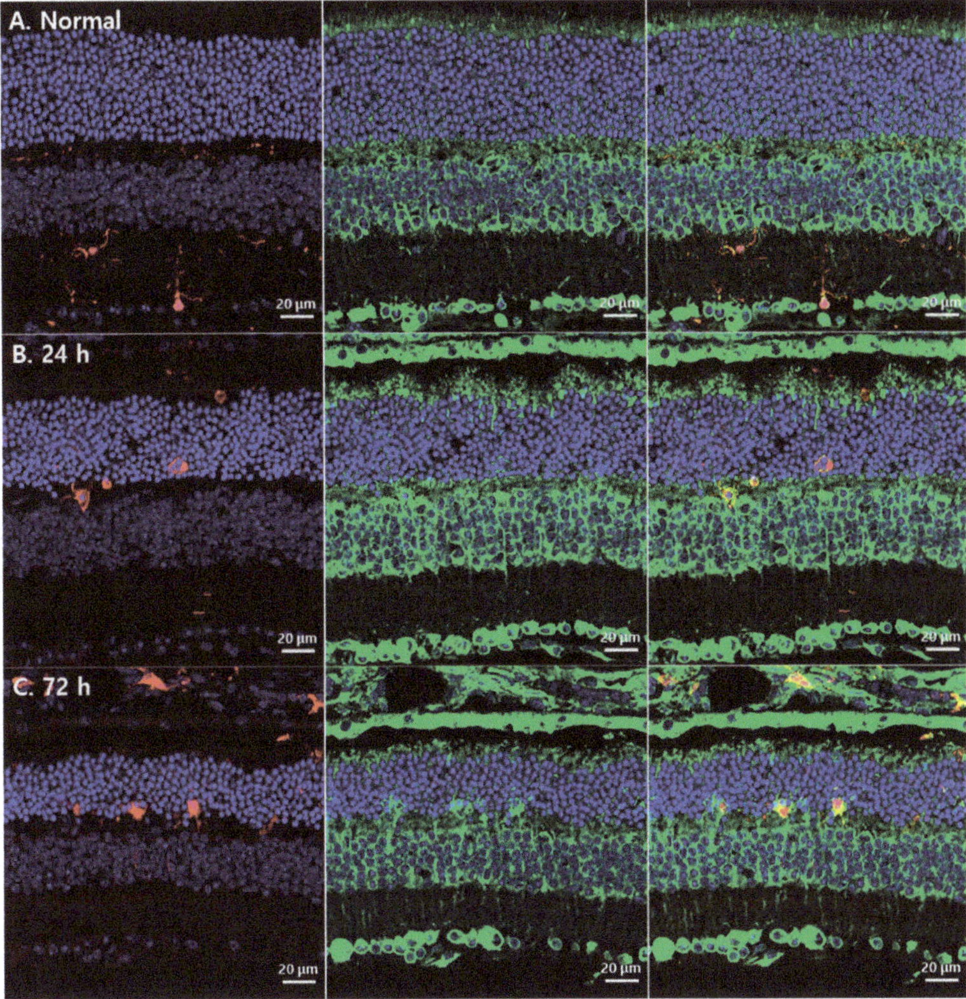

Figure 3. GRP78 expression in microglial cells in blue LED-induced RD retinas. (**A**) A representative confocal image of the normal retina labeled by DAPI (blue), GRP78 (green) and IBA1 (red). IBA1-labeled microglial cells were weakly labeled by GRP78 in the normal retina. (**B**). After 24 h of light exposure, microglial cells were detected in the ONL with an enlarged shape and increased GRP78 compared with the normal retina. (**C**) After 72 h of light exposure, microglial cells were markedly increased in the ONL and GRP78 was prominently increased in microglial cells compared with those after 24 h.

3.3. Subcellular Localization of GRP78 in Glial Cells of the Retina

GRP78 is known to localize various subcellular components including the cell membrane, nuclear membrane and cytosol [26] according to cell type. To elucidate the functional role of GRP78 in two retinal glial cells, Müller cells and microglia, we employed CLEM, an advanced EM technique combined with immunohistochemistry to determine the subcellular localization (Figure 4). As it is difficult to delineate the Müller cell processes in the INL and IPL without markers, we performed double-labeling with an anti-GS/Cy3-conjugated secondary antibody and an anti-GRP78/Alexa488-FluoroNanogold-conjugated secondary antibody in semi-thin cryosections of the RD retina. Under confocal microscopy, GS/Cy3-labeled Müller cell processes that were apparently co-labeled with GRP78/Alexa 488-FluoroNanogold were identified and selected for EM observation.

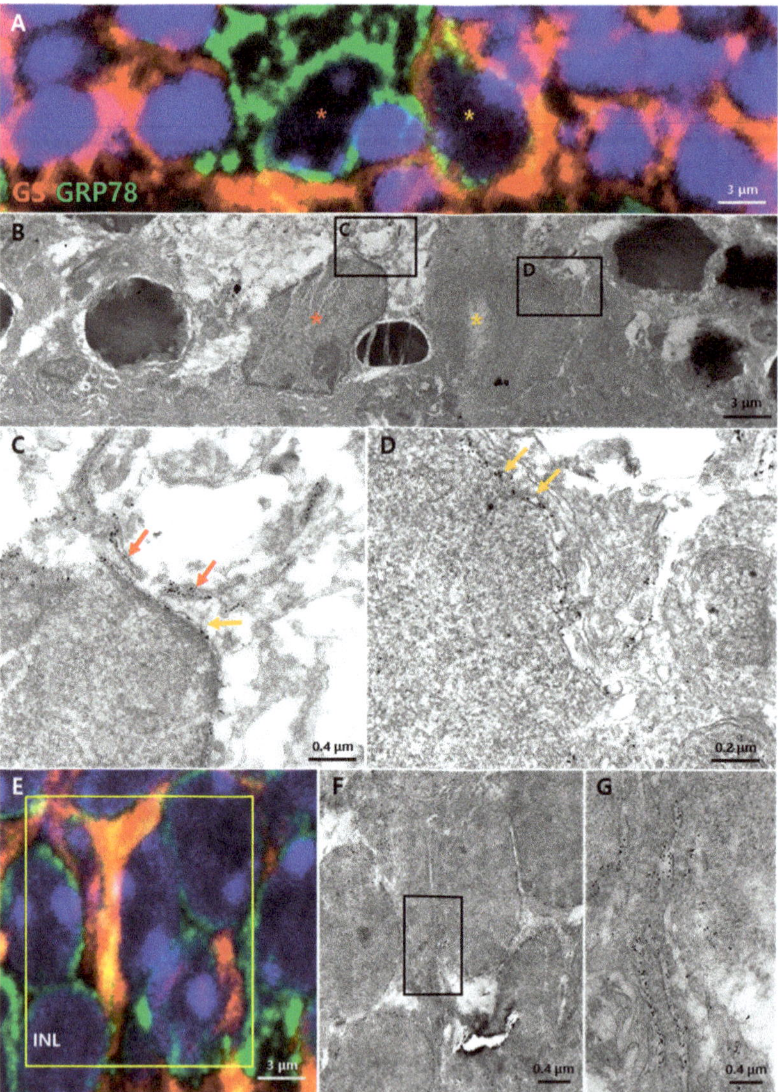

Figure 4. Subcellular GRP78 localization in the injured retina 72 h after light exposure. Semi-thin sections of the retina stained by GRP78 (green, gold), GS (red) and DAPI (blue). (**A**) Microglia and migrated Müller glia in the ONL. Microglia (red asterisk) were more strongly labeled by GRP78 than Müller glia without GS while Müller glia (orange asterisk) were labeled by both GRP78 and GS. (**B**) EM images of microglia (red asterisk) and migrated Müller glia (orange asterisk) in the same region as (**A**). (**C**) Immuno-gold-labeled GRP78 was detected in the perinuclear space (orange arrow) and lumen of the ER structure (red arrows) of the microglia. (**D**) Immuno-gold-labeled GRP78 was detected in the perinuclear space of the Müller glial cell (orange arrows). (**E**) Confocal images in the INL of an injured retina with Müller glia processes. (**F**) EM images matched with the yellow box in (**E**). Processes of the Müller glia in the INL labeled by GRP78. (**G**) A higher magnification of the Müller glia process in (**F**). Membranous structures containing immuno-gold-labeled GRP78 located in Müller glia processes.

In the confocal (Figure 4A) and its correlated EM (Figure 4B) images taken from the ONL in a RD retina, two types of retinal glial cells were identified. One type was GRP78 single-labeled microglial cells without GS immunoreactivity (red asterisk in Figure 4A), which showed typical microglial cell morphology characterized by irregular nuclei with a characteristic peripheral heterochromatin and heterochromatin net (red asterisk in Figure 4B). The other was GRP78/GS double-labeled Müller glial cells (orange asterisk in Figure 4A), which had irregular nuclei in shape that were apparently distinguished from neighboring photoreceptors and microglial cells (orange asterisk in Figure 4B). In higher magnification EM views of the microglial cells (Figure 4C) and Müller cells (Figure 4D) in Figure 4B, the immuno-gold particles of GRP78 were localized to the nuclear membrane, membranous structures in the perinuclear region (orange arrows) and ER (red arrows) in the perinuclear region.

In the INL, GRP78/GS double-labeled Müller cell processes were frequently observed by confocal microscopy (Figure 4E). In correlative EM (Figure 4F,G), the immuno-gold particles of GRP78 were found in an elongated membranous structure suggesting that GRP78 was expressed in newly formed ER in hypertrophied Müller glial cells in the RD retina.

4. Discussion

We determined the cellular and subcellular localization of GRP78 in normal and blue LED-induced RD retinas. GRP78 is expressed in various types of retinal neurons and glial cells. In RD, GRP78 was not increased in the damaged photoreceptors, resulting in photoreceptor apoptosis while it was increased in Müller and microglial cells together with the development of a new ER.

UPR related to ER stress is expected to be a promising target for the treatment of various degenerative diseases because it is involved in cell survival and death and thus controls cell fate [27–29]. Among all UPR products, GRP78 belongs to the heat shock protein 70 family (HSP70), which deals with damaged or misfolded proteins to protect cells [17,18]. In previous studies [13,19–21], the protective role of GRP78 in response to various retinal injuries was proposed. For instance, that role against ganglion cell injury has been relatively well described [21] while GRP78's role in photoreceptor degeneration remains unclear. Moreover, it is not clear whether photoreceptors in the normal retina express GRP78. Previous studies have shown GRP78 expression in the normal 661 W cell line, which was derived from mouse cone cells [13] while other studies have shown that GRP78 is rarely expressed in the ONL in the retinas [19] of normal and P23H rats [30] and ER stress-activated indicator (ERAI) transgenic mice [20]. We therefore wanted to describe the exact cellular and subcellular localization of GRP78 in photoreceptors.

In this study, we confirmed GRP78 expression in photoreceptors in the normal retina, albeit at low levels. At the subcellular level, it was localized to membranous structures. This expression and localization pattern was observed in RD photoreceptors but did not change (Figure 1 and Supplementary Figure S1). These results appeared to be inconsistent with previous studies showing that GRP78 expression was increased in the photoreceptor cells in RD suggesting a protective role in RD by the inhibition of photoreceptor apoptosis [13,19]. First, this discrepancy might be caused by in vitro RD conditions [13,19] in which only a 661 W cell, a transformed cell line derived from the cone cell, exists and it cannot interact with the retinal pigment epithelium and/or rod and/or Müller glial cells. In addition, the finding of increased GRP78 in the ONL being due to photoreceptors might be falsely attributed [13,19]. Actually, increased GRP78 expression occurred in Müller cell processes that occupied a narrow interphotoreceptor space (Figures 2 and 4), not in the photoreceptors in RD (Supplementary Figure S1) as demonstrated in this study. These findings suggested that GRP78 was not directly involved in photoreceptor cell death. Instead of GRP78, apoptotic UPR pathways might be activated during photoreceptor apoptosis in RD. CHOP may be a strong candidate for cell death and is already known to be expressed in photoreceptors and plays an essential role in photoreceptor apoptosis via ER stress in RD [31,32].

Glial cells play an important role in various types of brain injury. In particular, two glial cells, microglia and astrocytes, are involved in neuroinflammation. Each type is divided into two subpopulations (M1/M2 and A1/A2) depending on its function and with an opposing role: pro-inflammatory vs. anti-inflammatory [33,34], which eventually results in neurodegeneration or neuroprotection. In neuroinflammation related to various brain pathologies, ER stress contributes to determining the phenotypes of microglia [35,36] and astrocytes [37,38]. It has been reported that GRP78 induces phagocytic function and cytokine production in microglial cells [39] and activates astrocytes to promote neuroprotective cytokines [40]. Moreover, it protects astrocytes themselves under stress conditions [41].

In this study, we demonstrated that GRP78 expression was increased in two types of retinal glial cells, Müller cells and microglial cells, in a blue LED-induced RD model. These results were consistent with the increase in GRP78 expression in glial cells in response to several types of brain and spinal cord injuries [25,42–44]. Müller cells are the main glial cells in the retina, corresponding to astrocytes in the brain and spinal cord. As GRP78 could activate astrocytes to promote neuroprotective cytokines [40], increased GRP78 in Müller cells might have had a neuroprotective role in this RD model. Moreover, Müller cells are responsible for retinal metabolism through the UPR as demonstrated in diabetic retinopathy [45–48]. Considering that GRP78 belongs to the HSP70 family that deals with damaged or misfolded proteins to protect cells [17,18], GRP78 in Müller cells may be involved in retinal protection as a key component of UPR. In addition, a prominent GRP78 increase in the microglia occurred 72 h after blue LED exposure when microglial activation peaked [22,23]. This suggested that GRP78 was closely associated with microglia activation, which might phagocytose and clear degenerating photoreceptors [39]. However, the exact role of GRP78 and its mechanism of action in retinal glial cells should be further investigated in a study on GRP78 in Müller cells or microglia conditional knockout animals.

GRP78 is an ER chaperone and thus was used as a representative ER stress marker. At the subcellular level, it was localized to membranous structures including the nuclear membrane and putative ER in retinal neurons and glial cells (Figures 1 and 4, Supplementary Figure S1). In particular, activated Müller cell processes in blue LED-induced RD retinas showed an increased GRP78 expression with ER development (Figure 4). We confirmed this by double-labeling immunofluorescence with anti-GFAP, an activated Müller cell marker [49,50], and anti-calreticulin, a representative ER marker [51,52]. The expression of calreticulin became evident in GFAP-labeled Müller cell processes in the IPL and INL (Supplementary Figure S2). Based on the fact that Müller cells are responsible for metabolic modulation in the degenerative retina [45] and that it engulfs and clears the damaged photoreceptor [53,54], increased GRP78 within newly developed ER in Müller cells might therefore be essential to resolve the overloaded metabolic needs by preventing the accumulation of misfolded proteins. In addition, activated microglial cells showed higher GRP78 levels than inactivated ones (Figure 3) and calreticulin-immunoreactive lumps similar to GRP78-immunoreactive lumps in activated microglial cells in the ONL in RD retina were found at the same time and location (Supplementary Figure S2) suggesting the development of a new ER. This might be reasonable because microglia could be in an overloaded condition to produce the cytokines and phagocyte the damaged photoreceptors in RD.

Finally, we want to refer to GRP78 at the mitochondria and mitochondria-associated ER membrane. Recently, growing evidence has pointed towards GRP78 playing an important function in the mitochondria in association with co-chaperones known to be involved in calcium-mediated signaling between the ER and mitochondria, important for bioenergetics and cell survival [55]. As previous reports detected GRP78 in the mitochondria [56,57] and mitochondria-associated ER membrane [58], we tried to detect mitochondrial GRP78 localization in photoreceptors and glial cells by EM. However, immuno-gold-labeled GRP78 was rarely detected in the mitochondria either in photoreceptors or glial cells. There was also no distinguishable ER region of a strong GRP78 expression close to the mitochondria (data not shown). Nevertheless, this issue needs to be further investigated in the near future.

5. Conclusions

In summary, we determined the cellular and subcellular localization of GRP78 in normal and blue LED-induced RD retinas. GRP78 was expressed in various types of retinal neurons and glial cells. In RD, GRP78 was not increased in the damaged photoreceptors, resulting in photoreceptor apoptosis while it was increased in Müller and microglial cells together with the development of a new ER. These results suggested that increased GRP78 played a role in glial cell activation and neuroprotective function by modulating the UPR under stress conditions. Further studies to reveal the relationship between GRP78 and ER stress and glial cell responses in RD and its mechanism are needed, considering that the findings in this study were limited to the histology.

Supplementary Materials: The following are available online at https://www.mdpi.com/article/10.3390/cells10050995/s1, Figure S1: Subcellular localization of the GRP78 in the ONL of the normal retina and blue LED-induced RD retina. Figure S2: Expression of the calreticulin and GFAP in a blue LED-induced RD retina.

Author Contributions: Conceptualization, Y.S.P. and I.-B.K.; methodology, Y.S.P., H.-L.K., S.H.L. and Y.Z.; software, Y.S.P.; validation, Y.S.P. and I.-B.K.; formal analysis, Y.S.P. and I.-B.K.; investigation, Y.S.P., H.-L.K., S.H.L. and Y.Z. and I.-B.K.; resources, Y.S.P. and I.-B.K.; data curation, Y.S.P. and I.-B.K.; writing—original draft preparation, Y.S.P.; writing—review and editing, I.-B.K.; visualization, Y.S.P.; supervision, I.-B.K.; project administration, I.-B.K.; funding acquisition, I.-B.K. All authors have read and agreed to the published version of the manuscript.

Funding: This research was funded by the Basic Science Research Program through the National Research Foundation (NRF) of Korea funded by the Ministry of Education, Science and Technology (grant number 2017R1A2B2005309).

Institutional Review Board Statement: The study was conducted in accordance with the guidelines of the Declaration of Helsinki and approved by the Institutional Animal Care and Use Committee (IACUC) at the College of Medicine, The Catholic University of Korea (Approval number: CUMS-2017-0241-03).

Informed Consent Statement: Not applicable.

Data Availability Statement: Data is contained within the article or supplementary material.

Conflicts of Interest: The authors declare no conflict of interest.

References

1. Wright, A.F.; Chakarova, C.F.; Abd El-Aziz, M.M.; Bhattacharya, S.S. Photoreceptor degeneration: Genetic and mechanistic dissection of a complex trait. *Nat. Rev. Genet.* **2010**, *11*, 273–284. [CrossRef] [PubMed]
2. Handa, J.T.; Bowes Rickman, C.; Dick, A.D.; Gorin, M.B.; Miller, J.W.; Toth, C.A.; Ueffing, M.; Zarbin, M.; Farrer, L.A. A systems biology approach towards understanding and treating non-neovascular age-related macular degeneration. *Nat. Commun.* **2019**, *10*, 3347. [CrossRef]
3. Hernandez-Zimbron, L.F.; Zamora-Alvarado, R.; Ochoa-De la Paz, L.; Velez-Montoya, R.; Zenteno, E.; Gulias-Canizo, R.; Quiroz-Mercado, H.; Gonzalez-Salinas, R. Age-related macular degeneration: New paradigms for treatment and management of AMD. *Oxid. Med. Cell. Longev.* **2018**, *2018*, 8374647. [CrossRef] [PubMed]
4. Falavarjani, K.G.; Nguyen, Q.D. Adverse events and complications associated with intravitreal injection of anti-VEGF agents: A review of literature. *Eye* **2013**, *27*, 787–794. [CrossRef] [PubMed]
5. Schargus, M.; Frings, A. Issues with intravitreal administration of anti-VEGF drugs. *Clin. Ophthalmol.* **2020**, *14*, 897–904. [CrossRef] [PubMed]
6. Zając-Pytrus, H.M.; Pilecka, A.; Turno-Kręcicka, A.; Adamiec-Mroczek, J.; Misiuk-Hojło, M. The dry form of Age-Related Macular Degeneration (AMD): The Current concepts of pathogenesis and prospects for treatment. *Adv. Clin. Exp. Med.* **2015**, *24*, 1099–1104. [CrossRef]
7. Verkhratsky, A.; Toescu, E.C. Endoplasmic reticulum Ca^{2+} homeostasis and neuronal death. *J. Cell. Mol. Med.* **2003**, *7*, 351–361. [CrossRef]

8. Lindholm, D.; Wootz, H.; Korhonen, L. ER stress and neurodegenerative diseases. *Cell Death Differ.* **2006**, *13*, 385–392. [CrossRef]
9. Sano, R.; Reed, J.C. ER stress-induced cell death mechanisms. *Biochim. Biophys. Acta (BBA) Mol. Cell Res.* **2013**, *1833*, 3460–3470. [CrossRef]
10. Nishitoh, H. CHOP is a multifunctional transcription factor in the ER stress response. *J. Biochem.* **2012**, *151*, 217–219. [CrossRef]
11. Kroeger, H.; Chiang, W.-C.; Felden, J.; Nguyen, A.; Lin, J.H. ER stress and unfolded protein response in ocular health and disease. *FEBS J.* **2019**, *286*, 399–412. [CrossRef]
12. Libby, R.T.; Gould, D.B. Endoplasmic reticulum stress as a primary pathogenic mechanism leading to age-related macular degeneration. *Adv. Exp. Med. Biol.* **2010**, *664*, 403–409. [CrossRef]
13. Nakanishi, T.; Shimazawa, M.; Sugitani, S.; Kudo, T.; Imai, S.; Inokuchi, Y.; Tsuruma, K.; Hara, H. Role of endoplasmic reticulum stress in light-induced photoreceptor degeneration in mice. *J. Neurochem.* **2013**, *125*, 111–124. [CrossRef] [PubMed]
14. Kroeger, H.; Messah, C.; Ahern, K.; Gee, J.; Joseph, V.; Matthes, M.T.; Yasumura, D.; Gorbatyuk, M.S.; Chiang, W.C.; LaVail, M.M.; et al. Induction of endoplasmic reticulum stress genes, BiP and chop, in genetic and environmental models of retinal degeneration. *Investig. Ophthalmol. Vis. Sci.* **2012**, *53*, 7590–7599. [CrossRef] [PubMed]
15. Rana, T.; Shinde, V.M.; Starr, C.R.; Kruglov, A.A.; Boitet, E.R.; Kotla, P.; Zolotukhin, S.; Gross, A.K.; Gorbatyuk, M.S. An activated unfolded protein response promotes retinal degeneration and triggers an inflammatory response in the mouse retina. *Cell Death Dis.* **2014**, *5*, e1578. [CrossRef] [PubMed]
16. Mendes, C.S.; Levet, C.; Chatelain, G.; Dourlen, P.; Fouillet, A.; Dichtel-Danjoy, M.L.; Gambis, A.; Ryoo, H.D.; Steller, H.; Mollereau, B. ER stress protects from retinal degeneration. *EMBO J.* **2009**, *28*, 1296–1307. [CrossRef]
17. Lee, A.S. The ER chaperone and signaling regulator GRP78/BiP as a monitor of endoplasmic reticulum stress. *Methods* **2005**, *35*, 373–381. [CrossRef]
18. Pfaffenbach, K.T.; Lee, A.S. The critical role of GRP78 in physiologic and pathologic stress. *Curr. Opin. Cell Biol.* **2011**, *23*, 150–156. [CrossRef]
19. Liu, H.; Qian, J.; Wang, F.; Sun, X.; Xu, X.; Xu, W.; Zhang, X.; Zhang, X. Expression of two endoplasmic reticulum stress markers, GRP78 and GADD153, in rat retinal detachment model and its implication. *Eye* **2010**, *24*, 137–144. [CrossRef]
20. Shimazawa, M.; Inokuchi, Y.; Ito, Y.; Murata, H.; Aihara, M.; Miura, M.; Araie, M.; Hara, H. Involvement of ER stress in retinal cell death. *Mol. Vis.* **2007**, *13*, 578–587.
21. Inokuchi, Y.; Nakajima, Y.; Shimazawa, M.; Kurita, T.; Kubo, M.; Saito, A.; Sajiki, H.; Kudo, T.; Aihara, M.; Imaizumi, K.; et al. Effect of an inducer of BiP, a molecular chaperone, on endoplasmic reticulum (ER) stress-induced retinal cell death. *Investig. Ophthalmol. Vis. Sci.* **2009**, *50*, 334–344. [CrossRef]
22. Kim, G.H.; Paik, S.S.; Park, Y.S.; Kim, H.G.; Kim, I.B. Amelioration of mouse retinal degeneration after blue LED Exposure by glycyrrhizic acid-mediated inhibition of inflammation. *Front. Cell. Neurosci.* **2019**, *13*, 319. [CrossRef]
23. Kim, G.H.; Kim, H.I.; Paik, S.S.; Jung, S.W.; Kang, S.; Kim, I.B. Functional and morphological evaluation of blue light-emitting diode-induced retinal degeneration in mice. *Graefes Arch. Clin. Exp. Ophthalmol.* **2016**, *254*, 705–716. [CrossRef]
24. Chang, S.W.; Kim, H.I.; Kim, G.H.; Park, S.J.; Kim, I.B. Increased expression of osteopontin in retinal degeneration induced by blue light-emitting diode exposure in mice. *Front. Mol. Neurosci.* **2016**, *9*, 58. [CrossRef]
25. Jin, X.; Riew, T.-R.; Kim, H.L.; Kim, S.; Lee, M.-Y. Spatiotemporal expression of GRP78 in the blood vessels of rats treated with 3-nitropropionic acid correlates with blood–brain barrier disruption. *Front. Cell. Neurosci.* **2018**, *12*. [CrossRef]
26. Ibrahim, I.M.; Abdelmalek, D.H.; Elfiky, A.A. GRP78: A cell's response to stress. *Life Sci.* **2019**, *226*, 156–163. [CrossRef]
27. Hiramatsu, N.; Chiang, W.-C.; Kurt, T.D.; Sigurdson, C.J.; Lin, J.H. Multiple mechanisms of unfolded protein response–induced cell death. *Am. J. Pathol.* **2015**, *185*, 1800–1808. [CrossRef] [PubMed]
28. Lai, E.; Teodoro, T.; Volchuk, A. Endoplasmic reticulum stress: Signaling the unfolded protein response. *Physiology* **2007**, *22*, 193–201. [CrossRef]
29. Hetz, C.; Papa, F.R. The unfolded protein response and cell fate control. *Mol. Cell* **2018**, *69*, 169–181. [CrossRef]
30. Gorbatyuk, M.S.; Knox, T.; LaVail, M.M.; Gorbatyuk, O.S.; Noorwez, S.M.; Hauswirth, W.W.; Lin, J.H.; Muzyczka, N.; Lewin, A.S. Restoration of visual function in P23H rhodopsin transgenic rats by gene delivery of BiP/Grp78. *Proc. Nat. Acad. Sci. USA* **2010**, *107*, 5961. [CrossRef] [PubMed]
31. Wang, S.; Liu, Y.; Tan, J.W.; Hu, T.; Zhang, H.F.; Sorenson, C.M.; Smith, J.A.; Sheibani, N. Tunicamycin-induced photoreceptor atrophy precedes degeneration of retinal capillaries with minimal effects on retinal ganglion and pigment epithelium cells. *Exp. Eye Res.* **2019**, *187*, 107756. [CrossRef]
32. Zhu, H.; Qian, J.; Wang, W.; Yan, Q.; Xu, Y.; Jiang, Y.; Zhang, L.; Lu, F.; Hu, W.; Zhang, X.; et al. RNA interference of GADD153 protects photoreceptors from endoplasmic reticulum stress-mediated apoptosis after retinal detachment. *PLoS ONE* **2013**, *8*, e59339. [CrossRef]
33. Orihuela, R.; McPherson, C.A.; Harry, G.J. Microglial M1/M2 polarization and metabolic states. *Br. J. Pharmacol.* **2016**, *173*, 649–665. [CrossRef]
34. Liddelow, S.A.; Guttenplan, K.A.; Clarke, L.E.; Bennett, F.C.; Bohlen, C.J.; Schirmer, L.; Bennett, M.L.; Münch, A.E.; Chung, W.S.; Peterson, T.C.; et al. Neurotoxic reactive astrocytes are induced by activated microglia. *Nature* **2017**, *541*, 481–487. [CrossRef]
35. Salminen, A.; Kaarniranta, K.; Kauppinen, A. ER stress activates immunosuppressive network: Implications for aging and Alzheimer's disease. *J. Mol. Med.* **2020**, *98*, 633–650. [CrossRef]

36. Wang, Y.-W.; Zhou, Q.; Zhang, X.; Qian, Q.-Q.; Xu, J.-W.; Ni, P.-F.; Qian, Y.-N. Mild endoplasmic reticulum stress ameliorates lipopolysaccharide-induced neuroinflammation and cognitive impairment via regulation of microglial polarization. *J. Neuroinflamm.* **2017**, *14*, 233. [CrossRef]
37. Smith, H.L.; Freeman, O.J.; Butcher, A.J.; Holmqvist, S.; Humoud, I.; Schätzl, T.; Hughes, D.T.; Verity, N.C.; Swinden, D.P.; Hayes, J.; et al. Astrocyte unfolded protein response induces a specific reactivity state that causes non-cell-autonomous neuronal degeneration. *Neuron* **2020**, *105*, 855–866.e5. [CrossRef]
38. Meares, G.P.; Liu, Y.; Rajbhandari, R.; Qin, H.; Nozell, S.E.; Mobley, J.A.; Corbett, J.A.; Benveniste, E.N. PERK-dependent activation of JAK1 and STAT3 contributes to endoplasmic reticulum stress-induced inflammation. *Mol. Cell Biol.* **2014**, *34*, 3911–3925. [CrossRef]
39. Kakimura, J.; Kitamura, Y.; Taniguchi, T.; Shimohama, S.; Gebicke-Haerter, P.J. Bip/GRP78-induced production of cytokines and uptake of amyloid-beta(1-42) peptide in microglia. *Biochem. Biophys. Res. Commun.* **2001**, *281*, 6–10. [CrossRef] [PubMed]
40. Qian, Y.; Zheng, Y.; Weber, D.; Tiffany-Castiglioni, E. A 78-kDa glucose-regulated protein is involved in the decrease of interleukin-6 secretion by lead treatment from astrocytes. *Am. J. Physiol. Cell Physiol.* **2007**, *293*, C897–C905. [CrossRef] [PubMed]
41. Ouyang, Y.-B.; Xu, L.-J.; Emery, J.F.; Lee, A.S.; Giffard, R.G. Overexpressing GRP78 influences Ca2+ handling and function of mitochondria in astrocytes after ischemia-like stress. *Mitochondrion* **2011**, *11*, 279–286. [CrossRef] [PubMed]
42. Jin, X.; Kim, D.K.; Riew, T.R.; Kim, H.L.; Lee, M.Y. Cellular and subcellular localization of endoplasmic reticulum chaperone GRP78 following transient focal cerebral ischemia in rats. *Neurochem. Res.* **2018**, *43*, 1348–1362. [CrossRef] [PubMed]
43. Alberdi, E.; Wyssenbach, A.; Alberdi, M.; Sánchez-Gómez, M.V.; Cavaliere, F.; Rodríguez, J.J.; Verkhratsky, A.; Matute, C. Ca(2+)-dependent endoplasmic reticulum stress correlates with astrogliosis in oligomeric amyloid β-treated astrocytes and in a model of Alzheimer's disease. *Aging. Cell* **2013**, *12*, 292–302. [CrossRef] [PubMed]
44. Fan, H.; Tang, H.B.; Kang, J.; Shan, L.; Song, H.; Zhu, K.; Wang, J.; Ju, G.; Wang, Y.Z. Involvement of endoplasmic reticulum stress in the necroptosis of microglia/macrophages after spinal cord injury. *Neuroscience* **2015**, *311*, 362–373. [CrossRef] [PubMed]
45. Kelly, K.; Wang, J.J.; Zhang, S.X. The unfolded protein response signaling and retinal Müller cell metabolism. *Neural. Regen. Res.* **2018**, *13*, 1861–1870. [CrossRef] [PubMed]
46. Devi, T.S.; Lee, I.; Hüttemann, M.; Kumar, A.; Nantwi, K.D.; Singh, L.P. TXNIP links innate host defense mechanisms to oxidative stress and inflammation in retinal Muller glia under chronic hyperglycemia: Implications for diabetic retinopathy. *Exp. Diabetes Res.* **2012**, *2012*, 438238. [CrossRef]
47. Wu, M.; Yang, S.; Elliott, M.H.; Fu, D.; Wilson, K.; Zhang, J.; Du, M.; Chen, J.; Lyons, T. Oxidative and endoplasmic reticulum stresses mediate apoptosis induced by modified LDL in human retinal Müller cells. *Investig. Ophthalmol. Vis. Sci.* **2012**, *53*, 4595–4604. [CrossRef]
48. Zhong, Y.; Li, J.; Chen, Y.; Wang, J.J.; Ratan, R.; Zhang, S.X. Activation of endoplasmic reticulum stress by hyperglycemia is essential for Müller cell-derived inflammatory cytokine production in diabetes. *Diabetes* **2012**, *61*, 492–504. [CrossRef]
49. Eisenfeld, A.J.; Bunt-Milam, A.H.; Sarthy, P.V. Müller cell expression of glial fibrillary acidic protein after genetic and experimental photoreceptor degeneration in the rat retina. *Investig. Ophthalmol. Vis. Sci.* **1984**, *25*, 1321–1328.
50. Chang, M.-L.; Wu, C.-H.; Jiang-Shieh, Y.-F.; Shieh, J.-Y.; Wen, C.-Y. Reactive changes of retinal astrocytes and Müller glial cells in kainate-induced neuroexcitotoxicity. *J. Anat.* **2007**, *210*, 54–65. [CrossRef]
51. Müller-Taubenberger, A.; Lupas, A.N.; Li, H.; Ecke, M.; Simmeth, E.; Gerisch, G. Calreticulin and calnexin in the endoplasmic reticulum are important for phagocytosis. *EMBO J.* **2001**, *20*, 6772–6782. [CrossRef] [PubMed]
52. Johnson, S.; Michalak, M.; Opas, M.; Eggleton, P. The ins and outs of calreticulin: From the ER lumen to the extracellular space. *Trends Cell Biol.* **2001**, *11*, 122–129. [CrossRef]
53. Bejarano-Escobar, R.; Sánchez-Calderón, H.; Otero-Arenas, J.; Martín-Partido, G.; Francisco-Morcillo, J. Müller glia and phagocytosis of cell debris in retinal tissue. *J. Anat.* **2017**, *231*, 471–483. [CrossRef] [PubMed]
54. Sakami, S.; Imanishi, Y.; Palczewski, K. Müller glia phagocytose dead photoreceptor cells in a mouse model of retinal degenerative disease. *FASEB J.* **2019**, *33*, 3680–3692. [CrossRef] [PubMed]
55. Casas, C. GRP78 at the centre of the stage in cancer and neuroprotection. *Front. Neurosci.* **2017**, *11*. [CrossRef] [PubMed]
56. Suzuki, C.K.; Bonifacino, J.S.; Lin, A.Y.; Davis, M.M.; Klausner, R.D. Regulating the retention of T-cell receptor alpha chain variants within the endoplasmic reticulum: Ca(2+)-dependent association with BiP. *J. Cell Biol.* **1991**, *114*, 189–205. [CrossRef]
57. Sun, F.C.; Wei, S.; Li, C.W.; Chang, Y.S.; Chao, C.C.; Lai, Y.K. Localization of GRP78 to mitochondria under the unfolded protein response. *Biochem. J.* **2006**, *396*, 31–39. [CrossRef] [PubMed]
58. Prasad, M.; Pawlak, K.J.; Burak, W.E.; Perry, E.E.; Marshall, B.; Whittal, R.M.; Bose, H.S. Mitochondrial metabolic regulation by GRP78. *Sci. Adv.* **2017**, *3*, e1602038. [CrossRef] [PubMed]

cells

MDPI

Article

Long-Term Transplant Effects of iPSC-RPE Monolayer in Immunodeficient RCS Rats

Deepthi S. Rajendran Nair [1], Danhong Zhu [2], Ruchi Sharma [3], Juan Carlos Martinez Camarillo [1,4], Kapil Bharti [3], David R. Hinton [2], Mark S. Humayun [1,4] and Biju B. Thomas [1,4,*]

[1] Department of Ophthalmology, Roski Eye Institute, Keck School of Medicine, University of Southern California, Los Angeles, CA 90033, USA; deepthir@usc.edu (D.S.R.N.); juan.martinez@med.usc.edu (J.C.M.C.); humayun@med.usc.edu (M.S.H.)

[2] Department of Pathology and Ophthalmology, USC Roski Eye Institute, Keck School of Medicine, University of Southern California, Los Angeles, CA 90033, USA; dzhu@usc.edu (D.Z.); dhinton@usc.edu (D.R.H.)

[3] Unit on Ocular and Stem Cell Translational Research, National Eye Institute, NIH, Bethesda, MD 20892, USA; fnu.ruchi2@nih.gov (R.S.); kapil.bharti@nih.gov (K.B.)

[4] USC Ginsburg Institute for Biomedical Therapeutics, University of Southern California, Los Angeles, CA 90033, USA

* Correspondence: biju.thomas@med.usc.edu; Tel.: +1-323-442-5593

Citation: Rajendran Nair, D.S.; Zhu, D.; Sharma, R.; Martinez Camarillo, J.C.; Bharti, K.; Hinton, D.R.; Humayun, M.S.; Thomas, B.B. Long-Term Transplant Effects of iPSC-RPE Monolayer in Immunodeficient RCS Rats. *Cells* **2021**, *10*, 2951. https://doi.org/10.3390/cells10112951

Academic Editors: Maurice Ptito and Joseph Bouskila

Received: 18 September 2021
Accepted: 25 October 2021
Published: 29 October 2021

Publisher's Note: MDPI stays neutral with regard to jurisdictional claims in published maps and institutional affiliations.

Abstract: Retinal pigment epithelium (RPE) replacement therapy is evolving as a feasible approach to treat age-related macular degeneration (AMD). In many preclinical studies, RPE cells are transplanted as a cell suspension into immunosuppressed animal eyes and transplant effects have been monitored only short-term. We investigated the long-term effects of human Induced pluripotent stem-cell-derived RPE (iPSC-RPE) transplants in an immunodeficient Royal College of Surgeons (RCS) rat model, in which RPE dysfunction led to photoreceptor degeneration. iPSC-RPE cultured as a polarized monolayer on a nanoengineered ultrathin parylene C scaffold was transplanted into the subretinal space of 28-day-old immunodeficient RCS rat pups and evaluated after 1, 4, and 11 months. Assessment at early time points showed good iPSC-RPE survival. The transplants remained as a monolayer, expressed RPE-specific markers, performed phagocytic function, and contributed to vision preservation. At 11-months post-implantation, RPE survival was observed in only 50% of the eyes that were concomitant with vision preservation. Loss of RPE monolayer characteristics at the 11-month time point was associated with peri-membrane fibrosis, immune reaction through the activation of macrophages (CD 68 expression), and the transition of cell fate (expression of mesenchymal markers). The overall study outcome supports the therapeutic potential of RPE grafts despite the loss of some transplant benefits during long-term observations.

Keywords: iPSC-RPE; retinal pigment epithelium; immunodeficient RCS rat; ultrathin parylene; retinal degeneration; retinal transplantation

1. Introduction

Age-related macular degeneration (AMD), one of the most common causes of blindness in the developed world, is a degenerative disease of the retina often leading to progressive vision loss. Geographic atrophy, the advanced form of AMD, is characterized by dysfunction of retinal pigmented epithelium cells (RPEs) followed by degeneration of overlying photoreceptors leading to the loss of central vision. At present, no proven clinical treatments exist for the preservation or replacement of vulnerable RPE cells; however, RPE cell transplantation is perhaps the most obvious therapeutic option and has garnered significant interest. In the early stages of AMD, although the RPE cells are dysfunctional, surviving photoreceptors and the inner retina that transmit visual signals to the brain remain functional, rendering a realistic possibility that replacing the degenerating RPE with functional young RPE will restore vision.

Potential sources of healthy RPEs are pluripotent cells derived from embryonic [1–4] or adult cell sources [5–9], which are differentiated into RPE cells by employing spontaneous or directed differentiation methods. Early-phase clinical trials by various research groups used embryonic stem-cell-derived RPE (ESC-RPE) for cell replacement, which has already shown early signals of safety and potential efficacy [10–14]. Our team has demonstrated that human embryonic stem-cell-derived RPE (hESC-RPE) grown as a polarized monolayer on ultrathin parylene substrates can remain functional after transplantation in athymic nude rats [15] and in Royal College of Surgeons (RCS) rats [16,17], a model for RPE dysfunction. The product, termed the California Project to Cure Blindness-RPE (CPCB-RPE1), is being assessed in an FDA-approved phase1/2a clinical trial (NCT 02590692) for advanced dry AMD and exhibits promising outcomes for improving visual activity [12].

The autologous induced pluripotent stem-cell-derived RPE (iPSC-RPE) transplantation is considered more advantageous as the chance of graft rejection issues can be minimized. Recent research focuses on the generation of iPSC lines from adult cell sources, such as skin fibroblasts or peripheral blood mononuclear cells [5–9]. The four-year report of iPSC-RPE sheet transplant surgery for CNV (choroidal neovascularization—the wet form of AMD) in one patient has been published recently [18]. Another major step forward is the allogeneic transplantation of off-the-shelf available iPSC-RPE. Due to concerns regarding possible oncogenic mutations in cell preparation, the attention is now focused on personalized screening for mutations and the development of autologous iPSC-RPE therapies including HLA matching [19,20]. A study by Sugita et al. [19] aimed at examining the safety of six-loci HLA-matched allogeneic iPSC-RPE transplantation under local steroids. RPE cells grafted as a suspension into the patient's subretinal space survived in all five cases for more than one year [19]. These observations suggest that it is possible to manage the survival of iPSC-RPE, under immunosuppression. However, regenerative medicine is still in its infancy and the cells may behave differently in each individual. In the first iPSC-RPE transplantation clinical study [14,21], three aberrations in the deoxyribonucleic acid (DNA) copy number (deletions) were observed in the cell preparation of the second patient, which caused the study to end due to possible adverse effects.

Existing evidence indicates that the delivery of cells as a suspension may not consistently develop into a monolayer of RPE and that their long-term survival rate will be low compared to RPE cells transplanted as a monolayer [15]. In many preclinical studies for geographic atrophy, the iPSC-RPEs were delivered into the subretinal space as a bolus injection [22–25] and the animals were monitored for survival under immunosuppression for only a short period. In other studies, iPSC-RPEs transplanted as a monolayer and maintained under immunosuppression regimes were followed for up to 5 months [9,18]. Published data suggest that a confluent polarized monolayer of iPSC-RPEs transplanted as a patch rather than as a cell suspension can perform several basic functions of RPEs including phagocytosis of photoreceptor outer segments, the renewal of visual pigment, and the transport of metabolites [9,14]

Although transplantation of iPSC-RPE cells to replace the diseased RPE has been tested by several groups through preclinical studies and there are preliminary reports of ongoing preclinical studies, no major attempt has been made to perform an in-depth analysis of the long-term viability and fate of the transplanted RPE. Based on previous reports from our group [17], the progressive deterioration of visual function after the transplantation of ESC- RPE was evident during long-term observations. Monitoring cell survival and assessing the long-term functional benefits of transplants are significant since the transplanted cells are exposed to a progressively degenerating environment and there may be immunological factors that can cause adverse effects.

In preclinical studies of RPE transplantation, one of the major factors that influence the long-term benefits is the immune reaction and associated xenograft rejection. Sharma et al. [9] showed 70% survival of subretinally transplanted human iPSC-RPE (hiPSC-RPE) cells up to 2.5 months post-implantation in immunosuppressed rodents. In the above study, systemic and resident innate immune responses in animal models were suppressed by using prednisone, doxycycline, and minocycline whereas the adaptive immune responses were suppressed using tacrolimus and sirolimus [9]. Del Priore et al., in 2003 [26], demonstrated 'triple systemic' therapy with anti-inflammatory antibiotics to increase the survival of grafted RPE at four weeks post-implantation. However, based on a previous report, immunosuppressants can alter the visual function in RCS rats with depressed scores on behavioral and electrophysiological testing [27]. Hence, to minimize the complications associated with immunosuppressants, we used a newly developed immunodeficient RCS rat model characterized by an absence of T cells and a lack of natural cell-mediated cytotoxicity [28]. The iPSC-RPE cells grown as a polarized monolayer on ultrathin perylene substrates were transplanted into the subretinal space of immunodeficient RCS rats. The transplant effects were assessed at various post-implantation time points (1 to 11 months after transplantation).

2. Materials and Methods

2.1. Human Pluripotent Stem Cells Generated from iPSCs

iPSC-RPE (frozen, passage 2 cells) generated from iPSCs, reprogrammed from healthy adult fibroblasts, was obtained from Dr. Kapil Bharti, Unit on Ocular and Stem Cell Translational Research, National Eye Institute, NIH, Bethesda, USA [29]. Briefly, hiPSC were seeded at 20,000 cells per cm^2 on Matrigel and grown in mTeSR1 in a 10% CO_2/5% O_2 incubator for 5 days. Afterward, they were transferred to a 5% CO_2/20% O_2 incubator and cultured for 5 additional days. At this point, the culture medium was switched to a differentiation medium (DM) [29]. After 10–15 days, cells were maintained in DM for 3 more weeks and then switched to an RPE maintenance medium (RPEM) [29]. Differentiated cells were dissociated in Accumax (Sigma, Saint Louis, MO, USA), plated at 250–300,000 cells per cm^2, and grown in RPEM. The passage 3 cells were used for transplantation experiments. These cells are extensively characterized for clinical applications in Dr. Bharti's lab [9,30,31].

2.2. Preparation of Polarized hESC-RPE Implant on Parylene Membranes

Ultrathin parylene membranes (0.3 μm thickness supported on a 6.0 μm-thick mesh frame) made from parylene C were specially designed for implantation on rat retinas (1.0 × 0.4 mm) [15,17] and used successfully for culturing iPSC-RPEs. These ultrathin membranes were coated with Matrigel (AMS Biotechnology, Frankfurt, Germany) and seeded with iPSC-RPE based on our previously established protocol [17]. The cells were grown to confluence for approximately 4 weeks before implantation. The final density of each implant was kept as approximately 2700 cells/membrane [15].

2.3. Immunostaining of iPSC-RPE on Parylene Membrane

iPSC-RPE cells grown on Matrigel-coated parylene were stained for RPE specific markers zonula occludens protein 1 (ZO-1), and RPE65, based on established protocols [17]. Stained cells were mounted with an anti-fading mounting medium (Invitrogen) and images were captured by confocal microscopy (FV1000 Confocal Microscope, Olympus, Centre Valley, PA, USA).

2.4. Animals

The Royal College of Surgeons (RCS) rat is an established model of retinal degeneration, which has been mainly used for studying photoreceptor rescue with treatment at the age of 3–4 weeks. These rats develop a fully functional visual system, which degenerates secondarily due to their dysfunctional RPE (MertK mutation), resulting in the loss of most photoreceptors at the age of 3 months. Immunodeficient RCS rats were produced from a cross between female homozygous RCS (RCS-p+/RCS-p+) and male athymic nude rats (Hsd: RH-Foxn1mu, a mutation in the foxn1 gene; no T cells) as described previously [28]. All rats were maintained in an aseptic and temperature-controlled environment. All animals were included in accordance with the Association for Research in Vision and

Ophthalmology (ARVO) statement for the use of animals in research, and the Institutional Animal Care and Use Committee (IACUC) of USC.

2.5. Surgical Procedure

Animals underwent surgery at postnatal day (P) 28. Anesthesia was induced by intraperitoneal injection of ketamine (37.5 mg/kg) and xylazine (5 mg/kg). Only the left eyes were used for transplantation surgeries. All surgeries were performed by the same surgeon. Topical anesthesia was administered with a 0.5% proparacaine hydrochloride ophthalmic solution (Akorn, Inc., Lake Forest, IL, USA). Pupils were dilated using ophthalmic solutions of 2.5% phenylephrine hydrochloride and 0.5% tropicamide (Akorn, Inc., Lake Forest, IL, USA). Once the conjunctiva is removed, a scleral incision was performed in the temporal superior quadrant followed by an anterior chamber paracentesis to reduce intraocular pressure. A small incision (approximately 0.8–1.0 mm) was cut transsclerally at the temporal equator of the eye until the choroid was exposed with the help of a 32-gauge needle, and a 5 µL balanced salt solution (Alcon Laboratories, Inc., Fort Worth, TX, USA) was injected to create a local retinal detachment. The implant held by forceps was introduced through the sub scleral space into the subretinal bleb. Clinical assessment as well as retinal imaging by optical coherence tomography (OCT) using a Spectralis HRA + OCT device (Heidelberg Engineering, Heildeberg, Germany) were performed to confirm the placement of the implant. The rats were then allowed to recover from anesthesia in a thermal care incubator. Animals with surgical complications such as excessive bleeding, perforation of the retina, and implant delivery into the vitreous were immediately excluded from the study. Based on OCT images, 15 animals were selected for short-term experiments (1-month and 4-month study group) and 15 animals were selected for the 11-month study group.

2.6. Histopathology

Cohorts of rats were euthanized by intracardiac injection of euthasol (Virbac AH, Inc., Fort Worth, TX, USA) at 1, 4-, and 11-months post-surgery, and eyes were processed for histology. Contralateral eyes were considered as controls. Whole eyes were fixed in Davidson's solution overnight, and the cornea and lens were removed. Finally, the eye cups were embedded in paraffin and processed for sectioning (5-µm sections). Groups of consecutive slides were stained with hematoxylin and eosin (HE) for light microscopy. The HE-stained slides were scanned and photographed using an Aperio Scanscope CS (Aperio Technologies, INL., Vista, CA, USA) microscope. Histological sections of cell-seeded membranes were evaluated to assess iPSC-RPE survival. The surgical placement was considered acceptable if more than 70% of the implant was located inside the subretinal area. Transplant survival was confirmed only if iPSC-RPE were observed in at least three consecutive sections based on light microscopy and immunostaining evaluations. Cell migration or dead cell aggregation was considered when pigmented cells or cell clumps were seen adjacent to the substrate and confirmed by immunohistochemistry. If no human/RPE marker was found, the specimen was considered as non-surviving RPE clumps. The outer nuclear layer (ONL) integrity was evaluated for photoreceptor preservation in the transplanted area. Cellular reaction around the implants, observed by light microscopy, was assessed for the presence of macrophages or the expression of glial cells. Adjacent sections of the implanted eye were processed for immunohistochemical analysis using the following antibodies as needed: Human-specific cell surface marker (anti-TRA-1-85), a marker of differentiated RPE cells (anti-RPE65), a macrophage marker (anti-CD68), an astrocyte/Müller cell marker (anti-GFAP), mesenchymal markers (α Smooth muscle actin and Vimentin), photoreceptor phagocytosis marker (Rhodopsin), and RPE binding protein (RBP1).

Details of the antibodies used are included in Table 1. For immunostaining, all slides were deparaffinized, rehydrated, and antigen retrieved (sodium citrate, pH 6.0). After staining, the slides were mounted with fluorescent-enhanced mounting medium with 4′,6-

diamidino-2-phenylindole (DAPI) (Vector Laboratory, Burlingame, CA, USA). Images were taken using the Ultra viewer ERS dual-spinning disk confocal microscope (PerkinElmer, Waltham MA, USA) equipped with a C-Apochromat (Carl Zeiss, Thornwood, NY, USA) ×10 high dry lens, a C-Apochromat ×40 water immersion lens NA 1.2, an electron multiplier charge-coupled device cooled digital camera (Hamamatsu Orce_ERCC 12-bit camera]; PerkinElmer, Waltham, MA, USA) or by using a Keyence BZX-800 microscope. Images were captured and processed using PerkinElmer Velocity imaging software.

Table 1. List of antibodies used for immunostaining.

Antibodies	Purpose	Manufacturer	Catalog No	Dilution
TRA-1-85	Human marker	R&D Systems, Minneapolis, MN, USA	MAB3195	1:100
RPE65	RPE marker	Abcam	Ab231782	1:200
Rhodopsin	Rods	Abcam	Ab3267	1:100
CD68	Microglia	Abcam	ab201340	1:300
Vimentin	Mesenchymal marker	Abcam	ab137321	1:300
GFAP	Reactive glial cells	Invitrogen	MA5-12023	1:500
α Smooth muscle actin	Mesenchymal marker	Abcam	ab5694	1:250
Goat anti-mouse IgG conjugated with Rhodamine	Secondary antibody	Jackson Immuno Research, West Grove, PA, USA	115-025-146	1:500
Goat anti-rabbit IgG conjugated with FITC	Secondary antibody	Abcam	Ab150081	1:500
Ki67	Proliferation marker	Abcam	Ab16667	1:500
Donkey Anti-Mouse lgG H&L	Secondary Antibody	Abcam	Ab7003	1:500
Donkey Anti-Rabbit lgG H&L	Secondary Antibody	Abcam	Ab150063	1:500

2.7. Superior Colliculus Electrophysiology

Electrophysiological mapping of the superior colliculus (SC) was performed at approximately 11-months post-surgery based on an established protocol followed in our laboratory [7,17,28]. Based on OCT screening, 8 rats (8/15) were selected for SC experiments. Rats dark-adapted overnight were anesthetized by an intraperitoneal injection of xylazine/ketamine. The gas-inhalant anesthetic (1–2.0% isoflurane) was administered via an anesthetic mask (Stoelting Company, Wood Dale, IL, USA). Rats were mounted in a stereotactic apparatus; a craniotomy was performed, and the SC was exposed. Multi-unit visual responses were recorded extracellularly from the superficial laminae of the SC using custom-made tungsten microelectrodes. For SC mapping, the responses were recorded from approximately 30 different SC locations. At each recording location, approximately 10 presentations of a full-field strobe flash (1300 cd m22, Grass model PS 33 Photic stimulator, W. Warwick, RI, USA), positioned 30 cm in front of the rat's eye, were delivered to the contralateral eye. An interstimulus interval of 5 s was used. The neural activities were recorded using a digital data acquisition system (Power lab; ADI Instruments, Mountain View, CA, USA) 100 milliseconds before and 500 milliseconds after the onset of the stimulus. All responses at each site were averaged. Blank trials, in which the illumination of the eye was blocked with an opaque filter, were also recorded at each site.

2.8. Optokinetic Testing

Optokinetic (OKN) testing was performed at 4 months and 11 months post transplantation using a previously described protocol [17]. To record OKN responses, two tablet screens were used to display the OKN stimuli consisting of high-contrast black and white stripes generated using 'OKN Stripes Visualization Web Application', a freely available software (http://mdds.nyc/okn-stripes-visualization, accessed on 12 April 2021). A clear plexiglass holder was used to restrain the rat and keep its head continuously exposed to the tablet screen. A micro camera attached to the top of the rat holder recorded the headtracking responses during clockwise (1 min) and anticlockwise (1 min) stripe rotations. Visual acuity was tested by changing the stripe width at decrements of 0.5. Video recordings were evaluated to compute the head-tracking scores by two separate investigators who were both masked to the experimental condition. The OKN responses at various spatial frequencies were assessed based on the presence or absence of clear head-tracking and based on the duration of head-tracking.

2.9. Statistical Analysis

Statistical comparisons were made using GraphPad Prism software (GraphPad Software Inc., La Jolla, CA, USA). The Paired *t*-test was used for analyzing the OKN data. The remaining data were analyzed using Student's *t*-test or by the Analysis of Variance (ANOVA) followed by the appropriate post hoc test. For all comparisons, the significance level was determined at $p < 0.05$.

3. Results

3.1. Human iPSC-RPE Cells Can Grow as a Polarized Monolayer over Ultrathin Parylene Membrane and Demonstrate High-Purity and RPE Marker Expression

iPSC-RPE cells (Figure 1a,b) cultured on a Matrigel-coated ultrathin parylene substrate were expanded as a polarized confluent monolayer (Figure 1d) and expressed RPE 65 and ZO-1 as evidenced by immunocytochemistry (Figure 1e–h). This study demonstrated that iPSC-RPE can be grown as a polarized monolayer on ultrathin parylene, similar to our previous hESC-RPE implants [17].

3.2. iPSC-RPE Implant Survival and Functionality Assessed by Short-Term in Vivo Experiments in Immunodeficient RCS Rats (1- and 4-Month Study)

After the transplantation surgery, OCT imaging was performed to screen the animals for proper implant placement (Figure 2a–c). Animals with the implant placed as a flat sheet adjacent to the Bruch's membrane were selected for further analysis. Histological analysis at 1-month post-implantation showed the presence of a well-pigmented, intact iPSC-RPE cell layer attached to the parylene substrate in all transplanted eyes. No major signs of inflammation were observed in any of the implanted animals. A majority of the retinas (92.0%) maintained the basic retinal architecture without noticeable structural changes. The histological analysis revealed that transplanted cells survived very well, evidenced by TRA-1-85 (human specific marker) expression and retinol-binding protein expression (Figure 2e,f). Transplants in which iPSC-RPEs were present on the lower surface of the parylene membrane also showed good survival (Figure 2e). At the 1-month time point, the expression of CD68 (macrophage marker) and GFAP (retinal glial marker) was not observed in the transplant area (Supplementary Figure S1) or in the area outside the transplant (data not shown). The cells retained RPE 65 expression and human marker expression (Tra-1-85) without any evidence of mesenchymal marker expression (vimentin and α-smooth muscle actin, see Supplementary Figure S2). Good survival of iPSC-RPE implants was also observed at 4 months post-implantation (Figure 2g–i). Based on rhodopsin staining, the surviving cells performed a phagocytic function (Figure 2i). The absence of immunological markers comparable to the control group indicate the absence of detectable chronic inflammation induced by xenografts.

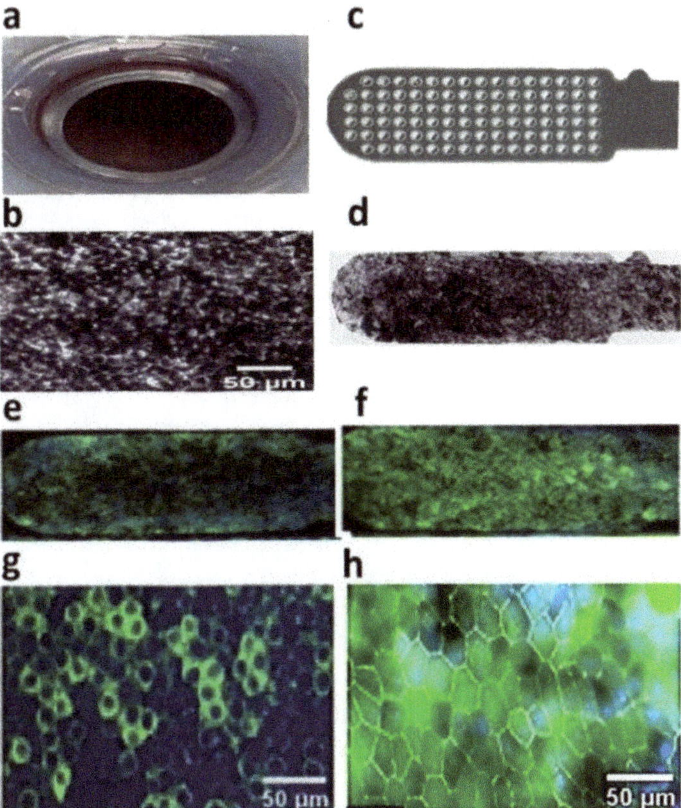

Figure 1. iPSC-RPE grown as polarized monolayer over parylene substrate. (**a**) iPSC-RPE polarized monolayer cultured on 24-well transwell insert, (**b**) enlarged view of iPSC-RPE monolayer, (**c**) ultrathin parylene membrane without cells, (**d**) iPSC-RPE grown as polarized monolayer on parylene membrane, low magnification (10×) image showing the whole implant stained for (**e**) RPE 65 (**f**) ZO-1 expression, enlarged view of expression of (**g**) RPE 65 and (**h**) ZO-1 on parylene membrane.

3.3. In Vivo Assessment of Long-Term Transplant Effects in Immunodeficient RCS Rats (11 Month Study)

Evaluation of histology images from serial sections at 11 months post-implantation showed the presence of transplanted RPE in seven eyes (7/15). Out of these seven eyes, four eyes retained an intact RPE monolayer structure. Immunostaining showed rhodopsin-containing phagosomes in the transplanted RPE. This was more prominent in eyes in which better preservation of the iPSC-RPE monolayer structure was observed (Figure 3c). In the remaining three eyes, the cells appeared as clumps (Figure 3e,f) out of which only two eyes retained RPE65 expression. There was no Ki 67 expression in the implanted areas, suggesting an absence of proliferative cells. Photoreceptor outer nuclear layer (ONL) preservation was evident in almost all eyes in which strong RPE65 expression was noticed (Figure 3c,f). A complete loss of transplanted cells was noticed in eight (8/15) eyes.

ONL preservation was not observed in the eyes in which iPSC-RPE survival was absent. The presence of fibrosis was noticed in the majority of the above eyes (Figure 4a,b). A summary of the histological result of the 11-month post-implantation study is given in Table 2.

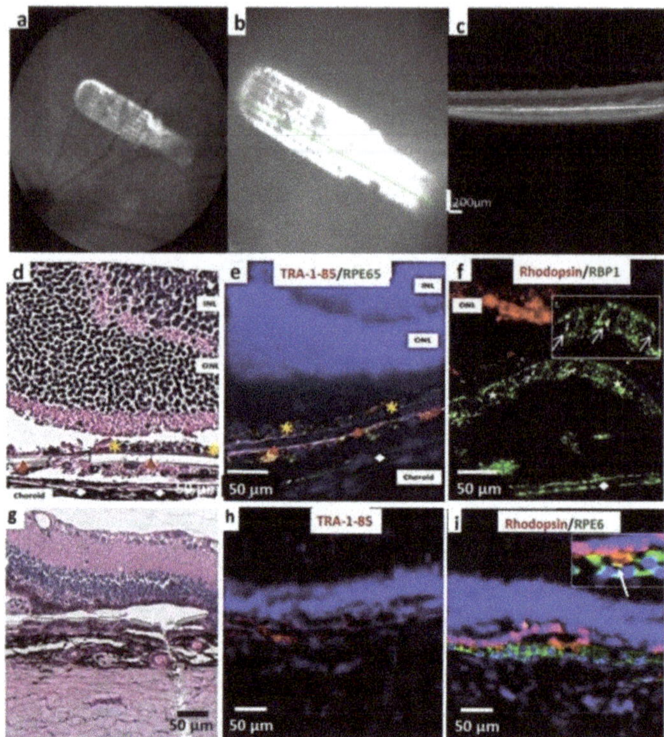

Figure 2. Short-term assessment of iPSC-RPE implant survival and functionality, in immunodeficient RCS rats. (**a**) iPSC-RPE implants observed during fundus examination of immunodeficient RCS rats at 1 month post-implantation, (**b**) enlarged view, (**c**) vertical OCT b-scan image through the transplant area, (**d**) HE image showing subretinal implant placement. The choroidal layer that appears to be separated from the remaining retina is considered a histologic artifact. Yellow asterisk indicates iPSC-RPE cells; Red arrow heads indicate the parylene membrane (**e**) transplant is identified by TRA-1-85 (human specific marker, red) and RPE65 (green) expression; Red arrowhead indicates the parylene substrate; white rhombus represents endogenous rat RPE, yellow asterisk indicates RPE on Parylene membrane (**f**) Rhodopsin (red) and retinol-binding protein (RBP1, green) staining to demonstrate that implanted iPSC-RPE can phagocytose photoreceptor outer segments (white arrows). Inset is a higher magnification of the above area. Red arrowhead indicates the parylene substrate; white rhombus represents endogenous rat RPE; (**g**) HE image showing subretinal implant 4 months after transplantation; (**h**) transplant at 4 months is identified by TRA-1-85 (human specific marker, red) (**i**) Rhodopsin (red) and RPE 65 (green) staining were used to show that implanted iPSC-RPE can phagocytose photoreceptor outer segments at 4 months after transplantation. Inset is a higher magnification of the transplant area indicating phagocytosis (white arrow).

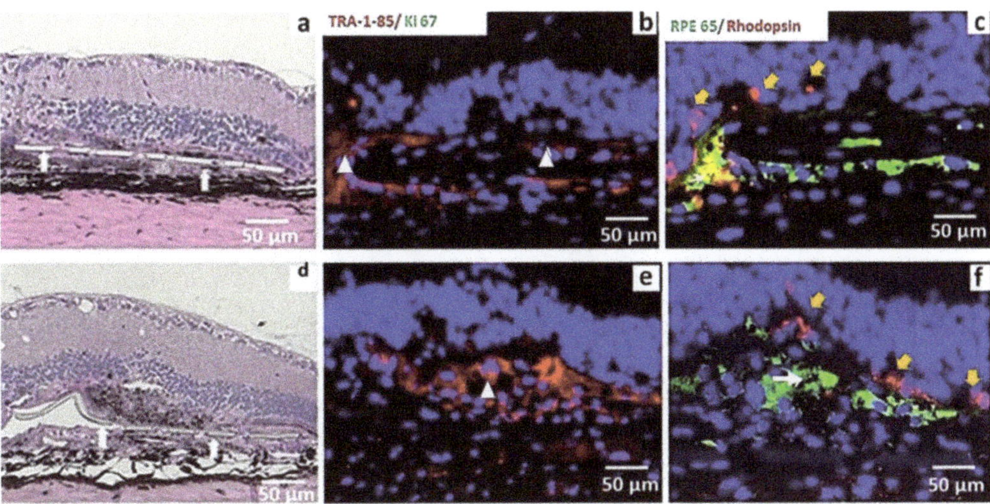

Figure 3. Representative HE and immunostaining images of immunodeficient RCS rats implanted with iPSC-RPE monolayer assessed at 11 months post implantation. Large white arrows (**a,d**) point to the parylene membrane. (**a–c**) Retina containing iPSC-RPE monolayer, (**d–f**) retina with iPSC-RPE appeared as multiple cell layers or cell clumps. TRA-1-85 (white triangle in (**b,e**) and RPE65 expression were used for identifying iPSC-RPE. Absence of Ki67 expression indicates absence of proliferative cells (**b,e**). Rhodopsin immunostaining is used to identify photoreceptor survival (yellow arrows in figure (**c,f**). Rhodopsin-containing phagosomes are found in the transplanted iPSC-RPE denoted by white arrows (**c,f**). Phagocytic activities were prominent in eyes in which monolayer structure was better preserved (**c**).

Table 2. Summary of the histological result for iPSC-RPE implantation in immunodeficient RCS rats (11-month post implantation).

iPSC-RPE Implant Status			RPE65			Phagocytosis			Fibrosis/inflammation		
No cells or cells died	Presence of intact monolayer	Cells developed into clumps, no intact monolayer	++	+	−	++	+	−	++	+	−
8	4	3	4	2	9	0	4	11	2	4	9

The expression of CD68 and GFAP was used to analyze inflammatory and glial reactions to donor tissues. GFAP was strongly expressed in the ganglion cell layer, the inner nuclear layer, and the choroid area, but was absent in the transplant area (Figure 4c). CD68 positivity observed in some of the implanted eyes suggests inflammatory reactions associated with transplantation (Figure 4d). To validate the cell loss associated with the loss of tight junctions and consequent loss of cell–cell contact and cell–matrix contact, the tissue was tested for classic EMT markers—α smooth muscle actin (α SMA) and vimentin. Interestingly, in the implants in which RPE expression was absent or feeble, there was a strong expression of α smooth muscle actin (Figure 4e), and vimentin (Figure 4f) was also observed.

3.4. Preservation of Low Light Level Visual Responses in the Superior Colliculus (SC) of iPSC-RPE-Implanted Rats at 11-Month Post-Implantation

iPSC-RPE-implanted immunodeficient RCS rats were subjected to SC luminance threshold mapping (Figure 5). Electrophysiological mapping of the SC allowed for correlation of the response area in the SC with the location of the implant placement in the eye based on the established retinocollicular map properties [32]. Age-matched normal

Long Evans (LE) rats showed visual activity from all over the SC (Figure 5a). Among transplanted rats, visual preservation was observed in only five rats (5/8).

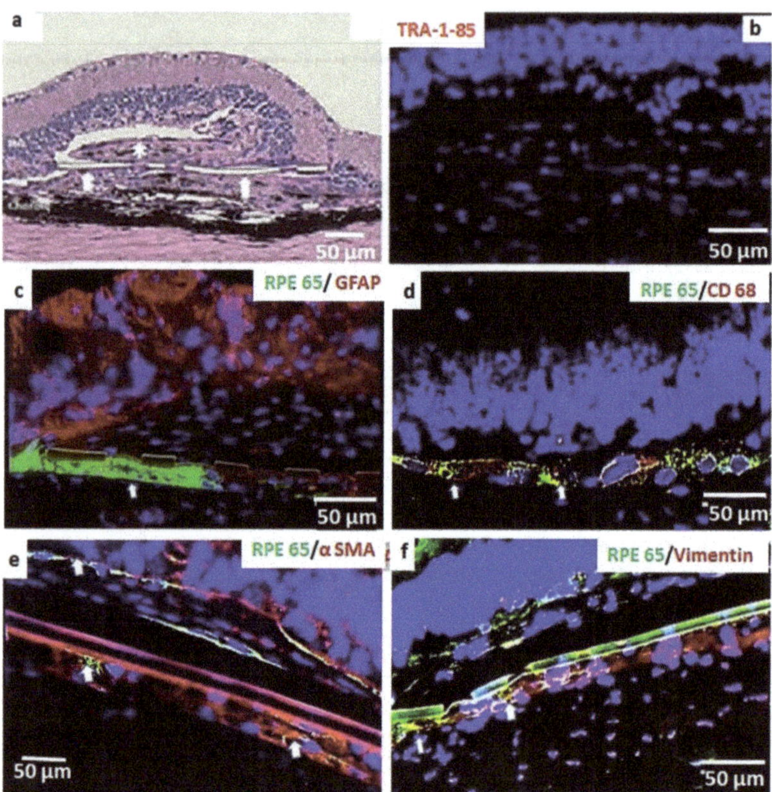

Figure 4. Representative HE and immunostaining images of immunodeficient RCS rat retinas implanted with iPSC-RPE monolayer assessed at 11 months post-implantation. Presence of fibrosis, immunoreactivity, and epithelial–mesenchymal transition (EMT) was assessed. (**a**) Retina with no surviving iPSC-RPE showing signs of inflammation and peri-membrane fibrosis indicated by white asterisk. (**b**) Absence of TRA-I-85 staining. (**c**) Retinas showing RPE65 expressing iPSC-RPE cells (white arrows) labelled for GFAP (glial cells), (**d**) CD68 (macrophages/microglia), (**e**) expression of classical mesenchymal markers α smooth muscle actin α SMA and vimentin. (**f**) Images in which iPSC-RPE monolayer appears to be present below the parylene membrane are either due to orientation difference in the implant placement or due to the survival of the RPE on the lower side of the parylene membrane.

Visually evoked activities in these rats were observed only in a small SC area corresponding to the implant placement in the eye. Visual activities were robust (higher light sensitivity) in two of the above rats (Figure 5b) whereas only weak visual activity (lower light sensitivity) was recorded in the remaining three rats (Figure 5c). No light-evoked visual activity was observed in the SC of age-matched non-transplanted rats (Figure 5d). All five rats that showed SC visual activity showed a presence of transplanted RPE in their eyes (Table 2). No correlation was observed between the light-sensitivity threshold and the degree of transplant survival.

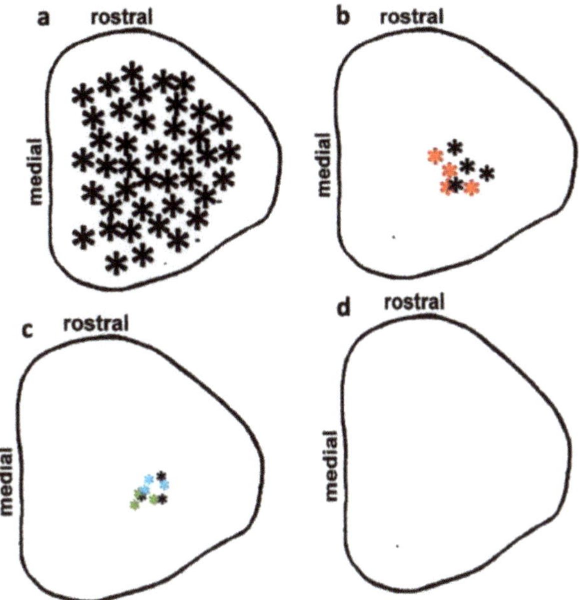

Figure 5. Visual activities recorded from the SC of 11-month-old immunodeficient RCS rats. Map properties of SC-evoked responses from individual rats are represented by colored asterisks. Larger asterisks show higher light sensitivity in the SC. (**a**) Age-matched normal rat. (**b**) * **Rat # 6005 and** * **Rat # 6012**. (**c**) * **Rat # 6001**, * **Rat # 6006**, and * Rat # 6011. Based on morphological examination, all these rats showed surviving iPSC-RPE in the retina (see Table 2). (**d**) No light-evoked visual activity was observed in the remaining transplanted rats and age-matched control RD rats.

3.5. Optokinetic (OKN) Responses in iPSC-RPE-Implanted Rats

Based on OKN data, visual improvement in the iPSC-RPE-implanted eyes was observed at 4 months post-transplantation (Figure 6). However, when tested at the 11-month time point, no measurable OKN responses were observed in any of the iPSC-RPE-implanted rats (data not shown).

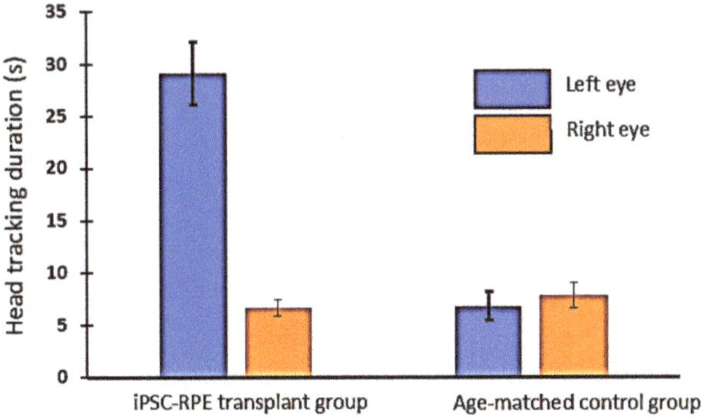

Figure 6. OKN testing data based on the duration of head-tracking recorded from 4-month-old immunodeficient RCS rats (±SE). The data show improved head-tracking response in the iPSC-RPE transplanted left eyes (n = 12) compared to the non-transplanted eyes and age-matched control rats (n = 5).

4. Discussion

Ongoing multicentral clinical studies have proven that stem-cell-derived RPE transplantation is a practical option to restore failing vision in retinal dystrophies [12,13,19,24]. Previous animal studies and pilot data from our Phase I/II clinical studies demonstrated the feasibility of using ultrathin parylene as a bio membrane for hESC-RPE growth and subretinal implantation [12,15,17,33]. Compared to hESC-RPE, using iPSC-derived RPE is considered more advantageous due to the possibility of generating a sufficient number of autologous RPE cells that strongly resemble primary human RPEs and its potential to minimize issues associated with immune rejection. Based on this, the present study evaluated the long-term benefits of polarized iPSC-RPE cells grown on an ultrathin parylene membrane that can act as an artificial Bruch's membrane. The ability of such implants to support transplant survival and viability of the host photoreceptors to preserve visual function is demonstrated in a new immunodeficient RCS rat model.

The majority of the previous iPSC-RPE transplantation studies were conducted in immunosuppressed animal models, and assessments were made for a short duration only, which may not be sufficient to extrapolate into long-term implications [9,34–36]. Hence, in the present study, transplant effects were analyzed in a new immunodeficient RCS rat model, and assessments conducted up to one year after implantation. The results from this study demonstrated the safety and potential bioactivity of the iPSC-RPE implant both during the short-term (1–4 months) investigation and long-term investigation (11-month study). Based on histology assessments, good coverage of the implanted iPSC-RPE on the parylene membrane was observed in the majority of eyes up to 4 months post-implantation along with improvement in visual function confirmed by OKN testing. In our long-term studies (11 months post-transplantation), iPSC-RPE survival and phagocytic function were only observed in less than 50% of the transplanted rats (7/15). In another report, a loss of transplanted iPSC- RPE in RCS rats, immunosuppressed by oral administration of cyclosporin, was observed at 13 weeks post-transplantation [36]. Based on the available data, the loss of transplanted RPE over the course of time and alteration in the monolayer structure can be related to the immune reaction to xenografts [19,37,38].

Previous studies have shown that the survival and integration of transplanted ESC-RPEs in the pathologic environment of a diseased retina is challenging due to it being prone to attack by macrophages [15]. CD 68 expression has been reported in studies using hESC-RPE transplantation and hESC-RPE cell suspension injections in immunosuppressed RCS rats [15,38,39]. In the present investigation, we used immunodeficient RCS rats to reduce immunological issues. However, signs of inflammation and microglia activation have been previously reported in immunodeficient RCS rats [28]. Hence, we used CD68 as a marker for assessing reactive microglia in the transplanted eyes. In our long-term studies, some CD 68 expression was observed in the implants and in areas adjacent to them (Figure 5d). However, this phenomenon was not observed at the early timepoint (1-month post-implantation, see Supplementary Figure S1). This suggests that reactive microglia/macrophages can play a role in the loss of transplanted RPEs at a later time point. In contrast to the above reports, a recent study by Zhu et.al suggested that an iPSC-PRE cell suspension injection can lower the microglial activation (CD68 expression) in rd10 mice [34]. The discrepancies in the study outcomes may be related to the differences in the animal models used, the time points in which CD68 staining was conducted, and the cell types used for transplantation experiments.

Glial fibrillary acid protein (GFAP) expression, which is known to occur in response to retinal injuries [40], can be also suggested to play a role in transplant loss in the 11-month study group. However, GFAP expression in the above study group was found to be mostly adjacent to the inner nuclear region, choroid area, and ganglionic layer, which is far from the implant area. Since this GFAP expression pattern is comparable to that of the non-transplanted control eyes [41], the presence of GFAP cannot be correlated to the loss of transplanted iPSC-RPEs.

In some of our transplanted eyes, the iPSC-RPEs developed into cell clumps on the surface of the parylene membrane (Figure 3d). Previous studies suggested that when the RPE transplant fails to establish a monolayer and form cell clumps, its survival will be poor and the cells will not be capable of performing normal RPE functions [42,43]. Based on this finding, we suggest that implants may need to be microscopically examined for potential signs of clumping prior to subretinal implantation.

The cell clump formation observed in about 20% of the implanted eyes (long-term study group) might have occurred even after transplantation due to cell migration. Cell migration is mainly attributed to the loss of RPE tight junctions [44]. This can lead to a loss of cell-to-cell contact and anchorage dependence, which are critical for RPE survival and functionality [45]. Emerging evidence demonstrates that RPE cells can be less differentiated and undergo the epithelial–mesenchymal transition (EMT) and enhanced migration in retinal degenerative diseases, including macular degenerations and proliferative vitre-oretinopathy [46–49]. Such a transition is also reported in higher-passage RPEs during in vitro observations [50,51]. Immunostaining of transplanted eyes from the 11-month study group revealed the presence of two classic mesenchymal cell markers, namely α SMA and vimentin, especially in areas of the parylene membrane where a loss of RPE expression was noticed. The transition to a mesenchymal fate may cause a loss of tight junctions and reduced cell adherence to the parylene membrane that can lead to a fibrob-lastic phenotype [44,45]. According to Zhou et al. [52], RPE cells retain the reprogramming capacity to move along a continuum between polarized epithelial cells and mesenchymal cells. This shift towards a mesenchymal phenotype can be defined as RPE dysfunction [52]. This change of RPE characteristics can cause senescence/fibrosis, eventually resulting in a loss of transplanted cells. In our transplanted rats, the expression of EMT markers was not evident at the earlier time point (1-month study group) when RPE survival was more robust (Supplementary Figure S2). Further studies are needed to identify the exact time point at which the EMT markers are expressed to determine whether changes in the iPSC-RPEs take place only in the long-term post-implantation period.

The RCS retina is widely known for its acute reactions to surgical interventions. Surgical trauma in rat eyes of a severe nature as a result of their small size can lead to increased tissue reactions in the implanted area. In support of this, mild inflammation and peri-membrane fibrosis were visible around the majority of the implants, in which the RPE monolayer was lost. It may be noted that, although the immunodeficient RCS rat is T-cell deficient, they possess bone-marrow-dependent B cells and natural killer (NK) cells. All of the above factors can contribute to the loss of transplanted iPSC-RPE cells.

In the long-term study group (11-month post-implantation), the visual functional preservation (based on SC electrophysiology) was correlated to the survival of the trans-planted iPSC-RPE. However, no considerable OKN visual activities were observed in these rats at this time point. Previously, in hESC-RPE-implanted immunosuppressed RCS rats, a progressive loss of OKN responses has been reported [17]. Improved OKN visual activities observed at an earlier time point (4 month) may be attributed to residual photoreceptors present in the RCS retina. When the photoreceptor degeneration is more advanced, the transplant benefit can be limited to a very small area of the retina and its contribution may not be strong enough to evoke measurable head-tracking activities.

In conclusion, the present study demonstrated the survival and functionality of iPSC-RPE transplanted as a polarized monolayer on a non-degradable substrate containing similarities to an artificial Bruch's membrane. The transplant benefits are higher during the earlier post-implantation period. Progressive deterioration of the transplant benefits observed in this study was correlated with the loss of transplanted iPSC-RPEs. The im-mune reactions and subretinal fibrosis can be considered the major causes of the loss of transplanted iPSC-RPE. From a clinical perspective, many of these adverse effects can be less severe in humans due to the differences in the eye architecture, surgical procedures, and the nature of the disease microenvironment. Moreover, in human eyes, easy application of target-specific and effective immune suppressants can help to reduce potential immunological reactions.

Supplementary Materials: The following are available online at https://www.mdpi.com/article/10.3390/cells10112951/s1, Figure S1: Representative images of immunodeficient RCS rat retinas implanted with iPSC-RPE monolayer (shown by RPE65 expression) cultured on a parylene membrane assessed at 1-month post implantation, Additional labelling of the retina sections was performed for expression of (**a**) CD68 (macrophages/microglia) and (**b**) GFAP (glial cells), Figure S2: Representative images of immunodeficient RCS rat retinas implanted with iPSC-RPE monolayer (shown by RPE65 expression) cultured on a parylene membrane assessed at 1-month post-implantation. Additional labelling of the retina sections was performed for expression of classical mesenchymal markers (**a**) vimentin and (**b**) α smooth muscle actin (α SMA).

Author Contributions: Conceptualization, B.B.T.; methodology, D.Z., D.S.R.N., J.C.M.C. and B.B.T.; validation B.B.T., K.B., M.S.H., J.C.M.C. and D.S.R.N.; resources, B.B.T., K.B., M.S.H., D.R.H. and R.S.; writing—original draft preparation, D.S.R.N. and B.B.T., writing—review and editing, B.B.T., D.S.R.N., D.Z., J.C.M.C., K.B., D.R.H., R.S. and M.S.H.; supervision B.B.T., funding acquisition, B.B.T. All authors have read and agreed to the published version of the manuscript.

Funding: This study was funded by a grant from the Bright Focus Foundation (M2016186, Thomas, PI). Research reported in this publication was supported by the National Eye Institute of the National Institutes of Health under Award Number P30EY029220. CIRM (California Institute for Regenerative Medicine) grants (DISC1-09912 PI-Thomas, DR3-07438-PI-Humayun), Unrestricted Grant to the Department of Ophthalmology from Research to Prevent Blindness, New York, NY. The content is solely the responsibility of the authors and does not necessarily represent the official views of the National Institutes of Health.

Institutional Review Board Statement: All experiments were approved by the University of Southern California Animal Care and Use Committee and were performed in accordance with the National Institute of Health Guide for the Care and Use of Laboratory Animals and the ARVO Statement for the Use of Animals in Ophthalmic and Vision Research. Ethical approval codes: 21068, BUA-14-00059 and 2020-3.

Informed Consent Statement: Not Applicable.

Data Availability Statement: The data presented in this study are available on request from the corresponding author.

Acknowledgments: We thank Jane Lebkowski (Regenerative Patch Technologies, Portola Valley, CA) for critically reviewing the manuscript. The authors want to thank Xiaopeng Wang (USC) for histological processing of the tissue samples.

Conflicts of Interest: Regenerative Patch Technologies: MSH and DZ have proprietary interests in the culture of RPE on ultrathin parylene.

References

1. Foltz, L.P.; Clegg, D.O. Rapid, Directed Differentiation of Retinal Pigment Epithelial Cells from Human Embryonic or Induced Pluripotent Stem Cells. *J. Vis. Exp. JoVE* **2017**, *128*, e56274. [CrossRef] [PubMed]
2. Klimanskaya, I.; Hipp, J.; Rezai, K.A.; West, M.; Atala, A.; Lanza, R. Derivation and Comparative Assessment of Retinal Pigment Epithelium from Human Embryonic Stem Cells Using Transcriptomics. *Cloning Stem Cells* **2004**, *6*, 217–245. [CrossRef]
3. Lund, R.D.; Wang, S.; Klimanskaya, I.; Holmes, T.; Ramos-Kelsey, R.; Lu, B.; Girman, S.; Bischoff, N.; Sauvé, Y.; Lanza, R. Human Embryonic Stem Cell-Derived Cells Rescue Visual Function in Dystrophic RCS Rats. *Cloning Stem Cells* **2006**, *8*, 189–199. [CrossRef]
4. Idelson, M.; Alper, R.; Obolensky, A.; Ben-Shushan, E.; Hemo, I.; Yachimovich-Cohen, N.; Khaner, H.; Smith, Y.; Wiser, O.; Gropp, M.; et al. Directed Differentiation of Human Embryonic Stem Cells into Functional Retinal Pigment Epithelium Cells. *Cell Stem Cell* **2009**, *5*, 396–408. [CrossRef] [PubMed]
5. Rowland, T.J.; Blaschke, A.J.; Buchholz, D.E.; Hikita, S.T.; Johnson, L.V.; Clegg, D.O. Differentiation of Human Pluripotent Stem Cells to Retinal Pigmented Epithelium in Defined Conditions Using Purified Extracellular Matrix Proteins. *J. Tissue Eng. Regen. Med.* **2013**, *7*, 642–653. [CrossRef] [PubMed]
6. Hazim, R.A.; Karumbayaram, S.; Jiang, M.; Dimashkie, A.; Lopes, V.S.; Li, D.; Burgess, B.L.; Vijayaraj, P.; Alva-Ornelas, J.A.; Zack, J.A.; et al. Differentiation of RPE Cells from Integration-Free IPS Cells and Their Cell Biological Characterization. *Stem Cell Res. Ther.* **2017**, *8*, 1–17. [CrossRef] [PubMed]

7. D'Antonio-Chronowska, A.; D'Antonio, M.; Frazer, K.A. In Vitro Differentiation of Human IPSC-Derived Retinal Pigment Epithelium Cells (IPSC-RPE). *Bio-Protocol* **2019**, *9*, e3469. [CrossRef]

8. Buchholz, D.E.; Hikita, S.T.; Rowland, T.J.; Friedrich, A.M.; Hinman, C.R.; Johnson, L.V.; Clegg, D.O. Derivation of Functional Retinal Pigmented Epithelium from Induced Pluripotent Stem Cells. *Stem Cells* **2009**, *27*, 2427–2434. [CrossRef]

9. Sharma, R.; Khristov, V.; Rising, A.; Jha, B.S.; Dejene, R.; Hotaling, N.; Li, Y.; Stoddard, J.; Stankewicz, C.; Wan, Q.; et al. Clinical-Grade Stem Cell-Derived Retinal Pigment Epithelium Patch Rescues Retinal Degeneration in Rodents and Pigs. *Sci. Transl. Med.* **2019**, *11*, 475. [CrossRef]

10. Schwartz, S.D.; Tan, G.; Hosseini, H.; Nagiel, A. Subretinal Transplantation of Embryonic Stem Cell-Derived Retinal Pigment Epithelium for the Treatment of Macular Degeneration: An Assessment at 4 Years. *Invest. Ophthalmol. Vis. Sci.* **2016**, *57*, ORSFc1–ORSFc9. [CrossRef] [PubMed]

11. Schwartz, S.D.; Hubschman, J.-P.; Heilwell, G.; Franco-Cardenas, V.; Pan, C.K.; Ostrick, R.M.; Mickunas, E.; Gay, R.; Klimanskaya, I.; Lanza, R. Embryonic Stem Cell Trials for Macular Degeneration: A Preliminary Report. *Lancet Lond. Engl.* **2012**, *379*, 713–720. [CrossRef]

12. Kashani, A.H.; Lebkowski, J.S.; Rahhal, F.M.; Avery, R.L.; Salehi-Had, H.; Dang, W.; Lin, C.-M.; Mitra, D.; Zhu, D.; Thomas, B.B.; et al. A Bioengineered Retinal Pigment Epithelial Monolayer for Advanced, Dry Age-Related Macular Degeneration. *Sci. Transl. Med.* **2018**, *10*. [CrossRef] [PubMed]

13. Da Cruz, L.; Fynes, K.; Georgiadis, O.; Kerby, J.; Luo, Y.H.; Ahmado, A.; Vernon, A.; Daniels, J.T.; Nommiste, B.; Hasan, S.M.; et al. Phase 1 Clinical Study of an Embryonic Stem Cell-Derived Retinal Pigment Epithelium Patch in Age-Related Macular Degeneration. *Nat. Biotechnol.* **2018**, *36*, 328–337. [CrossRef] [PubMed]

14. Mandai, M.; Watanabe, A.; Kurimoto, Y.; Hirami, Y.; Morinaga, C.; Daimon, T.; Fujihara, M.; Akimaru, H.; Sakai, N.; Shibata, Y.; et al. Autologous Induced Stem-Cell–Derived Retinal Cells for Macular Degeneration. *N. Engl. J. Med.* **2017**, *376*, 1038–1046. [CrossRef]

15. Diniz, B.; Thomas, P.; Thomas, B.; Ribeiro, R.; Hu, Y.; Brant, R.; Ahuja, A.; Zhu, D.; Liu, L.; Koss, M.; et al. Subretinal Implantation of Retinal Pigment Epithelial Cells Derived from Human Embryonic Stem Cells: Improved Survival When Implanted as a Monolayer. *Invest. Ophthalmol. Vis. Sci.* **2013**, *54*, 5087–5096. [CrossRef] [PubMed]

16. Hu, Y.; Liu, L.; Lu, B.; Zhu, D.; Ribeiro, R.; Diniz, B.; Thomas, P.B.; Ahuja, A.K.; Hinton, D.R.; Tai, Y.-C.; et al. A Novel Approach for Subretinal Implantation of Ultrathin Substrates Containing Stem Cell-Derived Retinal Pigment Epithelium Monolayer. *Ophthalmic Res.* **2012**, *48*, 186–191. [CrossRef]

17. Thomas, B.B.; Zhu, D.; Zhang, L.; Thomas, P.B.; Hu, Y.; Nazari, H.; Stefanini, F.; Falabella, P.; Clegg, D.O.; Hinton, D.R.; et al. Survival and Functionality of HESC-Derived Retinal Pigment Epithelium Cells Cultured as a Monolayer on Polymer Substrates Transplanted in RCS Rats. *Invest. Ophthalmol. Vis. Sci.* **2016**, *57*, 2877–2887. [CrossRef]

18. Takagi, S.; Mandai, M.; Gocho, K.; Hirami, Y.; Yamamoto, M.; Fujihara, M.; Sugita, S.; Kurimoto, Y.; Takahashi, M. Evaluation of Transplanted Autologous Induced Pluripotent Stem Cell-Derived Retinal Pigment Epithelium in Exudative Age-Related Macular Degeneration. *Ophthalmol. Retina* **2019**, *3*, 850–859. [CrossRef] [PubMed]

19. Sugita, S.; Mandai, M.; Hirami, Y.; Takagi, S.; Maeda, T.; Fujihara, M.; Matsuzaki, M.; Yamamoto, M.; Iseki, K.; Hayashi, N.; et al. HLA-Matched Allogeneic IPS Cells-Derived RPE Transplantation for Macular Degeneration. *J. Clin. Med.* **2020**, *9*, 2217. [CrossRef] [PubMed]

20. Sugita, S.; Iwasaki, Y.; Makabe, K.; Kamao, H.; Mandai, M.; Shiina, T.; Ogasawara, K.; Hirami, Y.; Kurimoto, Y.; Takahashi, M. Successful Transplantation of Retinal Pigment Epithelial Cells from MHC Homozygote IPSCs in MHC-Matched Models. *Stem Cell Rep.* **2016**, *7*, 635–648. [CrossRef] [PubMed]

21. Garber, K. RIKEN Suspends First Clinical Trial Involving Induced Pluripotent Stem Cells. *Nat. Biotechnol.* **2015**, *33*, 890–891. [CrossRef] [PubMed]

22. Kanemura, H.; Go, M.J.; Shikamura, M.; Nishishita, N.; Sakai, N.; Kamao, H.; Mandai, M.; Morinaga, C.; Takahashi, M.; Kawamata, S. Tumorigenicity Studies of Induced Pluripotent Stem Cell (IPSC)-Derived Retinal Pigment Epithelium (RPE) for the Treatment of Age-Related Macular Degeneration. *PLoS ONE* **2014**, *9*, e85336. [CrossRef] [PubMed]

23. Zhang, H.; Su, B.; Jiao, L.; Xu, Z.-H.; Zhang, C.-J.; Nie, J.; Gao, M.-L.; Zhang, Y.V.; Jin, Z.-B. Transplantation of GMP-Grade Human IPSC-Derived Retinal Pigment Epithelial Cells in Rodent Model: The First Pre-Clinical Study for Safety and Efficacy in China. *Ann. Transl. Med.* **2021**, *9*, 245. [CrossRef] [PubMed]

24. Sohn, E.; Jiao, C.; Kaalberg, E.; Cranston, C.; Mullins, R.; Stone, E.; Tucker, B. Allogenic IPSC-Derived RPE Cell Transplants Induce Immune Response in Pigs: A Pilot Study. *Sci. Rep.* **2015**, *5*. [CrossRef] [PubMed]

25. Westenskow, P.D.; Bucher, F.; Bravo, S.; Kurihara, T.; Feitelberg, D.; Paris, L.P.; Aguilar, E.; Lin, J.H.; Friedlander, M. IPSC-Derived Retinal Pigment Epithelium Allografts Do Not Elicit Detrimental Effects in Rats: A Follow-Up Study. *Stem Cells Int.* **2016**, *2016*. [CrossRef] [PubMed]

26. Priore, L.V.D.; Ishida, O.; Johnson, E.W.; Sheng, Y.; Jacoby, D.B.; Geng, L.; Tezel, T.H.; Kaplan, H.J. Triple Immune Suppression Increases Short-Term Survival of Porcine Fetal Retinal Pigment Epithelium Xenografts. *Invest. Ophthalmol. Vis. Sci.* **2003**, *44*, 4044–4053. [CrossRef] [PubMed]

27. Cooper, A.E.; Cho, J.-H.; Menges, S.; Masood, S.; Xie, J.; Yang, J.; Klassen, H. Immunosuppressive Treatment Can Alter Visual Performance in the Royal College of Surgeons Rat. *J. Ocul. Pharmacol. Ther. Off. J. Assoc. Ocul. Pharmacol. Ther.* **2016**, *32*, 296–303. [CrossRef]

28. Thomas, B.B.; Zhu, D.; Lin, T.-C.; Kim, Y.C.; Seiler, M.J.; Martinez-Camarillo, J.C.; Lin, B.; Shad, Y.; Hinton, D.R.; Humayun, M.S. A New Immunodeficient Retinal Dystrophic Rat Model for Transplantation Studies Using Human-Derived Cells. *Graefes Arch. Clin. Exp. Ophthalmol.* **2018**, *256*, 2113–2125. [CrossRef] [PubMed]
29. Maruotti, J.; Sripathi, S.R.; Bharti, K.; Fuller, J.; Wahlin, K.J.; Ranganathan, V.; Sluch, V.M.; Berlinicke, C.A.; Davis, J.; Kim, C.; et al. 30-Molecule–Directed, Efficient Generation of Retinal Pigment Epithelium from Human Pluripotent Stem Cells. *Proc. Natl. Acad. Sci. USA* **2015**, *112*, 10950–10955. [CrossRef] [PubMed]
30. Miyagishima, K.J.; Wan, Q.; Corneo, B.; Sharma, R.; Lotfi, M.R.; Boles, N.C.; Hua, F.; Maminishkis, A.; Zhang, C.; Blenkinsop, T.; et al. In Pursuit of Authenticity: Induced Pluripotent Stem Cell-Derived Retinal Pigment Epithelium for Clinical Applications. *Stem Cells Transl. Med.* **2016**, *5*, 1562–1574. [CrossRef]
31. May-Simera, H.L.; Wan, Q.; Jha, B.S.; Hartford, J.; Khristov, V.; Dejene, R.; Chang, J.; Patnaik, S.; Lu, Q.; Banerjee, P.; et al. Primary Cilium-Mediated Retinal Pigment Epithelium Maturation Is Disrupted in Ciliopathy Patient Cells. *Cell Rep.* **2018**, *22*, 189–205. [CrossRef] [PubMed]
32. Siminoff, R.; Schwassmann, H.O.; Kruger, L. An Electrophysiological Study of the Visual Projection to the Superior Colliculus of the Rat. *J. Comp. Neurol.* **1966**, *127*, 435–444. [CrossRef] [PubMed]
33. Koss, M.J.; Falabella, P.; Stefanini, F.R.; Pfister, M.; Thomas, B.B.; Kashani, A.H.; Brant, R.; Zhu, D.; Clegg, D.O.; Hinton, D.R.; et al. Subretinal Implantation of a Monolayer of Human Embryonic Stem Cell-Derived Retinal Pigment Epithelium: A Feasibility and Safety Study in Yucatán Minipigs. *Graefes Arch. Clin. Exp. Ophthalmol. Albrecht Von Graefes Arch. Klin. Exp. Ophthalmol.* **2016**, *254*, 1553–1565. [CrossRef] [PubMed]
34. Zhu, D.; Xie, M.; Gademann, F.; Cao, J.; Wang, P.; Guo, Y.; Zhang, L.; Su, T.; Zhang, J.; Chen, J. Protective Effects of Human IPS-Derived Retinal Pigmented Epithelial Cells on Retinal Degenerative Disease. *Stem Cell Res. Ther.* **2020**, *11*, 98. [CrossRef] [PubMed]
35. Fujii, S.; Sugita, S.; Futatsugi, Y.; Ishida, M.; Edo, A.; Makabe, K.; Kamao, H.; Iwasaki, Y.; Sakaguchi, H.; Hirami, Y.; et al. A Strategy for Personalized Treatment of IPS-Retinal Immune Rejections Assessed in Cynomolgus Monkey Models. *Int. J. Mol. Sci.* **2020**, *21*, 3077. [CrossRef] [PubMed]
36. Carr, A.-J.; Vugler, A.A.; Hikita, S.T.; Lawrence, J.M.; Gias, C.; Chen, L.L.; Buchholz, D.E.; Ahmado, A.; Semo, M.; Smart, M.J.K.; et al. Protective Effects of Human IPS-Derived Retinal Pigment Epithelium Cell Transplantation in the Retinal Dystrophic Rat. *PLoS ONE* **2009**, *4*, e8152. [CrossRef] [PubMed]
37. McGill, T.J.; Stoddard, J.; Renner, L.M.; Messaoudi, I.; Bharti, K.; Mitalipov, S.; Lauer, A.; Wilson, D.J.; Neuringer, M. Allogeneic IPSC-Derived RPE Cell Graft Failure Following Transplantation into the Subretinal Space in Nonhuman Primates. *Invest. Ophthalmol. Vis. Sci.* **2018**, *59*, 1374–1383. [CrossRef]
38. Ilmarinen, T.; Hiidenmaa, H.; Kööbi, P.; Nymark, S.; Sorkio, A.; Wang, J.-H.; Stanzel, B.V.; Thieltges, F.; Alajuuma, P.; Oksala, O.; et al. Ultrathin Polyimide Membrane as Cell Carrier for Subretinal Transplantation of Human Embryonic Stem Cell Derived Retinal Pigment Epithelium. *PLoS ONE* **2015**, *10*, e0143669. [CrossRef]
39. Karlstetter, M.; Scholz, R.; Rutar, M.; Wong, W.T.; Provis, J.M.; Langmann, T. Retinal Microglia: Just Bystander or Target for Therapy? *Prog. Retin. Eye Res.* **2015**, *45*, 30–57. [CrossRef]
40. Ekström, P.; Sanyal, S.; Narfström, K.; Chader, G.; Veen, T. Accumulation of Glial Fibrillary Acidic Protein in Muller Radial Glia during Retinal Degeneration. *Invest. Ophthalmol. Vis. Sci.* **1988**, *29*, 1363–1371.
41. Di Pierdomenico, J.; García-Ayuso, D.; Pinilla, I.; Cuenca, N.; Vidal-Sanz, M.; Agudo-Barriuso, M.; Villegas-Pérez, M.P. Early Events in Retinal Degeneration Caused by Rhodopsin Mutation or Pigment Epithelium Malfunction: Differences and Similarities. *Front. Neuroanat.* **2017**. [CrossRef] [PubMed]
42. Algvere, P.V.; Gouras, P.; Dafgård Kopp, E. Long-Term Outcome of RPE Allografts in Non-Immunosuppressed Patients with AMD. *Eur. J. Ophthalmol.* **1999**, *9*, 217–230. [CrossRef]
43. Sheridan, C.M.; Mason, S.; Pattwell, D.M.; Kent, D.; Grierson, I.; Williams, R. Replacement of the RPE Monolayer. *Eye* **2009**, *23*, 1910–1915. [CrossRef] [PubMed]
44. Carlsson, E.; Supharattanasitthi, W.; Jackson, M.; Paraoan, L. Increased Rate of Retinal Pigment Epithelial Cell Migration and Pro-Angiogenic Potential Ensuing From Reduced Cystatin C Expression. *Invest. Ophthalmol. Vis. Sci.* **2020**, *61*, 9. [CrossRef]
45. White, C.; DiStefano, T.; Olabisi, R. The Influence of Substrate Modulus on Retinal Pigment Epithelial Cells. *J. Biomed. Mater. Res. A* **2017**, *105*, 1260–1266. [CrossRef] [PubMed]
46. Lamouille, S.; Xu, J.; Derynck, R. Molecular Mechanisms of Epithelial–Mesenchymal Transition. *Nat. Rev. Mol. Cell Biol.* **2014**, *15*, 178–196. [CrossRef] [PubMed]
47. Ferrer-vaquer, A.; Viotti, M.; Hadjantonakis, A.-K. Transitions between Epithelial and Mesenchymal States and the Morphogenesis of the Early Mouse Embryo. *Cell Adhes. Migr.* **2010**, *4*, 447–457. [CrossRef]
48. Tamiya, S.; Kaplan, H.J. Role of Epithelial-Mesenchymal Transition in Proliferative Vitreoretinopathy. *Exp. Eye Res.* **2016**, *142*, 26–31. [CrossRef]
49. Ghosh, S.; Shang, P.; Terasaki, H.; Stepicheva, N.; Hose, S.; Yazdankhah, M.; Weiss, J.; Sakamoto, T.; Bhutto, I.A.; Xia, S.; et al. A Role for BA3/A1-Crystallin in Type 2 EMT of RPE Cells Occurring in Dry Age-Related Macular Degeneration. *Invest. Ophthalmol. Vis. Sci.* **2018**, *59*, AMD104–AMD113. [CrossRef]
50. Zou, H.; Shan, C.; Ma, L.; Liu, J.; Yang, N.; Zhao, J. Polarity and Epithelial-Mesenchymal Transition of Retinal Pigment Epithelial Cells in Proliferative Vitreoretinopathy. *PeerJ* **2020**, *8*, e10136. [CrossRef]

51. Shu, D.Y.; Butcher, E.; Saint-Geniez, M. EMT and EndMT: Emerging Roles in Age-Related Macular Degeneration. *Int. J. Mol. Sci.* **2020**, *21*, 4271. [CrossRef]
52. Zhou, M.; Geathers, J.S.; Grillo, S.L.; Weber, S.R.; Wang, W.; Zhao, Y.; Sundstrom, J.M. Role of Epithelial-Mesenchymal Transition in Retinal Pigment Epithelium Dysfunction. *Front. Cell Dev. Biol.* **2020**. [CrossRef]

Article

Neuroprotective Effects of Fingolimod in a Cellular Model of Optic Neuritis

Amritha A. Candadai [1,2,3], Fang Liu [1,2,3], Arti Verma [1,2], Mir S. Adil [1,2], Moaddey Alfarhan [1,2,3], Susan C. Fagan [1,2], Payaningal R. Somanath [1,2] and S. Priya Narayanan [1,2,3,*]

[1] Clinical and Experimental Therapeutics Program, College of Pharmacy, University of Georgia, Augusta, GA 30912, USA; amritha@uga.edu (A.A.C.); fliu1@augusta.edu (F.L.); Av01776@uga.edu (A.V.); madil@augusta.edu (M.S.A.); malfarhan@augusta.edu (M.A.); sfagan@augusta.edu (S.C.F.); sshenoy@augusta.edu (P.R.S.)
[2] Charlie Norwood VA Medical Center, Augusta, GA 30912, USA
[3] Culver Vision Discovery Institute, Augusta University, Augusta, GA 30912, USA
*   Correspondence: pnarayanan@augusta.edu

Abstract: Visual dysfunction resulting from optic neuritis (ON) is one of the most common clinical manifestations of multiple sclerosis (MS), characterized by loss of retinal ganglion cells, thinning of the nerve fiber layer, and inflammation to the optic nerve. Current treatments available for ON or MS are only partially effective, specifically target the inflammatory phase, and have limited effects on long-term disability. Fingolimod (FTY) is an FDA-approved immunomodulatory agent for MS therapy. The objective of the current study was to evaluate the neuroprotective properties of FTY in the cellular model of ON-associated neuronal damage. R28 retinal neuronal cell damage was induced through treatment with tumor necrosis factor-α (TNFα). In our cell viability analysis, FTY treatment showed significantly reduced TNFα-induced neuronal death. Treatment with FTY attenuated the TNFα-induced changes in cell survival and cell stress signaling molecules. Furthermore, immunofluorescence studies performed using various markers indicated that FTY treatment protects the R28 cells against the TNFα-induced neurodegenerative changes by suppressing reactive oxygen species generation and promoting the expression of neuronal markers. In conclusion, our study suggests neuroprotective effects of FTY in an in vitro model of optic neuritis.

Keywords: optic neuritis; multiple sclerosis; oxidative stress; neuroprotection; fingolimod

Citation: Candadai, A.A.; Liu, F.; Verma, A.; Adil, M.S.; Alfarhan, M.; Fagan, S.C.; Somanath, P.R.; Narayanan, S.P. Neuroprotective Effects of Fingolimod in a Cellular Model of Optic Neuritis. *Cells* **2021**, *10*, 2938. https://doi.org/10.3390/cells10112938

Academic Editors: Maurice Ptito and Joseph Bouskila

Received: 10 September 2021
Accepted: 21 October 2021
Published: 28 October 2021

Publisher's Note: MDPI stays neutral with regard to jurisdictional claims in published maps and institutional affiliations.

1. Introduction

Multiple sclerosis (MS) is an autoimmune disease of the central nervous system (CNS) prevalent in about 400,000 people in the US and 2.1 million people worldwide [1–4]. Approximately 20% of MS patients present with vision deficits associated with optic neuritis (ON) [5,6], and neurodegeneration characterized by loss of retinal ganglion cells, thinning of the nerve fiber layer, and axonal damage [7,8]. Parameters of visual function are utilized as necessary outcome measures in MS studies [9]. Although the current MS therapies target the inflammatory pathology, effects on the long-term neurodegenerative phases of the disease have not been shown. A treatment that effectively targets both aspects of MS would likely achieve preferred status as a disease-modifying agent.

Fingolimod (FTY720 or FTY), a sphingosine analog that functions as a potent immuno-suppressive agent in the CNS, is approved for MS therapy, especially for highly active disease [10,11]. Once phosphorylated into its active form by the sphingosine-1-kinase, it acts as an agonist on sphingosine-1-phosphate (S1P) receptors [12–15] and induces internalization of S1P receptors after binding [16]. Pharmacologically FTY is known as an immunomodulatory drug. FTY exerts its immunosuppressive effect by lymphocyte sequestration and thus reduces the numbers of T and B cells in circulation [17,18]. FTY has been shown to exhibit neuroprotective properties in experimental models of Alzheimer's, stroke, and Parkinson's disease [19–24]. A recent study showed that FTY exerts neuroprotective and anti-inflammatory effects on the retina and optic nerve in a mouse model of MS, perhaps explaining the potent protective effects in patients [25]. However, the molecular mechanisms regulating the neuroprotective properties have not been studied in retinal neurons. The current study was performed to assess the neuroprotective potential of FTY in an in vitro model of optic neuritis.

The R28 rat neuro-retinal cell line treated with TNFα was standardized to mimic MS-mediated neuronal injury in vitro. The cellular model was chosen based on the studies by Seigel et al. demonstrating the activity and/or expression of neuronal markers at the mRNA, protein, and functional levels in response to various stimuli [26,27]. The expression of neuron-specific markers such as microtubule associated protein 2 (MAP2), Syntaxin, neuron-specific enolase [NSE], Nestin, and receptors for neurotransmitters such as dopamine, serotonin, acetylcholine, and glycine justify the use of these cells to study CNS function [28]. Utilizing the in vitro experimental model of optic neuritis standardized in our laboratory using the tumor necrosis factor-α (TNFα) as an insult, the current study investigated the neuroprotective properties of FTY in reducing the TNFα-induced injury in R28 cells.

2. Materials and Methods

2.1. Cell Culture

Immortalized R28 postnatal day 6 rat neuro-retinal cells (heterogeneous population of cells derived from the parent cell line) (Cat # E1A-NR.3, Kerafast, Inc., Boston, MA, USA) were maintained in low-glucose DMEM medium (Cat # SH30021.01, Hyclone, Logan, UT, USA) supplemented with 10% fetal calf serum (Cat # SH30073.02, Hyclone, Logan, UT, USA), 0.225% Sodium bicarbonate (Cat # S8761, Sigma, St. Louis, MO, USA), 1X MEM non-essential amino acids (Cat # 11140-050, GIBCO, Waltham, MA, USA), 1X MEM vitamins (Cat # 11120-052, GIBCO, Waltham, MA, USA), 0.5 mM l-glutamine (Cat # 25030-081, GIBCO, Waltham, MA, USA), and 50 μg gentamicin (Cat # 15750-060, GIBCO, Waltham, MA, USA). The cells were differentiated to a neuronal phenotype with the help of 25 μg/mL laminin (Cat # 11243217001, Sigma, St. Louis, MO, USA) and 250 mM modified cyclic adenosine monophosphate (pCPT-cAMP) (Cat # C3912, Sigma, St. Louis, MO, USA) treatments, according to the published methods [29,30]. All other chemicals were purchased from Fisher Scientific, Waltham, MA, USA, unless otherwise mentioned.

2.2. The In Vitro Model of Optic Neuritis

Dose-response experiments were conducted to standardize the in vitro treatment with TNFα (recombinant rat tumor necrosis factor-α) (Cat # 510-RT, R&D Systems, Minneapolis, MN, USA) to induce neuronal injury in R28 cells. On day 0, cells were differentiated on 6-well culture plates (24 h), as described above. Treatments with TNFα at doses of 5, 10, 25, 50 ng/mL were initiated on day 1, followed by a 24 h incubation period. Cell viability with various doses of TNFα was compared against the control group with no treatment that depicted average growth and differentiation.

2.3. Treatments with Fingolimod

Once the effective dose of TNFα was established, experiments were conducted to identify an appropriate treatment concentration of FTY (Cat # 11975, Cayman Chemicals, Ann Arbor, MI, USA). Cells were pre-treated with FTY at concentrations of 2.5, 5, 10, 25, 50, and 100 nM for a 1 h incubation period before the TNFα treatment. Cell viability differences were compared among control (no treatment), TNFα-treated group, and TNFα co-incubated with varying FTY concentrations. FTY treatment alone at higher doses of 100, 200, and 500 nM were performed to test its cytotoxicity.

2.4. Cell Viability

The degree of viability of R28 cells post-treatment with TNFα and/or FTY was determined using the Trypan blue method [31]. Cells were seeded on six well plates at a density of 0.5×10^6 per well. Following treatments, they were trypsinized and collected in labeled tubes respective to their grouping. Equal volumes of a sample of cell suspension and trypan blue dye were thoroughly mixed using a micropipette, from which 10 µL was injected into a cell counting chamber (Cat # 02-671-55A, Fischer Scientific, Waltham, MA, USA) for manual counting. Trypan blue dye stains dead cells blue, and the number of viable cells in all four 16-squared tiles of the chamber were counted. This was repeated in triplicates for each cell suspension sample, and cell viability was plotted as the percentage with respect to 100% control. All graphs are represented as Mean ± SEM.

2.5. Western Blot Analysis

R28 cells seeded at a density of differentiated on six well plates and treated with TNFα and/or FTY as described earlier, resulting in four groups: Control (no treatment), TNFα, TNFα + FTY, and Control + FTY groups. Cells were homogenized, and the lysate was collected in RIPA buffer (Cat # 20-188, EMD Millipore, Burlington, MA, USA) containing protease (Cat # 78430, Fischer Scientific, Waltham, MA, USA) and phosphatase inhibitors (Cat # 78428, Fischer Scientific, Waltham, MA, USA). Protein estimation was performed using the Bradford's protein assay kit (Cat # 5000201, Bio-Rad Laboratories, Hercules, CA, USA). Samples with an equal amount of protein were prepared by using 4X Laemmli buffer (Cat # 161-0747, Bio-Rad Laboratories, Hercules, CA, USA) containing β-mercaptoethanol (Cat # O3446I-100, Fischer Scientific, Waltham, MA, USA). Samples were separated on SDS-PAGE and transferred to nitrocellulose membranes (Cat # 1620112, Bio-Rad Laboratories, Hercules, CA, USA). Membranes were blocked in 5% milk (Cat # 1706404, Bio-Rad Laboratories, Hercules, CA, USA) in tris-buffered saline with tween-20 (TBS-T) and incubated with respective primary antibodies (Table A1) overnight at 4 °C. Membranes were washed with $1\times$ TBS-T and incubated in appropriate secondary antibodies. Signals were detected using enhanced chemiluminescence (ECL) (Cat # 32106, Fischer Scientific, Waltham, MA, USA). NIH Image J software was utilized to conduct densitometric analysis, and the intensity measurements were normalized to loading control. Experiments were repeated for a minimum of three times.

2.6. Immunofluorescence Staining

Cells seeded at a density of 15,000 to 20,000 cells per well on 8 well glass chamber slides (Cat # 154941, Fischer Scientific, Waltham, MA, USA) were treated according to the study design described previously. Following the 24 h incubation period, the culture media was removed, cells were washed with $1\times$ PBS and fixed with 2% paraformaldehyde for 10 min. This was followed by a wash with PBS, and the chamber slides were stored in humidified containers at 4 °C. Slides were brought to room temperature and washed with PBS before initiating the staining protocol. Permeabilization was achieved using 0.1% Triton X-100 in PBS for 5 min, followed by a PBS wash, and blocking with 10% donkey serum at room temperature for 1 h. Cells were washed and incubated with respective primary antibodies (Table A1) overnight. The next day, the cells were incubated with appropriate secondary antibodies for 2 h. The Slides were washed, and the chambers were separated from the glass slide. Cells were covered with a coverslip using a mounting medium containing DAPI and stored in 4 °C. Images were taken using a confocal microscope (LSM 780; Carl Zeiss, Thornwood, NY, USA) available at the Augusta University imaging core facility. For the Tuj1 and NSE quantitative analysis, similar thresholds were set for all the images. Three to five regions of interest (ROI) were randomly selected per chamber slide images, and the fluorescent intensity of immunoreactivity was measured (Integrated Density) using NIH Image J software. Along with the average fluorescence intensities of a given ROI, the average fluorescence intensity of areas without staining (background) was measured as well. The Corrected Fluorescence was calculated as Integrated Density—(Area of selected cells X Mean fluorescence of background readings). The values were then normalized relative to percentage of control group. Experiments were repeated a minimum of three times and details are provided under figure legends.

2.7. Cellular ROS Formation Using DCF Assay

CM-H2DCFDA (General Oxidative Stress Indicator, Thermofisher Scientific, cat# C6827) was used to measure reactive oxygen species (ROS) formation in TNFα treated cells and any changes in response to FTY treatment, as per the manufacturer's instructions. Briefly, cells were washed with ice-cold PBS (three times), and then incubated with 10 μm CM-H2DCFDA (working solution) at 37 °C for 30 min under dark. Washed with ice-cold PBS (three times), cover-slipped using the mounting medium with DAPI stain (Vector laboratories), and the images were immediately taken by confocal microscope (LSM 780; Carl Zeiss, Thornwood, NY, USA). NIH Image J software was used for the analysis of fluorescent intensity. For single-cell quantification, single-cell was delineated and sampled at 40× from a random start point. Only cells with precise neuronal shape and specific nuclear staining with DAPI were analyzed. Three to five chamber slides per treatment were examined and the experiments were repeated three times.

2.8. Mitochondrial ROS Measurement

The production of superoxide by mitochondria in response to TNFα treatment and the impact of FTY on the mitochondrial ROS generation was measured using the MitoSOX™ Red reagent (Thermo Fisher Scientific, cat# M36008), following the manufacturer's instructions. Briefly, cells on chamber slides were washed with ice-cold PBS (three times), incubated with 5 μM MitoSOX™ reagent working solution, and incubated for 10 min at 37 °C, protected from light. Chamber slides were then carefully washed with ice-cold PBS (three times), cover slipped, and images were immediately taken by confocal microscope (LSM 780; Carl Zeiss, Thornwood, NY, USA). NIH Image J software was used for the analysis of fluorescent intensity. Three to five chamber slides per treatment were examined and the experiments were repeated three times.

2.9. H_2O_2 Treatment and LDH Assay

Oxidative stress was induced on differentiated R28 cells using H_2O_2 (SigMA, USA) following the method of Song et al. [32] with some minor modifications. Briefly, R28 cells were treated with multiple concentrations of H_2O_2 (0.0, 0.2, 0.4, 0.6, 0.8, 1.0, and 1.5 mM) for 24 h and cellular cytotoxicity was measured. In order to determine the effect of FTY on H_2O_2-induced oxidative stress, cells were pre-treated with 25 nM FTY (as described previously) and changes in cytotoxicity was measured.

Lactate dehydrogenase (LDH) assay was used to determine cellular cytotoxicity. LDH released into the culture media from damaged cells was measured following the manufacturer's instructions (CytoTox 96 non-radioactive cytotoxicity assay kit; Promega Corporation, Madison, WI, USA). The level of LDH release was normalized to the total LDH content following cell lysis in a medium. The absorbance was determined at 490 nm using a Multimode Microplate Reader (Berthold Technologies, Bad Wildbad, Germany). LDH release was expressed as a percentage of the maximum LDH released after cell lysis.

2.10. Statistical Analysis

All statistical analyses were performed with GraphPad Prism 7 (GraphPad Software Inc., La Jolla, CA, USA). Student t-test or Two-way ANOVA followed by Tukey's multiple comparisons test was employed to analyze the groups. A p value less than 0.05 was considered as statistically significant. Results are presented as Mean $\pm$ SEM.

3. Results

3.1. Fingolimod Treatment Reduces TNFα-Induced Neuronal Injury

We found that TNFα treatment resulted in a significant reduction ($p < 0.05$) in cell viability at doses of 10 ng/mL ($44.6 \pm 24.8\%$), 25 ng/mL ($39.8 \pm 8.7\%$), and 50 ng/mL ($60.2 \pm 13.0\%$) compared to the untreated control (Figure 1A). Our findings suggested that TNFα at 10 ng/mL desirably reduced the percentage of viable cells by nearly half that of the control group. In the next step, cells were pre-treated with fingolimod at concentrations of 0, 2.5, 5, 10, 25, 50, and 100 nM for 1 h prior to TNFα induction (10 ng/mL). Our results showed that FTY concentrations at 25 nM ($79.7 \pm 17.7\%$) and 50 nM ($71.0 \pm 32.3\%$) significantly prevented the TNFα-induced injury ($p < 0.001$) (Figure 2B). Experiments with high-dose FTY treatment alone resulted in cytotoxicity at doses above 100 nM. Doses of 100, 200, and 500 nM showed a decreasing viable cell count of 14.5×10^4, 3.5×10^4, and 2×10^4, respectively, compared to an average viable cell count of 47.5×10^4 in control cells without FTY treatment (not shown). Based on our findings, 25 nM was chosen as the dose of FTY to be used in further studies (Figure 1C).

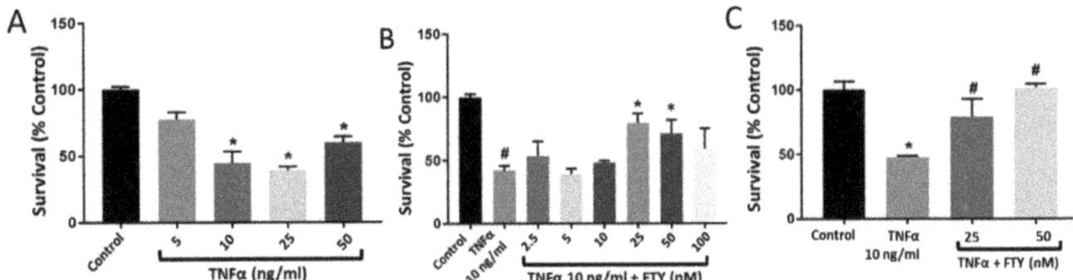

Figure 1. Dose-response effect of TNFα treatment on R28 cell survival and its reversal by co-treatment with fingolimod. (**A**) Neuronal damage was induced by treating R28 cells with various doses of TNFα for 24 h assessed by Trypan blue method. TNFα at 10ng/mL showed a marked reduction in cell survival and was chosen as the effective dose for further analysis. [$n = 3$; * $p < 0.05$ vs. dose 0]. (**B**) Cells were pre-treated with different doses of FTY, followed by TNFα (10 ng/mL), and cell survival was assessed at 24 h. Bar graph showing the effect of 25, 50, and 100 nM FTY on improving the rate of R28 cell survival [# $p < 0.005$ vs. Con; * $p < 0.05$ vs. TNFα]. (**C**) Bar graph indicating 25 nM as the optimal dose of FTY in protecting R28 cells from TNFα-induced injury in vitro [$n = 3$; * $p < 0.001$ vs. Con; # $p < 0.01$ vs. TNFα].

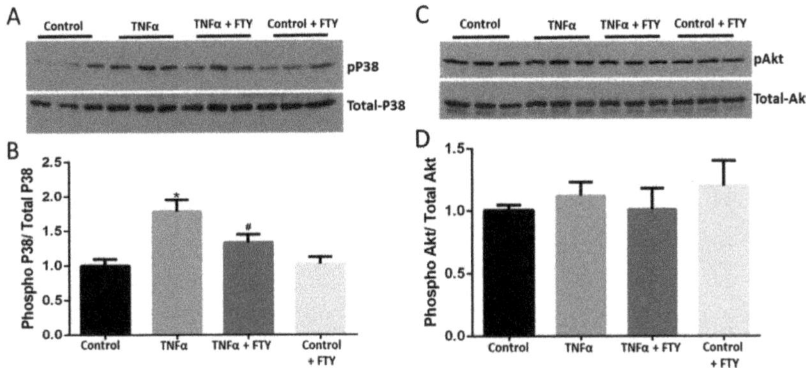

Figure 2. Fingolimod co-treatment mitigates TNFα-induced activation of cellular stress signaling. (**A**) Western blot images showing upregulation of phospho-p38 MAPK levels in response to TNFα (10 ng/mL) treatment in R28 cells, which was reduced in the presence of FTY. (**B**) Bar graph showing band densitometry quantification of Western blots indicating the effects of FTY in suppressing P38

MAPK activation induced by TNFα treatment [$n = 6$; * $p < 0.05$ vs. Con; # $p < 0.05$ vs. TNFα]. (**C**) Western blot images showing no changes in phospho-Akt (Ser473) levels in response to either TNFα (10 ng/mL) and/or FTY in R28 cells. (**D**) Bar graph showing band densitometry quantification of Western blots indicating the no changes in phospho-Akt (Ser473) levels in response to either TNFα (10 ng/mL) and/or FTY in R28 cells [$n = 6$; NS vs. TNFα].

3.2. Fingolimod Attenuates Cellular Stress and Survival Signaling

Changes in phosphorylated p38 MAP kinase expression were assessed to characterize cellular stress by Western blotting. Figure 2A shows that TNFα augmented the level of p-P38 MAPK, and this increase was markedly prevented in the presence of FTY. Moreover, we found that FTY treatment alone did not affect levels of p-P38 MAPK. Our quantification data demonstrated that in the presence of FTY, levels of p-P38/total-P38 were significantly reduced ($p < 0.05$) versus the TNFα group (Figure 2B). Changes in phosphorylated Akt levels were tested to assess cell survival using Western analysis. Figure 2C shows no changes in the expression of p-AKT with TNFα induction. Consistently, our quantification data also showed no significant decrease in levels of p-Akt/t-Akt in the presence of TNFα versus control (Figure 2D).

3.3. Effect of Fingolimod on Neuronal Cell Death

To evaluate apoptotic changes, Western blot analyses using apoptotic marker cleaved caspase-3, and anti-apoptotic marker Bcl-xL were performed. An upregulated expression on cleaved caspase-3 along with a decreased level of Bcl-xL in the presence of TNFα was observed. In the co-treatment group with TNFα and FTY, we observed a reversal in these changes (Figure 3). Quantification data showed a significant increase in the levels of cleaved caspase-3 ($p < 0.01$) versus control, while FTY treatment significantly reversed this effect ($p < 0.05$) versus the TNFα group (Figure 3B). TNFα caused a significant decrease in levels of Bcl-xL ($p < 0.05$) compared to the control group. However, the difference observed in response to FTY co-treatment was not statistically significant compared to the TNFα group (Figure 3D).

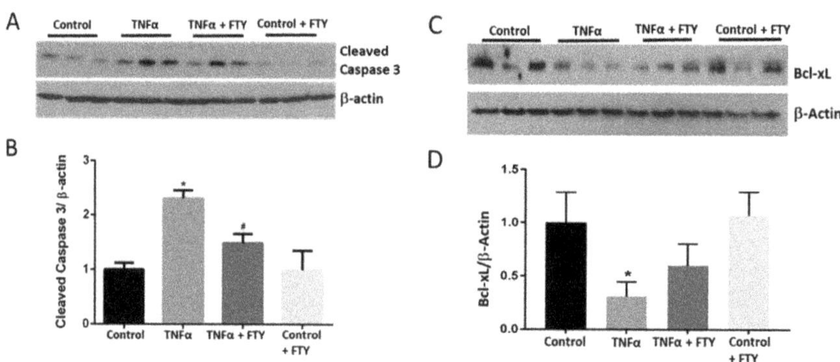

Figure 3. Fingolimod co-treatment blunted TNFα-induced activation of cleaved caspase-3 and expression of Bcl-xL. (**A**) Representative Western blot data showing increased expression of cleaved caspase-3 with TNFα treatment, which was reduced in the presence of FTY. (**B**) Bar graph showing band densitometry quantification indicating increased cleaved caspase-3 expression with TNFα treatment [* $p < 0.05$ vs. Con; $n = 3$] and its reversal by co-treatment with FTY [* $p < 0.005$ vs. control; # $p < 0.05$ vs. TNFα; $n = 4$]. (**C**) Western blot images showing reduced expression of Bcl-xL with TNFα treatment, which was increased in the presence of FTY. (**D**) Bar graph showing band densitometry quantification of Western blots indicating reduced Bcl-xL expression with TNFα treatment [* $p < 0.05$, $n = 3$]. The changes observed in response to FTY, however, were not statistically significant.

3.4. Effect of Fingolimod on TNFα-Induced Neuronal Damage

Immunofluorescence studies were conducted to study the neurodegenerative changes observed in R28 cells in response to the different treatments. β-tubulin III, also called Tuj1 contributes to microtubule formation in neuronal cell bodies and axons and plays important roles in axonal transport and cell differentiation. It is a useful marker for the detection of injury-related alterations [33]. Neuron specific enolase (NSE) is widely used and accepted as a neuronal marker, and is expressed by mature neurons and cells of neuronal origin [34]. In the present study, immunostaining with Tuj1 and NSE revealed the degenerative changes induced by TNFα, which was attenuated with FTY treatment (Figure 4A,C). Tuj1 expression was downregulated in TNFα-treated cells; however, we observed that FTY was able to protect the cells against neurofilament damage (Figure 4A lower panels). Consistently, the NSE marker was found to be reduced in the presence of TNFα, and the reduction was prevented by FTY treatment (Figure 4B,D).

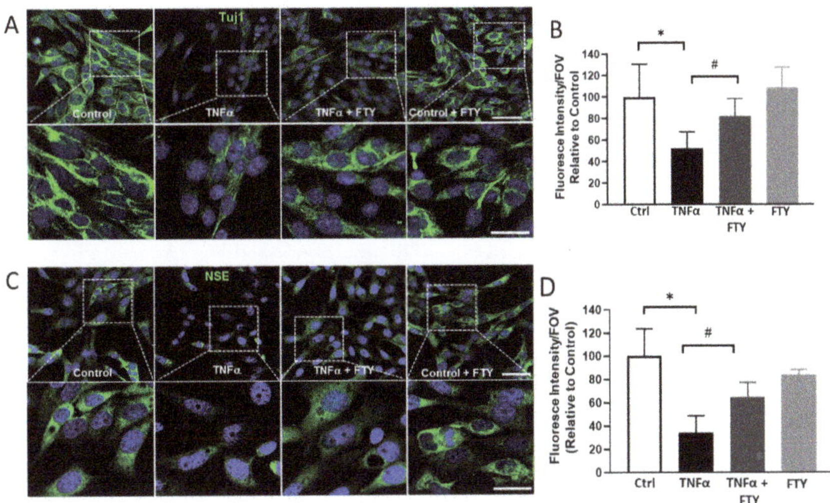

Figure 4. Fingolimod depicts protection against neuronal damage evidenced by immunofluorescence staining of marker proteins. (**A**) Representative confocal images showing the impact of TNFα treatment on neurofilament, Tuj1 (β-tubulin class III) indicating neurodegeneration, which was reduced by co-treatment with FTY. High magnification images of the boxed areas, indicating reduced Tuj1 expression, are presented in the lower panel. Scale bar 50 μm. (**B**) Bar graph showing the quantification of Tuj1 level in response to TNFα treatment, and the protective effect by co-treatment with FTY (**C**) Representative confocal images showing the changes in neuronal enolase (NSE) in response to TNFα and FTY treatments. Lower panel show high magnification images of the boxed areas, indicating reduced levels in TNFα treated group and the improved NSE expression in response to FTY co-treatment. Scale bar 50 μm. (**D**) Bar graph showing the quantification of Tuj1 and NSE levels in response to TNFα treatment, and the effect by co-treatment with FTY [* $p < 0.01$ vs. control; # $p < 0.01$ vs. TNFα, $n = 3$ per group].

3.5. Effect of Fingolimod on ROS Formation

Changes in ROS formation were studied using DCF assay. As shown in Figure 5A, ROS levels were observed to be markedly elevated in the neuronal cells in response to TNFα treatment. However, treatment with FTY downregulated the TNFα induced ROS generation. Our quantification studies demonstrate that the ROS levels are significantly higher in the TNFα group compared to control and are significantly reduced in response to FTY co-treatment (Figure 5B).

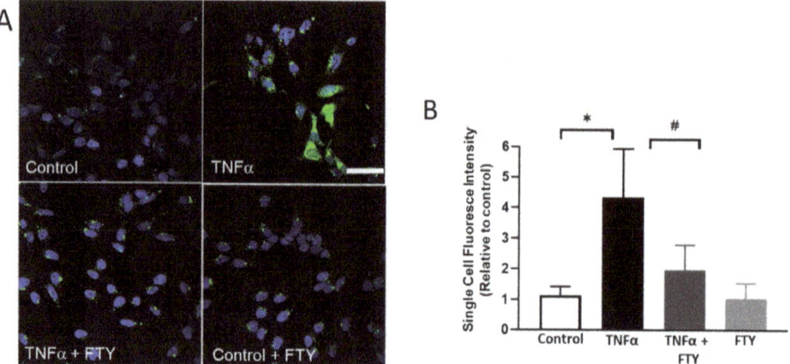

Figure 5. Fingolimod reduced TNFα-induced ROS formation. (**A**) Representative images showing the impact of FTY on reactive oxygen species (ROS) formation. H2DCFDA (DCF) assay was used to assess the generation of ROS. Scale bar 50 µm. (**B**) Quantification of single cell fluorescence intensity showing of the increased ROS formation in response to TNFα treatment, which was reduced in the presence of FTY [* $p < 0.01$ vs. control; # $p < 0.01$ vs. TNFα, n = 3 per group].

3.6. Effect of Fingolimod on Mitochondrial Dynamics and ROS Formation

In the present study, we investigated the effect of FTY treatment on the changes in proteins related to mitochondrial dynamics, including DRP-1 (dynamin related protein-1), Mitofusin 2 and OPA-1 (optic atrophy 1). As illustrated in Figure 6A,D, expression of Mitofusin 2 was significantly reduced in TNFα treated R28 cells, while FTY treatment normalized the level of Mitofusin 2 in TNFα treated cells, similar to control levels. An increase in the level of p-DRP1 was observed in TNFα treated group, while treatment with FTY reversed this effect. However, these changes were not statistically significant. No marked differences were seen in the other mitochondrial proteins studied. Further, we investigated the impact of FTY treatment on mitochondrial ROS formation using MitoSox assay (Figure 6G,H). TNFα treatment resulted in the generation of mitochondrial ROS as evidenced by elevated fluorescence indicator. FTY treatment markedly reduced the level of mitochondrial ROS formed in response to TNFα treatment. Quantification results demonstrate that the mitochondrial ROS level is significantly upregulated in TNFα treated group and the treatment with FTY significantly reduced the effect (Figure 6H).

3.7. Fingolimod Attenuates H_2O_2-Induced Cellular Damage and Stress Signaling in R28 Cells

To further assess the neuroprotective effect of FTY, experiments were performed using H_2O_2, another cellular stressor. Our results show that H_2O_2 induces cytotoxicity in differentiated R28 cells in a dose-dependent manner (Figure 7A). Treatment with FTY significantly reduced the cytotoxicity induced by TNFα at the various concentrations studied (Figure 7B). Our results indicate that H_2O_2 treatments significantly elevated p-P38 levels at all the concentrations studied, while FTY treatment significantly reduced the effect at two different contentions of H_2O_2. FTY treatment did not offer protection at a higher concentration, 1mM of H_2O_2 (Figure 7C).

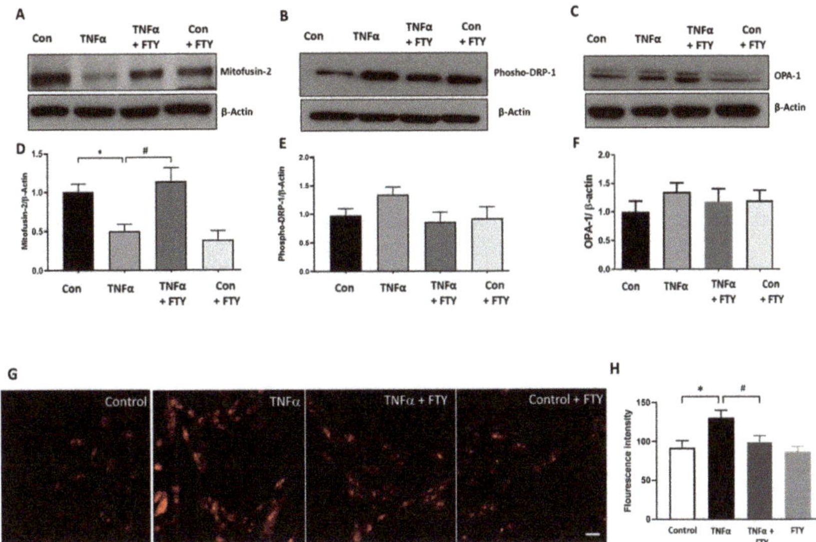

Figure 6. Changes in mitochondrial protein dynamics and ROS formation in response to Fingolimod treatment. (**A–C**) Representative Western blot data showing changes in expression of proteins associated with mitochondrial dynamics with TNFα treatment, and the effect of FTY cotreatment. (**D–F**) Respective bar graphs showing quantification of these proteins expression changes with TNFα treatment and any changes by co-treatment with FTY [* $p < 0.05$ vs. control; # $p < 0.05$ vs. TNFα; N = 3]. (**G**) Representative images of Mitosox Red staining showing the impact of TNFα on mitochondrial reactive oxygen species (ROS) formation and the effect of FTY on the treatment. Scale bar 20 μm. (**H**) Quantification of fluorescence intensity showing of the increased mitochondrial ROS formation in response to TNFα treatment, which was reduced in the presence of FTY [* $p < 0.01$ vs. control; # $p < 0.01$ vs. TNFα, $n = 3$ per group].

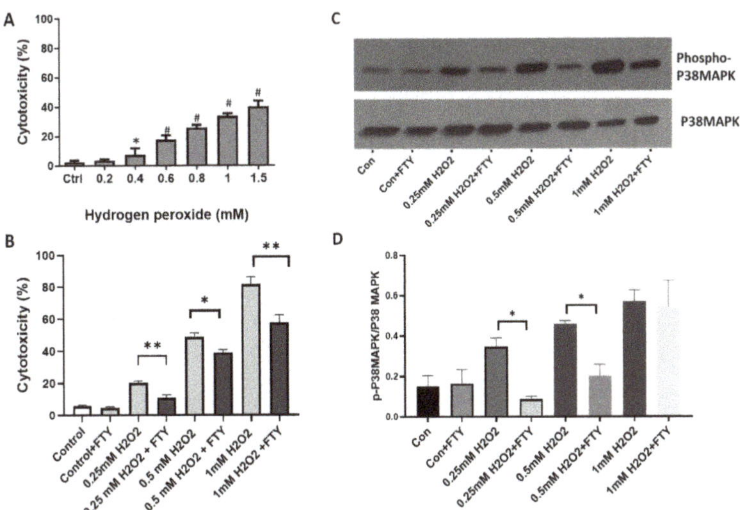

Figure 7. Fingolimod attenuates H_2O_2-induced cellular damage and stress signaling in R28 cells. (**A**) LDH assay results showing the dose-dependent effect of H_2O_2 treatment on R28 cells. * $p < 0.05$ vs control; # $p < 0.01$ vs. control and $n = 3$. (**B**) LDH assay showing the cytotoxic effect of H_2O_2 at

selected concentrations on R28 cells and the protective effect of FTY on the treatments. * $p < 0.05$; ** $p < 0.01$. (**C**) Representative Western blot data showing changes in p-P38 MAPK with H2O2 treatment, and the protective effect of FTY co-treatment on differentiated R28 cells. (**D**) Bar graph showing quantification of p-P38MAPK proteins expression changes with H_2O_2 and FTY treatments. * $p < 0.01$ and the experiment was repeated three times.

4. Discussion

The present study was conducted to assess the potential neuroprotective action of FTY in an in vitro model of optic neuritis. Lack of effective treatment strategies to reduce neurodegeneration continues to be a major problem in the field of MS research. It is vital to understand the underlying mechanisms of MS-induced neuronal damage and dysfunction. Even though MS research on the pathophysiology and associated molecular mechanisms have evolved over decades of research, including our laboratory [35,36], the field lacks reliable in vitro models to study neurodegeneration. Synthetic molecules such as trimethyltin [37], oxaliplatin [38], and cuprizone [39], although successful in creating a neurodegenerative environment, do not accurately represent the neuroinflammatory changes observed in an MS brain. In MS, inflammatory leukocytes are believed to infiltrate the CNS to mediate demyelination and neuronal degeneration via cytokines upon activation of T lymphocytes and antigen-presenting cells (APCs) [40]. tumor necrosis factor-α (TNFα) is one of the primary cytokines that are present in elevated levels in active MS lesions, serum, and cerebrospinal fluid of MS patients [41]. Studies conducted on BV-2 microglial cell lines [42] and primary mixed neuronal and glial cultures [43] show the effect of TNFα-induced damage and apoptosis. Hence, utilizing the R28 neuro-retinal cells, we standardized an in vitro experimental model of neurodegeneration to assess the impact of fingolimod on TNFα-induced neuronal damage. Using a set of functional and biochemical analyses, our study demonstrates the neuroprotective properties of fingolimod in MS-associated optic neuritis in vitro.

We first demonstrated the dose-dependent effects of TNFα on neuronal cell viability in R28 neuro-retinal cells in vitro and identified the optimal dose of TNFα for further molecular characterization. The R28 cells are immortalized, rat retinal origin, heterogenous, precursor cells with differentiation potential [26]. According to the published methods, R28 cells were differentiated to neuronal phenotype with the addition of a modified form of cAMP and laminin and grown using DMEM [44–46]. Studies by Seigel et al. utilizing R28 cells demonstrate the expression of neuron-specific markers such as MAP2, Syntaxin, NSE, and Nestin, along with neurotransmission receptors [26,28]. Our study characterized the expression of neurofilament marker Tuj1, along with NSE in these cells, thus demonstrating the reliability of this model in evaluating the impact of FTY on the neuronal injury. In MS, one of the major clinical presentations observed in patients is optic neuritis [47]. Studies have shown that approximately 20% of patients present with inflammation of the optic nerve as their first symptom of MS [5,6]. Another study conducted by the North American Research Committee on multiple sclerosis (NARCOMS) showed that of the 9107 patients participating in the study, 60% reported signs of vision impairment, and 14% of these depicted moderate/severe/very severe impairment of vision [9]. Based on the available research, visual dysfunction is a common component of MS disease progression and an important determinant of quality of life.

Current MS therapies function by suppressing the inflammatory pathways and have unknown impact on the long-term neuronal damage, causing a major knowledge gap and emphasizes the need to identify a neuroprotective therapeutic agent. Therefore, our study focused on assessing the neuroprotective effect of FTY in an in vitro model of MS-induced optic neuritis. Fingolimod, a sphingosine-1-phosphate (S1P) receptor modulator, has previously been shown to prevent neurodegenerative mechanisms targeting an inflammatory CNS state in vitro, in vivo, and clinical settings, as detailed below. Studies on Parkinson's disease models have shown a positive impact with FTY [19–21]. Mechanistically, it was found that the protective effects of FTY in Parkinson's disease were correlated with the activation of survival pathway mediated by Akt/ERK1/2 and increased expression of a neuron-specific brain-derived neurotrophic factor (BDNF). In a model of Alzheimer's, FTY was able to reverse the effect of damage by modulating the levels of different markers such as the Glial fibrillary acidic protein (astrogliosis marker), taurine (anti-inflammatory marker), and neuronal markers such as the N-acetyl aspartate and glutamate [23]. A meta-analysis was conducted by Liu et al. which included nine studies that focused on quantification of infarct volume and neurological deficit scoring in a transient middle cerebral artery occlusion model of ischemic stroke challenged with FTY [24]. The study concluded that FTY could be a possible candidate for stroke due to its protective effects on neurological deficit and infarct volume in eight of the nine included studies. Promising outcomes with FTY represent the need for further investigation to confirm the theories on its action as a neuroprotective agent. Utilizing the EAE (Experimental Autoimmune Encephalitis) model of MS, a recent study by Yang et al. investigated whether FTY is beneficial to the visual system [25]. Their results showed that FTY treatment offered neuroprotective and anti-inflammatory effects on the retina and optic nerve. FTY treatment alleviated EAE-induced gliosis, inflammation and reduced the apoptosis of RGCs and oligodendrocytes.

In response to FTY treatment, our studies found a reduction in the phosphorylation of p38 MAP kinase (a cellular stress signaling pathway) in TNFα-treated R28 cells. However, fingolimod treatment did not induce any changes in the levels of phosphorylated Akt, indicating its effect on cell survival is independent of Akt, with the conditions studied. TNFα-induced cell death was confirmed by the upregulation of cleaved caspase-3 (a cell death marker) expression along with reduced levels of Bcl-xL (an anti-apoptotic protein). These changes were reversed in response to FTY treatment, supporting its neuroprotective and anti-apoptotic function. FTY treatment also protected the retinal neurons against the TNFα-induced neuronal damage determined by the expression of neuronal markers.

Oxidative stress plays a critical part in the pathogenesis of various neurodegenerative diseases and neuroinflammation. There exists an increasing amount of data indicating that oxidative stress plays a major role in the pathogenesis of MS and optic neuritis [48]. Results from preclinical studies show that suppression of oxidative stress is a promising strategy for optic neuritis [49–53]. In the present study, we show that FTY mediated neuroprotection of R28 cells involves the regulation of ROS formation. A recent study performed on serum samples from patients with relapsing-remitting MS and healthy controls demonstrated that TFY treatment increased total antioxidant capacity [54]. Furthermore, FTY treatment reduced oxidative stress in experimental models of cardiomyopathy [55], multiple system atrophy [56], autism [57], and vitamin K-induced neurotoxicity [20].

Oxidative damages along with mitochondrial dysfunction are common characteristics of neurodegenerative diseases. Mitochondrial dysfunction is increasingly recognized as a major mechanism of MS associated pathologies [58–60]. The present study investigated the impact of FTY treatment on the changes in protein levels related to mitochondrial dynamics. Alterations in mitochondrial dynamics affect mitochondrial size and shape and impact mitochondrial metabolism and cell death. These events are controlled by mitochondrial dynamin-related GTPases, including mitofusin-1, mitofusin-2, OPA1, and DRP1. Our results show that one possible mechanism by which FTY offers neuroprotection of R28 neuronal cells via regulation of mitochondrial fusion. Altered levels of mitochondrial proteins were reported in retinal neurons in models of diabetic retinopathy [61] and glaucoma [62]. Results from our study are consistent with studies on neuroprotective properties of FTY in other models where FTY improved mitochondrial stability and restored mitochondrial dynamics under oxidative stress conditions [20,63–65]. Our present study did not investigate the changes in mitochondrial function.

Our study showed that FTY treatment reduces TNFα induced mitochondrial ROS formation in R28 cells. Other studies recently reported that FTY reduced mitochondrial dysfunction in a rat model of chronic cerebral hypoperfusion, reduced neuroinflammation and restored mitochondrial function in a model of Multiple System Atrophy [65], prevented mitochondrial dysfunction and protected neurons in prion protein-disease model [66].

Furthermore, we investigated the neuroprotective effect of FTY on R28 cells treated with H_2O_2, a cellular stressor known to induce oxidative stress. Our results reveal that FTY offers neuroprotection in response to oxidative stress-induced cellular damage. While FTY treatment rescued R28 cells at all concentrations of H_2O_2 studied, FTY did not significantly reduce phospho-P38 levels at higher concentration of H_2O_2, suggesting the possibility of other signaling pathways involved. Another possibility is that FTY at a different dose/time could reduce the phospho-P38 levels at higher concentrations of H_2O_2. Overall, these results indicate that FTY mediated neuroprotection could be offered through multiple mechanisms, including P38MAPK signaling. However, further studies are needed to confirm this observation.

Evidence as a regulator of oxidative stress, along with its immunomodulatory function, offers significant therapeutic potential to FTY in neuroinflammatory diseases such as optic neuritis. Figure 8 depicts the possible mechanisms of FTY mediated neuroprotection in response to TNFα-induced damage. Further studies are needed to define whether the effect of FTY on ROS level is direct or indirect and to delineate the molecular mechanisms of FTY mediated neuroprotection.

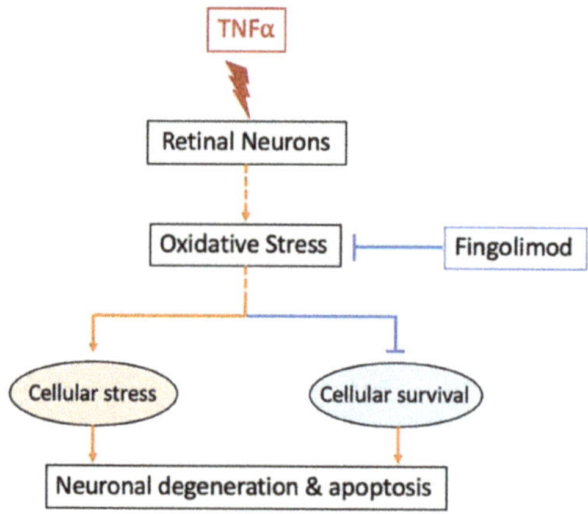

Figure 8. Proposed mechanism of FTY mediated neuroprotection. It is postulated that TNFα treatment induces cellular stress and cell death in retinal neurons by elevating oxidative stress. Treatment with FTY reduces TNFα-induced oxidative stress and improved neuronal survival.

One limitation of our study is that it did not elucidate the role of cell survival pathways that are directly associated with TNFα and FTY action. The concentration of FTY used in our study (25 nM) corresponds to 7.6875 ng/mL. However, as per the manufacturer's (Gilenya®) package insert, the concentration of active fingolimod phosphate in adult MS patients is 1.35 ng/mL. Lower concentrations used in our study, did not offer any neuroprotective effect. It is likely that the neurovascular unit could respond differently in response to other cytokines or injury mediators. This difference in human systemic concentration versus in vitro concentration in our models is an interesting aspect of the study and suggests that repurposing may be needed for the neuroprotective action of FTY in MS patients. Another limitation of the current study is that it is purely performed in a cellular model in vitro, which is not a true reflection of what happens in a complex in vivo set up. Nevertheless, the study provides reasonable optimism on the potential therapeutic benefits of FTY to treat ON. Studies performed on EAE model by Yang el tal demonstrated neuroprotective and anti-inflammatory effects on the retina and optic nerve, with no direct negative effects at the two different doses of FTY (0.3 and 1.0 mg/kg) utilized [25]. Results from our study complements the findings of Yang et al., where stronger neuroprotective effects on the visual system.

5. Conclusions

Overall, our study investigated the potential neuroprotective effects of FTY in an in vitro model of neurodegeneration. The R28 neuro-retinal cells are characterized as a successful platform for evaluating neuronal damage in the presence of TNFα, and its suppression with FTY. Furthermore, our studies demonstrated the antioxidant properties of FTY, a possible mechanism of neuroprotection. However, further studies are required to confirm the results. Based on the cellular and molecular analysis, FTY demonstrated the potential to be investigated as a novel neuroprotective strategy in conditions like MS and associated pathologies.

Author Contributions: Conceptualization, S.P.N. and S.C.F.; methodology, A.A.C., F.L., A.V., M.A. and M.S.A.; formal analysis, A.A.C., F.L., A.V. and M.S.A.; investigation, A.A.C., F.L., A.V. and M.A.; resources, S.P.N. and P.R.S.; writing—original draft preparation, A.A.C.; writing—review and editing, F.L., M.A., M.S.A., S.C.F., P.R.S. and S.P.N.; supervision, S.P.N.; project administration, S.P.N.; funding acquisition, S.P.N. All authors have read and agreed to the published version of the manuscript.

Funding: This study was supported in part by University of Georgia startup funds and the National Eye Institute to SPN (R01EY028569). This work has been accomplished using the resources and facilities at the Charlie Norwood VA Medical Center, Augusta, GA, and a core grant from the NIH/NEI to the Augusta University Vision Discovery Institute (P30EY031631). The funders had no role in study design, data collection, analysis, and decision to publish the data. The contents of the manuscript do not represent the views of the Department of Veteran Affairs or the United States.

Institutional Review Board Statement: Not applicable.

Informed Consent Statement: Not applicable.

Data Availability Statement: Not applicable.

Acknowledgments: The authors are thankful to Chithra Palani for her assistance with the R28 cell culture studies.

Conflicts of Interest: The authors declare no conflict of interest.

Appendix A

Table A1. List of antibodies used in the study.

Antibody Used	Catalogue No.	Company	Dilution
Phospho-P38 MAP Kinase	4511S	Cell Signaling	1:1000
Total P38 MAP Kinase	9212S	Cell Signaling	1:1000
Phospho-Akt	4060S	Cell Signaling	1:1000
Total Akt	9272S	Cell Signaling	1:1000
Bcl-xL	2764S	Cell Signaling	1:1000
Cleaved caspase-3	9664S	Cell Signaling	1:1000
β-Actin	A1978-200UL	Sigma	1:10,000
Tuj1	MAB1195	R&D Systems	1:500
Neuron Specific Enolase	NSE	Aves Lab	1:500
Mitofusin-2	74792	Cell Signaling	1:1000
p-DRP-1	4494	Cell Signaling	1:1000
OPA-1	80471S	Cell Signaling	1:1000

References

1. Filippi, M.; Rocca, M.A. Multiple sclerosis: New measures to monitor the disease. *Lancet Neurol.* **2013**, *12*, 12–13. [CrossRef]
2. Lassmann, H.; van Horssen, J. The molecular basis of neurodegeneration in multiple sclerosis. *FEBS Lett.* **2011**, *585*, 3715–3723. [CrossRef]
3. Zwibel, H.L.; Smrtka, J. Improving quality of life in multiple sclerosis: An unmet need. *Am. J. Manag. Care* **2011**, *17* (Suppl. 5), S139–S145. [PubMed]
4. Waubant, E.; Lucas, R.; Mowry, E.; Graves, J.; Olsson, T.; Alfredsson, L.; Langer-Gould, A. Environmental and genetic risk factors for MS: An integrated review. *Ann. Clin. Transl. Neurol.* **2019**, *6*, 1905–1922. [CrossRef] [PubMed]
5. Burman, J.; Raininko, R.; Fagius, J. Bilateral and recurrent optic neuritis in multiple sclerosis. *Acta Neurol. Scand.* **2011**, *123*, 207–210. [CrossRef] [PubMed]
6. Sorensen, T.L.; Frederiksen, J.L.; Bronnum-Hansen, H.; Petersen, H.C. Optic neuritis as onset manifestation of multiple sclerosis: A nationwide, long-term survey. *Neurology* **1999**, *53*, 473–478. [CrossRef] [PubMed]
7. Walter, S.D.; Ishikawa, H.; Galetta, K.M.; Sakai, R.E.; Feller, D.J.; Henderson, S.B.; Wilson, J.A.; Maguire, M.G.; Galetta, S.L.; Frohman, E.; et al. Ganglion cell loss in relation to visual disability in multiple sclerosis. *Ophthalmology* **2012**, *119*, 1250–1257. [CrossRef]
8. Trip, S.A.; Schlottmann, P.G.; Jones, S.J.; Altmann, D.R.; Garway-Heath, D.F.; Thompson, A.J.; Plant, G.T.; Miller, D.H. Retinal nerve fiber layer axonal loss and visual dysfunction in optic neuritis. *Ann. Neurol.* **2005**, *58*, 383–391. [CrossRef]
9. Salter, A.R.; Tyry, T.; Vollmer, T.; Cutter, G.R.; Marrie, R.A. "Seeing" in NARCOMS: A look at vision-related quality of life in the NARCOMS registry. *Mult. Scler.* **2013**, *19*, 953–960. [CrossRef]
10. Adachi, K.; Chiba, K. FTY720 story. Its discovery and the following accelerated development of sphingosine 1-phosphate receptor agonists as immunomodulators based on reverse pharmacology. *Perspect. Med. Chem.* **2007**, *1*, 11–23. [CrossRef]
11. Park, S.J.; Im, D.S. Sphingosine 1-Phosphate Receptor Modulators and Drug Discovery. *Biomol. Ther. (Seoul)* **2017**, *25*, 80–90. [CrossRef]
12. Brinkmann, V.; Billich, A.; Baumruker, T.; Heining, P.; Schmouder, R.; Francis, G.; Aradhye, S.; Burtin, P. Fingolimod (FTY720): Discovery and development of an oral drug to treat multiple sclerosis. *Nat. Rev. Drug Discov.* **2010**, *9*, 883–897. [CrossRef]
13. Brinkmann, V.; Davis, M.D.; Heise, C.E.; Albert, R.; Cottens, S.; Hof, R.; Bruns, C.; Prieschl, E.; Baumruker, T.; Hiestand, P.; et al. The immune modulator FTY720 targets sphingosine 1-phosphate receptors. *J. Biol. Chem.* **2002**, *277*, 21453–21457. [CrossRef]
14. Dyckman, A.J. Modulators of Sphingosine-1-phosphate Pathway Biology: Recent Advances of Sphingosine-1-phosphate Receptor 1 (S1P1) Agonists and Future Perspectives. *J. Med. Chem.* **2017**, *60*, 5267–5289. [CrossRef] [PubMed]
15. Miron, V.E.; Schubart, A.; Antel, J.P. Central nervous system-directed effects of FTY720 (fingolimod). *J. Neurol. Sci.* **2008**, *274*, 13–17. [CrossRef]
16. Matloubian, M.; Lo, C.G.; Cinamon, G.; Lesneski, M.J.; Xu, Y.; Brinkmann, V.; Allende, M.L.; Proia, R.L.; Cyster, J.G. Lymphocyte egress from thymus and peripheral lymphoid organs is dependent on S1P receptor 1. *Nature* **2004**, *427*, 355–360. [CrossRef] [PubMed]
17. Graler, M.H.; Goetzl, E.J. The immunosuppressant FTY720 down-regulates sphingosine 1-phosphate G-protein-coupled receptors. *FASEB J.* **2004**, *18*, 551–553. [CrossRef] [PubMed]
18. Chiba, K.; Yanagawa, Y.; Masubuchi, Y.; Kataoka, H.; Kawaguchi, T.; Ohtsuki, M.; Hoshino, Y. FTY720, a novel immunosuppressant, induces sequestration of circulating mature lymphocytes by acceleration of lymphocyte homing in rats. I. FTY720 selectively decreases the number of circulating mature lymphocytes by acceleration of lymphocyte homing. *J. Immunol.* **1998**, *160*, 5037–5044.
19. Zhao, P.; Yang, X.; Yang, L.; Li, M.; Wood, K.; Liu, Q.; Zhu, X. Neuroprotective effects of fingolimod in mouse models of Parkinson's disease. *FASEB J.* **2017**, *31*, 172–179. [CrossRef]
20. Martin-Montanez, E.; Pavia, J.; Valverde, N.; Boraldi, F.; Lara, E.; Oliver, B.; Hurtado-Guerrero, I.; Fernandez, O.; Garcia-Fernandez, M. The S1P mimetic fingolimod phosphate regulates mitochondrial oxidative stress in neuronal cells. *Free Radic. Biol. Med.* **2019**, *137*, 116–130. [CrossRef]
21. Ren, M.; Han, M.; Wei, X.; Guo, Y.; Shi, H.; Zhang, X.; Perez, R.G.; Lou, H. FTY720 Attenuates 6-OHDA-Associated Dopaminergic Degeneration in Cellular and Mouse Parkinsonian Models. *Neurochem. Res.* **2017**, *42*, 686–696. [CrossRef]
22. Vidal-Martinez, G.; Vargas-Medrano, J.; Gil-Tommee, C.; Medina, D.; Garza, N.T.; Yang, B.; Segura-Ulate, I.; Dominguez, S.J.; Perez, R.G. FTY720/Fingolimod Reduces Synucleinopathy and Improves Gut Motility in A53T Mice: Contributions of pro-brain-derived neurotrophic factor (pro-bdnf) and mature bdnf. *J. Biol. Chem.* **2016**, *291*, 20811–20821. [CrossRef]
23. Aytan, N.; Choi, J.K.; Carreras, I.; Brinkmann, V.; Kowall, N.W.; Jenkins, B.G.; Dedeoglu, A. Fingolimod modulates multiple neuroinflammatory markers in a mouse model of Alzheimer's disease. *Sci. Rep.* **2016**, *6*, 24939. [CrossRef] [PubMed]
24. Liu, J.; Zhang, C.; Tao, W.; Liu, M. Systematic review and meta-analysis of the efficacy of sphingosine-1-phosphate (S1P) receptor agonist FTY720 (fingolimod) in animal models of stroke. *Int. J. Neurosci.* **2013**, *123*, 163–169. [CrossRef]
25. Yang, T.; Zha, Z.; Yang, X.; Kang, Y.; Wang, X.; Tong, Y.; Zhao, X.; Wang, L.; Fan, Y. Neuroprotective Effects of Fingolimod Supplement on the Retina and Optic Nerve in the Mouse Model of Experimental Autoimmune Encephalomyelitis. *Front. Neurosci.* **2021**, *15*, 663541. [CrossRef] [PubMed]
26. Seigel, G.M. Review: R28 retinal precursor cells: The first 20 years. *Mol. Vis.* **2014**, *20*, 301–306. [PubMed]
27. Seigel, G.M.; Mutchler, A.L.; Imperato, E.L. Expression of glial markers in a retinal precursor cell line. *Mol. Vis.* **1996**, *2*, 9233983.

28. Seigel, G.M.; Sun, W.; Wang, J.; Hershberger, D.H.; Campbell, L.M.; Salvi, R.J. Neuronal gene expression and function in the growth-stimulated R28 retinal precursor cell line. *Curr. Eye Res.* **2004**, *28*, 257–269. [CrossRef]
29. Boriushkin, E.; Wang, J.J.; Li, J.; Jing, G.; Seigel, G.M.; Zhang, S.X. Identification of p58IPK as a novel neuroprotective factor for retinal neurons. *Investig. Ophthalmol. Vis. Sci.* **2015**, *56*, 1374–1386. [CrossRef]
30. Kong, D.; Gong, L.; Arnold, E.; Shanmugam, S.; Fort, P.E.; Gardner, T.W.; Abcouwer, S.F. Insulin-like growth factor 1 rescues R28 retinal neurons from apoptotic death through ERK-mediated BimEL phosphorylation independent of Akt. *Exp. Eye Res.* **2016**, *151*, 82–95. [CrossRef]
31. Goc, A.; Kochuparambil, S.T.; Al-Husein, B.; Al-Azayzih, A.; Mohammad, S.; Somanath, P.R. Simultaneous modulation of the intrinsic and extrinsic pathways by simvastatin in mediating prostate cancer cell apoptosis. *BMC Cancer* **2012**, *12*, 409. [CrossRef] [PubMed]
32. Song, Y.; Hong, S.; Iizuka, Y.; Kim, C.Y.; Seong, G.J. The neuroprotective effect of maltol against oxidative stress on rat retinal neuronal cells. *Korean J. Ophthalmol.* **2015**, *29*, 58–65. [CrossRef]
33. Geisert, E.E., Jr.; Frankfurter, A. The neuronal response to injury as visualized by immunostaining of class III beta-tubulin in the rat. *Neurosci. Lett.* **1989**, *102*, 137–141. [CrossRef]
34. Iwanaga, T.; Takahashi, Y.; Fujita, T. Immunohistochemistry of neuron-specific and glia-specific proteins. *Arch. Histol. Cytol.* **1989**, *52*, 13–24. [CrossRef]
35. Palani, C.D.; Fouda, A.Y.; Liu, F.; Xu, Z.; Mohamed, E.; Giri, S.; Smith, S.B.; Caldwell, R.B.; Narayanan, S.P. Deletion of Arginase 2 Ameliorates Retinal Neurodegeneration in a Mouse Model of Multiple Sclerosis. *Mol. Neurobiol.* **2019**, *56*, 8589–8602. [CrossRef] [PubMed]
36. Candadai, A.A.; Liu, F.; Fouda, A.Y.; Alfarhan, M.; Palani, C.D.; Xu, Z.; Caldwell, R.B.; Narayanan, S.P. Deletion of arginase 2 attenuates neuroinflammation in an experimental model of optic neuritis. *PLoS ONE* **2021**, *16*, e0247901. [CrossRef]
37. Long, J.; Wang, Q.; He, H.; Sui, X.; Lin, G.; Wang, S.; Yang, J.; You, P.; Luo, Y.; Wang, Y. NLRP3 inflammasome activation is involved in trimethyltin-induced neuroinflammation. *Brain Res.* **2019**, *1718*, 186–193. [CrossRef]
38. Miyagi, A.; Kawashiri, T.; Shimizu, S.; Shigematsu, N.; Kobayashi, D.; Shimazoe, T. Dimethyl Fumarate Attenuates Oxaliplatin-Induced Peripheral Neuropathy without Affecting the Anti-tumor Activity of Oxaliplatin in Rodents. *Biol. Pharm. Bull.* **2019**, *42*, 638–644. [CrossRef] [PubMed]
39. Clarner, T.; Janssen, K.; Nellessen, L.; Stangel, M.; Skripuletz, T.; Krauspe, B.; Hess, F.M.; Denecke, B.; Beutner, C.; Linnartz-Gerlach, B.; et al. CXCL10 triggers early microglial activation in the cuprizone model. *J. Immunol.* **2015**, *194*, 3400–3413. [CrossRef] [PubMed]
40. Gobel, K.; Ruck, T.; Meuth, S.G. Cytokine signaling in multiple sclerosis: Lost in translation. *Mult. Scler.* **2018**, *24*, 432–439. [CrossRef]
41. Sharief, M.K.; Hentges, R. Association between tumor necrosis factor-alpha and disease progression in patients with multiple sclerosis. *N. Engl. J. Med.* **1991**, *325*, 467–472. [CrossRef]
42. Lei, Q.; Tan, J.; Yi, S.; Wu, N.; Wang, Y.; Wu, H. Mitochonic acid 5 activates the MAPK-ERK-yap signaling pathways to protect mouse microglial BV-2 cells against TNFalpha-induced apoptosis via increased Bnip3-related mitophagy. *Cell Mol. Biol. Lett.* **2018**, *23*, 14. [CrossRef]
43. Neniskyte, U.; Vilalta, A.; Brown, G.C. Tumour necrosis factor alpha-induced neuronal loss is mediated by microglial phagocytosis. *FEBS Lett.* **2014**, *588*, 2952–2956. [CrossRef]
44. Pang, Y.; Qin, M.; Hu, P.; Ji, K.; Xiao, R.; Sun, N.; Pan, X.; Zhang, X. Resveratrol protects retinal ganglion cells against ischemia induced damage by increasing Opa1 expression. *Int. J. Mol. Med.* **2020**, *46*, 1707–1720. [CrossRef]
45. Mathew, B.; Chennakesavalu, M.; Sharma, M.; Torres, L.A.; Stelman, C.R.; Tran, S.; Patel, R.; Burg, N.; Salkovski, M.; Kadzielawa, K.; et al. Autophagy and post-ischemic conditioning in retinal ischemia. *Autophagy* **2020**, *17*, 1479–1499. [CrossRef]
46. Tan, J.; Digicaylioglu, M.; Wang, S.X.J.; Dresselhuis, J.; Dedhar, S.; Mills, J. Insulin attenuates apoptosis in neuronal cells by an integrin-linked kinase-dependent mechanism. *Heliyon* **2019**, *5*, e02294. [CrossRef] [PubMed]
47. Kale, N. Optic neuritis as an early sign of multiple sclerosis. *Eye Brain* **2016**, *8*, 195–202. [CrossRef] [PubMed]
48. Kimura, A.; Namekata, K.; Guo, X.; Noro, T.; Harada, C.; Harada, T. Targeting Oxidative Stress for Treatment of Glaucoma and Optic Neuritis. *Oxidative Med. Cell Longev.* **2017**, *2017*, 2817252. [CrossRef]
49. Gilgun-Sherki, Y.; Melamed, E.; Offen, D. The role of oxidative stress in the pathogenesis of multiple sclerosis: The need for effective antioxidant therapy. *J. Neurol.* **2004**, *251*, 261–268. [CrossRef] [PubMed]
50. Guo, X.; Harada, C.; Namekata, K.; Matsuzawa, A.; Camps, M.; Ji, H.; Swinnen, D.; Jorand-Lebrun, C.; Muzerelle, M.; Vitte, P.A.; et al. Regulation of the severity of neuroinflammation and demyelination by TLR-ASK1-p38 pathway. *EMBO Mol. Med.* **2010**, *2*, 504–515. [CrossRef]
51. Azuchi, Y.; Kimura, A.; Guo, X.; Akiyama, G.; Noro, T.; Harada, C.; Nishigaki, A.; Namekata, K.; Harada, T. Valproic acid and ASK1 deficiency ameliorate optic neuritis and neurodegeneration in an animal model of multiple sclerosis. *Neurosci. Lett.* **2017**, *639*, 82–87. [CrossRef]
52. Guo, X.; Harada, C.; Namekata, K.; Kimura, A.; Mitamura, Y.; Yoshida, H.; Matsumoto, Y.; Harada, T. Spermidine alleviates severity of murine experimental autoimmune encephalomyelitis. *Investig. Ophthalmol. Vis. Sci.* **2011**, *52*, 2696–2703. [CrossRef] [PubMed]

53. Guo, X.; Namekata, K.; Kimura, A.; Noro, T.; Azuchi, Y.; Semba, K.; Harada, C.; Yoshida, H.; Mitamura, Y.; Harada, T. Brimonidine suppresses loss of retinal neurons and visual function in a murine model of optic neuritis. *Neurosci. Lett.* **2015**, *592*, 27–31. [CrossRef] [PubMed]

54. Yevgi, R.; Demir, R. Oxidative stress activity of fingolimod in multiple sclerosis. *Clin. Neurol. Neurosurg.* **2021**, *202*, 106500. [CrossRef]

55. Ryba, D.M.; Warren, C.M.; Karam, C.N.; Davis, R.T., 3rd; Chowdhury, S.A.K.; Alvarez, M.G.; McCann, M.; Liew, C.W.; Wieczorek, D.F.; Varga, P.; et al. Sphingosine-1-Phosphate Receptor Modulator, FTY720, Improves Diastolic Dysfunction and Partially Reverses Atrial Remodeling in a Tm-E180G Mouse Model Linked to Hypertrophic Cardiomyopathy. *Circ. Heart Fail.* **2019**, *12*, e005835. [CrossRef] [PubMed]

56. Vargas-Medrano, J.; Segura-Ulate, I.; Yang, B.; Chinnasamy, R.; Arterburn, J.B.; Perez, R.G. FTY720-Mitoxy reduces toxicity associated with MSA-like alpha-synuclein and oxidative stress by increasing trophic factor expression and myelin protein in OLN-93 oligodendroglia cell cultures. *Neuropharmacology* **2019**, *158*, 107701. [CrossRef] [PubMed]

57. Wu, H.; Wang, X.; Gao, J.; Liang, S.; Hao, Y.; Sun, C.; Xia, W.; Cao, Y.; Wu, L. Fingolimod (FTY720) attenuates social deficits, learning and memory impairments, neuronal loss and neuroinflammation in the rat model of autism. *Life Sci.* **2017**, *173*, 43–54. [CrossRef]

58. Ng, X.; Sadeghian, M.; Heales, S.; Hargreaves, I.P. Assessment of Mitochondrial Dysfunction in Experimental Autoimmune Encephalomyelitis (EAE) Models of Multiple Sclerosis. *Int. J. Mol. Sci.* **2019**, *20*, 4975. [CrossRef]

59. Mancini, A.; Tantucci, M.; Mazzocchetti, P.; de Iure, A.; Durante, V.; Macchioni, L.; Giampa, C.; Alvino, A.; Gaetani, L.; Costa, C.; et al. Microglial activation and the nitric oxide/cGMP/PKG pathway underlie enhanced neuronal vulnerability to mitochondrial dysfunction in experimental multiple sclerosis. *Neurobiol. Dis.* **2018**, *113*, 97–108. [CrossRef]

60. Sadeghian, M.; Mastrolia, V.; Rezaei Haddad, A.; Mosley, A.; Mullali, G.; Schiza, D.; Sajic, M.; Hargreaves, I.; Heales, S.; Duchen, M.R.; et al. Mitochondrial dysfunction is an important cause of neurological deficits in an inflammatory model of multiple sclerosis. *Sci. Rep.* **2016**, *6*, 33249. [CrossRef]

61. Chang, J.Y.; Yu, F.; Shi, L.; Ko, M.L.; Ko, G.Y. Melatonin Affects Mitochondrial Fission/Fusion Dynamics in the Diabetic Retina. *J. Diabetes Res.* **2019**, *2019*, 8463125. [CrossRef] [PubMed]

62. Nivison, M.P.; Ericson, N.G.; Green, V.M.; Bielas, J.H.; Campbell, J.S.; Horner, P.J. Age-related accumulation of phosphorylated mitofusin 2 protein in retinal ganglion cells correlates with glaucoma progression. *Exp. Neurol.* **2017**, *296*, 49–61. [CrossRef] [PubMed]

63. Gil, A.; Martin-Montanez, E.; Valverde, N.; Lara, E.; Boraldi, F.; Claros, S.; Romero-Zerbo, S.Y.; Fernandez, O.; Pavia, J.; Garcia-Fernandez, M. Neuronal Metabolism and Neuroprotection: Neuroprotective Effect of Fingolimod on Menadione-Induced Mitochondrial Damage. *Cells* **2020**, *10*, 34. [CrossRef]

64. Gimenez-Molina, Y.; Garcia-Martinez, V.; Villanueva, J.; Davletov, B.; Gutierrez, L.M. Multiple sclerosis drug FTY-720 toxicity is mediated by the heterotypic fusion of organelles in neuroendocrine cells. *Sci. Rep.* **2019**, *9*, 18471. [CrossRef] [PubMed]

65. Vidal-Martinez, G.; Segura-Ulate, I.; Yang, B.; Diaz-Pacheco, V.; Barragan, J.A.; De-Leon Esquivel, J.; Chaparro, S.A.; Vargas-Medrano, J.; Perez, R.G. FTY720-Mitoxy reduces synucleinopathy and neuroinflammation, restores behavior and mitochondria function, and increases GDNF expression in Multiple System Atrophy mouse models. *Exp. Neurol.* **2020**, *325*, 113120. [CrossRef]

66. Moon, M.H.; Jeong, J.K.; Lee, Y.J.; Park, S.Y. FTY720 protects neuronal cells from damage induced by human prion protein by inactivating the JNK pathway. *Int. J. Mol. Med.* **2013**, *32*, 1387–1393. [CrossRef] [PubMed]

Article

Annexin A1 Mimetic Peptide and Piperlongumine: Anti-Inflammatory Profiles in Endotoxin-Induced Uveitis

Ana Paula Girol [1,2,3], Caroline de Freitas Zanon [2], Ícaro Putinhon Caruso [4], Sara de Souza Costa [1,2], Helena Ribeiro Souza [1,2], Marinônio Lopes Cornélio [4] and Sonia Maria Oliani [2,3,5,*]

[1] Department of Physical and Morphological Sciences, University Center Padre Albino (UNIFIPA), Catanduva 15809-144, SP, Brazil; anapaula.girol@unifipa.com.br (A.P.G.); sarah_sc_0705@hotmail.com (S.d.S.C.); helena.souza@unifipa.com.br (H.R.S.)
[2] Department of Biology, Institute of Biosciences, Humanities and Exact Sciences (Ibilce), São Paulo State University, (UNESP), São José do Rio Preto 15054-000, SP, Brazil; carolfzanon@yahoo.com.br
[3] Post Graduate Program in Structural and Functional Biology, Escola Paulista de Medicina (UNIFESP-EPM), Federal University of São Paulo, São Paulo 04023-062, SP, Brazil
[4] Department of Phisics, Institute of Biosciences, Humanities and Exact Sciences (Ibilce), São Paulo State University, (UNESP), São José do Rio Preto 15054-000, SP, Brazil; icaro.caruso@unesp.br (Í.P.C.); mariol.cornelio@gmail.com (M.L.C.)
[5] Advanced Research Center in Medicine (CEPAM), União das Faculdades dos Grandes Lagos (Unilago), São José do Rio Preto 15030-070, SP, Brazil
* Correspondence: sonia.oliani@unesp.br; Tel.: +55-17-32212381; Fax: +55-17-32212390

 **check for updates**

Citation: Girol, A.P.; de Freitas Zanon, C.; Caruso, Í.P.; de Souza Costa, S.; Souza, H.R.; Cornélio, M.L.; Oliani, S.M. Annexin A1 Mimetic Peptide and Piperlongumine: Anti-Inflammatory Profiles in Endotoxin-Induced Uveitis. *Cells* **2021**, *10*, 3170. https://doi.org/10.3390/cells10113170

Academic Editors: Maurice Ptito and Joseph Bouskila

Received: 12 September 2021
Accepted: 28 October 2021
Published: 15 November 2021

Publisher's Note: MDPI stays neutral with regard to jurisdictional claims in published maps and institutional affiliations.

Abstract: Uveitis is one of the main causes of blindness worldwide, and therapeutic alternatives are worthy of study. We investigated the effects of piperlongumine (PL) and/or annexin A1 (AnxA1) mimetic peptide Ac2-26 on endotoxin-induced uveitis (EIU). Rats were inoculated with lipopolysaccharide (LPS) and intraperitoneally treated with Ac2-26 (200 µg), PL (200 and 400 µg), or Ac2-26 + PL after 15 min. Then, 24 h after LPS inoculation, leukocytes in aqueous humor, mononuclear cells, AnxA1, formyl peptide receptor (fpr)1, fpr2, and cyclooxygenase (COX)-2 were evaluated in the ocular tissues, along with inflammatory mediators in the blood and macerated supernatant. Decreased leukocyte influx, levels of inflammatory mediators, and COX-2 expression confirmed the anti-inflammatory actions of the peptide and pointed to the protective effects of PL at higher dosage. However, when PL and Ac2-26 were administered in combination, the inflammatory potential was lost. AnxA1 expression was elevated among groups treated with PL or Ac2-26 + PL but reduced after treatment with Ac2-26. Fpr2 expression was increased only in untreated EIU and Ac2-26 groups. The interaction between Ac2-26 and PL negatively affected the anti-inflammatory action of Ac2-26 or PL. We emphasize that the anti-inflammatory effects of PL can be used as a therapeutic strategy to protect against uveitis.

Keywords: eye inflammation; lipopolysaccharide; natural bioactive extracts; Ac2-26; FPR receptor; inflammatory mediators

1. Introduction

Uveitis is an intraocular inflammation of different etiologies [1–5] characterized by leukocyte accumulation in ocular tissues and cytokine release. It is a painful condition and is associated with redness, photophobia, impaired vision, and blindness [6–10]

Pharmacological treatment for uveitis includes corticosteroids, chemotherapeutic agents, and tumor necrosis factor (TNF)-α inhibitors [3,7,8,11], but the use of these drugs is limited by their serious side effects, such as increased ocular pressure or cytotoxicity [9,12]. However, recent advances in the mechanisms of inflammation and the discovery of several endogenous anti-inflammatory mediators have provided new therapeutic possibilities for uveitis treatment [5,7,10,13–15].

In particular, the endogenous protein annexin A1 (AnxA1) may represent an alternative therapy for uveitis [16–20]. AnxA1 is an anti-inflammatory 37 kDa protein, which exhibits calcium and membrane phospholipid binding sites and is involved in the inhibition of glucocorticoid-induced eicosanoids and phospholipase A2 synthesis [21–24]. Structurally, AnxA1 comprises a specific small N-terminal region and a central domain formed by four to eight replicates of a highly conserved 70 to 80 amino acids sequence [25–27]. The N-terminal domain contains sites for post-translational processes, such as phosphorylation, glycosylation, and proteolysis [17,24,28,29].

Over the years, our research group has investigated the effect of AnxA1 on different ocular inflammatory conditions [17,19,30–35]. Positive modulation of endogenous AnxA1 in inflammatory cells in the eyes of mice and retinal pigment epithelial cells (ARPE-19) infected with *Toxoplasma gondii* suggests the protein can be used as a therapeutic target in ocular toxoplasmosis [30]. AnxA1 is involved in the signaling cascades of inflammatory processes, leading to decreased cell proliferation and increased migration by modulation of connective tissue growth factor (CTGF) and lecithin retinol acyltransferase (LRAT) gene expression in ARPE-19 cells [19]. In an experimental allergic conjunctivitis model using wild and AnxA1-null Balb/c mice, administration of the AnxA1-N-terminal region mimetic peptide (Ac2-26) was effective in reducing interleukin (IL)-2, IL-4, IL-10, IL-13, eotaxin, and regulated upon activation normal T cell expressed and presumably secreted (RANTES) [32]. In addition, the potential involvement of the formyl peptide receptor (fpr) family in the protective effect of Ac2-26 was investigated in the same allergic conjunctivitis model [35]. In a *Pseudomonas aeruginosa* keratitis model, the overexpression of AnxA1 and fpr2 occurred in the corneas of Balb/c mice and especially C57BL/6 mice, which is more susceptible to pathogens and infectious antigens [34]. Concerning the uveitis, the expression of AnxA1 in leukocytes and aqueous humor (AqH) was observed in endotoxin-induced uveitis (EIU) in rats [16], with this protein noted as one of the essential mediators in the inflammatory homeostasis process. Moreover, the mechanism of action and potential use of AnxA1 and Ac2-26 were demonstrated in EIU in rodents and lipopolysaccharide (LPS)-activated ARPE-19 cells [17]. The results of this investigation showed that following specific serine phosphorylation, AnxA1 can be translocated to the cell surface, where it interacts with fpr2 and inhibits the release of inflammatory mediators independent of the nuclear factor (NF)-kB signaling pathway and in a post-translational manner.

In recent years, another potent anti-inflammatory mediator, piperlongumine (PL) (5,6-dihydro-1-[(2E)-1-oxo-3-(3,4,5-trimethoxyphenyl)-2-propenyl]-2(1H)pyridinone), a biologically active component of *Piper* species (Piperaceae), has attracted the attention of our research group for its possible interaction with AnxA1 [36]. In particular, PL is the main alkaloid of long pepper (*Piper longum* L.), and its pharmacological actions include cytotoxic, antitumor, antiangiogenic, antiplatelet, antibacterial, antidiabetic, antianxiolytic, antiatherosclerotic, and antifungal effects [36–43]. PL induces apoptosis by interfering with redox and reactive oxygen species (ROS) homeostatic regulators, such as glutathione S-transferase pi 1 (GSTP1) and carbonyl reductase (CBR1) [44]. Moreover, on nonsmall cell lung cancer (NSCLC) in vivo and in vitro, *PL* suppressed lung cancer cell growth in a dose-dependent manner via inhibition of the NF-κB signaling pathway [45].

Regarding inflammation, LPS insult on PL protected the vascular barrier integrity by inhibiting hyperpermeability, expression of cellular adhesion molecules (CAMs), and adhesion and migration of leukocytes, thus endorsing its usefulness as a therapy for vascular inflammatory diseases [46]. Moreover, PL and derivatives reduced the amount of nitric oxide (NO) in LPS-stimulated RAW264.7 macrophages [47]. The protective effect of *P. longum* alkaloid extract containing piperine and PL on dopaminergic neurons against inflammatory reaction was observed in LPS-induced damage. The active extract attenuated the depletion of dopamine in the striatum, facilitated the survival of damaged neurons by inhibiting microglial activation, suppressed the release of neurotoxic factors, and improved LPS-induced behavioral dysfunctions [48]. For neuroinflammation caused by LPS in a model of amyloidogenesis, PL exhibited anti-inflammatory and antiamyloidogenic effects by inhibiting NF-κB [49].

Our investigations indicated that PL attenuated systemic and pulmonary inflammatory changes, partially by modulating the expression of the endogenous AnxA1, in lung inflammation induced by cigarette smoke [42]. The potential of PL as a therapeutic immunomodulator for cancer prevention and progression was reinforced by analyzing PL administration in human cancer cells from an epidermoid carcinoma of the larynx (Hep-2) and umbilical vein endothelial cells (HUVEC), in which PL modulated the expression of genes involved in inflammatory processes [36]. Although there are several publications related to PL, few studies have focused on its anti-inflammatory role, and the actions of PL in ocular inflammation are not known. Building upon these observations, we decided to investigate the role of PL as an alternative therapy for EIU.

Ac2-26 and PL interaction has been explored by our group through different molecular or computational screening techniques, such as phage display and molecular docking [36]. In silico analyses showed that, among other PL molecules, there was a terpene that appeared to interact with lysine 9 from AnxA1 in the region corresponding to the N-terminal peptide Ac2-26 [36]. However, the physiological reason for this interaction, whether positive or negative in vivo, with regard to the anti-inflammatory effects of AnxA1 is not yet understood, which opens up a new and stimulating field for research.

Given the above, we tested PL in EIU either alone or in coadministration with the peptide Ac2-26, followed by analyses of the leukocyte influx and inflammatory mediators, to verify the following hypotheses: (1) there is anti-inflammatory potential of PL in EIU; (2) Ac2-26 + PL coadministration may interfere with the anti-inflammatory response profile of Ac2-26, favoring or attenuating the effects of its administration on experimental uveitis.

2. Materials and Methods

2.1. Experimental Model of Uveitis and Treatment Protocols

Male Wistar rats, 6 to 8 weeks of age (200 g), were distributed in 7 groups (n = 10/group). The animals were kept in cages in a temperature-controlled environment (22 to 25 °C) and received water and food ad libitum. The experimental procedures were conducted according to the guidelines for biomedical research stated by the Brazilian Societies of Experimental Biology and also approved by the Ethic Committee on Animal Use of University Center Padre Albino (Certificate (No. 11/14). The experiments were designed to minimize the number of animals used and their suffering during the execution of the protocols. All animals were evaluated daily by the institution's veterinarian.

For the development of EIU, rats were anesthetized with isoflurane (1%) and inoculated subcutaneously in the right footpad with 200 μg (1 mg/kg) of lipopolysaccharide (LPS type *Escherichia coli* serotype 0127: B8, Sigma Chemical Co. Poole, Dorset, UK) diluted in 100 μL of phosphate-buffered saline (PBS) [16,17].

The anti-inflammatory effects of Ac2-26 peptide (Ac-AMVSEFLKQAWFIENEEQE-YVQTVK, Thermo Fisher Scientific, Grand Island, NY, USA) and PL ($C_{17}H_{19}NO_5$, CAS number: 20069-09-4, Sigma-Aldrich/Merck, Darmstadt, Hesse, Germany) administered singly or in combination were tested by intraperitoneal (ip) injection 15 min after LPS induction in five EIU groups (n = 10/group) (Figure 1) [17]. The dosage of Ac2-26 at 200 μg (1 mg/kg or 539 μM) diluted in 100 μL of PBS was based on previous studies [17]. For the selection of the PL dosage and for the purpose of comparison with the peptide, we chose two dosages. The lowest dosage of 200 μg (1 mg/kg or 1.57 mM) was equal to the one used for the peptide, while the highest dosage was based on another investigation by our group, which tested the therapeutic efficacy of PL by ip administration at a dosage of 400 μg (2 mg/kg or 3.15 mM) diluted in 100 μL of 10% dimethyl sulfoxide (DMSO, Gold Lab; Ribeirão Preto, São Paulo, Brazil) [42].

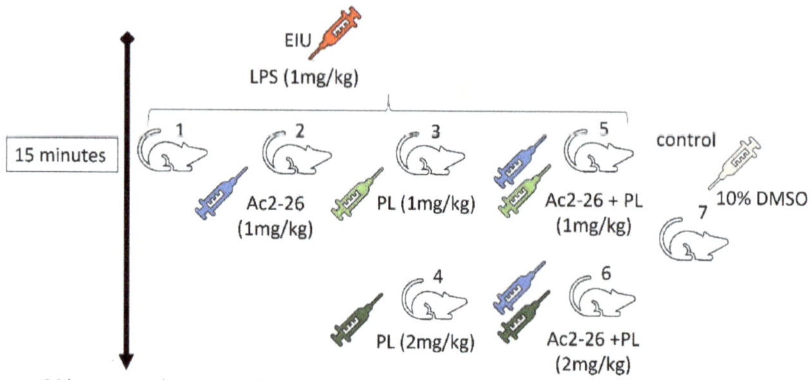

Figure 1. Schematic representation of experimental groups. Induced by (**1**) LPS, (**2**) LPS and treated with Ac2-26 200 µg diluted in PBS, (**3**) LPS and treated with PL 200 µg diluted in 10% DMSO, (**4**) LPS and treated with PL 400 µg diluted in 10% DMSO, (**5**) LPS and treated with Ac2-26 200 µg diluted in PBS and PL 200 µg diluted in 10% DMSO, (**6**) LPS and treated with Ac2-26 200 µg diluted in PBS and PL 400 µg diluted in 10% DMSO, and (**7**) uninduced, administered with 10% DMSO ($n = 5$/group).

Rats that received an intraperitoneal injection of 10% DMSO were used as the control group. The animals were anesthetized with isoflurane (1%) before each experimental treatment and euthanized 24 h after LPS inoculation by excessive dose of the anesthetic.

2.2. Histopathological and Quantitative Analyses

AqH was collected by puncturing the anterior chamber of the left eyes, and 10 µL samples were used and stained in Turk's solution (90 µL). Blood was collected by cardiac puncture, and 10 µL samples were diluted in 190 µL of Turk's solution. Neutrophils and monocytes were quantified in the Neubauer chamber. Values for quantification of AqH and blood leukocytes were expressed as mean $\pm$ standard error (SEM) of the average number of cells $\times 10^5$/mL in the AqH samples and the number of cells $\times 10^3$/mL in the blood samples [31].

After AqH collection, the left eyes were fixed in 4% formaldehyde, dehydrated in increasing order of alcohol content, and placed in paraffin for histopathological analyses and immunohistochemistry. These analyses were performed in the Leica DM500 microscope (Leica, Wetzlar, Hessen, Germany).

2.3. Immunohistochemical and Densitometric Studies

Detection of the AnxA1 protein, fpr1 and fpr2 receptors, cyclooxygenase (COX)-2 enzyme, and phagocytic mononuclear cells (macrophages and monocytes) were performed in 5 µm sections of the paraffin-embedded material. After an antigen retrieval step using citrate buffer (pH 6), the endogenous peroxide activity was blocked, and the sections were incubated overnight at 4 °C with the primary rabbit polyclonal antibodies anti-AnxA1 (1:2000) (Invitrogen, Camarillo, CA, USA), anti-fpr1 and anti-fpr2 (1:1000) (Bioss Inc., Woburn, MA, USA), anti-COX-2 (1:1000) (Bioss Inc, Wo-burn, MA, USA) and with monoclonal anti-ED-1 (monocytes and macrophages) (1:1000) (Millipore, Temecula, CA, USA) diluted in 1% BSA. Subsequently, the slides were incubated with biotinylated secondary antibody (Histostain kit, Invitrogen, Carlsbad, CA, USA). Positive staining was detected using a peroxidase-conjugated streptavidin complex, and color was developed using diaminobenzidine substrate (DAB Kit, Invitrogen, Carlsbad, CA, USA). The sections were counterstained with hematoxylin.

ED-1 positive cells were quantified in the anterior segment of the eyes of different groups with the aid of the Leica Image Analysis software, and the values obtained were expressed as number of cells per mm^2. For the protein densitometric analyses, 3 different slides from each animal were used, and 15 points were analyzed in 3 regions of the cornea, iris, and ciliary processes for an average related to the intensity of immunoreactivity. The values were obtained as arbitrary units [17].

2.4. Inflammatory Mediator Levels

The intact right eyes of all the studied groups were macerated in liquid nitrogen and placed in eppendorfs, which were added with 500 μL of protease (Protease Inhibitor Cocktail Set I, Cat. No. 53391, Millipore Corporation, CA, USA) and phosphatase (PhosphoSafe, Cat. No. 7,126-3-3, Novagen, Millipore Corporation, Billerica, CA, USA) inhibitor solution prepared according to the manufacturer's instructions. The material was incubated for 20 min at 4 °C under constant stirring and centrifuged at 14,000 RPM for 10 min at 4 °C. The supernatants were then collected and immediately frozen at −80 °C. The protein concentration in the supernatant was measured using a Bradford assay (Bio-Rad, Hemel Hempstead, UK).

IL-1β, IL-6, IL-10, monocyte chemoattractant protein (MCP)-1, and TNF-α inflammatory mediators were quantified in the eye macerate supernatant and in blood plasma using the rat cytokine MILLIPLEX MAP Kit (RECYTMAG-65K; Millipore Corporation, Billerica, CA, USA) according to the manufacturer's instructions and analyzed on the LUMINEX xMAP MAGPIX (Millipore Corporation, Billerica, CA, USA) equipment. The concentration of analytes was determined by MAGPIX xPONENT software (Millipore Corporation, Billerica, CA, USA). Results were expressed as mean ± SEM of cytokine concentrations (pg/mL).

2.5. Statistical Analyses

The results were first submitted to descriptive analysis and normality determination. As the samples presented normal distribution, the analysis of variance (ANOVA) was used, followed by the Bonferroni test. All values were expressed as mean ± SEM and *p* values less than 0.05 were considered statistically significant.

3. Results

3.1. Singly Administered, the Treatments Inhibited the Influx of Leukocytes, Indicating Protective Effects of PL, Especially at 400 µg Dosage, and Confirming the Anti-Inflammatory Action of Ac2-26, but These Effects Were Lost with Coadministration

Transmigrated leukocytes were absent in the control eyes (Figure 2A), but a high influx of these cells, mainly neutrophils, occurred 24 h after LPS inoculation without treatment (Figure 2B). The anterior eye segment was the most affected, and the inflammatory cells were observed in AqH, anterior and posterior chambers, and also in iris, ciliary body, and ciliary process stroma (Figure 2B). Except for Ac2-26 + PL 400 µg group (Figure 2G), fewer transmigrated leukocytes were presented after treatments (Figure 2C–F), especially in Ac2-26 (Figure 2C) and PL 400 µg (Figure 2E) groups.

Decreased neutrophil transmigration into AqH was verified after Ac2-26 ($p < 0.001$) and PL 400 µg ($p < 0.01$) administrations compared to the untreated LPS group (Figure 2H). The influx of monocytes into AqH was also reduced by treatments ($p < 0.001$), except in the Ac2-26 + PL 400 µg group (Figure 2I). Similarly, in blood quantifications, higher numbers of neutrophils and monocytes were observed, especially in the LPS and Ac2-26 + PL 400 µg groups, with a marked reduction after Ac2-26 and PL 400 µg treatments (Figure 3A,B).

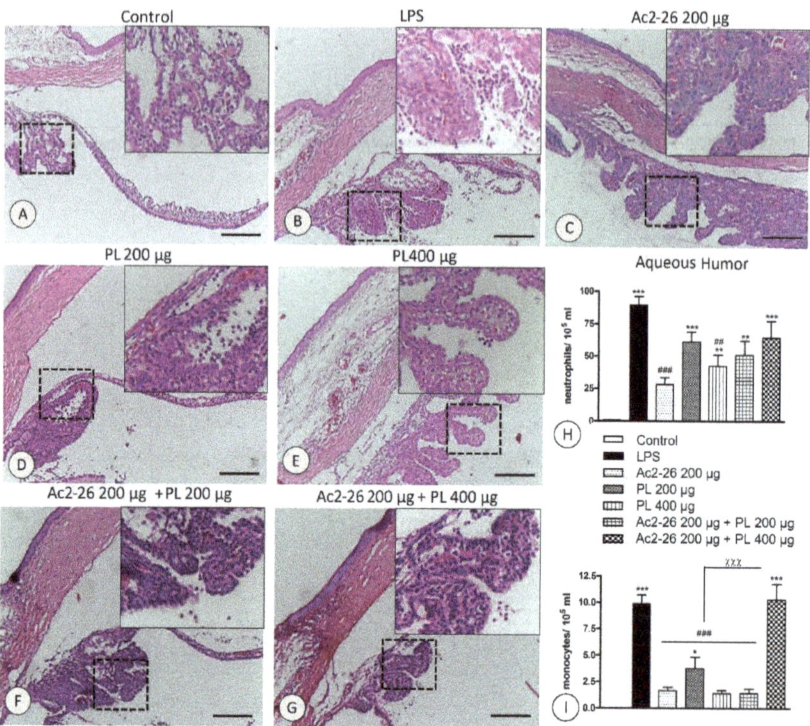

Figure 2. Histopathological analyses of ocular tissues in EIU. Absence of leukocytes in control tissues (**A**). Influx of neutrophils after 24 h LPS (**B**) and treated with Ac2-26 + PL 400 μg (**G**). Significant decrease in cellular extravasation after systemic treatments with Ac2-26 (**C**) and PL 400 μg (**E**) and moderate inflammatory influx reduction after treatments with PL 200 μg (**D**) and Ac2-26 + PL 200 μg (**F**). The details show enlargements of dashed areas. Sections: 5 μm, stain: HE, bars: 100 μm. Quantitative analyses of neutrophils (**H**) and monocytes (**I**) in the aqueous humor. The data show mean ± SEM of neutrophils and monocytes × 10^5 mL in the eyes of control, untreated (LPS), and treated (Ac2-26 200 μg, PL 200 μg, PL 400 μg, Ac2-26 + PL 200 μg, and Ac2-26+ PL 400 μg) rats ($n = 10$ animals/group). *** $p < 0.001$, ** $p < 0.01$, and * $p < 0.05$ versus control; ### $p < 0.001$ and ## $p < 0.01$ versus LPS; and χχχ $p < 0.001$ versus Ac2-26 200 μg, PL 200 μg, PL 400 μg, and Ac2-26 200 μg + PL 200 μg.

Phagocytic mononuclear cells were studied following immunohistochemical reaction with anti-ED-1 antibody in the ciliary body and iris (Figure 3D). Quantification showed a large number of ED-1 positive cells in the LPS and Ac2-26 + PL 400 μg (Figure 3C) groups ($p < 0.001$) compared to the control but significant decrease in the other groups.

Analyses related to influx of neutrophils and monocytes in AqH and blood and phagocytic mononuclear cells in the anterior eye segment showed anti-inflammatory action of Ac2-26 and dose-dependent effects of PL, which was more efficient at 400 μg dosage. In contrast, Ac2-26 + PL coadministration abrogated the peptide inhibitory action on neutrophil and monocyte extravasation, especially in the Ac2-26 + PL 400 μg group.

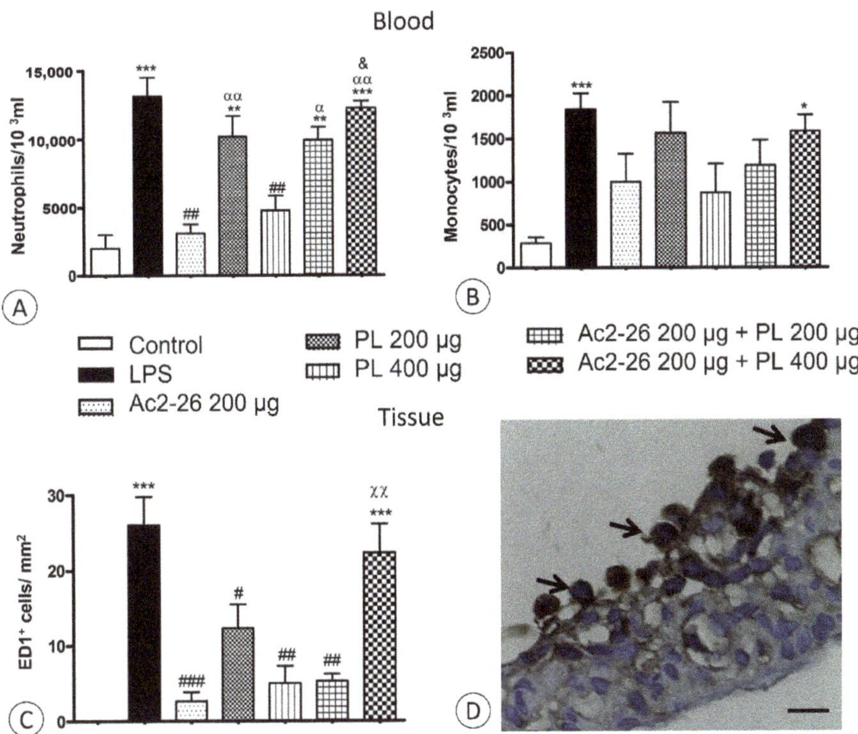

Figure 3. Quantitative analyses of neutrophils (**A**) and monocytes (**B**) in blood. Data show mean ± SEM of neutrophils and monocytes × 10^3 mL in the blood of control, untreated (LPS), and treated (Ac2-26 200 µg, PL 200 µg, PL 400 µg, Ac2-26 + PL 200 µg, and Ac2-26 + PL 400 µg) rats ($n = 10$ animals/group). *** $p < 0.001$, ** $p < 0.01$, and * $p < 0.05$ versus control; ## $p < 0.01$ versus LPS; αα $p < 0.01$ and α $p < 0.05$ versus Ac2-26 200 µg; and & $p < 0.05$ versus PL 400 µg. Quantification of phagocytic mononuclear cells in the anterior segment of the eye (**C**). Data show mean ± SEM of ED-1 positive cells per mm^2 in the eyes of control, untreated (LPS) and treated (Ac2-26 200 µg, PL 200 µg, PL 400 µg, Ac2-26 + PL 200 µg and Ac2-26 + PL 400 µg) rats. ($n = 10$ animals/group). *** $p < 0.001$ versus control; # $p < 0.05$, ## $p < 0.01$, and ### $p < 0.001$ versus LPS; and χχ $p < 0.01$ versus Ac2-26 200 µg, PL 400 µg, and Ac2-26 200 µg + PL 200 µg. ED-1 positive cells (arrows) on iris induced to uveitis and untreated (**D**). Sections: 5 µm, counterstain: hematoxylin, bars: 10 µm.

3.2. Ac2-26 and PL Singly Administered Reduced the Release of Proinflammatory Mediators in EIU, but These Effects Were Abrogated with Coadministration, Especially Ac2-26 + PL 400 µg

The supernatants of macerated ocular tissues of LPS animals and those treated with Ac2-26 + PL 400 µg showed significant increase in total protein levels ($p < 0.001$) compared to the control. In the other treated groups, the protein concentration was reduced in relation to the LPS group (Figure 4A).

The proinflammatory mediators IL-1β, IL-6, TNF-α, and MCP-1 and the anti-inflammatory cytokine IL-10, which are all multifunctional molecules that play important roles in host defense in acute phase inflammatory reactions, were analyzed in supernatants of the ocular tissues after maceration and in the blood plasma. The results indicated low levels of the proinflammatory cytokines in the control eyes and, as expected, significant increased levels in the untreated LPS group, both in the macerated supernatant (Figure 4B,D,F,H) and blood plasma (Figure 4C,E,G,I).

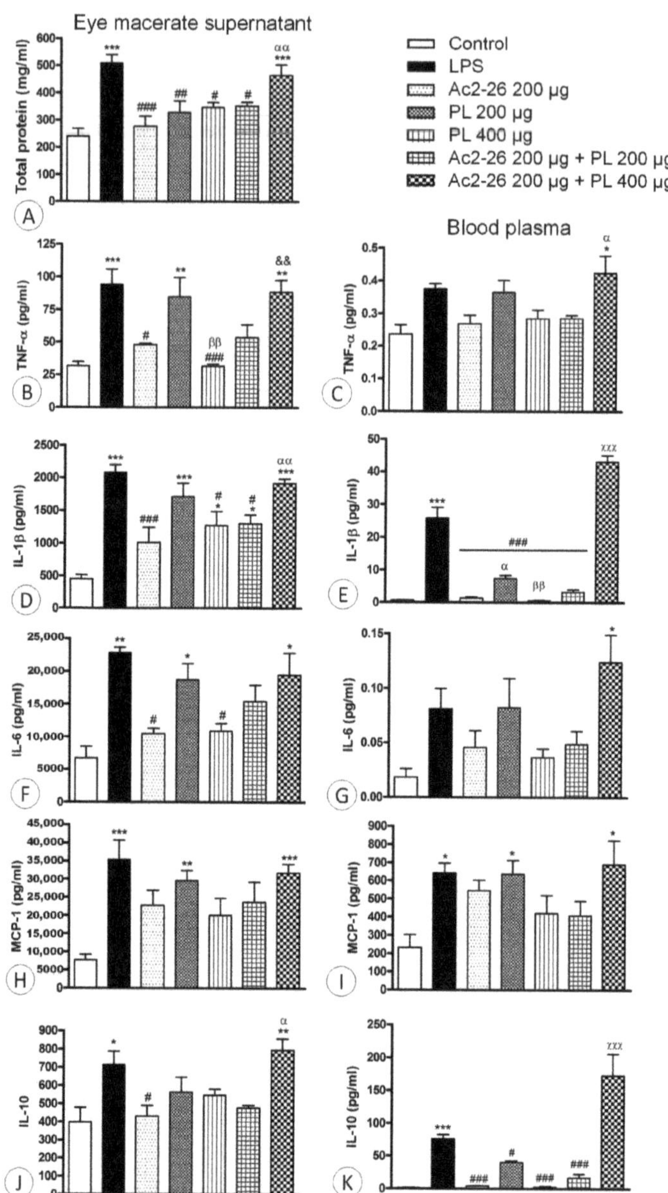

Figure 4. Effects of peptide Ac2-26 and PL, administered singly or in combination, on EIU. Levels of total proteins in supernatants after maceration of ocular tissues (**A**) and dosages of anti-inflammatory mediators TNF-α (**B,C**), IL-1β (**D,E**), IL-6 (**F,G**), and MCP-1 (**H,I**) and proinflammatory cytokine IL-10 (**J,K**) in the macerated eye supernatant and blood plasma. Data expressed as mean ± SEM of mg of proteins/mL and pg of cytokines/mL of control, untreated (LPS), and treated (Ac2-26 200 µg, PL 200 µg, PL 400 µg, Ac2-26 + PL 200 µg, and Ac2-26 + PL 400 µg) rats (n = 10 animals/group). *** $p < 0.001$, ** $p < 0.01$, and * $p < 0.05$ versus control; # $p < 0.05$, ## $p < 0.01$, and ### $p < 0.001$ versus LPS; αα $p < 0.01$ and α $p < 0.05$ versus Ac2-26 200 µg; ββ $p < 0.01$ versus PL 200 µg; && $p < 0.01$ versus PL 400 µg; and χχχ $p < 0.001$ versus all other groups.

Treatments with PL at 200 µg dosage singly administered or in combination with Ac2-26 peptide were able to reduce the levels of IL-1β only in the blood plasma (Figure 4E), while the combined administration of the AnxA1 mimetic peptide and PL at 400 µg dosage did not reduce the proinflammatory cytokine levels (Figure 4B–K). In contrast, when the peptide Ac2-26 or PL at 400 µg dosage were singly administered, IL-1β, IL-6, and TNF-α levels were reduced compared to the LPS group (Figure 4B–K), indicating resolution of the inflammatory process.

Regarding IL-10, increased levels were verified in the LPS and Ac2-26 + PL 400 µg groups in blood plasma and macerated supernatant (Figure 4J,K). Treatments with Ac2-26 peptide and PL (200 and 400 µg) reduced the cytokine concentration in blood plasma and eye supernatant, while Ac2-26 + PL 200 µg administration decreased IL-10 in blood plasma.

3.3. COX-2 Expression Is Not Inhibited after Treatments with PL 200 µg and Ac2-26 + PL 400 µg

In the anterior ocular segment of the control rats, especially in iris, ciliary body, and ciliary processes (Figure 5A), the expression COX-2 enzyme was not detected. However, in the same regions, in the untreated EIU animals (Figure 5B) and after systemic treatments with PL 200 µg and Ac2-26 + PL 400 µg (Figure 5D,G), strong COX-2 immunolabeling was observed. The Ac2-26, PL 400 µg, and Ac2-26 + PL 200 µg groups showed reduced enzyme expression (Figure 5C,E,F), mainly in the singly peptide treated group. There was no immunoreactivity for COX-2 in the reaction control (Figure 5H), confirming antibody specificity.

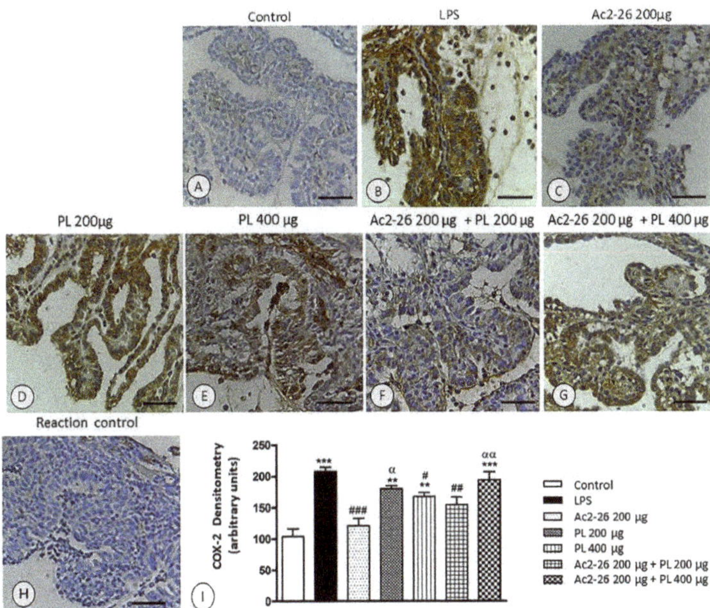

Figure 5. Expression of the COX-2 enzyme in ciliary processes in EIU. Absence of immunostaining in the control eyes (**A**). Strong immunostaining in the untreated, induced EIU animals (LPS) (**B**) and those treated with PL 200 µg (**D**) and Ac2-26 + PL 400 µg (**G**). Decreased expression after systemic treatments with Ac2-26 (**C**), PL 400 µg (**E**), and Ac2-26 + PL 200 µg (**F**). Absence of immunostaining in the control eyes (**H**). Counterstaining: hematoxylin, bars: 10 µm. Densitometric analysis of COX-2 (**I**). Results were obtained as mean ± SEM of the densitometric index of the eyes of control rats, untreated uveitis (LPS), and treated groups (Ac2-26 200 µg, PL 200 µg, PL 400 µg, Ac2-26 + PL 200 µg, and Ac2-26 + PL 400 µg) ($n = 10$ animals/group). *** $p < 0.001$ and ** $p < 0.01$ versus control; # $p < 0.05$, ## $p < 0.01$, and ### $p < 0.001$ versus LPS; αα $p < 0.01$ and α $p < 0.05$ versus Ac2-26 200 µg.

Densitometric analyses corroborated the immunohistochemical observations (Figure 5I), reinforced the anti-inflammatory activities of PL in a dose-dependent manner, and showed that the combination of Ac2-26 and PL, especially at the higher dosage, inhibited the protective action of the peptide and PL singly administered.

3.4. Endogenous AnxA1 Increased during Inflammation in Ocular Tissues, but Ac2-26 Administered Singly or in Combination with PL at Lower Dosage Reduced AnxA1 Immunoreactivity

The immunohistochemical and densitometric analyses of AnxA1 expression in the anterior eye segment showed significant increase in the endogenous protein, especially in the ciliary processes, 24 h after uveitis induction in the untreated group ($p < 0.001$) compared to the control (Figure 6A,B,I).

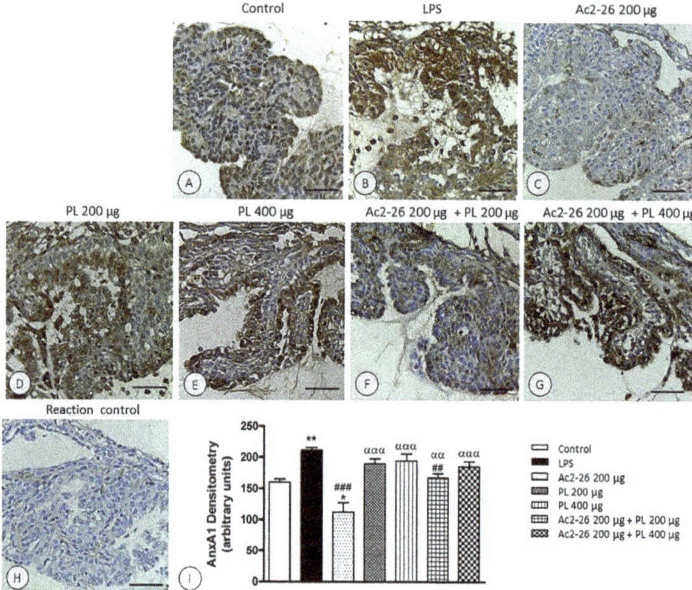

Figure 6. Expression of AnxA1 in ciliary processes in EIU. Increased expression after 24 h of induction of inflammation in the untreated (LPS) (**B**) and treated groups with PL (**D,E**) and Ac2-26 + PL 400 µg (**G**) relative to the control (**A**). Reduction of immunoreactivity after treatments with Ac2-26 (**C**) and Ac2-26 + PL 200 µg (**F**) compared to LPS. Absence of immunoreactivity in the control of the reaction (**H**). Counterstaining: hematoxylin, bars: 10 µm. Densitometric analysis of AnxA1 (**I**). Results were obtained as mean ± SEM of the densitometric index of the eyes of control rats, untreated uveitis (LPS), and treated groups (Ac2-26 200 µg, PL 200 µg, PL 400 µg, Ac2-26 + PL 200 µg, and Ac2-26+ PL 400 µg) ($n = 10$ animals/group). ** $p < 0.01$ and * $p < 0.05$ versus control; ## $p < 0.01$ and ### $p < 0.001$ versus LPS; $\alpha\alpha\alpha$ $p < 0.001$ and $\alpha\alpha$ $p < 0.01$ versus Ac2-262 00 µg.

AnxA1 immunolabeling remained increased after treatments with PL singly administered (Figure 6D,E,I) or in combination with Ac2-26 at 400 µg dosage (Figure 6G,I). In contrast, in the Ac2-26 (Figure 6C) and Ac2-26 + PL 200 µg groups (Figure 6F) decreased AnxA1 expression was observed relative to the LPS group (Figure 6I). In the group treated only with the peptide, the expression of the protein was also reduced compared to control group ($p < 0.05$). The specificity of the immunolabeling was confirmed by the reaction control (Figure 6H).

3.5. Expression of the fpr2 Receptor Is Modulated by Treatment with AnxA1 Peptide but Not with PL Singly Administered or Ac2-26 in Combination with PL 400 μg

The expression of fpr2 receptor in ocular tissues overlapped the location of the AnxA1 with increased labeling in the LPS group ($p < 0.05$) (Figure 7B,J) and mainly in animals treated with AnxA1 peptide ($p < 0.001$) (Figure 7C,J) compared to the control (Figure 7A,J). However, after other treatments, there was no significant change in receptor expression (Figure 7D–G,J). In contrast, we observed no immunoreactivity for fpr1 (Figure 7H) in all the studied groups. The control of the reaction indicated antibody specificity (Figure 7I).

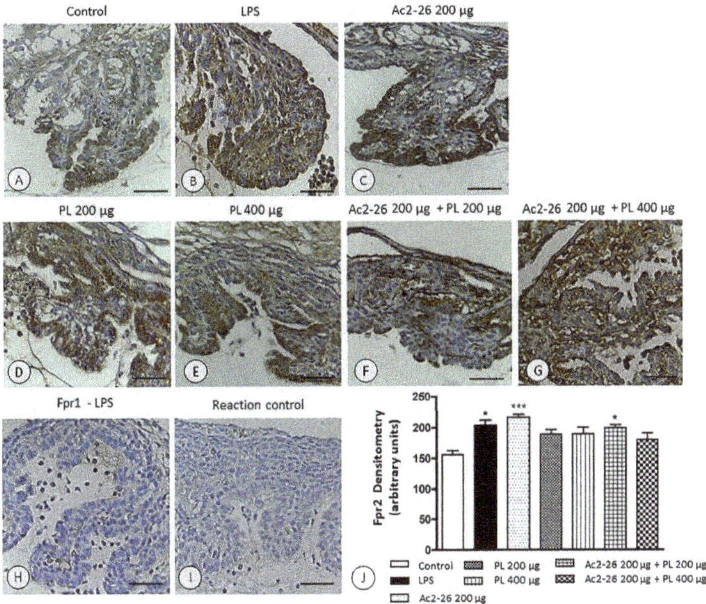

Figure 7. Specific expression of fpr2 in ocular tissues. Strong immunoreactivity for fpr2 in the LPS (**B**) and Ac2-26 (**C**) groups. Similar expressions were found among the control (**A**), PL (**D,E**), and Ac2-26 + PL (**F,G**) groups. Absence of labeling for fpr1 receptor (**H**) and in control of the reaction (**I**). Counterstaining: hematoxylin, bars: 10 μm. Densitometric analysis of fpr2 (**J**). Results were obtained as mean ± SEM of the densitometric index of the eyes of the control rats, untreated uveitis (LPS), and treated groups (Ac2-26 200 μg, PL 200 μg, PL 400 μg, Ac2-26 + PL 200 μg, and Ac2-26 + PL 400 μg) ($n = 10$ animals/group). *** $p < 0.001$ and * $p < 0.05$ versus control.

4. Discussion

Several investigations have explored the anti-inflammatory and protective activities of AnxA1 protein and its mimetic peptides, especially Ac2-26 [25,29]. In recent years, our group has researched the role of AnxA1 in the eye by means of in vivo and in vitro studies and highlighted the potential of the protein in controlling the ocular inflammatory process [16,17,19,34,35]. Recently, the understanding that AnxA1 could interact with PL [36] has opened up a new and exciting field of research. Thus, we proposed an investigation into the effects of PL administered alone in the EIU, and our findings indicate an important anti-inflammatory profile of PL in LPS-induced uveitis. Moreover, we evaluated the coadministration of Ac2-26 + PL in the experimental uveitis. We speculated whether their coadministration would have synergistic or antagonistic effects. It would be expected that the combination would enhance the action against uveitis, but the result proved to be the opposite. It follows from this result that both might also act in close proximity and that a possible peptide structure conformational change occurred.

Our initial analysis, as expected, showed that inflammatory stimuli induced by LPS released inflammatory mediators IL-1β, IL-6, TNF-α, and MCP-1 and increased expression of COX-2 by promoting disruption of the blood–ocular barrier and intense influx of leukocytes, reinforcing previous studies [17,31]. Neutrophils were the predominantly extravasated inflammatory cells, especially near ciliary processes. In EIU, neutrophil transmigration occurs at the base of the ciliary body, whereas the infiltrate of phagocytic mononuclear cells and lymphocytes occurs in the iris vessels [16,17]. The expression of *Toll-like receptor 4* (TLR4), preferably by cells in the anterior region of the eye, may explain the apparent susceptibility of anterior uvea to disruption of the blood–ocular barrier and development of uveitis [50,51].

Data obtained after systemic treatments with Ac2-26 and PL in EIU confirmed the anti-inflammatory action of AnxA1 mimetic peptide in experimental ocular inflammation [18,32], including uveitis, as previously reported by our research group [17]. However, as a novelty, our current results also indicated the protective effects of PL at 400 μg dosage and therefore a dose-dependent manner. Both Ac2-26 and PL at 400 μg dosage promoted decrease in the influx of neutrophils and monocytes into AqH and blood as well the reduction of the phagocytic mononuclear cells in the iris and ciliary body. Furthermore, the anti-inflammatory effects of the mimetic peptide and PL at higher dosage stimulated the reduction of IL-1β, IL-6, TNF-α, and MCP-1 levels, which are produced especially by neutrophils and phagocytic mononuclear cells [52,53].

Our findings in relation to PL in EIU are in agreement with other investigations that showed the anti-inflammatory potential of PL, such as in the reduction of ear edema induced by croton oil [54], analgesia and suppression of the stress response caused by pain in a dose-dependent manner [55], and decreased release of TNF-α and IL-1β in collagen-induced arthritis [56]. In particular, in LPS-induced-inflammation, PL suppressed leukocytes migration, TNF-α and IL-6 production, NF-κB and regulated extracellular kinases (ERK) 1 and 2 activation [46], and reduced mortality in sepsis [44]. Moreover, in LPS-induced neuroinflammation, PL protected dopaminergic neurons against inflammation by inhibiting microglial activation and decreasing levels of TNF-α, IL-1β, IL-6, and the production of ROS and NO [48] as well as inhibiting NF-κB and amyloidogenesis [49].

However, when Ac2-26 was administered in combination with PL, anti-inflammatory effects were abrogated, especially in combination with PL at 400 μg dosage. Interestingly, the groups that showed increased levels of the anti-inflammatory cytokine IL-10 were LPS and Ac2-26 + PL 400 μg, whereas treatment with peptide singly administered led to reduction of this cytokine level. In the model of allergic conjunctivitis, low levels of IL-10 were also observed after treatment with Ac2-26, and a significant increase in this cytokine occurred in AnxA1-null animals [32], indicating the importance of Th1/Th2 balance in the development of allergic inflammatory responses and suggesting that the protective role of AnxA1 in ocular allergy occurs through downregulation of both cytokine profiles. The same seems to happen in EIU.

The efficacy of Ac2-26 and PL at 400 μg dosage was also verified by the reduction in COX-2 proinflammatory enzyme expression. Again, the administration of PL at 200 μg dosage and the combination of Ac2-26 + PL 400 μg did not revert the inflammatory process. In a previous research, we had shown the exacerbated inflammatory response, characterized by COX-2 overexpression in the eyes of AnxA1-null mice, reinforcing the actions of AnxA1 in the resolution of ocular inflammation [17]. Concerning PL, the importance of its analogs in COX-2 inhibition after LPS induction was demonstrated in the RAW264.7 lineage of macrophages [38]. More recently, our research group showed that PL administration decreased expressions of COX-2, NF-κB, and neutrophil elastase and recovered lung tissues in a model of lung inflammation [42].

We studied the endogenous expression of AnxA1 in the different experimental groups, especially in the anterior eye segment (iris, ciliary body, and ciliary processes). The immunoreactivity for AnxA1 in the ocular tissues overlapped with the sites of TLR4 expression and production of inflammatory mediators [57,58]. Indeed, TLR4 in the eye is particularly expressed by epithelial cells (cornea and pigmented epithelia of the ciliary body), iris endothelial cells [50,51], and resident antigen presenting cells of the uvea [52,59]. Moreover, in uveitis, cytokines are produced mainly by inflammatory and endothelial cells as well as by corneal epithelial cells and retinal pigmented cells [57,58].

The immunohistochemical analysis of untreated EIU eyes showed an increase in the intensity of immunolabeling for AnxA1, corroborating our previous findings in ocular tissues [17,34] and ocular inflammatory cells [16]. The higher expression of AnxA1 was also observed in neutrophils in ocular toxoplasmosis in mice and culture of retinal pigmented cells infected with *Toxoplasma gondii* [30]. In contrast, in animals treated with Ac2-26 singly or in combination with PL at lower dosage, there was a decrease in immunoreactivity, probably due to a negative feedback mechanism, as hitherto observed in the uveitis [17] and allergic conjunctivitis [32] models.

The fact that endogenous AnxA1 is strongly induced by LPS has already been reported in other investigations and reinforces the action of AnxA1 as a proresolving mediator in inflammation [23,26,27]. In the systemic inflammatory reaction induced by LPS, higher AnxA1 expression was observed and associated with the combined actions of endogenous glucocorticoids IL-6 and TNF-α [60]. Intense increase in *AnxA1* gene activity and protein synthesis in hepatocytes, endothelial cells, and leukocytes at the beginning of the inflammatory process, followed by reduction in AnxA1 expression in the late phase of inflammation, was verified by means of the *LacZ* reporter gene in AnxA1-null mice after LPS endotoxemia [61]. Similar to our findings, AnxA1 expression increased during lung inflammation induced by LPS but decreased after peptide Ac2-26 treatment [62]. Moreover, in the model of LPS-induced pleurisy, glucocorticoid-induced leucine zipper (GILZ) deficiency was associated with an early increase in AnxA1, so the lack of endogenous GILZ during the resolution of inflammation was compensated by AnxA1 overexpression [63].

Although our results indicated that the combination of the peptide with PL promoted a decrease in the effects of Ac2-26, the dosage of 200 µg of PL still allowed the peptide actions in a moderate manner, which may explain the lower expression of the endogenous AnxA1 in this group. Modulation in the expression of AnxA1 after LPS inoculation found in this investigation reinforces the involvement of the protein in ocular tissue physiology during inflammatory [16,17], infectious [34], allergic [32,35], and autoimmune [18,20] processes in experimental models. Interestingly, the expression of AnxA1 remains increased after treatments with PL singly administered or at 400 µg dosage in combination with the peptide. This finding could reflect the results found by other researchers who used herbal medicines, as in rats induced with sepsis and treated with Xuebijing (XBJ) [64] and culture of lung tumor cells administered with *Camellia simensis* [65]. Similar to our results, in a model of lung inflammation induced by cigarette smoke, PL administration promoted increased expression of AnxA1, concomitant with the reduction in COX-2 [42].

Following the study, we investigated the expression of fpr1 and fpr2 receptors in all groups. In previous studies on inflamed ocular tissues, it was shown that the expression of fpr2 perfectly overlapped the distribution of AnxA1 and that it was increased after treatment with Ac2-26 [17,34]. Again, our results strengthen the possible specificity of the AnxA1/fpr2 interaction as the expression of the fpr1 receptor did not occur in any of the experimental groups. In contrast, intense expression of both fprs were detected in conjunctival epithelial cells in an allergic conjunctivitis model [35], but the lack of AnxA1 protein in the ovalbumin-sensitized mice produced a marked increase only in fpr2 expression.

In addition, we found that the fpr2 expression was not altered by PL singly administered or in combination with Ac2-26, suggesting that the attenuation of the protective effects of mimetic peptide by interaction with PL probably is not related to receptor expression changes. In the light of these data and our previous findings about the interaction between Ac2-26 and PL [36], we speculate that conformational alteration may have occurred in the Ac2-26 + PL complex and may have interfered with the fpr2 binding receptor, which impaired the anti-inflammatory actions of the peptide. This reasoning is supported by recently reported results as we demonstrated that the interaction of PL with the AnxA1-derived peptide Ac2-26 occurred spontaneously, was enthalpically driven, and that the forces governing the interaction were hydrophobic. Moreover, Ac2-26 peptide binds to PL via two hydrogen-bonding interactions at lysine 9 but not at tryptophan 12 [36]. Previous data from our group have allowed us to report that the anti-inflammatory activities of AnxA1 occur after a specific serine phosphorylation event in the N-terminal region [17]. Therefore, as the interaction between Ac2-26 and PL occurs on tyrosine, the serine site, which is an important post-translational modification related to the translocation of AnxA1 from cytoplasm to cell surface [28], remains free for phosphorylation.

At first, this could indicate that PL does not affect the action of endogenous AnxA1. This thought was supported by our current findings, which showed increased expression of AnxA1 when PL was administered alone. However, with coadministration in vivo, Ac2-26 and PL promoted the reversal of the anti-inflammatory effects when administrated alone (Figure 8). Moreover, the higher the concentration of PL, the greater the possibility of interaction of this molecule with Ac2-26 and, consequently, the lower the anti-inflammatory response. These findings indicate that the impairment of the anti-inflammatory action of Ac2-26 may be related to a conformational change, which prevents the peptide from binding to the fpr2 receptor. Relevantly, the PL sequestered by the complex is also prevented from entering the cells and performing its anti-inflammatory role. Thus, there is a competition between Ac2-26 and PL in the anti-inflammatory action against uveitis. This important result implies there are other factors to be considered, such as molecular weight, size of the molecules, charge distribution on the peptide, possible conformational structure adopted by the peptide during the action, and localization of the interaction site. Thus, a reasonable next step will be studies requiring new approaches, both from the point of view of action against uveitis as well as experimental evidence.

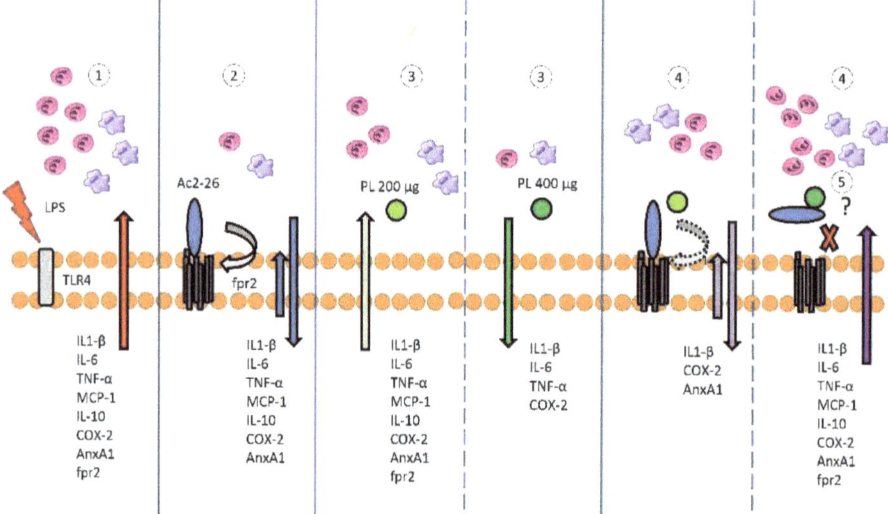

Figure 8. Schematic model of Ac2-26 and PL actions in EIU. (**1**) LPS triggers the release of cytokines (IL1-β, IL-6, IL-10, TNF-α, MCP-1), leukocyte influx (neutrophils and monocytes), and the production of endogenous COX-2, AnxA1, and fpr2. (**2**) These inflammatory responses are mitigated by Ac2-26, mediated by fpr2 receptor, and (**3**) PL, especially at a higher dosage. (**4**) The coadministration of Ac2-26 and PL abrogates the anti-inflammatory effects of the singly administered compounds. (**5**) Conformational changes on the Ac2-26 peptide due to its interaction with PL may impair its binding to the fpr2 receptor. Larger arrows directed upwards or downwards indicate an increase or decrease, respectively, of cytokines and endogenous proteins produced by the cell. Smaller arrows pointing upwards indicate increased expression of fpr2. Outside the cell, pink shape represents neutrophils, purple shape represents monocytes, light green circle depicts PL at 200 μg dosage, dark green circle depicts PL at 400 μg dosage, and blue oval represents Ac2-26.

5. Conclusions

Our study sheds light on the protective effects of PL, revealing it as a potential therapeutic target in ocular inflammation. Furthermore, the results show that Ac2-26 + PL combination abrogates the anti-inflammatory actions of Ac2-26 and PL singly administered.

Author Contributions: Conceptualization, A.P.G., M.L.C. and S.M.O.; methodology, A.P.G., C.d.F.Z., Í.P.C., S.d.S.C. and H.R.S.; formal analysis, A.P.G., C.d.F.Z., Í.P.C., S.d.S.C. and H.R.S.; investigation, A.P.G., C.d.F.Z., Í.P.C., S.d.S.C., H.R.S., M.L.C. and S.M.O.; resources, A.P.G., M.L.C. and S.M.O.; data curation, A.P.G.; writing—original draft preparation, A.P.G. and H.R.S.; writing—review and editing, M.L.C. and S.M.O.; supervision, A.P.G.; project administration, S.M.O.; funding acquisition, A.P.G. and S.M.O. All authors have read and agreed to the published version of the manuscript.

Funding: This research was funded by the Fundação de Amparo à Pesquisa do Estado de São Paulo (FAPESP), grant numbers 2015/03359-5 (APG) and 2019/19949-7 (SMO); Conselho Nacional de Desenvolvimento Científico e Tecnológico (CNPq), grant number 404190/2016-2 (APG); and Advanced Research Center in Medicine, CEPAM, Unilago, Brazil (56.569.197/0001-39) (SMO).

Institutional Review Board Statement: The rat experimental model was conducted according to the Brazilian Law 11.794 of 8 October 2008, Decree 6899 of 15 July 2009, and the rules issued by the National Council for Control of Animal Experimentation (CONCEA) and approved (Certificate No. 11/14) by the Ethic Committee on Animal Use at University Center Padre Albino (CEUA/UNIFIPA) in the meeting of 20 November 2014 (protocol code 14.11.03-13). The study did not involve humans.

Informed Consent Statement: Not applicable.

Data Availability Statement: Data available on request due to restrictions.

Conflicts of Interest: The authors declare no conflict of interest.

References

1. Majumder, P.D.; Ghosh, A.; Biswas, J. Infectious uveitis: An enigma. *Middle East Afr. J. Ophthalmol.* **2017**, *24*, 2–10. [PubMed]
2. Mohlin, C.; Sandholm, K.; Ekdahl, K.N.; Nilsson, B. The link between morphology and complement in ocular disease. *Mol. Immunol.* **2017**, *89*, 84–99. [CrossRef] [PubMed]
3. You, C.; Sahawneh, H.F.; Ma, L.; Kubaisi, B.; Schmidt, A.; Foster, C.S. A review and update on orphan drugs for the treatment of noninfectious uveitis. *Clin. Ophthalmol.* **2017**, *11*, 257–265. [CrossRef] [PubMed]
4. Oliver, G.F.; Carr, J.M.; Smith, J.R. Emerging infectious uveitis: Chikungunya, dengue, Zika and Ebola: A review. *Clin. Exp. Ophthalmol.* **2019**, *47*, 372–380. [CrossRef]
5. Leclercq, M.; Desbois, C.; Domont, F.; Maalouf, G.; Touhami, S.; Cacoub, P.; Bodaghi, B.; Saadoun, D. Biotherapies in Uveitis. *J. Clin. Med.* **2020**, *9*, 3599. [CrossRef]
6. Barry, R.J.; Nguyen, Q.D.; Lee, R.W.; Murray, P.I.; Denniston, A.K. Pharmacotherapy for uveitis: Current management and emerging therapy. *Clin. Ophthalmol.* **2014**, *8*, 1891–1911.
7. Chen, S.C.; Sheu, S.J. Recent advances in managing and understanding uveitis. *F1000Res* **2017**, *6*, 280. [CrossRef]
8. Trivedi, A.; Katelaris, C. The Use of Biologic Agents in the Management of Uveitis. *Intern. Med. J.* **2019**, *49*, 1352–1363. [CrossRef]
9. Hassan, M.; Nguyen, N.V.; Halim, M.S.; Afridi, R.; Sadiq, M.A.; Karkhur, S.; Vigil, E.; Karabekirogullari, S.; Nguyen, Q.D.; Do, D.V.; et al. Effect of vitreomacular adhesion on the treatment outcomes in the STOP-Uveitis clinical trial for non-infectious uveitis. *J. Ophthalmic Inflamm. Infect.* **2019**, *9*, 12. [CrossRef]
10. Pleyer, U.; Neri, P.; Deuter, C. New pharmacotherapy options for noninfectious posterior uveitis. *Int. Ophthalmol.* **2021**, *41*, 2265–2281. [CrossRef]
11. Touhami, S.; Diwo, E.; Sève, P.; Trad, S.; Bielefeld, P.; Sène, D.; Abad, S.; Brézin, A.; Quartier, P.; Paut, I.K.; et al. Expert opinion on the use of biological therapy in non-infectious uveitis. *Expert Opin. Biol. Ther.* **2019**, *19*, 477–490. [CrossRef]
12. Sheppard, J.D.; Comstock, T.L.; Cavet, M.E. Impact of the Topical Ophthalmic Corticosteroid Loteprednol Etabonate on Intraocular Pressure. *Adv. Ther.* **2016**, *33*, 532–552. [CrossRef]
13. Fang, C.; Zhou, D.; Zhang, S.; He, Y.; Lin, Z.; Huang, C.; Li, J. Amelioration of experimental autoimmune uveitis by leflunomide in Lewis rats. *PLoS ONE* **2013**, *8*, e62071. [CrossRef]
14. Bhuyan, Z.A.; Asanoma, M.; Iwata, A.; Ishifune, C.; Maekawa, Y.; Shimada, M.; Yasutomo, K. Abrogation of Rbpj Attenuates Experimental Autoimmune Uveoretinitis by Inhibiting IL-22-Producing CD4(+) T Cells. *PLoS ONE* **2014**, *9*, e89266. [CrossRef]
15. Harthan, J.S.; Opitz, D.L.; Fromstein, S.R.; Morettin, C.E. Diagnosis and treatment of anterior uveitis: Optometric management. *Clin. Optom.* **2016**, *8*, 23–35. [CrossRef]
16. Da Silva, P.S.; Girol, A.P.; Oliani, S.M. Mast cells modulate the inflammatory process in endotoxin-induced uveitis. *Mol. Vis.* **2011**, *17*, 1310–1319.

17. Girol, A.P.; Mimura, K.K.O.; Drewes, C.C.; Bolonheis, S.M.; Solito, E.; Farsky, S.H.P.; Gil, C.D.; Oliani, S.M. Anti-inflammatory mechanisms of the annexin A1 protein and its mimetic peptide Ac2-26 in models of ocular inflammation in vivo and *in vitro*. *J. Immunol.* **2013**, *190*, 5689–5701. [CrossRef]
18. Yazid, S.; Gardner, P.J.; Carvalho, L.; Chu, C.J.; Flower, R.J.; Solito, E.; Lee, R.W.J.; Ali, R.R.; Dick, A.D. Annexin-A1 restricts Th17 cells and attenuates the severity of autoimmune disease. *J. Autoimmun.* **2015**, *58*, 1–11. [CrossRef]
19. Cardin, L.T.; Sonehara, N.M.; Mimura, K.K.O.; Dos Santos, A.R.D.; da Silva Junior, W.A.; Sobral, L.M.; Leopoldino, A.M.; Cunha, B.R.; Tajara, E.H.; Oliani, S.M.; et al. ANXA1Ac2-26 peptide, a possible therapeutic approach in inflammatory ocular diseases. *Gene* **2017**, *614*, 26–36. [CrossRef]
20. Gardner, P.J.; Yazid, S.; Ribeiro, J.; Ali, R.R.; Dick, A.D. Augmenting Endogenous Levels of Retinal Annexin A1 Suppresses Uveitis in Mice. *Transl. Vis. Sci. Technol.* **2017**, *6*, 10. [CrossRef]
21. Flower, R.; Blackwell, G. Anti-inflammatory steroids induce biosynthesis of a phospholipase A2 inhibitor which prevents prostaglandin generation. *Nature* **1979**, *278*, 456–459. [CrossRef] [PubMed]
22. Flower, R.; Gaddum, E. Lipocortin and the mechanism of action of the glucocorticoids. *Br. J. Pharmacol.* **1988**, *94*, 987–1015. [CrossRef] [PubMed]
23. Sheikh, M.H.; Solito, E. Annexin A1: Uncovering the Many Talents of an Old Protein. *Int. J. Mol. Sci.* **2018**, *19*, 1045. [CrossRef] [PubMed]
24. Shao, G.; Zhou, H.; Zhang, Q.; Jin, Y.; Fu, C. Advancements of Annexin A1 in inflammation and tumorigenesis. *Onco Targets Ther.* **2019**, *12*, 3245–3254. [CrossRef]
25. Lizarbe, M.A.; Barrasa, J.I.; Olmo, N.; Gavilanes, F.; Turnay, J. Annexin-phospholipid interactions. Functional implications. *Int. J. Mol. Sci.* **2013**, *14*, 2652–2683. [CrossRef]
26. Gobbetti, T.; Cooray, S.N. Annexin A1 and resolution of inflammation: Tissue repairing properties and signalling signature. *Biol. Chem.* **2016**, *397*, 981–993. [CrossRef]
27. Grewal, T.; Wason, S.J.; Enrich, C.; Rentero, C. Annexins-insights from knockout mice. *Biol. Chem.* **2016**, *397*, 1031–1053. [CrossRef]
28. Solito, E.; Christian, H.C.; Festa, M.; Mulla, A.; Tierney, T.; Flower, R.J.; Buckingham, J.C. Post-translational modification plays an essential role in the translocation of annexin A1 from the cytoplasm to the cell surface. *FASEB J.* **2006**, *20*, 1498–1500. [CrossRef]
29. Leoni, G.; Nusrat, A. Annexin A1: Shifting the balance towards resolution and repair. *Biol. Chem.* **2016**, *397*, 971–979. [CrossRef]
30. Mimura, K.K.; Tedesco, R.C.; Calabrese, K.S.; Gil, C.D.; Oliani, S.M. The involvement of anti-inflammatory protein, annexin A1, in ocular toxoplasmosis. *Mol. Vis.* **2012**, *18*, 1583–1593.
31. Zanon, C.F.; Sonehara, N.M.; Girol, A.P.; Gil, C.D.; Oliani, S.M. Protective effects of the galectin-1 protein on in vivo and in vitro models of ocular inflammation. *Mol. Vis.* **2015**, *21*, 1036–1050.
32. Gimenes, A.D.; Andrade, T.R.M.; Mello, C.B.; Ramos, L.; Gil, C.D.; Oliani, S.M. Beneficial effect of annexin A1 in a model of experimental allergic conjunctivitis. *Exp. Eye Res.* **2015**, *134*, 24–32. [CrossRef]
33. Jorge, Y.C.; Mataruco, M.M.; Araújo, L.P.; Rossi, A.F.T.; Oliveira, J.G.; Valsechi, M.C.; Caetano, A.; Miyazaki, K.; Fazzio, C.S.J.; Thomé, J.A.; et al. Expression of annexin-A1 and galectin-1 anti-inflammatory proteins and mRNA in chronic gastritis and gastric cancer. *Mediat. Inflamm.* **2013**, *2013*, 152860. [CrossRef]
34. Da Silva, R.A.; Hamade, A.M.A.; Silva, G.A.S.; Pereira, G.H.; De Oliveira, F.F.J.; Costa, S.S.; Iyomasa-Pilon, M.M.; Souza, H.R.; Possebon, L.; Girol, A.P. Evaluation of Annexin A1 Protein in an Infectious Keratitis Model: Therapeutic Perspectives. *Curr. Trends Ophthal.* **2019**, *2*, 104–112. [CrossRef]
35. Marmorato, M.P.; Gimenes, A.D.; Andrade, F.E.C.; Oliani, S.M.; Gil, C.D. Involvement of the annexin A1-Fpr anti-inflammatory system in the ocular allergy. *Eur. J. Pharmacol.* **2019**, *842*, 298–305. [CrossRef]
36. Henrique, T.; Zanon, C.F.; Girol, A.P.; Stefanini, A.C.B.; Contessoto, N.S.A.; Silveira, N.J.F.; Bezerra, D.P.; Silveira, E.R.; Barbosa-Filho, J.M.; Cornélio, M.L.; et al. Biological and physical approaches on the role of piplartine (piperlongumine) in cancer. *Sci. Rep.* **2020**, *10*, 22283. [CrossRef]
37. Bezerra, D.P.; Pessoa, C.; Moraes, M.O.; Saker-Neto, N.; Silveira, E.R.; Costa-Lotufo, L.V. Overview of the therapeutic potential of piplartine (piperlongumine). *Eur. J. Pharm. Sci.* **2013**, *48*, 453–463. [CrossRef]
38. Sun, L.D.; Wang, F.; Dai, F.; Wang, Y.H.; Lin, D.; Zhou, B. Development and mechanism investigation of a new piperlongumine derivative as a potent anti-inflammatory agent. *Biochem. Pharmacol.* **2015**, *95*, 156–169. [CrossRef]
39. Prasad, S.; Tyagi, A.K. Historical Spice as a Future Drug: Therapeutic Potential of Piperlongumine. *Curr. Pharm. Des.* **2016**, *22*, 4151–4159. [CrossRef]
40. Meegan, M.J.; Nathwani, S.; Twamley, B.; Zisterer, D.M.; O'Boyle, N.M. Piperlongumine (piplartine) and analogues: Antiproliferative microtubule-destabilising agents. *Eur. J. Med. Chem.* **2017**, *125*, 453–463. [CrossRef]
41. Srivastava, A.; Karthick, T.; Joshi, B.D.; Mishra, R.; Tandon, P.; Ayala, A.P.; Ellena, J. Spectroscopic (far or terahertz, mid-infrared and Raman) investigation, thermal analysis and biological activity of piplartine. *Spectrochim. Acta A Mol. Biomol. Spectrosc.* **2017**, *184*, 368–381. [CrossRef]
42. Sant'Ana, M.; Souza, H.R.; Possebon, L.; Cornélio, M.L.; Riffo-Vasquez, Y.; Girol, A.P.; Oliani, S.M. Effect of piperlongumine during exposure to cigarette smoke reduces inflammation and lung injury. *Pulm. Pharmacol. Ther.* **2020**, *61*, 10–16. [CrossRef]
43. Bezerra, D.P.; Castro, F.O.; Alves, A.P.N.N.; Pessoa, C.; Moraes, M.O.; Silveira, E.R.; Lima, M.A.S.; Elmiro, F.J.M.; Alencar, N.M.N.; Mesquita, R.O.; et al. In vitro and in vivo antitumor effect of 5-FU combined with piplartine and piperine. *J. Appl. Toxicol.* **2008**, *28*, 156–163. [CrossRef]

44. Raj, L.; Ide, T.; Gurkar, A.U.; Foley, M.; Schenone, M.; Li, X.; Tolliday, N.J.; Golub, T.R.; Carr, S.A.; Shamji, A.F.; et al. Selective killing of cancer cells by a small molecule targeting the stress response to ROS. *Nature* **2011**, *475*, 231–234. [CrossRef]

45. Zheng, J.; Son, D.J.; Gu, S.M.; Woo, J.R.; Ham, Y.W.; Lee, H.P.; Kim, W.J.; Jung, J.K.; Hong, J.T. Piperlongumine inhibits lung tumor growth via inhibition of nuclear factor kappa B signaling pathway. *Sci. Rep.* **2016**, *6*, 26357. [CrossRef]

46. Lee, W.; Yoo, H.; Kim, J.A.; Lee, S.; Jee, J.G.; Lee, M.Y.; Lee, Y.M.; Bae, J.S. Barrier protective effects of piperlonguminine in LPS-induced inflammation in vitro and *in vivo*. *Food Chem. Toxicol.* **2013**, *58*, 149–157. [CrossRef]

47. Seo, Y.H.; Kim, J.K.; Jun, J.G. Synthesis and biological evaluation of piperlongumine derivatives as potent anti-inflammatory agents. *Bioorg. Med. Chem. Lett.* **2014**, *24*, 5727–5730. [CrossRef]

48. He, H.; Guo, W.W.; Xu, R.R.; Chen, X.Q.; Zhang, N.; Wu, X.; Wang, X.M. Alkaloids from piper longum protect dopaminergic neurons against inflammation-mediated damage induced by intranigral injection of lipopolysaccharide. *BMC Complement. Altern. Med.* **2016**, *16*, 412. [CrossRef]

49. Gu, S.M.; Lee, H.P.; Ham, Y.W.; Son, D.J.; Kim, H.Y.; Oh, K.W.; Han, S.; Yun, J.; Hong, J.T. Piperlongumine Improves Lipopolysaccharide-Induced Amyloidogenesis by Suppressing NF-KappaB Pathway. *Neuromol. Med.* **2018**, *20*, 312–327. [CrossRef]

50. Brito, B.; Zamora, D.O.; Bonnah, R.A.; Pan, Y.; Planck, S.R.; Rosenbaum, J.T. Toll-like receptor 4 and CD14 expression in human ciliary body and TLR-4 in human iris endothelial cells. *Exp. Eye Res.* **2004**, *79*, 203–208. [CrossRef]

51. Chang, J.; McCluskey, P.; Wakefield, D. Toll-like receptors in ocular immunity and the immunopathogenesis of inflammatory eye disease. *Br. J. Ophthalmol.* **2006**, *90*, 103–108. [CrossRef] [PubMed]

52. Li, S.; Lu, H.; Hu, X.; Chen, W.; Xu, Y.; Wang, J. Expression of TLR4-MyD88 and NF-κB in the iris during endotoxin-induced uveitis. *Mediat. Inflamm.* **2010**, *2010*, 748218. [CrossRef] [PubMed]

53. Yang, S.; Lu, H.; Wang, J.; Qi, X.; Liu, X.; Zhang, X. The effect of toll-like receptor 4 on macrophage cytokines during endotoxin induced uveitis. *Int. J. Mol. Sci.* **2012**, *13*, 7508–7520. [CrossRef] [PubMed]

54. Rodrigues Silva, D.; Baroni, S.; Svidzinski, A.E.; Bersani-Amado, C.A.; Cortez, D.A.G. Anti-inflammatory activity of the extract, fractions and amides from the leaves of Piper ovatum Vahl (Piperaceae). *J. Ethnopharmacol.* **2008**, *116*, 569–573. [CrossRef]

55. Yadav, V.; Chatterjee, S.S.; Majeed, M.; Kumar, V. Preventive potentials of piperlongumine and a *Piper longum* extract against stress responses and pain. *J. Tradit. Complement. Med.* **2015**, *6*, 413–423.

56. Sun, J.; Xu, P.; Du, X.; Zhang, Q.; Zhu, Y. Piperlongumine attenuates collagen-induced arthritis via expansion of myeloid-derived suppressor cells and inhibition of the activation of fibroblast-like synoviocytes. *Mol. Med. Rep.* **2015**, *11*, 2689–2694. [CrossRef]

57. Weinstein, J.E.; Pepple, K.L. Cytokines in uveitis. *Curr. Opin. Ophthalmol.* **2018**, *29*, 267–274. [CrossRef]

58. Balamurugan, S.; Das, D.; Hasanreisoglu, M.; Toy, B.C.; Akhter, M.; Anuradha, V.K.; Anthony, E.; Gurnani, B.; Kaur, K. Interleukins and cytokine biomarkers in uveitis. *Indian J. Ophthalmol.* **2020**, *68*, 1750–1763. [CrossRef]

59. Chen, W.; Hu, W.F.; Zhao, L.; Li, S.; Lu, H. Toll-like receptor 4 expression in macrophages in endotoxin-induced uveitis in Wistar rats. *Chin. J. Ophthalmol.* **2010**, *46*, 355–361.

60. de Coupade, C.; Ajuebor, M.N.; Russo-Marie, F.; Perretti, M.; Solito, E. Cytokine modulation of liver annexin 1 expression during experimental endotoxemia. *Am. J. Pathol.* **2001**, *159*, 1435–1443. [CrossRef]

61. Damazo, A.; Flower, R.J.; Solito, E.; Oliani, S.M. Annexin-A1 gene expression during liver development and post-translation modification after experimental endotoxemia. *Inflamm. Res.* **2008**, *57*, 97–103. [CrossRef]

62. Da Cunha, E.E.; OlianI, S.M.; Damazo, A.S. Effect of annexin-A1 peptide treatment during lung inflammation induced by lipopolysaccharide. *Pulm. Pharmacol. Ther.* **2012**, *25*, 303–311. [CrossRef]

63. Vago, J.P.; Tavares, L.P.; Garcia, C.C.; Lima, K.M.; Perucci, L.O.; Vieira, É.L.; Nogueira, C.R.; Soriani, F.M.; Martins, J.O.; Silva, P.M.; et al. The role and effects of glucocorticoid-induced leucine zipper in the context of inflammation resolution. *J. Immunol.* **2015**, *194*, 4940–4950. [CrossRef]

64. He, X.D.; Wang, Y.; Wu, Q.; Wang, H.X.; Chen, Z.D.; Zheng, R.S.; Wang, Z.S.; Wang, J.B.; Yang, Y. Xuebijing Protects Rats from Sepsis Challenged with Acinetobacter baumannii by Promoting Annexin A1 Expression and Inhibiting Proinflammatory Cytokines Secretion. *Evid. Based Complement. Altern. Med.* **2013**, *2013*, 804940. [CrossRef]

65. Lu, Q.Y.; Jin, Y.; Mao, J.T.; Zhang, Z.F.; Heber, D.; Dubinett, S.M.; Rao, J. Green tea inhibits cycolooxygenase-2 in non-small cell lung cancer cells through the induction of Annexin-1. *Biochem. Biophys. Res. Commun.* **2012**, *427*, 725–730. [CrossRef]

Article

Differential Response of Müller Cells and Microglia in a Mouse Retinal Detachment Model and Its Implications in Detached and Non-Detached Regions

Seung-Hee Lee [1,2], **Yong-Soo Park** [1,2], **Sun-Sook Paik** [1,2] and **In-Beom Kim** [1,2,3,4,*]

[1] Department of Anatomy, College of Medicine, The Catholic University of Korea, 222 Banpo-daero, Seocho-gu, Seoul 06591, Korea; seunghui6310@daum.net (S.-H.L.); yongsoopark88@gmail.com (Y.-S.P.); paikss@catholic.ac.kr (S.-S.P.)

[2] Catholic Neuroscience Institute, College of Medicine, The Catholic University of Korea, 222 Banpo-daero, Seocho-gu, Seoul 06591, Korea

[3] Department of Biomedicine & Health Sciences, Graduate School, The Catholic University of Korea, 222 Banpo-daero, Seocho-gu, Seoul 06591, Korea

[4] Catholic Institute for Applied Anatomy, College of Medicine, The Catholic University of Korea, 222 Banpo-daero, Seocho-gu, Seoul 06591, Korea

[*] Correspondence: ibkimmd@catholic.ac.kr; Tel.: +82-2-2258-7263

Citation: Lee, S.-H.; Park, Y.-S.; Paik, S.-S.; Kim, I.-B. Differential Response of Müller Cells and Microglia in a Mouse Retinal Detachment Model and Its Implications in Detached and Non-Detached Regions. *Cells* **2021**, *10*, 1972. https://doi.org/10.3390/cells 10081972

Academic Editors: Maurice Ptito and Joseph Bouskila

Received: 12 June 2021
Accepted: 29 July 2021
Published: 3 August 2021

Publisher's Note: MDPI stays neutral with regard to jurisdictional claims in published maps and institutional affiliations.

Abstract: Retinal detachment (RD) is a sight-threatening condition, leading to photoreceptor cell death; however, only a few studies provide insight into its effects on the entire retinal region. We examined the spatiotemporal changes in glial responses in a mouse RD model. In electroretinography, a- and b-waves were reduced in a time-dependent manner. Hematoxylin and eosin staining revealed a gradual decrease in the outer nuclear layer throughout the retinal region. Terminal deoxynucleotidyltransferase dUTP nick end labeling (TUNEL) assay showed that TUNEL-positive photoreceptors increased 5 days after RD and decreased by 14 days. Glial response was evaluated by immunohistochemistry using antibodies against glial fibrillary acidic protein (GFAP, Müller glial marker) and Iba-1 (microglial marker) and osteopontin (OPN, activated microglial marker). GFAP immunoreactivity increased after 7 days in complete RD, and was retained for 14 days. OPN expression increased in microglial cells 3–7 days after RD, and decreased by 14 days in the detached and border regions. Although OPN was not expressed in the intact region, morphologically activated microglial cells were observed. These retinal glial cell responses and photoreceptor degeneration in the border and intact regions suggest that the effects of RD in the border and intact retinal regions need to be understood further.

Keywords: GFAP; osteopontin; Müller cells; microglia; retinal detachment

1. Introduction

Retinal detachment (RD) is a sight-threatening condition, in which the outer segments of the photoreceptors physically separate from the underlying retinal pigment epithelium (RPE), which acts as a photoreceptor nourishment source. Photoreceptor cell death continues during RD, causing progressive visual impairment. RD can be classified into three types: rhegmatogenous, tractional, and exudative. Rhegmatogenous detachment is the most common cause of RD, which is characterized by retinal tear-inducing vitreous penetration under the retina [1], while tractional detachment and exudative detachment are rare types of RD caused by a complication of various retinal disorders, including age-related macular degeneration (AMD) [2] and diabetic retinopathy (DR) [3].

Many studies have used various animal RD models to understand its pathogenesis [4,5]. Progressive photoreceptor cell death occurs in multiple ways in the detached region of the retina, including via apoptosis [6,7], necroptosis [8–10], and autophagy [11,12].

Based on the findings of microglial activation and macrophage infiltration into the subretinal space [13,14], local inflammation is thought to be involved in RD progression. However, there have been only a few studies providing an overall insight into the entire retinal region that includes intact (undetached), borderline, and detached regions.

Previously, we established and characterized animal retinal degeneration models, such as N-methyl-nitrosourea (MNU)-induced rat model [15] and a blue light-emitting diode (LED)-induced mouse model [16]. In these retinal degeneration models, we found that microglial cells and Müller glial cells were activated, and an increase in pro-inflammatory cytokines was seen [15–19], suggesting that inflammation is a critical mechanism in its pathogenesis. In addition, we introduced the differential activation patterns of two retinal glial cells, Müller glial cells, and microglial cells. In other words, Müller glial cells gradually express the glial fibrillary acidic protein (GFAP), an activation marker of Müller glial cells, from their foot process in the nerve fiber layer to the outer retina in the entire retina, regardless of injury location and type [15–17]. On the other hand, microglial cells express osteopontin (OPN), a newly identified marker for activated microglial cell marker/retinal inflammation, near the injury location, but not far from the injury region [17].

In this study, we examined the morphological characteristics of a mouse RD model with deconstructed photoreceptor outer segments and apoptotic photoreceptors [20]. In particular, we focused on retinal glia located in different regions (detached, border, and intact) of the detached retina by using three markers for retinal glial cells and inflammation: GFAP for Müller glia and retinal stress, Iba-1 for microglia, and OPN for microglia and neuroinflammation. Additionally, we tested whether OPN is a lesion-specific marker of retinal inflammation in RD, as shown in a blue LED exposure-induced retinal degeneration model [16].

2. Materials and Methods

2.1. Animals and RD Generation

All mice-related experiments were performed according to the regulations of the Catholic ethics committee of the Catholic University of Korea, Seoul, which are based on the National Institute of Health (NIH) guidelines for the Care and Use of Laboratory Animals (NIH Publication NO. 80-23), as revised in 1996. The experimental procedures were approved by the Institutional Animal Care and Use Committee of the College of Medicine, The Catholic University of Korea (approval number: CUMC-2019-0266-06).

A total of 50 male C57BL6 mice which were 6–7 weeks old were used in this study. The animals were kept in a 12-h light/12-h dark cycle in a climate-controlled laboratory. Before RD, the mice were anesthetized by intraperitoneal injection of 20 mg/kg zolazepam and 7.5 mg/kg xylazine, and the pupils were dilated using 0.5% topical tropicamide and 0.5% phenylephrine hydrochloride (Mydrin-P; Santen Pharmaceutical Co. Ltd.; Osaka, Japan). RD was induced by subretinally injecting 2 µL PBS through the sclera, by using Nanofil Application Kits (World Precision Instruments, Sarasota, FL, USA) with a 34-gauge blunt needle. This administration resulted in a detached lesion approximately 700 µm in diameter. Electroretinography (ERG) recordings were performed 1, 3, 5, 7, and 14 days after RD, and the mice were subsequently sacrificed.

2.2. ERG

ERG recordings were performed following the procedures described in our previous study [16]. All animals were kept in a completely dark room for 16 h before the ERG recording. All procedures were performed under dim red light ($\lambda > 600$ nm). The mice were anesthetized by intraperitoneal injection of 20 mg/kg zolazepam and 7.5 mg/kg xylazine. The corneas were coated with hydroxypropyl methylcellulose gel and covered with gold ring contact electrodes. A ground electrode and reference electrode were placed subcutaneously in the tail and ear, respectively. Stimuli were short-term white flashes, delivered via a Ganzfeld stimulator (UTAS-3000; LKC Technologies, Gaithersburg, MD, USA).

The signals were amplified and filtered through a digital band-pass filter, ranging from 5 Hz to 300 Hz to yield a- and b-waves. All scotopic ERG, rod-mediated responses were obtained at the following increasing light intensities: 0.99 and 3.96 cd/s.m^2. Each recording was the average of all responses obtained within a 15 s inter-stimulus interval. The amplitude of the a-wave was measured from the baseline to the maximum a-wave peak, and the b-wave was measured from the maximum a-wave peak to the maximum b-wave peak.

2.3. Hematoxylin and Eosin (H&E) Staining

The eyecup was fixed by immersion in 4% paraformaldehyde in the 0.1 M phosphate buffer (PB; pH 7.4) for 2 h. Then, the tissue was rinsed in PB, transferred to a 30% sucrose solution in PB, infiltrated overnight, and embedded the next day in a supporting medium for frozen tissue specimens (Tissue-Tek O.C.T compound; Sakura; Alphen aan den Rijn, The Netherlands). Vertical retinal sections (8 μm in thickness) were prepared using a cryostat at −25 °C, stored at −20 °C, and then stained with H&E.

2.4. Terminal Deoxynucleotidyl Transferase dUTP Nick End Labeling (TUNEL) Assay

As described in our previous report [16], the TUNEL assay was performed following the manufacturer's protocols (In Situ Cell Death Detection Kit; Roche Biochemical; Mannheim, Germany) to detect retinal cell death. In cryosections of the eyecup preparations, cell nuclei were counterstained with 4,6-diamidino-2-phenylindole (DAPI; dilution, 1:1500; Roche Biochemical). The number of TUNEL-positive photoreceptors in the outer nuclear layer (ONL) was counted in two sections of the detached, border, and intact retinal regions.

2.5. Immunohistochemistry

Immunohistochemical analysis was performed as described in our previous study [17]. Cryostat retinal sections of 8 μm thickness were washed three times with 0.1 M PB for 10 min each, and incubated in 10% normal donkey serum for 1 h at room temperature. The tissues were then incubated with primary antibodies, including rabbit anti-Iba-1 (dilution, 1:500; Wako Pure Chemical Industries; Osaka, Japan), mouse anti-GFAP (1:1500; Chemicon; Temecula, CA, USA), and goat anti-OPN (dilution, 1:1000; R&D Systems, Minneapolis, MN, USA) antibodies overnight at 4 °C. Then, the tissues were thoroughly rinsed in PB and incubated with appropriate secondary antibodies conjugated with Cy3 (dilution, 1:2000; Jackson ImmunoResearch; West Grove, PA, USA) or Alexa 488 (dilution, 1:1000; Molecular Probes; Eugene, OR, USA) for 2 h at room temperature. After rinsing several times in PB, the cell nuclei were counterstained with DAPI for 10 min and mounted with anti-fading mounting media (Vector Laboratories; Burlingame, CA, USA).

A Zeiss LSM 800 confocal microscope (Carl Zeiss Co. Ltd., Oberkochen, Germany) was used for observation and image acquisition. Quantitative image analysis was performed using Zen 2.3 software (Blue edition; Carl Zeiss), as described in our previous study [18]. The region of interest (ROI) was selected in the detached, border, and intact regions (480 μm length) of each retinal section, and the intensity of GFAP immunoreactivity was automatically measured.

2.6. Statistical Analysis

All statistical analyses for ERG amplitude, histology image analysis, TUNEL-positive quantification, and immunohistochemistry were performed using GraphPad Prism 8.0 (GraphPad Software; San Diego, CA, USA) by one-way ANOVA with Bonferroni's multiple comparisons test. Differences were considered statistically significant at $p < 0.05$.

3. Results

3.1. Functional and Histological Changes in Experimental RD

First, we evaluated the functional and histological changes in detached retinas. Functional changes in RD mice were investigated using ERG recordings. Figure 1A shows the scotopic ERG response in normal and RD mice at 1, 3, 5, 7, and 14 days after RD as representative waveforms of scotopic 0 dB flash at 0.99 cd·s/m^2. The amplitudes of the ERG responses were significantly reduced in a time-dependent manner, compared to those in the normal group (Figure 1B, $p < 0.001$, $n = 5$ in each group). The ERG responses decreased abruptly in the early phase (within five days after RD). Afterward, the responses slightly decreased or were sustained until the last day of the experiment (14 days after RD).

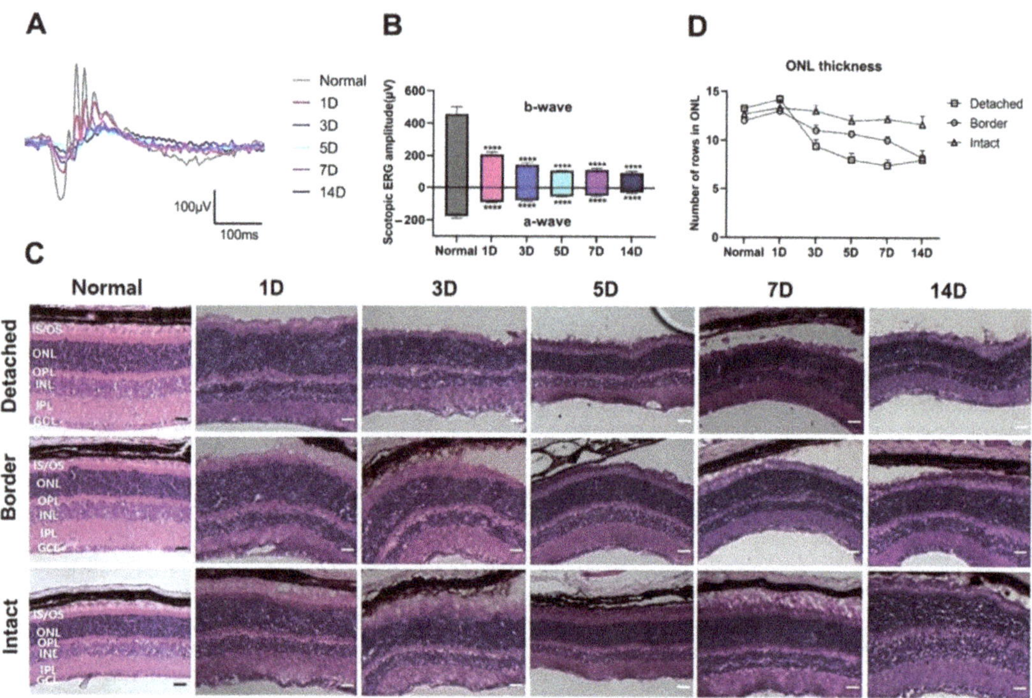

Figure 1. Functional and histological evaluation of retinal detachment (RD) at different time points. Electroretinograms (ERGs) were recorded in RD mice. (**A**) Representative scotopic ERG responses in normal (gray) and RD mice after 1 (pink), 3 (blue), 5 (cyan), 7 (purple), and 14 days (navy). Based on the functional changes, the amplitudes of both scotopic a- and b-waves significantly reduced in a time-dependent manner, compared to those in the normal group. (**B**) Values are represented as the mean ± SEM ($n = 5$, $p < 0.001$, one-way ANOVA). (**C**) Hematoxylin and eosin (H&E) staining of representative vertical sections from normal control and the detached retina at different time points (1, 3, 5, 7, and 14 days). Retinal thickness of the outer nuclear layer (ONL), where the photoreceptor residue gradually decreased. Scale bars, 20 µm. (**D**) ONL thickness was measured manually at each time point. According to histological changes in the detached region, ONL thickness significantly reduced in a time-dependent manner, and no significant changes were observed in the intact region. IS/OS, inner segment and outer segment; OPL, outer plexiform layer; INL, inner nuclear layer; IPL, inner plexiform layer; GCL, ganglion cell layer.

Next, to assess the histology, retinal vertical sections collected from the mice after ERG recording were stained with H&E (Figure 1C). Consistent with the results of ERG, ONL thickness gradually decreased in the detached retinal regions with photoreceptor degeneration. Specifically, ONL thickness in the detached region abruptly decreased (~14 to ~7 rows of photoreceptors) in the early phase, and then remained steady (Figure 1C,D). However, ONL thickness decreased in the border and intact regions at a relatively uniform pace, although the steepness in the border region was sharp, while that in the intact region was gradual (Figure 1D).

3.2. Photoreceptor Degeneration in Experimental RD

We evaluated the spatiotemporal pattern of photoreceptor degeneration after RD, using the TUNEL assay at each time point. During the entire experimental period, TUNEL-positive cells were exclusively observed in the ONL, indicating that they corresponded to photoreceptors (Figure 2A).

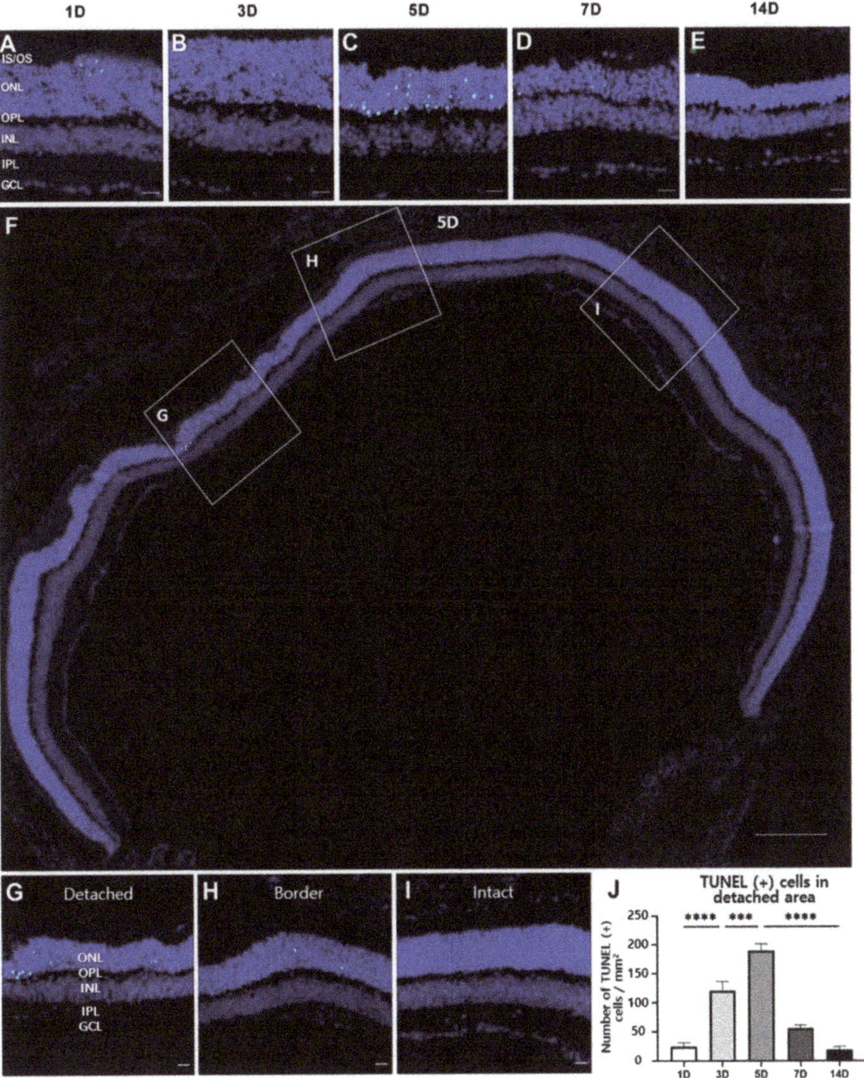

Figure 2. TUNEL assay to evaluate the apoptotic cell death in RD. (**A–E**) TUNEL-positive cells mostly observed in ONL of the detached region were significantly increased 5 days after RD, and, thereafter, decreased by 14 days. Scale bars, 20 µm. (**F**) A low-magnification view 5 days after RD. Scale bar, 200 µm. (**G–I**) TUNEL-positive cells located at photoreceptors were present in the (**G,H**) detached and border regions, but not in the (**I**) intact region. Scale bars, 20 µm. (**J**) Quantitative analysis of the number of TUNEL-positive cells was manually conducted ($n = 5$, 5 fields per time point). Data are shown as mean $\pm$ SEM. *** $p < 0.001$ and **** $p < 0.0001$ based on one-way ANOVA followed by Bonferroni's multiple comparisons test.

As expected from the histological findings shown in Figure 1C,D, most TUNEL-positive photoreceptors were observed in the detached region in the retinas (Figure 2A–C). They occurred at day 1, reached a peak at day 5, and rarely existed 14 days after RD. However, fewer TUNEL-positive photoreceptors were found in the border retinal region than in the detached retinal region, and only a few TUNEL-positive cells were found in the intact region 5 days after RD (Figure 2F–I). Thereafter, TUNEL-positive cells were manually counted at different time points (Figure 2J). TUNEL-positive cells gradually increased until 5 days after RD, after which they significantly decreased ($p < 0.05$). Together with histological results, these results demonstrate that the degeneration of photoreceptors occurs in early RD, mainly in the detached region, while progressing slowly in the border and intact regions.

3.3. GFAP Expression in Müller Glial Cells in RD

To assess retinal injury or stress in RD, immunohistochemical assays with anti-GFAP, a widely used marker for retinal injury and/or reactive Müller glial cells [15,17,21], were performed.

As shown in Figure 3A, GFAP immunoreactivity was observed in the inner limiting membrane and in few thin Müller cell processes in the inner plexiform layer (IPL) and inner nuclear layer (INL) of the detached retinal region, as well as in border and intact regions at 1 day after RD, similar to a normal retina (data not shown). However, GFAP immunoreactivity was observed in many Müller cell processes in the IPL, INL, outer plexiform layer (OPL), and ONL in the entire retina 3 days after RD. GFAP expression peaked 7 days after RD in the detached region, where many GFAP-immunoreactive Müller cell processes were distributed throughout all retinal layers, and some even appeared stout. In contrast, GFAP expression was significantly reduced in the border and intact regions, compared to that in the detached region. GFAP immunoreactivity in the detached region decreased 14 days after RD. However, GFAP immunoreactivity in the border region was similar to that at 7 days after RD, while that in the intact region sometimes appeared to be stronger than that at 7 days after RD.

The images captured after immunohistochemical analysis for GFAP were quantified as the GFAP profile per ONL, where photoreceptors degenerate (Figure 3B). Quantification results demonstrated that GFAP expression in the detached region gradually increased until 7 days after RD, after which it significantly decreased ($p < 0.05$). Moreover, GFAP expression slightly increased or was sustained in both the border and intact regions 14 days after RD. These results suggest that GFAP expression may be a general, but not a lesion-specific, marker for detecting retinal injury or stress in RD.

3.4. OPN Expression in Iba-1-Immunolabeled Microglial Cell in RD

We also evaluated retinal inflammation, and, thus, performed double-labeling immunofluorescence with anti-Iba-1, a microglial cell marker [17,22,23], and anti-OPN, an activated microglial cell marker [24–26], which was used for a new marker for retinal injury [17,27].

Few Iba-1-labeled microglial cells were found in the OPL and IPL of the entire retinal regions, including detached, border, and intact (normal data not shown) regions at 1 day after RD, and OPN immunoreactivity was rarely observed in normal retinas at 1 day after RD (Figure 4A). The number of Iba-1-labeled microglial cells was significantly increased in the detached and border regions, but not in the intact regions ($p < 0.05$) 3 days after RD. In the detached and border regions, Iba-1-labeled microglial cells were found throughout the retinal layer from the IPL to the subretinal space. OPN immunoreactivity was observed in Iba-1-labeled microglial cells in the subretinal space, but not in the retina. The number of Iba-1-labeled and OPN co-labeled microglial cells increased in both detached and border regions 5 days after RD. In addition, OPN immunoreactivity was observed in the rod and cone layers, which may be fragmented or degenerating photoreceptor outer and inner segments, as previously reported [17]. The number of Iba-1-labeled microglial cells in the detached and border regions peaked 7 days after RD. Interestingly, Iba-1-labeled microglial cells decreased in the subretinal space, while those within the outer and inner retina increased. The number of OPN co-labeled microglial cells decreased because of the decrease in Iba-1-labeled microglial cells in the subretinal space. The number of Iba-1-labeled microglial cells markedly decreased, particularly in the subretinal space, 14 days after RD; thus, OPN co-labeled microglial cells were rarely observed. In the intact region, throughout the entire experimental period, Iba-1-labeled microglial cells were only observed within the retina but not in the subretinal space; thus, OPN immunoreactivity was not observed. Iba-1-labeled microglial cells within the retina slightly increased in number and changed in size and shape. That is, they were stout and their processes were thick and long compared to the Iba-1-labeled microglial cells in the normal retina (Figure 4B).

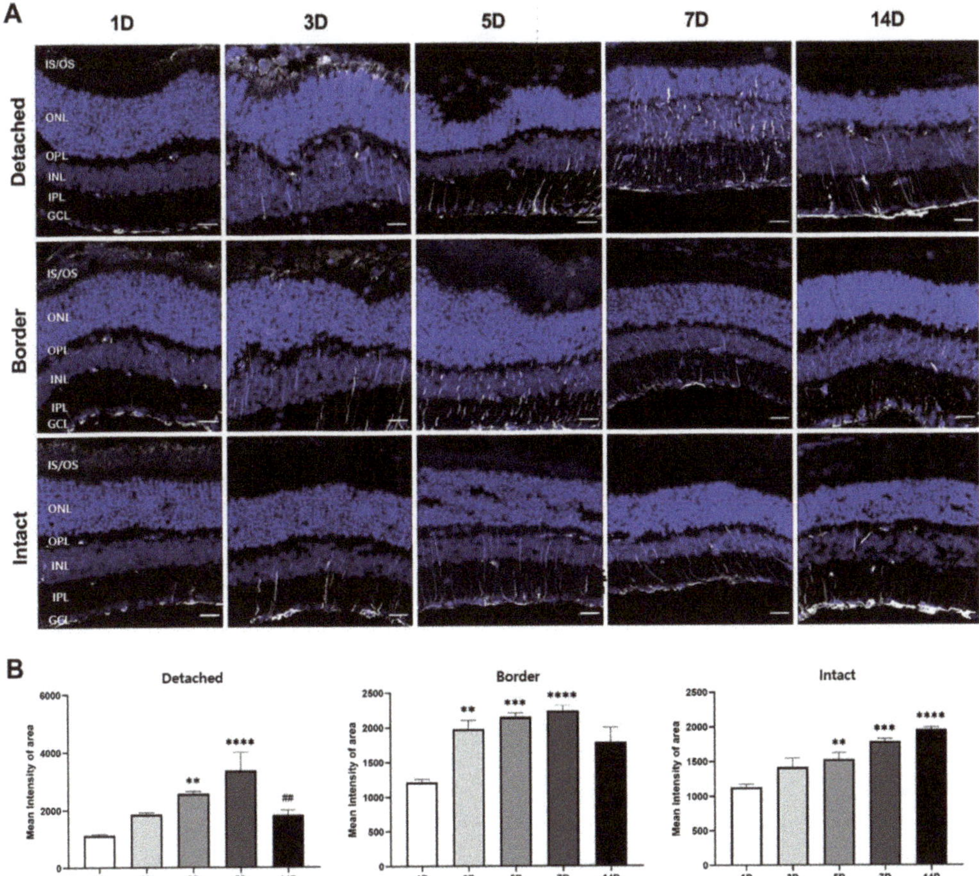

Figure 3. GFAP expression in mouse detached retinas. (**A**) Confocal micrographs of vertical cryosections collected 1, 3, 5, 7, and 14 days after RD. The sections were immunostained with anti-GFAP (white), an activated Müller glial cell marker. In the detached region, GFAP immunoreactivity gradually increased until 7 days after RD, and then decreased. In the border and intact regions, GFAP expression was significantly weak. Scale bars, 20 µm. (**B**) In each region, ROIs were positioned manually per outer nuclear layer (ONL; $n = 6$, 15 fields per time point). In the detached region, GFAP expression gradually increased until 7 days after RD ($p < 0.05$), after which it significantly decreased ($p < 0.05$). In border regions, GFAP levels tended to increase or at least be sustained by 14 days after RD. In the intact region, GFAP expression steadily increased time-dependently ($p < 0.05$). Data are shown as mean ± SEM. ** $p < 0.01$, *** $p < 0.001$, and **** $p < 0.0001$ versus 1 day after RD, and ## $p < 0.01$ versus 7 days after RD, based on one-way ANOVA followed by Bonferroni's multiple comparisons test.

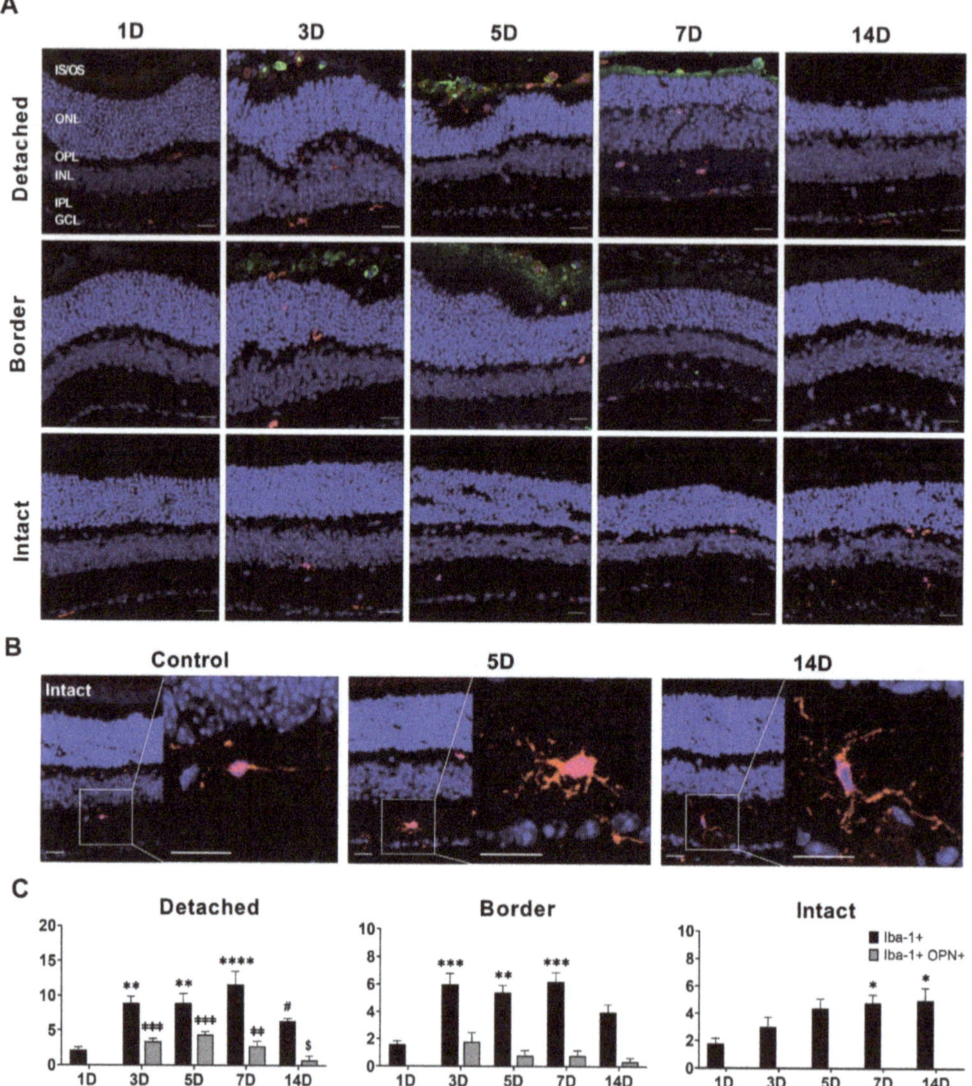

Figure 4. Iba-1 and OPN expression in mouse RD model. (**A**) Confocal micrographs of vertical cryosections collected 1, 3, 5, 7, and 14 days after RD. The sections were double immunostained with anti-Iba-1 (red) as a microglial cell marker and anti-OPN (green) as an inflammation marker. Iba-1 and OPN co-labeled microglial cells were mainly detected from 3 to 7 days after RD, and, thereafter, decreased by 14 days after RD in the detached and border regions. In the intact region, the number of Iba-1-labeled microglial cells was lower than that in the detached region, and OPN co-labeled microglial cells were undetected. OPN immunoreactivity was detected only in the subretinal space, but Iba-1 immunoreactivity was observed in all the layers. Scale bars, 20 μm. (**B**) Representative Iba-1-labeled microglial cells at the normal control, peak, and end of the experiment in the intact region. Iba-1-labeled microglial cells showed changes in their morphological characteristics. Scale bars, 20 μm. (**C**) Quantitative analysis of the number of Iba-1 and OPN co-labeled cells was conducted. In each region, the number of Iba-1- and OPN co-labeled cells were measured for all layers. Iba-1-labeled, and not OPN co-labeled, microglial cells were observed in each region. Most OPN co-labeled microglial cells were detected in the detached and border regions, but not in the intact region ($n = 6$, 15 fields per time point). Data are shown as mean ± SEM. * $p < 0.05$, ** $p < 0.01$, *** $p < 0.001$, **** $p < 0.0001$, ++ $p < 0.01$ and +++ $p < 0.001$ versus 1 day after RD, and # $p < 0.05$ and $ $p < 0.05$ versus 7 days after RD based on one-way ANOVA followed by Bonferroni's multiple comparisons test.

The images captured after immunohistochemical analysis for Iba-1 and OPN were measured for all of the layers (Figure 4C). In the detached region, Iba-1-labeled cells significantly increased until 7 days after RD ($p < 0.05$) and decreased by 14 days after RD in the detached and border regions ($p < 0.05$). OPN co-labeled microglial cells were detected from 3 days after RD and were rarely observed 14 days after RD. In the intact region, the number of Iba-1-labeled cells increased from 3 to 7 days after RD, and then sustained for 14 days, unlike those in detached and border regions. OPN co-labeled microglial cells were not observed during the experimental period.

4. Discussion

RD is an eye disorder in which photoreceptors separate from the choroidal blood vessels that provide oxygen and nourishment. It can occur spontaneously as a complication from various retinal disorders, such as AMD and DR. Untreated RD can lead to permanent vision loss in the affected eye. Nevertheless, RD pathogenesis is little known, compared to other retinal diseases, including AMD and DR. In particular, events that occur in non-detached areas, including the penumbra, have not been extensively studied. This might be because surgery is almost always performed for the treatment of RD, and its outcome is generally good [28].

The study demonstrated progressive photoreceptor loss in the detached region with a decreased ERG response (Figure 1). TUNEL-positive cells were first observed in the detached region 1 day after RD, prominently increased from 3 days after RD, and peaked at 5 days after RD (Figure 2). These findings reveal that photoreceptor degeneration starts early (<24 h), accelerates, and rapidly subsides within a week, suggesting that surgical intervention is recommended in early RD [29]. In addition, we observed an indispensable number of TUNEL-positive cells in the border region near the detached region, while histological findings showed thinned ONL. Additionally, in a few cases, the ONL thickness in the intact region decreased, albeit statistically insignificant ($p > 0.05$). Inflammation in the border region was comparable to that in the detached region. These findings also suggest that early intervention in RD is important for preventing its progression into the neighboring regions of the retina.

Many studies have been conducted in order to understand the consequences of induced RD in animal models, and have strongly implicated inflammatory factors [13,14,30–32]. Two retinal glial cells, Müller glia and microglia, play a key role in the inflammatory response in the retina [33,34]. Thus, characterizing the glial response activation pattern is important for interpreting the inflammatory processes. Previously, we reported the differential activation patterns of two retinal glial cells, Müller glial cells and microglial cells, in two different RD models. Müller glial cells were gradually activated and slowly subsided in whole retinal regions, regardless of the injury site and type [16,17], whereas microglial cells were abruptly activated and deactivated at the injury site, but not far from the injury region [17]. However, there have been few reports of glial responses throughout the retina in the RD model.

In this study, GFAP expression, the activation marker of Müller glial cells, increased from the inner limiting membrane of the detached region to the ONL until 7 days after RD, and then decreased slightly. GFAP expression increased similarly in the entire retina and was not limited to the detached region (Figure 3). This GFAP expression pattern has been previously reported in light-induced photoreceptor degeneration [16,35] and RD models [36,37]. Interestingly, GFAP expression increased in the intact region, regardless of RD. This finding was also reported in our previous studies in the blue LED-induced retinal degeneration model [16,17]. Taken together, these findings suggest that regardless of the injury size and type, Müller cells in the entire retina respond to injury.

Microglial distribution was more distinct than Müller glial distribution in RD. Iba-1-labeled microglial cells were detected in the subretinal space and ONL in the detached and border regions from 3 to 14 days after RD, with a peak at 7 days, but not in the intact region. Interestingly, Iba-1-labeled microglial cells in the subretinal space peaked 5 days after RD and showed OPN immunoreactivity (Figure 4), and also expressed major histocompatibility class II (MHC-II) (Supplementary Figure S1). This is consistent with previous studies, in which MHC-II-positive microglial cells were located in the subretinal space in the retinal degeneration model, suggesting that they play a pro-inflammatory and phagocytic role during photoreceptor degeneration in a region-specific manner [38,39]. In addition, they express OPN, which is secreted by microglia and acts as a pro-inflammatory cytokine [24–26,40]. In contrast, within the retina, Iba-1-labeled microglial cells peaked 7 days after RD, followed by a slight decrease at 14 days after RD in the detached region. Moreover, they tended to proliferate in the intact region. In addition, some Iba-1-labeled microglial cells showed activated microglial morphology characterized by a stout cell body and thick and extended processes in the intact region at 14 days after RD, compared to those in normal retinas (Figure 4B). These findings suggest that microglial cells can be activated by the propagation of alarm signals from the injury site, and are involved in systemic inflammation in the retina. This inference is supported by previous studies, showing that unilateral ocular hypertension and axotomy induce contralateral microglial cell activation (e.g., the advent of microglial cells with activated form, MHC-II upregulation, and increase in Iba-1-labeled microglial cell number) [41] and retinal neuronal degeneration in the contralateral eye [42]. Taken together, inflammation during RD in the detached region could be propagated to the intact region, eventually leading to photoreceptor degeneration. Further studies are needed to elucidate the signaling molecules that mediate alarm signals from detached subretinal microglial cells to these cells in the intact region.

Lastly, this study aimed to confirm that OPN is a good lesion-specific marker for retinal inflammation in RD. OPN is upregulated or secreted by activated microglial cells in various types of brain injury [43–46] and RD [16,47]. Our findings demonstrated that OPN is expressed in activated microglial cells in the detached and border regions, but not in the intact region. Moreover, the temporal profile of OPN expression in microglial cells is more closely related to photoreceptor degeneration in RD than GFAP. This spatiotemporal correlation between OPN expression and cell death in retinal lesions has been previously reported in RD [17] and glaucoma [27]. Therefore, OPN can be used as a marker of retinal inflammation, which is accompanied by retinal injury.

Supplementary Materials: The following are available online at https://www.mdpi.com/article/10.3390/cells10081972/s1, Figure S1: MHC-II expression in a mouse RD model.

Author Contributions: Conceptualization, S.-H.L. and I.-B.K.; methodology, S.-H.L., Y.-S.P. and S.-S.P.; software, S.-H.L.; validation, Y.-S.P., S.-S.P. and I.-B.K.; formal analysis, S.-H.L. and I.-B.K.; investigation, S.-H.L., Y.-S.P., S.-S.P. and I.-B.K.; resources, S.-H.L. and I.-B.K.; data curation, S.-H.L. and I.-B.K.; writing—original draft preparation, S.-H.L.; writing—review and editing, Y.-S.P. and I.-B.K.; visualization, S.-H.L.; supervision, I.-B.K.; project administration, I.-B.K.; funding acquisition, I.-B.K. All authors have read and agreed to the published version of the manuscript.

Funding: This research was funded by the Basic Science Research Program through the National Research Foundation (NRF) of Korea, funded by the Ministry of Education, Science, and Technology (grant number 2017R1A2B2005309).

Institutional Review Board Statement: The study was conducted under the guidelines of the Declaration of Helsinki and approved by the Institutional Animal Care and Use Committee (IACUC) at the College of Medicine, The Catholic University of Korea (Approval number: CUMS-2017-0241-03).

Informed Consent Statement: Not applicable.

Data Availability Statement: Data is contained within the article or supplementary material.

Conflicts of Interest: The authors declare no conflict of interest.

References

1. Feltgen, N.; Walter, P. Rhegmatogenous retinal detachment–an ophthalmologic emergency. *Dtsch Arztebl Int.* **2014**, *111*, 12–21; quiz 22. [CrossRef]
2. Dunaief, J.L.; Dentchev, T.; Ying, G.S.; Milam, A.H. The role of apoptosis in age-related macular degeneration. *Cinical. Sci.* **2002**, *20*, 1435–1442. [CrossRef] [PubMed]

3. Barber, A.J.; Lieth, E.; Khin, S.A.; Antonetti, D.A.; Buchanan, A.G.; Gardner, T.W. Neural apoptosis in the retina during experimental and human diabetes. Early onset and effect of insulin. *J. Clin. Investig.* **1998**, *102*, 783–791. [CrossRef]
4. Lewis, G.P.; Sethi, C.S.; Linberg, K.A.; Charteris, D.G.; Fisher, S.K. Experimental retinal reattachment: A new perspective. *Mol. Neurobiol.* **2003**, *8*, 159–175. [CrossRef]
5. Fisher, S.K.; Lewis, G.P.; Linberg, K.A.; Verardo, M.R. Cellular remodeling in mammalian retina: Results from studies of experimental retinal detachment. *Prog. Retin. Eye Res.* **2005**, *24*, 395–431. [CrossRef] [PubMed]
6. Arroyo, J.G.; Yang, L.; Bula, D.; Chen, D.F. Photoreceptor apoptosis in human retinal detachment. *Am. J. Ophthalmol.* **2005**, *139*, 605–610. [CrossRef]
7. Cook, B.; Lewis, G.P.; Fisher, S.K.; Adler, R. Apoptotic photoreceptor degeneration in experimental retinal detachment. *Investig. Ophthalmol. Vis. Sci.* **1995**, *36*, 990–996.
8. Dong, K.; Zhu, H.; Song, Z.; Gong, Y.; Wang, F.; Wang, W.; Zheng, Z.; Yu, Z.; Gu, Q.; Xu, X.; et al. Necrostatin-1 protects photoreceptors from cell death and improves functional outcome after experimental retinal detachment. *Am. J. Pathol.* **2012**, *181*, 1634–1641. [CrossRef]
9. Lo, A.C.; Woo, T.T.; Wong, R.L.; Wong, D. Apoptosis and other cell death mechanisms after retinal detachment: Implications for photoreceptor rescue. *Ophthalmologica* **2011**, *226* (Suppl. 1), 10–17. [CrossRef]
10. Murakami, Y.; Notomi, S.; Hisatomi, T.; Nakazawa, T.; Ishibashi, T.; Miller, J.W.; Vavvas, D.G. Photoreceptor cell death and rescue in retinal detachment and degenerations. *Prog. Retin. Eye Res.* **2013**, *37*, 114–140. [CrossRef]
11. Kunchithapautham, K.; Rohrer, B. Autophagy is one of the multiple mechanisms active in photoreceptor degeneration. *Autophagy* **2007**, *3*, 65–66. [CrossRef]
12. Chinskey, N.D.; Zheng, Q.D.; Zacks, D.N. Control of photoreceptor autophagy after retinal detachment: The switch from survival to death. *Investig. Ophthalmol. Vis. Sci.* **2014**, *55*, 688–695. [CrossRef] [PubMed]
13. Iandiev, I.; Uckermann, O.; Pannicke, T.; Wurm, A.; Tenckhoff, S.; Pietsch, U.C.; Reichenbach, A.; Wiedemann, P.; Bringmann, A.; Uhlmann, S. Glial cell reactivity in a porcine model of retinal detachment. *Investig. Ophthalmol. Vis. Sci.* **2006**, *47*, 2161–2171. [CrossRef] [PubMed]
14. Lewis, G.P.; Sethi, C.S.; Carter, K.M.; Charteris, D.G.; Fisher, S.K. Microglial cell activation following retinal detachment: A comparison between species. *Mol. Vis.* **2005**, *11*, 491–500.
15. Jeong, E.; Paik, S.S.; Jung, S.W.; Chun, M.H.; Kim, I.B. Morphological and functional evaluation of an animal model for the retinal degeneration induced by N-methyl-N-nitrosourea. *Anat. Cell Biol.* **2011**, *44*, 314–323. [CrossRef] [PubMed]
16. Kim, G.H.; Kim, H.I.; Paik, S.S.; Jung, S.W.; Kang, S.; Kim, I.B. Functional and morphological evaluation of blue light-emitting diode-induced retinal degeneration in mice. *Graefe's Arch. Clin. Exp. Ophthalmol.* **2016**, *254*, 705–716. [CrossRef]
17. Chang, S.W.; Kim, H.I.; Kim, G.H.; Park, S.J.; Kim, I.B. Increased Expression of Osteopontin in Retinal Degeneration Induced by Blue Light-Emitting Diode Exposure in Mice. *Front. Mol. NeuroSci.* **2016**, *9*, 58. [CrossRef]
18. Kim, G.H.; Paik, S.S.; Park, Y.S.; Kim, H.G.; Kim, I.B. Amelioration of Mouse Retinal Degeneration After Blue LED Exposure by Glycyrrhizic Acid-Mediated Inhibition of Inflammation. *Front. Cell NeuroSci.* **2019**, *13*, 319. [CrossRef]
19. Paik, S.S.; Jeong, E.; Jung, S.W.; Ha, T.J.; Kang, S.; Sim, S.; Jeon, J.H.; Chun, M.H.; Kim, I.B. Anthocyanins from the seed coat of black soybean reduce retinal degeneration induced by N-methyl-N-nitrosourea. *Exp. Eye Res.* **2012**, *97*, 55–62. [CrossRef]
20. Fisher, S.K.; Lewis, G.P. Müller cell and neuronal remodeling in retinal detachment and reattachment and their potential consequences for visual recovery: A review and reconsideration of recent data. *Vis. Res.* **2003**, *43*, 887–897. [CrossRef]
21. Kim, I.B.; Kim, K.Y.; Joo, C.K.; Lee, M.Y.; Oh, S.J.; Chung, J.W.; Chun, M.H. Reaction of Müller cells after increased intraocular pressure in the rat retina. *Exp. Brain Res.* **1998**, *121*, 419–424. [CrossRef] [PubMed]
22. Imai, Y.; Ibata, I.; Ito, D.; Ohsawa, K.; Kohsaka, S. A novel gene iba1 in the major histocompatibility complex class III region encoding an EF hand protein expressed in a monocytic lineage. *Biochem. Biophys. Res. Commun.* **1996**, *224*, 855–862. [CrossRef]
23. Hikita, S.T.; Vistica, B.P.; Jones, H.R.; Keswani, J.R.; Watson, M.M.; Ericson, V.R.; Ayoub, G.S.; Gery, I.; Clegg, D.O. Osteopontin is proinflammatory in experimental autoimmune uveitis. *Investig. Ophthalmol. Vis. Sci.* **2006**, *47*, 4435–4443. [CrossRef]
24. Kim, M.D.; Cho, H.J.; Shin, T. Expression of osteopontin and its ligand, CD44, in the spinal cords of Lewis rats with experimental autoimmune encephalomyelitis. *J. Neuroimmunol.* **2004**, *151*, 78–84. [CrossRef]
25. Choi, J.S.; Kim, H.Y.; Cha, J.H.; Choi, J.Y.; Lee, M.Y. Transient microglial and prolonged astroglial upregulation of osteopontin following transient forebrain ischemia in rats. *Brain Res.* **2007**, *1151*, 195–202. [CrossRef] [PubMed]
26. Schroeter, M.; Zickler, P.; Denhardt, D.T.; Hartung, H.P.; Jander, S. Increased thalamic neurodegeneration following ischaemic cortical stroke in osteopontin-deficient mice. *Brain* **2006**, *129*, 1426–1437. [CrossRef] [PubMed]
27. Yu, H.; Zhong, H.; Li, N.; Chen, K.; Chen, J.; Sun, J.; Xu, L.; Wang, J.; Zhang, M.; Liu, X.; et al. Osteopontin activates retinal microglia causing retinal ganglion cells loss via p38 MAPK signaling pathway in glaucoma. *FASEB J.* **2021**, *35*, e21405. [CrossRef]
28. Heiss, W.D. The ischemic penumbra: Correlates in imaging and implications for treatment of ischemic stroke. The Johann Jacob Wepfer award 2011. *Cereb. Dis.* **2011**, *32*, 307–320. [CrossRef]
29. Angermann, R.; Bechrakis, N.E.; Rauchegger, T.; Casazza, M.; Nowosielski, Y.; Zehetner, C. Effect of Timing on Visual Outcomes in Fovea-Involving Retinal Detachments Verified by SD-OCT. *J. Ophthalmol.* **2020**, *2020*, 2307935. [CrossRef]
30. Kiang, L.; Ross, B.X.; Yao, J.; Shanmugam, S.; Andrews, C.A.; Hansen, S.; Besirli, C.G.; Zacks, D.N.; Abcouwer, S.F. Vitreous Cytokine Expression and a Murine Model Suggest a Key Role of Microglia in the Inflammatory Response to Retinal Detachment. *Invest Ophthalmol. Vis. Sci.* **2018**, *59*, 3767–3778. [CrossRef]

31. Wurm, A.; Pannicke, T.; Iandiev, I.; Bühner, E.; Pietsch, U.C.; Reichenbach, A.; Wiedemann, P.; Uhlmann, S.; Bringmann, A. Changes in membrane conductance play a pathogenic role in osmotic glial cell swelling in detached retinas. *Am. J. Pathol.* **2006**, *169*, 1990–1998. [CrossRef] [PubMed]
32. Hollborn, M.; Francke, M.; Iandiev, I.; Bühner, E.; Foja, C.; Kohen, L.; Reichenbach, A.; Wiedemann, P.; Bringmann, A.; Uhlmann, S. Early activation of inflammation- and immune response-related genes after experimental detachment of the porcine retina. *Investig. Ophthalmol. Vis. Sci.* **2008**, *49*, 1262–1273. [CrossRef] [PubMed]
33. Eastlake, K.; Banerjee, P.J.; Angbohang, A.; Charteris, D.G.; Khaw, P.T.; Limb, G.A. Müller glia as an important source of cytokines and inflammatory factors present in the gliotic retina during proliferative vitreoretinopathy. *Glia* **2016**, *64*, 495–506. [CrossRef]
34. Langmann, T. Microglia activation in retinal degeneration. *J. Leukoc Biol.* **2007**, *81*, 1345–1351. [CrossRef]
35. Torbidoni, V.; Iribarne, M.; Ogawa, L.; Prasanna, G.; Suburo, A.M. Endothelin-1 and endothelin receptors in light-induced retinal degeneration. *Exp. Eye Res.* **2005**, *81*, 265–275. [CrossRef] [PubMed]
36. Lewis, G.P.; Fisher, S.K. Up-regulation of glial fibrillary acidic protein in response to retinal injury: Its potential role in glial remodeling and a comparison to vimentin expression. *Int. Rev. Cytol.* **2003**, *230*, 263–290. [CrossRef]
37. Luna, G.; Lewis, G.P.; Banna, C.D.; Skalli, O.; Fisher, S.K. Expression profiles of nestin and synemin in reactive astrocytes and Müller cells following retinal injury: A comparison with glial fibrillar acidic protein and vimentin. *Mol. Vis.* **2010**, *16*, 2511–2523. [PubMed]
38. Ma, W.; Zhang, Y.; Gao, C.; Fariss, R.N.; Tam, J.; Wong, W.T. Monocyte infiltration and proliferation reestablish myeloid cell homeostasis in the mouse retina following retinal pigment epithelial cell injury. *Sci. Rep.* **2017**, *7*, 8433. [CrossRef] [PubMed]
39. Aredo, B.; Zhang, K.; Chen, X.; Wang, C.X.; Li, T.; Ufret-Vincenty, R.L. Differences in the distribution, phenotype and gene expression of subretinal microglia/macrophages in C57BL/6N (Crb1 rd8/rd8) versus C57BL6/J (Crb1 wt/wt) mice. *J. Neuroinflamm.* **2015**, *12*, 6. [CrossRef]
40. Wirths, O.; Breyhan, H.; Marcello, A.; Cotel, M.C.; Brück, W.; Bayer, T.A. Inflammatory changes are tightly associated with neurodegeneration in the brain and spinal cord of the APP/PS1KI mouse model of Alzheimer's disease. *Neurobiol. Aging* **2010**, *31*, 747–757. [CrossRef]
41. Rojas, B.; Gallego, B.I.; Ramírez, A.I.; Salazar, J.J.; de Hoz, R.; Valiente-Soriano, F.J.; Avilés-Trigueros, M.; Villegas-Perez, M.P.; Vidal-Sanz, M.; Triviño, A.; et al. Microglia in mouse retina contralateral to experimental glaucoma exhibit multiple signs of activation in all retinal layers. *J. Neuroinflamm.* **2014**, *11*, 133. [CrossRef]
42. Lucas-Ruiz, F.; Galindo-Romero, C.; Rodríguez-Ramírez, K.T.; Vidal-Sanz, M.; Agudo-Barriuso, M. Neuronal Death in the Contralateral Un-Injured Retina after Unilateral Axotomy: Role of Microglial Cells. *Int. J. Mol. Sci.* **2019**, *20*, 5733. [CrossRef]
43. Jarrett, W.H., 2nd. Rhematogenous retinal detachment complicated by severe intraocular inflammation, hypotony, and choroidal detachment. *Trans. Am. Ophthalmol. Soc.* **1981**, *79*, 664–683. [PubMed]
44. Shah, D.N.; Al-Moujahed, A.; Newcomb, C.W.; Kaçmaz, R.O.; Daniel, E.; Thorne, J.E.; Foster, C.S.; Jabs, D.A.; Levy-Clarke, G.A.; Nussenblatt, R.B.; et al. Exudative Retinal Detachment in Ocular Inflammatory Diseases: Risk and Predictive Factors. *Am. J. Ophthalmol.* **2020**, *218*, 279–287. [CrossRef] [PubMed]
45. Lv, Z.; Li, Y.; Wu, Y.; Qu, Y. Surgical complications of primary rhegmatogenous retinal detachment: A meta-analysis. *PLoS ONE* **2015**, *10*, e0116493. [CrossRef] [PubMed]
46. Ellison, J.A.; Velier, J.J.; Spera, P.; Jonak, Z.L.; Wang, X.; Barone, F.C.; Feuerstein, G.Z. Osteopontin and its integrin receptor alpha(v)beta3 are upregulated during formation of the glial scar after focal stroke. *Stroke* **1998**, *29*, 1698–1706; discussion 1707. [CrossRef] [PubMed]
47. Hedtjärn, M.; Mallard, C.; Hagberg, H. Inflammatory gene profiling in the developing mouse brain after hypoxia-ischemia. *J. Cereb. Blood Flow Metab.* **2004**, *24*, 1333–1351. [CrossRef]

Review

Retinal and Brain Microglia in Multiple Sclerosis and Neurodegeneration

Soyoung Choi [1], Li Guo [1] and Maria Francesca Cordeiro [1,2,*]

1 UCL Institute of Ophthalmology, London EC1V 9EL, UK; skgthoi@ucl.ac.uk (S.C.); l.guo@ucl.ac.uk (L.G.)
2 ICORG, Imperial College London, London NW1 5QH, UK
* Correspondence: M.Cordeiro@ucl.ac.uk

Abstract: Microglia are the resident immune cells of the central nervous system (CNS), including the retina. Similar to brain microglia, retinal microglia are responsible for retinal surveillance, rapidly responding to changes in the environment by altering morphotype and function. Microglia become activated in inflammatory responses in neurodegenerative diseases, including multiple sclerosis (MS). When activated by stress stimuli, retinal microglia change their morphology and activity, with either beneficial or harmful consequences. In this review, we describe characteristics of CNS microglia, including those in the retina, with a focus on their morphology, activation states and function in health, ageing, MS and other neurodegenerative diseases such as Alzheimer's disease, Parkinson's disease, glaucoma and retinitis pigmentosa, to highlight their activity in disease. We also discuss contradictory findings in the literature and the potential ways of reducing inconsistencies in future by using standardised methodology, e.g., automated algorithms, to enable a more comprehensive understanding of this exciting area of research.

Keywords: retina; microglia; neurodegeneration; multiple sclerosis; retinal microglia; microglia morphotype

Citation: Choi, S.; Guo, L.; Cordeiro, M.F. Retinal and Brain Microglia in Multiple Sclerosis and Neurodegeneration. *Cells* **2021**, *10*, 1507. https://doi.org/10.3390/cells10061507

Academic Editor: Maurice Ptito

Received: 1 May 2021
Accepted: 11 June 2021
Published: 15 June 2021

Publisher's Note: MDPI stays neutral with regard to jurisdictional claims in published maps and institutional affiliations.

1. Introduction

Microglia are resident immune glial cells of the central nervous system (CNS). They dynamically shift into various different morphologies, which have also been associated with specific activation states that may be related to neuroprotective and/or neurotoxic functions in response to stimuli, injury or insult [1–6]. These morphological and functional changes are essential to support a healthy CNS by contributing to homeostasis [2,7,8]. However, emerging evidence has started to show the involvement of microglia in disease whereby microglial dysfunction may be caused by disease and/or actually cause augmented disease-associated pathologies [2]. Nevertheless, the exact degree of involvement and mechanisms of how microglia may influence health and disease is unknown and currently being investigated. The purpose of this review was to compile information on the characteristics of CNS microglia, including those in the retina, with a focus on their activation states, morphology and function, in relation to health and disease. In the first section, we have compiled an extensive description of the main features of microglia in the general CNS and in the retina. We also explain some of the microglial changes that occur throughout normal development and the ageing process. In the next section, we discuss the current understanding of multiple sclerosis (MS) as an autoimmune disease, the surprisingly common ocular manifestations of MS, microglia in MS and finally, microglia in other neurodegenerative diseases. Finally, we provide insight as to why there may be contradicting findings in relation to the characteristics of CNS and retinal microglia. Here, we suggest different methods of experimentation such as using automated algorithms in light of producing more conclusive and consistent results.

2. Microglia

The CNS is made of several different types of cells, 5–10% of which are microglia, the resident immune cells [9]. Microglia were originally thought to exist as quiescent or "resting" cells that continuously survey their microenvironment for any stimuli or injury that may be harmful [5]. Conversely, more recent findings have suggested explanations for their dynamic properties. Microglia can shift into different morphological states, which will be referred to as morphotypes. Each morphotype has been correlated to different activation states, which have also been associated with unique functions required to preserve a physiologically "normal" environment [5].

Once microglia are activated from their resting state, they can further differentiate into two main phenotypes: M1 and M2 [6]. Although there is a lack of understanding of the specific mechanisms that induce this differentiation, M1 and M2 microglia have also been associated with distinct cytokines, chemokines and trophic factors [6]. Pro-inflammatory responses are associated with the "classically" activated M1 microglia, which encourage neuroinflammation as a response to insult or injury, creating a neuro-toxic environment and removing dysfunctional fragments of cellular debris [6]. This may occur as a result of inflammatory factors such as interleukin-1ß (IL-1ß), tumour necrosis factor-alpha (TNF-alpha) and inducible nitric oxide synthase (iNOS) [6,10]. Conversely, the "alternatively" activated M2 microglia are notoriously responsible for anti-inflammatory responses that encourage neuroprotective and restorative processes [6,10]. More recently, at least three more sub-phenotypes of the M2 type (M2a-c) have been found [11] whereby, more specifically, the M2a type secrete anti-inflammatory factors such as IL-10 and insulin-such as growth factor-1 (IGF-1), promoting cell debris removal and neuroprotection [6,11,12]. The M2b is said to be stimulated by inflammatory factors such as IL-1ß and lipopolysaccharides (LPS), which may also increase the expression of IL-10 [11]. These M2b microglia have been found to have phagocytic properties in brains modelled for Alzheimer's disease (AD) and expressing high levels of CD64 [11]. The M2c "acquire deactivation" by IL-10 or glucocorticoids, in turn increasing expression of growth factors such as TGFß [11]. Despite these differences, M1 and M2 activation are functionally required to guarantee the removal of dysfunctional cells or noxious aggregates of cellular debris [6]. M1 microglia are usually involved in the clearance of cell debris, and this inflammatory response must be controlled by M2 microglia to avoid needlessly prolonged inflammation [6]. Often in pathological processes, the typical balance of M1 and M2 polarisation seen in normal conditions may be affected [6]. This can result in the clearance of healthy cells due to excessive M1 inflammation and the M2's dampening effect of M1s may be overwhelmed causing further damage [6]. This often occurs in neurodegenerative illnesses and therefore, some therapeutic candidates that target M1 and M2 polarisation have been proposed [13]. Despite this, there is also emerging evidence that suggests that the M1/M2 polarisation may be outdated. It was first introduced and used to accommodate easier methods of data interpretation [14]. However, recent advances in technology have revealed overlapping morphological and genetic characteristics between M1/M2 types, suggesting a need to re-evaluate microglia types [14]. More recently, disease-associated microglia (DAM) have also been recognised as a unique microglia type seen in disease [14]. DAMs are characterised by microglia that express low levels of surveillant and homeostatic genes and high levels of markers associated with degeneration such as triggering receptor expressed on myeloid cells 2 (TREM2) [14,15].

With regards to morphology, there about five main microglial morphotypes that have been recognised, including the ramified, hyper-ramified, activated, amoeboid and rod types. Under non-primed or "inactive" conditions, microglia appear "ramified". They are distributed evenly like a "mosaic", with each consisting of a small and round cell body to which are attached several thin and long processes that constantly extend and retract to facilitate their surveillant functions (Figure 1) [2,5,16]. Experimental in vivo imaging of the brain has shown that these dynamic processes come into close proximity with neurons, glia and blood vessels, suggesting that microglia actively co-operate with other parts of the cortex to sustain a physiologically normal CNS environment [17]. Sometimes, ramified microglia may recognise microenvironmental changes and respond by changing into "hyper-ramified" microglia typically defined by more abundant processes that are longer and thicker, attached to larger, lobular and irregularly shaped cell bodies (Figure 1) [16,18]. Most types of primed "non-ramified" microglia, including hyper-ramified cells, are scattered in the CNS in an irregular and "clustered" distribution [2]. Hyper-ramified microglia may also shift into the activated form upon exposure to significantly noxious stimuli, which also have similar cell bodies to hyper-ramified cells whilst having much fewer processes that are thicker and shorter (Figure 1) [2,16,19]. When noxious stimuli are extensively prolonged, activated microglia may morph into the "amoeboid" state with a rounder, larger and more regularly shaped cell and very few or no processes (Figure 1) [16,18]. A recently re-discovered morphotype are "rod" microglia characterised by a long, sausage-shaped cell body with a few processes that may not always extend beyond the length of ramified microglia (Figure 1) [20,21]. Although emerging evidence has shown rod microglia localising near neurons and aligning itself along the nerve fibres, their exact function is still yet to be discovered [20–23]. The final type is the amoeboid microglia, often referred to as the phagocytic type that moves to the site of damage and phagocytose dead or dying neurons and cell debris (Figure 1) [5,24]. Recently emerging evidence has led to a theory that the hyper-ramified, activated and rod morphotypes may be "transitioning" forms that exist between the ramified and amoeboid states [5,25].

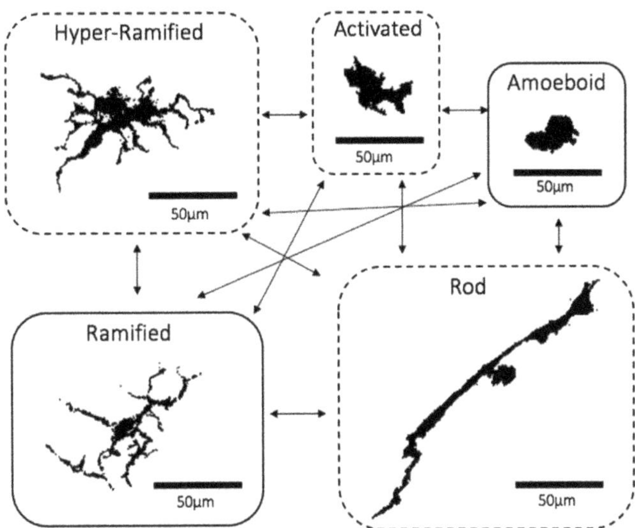

Figure 1. Microglia Morphotypes. A diagram showing microglia morphotypes with arrows showing potential orders of morphotype modifications. Dashed boxes represent the "transitioning" microglia. Microglia are from whole-mount retinal images of iba-1-stained C57BL6 mouse (28 months old) retinal microglia. Scale bar = 50 μm. Images were obtained from the Cordeiro Laboratory following similar protocols mentioned in Davis et al. [24].

2.1. Retinal Microglia

The retina is an essential part of the CNS. Due to its transparent nature to light, it is possible to use less invasive and high-resolution imaging modalities to visualise the retina [26,27]. It is predominantly known for its involvement in converting light energy into electrical signals [28]. In humans, the retina develops from the first month embryologically to the end of the first year, originating from the neuroectoderm [28,29]. The retina consists of several distinct layers with the innermost layer being the retinal nerve fibre layer (RNFL), then the ganglion cell layer (GCL), the inner plexiform layer (IPL), the inner nuclear layer

(INL), the outer plexiform layer (OPL), the outer nuclear layer (ONL) and the final layer the retinal pigment epithelium (RPE) (Figure 2) [29]. Throughout these layers exist several types of cells, including amacrine cells, Müller cells, astrocytes, horizontal cells, rod and cone photoreceptors and bipolar cells, all of which may also be found in the rest of the CNS [29].

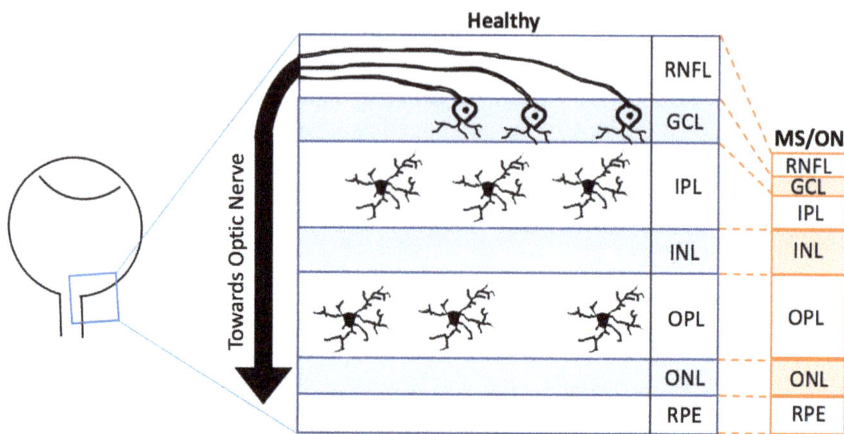

Figure 2. Comparison of retinal changes in Healthy and Diseased Retinas. A schematic diagram showing changes in the main layers in the healthy (blue) and MS/ON (multiple sclerosis/optic neuritis, orange) retinae. The diagram is not to scale and a hypothetical representation. The ganglion cell axons are found in the RNFL whilst their cell bodies are in the GCL. Microglia are said to reside in the IPL and OPL. There are also other retinal neurons and glia that are not specified in this diagram. RNFL = retinal nerve fibre layer, GCL = ganglion cell layer, IPL = inner plexiform layer, INL = inner nuclear layer, OPL = outer plexiform layer, ONL = outer nuclear layer, RPE = retinal pigment epithelium, MS = multiple sclerosis, ON = optic neuritis.

Approximately 0.2% of the total retinal cells are made of microglia, of which 50% usually reside in the IPL whilst the rest reside in the OPL (Figure 2) [2]. Through development and homeostasis, microglia dynamically move through the different layers of the retina, although avoiding the ONL [2]. Furthermore, many retinal diseases and retinal injury models have shown that microglia may migrate towards the region of degeneration, become activated and proliferate [30,31], as described in more detail later. Although both retinal and brain microglia are developed from the primitive yolk sac, the appearance of each morphotype may vary depending on the region of the CNS [10]. For example, microglia in the striatum, hippocampus and frontal cortex have larger cell bodies with more processes compared to those in the cerebellum [32]. Additionally, there is evidence that distinct layers of the cerebral cortex contain microglia of different sizes [32]. Morphotype appearances may also vary depending on the methods used or the axes of dissection. For instance, ramified microglia may appear horizontally ramified, which, in a cross-sectional observation, appear as one horizontally long cell whilst in a whole-mount observation, appear as and like the previously mentioned standard morphological description of ramified microglia [2,33]. When microglia are primed and become hyper-ramified, their processes extend radially, reaching across different layers, which is more visualisable in cross-sectional dissection compared to whole-mount observations [2,33]. Despite these differences, retinal and brain microglia have been found to share the expression of several transcription factors [2]. Numerous studies have also been able to observe each morphotype in both the brain and the retina [2,4,5,22,33,34]. However, it is still not clear whether or not each morphotype in both CNS regions share the same features [2].

2.2. Molecular Markers and Stimuli Affecting Microglial Morphology

Similar to the brain, individual microglia morphotypes in the retina may result from microglial responses to various cytokines, chemokines or damage-associated molecular patterns present in its microenvironment, suggesting heterogeneity in microglial genetic expression [9]. For instance, ramified microglia are said to have high expressions of P2RY12, which is often associated with surveillant functions, whilst amoeboid microglia have been found to express high levels of CD68—a notorious marker of phagocytosis [35,36]. Differential expression levels of microglial ion channels and surface receptors can also interact with such molecules (e.g., cytokines) in the microglial microenvironment, resulting in microglial changes including density, spatial distribution, activation state and morphotype and disease pathogenesis [9,37]. For instance, transforming growth factor beta (TGFß) is an important microglial cytokine, pleiotropically involved in the physiological development of retinal neurons and vessels [37]. Ma et al., observed iba-1 positive microglia, from whole-mounted retinas of tamoxifen-induced ablation of TGFßR2 (TGFß receptor) in 2-month-old Cx3cr1$^{CreER/+}$, Tgfbr2$^{flox/flox}$ mice [37]. The microglial morphology appeared ramified at 1 day post TGFßR2 ablation (PTA), which then became less ramified with "stubby" processes by 2–5 days PTA and finally appeared with lengthy processes that were aligned along the retinal blood vessels by 3–10 weeks PTA [37]. Real-time polymerase chain reaction (RT-PCR) analyses of TGFßR2 ablated retinal microglia revealed decreased expression of growth factors (e.g., BDNF, PDGFA) and increased expression of inflammatory activation markers (e.g., MHCII, CD68). This was not observable in healthy microglia. Additionally, whilst the TGFßR2 ablated animals showed no effects of the retinal vasculature, they experienced retinal thinning and amplified rates of pathological choroidal neovascularisation in response to injury [37].

2.3. Extracellular Vesicles: Effects on Microglia

Retinal (and brain) microglia also secrete extracellular vesicles (EVs), which are membrane-bound particles composed of mRNA, miRNA, DNA, cytokines, lipids and proteins, regarded as a biopsy of its origin cell [6,38,39]. EVs are involved in cell-to-cell communication by transporting their neuro-protective/toxic components in response to intracellular and extracellular cues as they move through the bloodstream and cerebrospinal fluid (CSF) to reach other cells within close and distant proximity [6,38,39]. EVs include exosomes and ectosomes, which consist of apoptotic bodies and microvesicles [39]. Firstly, many endosomal vesicular bodies fuse to form exosomes (40–160 nm in size) which are then released for intercellular communication [39]. Secondly, microvesicles instead originate from the plasma membrane undergoing outward budding (100–1000 nm) [38]. Finally, apoptotic bodies are >1000 nm in size and form through membrane blebbing of disintegrating, e.g., retinal microglia [38]. In fact, components of apoptotic bodies such as phosphatidyl serine modulate microglial phagocytosis [38,39]. As a result of these diverse features, microglial EVs have recently started to be investigated in relation to neurodegenerative diseases.

2.4. Microglia and Ageing

Microglial cells in the retina of newborn and postnatal rats have a round or amoeboid form and show pseudopodal processes involved in cell debris phagocytosis and developmental synapse remodelling [2]. Steadily, as the second and third week of the postnatal period approaches, microglial cell bodies become smaller with fine ramifications [2], assuming a highly ramified phenotype with the progress in brain development [40]. This progression is also reflected in brain regions, including the cerebellum, which suggests that microglia are actively involved through the maturation of the CNS.

A study investigated cortical microglia characteristics in old adult (24 months) and young adult mice (6 months) to find decreased microglial density and irregularly distributed clusters [41]. However, retinal microglia studies have revealed some contradictory results. Firstly, similar age groups of mice as the previous study were examined

(18–24 months vs. 3–4 months) to find that the older adults had significantly elevated retinal microglia densities compared to that of younger adult mice [42]. This suggests that the microglial responses to ageing may be region-specific. Further inspection of real-time retinal imaging showed that most of the old adult microglia appeared to have fewer and shorter processes, suggestive of activated or amoeboid microglia, than those in the young mice [42]. These were found to be arranged in a mosaic distribution. A more recent study looked at the morphological and marker expression differences of whole-mounted retinal microglia in old (15 months) and young (age not specified) mice [43]. Unlike the results from Damani et al., this study found no significant differences in iba-1 positive retinal microglia cell density and area covered by the processes between the two age groups, whilst the cell soma area in the OPL, IPL, NFL and GCL and the number of vertical processes had significantly increased [42,43]. Additionally, the young microglia cells were mostly expressing P2RY12 and no CD68, whilst the aged microglia were CD68+ and appeared amoeboid [43]. These results, instead, imply that with ageing, microglia may shift their morphologies and genetic profile to accommodate the ageing process [43]. Fernández-Albarral et al., however, do not disclose the exact ages, species of mice or the quantitation methods used. Such differences between studies may impact the results.

Aged microglia become less dynamic, showing significantly slower process motilities compared to those in their younger counterparts, which likely compromise their ability to continuously survey and interact with their environment [42]. Post-mortem hippocampal and cortical investigations in "young", "middle", or "old" (20–69 vs. 70+ vs. 90+ years old) adults revealed more region-specific microglial responses to ageing [4]. Rod-shaped microglia were found to be significantly more abundant in the "middle" compared to "young" adults, although this was only seen in the hippocampus whilst the "old" adults only had significant rod increases in the hippocampus [4]. Other cortical and hippocampal studies found similar microglial trends with age, which suggests that aged microglia may have diminished surveillant capabilities required for maintaining a healthy CNS, which in turn may increase the risk of developing neurodegenerative diseases [44–46].

Myelinated axons of neurons are an essential feature of the CNS that enables efficient action potential conductance [47]. Whilst myelin is produced by many oligodendrocyte cells, it undergoes constant renewal, with myelin debris being directly removed and indirectly replaced by microglia [47,48]. However, it has been proposed that changes in myelin debris formation with ageing may also result in age-related dysfunction of CNS immune cells such as microglia [49]. Hence with age, there are more impaired myelin-associated molecules, and a higher myelin protein turnover rate is required [49]. This leads to an increased rate of myelin breakdown, causing burdensome myelin accumulation, which then forms insoluble lysosomal aggregates within microglia cells [49]. TREM2 are expressed on microglial cell surface membranes; however, its deficiency or mutation can result in disease caused by excessive demyelination [50]. Poliani et al. found "aged" TREM2 deficient brains had demyelination with dystrophic and amoeboid looking microglia [50]. It is also said that TREM2 positive amoeboid microglia may then morph into different shapes as it starts to produce factors such as TNF and IL-1, which are considered to be "pro-regenerative" factors [48]. Thus, the extracellular matrix is modified to trigger modifying the extracellular matrix to attract and activate oligodendrocyte precursor cells (OPCs) which then remyelinate the axon [48]. Furthermore, healthy microglial responses to demyelination were found to increase expression of genes associated with activation, phagocytosis and lipid metabolism, whilst that of TREM2 deficient microglia were found not to [50]. Other studies have shown that although debris clearance by microglial phagocytosis is greater with the increase in age, "younger" myelin phagocytosis was more proficient than with "older" myelin [51]. These age-related changes were also correlated with the more frequent appearance of "dystrophic" non-ramified microglia [51]. There are not yet many investigations of myelin debris related retinal microglial phagocytosis, possibly due to the lack of myelin in the retina itself.

With ageing, the microglial populations become dystrophic and undergo structural and morphological changes. The cytoplasm starts to fragment, their cell processes gradually lose the fine ramifications and show spheroidal swellings [52]. In addition, their constitutive microglial function starts to decline and show abnormal microglial injury responses. These alterations, combined with molecular and gene expression ageing changes within microglia, resulting in their reduced capability of maintaining homeostasis in the immune environment, and this may contribute to neuronal impairments, cognitive decline and age-related diseases [5,27,53–55].

2.5. Genetic Factors: Effects on Ageing Microglia

Age is a well-known risk factor for neurodegenerative diseases [56–58]. Therefore, the change of gene expression in the ageing retina has become of significant interest. For instance, Chen et al. investigated the total retinal RNA of 3-month- and 20-month-old C57BL/6 mice [57]. With the increase in age, 298 genes, including those related to stress response and glycoprotein synthesis, were upregulated more than two-fold whilst 137 genes, including those related to immune and defence responses, had also been downregulated more than two-fold [57]. Additionally, RT-PCR analyses showed increases in inflammatory cytokine-, chemokine-, or complement activation-associated genes, e.g., chemokine (C-C motif) ligand 2 (CCL2), CCL12 or complement component 3 (C3) [57]. The authors then hypothesised that this might reflect microglial activation, which was supported by the immunohistological observation of isolectin B4$^+$ amoeboid microglia in the IPL of aged mice only [57]. Another study was able to specifically investigate transcriptional changes in the ageing retinal microglia by comparing the RNA extracted from isolated retinal microglia of 3-, 12-, 18-, and 24-month-old C57BL/6 mice [58]. A total of 719 differentially expressed genes were identified and were functionally associated with microglial immune regulation, e.g., IL3 and IL7, angiogenesis, e.g., vascular endothelial growth factors and trophic growth factors, e.g., neurotrophin [58]. Interestingly, like Chen et al., expression of C3, a gene associated with age-related macular degeneration (AMD), increased with age which implied that senescence-associated retinal microglia transcriptional changes might contribute to AMD pathogenesis [58].

3. Multiple Sclerosis

3.1. Multiple Sclerosis

Multiple Sclerosis (MS) is a chronic inflammatory disease of the CNS, affecting approximately 100,000 patients in the U.K. and 2,500,000 patients globally [59]. Interestingly, it can affect up to three times as many women as men, with an average age of onset in early adulthood [60]. Whilst the initial records of pathophysiological features related to MS date back to 1838, their pathological and clinical presentations were first identified as "MS" in 1863 by Jean-Martin Charcot [61]. Since then, the complex pathologies of MS have been revealed. Frequently occurring symptoms of MS involves the ocular pathway such as RNFL thinning, optic neuritis (ON) characterised by inflammatory damage to the optic nerve and uveitis characterised by intraocular inflammation of the vitreous body, retina, and uveal tract [62–66]. It has also been reported that such retinal changes may occur before changes in the rest of the CNS in many neurodegenerative diseases, including MS [26]. These recent findings have led to much interest in retinal research. Despite the complex processes involved in MS development, it is a disease defined by autoimmune responses through activation of immune cells such as T-lymphocytes, B-lymphocytes, microglia cells and macrophages, demyelination, remyelination and neurodegeneration [12,67–69].

A well-recognised hypothesis of MS is that it develops through two main phases. Firstly, T-cells and B-cells mediate inflammatory responses by releasing cytokines which induce activation of inflammatory cells such as microglia behind a "closed" blood-brain barrier (BBB) [68,70–72]. Chronic inflammation can then result in mitochondrial dysfunction, causing energy deficiency. Secondly, the neuroprotective signals are over-ridden, which impairs the ability to repair demyelination, damaged axons and neurodegeneration [68,70]. As a result, there are significant blockages of axonal conductance whereby eventually, the patient is left with irreversible lesions of the CNS such as the BBB, causing it to become "leaky" [68]. This, however, remains a theory due to recent emerging evidence showing that some patients respond better to therapeutics agents that target B-cells as opposed to those that target T-cells [70]. Although there have been multiple investigations using ever-evolving methods, there is a lack of consensus due to contradictory scientific results [73].

3.2. Sub-Types of Multiple Sclerosis

There is a spectrum of severity for symptoms and rates of progression, reflecting the heterogeneity of MS [70]. The earliest presentation of MS patients can be recognised as a clinically isolated syndrome (CIS) [67]. CIS patients have monophasic and monofocal symptomatic episodes, experiencing, e.g., ataxia, photophobia or areflexia, which lasts between 24 h at 3 weeks [67]. Primary progressive MS (PPMS) affects around 15% of MS patients who usually experience a steady and progressive decline in health [74]. Progressive relapsing MS (PRMS) is the least common form of MS, which only affects around 5% of patients that experience a steady decline in health with unexpected spikes of deterioration and recovery [74]. Relapse remitting MS (RRMS) is a more common form, affecting 80–90% of patients that experience unforeseen surges of disability [70]. Usually, these surges are succeeded by complete recoveries, but as patients age and progress to later stages, these recoveries become more partial [70,74]. This may be justified by many RRMS patients proceeding on to develop secondary progressive MS (SPMS), where patients may experience fewer spikes of impairment and start to mimic the disease trajectory of PPMS [70,74]. Recently, wider interests in genetic studies have enabled MS-specific alleles and gene variants to be recognised, especially through the genome-wide association study [70,75]. Large amounts of data have accommodated more sophisticated analyses that revealed that specific gene variants were correlated to the various types of MS [76]. These gene variants were predominantly found physically and functionally near to immunomodulatory genes associated with MS pathogenesis [70,75].

Despite this, the profiles of these sub-types are only descriptive as there is a lack of sufficient evidence to make accurate pathophysiological distinctions [77]. For instance, a large proportion of MS patients experience asymptomatic phases whereby MS-associated lesions and other pathophysiological changes may occur silently [78]. Some post-mortem brain studies have even shown that there were distinctive MS-associated pathologies seen in subjects who had been considered "healthy" during their lifetime [78]. More recently, references to these sub-types have therefore evolved, distinguishing between disease activity or no disease activity with details of "surges" or steady progression and observations of new CNS lesions [77].

3.3. Ocular Manifestations of Multiple Sclerosis

Post-mortem and retinal imaging studies have revealed MS-related pathologies such as optic nerve damage by demyelination and atrophy and GCL, IPL, and RNFL thinning (Figure 2) where the RGCs and microglia can be found [79,80]. It is theorised that retinal thinning is a downstream effect of the optic nerve and RGC degeneration following microglia-mediated immune responses [79,81]. Interestingly, these hallmarks are symptomatically present in more than 20% of pMS and asymptomatically present in more than 68% of pMS, making almost a staggering 90% of pMS who are estimated to have such retinal pathologies. The typical symptomatic features consist of photopsia, monocular loss of visual acuity and colour vision with mild eye pain, whilst some atypical features consist of photophobia, binocular prolonged vision loss and severe or absence of pain [62,67,82]. These symptoms may develop from a couple of hours to days, peaking at roughly 2–3 weeks from the initial signs followed by recovery [83].

Clinicians have been encouraged to use neuroimaging techniques such as magnetic resonance imaging (MRI) to confirm MS-associated diagnoses [63,79]. Such neuroimaging methods can be used to image the entire CNS, including the visual pathway. For example, an ON-patient MRI showed ventricular brain enlargement and atypical thickening of the optic nerve, which were indicative of brain atrophy and optic nerve demyelination [84]. Optical coherence tomography (OCT) and motion perception assessments, both of which examine the visual system, may also be used as adjuncts to MRIs [85]. The majority of neuroimaging techniques may be complex, low reproducibility, uncomfortable and extremely costly (e.g., MRI) [63,79,80]. In contrast, OCT retinal imaging offers a non-invasive, simpler and less expensive technique to provide high-resolution images with detailed analysis of distinct retinal layers [63,79,80,83]. Whilst sophisticated MRI techniques can be used to identify dynamic inflammation, typically seen in earlier stages of MS, standard MRI parameters can be correlated to OCT evidence indicative of irreversible neurodegeneration and, therefore, significant reductions to retinal thickness [79]. Despite the shortcomings of both methods, each provides unique information relating to distinct aspects of MS pathogenesis. This emphasises that both MRI and OCT should be used in conjunction to maximise vital information extraction, enabling in-depth monitoring of MS-associated atypical pathologies and making MS trajectory predictions.

3.4. Microglia and Multiple Sclerosis

Like other neurodegenerative disorders, changes in microglial activation has also been observed both in models of MS and in pMS. Experimental autoimmune encephalomyelitis (EAE) is often used in scientific investigations to model MS and ON [48]. Although all experimental models of diseases are not fully representative, studies have shown similarities between human and animal investigations. For instance, compared to healthy, EAE-induced mice were found to have more "de-ramified" amoeboid microglia in the optic nerve and spinal cord [86]. Horstmann et al. found in MS modelled mice by injection of MOG35-55 (encephalitogenic peptide which induces EAE) that retinal microglia were significantly more abundant compared to that of healthy mice injected with PBS [87]. The same investigation also found that, at 60 days post-immunisation, the MS-induced mice had non-significantly but slightly more amoeboid microglia but significantly more ramified microglia compared to the control mice [87]. Another investigation also used MOG35-55 to induce EAE in mice [88]. Within the first week of immunisation, retinal microglia density had increased significantly [88]. In this investigation, amoeboid microglia were also found to be more abundant at 7 and 28 days post-MOG35-55-immunisation; however, no statistical analyses were carried out. These variations of microglial morphotypes indicate that retinal microglia may behave dynamically, becoming "activated" and "de-activated" throughout the progress of EAE. This could correlate to changes in microglial activation, which may occur during fluctuations of demyelination and remyelination seen in MS. Additionally, more sophisticated retinal layer-specific analyses were performed on EAE-induced and healthy mice to find layer-specific microglial morphotypes [86]. Whilst both EAE and healthy mice consisted of ramified microglia in the IPL and OPL, amoeboid microglia were only found in the GCL layer of EAE-induced mice [86]. This interestingly correlates to the GCL and IPL thinning hallmarks of MS, although it has not yet been investigated if this thinning pathology is directly linked to microglial activation. Ultimately, this suggests that retinal microglia may respond and activate in a temporally specific and layer-specific manner [32,86].

Similar implications can also be seen in human studies whereby there was a significantly reduced expression of P2RY12 (homeostatic microglia marker) and significantly increased expression of CD68 in both normal-appearing white matter and areas of active lesions of MS brains compared to healthy brains [89]. Since these markers have been associated with specific functions and microglia morphotypes, the results from Zrzavy et al. suggest that MS brains have fewer surveillant ramified microglia whilst having more phagocytic amoeboid [32,36,89].

Myelin phagocytosis is an additional pathological feature of MS [51]. When Hendrickx et al. investigated the phagocytic capabilities of myelin by microglia through flow cytometry, microglia derived from normal-appearing white matter of post-mortem brains seemed to engulf more myelin from MS donors compared to that from healthy donors [51]. Furthermore, myelin debris processing has been correlated to microglia taking on other "dystrophic" morphologies [51,90]. Cuprizone (CPZ) is a toxic and systemic copper chelator that can be administered intravenously and orally to model MS and demyelinating diseases [90,91]. Cantoni et al. investigated demyelination and microglia characteristics in healthy or CPZ administered [90]. Histological observations revealed that healthy animals were unaffected with their myelin integrity whilst most microglia appeared ramified. Conversely, CPZ administered animals exhibited hippocampal and corpus callosum (CC) neuron demyelination that became more "profoundly" evident with longer exposure to CPZ [90]. Interestingly, amoeboid microglia were abundant up to 4 weeks of CPZ administration, but as this increased to 6 or 12 weeks, more ramified microglia were observed [90]. This suggests that microglial activation plays a more active role in the earlier stages of demyelination or demyelinating diseases such as MS. This is further supported by clinical evidence showing that younger MS patients in earlier stages of the disease are more responsive to therapeutic interventions such as Rituximab that can reduce microglial activation [92,93]. Conversely, transcriptional investigations in microglia from normal-appearing white (NAWM) and grey matter (NAGM) of MS and healthy volunteers showed some interesting differences [94]. Firstly, expression of surveillant microglia markers such as P2RY12 was found to be similar in NAGM and NAWM [94]. Secondly, compared to the NAGM, the NAWM had higher levels of genes involved in the NF-κB pathway, which is involved in the immunomodulatory responses in MS pathology [94]. Unlike the findings from the CPZ MS models [90], van der Poel et al. [94] suggest that surveillant properties of microglia may remain unaffected in the initial stages of MS whilst the transcriptional profile of human-derived microglia are region-specific. Although evidence from human studies would provide a valuable translational comparison, there are, to date, no reports that investigate the morphological characteristics of retinal microglia in patients with MS.

3.5. Immunological Markers

Retinal microglia are also unsurprisingly involved in uveitis, an infrequent manifestation of MS. Experimental autoimmune uveitis (EAU) is a commonly used model of uveitis, which can be induced by immunisation of interphotoreceptor retinoid-binding protein (IRBP) [95]. Whole-mounted retinas of EAU-induced mice were found with P2RY12 positive microglia that underwent morphological activation from ramified to amoeboid morphology within 7 days post-IRBP-immunisation, indicating the early involvement of retinal microglia in EAU pathology [95]. Interestingly, ablation of microglia with the antagonist PLX5622 in IRBP-immunised was found to suppressed EAU. This was observed by a decrease in infiltrating inflammatory cells, lack of photoreceptor folds and retinal granulomas compared to untreated animals [95]. Additionally, three-dimensional images were constructed from images of retino-vascular leukocytes, including T-cells and microglia, labelled with CD11b, CD4/CD8, MHCII and P2RY12 [95]. The expression of all markers could be found co-localised with lectin, a marker of blood vessels. Together, these results emphasise the active role of microglia activation in initiating the transport of leukocytes between the blood-retinal-barrier [95]. Whether the morphological shift had a direct correlation to EAU pathogenesis was not explored.

3.6. Extracellular Vesicles: Effects on Multiple Sclerosis

The EAE model and MS patients have been investigated in parallel to observe microglia-derived microvesicle (M-MV) EVs in MS/EAE brain pathology [96]. Firstly, M-MVs were found to be neuroinflammatory as evidenced by upregulation of inflammatory markers such as CD86 (T-cell receptor ligand), iNOs and IL1ß, by the transfer of the vesicular components from the origin microglial cell to the recipient [96]. Additionally, EAE mice injected with M-MV into the corpus callosum had many amoeboid microglia, which were not observable in control EAE mice (injected with saline, liposomes or mesenchymal stem cells). Mice with and without acid sphingomyelinase knockout (ASKO) were induced with EAE [96]. ASKO-EAE mice showed significantly reduced M-MVs in the CSF, and CD45^{low}CD11b$^+$ microglia and inflammatory T-cells in the spinal cord, whilst no ASKO-EAE mice had developed EAE pathologies [96]. In humans, healthy participants were found to have M-MVs in their CSF with increased concentrations in MS: significantly greater in CIS compared to RRMS. These results suggest that M-MVs are differentially present in MS patients and EAE models, which may cause microglial activation and morphotype changes whilst these responses may be reduced by preventing M-MV shedding [96]. This also reveals the potential of targeting M-MVs to monitor disease and its therapeutic strategies. Their role in the retina is still to be established, although if they are found to influence retinal microglia morphology, this could have significant implications for using the eye as a monitoring and treatment efficacy tool.

3.7. Genetic Factors: Effects on Retinal Microglia in Multiple Sclerosis

The MOG35-55 induced EAE model of MS was used to identify the genetic and morphological differences of retinal microglia between 8-week-old C57BL/6 mice and healthy controls [97]. Iba-1 staining of cross-sections of the retina revealed mostly ramified microglia in healthy controls whilst there were rounder and amoeboid-like microglia with higher expression of iNOS in early EAE mice (16 days post-MOG35-55 induction—16DPI) [97]. This iNOS expression was reduced by 41DPI, although still higher than that in healthy controls. Additionally, there was increased expression of C1q and TNF-alpha in EAE [97]. This could suggest that early transcriptional changes in retinal microglia reflected by morphological changes may be pathological features of optic neuropathies.

Additionally, Bell et al. investigated the retinal microglia transcriptional profile of Cx3cr1^{CreER}:R26-tdTomato C57BL/6 mice that were induced with endotoxin-induced uveitis (EIU) by a single intravitreal injection of LPS [98]. Isolated retinal microglia were analysed with flow cytometry, Fluorescence-Activated Cell Sorting (FACS), mRNA sequencing and quantitative PCR to discover a total of 1069 differentially expressed genes (DEGs) [98]. There were 613 DEGs found at 4 h post-EIU, 537 DEGS at 18 h and none at 2 weeks [98]. Initially, activation markers such as *Fas* and *CD44* were upregulated whilst homeostatic markers such as *P2rY12* and *Mertk* were downregulated in EIU induced samples whilst healthy controls did not show a similar profile, e.g., high P2RY12 levels throughout [98]. Additionally, confocal microscopy of tdTomato$^+$ microglial cells in naïve was ramified whilst EIU cells underwent morphological changes from ramified at 4 hours to amoeboid at 18 h, which returned to ramified by 2 weeks post-injection [98]. Together these results indicate that uveitis may be characterised by some initial transcriptional changes that may be reflected by retinal microglia morphology changes [98]. Similar investigations in patient cells would be useful to determine the clinical translatability of these findings.

4. Other Neurodegenerative Diseases

4.1. Alzheimer's Disease

AD is the most common cause of dementia. It is characterised by the progressive loss of neurons as a result of the build-up of extracellular amyloid-beta (Aß) plaques and intracellular neurofibrillary tangles composed of hyperphosphorylated tau proteins [99–101]. By the time the first cognitive symptoms start to appear, roughly 20 years of irreversible neurodegeneration would already have occurred [100]. Although the exact pathogenesis is still unknown, microglial activation in the brain and retina has been associated with AD pathology [4,13,21,99–103]. This highlights the unmet need for earlier detection of AD pathogenesis and that characterisation of microglia may be a potential answer to this dilemma.

Brain AD investigations have shown more activated microglia in the cortex and hippocampus compared to age-matched controls [99,103]. Specifically, Cao et al. found a significantly greater degree of microglial activation in these areas of AD modelled mice at 12 months old compared to other ages and compared to age-matched or older healthy controls [99]. Additionally, 4-month-old 5× familial AD modelled mice showed that Aß plaques were surrounded by activated microglia in the cortex [103]. These activated microglia were also found to co-localise with Aß aggregates, and as Aß accumulated, microglial cell death was observed [103]. Similarly, Aß plaques were found surrounded by retinal microglia with a large cell body and fewer and thicker ramifications, suggesting activated microglia observations could be seen in the AD modelled retina [100–102]. Some sources have found activated retinal microglia in 3× transgenic AD mice as young as 5 postnatal weeks old, suggesting an initial microglial response in AD pathology [102]. This suggests that some microglial responses to AD may occur in the retina before the brain. Despite these findings, it is important to note the ambiguous definition of activated microglia. For instance, some literature refers to all non-resting microglia as "activated microglia", and some may use the term interchangeably with "activation" of microglia, whilst more recent literature may use the term in reference to the unique morphological description of the activated microglia.

Rod microglia may have a region-specific involvement in AD pathology [4,21]. For instance, when three brain regions of AD patient autopsies were examined, rod microglia were only seen in the parietal cortex and not in the hippocampus or temporal cortex [4]. Additionally, there were no significant observations seen relating rod microglia in hippocampal and cortical brain regions of patients who had traumatic brain injury [4]. However, rod microglia were found to be significantly abundant in the parietal cortex but not the hippocampus of AD patients compared to healthy controls [4]. Rod microglia have also been observed in those with mutated *C9orf72* whilst they were absent in age-matched controls whilst additionally, a high density of rod microglia in the grey matter of those with Down's Syndrome AD correlated to more severe pathological hallmarks compared to those just with AD [21].

Recently, Grimaldi et al. investigated the retinal pathology in post-mortem AD patient retinas [15]. These samples also had Aß plaques and phosphorylated tau tangles, with signs of retino-neuronal degeneration evidenced by increased expression of caspase 3, a marker of AD-associated neurodegeneration [15]. Additionally, the AD retinal slices examined had a significantly higher density of iba-1 positive microglial cells which also expressed higher levels of DAM marker, IL-1ß, compared to healthy samples [15]. TREM2 is also a DAM associated marker, although this was not upregulated in the iba-1 positive AD microglia [15]. Despite this, the AD retinas expressed higher levels of TREM2 mRNA compared to healthy controls [15]. This unexpected trend in AD retinal microglia may be due to historic evidence of post-mortem retinal samples with Aß deposits expressing downregulated TREM2 [15]. Investigations that observe other retinal DAM markers in parallel to morphological parameters may provide useful insight to elucidating AD retino-pathologies.

4.2. Parkinson's Disease

Parkinson's disease (PD) is the second most common type of neurodegenerative disease, affecting over 10 million people worldwide [104,105]. Its pathological hallmark consists of Lewy body (LB) aggregates composed of alpha-synuclein (α-syn) that accumulate inside the neuronal cell bodies and axons [104,105]. This is followed by the gradual neurodegeneration of dopaminergic neurons in CNS regions such as the substantia niagra (SN) pars compacta [6,106,107]. By the time the initial clinical symptoms such as tremor, rigidity, akinesia and postural stability start to appear, 50–80% of dopaminergic neurons have already been lost [107,108]. Much like AD, PD research has revealed many potential mechanisms of pathogenesis, including neuroinflammation as a result of microglial activation [6,105,109]. Recent studies have also revealed that up to 80% of PD patients may experience some visual dysfunction at least once throughout their disease progression [105]. Additionally, some studies have reported these retinal changes to occur prior to that of the brain [110]. Considering these factors, exploring the retinal microglia characteristics may serve as an ideal starting point in identifying earlier biomarkers of PD.

One study investigated the characteristics of SN and hippocampal microglia in three participant groups: PD patients, matched healthy participants and prodromal PD (p-PD) participants, with age-matched analyses performed (age range from 56 to 96 years old) [111]. Whilst there were almost no α-syn and LBs found in both brain regions of healthy controls, there were some in that of the p-PD patient specimens and significantly more in PD [111]. Additionally, healthy controls had significantly more ramified microglia compared to p-PD and PD groups, p-PD had significantly more amoeboid microglia compared to healthy participants, and PD patients also had significantly more than p-PD [111]. "Primed" or activated microglia were prominently seen in p-PD compared to any other participant group. This suggests that activated microglia may be an early indicator whilst amoeboid microglia are associated with PD pathology.

Similarly, healthy retinas were free of α-syn whilst there were some in 5-month-old retinas of transgenic mice (TgM83) modelled for PD and significantly more in the OPL of 8-month-old TgM83 retinal samples [112]. Only 8-month-old mouse retinas indicated microglial activation due to significant increases of CD68 expression (microglial activation marker) and visual observation of amoeboid microglia, whilst this was not seen in other experimental groups [112]. This suggests that there may be a differential expression of activated microglial markers and morphological changes through the early to intermediate stages of PD pathology. However, there are other studies that show that whilst there was OPL α-syn accumulation in mice modelled for α-syn-dependent diseases (Plp-α-syn), there were mostly ramified microglia found in whole-mounted retinal samples with no sign of microglial activation for all age groups investigated [113]. This may have been due to differences in disease models used, age groups investigated, and methods of analysis.

4.3. Glaucoma

Glaucoma is one of the main causes of blindness worldwide, which is estimated to affect more than 120 million people by 2040 [114]. It is characterised by the gradual loss of RGCs and, like AD and PD, significant neurodegeneration would have already occurred before clinical symptoms start to emerge [114]. Numerous methods for detecting the activity of glaucoma have been developed, including elevated intraocular pressure and an increased numbers of dying retinal cells, which is measured by detecting apoptosing retinal cells (DARC) [115,116]. Additionally, microglial activation may also play a role in glaucoma pathogenesis.

Microglia activation has been identified as one of the first events in glaucomatous neural damage, which may occur before RGC death [117,118]. For instance, when Swiss albino mice were modelled with the unilateral ocular hypertension (OHT) model of glaucoma, there was microglial activation [108]. This was seen in the form of increased cell body size and retraction of cell processes by as early as 1 day post-surgery (PS) and increased cell density by 3 days PS. Ramírez et al. additionally found that retinal microglia were P2RY12+ at 1 day PS [108]. This expression was decreased at 3 and 5 days PS, which then returned to normal levels by 15 days PS [108]. Moreover, the extent of neurodegeneration could also be correlated with early microglial changes, including microgliosis or microglial activation, as evidenced in transgenic $Cx3cr1^{GFP/+}$ DBA/2J mouse retinas modelled for pigmentary glaucoma [117]. In addition, in another experimental glaucoma model, treatment with minocycline or with a high dose of irradiation reduced microglial activation, resulting in lower RGC death [119,120].

Rod microglia have also been associated with several disease states. In a model of progressive RGC degeneration in rats, it has been shown that optic nerve transection (ONT) induced retinal rod microglia formation [23]. This was observable within the first 7 days, peaking at 14 days and disappearing after 2 months [23]. These periodic trends were interestingly reflected by the trends of ONT-induced RGC death. Additionally, these retinal rod microglia were found to orient themselves along the axon and cell body of RGCs, suggesting their active role in ONT-associated neurodegeneration [23].

4.4. Inherited Eye Disease: Retinitis Pigmentosa

Another major cause of inherited irreversible blindness is retinitis pigmentosa (RP). It affects roughly 1 in 4000 people in the world. Typically, patients may undergo loss of retinal rod cells, whilst more advanced cases may be caused by loss of cone cells. Although the exact mechanism of RP is still unclear, there is evidence of early retinal microglia activation occurring initially in the process of photoreceptor degeneration [121]. For instance, Di Pierdominico et al. observed P23H-1 rats, a model of inherited photoreceptor degeneration [121]. A reduction in rows of photoreceptor nuclei occurred between P10-21 with a greater reduction between P15-21 [121]. This occurred simultaneously with microglial morphology changes to amoeboid microglia, observed through retinal cross-section histology analyses showing that iba-1 positive microglia with shorter processes by P15 whilst the total retinal microglia density was significantly greater in P23H-1 rats compared to controls between P15-45 [121]. The same study investigated Royal College of Surgeons (RCS) rats, another model for inherited photoreceptor degeneration, to find a similar trend. The most significant photoreceptor degeneration occurred between P33-60 whilst retinal microglia took on the amoeboid appearance by P21-33 [121]. These microglia also migrated to the outer retinal layers between P21-60. RCS retinas had a significantly higher density of microglia compared to controls P21-60, whilst there was a significant decline between P45-60 [121].

5. Perivascular Microglia

Investigating microglia that reside in or around the perivascular region (PM) has become a popular field of research because of the associations that have been made between immune responses to neurological conditions and the CNS vascular network [122]. These associations are seen where patients with diabetic retinopathy had a significantly elevated number of hyper-ramified retinal PM microglia cells in comparison to that of normal subjects [34]. Additionally, PM microglia are said to closely track any incoming compounds from outside the blood-brain barrier through the vessels and into the CNS [95]. For instance, drug-induced depletion of PM microglia in mice with experimental AD resulted in corticovascular accumulation of amyloid plaques [122]. Other investigations also found that glaucoma modelled mice (ocular hypertension model induced by laser) presented with rod microglia both in the retinal PM and adjacent to the retinal axons [22]. This indicates that microglia morphotypes are not specific to perivascular regions.

6. Imaging Retinal Microglia

The establishment of a robust method to image retinal microglia both non-invasively and in vivo would provide an invaluable avenue to tracking disease-associated retinal microglia pathologies. There are indeed transgenic, e.g., CX3CR-1^{GFP} mice, which are specially bred to have enhanced expression of the green fluorescent protein (GFP) in the chemokine receptor 1 (CX3CR-1) gene [123–125]. As a result, scanning laser ophthalmoscopes (SLOs) and OCT systems may be used to non-invasively image the fluorescing retinal microglia at a cellular resolution, to investigate their shape, spatial distribution and density in real time [37,124–126]. Each of these studies could use distinctive methods with unique SLO/OCT systems including: multi-colour confocal SLO to detect multiple fluorescent markers [124], widefield autofluorescence imaging [125] and 488 nm fluorescence imaging using commercial SLO/OCT devices [126]. Although this reveals the imaging versatility of the Cx3CR-1^{GFP} mice, genetic modifications often have undesirable adverse effects and are not currently clinically translatable.

Despite this, there have also been reports of attempts of in vivo and non-invasive imaging methods in humans [127,128]. For example, Liu et al. developed a multimodal adaptive optics (AO) system which combined SLO and OCT systems [127]. Superluminescent diodes were also used to produce the imaging beams from each of the AOSLO and AOOCT with the addition of mirrors positioned precisely for it to be directed into the eye, enabling the retina to be imaged. Although they were able to image microglia in the inner limiting membrane (ILM), these were identified by subjective visual interpretation, and the authors emphasise the need for improvement due to the lengthy process (1 hour to image) [127]. Although AOOCTs can produce high-resolution and cellular levels of imaging, this is compromised by ~6-fold reduction in the field of view compared to a commercial machine [128]. For instance, Castanos et al. used spectral-domain en face OCT imaging in the reflectance mode (OCT-R) to obtain larger images of the retina [128]. On examination of the ILM in healthy participants, uniformly distributed ramified macrophage-like cells were observable [128]. In patients with diabetic retinopathy, central retinal vein occlusion and open-angle glaucoma, these cells appeared amoeboid and not uniform [128]. Through these morphological and distributional characteristics, the authors interpret these cells as microglia or hyalocytes. Although this reveals the possibility of non-invasive in vivo imaging of the human retina, the authors recognise limitations to their technology, including manual subjective interpretations of cell morphology and the indefinite cell type distinction [128].

7. Contradictory Findings: Reasons and Limitations

Although the aforementioned studies had similar objectives of finding densitometric or morphological features of microglia, there were clearly some which produced contradictory results. This may have been because many of the methods of analyses used were subjective, varied and even ambiguous. These aspects could have been amended to allow for more consistent, representative and less biased results. For example, when specific regions of the CNS, such as the striatum, SN, hippocampus or the retina, have been investigated, randomly chosen "sections" or "slices" are analysed [86,111,129]. Additionally, some investigations took subjective measures to distinguish "ramified" and "amoeboid" cells whilst also using manual counting methods [99]. There are indeed other methods that can certainly be used to reduce the impact of these subjective and non-representative techniques. For instance, more extensive and efficient imaging methods can be achieved by scanning the whole retina rather than sectors and including z-stacks of all the layers and maximum intensity projections [24]. In this way, observations of the whole retina and all its layers can be comprehensively analysed. Furthermore, algorithms that automate cell counting can be used to quantify densitometric parameters [24]. In light of encouraging future investigations, avoidance of subjective categorisation of microglia morphotypes has successfully already been performed using changes in 15 parameters defined through the course of microglial inflammation enabling morphometric analyses output by automated software (Fraclac; ImageJ) [130]. However, it is important to note that Fernández-Arjona et al., manually selected the microglia cells to be analysed based on a set of criterion which consisted of: fully visible cell body and processes which are nonoverlapping with parts of other adjacent cells [130]. Therefore, the resulting data were not fully representative or free from bias, although this clearly demonstrates the principle of using quantitative data to determine the different morphotypes.

Clinical and preclinical trials often have only male subjects [60,91]. This may be in order to reduce the effects of physiological differences such as the ovarian cycle, confounding the results of the trial or experiment. However, are known sexual dimorphisms for the structural and functional differences in microglial responses to ageing and disease, including MS [131]. This may be due to X chromosomes that express the greatest number of immune-related genes in the whole human genome and epigenetic and environmental influences [131,132]. Han et al. also suggest that hormonal differences may influence males having more activated microglia in development stages, putting them at higher risk of disease, whilst for females, this occurs during adulthood [131]. Other investigations on mediating hypersensitivities to pain have shown the involvement of microglia in male mice whilst other immune cells such as T-cells are required in female mice [133,134]. Although there are no reports which investigate the sexual differences in the morphological characteristics of retinal microglia in MS, transcriptome microarray analyses of male and female mice of two age groups (3 or 24 months) have shown that age-associated changes of retinal microglia cell-specific gene expression are sexually dimorphic [135]. Designing trial/experimental studies based on one sex may limit the ability to identify sex-associated pathological differences in MS, a disease more common in females [60]. Using stratified analyses in investigations that are inclusive of both sexes instead would benefit understanding pathological differences between different sub-sets of patients within the heterogeneous MS population. This would also enable larger patient group sizes to be investigated in larger multicentre research collaborations with consequently greater impact and generalisability.

There has also been a recent interest in the use of computer vision and machine learning algorithms to enable automated recognition and/or categorisation of "cells" or "regions of interests" in neuroimaging data. Commonly used in biomedical research are support vector machines (SVMs), regarded as supervised machine learning techniques that can categorise data and perform regression analyses [136]. Linear SVMs can use the input "training" data for each category to extract features that may be used to define each category [137]. Then, it uses the input "training" data to effectively categorise each data point by examining said defining features [137]. This is achieved by automatically identifying the optimal hyperplane that allows each data point to be distinguished in one category from those of another. These machine learning algorithms have already been implemented in research in efforts to predict dementia in patients using data from high-resolution MRIs and even mental state assessment scores [136]. Although these studies do not yet produce algorithms that have 100% accuracy and precision, the concept of using shareable automated algorithms presents as an excellent starting point to improving biomedical data processing and analysis [136].

8. Concluding remarks

In conclusion, microglia offer an exciting avenue for research in neurodegenerative disorders, including MS. There is also mounting evidence to show some early retinal microglial changes in response to diseases, further supporting the idea of using retinal biomarkers as an earlier method of disease detection. Additionally, there is a growing number of studies that investigate retinal and brain microglia with respect to its morphological, genetic and molecular characteristics. However, the key to our understanding of their role is the ability to perform accurate and consistent quantitative analyses of the dynamic changes in structure, as morphotypes may be closely linked to microglia function. We would encourage the development of standardised experimental designs and validated data analysis algorithms, as these are necessary to comprehensively delineate the role that these important cells have in disease pathogenesis and severity.

Author Contributions: Writing—planning, S.C., L.G. and M.F.C.; writing—original draft preparation, S.C.; writing—review and editing, S.C., L.G. and M.F.C.; visualisation—S.C.; supervision—L.G. and M.F.C. All authors have read and agreed to the published version of the manuscript.

Funding: This research received no external funding.

Conflicts of Interest: The authors declare no conflict of interest. The funders had no role in the writing of the manuscript.

References

1. Au, N.P.B.; Ma, C.H.E. Recent Advances in the study of bipolar/rod-shaped microglia and their roles in neurodegeneration. *Front. Aging Neurosci.* **2017**, *9*, 128. [CrossRef]
2. Li, F.; Jiang, D.; Samuel, M.A. Microglia in the developing retina. *Neural Dev.* **2019**, *14*, 1–13. [CrossRef]
3. Silverman, S.M.; Wong, W.T. Microglia in the retina: Roles in development, maturity, and disease. *Annu. Rev. Vis. Sci.* **2018**, *4*, 45–77. [CrossRef]
4. Bachstetter, A.D.; Ighodaro, E.T.; Hassoun, Y.; Aldeiri, D.; Neltner, J.H.; Patel, E.; Abner, E.L.; Nelson, P.T. Rod-shaped microglia morphology is associated with aging in 2 human autopsy series. *Neurobiol. Aging* **2017**, *52*, 98–105. [CrossRef]
5. Holloway, O.G.; Canty, A.J.; King, A.E.; Ziebell, J.M. Rod microglia and their role in neurological diseases. *Semin. Cell Dev. Biol.* **2019**, *94*, 96–103. [CrossRef]

6. Ramirez, A.I.; De Hoz, R.; Salobrar-Garcia, E.; Salazar, J.J.; Rojas, B.; Ajoy, D.; López-Cuenca, I.; Rojas, P.; Triviño, A.; Ramírez, J.M. The role of microglia in retinal neurodegeneration: Alzheimer's disease, parkinson, and glaucoma. *Front. Aging Neurosci.* **2017**, *9*, 214. [CrossRef] [PubMed]

7. Rashid, K.; Akhtar-Schaefer, I.; Langmann, T. Microglia in Retinal Degeneration. *Front. Immunol.* **2019**, *10*, 1975. [CrossRef] [PubMed]

8. Laprell, L.; Schulze, C.; Brehme, M.L.; Oertner, T.G. The role of microglia membrane potential in chemotaxis. *J. Neuroinflammation* **2021**, *18*, 1–10.

9. Karlstetter, M.; Scholz, R.; Rutar, M.; Wong, W.T.; Provis, J.M.; Langmann, T. Retinal microglia: Just bystander or target for therapy? *Prog. Retin. Eye Res.* **2015**, *45*, 30–57. [CrossRef] [PubMed]

10. Wang, J.; Wang, J.; Wang, J.; Yang, B.; Weng, Q.; He, Q. Targeting Microglia and Macrophages: A potential treatment strategy for multiple sclerosis. *Front. Pharmacol.* **2019**, *10*, 286. [CrossRef]

11. Walker, D.G.; Lue, L.-F. Immune phenotypes of microglia in human neurodegenerative disease: Challenges to detecting microglial polarization in human brains. *Alzheimer's Res. Ther.* **2015**, *7*, 1–9. [CrossRef]

12. Luo, C.; Jian, C.; Liao, Y.; Huang, Q.; Wu, Y.; Liu, X.; Zou, D.; Wu, Y. The role of microglia in multiple sclerosis. *Neuropsychiatr. Dis. Treat.* **2017**, *13*, 1661–1667. [CrossRef]

13. Yao, K.; Zu, H.-B. Microglial polarization: Novel therapeutic mechanism against Alzheimer's disease. *Inflammopharmacology* **2020**, *28*, 95–110. [CrossRef]

14. Ransohoff, R.M. A polarizing question: Do M1 and M2 microglia exist? *Nat. Neurosci.* **2016**, *19*, 987–991. [CrossRef]

15. Grimaldi, A.; Pediconi, N.; Oieni, F.; Pizzarelli, R.; Rosito, M.; Giubettini, M.; Santini, T.; Limatola, C.; Ruocco, G.; Ragozzino, D.; et al. Neuroinflammatory processes, A1 astrocyte activation and protein aggregation in the retina of Alzheimer's disease patients, possible biomarkers for early diagnosis. *Front. Neurosci.* **2019**, *13*, 925. [CrossRef]

16. Torres-Platas, S.G.; Comeau, S.; Rachalski, A.; Bo, G.D.; Cruceanu, C.; Turecki, G.; Giros, B.; Mechawar, N. Morphometric characterization of microglial phenotypes in human cerebral cortex. *J. Neuroinflammation* **2014**, *11*, 12. [CrossRef]

17. Nimmerjahn, A.; Helmchen, F.; Kirchhoff, F. Resting microglial cells are highly dynamic surveillants of brain parenchyma In Vivo. *Neuroforum* **2005**, *11*, 95. [CrossRef]

18. Ekarperien, A.; Eahammer, H.; Jelinek, H.F. Quantitating the subtleties of microglial morphology with fractal analysis. *Front. Cell. Neurosci.* **2013**, *7*, 3. [CrossRef]

19. Sasaki, A. Microglia and brain macrophages: An update. *Neuropathology* **2016**, *37*, 452–464. [CrossRef]

20. Taylor, S.E.; Morganti-Kossmann, C.; Lifshitz, J.; Ziebell, J.M. Rod microglia: A morphological definition. *PLoS ONE* **2014**, *9*, e97096. [CrossRef]

21. Giordano, K.R.; Denman, C.R.; Dubisch, P.S.; Akhter, M.; Lifshitz, J. An update on the rod microglia variant in experimental and clinical brain injury and disease. *Brain Commun.* **2021**, *3*, fcaa227. [CrossRef]

22. De Hoz, R.; Gallego, B.I.; Ramírez, A.I.; Rojas, B.; Salazar, J.J.; Soriano, F.J.V.; Avilés-Trigueros, M.; Pérez, M.P.V.; Vidal-Sanz, M.; Triviño, A.; et al. Rod-like microglia are restricted to eyes with laser-induced ocular hypertension but absent from the microglial changes in the contralateral untreated eye. *PLoS ONE* **2013**, *8*, e83733. [CrossRef]

23. Yuan, T.-F.; Liang, Y.-X.; Peng, B.; Lin, B.; So, K.-F. Local proliferation is the main source of rod microglia after optic nerve transection. *Sci. Rep.* **2015**, *5*, 10788. [CrossRef]

24. Davis, B.M.; Salinas-Navarro, M.; Cordeiro, M.F.; Moons, L.; De Groef, L. Characterizing microglia activation: A spatial statistics approach to maximize information extraction. *Sci. Rep.* **2017**, *7*, 1–12. [CrossRef]

25. Walker, F.R.; Beynon, S.B.; Jones, K.A.; Zhao, Z.; Kongsui, R.; Cairns, M.; Nilsson, M. Dynamic structural remodelling of microglia in health and disease: A review of the models, the signals and the mechanisms. *Brain Behav. Immun.* **2014**, *37*, 1–14. [CrossRef]

26. Snyder, P.J.; Alber, J.; Alt, C.; Bain, L.J.; Bouma, B.E.; Bouwman, F.H.; DeBuc, D.C.; Campbell, M.C.; Carrillo, M.C.; Chew, E.Y.; et al. Retinal imaging in Alzheimer's and neurodegenerative diseases. *Alzheimer's Dement.* **2021**, *17*, 103–111. [CrossRef]

27. Kashani, A.H.; Asanad, S.; Chan, J.W.; Singer, M.B.; Zhang, J.; Sharifi, M.; Khansari, M.M.; Abdolahi, F.; Shi, Y.; Biffi, A.; et al. Past, present and future role of retinal imaging in neurodegenerative disease. *Prog. Retin. Eye Res.* **2021**, *100938*, 100938. [CrossRef]

28. Alves, C.H.; Fernandes, R.; Santiago, A.R.; Ambrósio, A.F. Microglia contribution to the regulation of the retinal and choroidal vasculature in age-related macular degeneration. *Cells* **2020**, *9*, 1217. [CrossRef]

29. Madeira, M.H.; Boia, R.; Santos, P.F.; Ambrósio, A.F.; Santiago, A.R. Contribution of microglia-mediated neuroinflammation to retinal degenerative diseases. *Mediat. Inflamm.* **2015**, *2015*, 1–15. [CrossRef]

30. Di Pierdomenico, J.; Martínez-Vacas, A.; Hernández-Muñoz, D.; Gómez-Ramírez, A.M.; Valiente-Soriano, F.J.; Agudo-Barriuso, M.; Vidal-Sanz, M.; Villegas-Pérez, M.P.; García-Ayuso, D. Coordinated intervention of microglial and Müller cells in light-induced retinal degeneration. *Investig. Opthalmol. Vis. Sci.* **2020**, *61*, 47. [CrossRef]

31. Sobrado-Calvo, P.; Vidal-Sanz, M.; Villegas-Pérez, M.P. Rat retinal microglial cells under normal conditions, after optic nerve section, and after optic nerve section and intravitreal injection of trophic factors or macrophage inhibitory factor. *J. Comp. Neurol.* **2007**, *501*, 866–878. [CrossRef]

32. Tan, Y.-L.; Yuan, Y.; Tian, L. Microglial regional heterogeneity and its role in the brain. *Mol. Psychiatry* **2020**, *25*, 351–367. [CrossRef] [PubMed]

33. Zhao, L.; Zabel, M.; Wang, X.; Ma, W.; Shah, P.; Fariss, R.; Qian, H.; Parkhurst, C.N.; Gan, W.; Wong, W.T. Microglial phagocytosis of living photoreceptors contributes to inherited retinal degeneration. *EMBO Mol. Med.* **2015**, *7*, 1179–1197. [CrossRef] [PubMed]

34. Zeng, H.-Y.; Green, W.R.; Tso, M.O.M. Microglial activation in human diabetic retinopathy. *Arch. Ophthalmol.* **2008**, *126*, 227–232. [CrossRef]
35. Waller, R.; Baxter, L.; Fillingham, D.J.; Coelho, S.; Pozo, J.M.; Mozumder, M.; Frangi, A.; Ince, P.G.; Simpson, J.E.; Highley, J.R. Iba-1-/CD68+ microglia are a prominent feature of age-associated deep subcortical white matter lesions. *PLoS ONE* **2019**, *14*, e0210888. [CrossRef]
36. Dubbelaar, M.L.; Kracht, L.; Eggen, B.J.L.; Boddeke, E.W.G.M. The kaleidoscope of microglial phenotypes. *Front. Immunol.* **2018**, *9*, 1753. [CrossRef] [PubMed]
37. Ma, W.; Silverman, S.M.; Zhao, L.; Villasmil, R.; Campos, M.M.; Amaral, J.; Wong, W.T. Absence of TGFb signaling in retinal microglia induces retinal degeneration and exacerbates choroidal neovascularization. *eLife* **2019**, *8*, e42049. [CrossRef]
38. Mead, B.; Tomarev, S. Extracellular vesicle therapy for retinal diseases. *Prog. Retin. Eye Res.* **2020**, *79*, 100849. [CrossRef]
39. Aires, D.I.; Ribeiro-Rodrigues, T.; Boia, R.; Ferreira-Rodrigues, M.; Girão, H.; Ambrósio, A.; Santiago, A. Microglial extracellular vesicles as vehicles for neurodegeneration spreading. *Biomolecules* **2021**, *11*, 770. [CrossRef]
40. Rathnasamy, G.; Foulds, W.S.; Ling, E.-A.; Kaur, C. Retinal microglia—A key player in healthy and diseased retina. *Prog. Neurobiol.* **2019**, *173*, 18–40. [CrossRef]
41. Zöller, T.; Attaai, A.; Potru, P.S.; Ruß, T.; Spittau, B. Aged mouse cortical microglia display an activation profile suggesting immunotolerogenic functions. *Int. J. Mol. Sci.* **2018**, *19*, 706. [CrossRef] [PubMed]
42. Damani, M.R.; Zhao, L.; Fontainhas, A.M.; Amaral, J.; Fariss, R.N.; Wong, W.T. Age-related alterations in the dynamic behavior of microglia. *Aging Cell* **2010**, *10*, 263–276. [CrossRef]
43. Fernández-Albarral, J.A.; Salobrar-Garcia, E.; López-Cuenca, I.; Lozano, M.P.R.; Corral, J.J.S.; De Hoz, R.; Aviles-Trigueros, M.; Vidal-Sanz, M.; Ramirez, J.M.; Ramírez, A.I. Microglial changes in healthy mice retina in an early aging stage. *Acta Ophthalmol.* **2021**, *99*. [CrossRef]
44. Hefendehl, J.K.; Neher, J.J.; Sühs, R.B.; Kohsaka, S.; Skodras, A.; Jucker, M. Homeostatic and injury-induced microglia behavior in the aging brain. *Aging Cell* **2014**, *13*, 60–69. [CrossRef]
45. Mouton, P.R.; Long, J.M.; Lei, D.-L.; Howard, V.; Jucker, M.; Calhoun, M.E.; Ingram, D.K. Age and gender effects on microglia and astrocyte numbers in brains of mice. *Brain Res.* **2002**, *956*, 30–35. [CrossRef]
46. Streit, W.J.; Xue, Q.-S. Life and Death of Microglia. *J. Neuroimmune Pharmacol.* **2009**, *4*, 371–379. [CrossRef] [PubMed]
47. Williamson, J.M.; Lyons, D.A. Myelin dynamics throughout life: An ever-changing landscape? *Front. Cell. Neurosci.* **2018**, *12*, 424. [CrossRef]
48. Lloyd, A.F.; Miron, V.E. The pro-remyelination properties of microglia in the central nervous system. *Nat. Rev. Neurol.* **2019**, *15*, 447–458. [CrossRef]
49. Safaiyan, S.; Kannaiyan, N.; Snaidero, N.; Brioschi, S.; Biber, K.; Yona, S.; Edinger, A.L.; Jung, S.; Rossner, M.J.; Simons, M. Age-related myelin degradation burdens the clearance function of microglia during aging. *Nat. Neurosci.* **2016**, *19*, 995–998. [CrossRef]
50. Poliani, P.L.; Wang, Y.; Fontana, E.; Robinette, M.L.; Yamanishi, Y.; Gilfillan, S.; Colonna, M. TREM2 sustains microglial expansion during aging and response to demyelination. *J. Clin. Investig.* **2015**, *125*, 2161–2170. [CrossRef] [PubMed]
51. Hendrickx, D.A.E.; Schuurman, K.G.; Van Draanen, M.; Hamann, J.; Huitinga, I. Enhanced uptake of multiple sclerosis-derived myelin by THP-1 macrophages and primary human microglia. *J. Neuroinflammation* **2014**, *11*, 64. [CrossRef] [PubMed]
52. Streit, W.J.; Sammons, N.W.; Kuhns, A.J.; Sparks, D.L. Dystrophic microglia in the aging human brain. *Glia* **2004**, *45*, 208–212. [CrossRef]
53. Ma, W.; Wong, W.T. Aging changes in retinal microglia and their relevance to age-related retinal disease. In *Retinal Degenerative Diseases. Mechanisms and Experimental Therapy, Proceedings of the XVI International Symposium on Retinal Degeneration (RD2014), Pacific Grove, California, USA, 13–18 July 2014*; Bowes Rickman, C., LaVail, M., Anderson, R., Grimm, C., Hollyfield, J., Ash, J., Eds.; Springer: Cham, Switzerland, 2016; Volume 854. [CrossRef]
54. Udeochu, J.C.; Shea, J.M.; Villeda, S.A. Microglia communication: Parallels between aging and Alzheimer's disease. *Clin. Exp. Neuroimmunol.* **2016**, *7*, 114–125. [CrossRef]
55. Angelova, D.M.; Brown, D.R. Microglia and the aging brain: Are senescent microglia the key to neurodegeneration? *J. Neurochem.* **2019**, *151*, 676–688. [CrossRef]
56. Spittau, B. Aging Microglia—Phenotypes, functions and implications for age-related neurodegenerative diseases. *Front. Aging Neurosci.* **2017**, *9*, 194. [CrossRef]
57. Chen, M.; Muckersie, E.; Forrester, J.V.; Xu, H. Immune activation in retinal aging: A gene expression study. *Investig. Opthalmol. Vis. Sci.* **2010**, *51*, 5888–5896. [CrossRef]
58. Ma, W.; Cojocaru, R.; Gotoh, N.; Gieser, L.; Villasmil, R.; Cogliati, T.; Swaroop, A.; Wong, W.T. Gene expression changes in aging retinal microglia: Relationship to microglial support functions and regulation of activation. *Neurobiol. Aging* **2013**, *34*, 2310–2321. [CrossRef]
59. Nazari, F.; Shaygannejad, V.; Sichani, M.M.; Mansourian, M.; Hajhashemi, V. The prevalence of lower urinary tract symptoms based on individual and clinical parameters in patients with multiple sclerosis. *BMC Neurol.* **2020**, *20*, 24. [CrossRef]
60. Confavreux, C.; Vukusic, S. Natural history of multiple sclerosis: A unifying concept. *Brain* **2006**, *129*, 606–616. [CrossRef] [PubMed]
61. Compston, A. The 150th anniversary of the first depiction of the lesions of multiple sclerosis. *J. Neurol. Neurosurg. Psychiatry* **1988**, *51*, 1249–1252. [CrossRef]
62. Kale, N. Optic neuritis as an early sign of multiple sclerosis. *Eye Brain* **2016**, *8*, 195–202. [CrossRef] [PubMed]

63. Aboulenein-Djamshidian, F.; Serbecic, N. OCT is an alternative to MRI for monitoring MS–NO. *Mult. Scler. J.* **2018**, *24*, 703–705. [CrossRef] [PubMed]

64. Henderson, A.P.D.; Trip, S.A.; Schlottmann, P.G.; Altmann, D.R.; Garway-Heath, D.F.; Plant, G.T.; Miller, D.H. An investigation of the retinal nerve fibre layer in progressive multiple sclerosis using optical coherence tomography. *Brain* **2007**, *131*, 277–287. [CrossRef] [PubMed]

65. Britze, J.; Frederiksen, J.L. Optical coherence tomography in multiple sclerosis. *Eye* **2018**, *32*, 884–888. [CrossRef] [PubMed]

66. Kaya, D.; Kaya, M.; Özakbaş, S.; Idiman, E. Uveitis associated with multiple sclerosis: Complications and visual prognosis. *Int. J. Ophthalmol.* **2014**, *7*, 1010–1013. [CrossRef]

67. Miller, D.H.; Chard, D.T.; Ciccarelli, O. Clinically isolated syndromes. *Lancet Neurol.* **2012**, *11*, 157–169. [CrossRef]

68. Faissner, S.; Plemel, J.R.; Gold, R.; Yong, V.W. Progressive multiple sclerosis: From pathophysiology to therapeutic strategies. *Nat. Rev. Drug Discov.* **2019**, *18*, 905–922. [CrossRef]

69. Nazari, F.; Shaygannejad, V.; Sichani, M.M.; Mansourian, M.; Hajhashemi, V. Quality of life among patients with multiple sclerosis and voiding dysfunction: A cross-sectional study. *BMC Urol.* **2020**, *20*, 1–10. [CrossRef]

70. Dobson, R.; Giovannoni, G. Multiple sclerosis—A review. *Eur. J. Neurol.* **2019**, *26*, 27–40. [CrossRef]

71. Fletcher, J.M.; Lalor, S.; Sweeney, C.M.; Tubridy, N.; Mills, K.H.G. T cells in multiple sclerosis and experimental autoimmune encephalomyelitis. *Clin. Exp. Immunol.* **2010**, *162*, 1–11. [CrossRef]

72. Cavallo, S. Immune-mediated genesis of multiple sclerosis. *J. Transl. Autoimmun.* **2020**, *3*, 100039. [CrossRef]

73. Hemmer, B.; Archelos, J.J.; Hartung, H.-P. New concepts in the immunopathogenesis of multiple sclerosis. *Nat. Rev. Neurosci.* **2002**, *3*, 291–301. [CrossRef]

74. Lublin, F.D.; Reingold, S.C.; Cohen, J.A.; Cutter, G.R.; Sørensen, P.S.; Thompson, A.J.; Wolinsky, J.S.; Balcer, L.J.; Banwell, B.; Barkhof, F.; et al. Defining the clinical course of multiple sclerosis: The 2013 revisions. *Neurology* **2014**, *83*, 278–286. [CrossRef] [PubMed]

75. Bashinskaya, V.V.; Kulakova, O.; Boyko, A.N.; Favorov, A.V.; Favorova, O.O. A review of genome-wide association studies for multiple sclerosis: Classical and hypothesis-driven approaches. *Hum. Genet.* **2015**, *134*, 1143–1162. [CrossRef] [PubMed]

76. Jia, X.; Madireddy, L.; Caillier, S.; Santaniello, A.; Esposito, F.; Comi, G.; Stuve, O.; Zhou, Y.; Taylor, B.; Kilpatrick, T.; et al. Genome sequencing uncovers phenocopies in primary progressive multiple sclerosis. *Ann. Neurol.* **2018**, *84*, 51–63. [CrossRef]

77. Thompson, A.J.; Baranzini, S.E.; Geurts, J.; Hemmer, B.; Ciccarelli, O. Multiple sclerosis. *Lancet* **2018**, *391*, 1622–1636. [CrossRef]

78. Siva, A. Asymptomatic MS. *Clin. Neurol. Neurosurg.* **2013**, *115*, S1–S5. [CrossRef] [PubMed]

79. Saidha, S.; Al-Louzi, O.; Ratchford, J.N.; Bhargava, P.; Oh, J.; Newsome, S.D.; Prince, J.L.; Pham, D.; Roy, S.; Van Zijl, P.; et al. Optical coherence tomography reflects brain atrophy in multiple sclerosis: A four-year study. *Ann. Neurol.* **2015**, *78*, 801–813. [CrossRef] [PubMed]

80. Sotirchos, E.S.; Saidha, S. OCT is an alternative to MRI for monitoring MS–YES. *Mult. Scler. J.* **2018**, *24*, 701–703. [CrossRef]

81. Pawlitzki, M.; Horbrügger, M.; Loewe, K.; Kaufmann, J.; Opfer, R.; Wagner, M.; Al-Nosairy, K.O.; Meuth, S.G.; Hoffmann, M.B.; Schippling, S. MS optic neuritis-induced long-term structural changes within the visual pathway. *Neurol. Neuroimmunol. Neuroinflammation* **2020**, *7*, e665. [CrossRef]

82. Beck, R.W. The Clinical profile of optic neuritis. *Arch. Ophthalmol.* **1991**, *109*, 1673–1678. [CrossRef]

83. Toosy, A.T.; Mason, D.F.; Miller, D.H. Optic neuritis. *Lancet Neurol.* **2014**, *13*, 83–99. [CrossRef]

84. Garrett, B.; Dmytriw, A.A.; Maxner, C. Acute optic neuritis in multiple sclerosis. *Can. Med Assoc. J.* **2016**, *188*, E199. [CrossRef] [PubMed]

85. Backner, Y.; Petrou, P.; Glick-Shames, H.; Raz, N.; Zimmermann, H.; Jost, R.; Scheel, M.; Paul, F.; Karussis, D.; Levin, N. Vision and vision-related measures in progressive multiple sclerosis. *Front. Neurol.* **2019**, *10*. [CrossRef] [PubMed]

86. Namekata, K.; Guo, X.; Kimura, A.; Arai, N.; Harada, C.; Harada, T. DOCK8 is expressed in microglia, and it regulates microglial activity during neurodegeneration in murine disease models. *J. Biol. Chem.* **2019**, *294*, 13421–13433. [CrossRef]

87. Horstmann, L.; Kuehn, S.; Pedreiturria, X.; Haak, K.; Pfarrer, C.; Dick, H.B.; Kleiter, I.; Joachim, S.C. Microglia response in retina and optic nerve in chronic experimental autoimmune encephalomyelitis. *J. Neuroimmunol.* **2016**, *298*, 32–41. [CrossRef] [PubMed]

88. Manogaran, P.; Samardzija, M.; Schad, A.N.; Wicki, C.A.; Walker-Egger, C.; Rudin, M.; Grimm, C.; Schippling, S. Retinal pathology in experimental optic neuritis is characterized by retrograde degeneration and gliosis. *Acta Neuropathol. Commun.* **2019**, *7*, 1–22. [CrossRef]

89. Zrzavy, T.; Hametner, S.; Wimmer, I.; Butovsky, O.; Weiner, H.L.; Lassmann, H. Loss of 'homeostatic' microglia and patterns of their activation in active multiple sclerosis. *Brain* **2017**, *140*, 1900–1913. [CrossRef]

90. Cantoni, C.; Bollman, B.; Licastro, D.; Xie, M.; Mikesell, R.; Schmidt, R.; Yuede, C.M.; Galimberti, D.; Olivecrona, G.; Klein, R.S.; et al. TREM2 regulates microglial cell activation in response to demyelination in vivo. *Acta Neuropathol.* **2015**, *129*, 429–447. [CrossRef]

91. Vega-Riquer, J.M.; Mendez-Victoriano, G.; Morales-Luckie, R.A.; Gonzalez-Perez, O. Five Decades of Cuprizone, an Updated Model to Replicate Demyelinating Diseases. *Curr. Neuropharmacol.* **2019**, *17*, 129–141. [CrossRef]

92. Hawker, K.; O'Connor, P.; Freedman, M.S.; Calabresi, P.A.; Antel, J.; Simon, J.; Hauser, S.; Waubant, E.; Vollmer, T.; Panitch, H.; et al. Rituximab in patients with primary progressive multiple sclerosis: Results of a randomized double-blind placebo-controlled multicenter trial. *Ann. Neurol.* **2009**, *66*, 460–471. [CrossRef] [PubMed]

93. Baecher-Allan, C.; Kaskow, B.J.; Weiner, H.L. Multiple sclerosis: Mechanisms and immunotherapy. *Neuron* **2018**, *97*, 742–768. [CrossRef]

94. van Der Poel, M.; Ulas, T.; Mizee, M.R.; Hsiao, C.-C.; Miedema, S.; Adelia; Schuurman, K.G.; Helder, B.; Tas, S.W.; Schultze, J.L.; et al. Transcriptional profiling of human microglia reveals grey-white matter heterogeneity and multiple sclerosis-associated changes. *Nat. Commun.* **2019**, *10*, 1–13. [CrossRef] [PubMed]

95. Okunuki, Y.; Mukai, R.; Nakao, T.; Tabor, S.J.; Butovsky, O.; Dana, R.; Ksander, B.R.; Connor, K.M. Retinal microglia initiate neuroinflammation in ocular autoimmunity. *Proc. Natl. Acad. Sci. USA* **2019**, *116*, 9989–9998. [CrossRef]

96. Verderio, C.; Muzio, L.; Turola, E.; Bergami, A.; Novellino, L.; Ruffini, F.; Riganti, L.; Corradini, I.; Francolini, M.; Garzetti, L.; et al. Myeloid microvesicles are a marker and therapeutic target for neuroinflammation. *Ann. Neurol.* **2012**, *72*, 610–624. [CrossRef]

97. Jin, J.; Smith, M.D.; Kersbergen, C.J.; Kam, T.-I.; Viswanathan, M.; Martin, K.; Dawson, T.M.; Dawson, V.L.; Zack, D.J.; Whartenby, K.; et al. Glial pathology and retinal neurotoxicity in the anterior visual pathway in experimental autoimmune encephalomyelitis. *Acta Neuropathol. Commun.* **2019**, *7*, 1–17. [CrossRef]

98. Bell, O.H.; Copland, D.A.; Ward, A.; Nicholson, L.B.; Lange, C.A.; Chu, C.J.; Dick, A.D. Single eye mRNA-seq reveals normalisation of the retinal microglial transcriptome following acute inflammation. *Front. Immunol.* **2020**, *10*, 3033. [CrossRef]

99. Cao, S.; Fisher, D.W.; Rodriguez, G.; Yu, T.; Dong, H. Comparisons of neuroinflammation, microglial activation, and degeneration of the locus coeruleus-norepinephrine system in APP/PS1 and aging mice. *J. Neuroinflammation* **2021**, *18*, 1–16. [CrossRef]

100. Vandenabeele, M.; Veys, L.; Lemmens, S.; Hadoux, X.; Gelders, G.; Masin, L.; Serneels, L.; Theunis, J.; Saito, T.; Saido, T.C.; et al. The AppNL-G-F mouse retina is a site for preclinical Alzheimer's disease diagnosis and research. *Acta Neuropathol. Commun.* **2021**, *9*, 1–16. [CrossRef]

101. Salobrar-García, E.; Rodrigues-Neves, A.C.; Ramírez, A.I.; de Hoz, R.; Fernández-Albarral, J.A.; López-Cuenca, I.; Ramírez, J.M.; Ambrósio, A.F.; Salazar, J.J. Microglial activation in the retina of a triple-transgenic alzheimer's disease mouse model (3xTg-AD). *Int. J. Mol. Sci.* **2020**, *21*, 816. [CrossRef]

102. Grimaldi, A.; Brighi, C.; Peruzzi, G.; Ragozzino, D.; Bonanni, V.; Limatola, C.; Ruocco, G.; Di Angelantonio, S. Inflammation, neurodegeneration and protein aggregation in the retina as ocular biomarkers for Alzheimer's disease in the 3xTg-AD mouse model. *Cell Death Dis.* **2018**, *9*, 1–10. [CrossRef]

103. Baik, S.H.; Kang, S.; Son, S.M.; Mook-Jung, I. Microglia contributes to plaque growth by cell death due to uptake of amyloid β in the brain of Alzheimer's disease mouse model. *Glia* **2016**, *64*, 2274–2290. [CrossRef]

104. Ball, N.; Teo, W.P.; Chandra, S.; Chapman, J. Parkinson's Disease and the Environment. *Front. Neurol.* **2019**, *10*, 218. [CrossRef] [PubMed]

105. Veys, L.; Vandenabeele, M.; Ortuño-Lizarán, I.; Baekelandt, V.; Cuenca, N.; Moons, L.; De Groef, L. Retinal α-synuclein deposits in Parkinson's disease patients and animal models. *Acta Neuropathol.* **2019**, *137*, 379–395. [CrossRef] [PubMed]

106. Cheng, H.-C.; Ulane, C.M.; Burke, R.E. Clinical progression in Parkinson disease and the neurobiology of axons. *Ann. Neurol.* **2010**, *67*, 715–725. [CrossRef]

107. Emamzadeh, F.N.; Surguchov, A. Parkinson's disease: Biomarkers, treatment, and risk factors. *Front. Neurosci.* **2018**, *12*, 612. [CrossRef] [PubMed]

108. Guo, L.; Normando, E.M.; Shah, P.A.; De Groef, L.; Cordeiro, M.F. Oculo-visual abnormalities in Parkinson's disease: Possible value as biomarkers. *Mov. Disord.* **2018**, *33*, 1390–1406. [CrossRef] [PubMed]

109. Maiti, P.; Manna, J.; Dunbar, G.L. Current understanding of the molecular mechanisms in Parkinson's disease: Targets for potential treatments. *Transl. Neurodegener.* **2017**, *6*, 1–35. [CrossRef]

110. Normando, E.M.; Davis, B.M.; De Groef, L.; Nizari, S.; Turner, L.A.; Ravindran, N.; Pahlitzsch, M.; Brenton, J.; Malaguarnera, G.; Guo, L.; et al. The retina as an early biomarker of neurodegeneration in a rotenone-induced model of Parkinson's disease: Evidence for a neuroprotective effect of rosiglitazone in the eye and brain. *Acta Neuropathol. Commun.* **2016**, *4*, 1–15. [CrossRef]

111. Doorn, K.J.; Moors, T.; Drukarch, B.; Van De Berg, W.D.; Lucassen, P.J.; Van Dam, A.-M. Microglial phenotypes and toll-like receptor 2 in the substantia nigra and hippocampus of incidental Lewy body disease cases and Parkinson's disease patients. *Acta Neuropathol. Commun.* **2014**, *2*, 1–17. [CrossRef]

112. Mammadova, N.; Summers, C.M.; Kokemuller, R.D.; He, Q.; Ding, S.; Baron, T.; Yu, C.; Valentine, R.J.; Sakaguchi, D.S.; Kanthasamy, A.G.; et al. Accelerated accumulation of retinal α-synuclein (pSer129) and tau, neuroinflammation, and autophagic dysregulation in a seeded mouse model of Parkinson's disease. *Neurobiol. Dis.* **2019**, *121*, 1–16. [CrossRef] [PubMed]

113. Kaehler, K.; Seitter, H.; Sandbichler, A.M.; Tschugg, B.; Obermair, G.J.; Stefanova, N.; Koschak, A. Assessment of the retina of Plp-α-Syn mice as a model for studying synuclein-dependent diseases. *Investig. Opthalmol. Vis. Sci.* **2020**, *61*, 12. [CrossRef] [PubMed]

114. García-Bermúdez, M.Y.; Freude, K.K.; Mouhammad, Z.A.; van Wijngaarden, P.; Martin, K.K.; Kolko, M. Glial cells in glaucoma: Friends, foes, and potential therapeutic targets. *Front. Neurol.* **2021**, *12*, 624983. [CrossRef]

115. Normando, E.M.; Yap, T.E.; Maddison, J.; Miodragovic, S.; Bonetti, P.; Almonte, M.; Mohammad, N.G.; Ameen, S.; Crawley, L.; Ahmed, F.; et al. A CNN-aided method to predict glaucoma progression using DARC (Detection of Apoptosing Retinal Cells). *Expert Rev. Mol. Diagn.* **2020**, *20*, 737–748. [CrossRef] [PubMed]

116. Cordeiro, M.F. DARC: A new method for detecting progressive neuronal death. *Eye* **2007**, *21*, S15–S17. [CrossRef]

117. Bosco, A.; Romero, C.O.; Breen, K.T.; Chagovetz, A.A.; Steele, M.R.; Ambati, B.K.; Vetter, M.L. Neurodegeneration severity can be predicted from early microglia alterations monitored In Vivo in a mouse model of chronic glaucoma. *Dis. Model. Mech.* **2015**, *8*, 443–455. [CrossRef]

118. Ramírez, A.I.; de Hoz, R.; Fernández-Albarral, J.A.; Salobrar-Garcia, E.; Rojas, B.; Valiente-Soriano, F.J.; Avilés-Trigueros, M.; Villegas-Pérez, M.P.; Vidal-Sanz, M.; Triviño, A.; et al. Time course of bilateral microglial activation in a mouse model of laser-induced glaucoma. *Sci. Rep.* **2020**, *10*, 1–17. [CrossRef]

119. Bosco, A.; Inman, D.M.; Steele, M.R.; Wu, G.; Soto, I.; Marsh-Armstrong, N.; Hubbard, W.C.; Calkins, D.J.; Horner, P.J.; Vetter, M.L. Reduced retina microglial activation and improved optic nerve integrity with minocycline treatment in the DBA/2J mouse model of glaucoma. *Investig. Ophthalmol. Vis. Sci.* **2008**, *49*, 1437–1446. [CrossRef] [PubMed]

120. Bosco, A.; Crish, S.D.; Steele, M.R.; Romero, C.O.; Inman, D.M.; Horner, P.J.; Calkins, D.J.; Vetter, M.L. Early reduction of microglia activation by irradiation in a model of chronic glaucoma. *PLoS ONE* **2012**, *7*, e43602. [CrossRef]

121. García-Ayuso, D.; Villegas-Pérez, M.P.; Di Pierdomenico, J.; Agudo-Barriuso, M.; Vidal-Sanz, M. Role of microglial cells in photoreceptor degeneration. *Neural Regen. Res.* **2019**, *14*, 1186–1190. [CrossRef]

122. Koizumi, T.; Kerkhofs, D.; Mizuno, T.; Steinbusch, H.W.M.; Foulquier, S. Vessel-associated immune cells in cerebrovascular diseases: From perivascular macrophages to vessel-associated microglia. *Front. Neurosci.* **2019**, *13*, 1291. [CrossRef] [PubMed]

123. The Jackson Laboratory. CX3CR-1GFP. Available online: https://www.jax.org/strain/005582 (accessed on 26 May 2021).

124. Alt, C.; Runnels, J.M.; Mortensen, L.; Zaher, W.; Lin, C.P. In Vivo imaging of microglia turnover in the mouse retina after ionizing radiation and dexamethasone treatment. *Investig. Opthalmol. Vis. Sci.* **2014**, *55*, 5314–5319. [CrossRef]

125. Kokona, D.; Schneider, N.; Giannakaki-Zimmermann, H.; Jovanovic, J.; Ebneter, A.; Zinkernagel, M. Noninvasive Quantification of retinal microglia using widefield autofluorescence imaging. *Investig. Opthalmol. Vis. Sci.* **2017**, *58*, 2160–2165. [CrossRef] [PubMed]

126. Bosco, A.; Romero, C.O.; Ambati, B.K.; Vetter, M.L. In Vivo dynamics of retinal microglial activation during neurodegeneration: Confocal Ophthalmoscopic imaging and cell morphometry in mouse glaucoma. *J. Vis. Exp.* **2015**, *52731*, e52731. [CrossRef] [PubMed]

127. Liu, Z.; Tam, J.; Saeedi, O.; Hammer, D.X. Trans-retinal cellular imaging with multimodal adaptive optics. *Biomed. Opt. Express* **2018**, *9*, 4246–4262. [CrossRef]

128. Castanos, M.V.; Zhou, D.; Linderman, R.; Allison, R.; Milman, T.; Carroll, J.; Migacz, J.; Rosen, R.B.; Chui, T.Y. Imaging of macrophage-like cells in living human retina using clinical OCT. *Investig. Opthalmol. Vis. Sci.* **2020**, *61*, 48. [CrossRef] [PubMed]

129. Savage, J.C.; St-Pierre, M.-K.; Carrier, M.; El Hajj, H.; Novak, S.W.; Sanchez, M.G.; Cicchetti, F.; Tremblay, M.-È. Microglial physiological properties and interactions with synapses are altered at presymptomatic stages in a mouse model of Huntington's disease pathology. *J. Neuroinflammation* **2020**, *17*, 1–18. [CrossRef]

130. Fernández-Arjona, M.D.M.; Grondona, J.M.; Granados-Durán, P.; Fernández-Llebrez, P.; López-Ávalos, M.D. Microglia morphological categorization in a rat model of neuroinflammation by hierarchical cluster and principal components analysis. *Front. Cell. Neurosci.* **2017**, *11*, 235. [CrossRef]

131. Han, J.; Fan, Y.; Zhou, K.; Blomgren, K.; Harris, R.A. Uncovering sex differences of rodent microglia. *J. Neuroinflammation* **2021**, *18*, 1–11. [CrossRef]

132. Bianchi, I.; Lleo, A.; Gershwin, M.E.; Invernizzi, P. The X chromosome and immune associated genes. *J. Autoimmun.* **2012**, *38*, J187–J192. [CrossRef]

133. Sorge, R.E.; Mapplebeck, J.C.S.; Rosen, S.; Beggs, S.; Taves, S.; Alexander, J.K.; Martin, L.J.; Austin, J.-S.; Sotocinal, S.G.; Chen, D.; et al. Different immune cells mediate mechanical pain hypersensitivity in male and female mice. *Nat. Neurosci.* **2015**, *18*, 1081–1083. [CrossRef] [PubMed]

134. Coraggio, V.; Guida, F.; Boccella, S.; Scafuro, M.; Paino, S.; Romano, D.; Maione, S.; Luongo, L. Neuroimmune-driven neuropathic pain establishment: A focus on gender differences. *Int. J. Mol. Sci.* **2018**, *19*, 281. [CrossRef]

135. Du, M.; Mangold, C.A.; Bixler, G.V.; Brucklacher, R.M.; Masser, D.R.; Stout, M.B.; Elliott, M.H.; Freeman, W.M. Retinal gene expression responses to aging are sexually divergent. *Mol. Vis.* **2017**, *23*, 707–717. [PubMed]

136. Battineni, G.; Chintalapudi, N.; Amenta, F. Machine learning in medicine: Performance calculation of dementia prediction by support vector machines (SVM). *Inform. Med. Unlocked* **2019**, *16*, 100200. [CrossRef]

137. Wang, M.; Li, C.; Zhang, W.; Wang, Y.; Feng, Y.; Liang, Y.; Wei, J.; Zhang, X.; Li, X.; Chen, R. Support Vector machine for analyzing contributions of brain regions during task-state fMRI. *Front. Neuroinform.* **2019**, *13*, 10. [CrossRef] [PubMed]

Review

Harnessing Astrocytes and Müller Glial Cells in the Retina for Survival and Regeneration of Retinal Ganglion Cells

Hyung-Suk Yoo, Ushananthini Shanmugalingam [iD] and Patrice D. Smith *

Department of Neuroscience, Carleton University, Ottawa, ON K1S 5B6, Canada;
hyungyoo@cmail.carleton.ca (H.-S.Y.); UshananthiniShanmuga@cmail.carleton.ca (U.S.)
* Correspondence: patrice.smith@carleton.ca

Abstract: Astrocytes have been associated with the failure of axon regeneration in the central nervous system (CNS), as it undergoes reactive gliosis in response to damages to the CNS and functions as a chemical and physical barrier to axon regeneration. However, beneficial roles of astrocytes have been extensively studied in the spinal cord over the years, and a growing body of evidence now suggests that inducing astrocytes to become more growth-supportive can promote axon regeneration after spinal cord injury (SCI). In retina, astrocytes and Müller cells are known to undergo reactive gliosis after damage to retina and/or optic nerve and are hypothesized to be either detrimental or beneficial to survival and axon regeneration of retinal ganglion cells (RGCs). Whether they can be induced to become more growth-supportive after retinal and optic nerve injury has yet to be determined. In this review, we pinpoint the potential molecular pathways involved in the induction of growth-supportive astrocytes in the spinal cord and suggest that stimulating the activation of these pathways in the retina could represent a new therapeutic approach to promoting survival and axon regeneration of RGCs in retinal degenerative diseases.

Keywords: macroglia; astrocytes; Müller cells; optic nerve crush; retinal ganglion cells; spinal cord injury; signal transducer and activator of transcription 3; epidermal growth factor

Citation: Yoo, H.-S.; Shanmugalingam, U.; Smith, P.D. Harnessing Astrocytes and Müller Glial Cells in the Retina for Survival and Regeneration of Retinal Ganglion Cells. *Cells* **2021**, *10*, 1339. https://doi.org/10.3390/cells10061339

Academic Editors: Maurice Ptito and Joseph Bouskila

Received: 27 April 2021
Accepted: 26 May 2021
Published: 28 May 2021

Publisher's Note: MDPI stays neutral with regard to jurisdictional claims in published maps and institutional affiliations.

1. Introduction

The retina originates from the CNS during embryonic development [1]. The inner most layer of the retina harbors RGCs whose axons form the optic nerve that directly relays visual information to the brain [2]. RGCs can be considered CNS neurons because the optic nerve is myelinated by oligodendrocytes and do not regenerate spontaneously after injury [2,3]. Hence, the optic nerve crush (ONC) model has been widely used to determine the molecular mechanisms of neuronal survival and axon regeneration in the CNS [3–9]. While genetic and pharmacologic manipulations have been shown to activate neural repair mechanisms in RGCs after ONC, activation of macroglia in the retina, particularly astrocytes and Müller cells, has been shown to exert either detrimental or beneficial effects on survival and regeneration of RGCs after the injury [3,10–15].

Astrocytes in the CNS become activated in response to neuronal damage and neuroinflammation and form a dense network encapsulating the lesion site, known as the glial scar [15,16]. Although the glial scar functions as a physical and chemical barrier against further exposure to inflammatory agents, it also prevents growth of axons into the lesion site [16,17]. Astrocytes, therefore, have long been associated with the failure of axon regeneration in the CNS, and studies have attempted to eliminate or inhibit astrocytes to promote axon regeneration after CNS injury [16,18–20]. However, accumulating evidence now supports the concept that astrocytes are required for successful neuronal survival and axon regeneration in the CNS [12,13,16,19,21–25]. In this review, we will outline the evidence that reactive gliosis is required for successful neural repair in the CNS and suggest that harnessing the function of macroglia in the retina could promote survival and axon regeneration of RGCs.

2. Macroglia in the Retina

There are two types of macroglia in the retina: astrocytes and Müller cells. During retinal development, astrocytes from the brain enter the retina along the developing optic nerve [26,27]. In the mature retina, they are confined to the nerve fiber and ganglion cell layers [28]. On the contrary, Müller cells, the largest glial cell in the retina, originate from the retinal epithelium and span the entire retinal thickness [29,30]. The somata of Müller cells are located at the inner nuclear layer, and they extend their processes toward the outer and inner limiting membranes [31]. Müller cells ensheath retinal neurons and blood vessels in the plexiform and nerve fiber layers, allowing metabolic exchange between retinal vasculature and RGCs [32]. Astrocytes play a vital role in the development of the vascular system in the retina and contributes to the formation of the blood-retinal barrier [33,34]. Unlike Müller cells, astrocytes only envelop blood vessels in the nerve fiber and ganglion cell layers [32]. Astrocytes and Müller cells together maintain the integrity of the blood-retinal barrier by stabilizing tight junctions between endothelial cells and ensure the immune privilege of the eye [15,30]. They also provide essential nutrients, such as lactate and amino acids, from the circulation to neurons while participating in the retinal regulation of neurotransmitters, glucose metabolism and blood flow [10,15,30,35].

Astrocytes and Müller cells can provide RGCs with neurotrophic factors and antioxidants to maintain their viability [15,36–38]. Astrocytes can produce and release ciliary neurotrophic factor (CNTF), while Müller cells are known to be the major source of retinal brain derived neurotrophic factor (BDNF) and are capable of producing other well-known neurotrophic factors such as nerve growth factor (NGF) and glial cell line derived neurotrophic factor (GDNF) [14,39,40]. Increasing the retinal expression of these neurotrophic factors has been shown to promote survival and/or axon regeneration of RGCs. Moderate overexpression of BDNF in glaucomatous eye can result in long-term RGC survival while daily topical application of NGF can promote both survival and axon regeneration of RGCs after ONC [41,42]. Additionally, intravitreal co-administration of GDNF and CNTF can lead to survival and axon regeneration of RGCs after ONC possibly by directly binding to their respective receptors expressed by RGCs and/or inducing Müller cells to release additional neuroprotective factors, including BDNF and osteopontin [43]. Both macroglia can synthesize glutathione and counteract reactive oxygen species produced in the retina [11,15,36,37].

In response to the retinal injury, both astrocytes and Müller cells undergo reactive gliosis, up-regulating intermediate filaments, namely glial fibrillary acidic protein (GFAP), vimentin and nestin, and becoming more rigid [35]. This increased rigidness of both macroglia is mediated by signal transducer and activator of transcription 3 (STAT3), which is known as the master regulator of glial scar formation [18,35]. The rigidness allows for the glial scar formation, establishing the physical and chemical barrier to RGC axon regeneration [35]. Despite their proposed roles in inhibiting axon regeneration, studies show that these macroglia are rather essential in axon regeneration of RGCs. Astrocytes are known to release CNTF after lens injury and transform mature RGCs into a regenerative state, and the up-regulation of CNTF in astrocytes is also mediated by STAT3 [14,25]. Müller cells are also known to express CNTF after lens injury and hence may be involved in promoting the regeneration of RGCs in cooperation with astrocytes [44].

Since lens injury leads to intraocular inflammation, activation of inflammatory responses has been proposed to contribute to RGC axon regeneration [3,45]. Indeed, injecting the yeast wall extract zymosan can reproduce the regenerative effects of lens injury [3,14,44,45]. However, bacterial membrane component lipopolysaccharide (LPS) could not yield the same RGC axon regeneration, and this is due to zymosan's unique ability to stimulate dectin-1 receptors on leukocytes that invade the eye in response to the intraocular inflammation [46]. This finding suggests that infiltrating immune cells may secret neurotrophic factors that ultimately stimulate RGCs to regenerate after ONC [46]. Indeed, both macroglia and macrophages have been proposed to be the source of neurotrophic factors that promote RGC axon regeneration [44]. However, studies have shown that depletion of macrophages from the eye does not reduce the regenerative effects of lens injury whereas reduced number of reactive macroglia compromised the beneficial effects of zymosan [44,47]. This suggests that macroglia may be the major mediators of the regenerative effects in response to the intraocular inflammation [44]. Considering the dual role of macroglia in RGC axon regeneration, they may exist in two reactive states: neurotoxic state and growth-supportive state. Since RGC axon regeneration could be induced specifically by lens injury, the reactive state of macroglia may depend on the type of injury.

3. Two Distinct Reactive States of Astrocytes in the CNS

Pioneering studies by Sofroniew and colleagues established the concept that reactive astrocytes are necessary for protecting neurons from further damages after CNS injury and helping them survive and regenerate [17,20,23,48–52]. Additionally, accumulating evidence now posits the idea of phenotypical heterogeneity among reactive astrocytes [18,53–56]. Zamanian et al. have identified two distinct reactive states of astrocytes in the CNS using a transcriptome; they genetically profiled astrocytes after a systemic injection of LPS or cerebral ischemia [55]. LPS-induced neuroinflammation resulted in astrocytes expressing the components of classical complement cascade that are hypothesized to drive the loss of synapses and subsequently neurodegeneration [55]. On the contrary, the ischemic injury induced astrocytes to express neurotrophic factors and cytokines that can promote neural repair in the CNS [55]. The two types of reactive astrocytes induced by neuroinflammation and ischemia are known as A1 and A2, respectively [56]. Studies showed that the A1 phenotype is neurotoxic while the A2 phenotype is beneficial for neuronal survival and axon regeneration [56–60]. Figure 1 depicts the two distinct molecular pathways leading to changes in phenotype and function of reactive astrocytes.

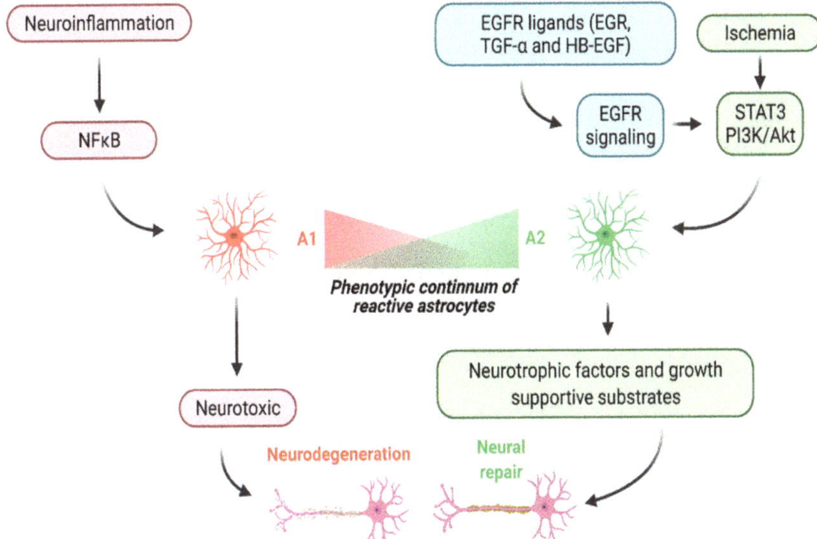

Figure 1. Neuroinflammation and ischemia lead to the generation of A1 and A2 astrocytes, respectively, in the spinal cord. Neuroinflammation may activate the NFκB pathway in reactive astrocytes and induce the A1 phenotype; on the other hand, ischemia may activate the STAT3 and/or PI3K/Akt pathways and induce the A2 phenotype. A1 and A2 astrocytes may exist on a phenotypic continuum. A1 astrocytes may be neurotoxic and promote neurodegeneration whereas A2 astrocytes may release neurotrophic factors and growth-supportive substrates and promote both survival and axon regeneration of CNS neurons. EGFR ligands may induce the generation of A2 astrocytes via activation of STAT3 and PI3K/Akt pathways (created with BioRender.com). Abbreviations: EGF, epidermal growth factor, EGFR, epidermal growth factor receptor, TGF-α, transforming growth factor-α, HB-EGF, heparin-binding EGF-like growth factor, NFκB, nuclear factor κB, STAT3, signal transducer and activator of transcription 3, PI3K/Akt, phosphoinositide 3-kinase/protein kinase B.

It should be noted that ONC results in generation of A1 astrocytes [57]. Neutralizing the factors that induce the A1 phenotype, such as interleukin-1α, tumor necrosis factor α and complement component 1q, could prevent the A1 formation and RGC death up to 14 days after injury [57]. Currently, it is unknown whether lens injury can generate A2 astrocytes in the retina and whether this phenotypic division also exists among Müller cells. However, identifying growth factors and downstream effectors involved in the induction of the A2 phenotype would allow for developing therapeutic strategies for protecting RGCs and promoting axon regeneration. Considering that findings from the ONC model generally have been found to hold true for spinal cord injury (SCI) [3], molecular mechanisms of inducing the A2 phenotype in the spinal cord could be applied to the retina and may hold therapeutic potential for retinal degenerative diseases.

4. Harnessing Astrocytes to Promote Neural Repair in the Spinal Cord

Astrogliosis has been extensively studied in the spinal cord, and accumulating evidence now suggests that it may have both beneficial and detrimental roles in the pathophysiology of SCI [18,23,51,52,56,61]. Although the signaling pathways leading to A1 and A2 phenotypes following SCI have yet to be determined, the nuclear factor κB (NFκB) and STAT3 pathways are believed to transform astrocytes into A1 and A2, respectively, because the roles of these two pathways seem to coincide with the hypothetical functions of A1 and A2 astrocytes (Figure 1) [61]. The NFκB pathway has a pivotal role in inducing neuroinflammation, and the inactivation of the pathway can reduce the expression of proinflammatory cytokines and significantly enhance the function of the spinal cord after SCI [62,63]. On the contrary, the inactivation of STAT3 pathway results in widespread infiltration of inflammatory cells and demyelination after SCI, whereas the activation of this pathway leads to rapid migration of reactive astrocytes to the lesion site to establish a physical barrier against inflammatory cells and promote significant improvement in functional recovery [64]. In further support of the idea that the STAT3 pathway may be involved in the generation of A2 astrocytes, Su et al. recently showed that down-regulation of microRNA-21 (miR-21) in astrocytes leads to STAT3-mediated conversion of A1 to A2 while up-regulation of miR-21 in astrocytes suppresses STAT3 activation and reverses the conversion process [60]. They also showed that these A2 astrocytes can promote axonal growth of neurons through the STAT3 pathway in vitro, suggesting that they may be beneficial to axon regeneration [60]. In addition to the STAT3 pathway, the phosphoinositide 3-kinase/protein kinase B (PI3K/Akt) pathway may also contribute to the generation of A2 astrocytes, as Xu et al. have shown that up-regulation of PI3K/Akt pathway and down-regulation of NFκB pathway are involved in counteracting A1 formation and promoting A2 formation [65]. Considering the evidence that the activation of STAT3 and PI3K/Akt pathways may be involved in the generation of A2 astrocytes, pharmacological stimulation of these pathways may lead to the increased population of A2 astrocytes that can promote neuronal survival and axon regeneration after CNS injury. The members of epidermal growth factor (EGF) family are known to activate these pathways via activation of epidermal growth factor receptors (EGFRs), and there is growing evidence that EGFR signaling can harness astrocytes to promote neuronal survival and axon regeneration in the CNS [21,22,66–70].

5. Manipulating Epidermal Growth Factor Signaling to Promote Neural Repair in the CNS

The EGF family is a group of related growth factors that are involved in a wide range of developmental processes, including proliferation, differentiation, and migration; the most notable members are EGF, transforming growth factor-α (TGF-α) and heparin-binding EGF-like growth factor (HB-EGF) [70]. These ligands signal through EGFR and three other homologous receptors, ErbB2, ErbB3 and ErbB4 [71]. Upon ligand binding, EGFRs undergo either homodimerization or heterodimerization; as an example, EGF can induce EGFR-EGFR homodimerization or EGFR-ErbB2 heterodimerization [70]. After this dimerization process, phosphorylated tyrosine residues function as docking sites for signaling protein complexes that are involved in PI3K/Akt and STAT3 pathways [70].

Members of the EGF and EGFR families are widely expressed in various regions of the CNS, including spinal cord, brainstem, cerebellum, diencephalon, telencephalon and hippocampus [70]. Their main function in the developing and adult CNS is to stimulate the proliferation and differentiation of neural progenitors; as an example, EGF and TGF-α stimulate both embryonic and adult striatal progenitors to proliferate and then differentiate into neurons and astrocytes [72–74]. The EGF and EGFR families also have been shown to be involved in neural repair after CNS injury. Studies showed not only that EGFR expression increases in subventricular zone (SVZ) after ischemic injury but also that intraventricular infusion of EGF promotes proliferation of neural stem cells in SVZ after cerebral ischemia and eventually leads to neuronal replacement in the injured striatum [75,76]. EGFR ligands can also exert neuroprotective effects against neurodegeneration; studies have shown that EGF and HB-EGF can promote dopaminergic neuronal survival in animal models of Parkinson's disease [77,78].

Although EGF and EGFR have been shown to promote neurite outgrowth of cultured CNS neurons, retinal studies have shown that EGFR signaling is activated by growth-inhibitory molecules, and inhibition of the EGFR signaling can promote axon regeneration of RGCs [79–81]. Koprivica et al. showed that myelin-derived proteins Nogo-66 and oligodendrocyte myelin glycoprotein can trigger indirect phosphorylation of EGFR in cultured postnatal cerebellar granule cells by activating their common receptor complexes that consist of Nogo-66 receptor (NgR) and its co-receptors p75/TROY and Lingo-1 [81]. Blocking the NgR-induced phosphorylation of EGFR using irreversible EGFR inhibitor PD168393 promoted neurite growth in retinal explant and axon regeneration in the ONC model [81]. The authors suggested that the EGFR inhibitor acts directly on RGCs to block their growth-inhibitory responses to myelin-derived proteins [81]. However, Douglas et al. later showed that the EGFR inhibitor has no direct impact on RGCs and, surprisingly, EGFR [82]. They reported that EGFR is only activated in glial cells, such as astrocytes and oligodendrocytes, in the retina and optic nerve, 14 days after optic nerve injury [82]. Furthermore, siRNA-mediated knockdown of EGFR in retinal culture could not promote neurite growth, but addition of competitive EGFR inhibitor AG1478 could restore neurite growth to the siRNA-treated cultures, suggesting that EGFR itself does not mediate the inhibition of axon regeneration [82]. Although the authors could not pinpoint the exact target of the EGFR inhibitor, they provided in vitro evidence that the inhibitor stimulates the release of neurotrophins, such as BDNF and NGF, from RGCs and retinal glia, and increases cyclic adenosine monophosphate, a second messenger involved in axon regeneration [82].

Currently, it is unclear whether EGFR signaling is involved in inhibition of axon regeneration. However, in support of the possibility that EGFR signaling could support axon regeneration in the CNS, a couple of studies have shown that TGF-α can promote axon regeneration after SCI [21,22]. Based on previous evidence that astrocytes can promote neuroprotection and may support axonal growth after injury, White et al. hypothesized that endogenous astrocytes could be harnessed to support axon regeneration with proper stimulation after SCI, and they intrathecally administered TGF-α in adult mice for two weeks following the injury [21]. TGF-α was able to stimulate proliferation and migration of astrocytes toward the lesion center and promote axon regeneration within the lesion [21]. As EGFR immunoreactivity was the strongest in GFAP-positive cells, the authors suggested that TGF-α acts directly on astrocytes [21]. Interestingly, TGF-α treatment not only increased the expression of neurocan, which is associated with inhibition of axonal growth, but also increased the expression of laminin throughout the lesion site [21]. As laminin immunoreactivity was co-localized with axons, the authors suggested that astrocytes may contribute to the formation of basal lamina structures throughout the lesion and provide a permissive substrate for axon elongation [21]. White and colleagues later published another study showing not only that TGF-α treatment can transform astrocytes into a growth-supportive phenotype that supports robust neurite outgrowth of dorsal root ganglion cells in vitro but also that overexpression of TGF-α in vivo by intraparenchymal adeno-associated virus injection adjacent to the injury site increased axon regeneration at the rostral lesion border [22]. In support of these findings, Sofroniew and colleagues included EGF as a part of the combinatorial pharmacological treatment for CNS axon regeneration, showing that EGF can increase the release of axon growth-supportive substrates, including laminin, fibronectin and collagen, and contribute to the overall axon regeneration after SCI [68]. They also noted that although EGF significantly increased astrocyte proliferation and density, axons were able to grow through and beyond glial scar formation [68]. Recently, Chen et al. showed that EGF can generate A2 astrocytes in vitro by down-regulating A1-like genes and up-regulating A2-like genes, further supporting the previous findings [66]. Overall, it is evident that manipulation of EGFR signaling can aid CNS axon regeneration by inducing the A2 phenotype that can provide axon growth-supportive substrates.

6. Manipulating EGFR Signaling in the Retina

Although EGFR signaling seems to induce the A2 phenotype in the spinal cord, it is currently unknown whether EGFR signaling can also generate A2 astrocytes in the retina and promote RGC survival and axon regeneration after injury. Ever since the discovery of the intrinsic ability of mature RGCs to regenerate their axons, the major focus of ONC studies has been on further deciphering how to activate the intrinsic growth ability of RGCs [4–9]. Despite the evidence that retinal astrocytes and Müller cells can provide neurotrophic support to RGCs and maintain their viability, the post-injury reparative roles of these retinal macroglia still need to be elucidated. Additionally, no studies have shown whether EGFR ligand-induced EGFR signaling can lead to survival and axon regeneration of RGCs. Figure 2 depicts three molecular pathways that may be induced by manipulating EGFR signaling in the retina after ONC.

EGFR signaling is involved in proliferation of retinal progenitor cells and macroglia during retina development, and retinal EGFR expression has been shown to decrease as the retina matures and loses its mitogenic response to EGFR ligands [83]. However, retinal injury can increase EGFR expression in adult retina, suggesting that the retina becomes more responsive to EGFR ligands after injury [83]. Whether intravitreal injection of EGFR ligands after ONC can induce A2 formation and promote RGC survival and axon regeneration has yet to be investigated. However, as Harder et al. recently suggested that astrocytes may exert protective effects on RGCs through EGFR signaling [84], it may be possible that activation of EGFR signaling in astrocytes may at least promote RGC survival (Figure 2). Interestingly, they also showed that the up-regulation of complement C3 in astrocytes is mediated by EGFR signaling and responsible for protecting RGCs against high intraocular pressure [84]. Because C3 is a known marker of A1 astrocytes [57], this finding does not support the idea that A1 phenotype is neurotoxic and raises the question whether so-called neurotoxic astrocytes are indeed detrimental to neurons. However, this could also imply that astrocytes perhaps exist as a continuum, with a heterogeneous population of A1 and A2, as previously suggested by Liddelow and Barres (Figure 1) [56]. Therefore, it is possible that those neuro-supportive astrocytes with increased C3 expression could have exhibited a genetic profile that may lie somewhere in the middle of the phenotypic continuum. Overall, future research could consider investigating: (1) whether EGFR ligands can promote RGC survival and/or axon regeneration after injury via A2 formation, (2) whether downstream targets of EGFR signaling are involved in the potential neuroprotective and regenerative effects of EGFR ligands, and (3) the phenotypic ratio between retinal A1 and A2 astrocytes after injury and after post-injury treatment with EGFR ligands.

Since the retina also contains Müller cells that undergo reactive gliosis along with astrocytes, dissecting the reparative roles of retinal gliosis would require understanding the post-injury molecular changes in Müller cells. Unlike reactive astrocytes, the phenotypic dichotomy of reactive Müller cells has yet to be defined although studies seem to suggest that Müller cells may also exhibit a similar continuum of reactive states [35,85]. Because these two macroglia both express EGFR and respond to EGFR ligands [83,86,87], they may undergo similar molecular changes after injury (Figure 2). However, what makes Müller cells unique is that they can dedifferentiate into retinal progenitor-like cells and become pluripotent in response to EGFR signaling (Figure 2) [35,87]. Although how this dedifferentiation of Müller cells may contribute to the overall reactive gliosis in the retina remains unknown, studies have suggested that inducing Müller cells to become pluripotent may be a therapeutic strategy for retina regeneration in retinal degenerative diseases, such as age-related macular degeneration and glaucoma [35,87–89]. Future studies could try to determine the continuum of reactive states of Müller cells and investigate: (1) whether EGFR ligand treatments can harness Müller cells to become like A2 astrocytes and/or dedifferentiate into progenitor-like cells that may also contribute to survival and axon regeneration of RGCs, (2) the potential interaction between Müller cell gliosis and Müller cell-derived progenitor-like cells, and (3) how this interaction may affect retinal astrogliosis and the overall survival and axon regeneration of RGCs.

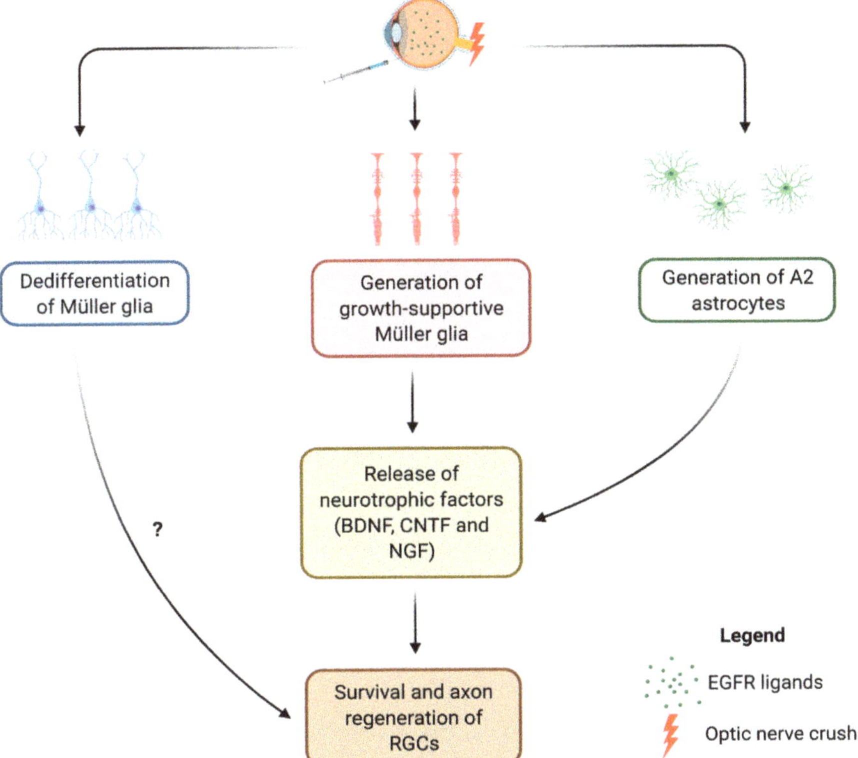

Figure 2. Intravitreal injection of EGFR ligands after ONC may lead to generation of A2 astrocytes and promote survival and axon regeneration of RGCs. It may also induce Müller cells to become more growth-supportive and help maintaining survival of RGCs and promoting optic nerve regeneration in cooperation with A2 astrocytes. Alternatively, Müller cells may dedifferentiate into retinal progenitor-like cells and contribute to survival and axon regeneration of RGCs (created with Biorender.com). Abbreviations: BDNF, brain derived neurotrophic factor, CNTF, ciliary neurotrophic factor, NGF, nerve growth factor, RGC, retinal ganglion cell, EGFR, epidermal growth factor.

7. Conclusions

Despite having an overall negative connotation for over a decade, reactive gliosis is now being recognized as an essential contributor to CNS repair [20,23,68]. Although its positive contributions to neuronal survival and axon regeneration in the spinal cord has been established, its potential roles in promoting RGC survival and axon regeneration have yet to be determined. However, a growing body of evidence suggests that retinal macroglia may be essential in aiding post-injury survival and axon regeneration of RGCs [14,15,44,45,47,84]. Considering that EGFR signaling can contribute to axon regeneration in the spinal cord and activate the STAT3 pathway, which is critical to generation of A2 astrocytes, future research should investigate whether EGFR ligand-mediated EGFR signaling can induce A2 formation and promote survival and axon regeneration of RGCs after injury. It is also critical to understand whether EGFR signaling can induce Müller cells to become like A2 astrocytes, as they undergo reactive gliosis along with retinal astrocytes in response to injury. Additionally, there has been increasing attention to the possibility to deliver neurotrophic factors by eye drops, and clinical studies suggest that this could represent a safe and effective strategy for treating retinal degenerative diseases [42,90–94]. Hence, future research should evaluate the efficacy of EGFR ligand-based eye drops after ONC and investigate whether the topical delivery of EGFR ligands could generate growth-supportive phenotypes of retinal astrocytes and Müller cells. Harnessing retinal macroglia to promote survival and axon regeneration of RGCs would contribute to designing and improving therapeutic strategies for degenerative retinal diseases.

Funding: This research received no external funding.

Institutional Review Board Statement: Not applicable.

Informed Consent Statement: Not applicable.

Conflicts of Interest: The authors declare no conflict of interest.

References

1. London, A.; Benhar, I.; Schwartz, M. The retina as a window to the brain-from eye research to CNS disorders. *Nat. Rev. Neurol.* **2013**, *9*, 44–53. [CrossRef]
2. Berry, M.; Ahmed, Z.; Lorber, B.; Douglas, M.; Logan, A. Regeneration of axons in the visual system. *Restor. Neurol. Neurosci.* **2008**, *26*, 147–174.
3. Williams, P.R.; Benowitz, L.I.; Goldberg, J.L.; He, Z. Axon Regeneration in the Mammalian Optic Nerve. *Annu. Rev. Vis. Sci.* **2020**, *6*, 195–213. [CrossRef] [PubMed]
4. Park, K.K.; Liu, K.; Hu, Y.; Smith, P.D.; Wang, C.; Cai, B.; Xu, B.; Connolly, L.; Kramvis, I.; Sahin, M.; et al. Promoting axon regeneration in the adult CNS by modulation of the PTEN/mTOR pathway. *Science* **2008**, *322*, 963–966. [CrossRef] [PubMed]
5. Smith, P.D.; Sun, F.; Park, K.K.; Cai, B.; Wang, C.; Kuwako, K.; Martinez-Carrasco, I.; Connolly, L.; He, Z. SOCS3 deletion promotes optic nerve regeneration in vivo. *Neuron* **2009**, *64*, 617–623. [CrossRef] [PubMed]
6. Li, Y.; Andereggen, L.; Yuki, K.; Omura, K.; Yin, Y.; Gilbert, H.Y.; Erdogan, B.; Asdourian, M.S.; Shrock, C.; de Lima, S.; et al. Mobile zinc increases rapidly in the retina after optic nerve injury and regulates ganglion cell survival and optic nerve regeneration. *Proc. Natl. Acad. Sci. USA* **2017**, *114*, E209–E218. [CrossRef] [PubMed]
7. Norsworthy, M.W.; Bei, F.; Kawaguchi, R.; Wang, Q.; Tran, N.M.; Li, Y.; Brommer, B.; Zhang, Y.; Wang, C.; Sanes, J.R.; et al. Sox11 Expression Promotes Regeneration of Some Retinal Ganglion Cell Types but Kills Others. *Neuron* **2017**, *94*, 1112–1120. [CrossRef]
8. Trakhtenberg, E.F.; Li, Y.; Feng, Q.; Tso, J.; Rosenberg, P.A.; Goldberg, J.L.; Benowitz, L.I. Zinc chelation and Klf9 knockdown cooperatively promote axon regeneration after optic nerve injury. *Exp. Neurol.* **2018**, *300*, 22–29. [CrossRef] [PubMed]
9. Thompson, A.; Berry, M.; Logan, A.; Ahmed, Z. Activation of the BMP4/Smad1 Pathway Promotes Retinal Ganglion Cell Survival and Axon Regeneration. *Investig. Ophthalmol. Vis. Sci.* **2019**, *60*, 1748–1759. [CrossRef]
10. Boia, R.; Ruzafa, N.; Aires, I.D.; Pereiro, X.; Ambrosio, A.F.; Vecino, E.; Santiago, A.R. Neuroprotective Strategies for Retinal Ganglion Cell Degeneration: Current Status and Challenges Ahead. *Int. J. Mol. Sci.* **2020**, *21*, 2262. [CrossRef]
11. Garcia, M.; Vecino, E. Role of Muller glia in neuroprotection and regeneration in the retina. *Histol. Histopathol.* **2003**, *18*, 1205–1218. [CrossRef]
12. Sun, D.; Moore, S.; Jakobs, T.C. Optic nerve astrocyte reactivity protects function in experimental glaucoma and other nerve injuries. *J. Exp. Med.* **2017**, *214*, 1411–1430. [CrossRef]
13. Harada, C.; Azuchi, Y.; Noro, T.; Guo, X.; Kimura, A.; Namekata, K.; Harada, T. TrkB Signaling in Retinal Glia Stimulates Neuroprotection after Optic Nerve Injury. *Am. J. Pathol.* **2015**, *185*, 3238–3247. [CrossRef]
14. Muller, A.; Hauk, T.G.; Fischer, D. Astrocyte-derived CNTF switches mature RGCs to a regenerative state following inflammatory stimulation. *Brain* **2007**, *130*, 3308–3320. [CrossRef] [PubMed]
15. de Hoz, R.; Rojas, B.; Ramirez, A.I.; Salazar, J.J.; Gallego, B.I.; Trivino, A.; Ramirez, J.M. Retinal Macroglial Responses in Health and Disease. *BioMed Res. Int.* **2016**, *2016*, 2954721. [CrossRef] [PubMed]

16. White, R.E.; Jakeman, L.B. Don't fence me in: Harnessing the beneficial roles of astrocytes for spinal cord repair. *Restor. Neurol. Neurosci.* **2008**, *26*, 197–214. [PubMed]

17. Sofroniew, M.V.; Vinters, H.V. Astrocytes: Biology and pathology. *Acta Neuropathol.* **2010**, *119*, 7–35. [CrossRef]

18. Yang, T.; Dai, Y.; Chen, G.; Cui, S. Dissecting the Dual Role of the Glial Scar and Scar-Forming Astrocytes in Spinal Cord Injury. *Front. Cell. Neurosci.* **2020**, *14*, 78. [CrossRef]

19. Tang, B.L. The astrocyte scar—Not so inhibitory after all? *Neural Regen. Res.* **2016**, *11*, 1054–1055. [CrossRef] [PubMed]

20. Sofroniew, M.V. Molecular dissection of reactive astrogliosis and glial scar formation. *Trends Neurosci.* **2009**, *32*, 638–647. [CrossRef]

21. White, R.E.; Yin, F.Q.; Jakeman, L.B. TGF-alpha increases astrocyte invasion and promotes axonal growth into the lesion following spinal cord injury in mice. *Exp. Neurol.* **2008**, *214*, 10–24. [CrossRef] [PubMed]

22. White, R.E.; Rao, M.; Gensel, J.C.; McTigue, D.M.; Kaspar, B.K.; Jakeman, L.B. Transforming growth factor alpha transforms astrocytes to a growth-supportive phenotype after spinal cord injury. *J. Neurosci.* **2011**, *31*, 15173–15187. [CrossRef]

23. Anderson, M.A.; Burda, J.E.; Ren, Y.; Ao, Y.; O'Shea, T.M.; Kawaguchi, R.; Coppola, G.; Khakh, B.S.; Deming, T.J.; Sofroniew, M.V. Astrocyte scar formation aids central nervous system axon regeneration. *Nature* **2016**, *532*, 195–200. [CrossRef]

24. Xie, C.; Shen, X.; Xu, X.; Liu, H.; Li, F.; Lu, S.; Gao, Z.; Zhang, J.; Wu, Q.; Yang, D.; et al. Astrocytic YAP Promotes the Formation of Glia Scars and Neural Regeneration after Spinal Cord Injury. *J. Neurosci.* **2020**, *40*, 2644–2662. [CrossRef] [PubMed]

25. Linnerbauer, M.; Rothhammer, V. Protective Functions of Reactive Astrocytes Following Central Nervous System Insult. *Front. Immunol.* **2020**, *11*, 573256. [CrossRef] [PubMed]

26. Ling, T.L.; Mitrofanis, J.; Stone, J. Origin of retinal astrocytes in the rat: Evidence of migration from the optic nerve. *J. Comp. Neurol.* **1989**, *286*, 345–352. [CrossRef] [PubMed]

27. Huxlin, K.R.; Sefton, A.J.; Furby, J.H. The origin and development of retinal astrocytes in the mouse. *J. Neurocytol.* **1992**, *21*, 530–544. [CrossRef]

28. Chang, M.L.; Wu, C.H.; Jiang-Shieh, Y.F.; Shieh, J.Y.; Wen, C.Y. Reactive changes of retinal astrocytes and Muller glial cells in kainate-induced neuroexcitotoxicity. *J. Anat.* **2007**, *210*, 54–65. [CrossRef] [PubMed]

29. Hamon, A.; Roger, J.E.; Yang, X.J.; Perron, M. Muller glial cell-dependent regeneration of the neural retina: An overview across vertebrate model systems. *Dev. Dyn.* **2016**, *245*, 727–738. [CrossRef]

30. Fernandez-Sanchez, L.; Lax, P.; Campello, L.; Pinilla, I.; Cuenca, N. Astrocytes and Muller Cell Alterations During Retinal Degeneration in a Transgenic Rat Model of Retinitis Pigmentosa. *Front. Cell. Neurosci.* **2015**, *9*, 484. [CrossRef] [PubMed]

31. Lindqvist, N.; Liu, Q.; Zajadacz, J.; Franze, K.; Reichenbach, A. Retinal glial (Muller) cells: Sensing and responding to tissue stretch. *Investig. Ophthalmol. Vis. Sci.* **2010**, *51*, 1683–1690. [CrossRef] [PubMed]

32. Antonetti, D.A.; Barber, A.J.; Bronson, S.K.; Freeman, W.M.; Gardner, T.W.; Jefferson, L.S.; Kester, M.; Kimball, S.R.; Krady, J.K.; LaNoue, K.F.; et al. Diabetic retinopathy: Seeing beyond glucose-induced microvascular disease. *Diabetes* **2006**, *55*, 2401–2411. [CrossRef]

33. Tao, C.; Zhang, X. Development of astrocytes in the vertebrate eye. *Dev. Dyn.* **2014**, *243*, 1501–1510. [CrossRef]

34. Yao, H.; Wang, T.; Deng, J.; Liu, D.; Li, X.; Deng, J. The development of blood-retinal barrier during the interaction of astrocytes with vascular wall cells. *Neural Regen. Res.* **2014**, *9*, 1047–1054. [CrossRef]

35. Vecino, E.; Rodriguez, F.D.; Ruzafa, N.; Pereiro, X.; Sharma, S.C. Glia-neuron interactions in the mammalian retina. *Prog. Retin. Eye Res.* **2016**, *51*, 1–40. [CrossRef]

36. Zhang, T.; Gillies, M.C.; Madigan, M.C.; Shen, W.; Du, J.; Grunert, U.; Zhou, F.; Yam, M.; Zhu, L. Disruption of De Novo Serine Synthesis in Muller Cells Induced Mitochondrial Dysfunction and Aggravated Oxidative Damage. *Mol. Neurobiol.* **2018**, *55*, 7025–7037. [CrossRef] [PubMed]

37. Guo, X.; Jiang, Q.; Tuccitto, A.; Chan, D.; Alqawlaq, S.; Won, G.J.; Sivak, J.M. The AMPK-PGC-1alpha signaling axis regulates the astrocyte glutathione system to protect against oxidative and metabolic injury. *Neurobiol. Dis.* **2018**, *113*, 59–69. [CrossRef] [PubMed]

38. Balzamino, B.O.; Esposito, G.; Marino, R.; Keller, F.; Micera, A. Changes in vitreal protein profile and retina mRNAs in Reeler mice: NGF, IL33 and Muller cell activation. *PLoS ONE* **2019**, *14*, e0212732. [CrossRef]

39. Taylor, S.; Srinivasan, B.; Wordinger, R.J.; Roque, R.S. Glutamate stimulates neurotrophin expression in cultured Muller cells. *Brain Res. Mol. Brain Res.* **2003**, *111*, 189–197. [CrossRef]

40. Seki, M.; Tanaka, T.; Sakai, Y.; Fukuchi, T.; Abe, H.; Nawa, H.; Takei, N. Muller Cells as a source of brain-derived neurotrophic factor in the retina: Noradrenaline upregulates brain-derived neurotrophic factor levels in cultured rat Muller cells. *Neurochem. Res.* **2005**, *30*, 1163–1170. [CrossRef]

41. Wojcik-Gryciuk, A.; Gajewska-Wozniak, O.; Kordecka, K.; Boguszewski, P.M.; Waleszczyk, W.; Skup, M. Neuroprotection of Retinal Ganglion Cells with AAV2-BDNF Pretreatment Restoring Normal TrkB Receptor Protein Levels in Glaucoma. *Int. J. Mol. Sci.* **2020**, *21*, 6262. [CrossRef] [PubMed]

42. Mesentier-Louro, L.A.; Rosso, P.; Carito, V.; Mendez-Otero, R.; Santiago, M.F.; Rama, P.; Lambiase, A.; Tirassa, P. Nerve Growth Factor Role on Retinal Ganglion Cell Survival and Axon Regrowth: Effects of Ocular Administration in Experimental Model of Optic Nerve Injury. *Mol. Neurobiol.* **2019**, *56*, 1056–1069. [CrossRef] [PubMed]

43. Dulz, S.; Bassal, M.; Flachsbarth, K.; Riecken, K.; Fehse, B.; Schlichting, S.; Bartsch, S.; Bartsch, U. Intravitreal Co-Administration of GDNF and CNTF Confers Synergistic and Long-Lasting Protection against Injury-Induced Cell Death of Retinal Ganglion Cells in Mice. *Cells* **2020**, *9*, 2082. [CrossRef] [PubMed]

44. Hauk, T.G.; Leibinger, M.; Muller, A.; Andreadaki, A.; Knippschild, U.; Fischer, D. Stimulation of axon regeneration in the mature optic nerve by intravitreal application of the toll-like receptor 2 agonist Pam3Cys. *Investig. Ophthalmol. Vis. Sci.* **2010**, *51*, 459–464. [CrossRef]

45. Yin, Y.; Cui, Q.; Gilbert, H.Y.; Yang, Y.; Yang, Z.; Berlinicke, C.; Li, Z.; Zaverucha-do-Valle, C.; He, H.; Petkova, V.; et al. Oncomodulin links inflammation to optic nerve regeneration. *Proc. Natl. Acad. Sci. USA* **2009**, *106*, 19587–19592. [CrossRef] [PubMed]

46. Baldwin, K.T.; Carbajal, K.S.; Segal, B.M.; Giger, R.J. Neuroinflammation triggered by beta-glucan/dectin-1 signaling enables CNS axon regeneration. *Proc. Natl. Acad. Sci. USA* **2015**, *112*, 2581–2586. [CrossRef]

47. Hauk, T.G.; Muller, A.; Lee, J.; Schwendener, R.; Fischer, D. Neuroprotective and axon growth promoting effects of intraocular inflammation do not depend on oncomodulin or the presence of large numbers of activated macrophages. *Exp. Neurol.* **2008**, *209*, 469–482. [CrossRef]

48. Tong, X.; Ao, Y.; Faas, G.C.; Nwaobi, S.E.; Xu, J.; Haustein, M.D.; Anderson, M.A.; Mody, I.; Olsen, M.L.; Sofroniew, M.V.; et al. Astrocyte Kir4.1 ion channel deficits contribute to neuronal dysfunction in Huntington's disease model mice. *Nat. Neurosci.* **2014**, *17*, 694–703. [CrossRef]

49. Pekny, M.; Pekna, M.; Messing, A.; Steinhauser, C.; Lee, J.M.; Parpura, V.; Hol, E.M.; Sofroniew, M.V.; Verkhratsky, A. Astrocytes: A central element in neurological diseases. *Acta Neuropathol.* **2016**, *131*, 323–345. [CrossRef]

50. Burda, J.E.; Bernstein, A.M.; Sofroniew, M.V. Astrocyte roles in traumatic brain injury. *Exp. Neurol.* **2016**, *275*, 305–315. [CrossRef]

51. Faulkner, J.R.; Herrmann, J.E.; Woo, M.J.; Tansey, K.E.; Doan, N.B.; Sofroniew, M.V. Reactive astrocytes protect tissue and preserve function after spinal cord injury. *J. Neurosci.* **2004**, *24*, 2143–2155. [CrossRef]

52. Herrmann, J.E.; Imura, T.; Song, B.; Qi, J.; Ao, Y.; Nguyen, T.K.; Korsak, R.A.; Takeda, K.; Akira, S.; Sofroniew, M.V. STAT3 is a critical regulator of astrogliosis and scar formation after spinal cord injury. *J. Neurosci.* **2008**, *28*, 7231–7243. [CrossRef] [PubMed]

53. Adams, K.L.; Gallo, V. The diversity and disparity of the glial scar. *Nat. Neurosci.* **2018**, *21*, 9–15. [CrossRef] [PubMed]

54. Anderson, M.A.; Ao, Y.; Sofroniew, M.V. Heterogeneity of reactive astrocytes. *Neurosci. Lett.* **2014**, *565*, 23–29. [CrossRef] [PubMed]

55. Zamanian, J.L.; Xu, L.; Foo, L.C.; Nouri, N.; Zhou, L.; Giffard, R.G.; Barres, B.A. Genomic analysis of reactive astrogliosis. *J. Neurosci.* **2012**, *32*, 6391–6410. [CrossRef]

56. Liddelow, S.A.; Barres, B.A. Reactive Astrocytes: Production, Function, and Therapeutic Potential. *Immunity* **2017**, *46*, 957–967. [CrossRef]

57. Liddelow, S.A.; Guttenplan, K.A.; Clarke, L.E.; Bennett, F.C.; Bohlen, C.J.; Schirmer, L.; Bennett, M.L.; Munch, A.E.; Chung, W.S.; Peterson, T.C.; et al. Neurotoxic reactive astrocytes are induced by activated microglia. *Nature* **2017**, *541*, 481–487. [CrossRef]

58. Morizawa, Y.M.; Hirayama, Y.; Ohno, N.; Shibata, S.; Shigetomi, E.; Sui, Y.; Nabekura, J.; Sato, K.; Okajima, F.; Takebayashi, H.; et al. Reactive astrocytes function as phagocytes after brain ischemia via ABCA1-mediated pathway. *Nat. Commun.* **2017**, *8*, 28. [CrossRef]

59. Yun, S.P.; Kam, T.I.; Panicker, N.; Kim, S.; Oh, Y.; Park, J.S.; Kwon, S.H.; Park, Y.J.; Karuppagounder, S.S.; Park, H.; et al. Block of A1 astrocyte conversion by microglia is neuroprotective in models of Parkinson's disease. *Nat. Med.* **2018**, *24*, 931–938. [CrossRef]

60. Su, Y.; Chen, Z.; Du, H.; Liu, R.; Wang, W.; Li, H.; Ning, B. Silencing miR-21 induces polarization of astrocytes to the A2 phenotype and improves the formation of synapses by targeting glypican 6 via the signal transducer and activator of transcription-3 pathway after acute ischemic spinal cord injury. *FASEB J.* **2019**, *33*, 10859–10871. [CrossRef] [PubMed]

61. Li, X.; Li, M.; Tian, L.; Chen, J.; Liu, R.; Ning, B. Reactive Astrogliosis: Implications in Spinal Cord Injury Progression and Therapy. *Oxid. Med. Cell. Longev.* **2020**, *2020*, 9494352. [CrossRef]

62. Brambilla, R.; Bracchi-Ricard, V.; Hu, W.H.; Frydel, B.; Bramwell, A.; Karmally, S.; Green, E.J.; Bethea, J.R. Inhibition of astroglial nuclear factor kappaB reduces inflammation and improves functional recovery after spinal cord injury. *J. Exp. Med.* **2005**, *202*, 145–156. [CrossRef]

63. Liu, T.; Zhang, L.; Joo, D.; Sun, S.C. NF-kappaB signaling in inflammation. *Signal Transduct. Target. Ther.* **2017**, *2*. [CrossRef] [PubMed]

64. Okada, S.; Nakamura, M.; Katoh, H.; Miyao, T.; Shimazaki, T.; Ishii, K.; Yamane, J.; Yoshimura, A.; Iwamoto, Y.; Toyama, Y.; et al. Conditional ablation of Stat3 or Socs3 discloses a dual role for reactive astrocytes after spinal cord injury. *Nat. Med.* **2006**, *12*, 829–834. [CrossRef] [PubMed]

65. Xu, X.; Zhang, A.; Zhu, Y.; He, W.; Di, W.; Fang, Y.; Shi, X. MFG-E8 reverses microglial-induced neurotoxic astrocyte (A1) via NF-kappaB and PI3K-Akt pathways. *J. Cell. Physiol.* **2018**, *234*, 904–914. [CrossRef] [PubMed]

66. Chan, S.J.; Niu, W.; Hayakawa, K.; Hamanaka, G.; Wang, X.; Cheah, P.S.; Guo, S.; Yu, Z.; Arai, K.; Selim, M.H.; et al. Promoting Neuro-Supportive Properties of Astrocytes with Epidermal Growth Factor Hydrogels. *Stem Cells Transl. Med.* **2019**, *8*, 1242–1248. [CrossRef]

67. Ito, M.; Komai, K.; Mise-Omata, S.; Iizuka-Koga, M.; Noguchi, Y.; Kondo, T.; Sakai, R.; Matsuo, K.; Nakayama, T.; Yoshie, O.; et al. Brain regulatory T cells suppress astrogliosis and potentiate neurological recovery. *Nature* **2019**, *565*, 246–250. [CrossRef]

68. Anderson, M.A.; O'Shea, T.M.; Burda, J.E.; Ao, Y.; Barlatey, S.L.; Bernstein, A.M.; Kim, J.H.; James, N.D.; Rogers, A.; Kato, B.; et al. Required growth facilitators propel axon regeneration across complete spinal cord injury. *Nature* **2018**, *561*, 396–400. [CrossRef] [PubMed]

69. Zhang, C.; Xu, H.; Zhou, Z.; Tian, Y.; Cao, X.; Cheng, G.; Liu, Q. Blocking of the EGFR-STAT3 signaling pathway through afatinib treatment inhibited the intrahepatic cholangiocarcinoma. *Exp. Ther. Med.* **2018**, *15*, 4995–5000. [CrossRef]

70. Chen, J.; Zeng, F.; Forrester, S.J.; Eguchi, S.; Zhang, M.Z.; Harris, R.C. Expression and Function of the Epidermal Growth Factor Receptor in Physiology and Disease. *Physiol. Rev.* **2016**, *96*, 1025–1069. [CrossRef]

71. Zeng, F.; Singh, A.B.; Harris, R.C. The role of the EGF family of ligands and receptors in renal development, physiology and pathophysiology. *Exp. Cell. Res.* **2009**, *315*, 602–610. [CrossRef]

72. Reynolds, B.A.; Tetzlaff, W.; Weiss, S. A multipotent EGF-responsive striatal embryonic progenitor cell produces neurons and astrocytes. *J. Neurosci.* **1992**, *12*, 4565–4574. [CrossRef]

73. Fallon, J.; Reid, S.; Kinyamu, R.; Opole, I.; Opole, R.; Baratta, J.; Korc, M.; Endo, T.L.; Duong, A.; Nguyen, G.; et al. In vivo induction of massive proliferation, directed migration, and differentiation of neural cells in the adult mammalian brain. *Proc. Natl. Acad. Sci. USA* **2000**, *97*, 14686–14691. [CrossRef]

74. McGinn, M.J.; Sun, D.; Schneider, S.L.; Alexander, J.K.; Colello, R.J. Epidermal growth factor-induced cell proliferation in the adult rat striatum. *Brain Res.* **2004**, *1007*, 29–38. [CrossRef] [PubMed]

75. Teramoto, T.; Qiu, J.; Plumier, J.C.; Moskowitz, M.A. EGF amplifies the replacement of parvalbumin-expressing striatal interneurons after ischemia. *J. Clin. Investig.* **2003**, *111*, 1125–1132. [CrossRef] [PubMed]

76. Sun, D.; Bullock, M.R.; Altememi, N.; Zhou, Z.; Hagood, S.; Rolfe, A.; McGinn, M.J.; Hamm, R.; Colello, R.J. The effect of epidermal growth factor in the injured brain after trauma in rats. *J. Neurotrauma* **2010**, *27*, 923–938. [CrossRef] [PubMed]

77. Iwakura, Y.; Piao, Y.S.; Mizuno, M.; Takei, N.; Kakita, A.; Takahashi, H.; Nawa, H. Influences of dopaminergic lesion on epidermal growth factor-ErbB signals in Parkinson's disease and its model: Neurotrophic implication in nigrostriatal neurons. *J. Neurochem.* **2005**, *93*, 974–983. [CrossRef] [PubMed]

78. Hanke, M.; Farkas, L.M.; Jakob, M.; Ries, R.; Pohl, J.; Sullivan, A.M. Heparin-binding epidermal growth factor-like growth factor: A component in chromaffin granules which promotes the survival of nigrostriatal dopaminergic neurones in vitro and in vivo. *Neuroscience* **2004**, *124*, 757–766. [CrossRef]

79. Chalazonitis, A.; Kessler, J.A.; Twardzik, D.R.; Morrison, R.S. Transforming growth factor alpha, but not epidermal growth factor, promotes the survival of sensory neurons in vitro. *J. Neurosci.* **1992**, *12*, 583–594. [CrossRef]

80. Peng, H.; Wen, T.C.; Tanaka, J.; Maeda, N.; Matsuda, S.; Desaki, J.; Sudo, S.; Zhang, B.; Sakanaka, M. Epidermal growth factor protects neuronal cells in vivo and in vitro against transient forebrain ischemia- and free radical-induced injuries. *J. Cereb. Blood Flow Metab.* **1998**, *18*, 349–360. [CrossRef]

81. Koprivica, V.; Cho, K.S.; Park, J.B.; Yiu, G.; Atwal, J.; Gore, B.; Kim, J.A.; Lin, E.; Tessier-Lavigne, M.; Chen, D.F.; et al. EGFR activation mediates inhibition of axon regeneration by myelin and chondroitin sulfate proteoglycans. *Science* **2005**, *310*, 106–110. [CrossRef]

82. Douglas, M.R.; Morrison, K.C.; Jacques, S.J.; Leadbeater, W.E.; Gonzalez, A.M.; Berry, M.; Logan, A.; Ahmed, Z. Off-target effects of epidermal growth factor receptor antagonists mediate retinal ganglion cell disinhibited axon growth. *Brain* **2009**, *132*, 3102–3121. [CrossRef]

83. Close, J.L.; Liu, J.; Gumuscu, B.; Reh, T.A. Epidermal growth factor receptor expression regulates proliferation in the postnatal rat retina. *Glia* **2006**, *54*, 94–104. [CrossRef]

84. Harder, J.M.; Braine, C.E.; Williams, P.A.; Zhu, X.; MacNicoll, K.H.; Sousa, G.L.; Buchanan, R.A.; Smith, R.S.; Libby, R.T.; Howell, G.R.; et al. Early immune responses are independent of RGC dysfunction in glaucoma with complement component C3 being protective. *Proc. Natl. Acad. Sci. USA* **2017**, *114*, E3839–E3848. [CrossRef] [PubMed]

85. Bringmann, A.; Iandiev, I.; Pannicke, T.; Wurm, A.; Hollborn, M.; Wiedemann, P.; Osborne, N.N.; Reichenbach, A. Cellular signaling and factors involved in Muller cell gliosis: Neuroprotective and detrimental effects. *Prog. Retin. Eye Res.* **2009**, *28*, 423–451. [CrossRef] [PubMed]

86. Liu, B.; Chen, H.; Johns, T.G.; Neufeld, A.H. Epidermal growth factor receptor activation: An upstream signal for transition of quiescent astrocytes into reactive astrocytes after neural injury. *J. Neurosci.* **2006**, *26*, 7532–7540. [CrossRef]

87. Wan, J.; Ramachandran, R.; Goldman, D. HB-EGF is necessary and sufficient for Muller glia dedifferentiation and retina regeneration. *Dev. Cell.* **2012**, *22*, 334–347. [CrossRef] [PubMed]

88. Lahne, M.; Nagashima, M.; Hyde, D.R.; Hitchcock, P.F. Reprogramming Muller Glia to Regenerate Retinal Neurons. *Annu. Rev. Vis. Sci.* **2020**, *6*, 171–193. [CrossRef]

89. Goldman, D. Muller glial cell reprogramming and retina regeneration. *Nat. Rev. Neurosci.* **2014**, *15*, 431–442. [CrossRef]

90. Domenici, L.; Origlia, N.; Falsini, B.; Cerri, E.; Barloscio, D.; Fabiani, C.; Sanso, M.; Giovannini, L. Rescue of retinal function by BDNF in a mouse model of glaucoma. *PLoS ONE* **2014**, *9*, e115579. [CrossRef] [PubMed]

91. Falsini, B.; Iarossi, G.; Chiaretti, A.; Ruggiero, A.; Manni, L.; Galli-Resta, L.; Corbo, G.; Abed, E. NGF eye-drops topical administration in patients with retinitis pigmentosa, a pilot study. *J. Transl. Med.* **2016**, *14*, 8. [CrossRef] [PubMed]

92. Lambiase, A.; Coassin, M.; Tirassa, P.; Mantelli, F.; Aloe, L. Nerve growth factor eye drops improve visual acuity and electrofunctional activity in age-related macular degeneration: A case report. *Ann. Ist. Super. Sanita* **2009**, *45*, 439–442. [CrossRef] [PubMed]

93. Ciavarella, C.; Buzzi, M.; Bergantin, E.; Di Marco, S.; Giannaccare, G.; Campos, E.; Bisti, S.; Versura, P. Effects of Cord Blood Serum (CBS) on viability of retinal Muller glial cells under in vitro injury. *PLoS ONE* **2020**, *15*, e0234145. [CrossRef] [PubMed]
94. Tirassa, P.; Rosso, P.; Iannitelli, A. Ocular Nerve Growth Factor (NGF) and NGF Eye Drop Application as Paradigms to Investigate NGF Neuroprotective and Reparative Actions. *Methods Mol. Biol.* **2018**, *1727*, 19–38. [CrossRef] [PubMed]

Article

Loss of Motor Protein MYO1C Causes Rhodopsin Mislocalization and Results in Impaired Visual Function

Ashish K. Solanki [1,†](ID), Manas R. Biswal [2,†](ID), Stephen Walterhouse [1,†], René Martin [3], Altaf A. Kondkar [4](ID), Hans-Joachim Knölker [3](ID), Bushra Rahman [1], Ehtesham Arif [1], Shahid Husain [5](ID), Sandra R. Montezuma [6], Deepak Nihalani [7,*] and Glenn Prazere Lobo [1,5,8,*]

1 Department of Medicine, Medical University of South Carolina, Charleston, SC 29425, USA; solankia@musc.edu (A.K.S.); walterho@musc.edu (S.W.); rahman@musc.edu (B.R.); arif@musc.edu (E.A.)
2 Department of Pharmaceutical Sciences, Taneja College of Pharmacy, University of South Florida, Tampa, FL 33612, USA; biswal@usf.edu
3 Faculty of Chemistry, Technische Universität Dresden, Bergstraße 66, 01069 Dresden, Germany; rene.martin@gmx.de (R.M.); hans-joachim.knoelker@tu-dresden.de (H.-J.K.)
4 Department of Ophthalmology, College of Medicine, King Saud University, Riyadh 11411, Saudi Arabia; akondkar@gmail.com
5 Department of Ophthalmology, Medical University of South Carolina, Charleston, SC 29425, USA; husain@musc.edu
6 Department of Ophthalmology and Visual Neurosciences, University of Minnesota, 516 Delaware Street S.E., 9th Floor, Minneapolis, MN 55455, USA; smontezu@umn.edu
7 National Institute of Diabetes and Digestive and Kidney Diseases (NIDDK), National Institutes of Health, Bldg. 2DEM, Room 6085, 6707 Democracy Blvd., Bethesda, MD 20817, USA
8 Department of Ophthalmology and Visual Neurosciences, Lions Research Building, University of Minnesota, 2001 6th Street S.E., Room 225, Minneapolis, MN 55455, USA
* Correspondence: nihalanideepak@hotmail.com (D.N.); lobo0023@umn.edu (G.P.L.)
† These authors contributed equally to this work.

Citation: Solanki, A.K.; Biswal, M.R.; Walterhouse, S.; Martin, R.; Kondkar, A.A.; Knölker, H.-J.; Rahman, B.; Arif, E.; Husain, S.; Montezuma, S.R.; et al. Loss of Motor Protein MYO1C Causes Rhodopsin Mislocalization and Results in Impaired Visual Function. *Cells* **2021**, *10*, 1322. https://doi.org/10.3390/cells10061322

Academic Editors: Maurice Ptito and Joseph Bouskila

Received: 4 May 2021
Accepted: 25 May 2021
Published: 26 May 2021

check for **updates**

Abstract: Unconventional myosins, linked to deafness, are also proposed to play a role in retinal cell physiology. However, their direct role in photoreceptor function remains unclear. We demonstrate that systemic loss of the unconventional myosin MYO1C in mice, specifically causes rhodopsin mislocalization, leading to impaired visual function. Electroretinogram analysis of *Myo1c* knockout (*Myo1c*-KO) mice showed a progressive loss of photoreceptor function. Immunohistochemistry and binding assays demonstrated MYO1C localization to photoreceptor inner and outer segments (OS) and identified a direct interaction of rhodopsin with MYO1C. In *Myo1c*-KO retinas, rhodopsin mislocalized to rod inner segments (IS) and cell bodies, while cone opsins in OS showed punctate staining. In aged mice, the histological and ultrastructural examination of the phenotype of *Myo1c*-KO retinas showed progressively shorter photoreceptor OS. These results demonstrate that MYO1C is important for rhodopsin localization to the photoreceptor OS, and for normal visual function.

Keywords: motor protein; myosin 1C; photoreceptor; rhodopsin; retina; outer segments; visual function

1. Introduction

Protein trafficking and proper localization within the photoreceptors must occur efficiently and at high fidelity for photoreception, photoreceptor structural maintenance, and overall retinal cell homeostasis. Additionally, it is well-known that proper opsin localization is tightly coupled to photoreceptor cell survival and function [1–9]. However, the cellular events that participate in retinal injuries due to improper signalling and protein localization to the photoreceptor outer segments (OS) are not yet fully understood. While many proteins are known to play essential roles in retinal cell development and function, the involvement of motor proteins in eye biology is less understood. Identification of genetic mutations in the *Myo7a* gene, associated with retinal degeneration in Usher syndrome, suggests that unconventional myosins play a critical role in retinal pigmented epithelium (RPE) and photoreceptor cell function [10,11]. Unconventional myosins are motor proteins that are proposed to transport membranous organelles along the actin filaments in an adenosine triphosphate (ATP)-dependent manner, and additional roles are currently being discovered [11–13]. The loss of *Myo7a* primarily affects RPE and OS phagocytosis, leading to retinal cell degeneration [10,11]. However, it is believed that other yet unidentified class I myosins may participate more directly in photoreceptor cell function. Here, we present compelling evidence for another unconventional actin-binding motor protein, MYO1C, which plays an important role in retinal cell structure and function via opsin localization to the photoreceptor OS.

Rhodopsin and cone pigments in photoreceptor OS mediate scotopic and photopic vision, respectively. The visual pigment rhodopsin is a prototypical G-protein-coupled receptor (GPCR), expressed by retinal rods for photon absorption. Light sensitivity is conferred by 11-*cis* retinaldehyde, a chromophore that is covalently linked to the K296 residue of the opsin protein [14–18]. Photon absorption causes a cis-to-trans conformational shift in the retinaldehyde, leading to structural changes in the opsin protein moiety [6,15]. This initiates a GPCR signalling pathway/phototransduction cascade, signalling the presence of light. Each photoreceptor cell contains an OS housing the phototransduction machinery, an inner segment (IS) where proteins are biosynthesized, and a synaptic terminal for signal transmission. One of the fundamental steps in vision is the proper assembly of signal-transducing membranes, including the transport and sorting of protein components. A major cause of neurodegenerative and other inherited retinal disorders is the improper localization of proteins. Mislocalization of the dim-light photoreceptor protein, rhodopsin, is a phenotype observed in many forms of blinding diseases, including retinitis pigmentosa (RP) [3,16]. The proteins that participate in phototransduction (including rhodopsin, transducin, phosphodiesterase (PDE6), or the cyclic nucleotide-gated channels (CNG)) are synthesized in the IS and must be transported through the connecting cilium to the OS. These proteins are either transmembrane or peripherally associated membrane, which are attached to the membrane surface [1–9]. How the transmembrane proteins (e.g., rhodopsin and CNG) and peripherally associated proteins (e.g., transducin and PDE6) traffic through the IS to incorporate eventually in the nascent disc membrane, or the photoreceptor outer membrane, is not fully understood and constitutes an area of intense research, as the mislocalization of these proteins causes retinal cell degeneration and can lead to blindness [1–9].

The myosin-1 family of molecular motors consists of eight different isoforms that participate in a wide range of cell biological processes that require generation or regulation of membrane tension, angiogenesis, formation of cell adhesions, and changes in the actin architecture [19–22]. Additionally, myosin-1 motors affect intracellular trafficking; function as tension-sensitive docks, phagocytosis, or tethers; and power membrane deformation [19–22]. Unconventional myosins are also proposed to be involved in the light-induced translocation of mitochondria in photoreceptors and in human non-syndromic deafness [23–28]. Genetic mutations in myosins that lead to hearing loss have also been associated with retinal degeneration [29–33]. Some of the essential genes involved in either or both of these functions belong to a family of unconventional motor proteins and include MYO3A [29], MYO7A, MYO6, MYO15 [29–31], and MYO5. Recently, it was reported that mutations affected the nucleotide-binding pocket and calcium binding ability of another unconventional myosin, MYO1C, and these were associated with deafness [32,33]. Importantly, MYO1C was identified in proteomic analysis of the retina and vitreous fluid as part of a protein hub involved in oxidative stress [34]. MYO1C is an actin-binding motor protein that is widely expressed in multiple cell types. It participates in a variety of cellular functions, including protein trafficking and translocation [12,35–37]. As MYO1C has low tissue specificity based on mRNA and protein expression, it remains unclear which cell type is most dependent on MYO1C function and is affected by the loss of MYO1C.

In this study, we systematically analysed the function of the unconventional motor protein, MYO1C, in proper protein localization in photoreceptors. We found that a global genetic deletion of *Myo1c* resulted in a retinal phenotype only, which manifested as a progressive mislocalization of opsins to the OS. Using retinal lysate from wild-type (WT) mice in co-immunoprecipitation assays, we showed that MYO1C and rhodopsin directly interact, indicating that opsin is a cargo for MYO1C. Loss of MYO1C promoted a progressive shortening of OS that was concomitant with a reduction in photoreceptor function, suggesting that MYO1C is critical for maintenance of photoreceptor cell structure and for visual function. Our findings have significant clinical implications for degenerative rod and cone diseases, as mutations in MYO1C or its interacting partners are predicted to affect retinal health and visual function by altering opsin localization to the photoreceptor OS, a fundamental step for maintaining visual function in humans.

2. Experimental Procedures

2.1. Materials

All chemicals, unless stated otherwise, were purchased from Sigma-Aldrich (St. Louis, MO, USA) and were of molecular or cell culture grade quality.

2.2. Myo1c-Knockout (Myo1c-KO) Mouse Model

Mice were kept with ad libitum access to food and water at 24 °C in a 12:12 h light–dark cycle. All mice experiments were approved by the Institutional Animal Care and Use Committee (IACUC protocol #00780; G.P.L.) of the Medical University of South Carolina and performed in compliance with ARVO Statement for the use of Animals in Ophthalmic and Vision Research. We previously generated *Myo1c* transgenic mice (*Myo1cfl/fl*) in C57BL/6N-derived embryonic stem cells, flanking exons 5 to 13 of the mouse *Myo1c* gene, which allowed us to specifically delete all *Myo1c* isoforms in a cell-specific manner [28]. Here, a complete *Myo1c*-knockout was generated by crossing *Myo1cfl/fl* mice with an F-actin Cre mouse strain (B6N.FVB-Tmem163Tg (ACTB-cre)2Mrt/CjDswJ) obtained from Jackson Labs. We refer to the *Myo1cfl/fl* × f-actin Cre cross as *Myo1c* knockout (*Myo1c*-KO) mice. Since the role of Myo1c has not been investigated, the F-actin Cre+ mice gave us an opportunity to study MYO1C function in an un-biased fashion in various cell types/tissues. For this study, the *Myo1c*-KO mice were crossed onto a C57BL/6J background to avoid potential problems with the *Rd8* mutation (found in C57BL/6N lines), and all breeding pairs were sequenced and were negative for *Rd8* and *Rd1* mutations [38]. Equal numbers of male and female mice (50:50 ratio) were used per group and timepoint.

2.3. Immunohistochemistry and Fluorescence Imaging

Light-adapted mice were euthanized, and their eyes were immediately enucleated. The eyes were fixed in 4% paraformaldehyde and buffered with 1X PBS for 2 h at 4 °C, using established protocols [39]. After fixation, samples were washed in 1X PBS and embedded in paraffin and processed (MUSC Histology core facility). Sections (10 μm) were cut and transferred onto frost-free slides. Slide edges were lined with a hydrophobic marker (PAP pen), deparaffinized using xylene, and processed through ethanol washes before blocking for 1–2 h at RT. Blocking solution (1% BSA, 5% normal goat serum, 0.2% Triton-X-100, 0.1% Tween-20 in 1X PBS) was applied for 2 h in a humidified chamber. Primary antibodies were diluted in blocking solution as follows: anti-rhodopsin 1D4 (1:500, Abcam, Cambridge, MA, USA), anti-Myo1c M2 (1:100) [40], cone-arrestin (1:250, Millipore-Sigma, St. Louis, MO, USA), conjugated PNA-488 (1:2000, Molecular Probes, Eugene, OR), anti-red/green cone opsin (M-opsin; 1:500; Millipore, St. Louis, MO, USA), anti S-opsin (1:500, Millipore-Sigma, St. Louis, MO), ZO1 (1:2000, Invitrogen, Waltham, MA, USA), Pde6b (1:300, ThermoFisher, Waltham, MA, USA), CNGA1 (1:250, Abcam), rod arrestin (1:250, Invitrogen), Stra6 (1:250, Millipore-Sigma), CRALBP (1:100, Invitrogen), rod transducin (1:250, Santa Cruz, Dallas, TX, USA), and 4′,6-diamidino-2-phenylendole (DAPI; 1:5000, Invitrogen) or Hoechst (1:10,000, Invitrogen), which were used to label nuclei. All secondary antibodies (Alexa 488 or Alexa 594) were used at 1:5000 concentrations (Molecular Probes, Eugene, OR, USA). Optical sections were obtained with a Leica SP8 confocal microscope (Leica, Wetzlar, Germany) and processed with the Leica Viewer software, or using a Keyence BZ-X800 scope. All fluorescently labelled retinal sections on slides were analysed by the BioQuant NOVA Prime Software (R & M Biometrics, Nashville, TN, USA) and fluorescence within individual retinal layers quantified using Image J or Fiji (NIH).

2.4. Measurement of Photoreceptor ONL Thickness and OS Lengths

The lengths of the photoreceptor OS in WT and *Myo1c*-KO animals (from H&E sections of retinas) were imaged (Keyence BZ-X800 microscope) and measured at 12 consecutive points (at 150 μm distances) from the optic nerve (ON). The OS length was measured from the base of the OS to the inner side of the retinal pigment epithelium. The total number of layers of nuclei in the ONL of retinal sections through the optic nerve (ON) was imaged (Keyence BZ-X800 microscope) and measured at 12 locations around the retina, six each in the superior and inferior hemispheres, starting at 150 μm from the ON. Retinal sections (n = 5–7 retinal sections per eye) from n = 8 mice for each genotype and timepoint were analysed. Two-way ANOVA with Bonferroni post-tests compared *Myo1c*-KO to WT mice at each segment measured.

2.5. ERG Analysis

Dark-adapted WT and *Myo1c*-KO mice (50:50 male–female ratio; n = 8 each genotype) at 2 months of age (young mice; early timepoint), and 6 months of age (end timepoint) were anesthetized by intraperitoneal injection of a ketamine/xylene anaesthetic cocktail (100 mg/kg and 20 mg/kg, respectively), and their pupils were dilated with 1% tropicamide and 2.5% phenylephrine HCl. ERGs were performed under dim red-light in the ERG rooms in the morning (8 a.m.–12 noon). Scotopic ERGs were recorded with a computerized system (UTASE-3000; LKC Technologies, Inc., Gaithersburg, MD, USA), as previously described [39,41,42].

2.6. TEM Analysis of Retinas

Eyecups at the indicated timepoints were harvested and fixed overnight at 4 °C in a solution containing 2% paraformaldehyde/2.5% glutaraldehyde (buffered in 0.1 M cacodylate buffer). Samples were rinsed in the buffer (0.1 M cacodylate buffer) and then placed in a post-fixative of 2% OsO_4/0.2 M cacodylate buffer for 1 h at 4 °C, followed by a 0.1 M cacodylate buffer wash. The samples were dehydrated through a graded ethanol series and then embedded in Epon (EMbed 812; EM Sciences, Hatfield, PA, USA). For TEM analysis, each eye (n = 6 individual eyes from n = 6 animals of each genotype) was cut in half before embedding in Epon blocks. Sections were parallel to the dorsoventral meridian and near the optic nerve (ON). The cured blocks were sectioned at 0.5 microns (semi-thin plastic sections) and stained with 1% toluidine blue to orient the blocks to the required specific cell types. The blocks were trimmed to the precise size needed for ultrathin sectioning. The blocks were cut at 70 nm and gathered on one-micron grids. The grids were air-dried, stained with uranyl acetate for 15 min and lead citrate for 5 min, and rinsed between each stain. They were allowed to dry and imaged with a JEOL 1010. Images were acquired with a Hamamatsu camera and software. All samples were processed by the Electron Microscopy Resource Laboratory at the Medical University of South Carolina, as previously described [39].

2.7. Western Blot Analysis and Densitometry

Total proteins from cells or mouse tissues (n = 3 per genotype) were extracted using the M-PER protein lysis buffer (ThermoScientific, Beverly, MA, USA) containing protease inhibitors (Roche, Indianapolis, IN, USA). Approximately 25 μg of total protein was electrophoresed on 4–12% SDS-PAGE gels and transferred to PVDF membranes. Membranes were probed with primary antibodies against anti-*Myo1c* (1:250), CRALBP (1:100, Invitrogen), rod transducin (1:250, Santa Cruz), PKCα (1:500, Novus Biologicals, Centennial, CO, USA), and β-Actin or Gapdh (1:10,000, Sigma) in antibody buffer (0.2% Triton X-100, 2% BSA, 1X PBS) [39,43]. HRP-conjugated secondary antibodies (BioRad, Hercules, CA, USA) were used at 1:10,000 dilution. Protein expression was detected using a LI-COR Odyssey system, and relative intensities of each band were quantified (densitometry) using Image *J* software version 1.49 and normalized to their respective loading controls. Each Western blot analysis was repeated thrice.

2.8. Co-Immunoprecipitation (Co-IP) Assays

Co-immunoprecipitation of endogenously expressed proteins (MYO1C and rhodopsin) was performed using mouse retinal extracts. Six retinas of each genotype (*n* = 3 animals of WT and *Myo1c*-KO) were used for extraction of retinal proteins in 250 μL of RIPA buffer (phosphate-buffered saline (PBS) containing 0.1% sodium dodecyl sulphate (SDS), 1% Nonidet P-40, 0.5% sodium deoxycholate, and 100 mM potassium iodide) with EDTA-free proteinase inhibitor mixture (Roche Molecular Biochemicals). Lysates were cleared by centrifugation at 10,000 rpm for 10 min at 4 °C. The prepared lysates were further incubated with anti-rhodopsin and mouse/rabbit IgG overnight at 4 °C and further with protein G-coupled agarose beads (ROCHE) for 1–2 h. Beads were then collected by centrifugation at 3000 rpm for 5 min at 4 °C, extensively washed in 1X PBS, and resuspended in SDS gel loading buffer. The proteins were separated on a 10% SDS-PAGE, transferred to a nitrocellulose membrane, and analysed by immunoblotting with the corresponding antibodies.

2.9. Overlay Direct Binding Assay

Rhodopsin protein was overexpressed in HEK293 cells using transient transfection (pcDNA3 rod opsin construct, a gift from Robert Lucas (Addgene plasmid #109361, http://n2t.net/addgene:109361, accessed on 25 May 2021; RRID:Addgene_109361) [44], and cell lysate with overexpressed rhodopsin was subjected to SDS-PAGE gel and transferred to PVDF membrane. The membrane was then probed by overlaying it with 5 μg of baculovirus-produced and purified recombinant full-length MYO1C FL [13] protein by incubating at 4 °C for 4 h. Following incubation, the membrane was Western blotted with MYO1C antibody to detect the direct binding of MYO1C to the rhodopsin bands. The location of rhodopsin on the membranes was marked by separately probing these membranes with an anti-rhodopsin (1:500, Millipore Sigma) antibody (Figure 7B).

2.10. ELISA

ELISA was performed as described previously, with minor modifications [44]. In total, 100 ng of mammalian-expressed and purified rhodopsin was coated on individual wells of a 96-well Maxisorp Immunoplate (Nunc, Rochester, NY, USA) and incubated at 4 °C overnight. The wells were blocked with 5% BSA (Sigma) in PBS for 4 h at 37 °C, and then washed with 1X PBS, 0.1% Tween 20 solution (PBS-T). The wells in the plates were incubated with 200 ng of MYO1C protein for 4 h at 37 °C. Following incubation, the wells were washed with PBS-T solution and incubated with MYO1C antibody for 4 h. Post incubation, secondary antibody (HRP-conjugated) against the Fc region of human IgG1 mAbs at a dilution of 1:5000 in PBS containing 5% BSA was added, and the plates were kept for 1 h at room temperature. The plates were then washed three times with PBS-T and twice with PBS and developed by adding 100 μL of substrate (3,3,5,5-tetramethylbenzidine) solution (Pierce, Hägersten, Sweden). Incubation was conducted at room temperature, the reaction was stopped as the colour developed by adding 100 μL of 2 M H_2SO_4, and absorbance at 450 nm was measured on a microplate reader (Biotek, Winuschi, VT, USA).

2.11. Quantitative Real-Time PCR

RNA was isolated from the retinas of WT and *Myo1c*-KO animals using Trizol reagent and processed as described previously [43]. One microgram of total RNA was reverse transcribed using the SuperScript II cDNA Synthesis Kit (Invitrogen, Eugene, OR, USA). Quantitative real-time PCR (qRT-PCR) was carried out using SYBR green 1 chemistry (BioRad, Hercules, CA, USA). Samples for qRT-PCR experiments were assayed in triplicate, using the BioRad CFX96 Q-PCR machine. Each experiment was repeated twice ($n = 6$ reactions for each gene), using newly synthesized cDNA.

2.12. Liver Function Tests Using Alanine Aminotransferase (ALT) Assays

To extract total protein, liver tissues from WT or *Myo1c*-KO mice (pooled livers $n = 4$ mice per genotype) were homogenized in RIPA buffer on ice and then centrifuged at 14,000 rpm at 4 °C for 10 min. Supernatant was collected, and the protein concentration was estimated using the Bio-Rad Protein Assay Dye Reagent (Sigma). A total of 10 μL of liver lysate was transferred to 96-well plate, and ALT was measured using a microplate-based ALT activity assay kit (Pointe Scientific, Cat. A7526, Irvine, CA, USA). Five biological replicates were used in the assay.

2.13. Heart Function Tests Using Echocardiographic (ECHO) Analyses

Echocardiographic (ECHO) analysis was performed on adult wild-type (WT) and *Myo1c*-KO animals ($n = 4$ per genotype) at the MUSC Cardiology Core Facility. For ECHO experiments, mutant and wild-type littermate controls were anesthetized in an induction chamber with 5% isoflurane in 100% oxygen. They were removed and placed on a warming table where anaesthesia was maintained via nose cone delivery of isoflurane (1% in 100% oxygen). They were placed in the supine position, and the thoracic area was shaved. The limbs were taped to the platform to restrict animal movement during echocardiography acquisition. This also provided a connection to ECG leads embedded in the platform. Sonography gel was applied to the chest and echocardiographic measurements of the peristernal long axis and short axis of the heart were acquired to derive the systolic and diastolic parameters of heart function. ECHO measurements were estimated using Vevo 2100 instrumentation.

2.14. Statistical Analysis

Data are expressed as means $\pm$ standard deviation by ANOVA in the Statistica 12 software (StatSoft Inc., Tulsa, OK, USA). Differences between means were assessed by Tukey's honestly significant difference (HSD) test. *P*-values below 0.05 ($p < 0.05$) were considered statistically significant. For Western blot analysis, relative intensities of each band were quantified (densitometry) using the Image *J* software version 1.49 and normalized to the loading control β-actin. The qRT-PCR analysis was normalized to 18S RNA, and the ΔΔCt method was employed to calculate fold changes. Data of qRT-PCR are expressed as mean $\pm$ standard error of mean (SEM). Statistical analysis was carried out using PRISM 8 software-GraphPad.

3. Results

3.1. Construction and Validation of Myo1c Null Mice

We previously generated *Myo1c* floxed mice using the standard knockout strategy [45] (Figure S1A). Systemic deletion of *Myo1c* was achieved by crossing *Myo1c* floxed (*Myo1cfl/fl*) mice with Actin Cre+ (ActCre+; JAX labs) mice to generate *Myo1cfl/fl-ActCre+/−* knockout mice (referred to as *Myo1c*-KO mice in this manuscript). Western blotting of protein lysates from various tissues, including kidney, heart, and liver of *Myo1c*-KO mice showed complete loss of MYO1C, thus confirming the systemic deletion of *Myo1c* (Figure S1B). Additionally, immunofluorescence expression analysis of these tissues further confirmed loss of MYO1C protein in *Myo1c*-KO mice (Figure S2A–C).

3.2. Genetic Deletion of Myo1c Induced Visual Impairment in Mice

Immunofluorescence analysis showed that MYO1C was enriched in the rod photoreceptor outer (OS) and inner segments (IS) (Figure 1A), as well as in the cone photoreceptor OS of wild-type (WT) mice (Figure 1B), but it was absent in the photoreceptors of *Myo1c*-KO animals (Figure 1A,C). Western blot analysis further confirmed that MYO1C protein was absent in the retinas of *Myo1c*-KO mice (Figure 1D). Since mutations or deletion of the motor protein, MYO7A, were associated with retinal degeneration in Usher syndrome and its animal model, we were prompted to investigate the effect of *Myo1c* in retinal function. Using electroretinograms (ERGs) [46,47], we tested photoreceptor cell function of *Myo1c*-KO and WT mice (n = 8 mice per genotype and age-group; 50:50 male–female ratio) under dark-adapted scotopic conditions. In contrast to WT animals, we observed reduced ERGs for *Myo1c*-KO mice at different ages. Two-month-old *Myo1c*-KO mice showed a significant reduction in the *a*-wave amplitudes, but not in *b*-wave amplitudes ($p < 0.0068$ and $p < 0.098$, respectively) (Figure 2A,C). Strikingly, ERG analysis of adult six-month-old *Myo1c*-KO mice showed loss of retinal function, in which a significant reduction in both *a*- and *b*-waves was observed (38–45% lower than WT animals (** $p < 0.005$; Figure 2B,D).

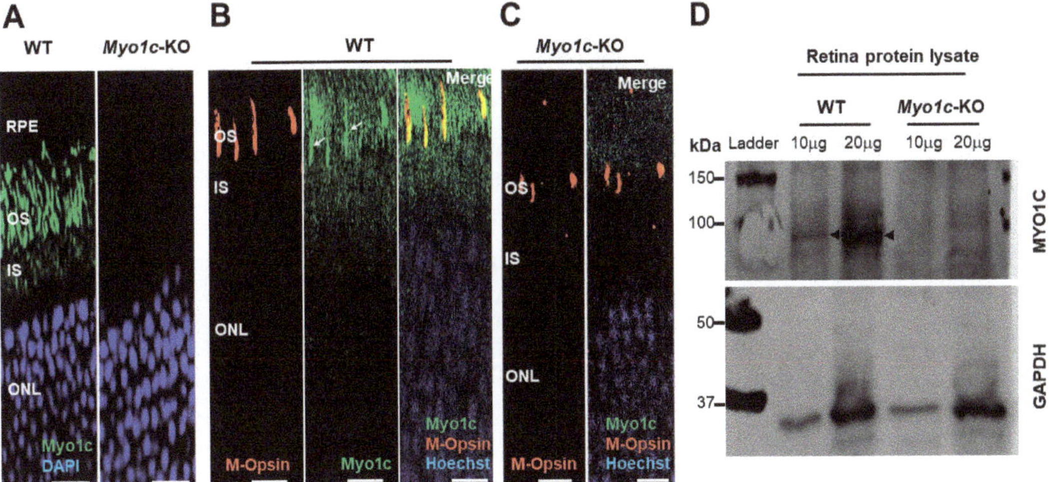

Figure 1. MYO1C localizes to photoreceptors in mouse retina. Eyes from adult wild-type (WT) and *Myo1c*-KO mice (n = 8 mice per genotype; 50:50 male–female ratio) were harvested, and retina sections (n = 5–7 sections per eye) were immunostained with an anti-MYO1C antibody (**A–C**) and M-opsin antibody (**B,C**), followed by secondary (Alexa 488 or Alexa 594) antibody staining. MYO1C (green fluorescence), M-Opsin (red fluorescence), and DAPI or Hoechst (blue fluorescence). Figures in (**A–C**) are representative of retinal sections (n = 5–7 sections per eye) imaged from n = 8 animals per genotype. (**B,C**) Merge (orange) represents co-localization of MYO1C-488 (green) with M-Opsin-594 (red). White arrows in *B* show cones. *RPE*, retinal pigmented epithelium; *OS*, outer segments; *IS*, inner segments; *ONL*, outer nuclear layer. (**A–C**) Scale bar = 50 µm. (**D**) Total proteins isolated from WT (n = 4) and *Myo1c*-KO (n = 4) mouse retinas were pooled sequentially and subjected to SDS-PAGE. Two different concentrations of protein (10 µg and 20 µg) were used. Blots were then probed with anti-Myo1c and Gapdh antibodies. Western blot analysis were repeated thrice. Arrows indicate MYO1C protein band in retinal lysates of WT mice.

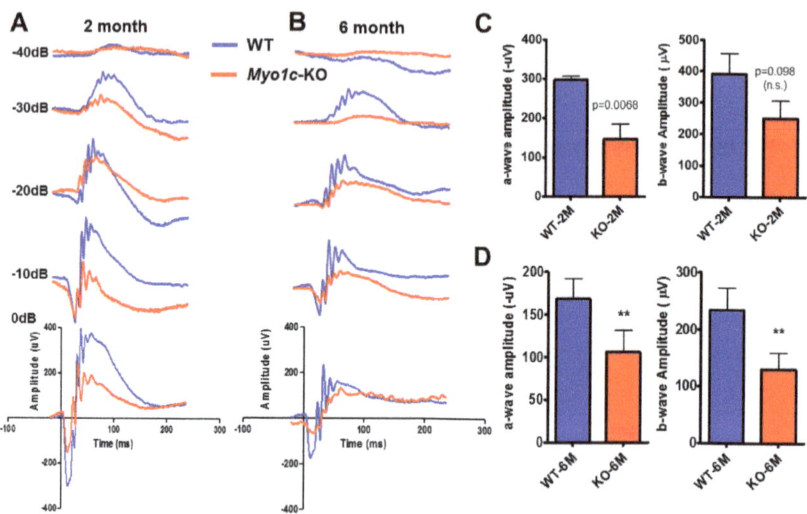

Figure 2. Genetic deletion of *Myo1c* in mice results in a decreased visual function. Dark-adapted scotopic ERGs were recorded in response to increasing light intensities in cohorts of control wild-type (WT) (blue bars, blue-traces) and *Myo1c*-KO (red bars, red-traces) mice, aged two months old (**A,C**), and six months old (**B,D**). Two-month-old *Myo1c*-KO mice had lower dark-adapted *a*- and *b*-wave amplitudes compared with those of controls (post-hoc ANOVA: *a*-waves, $p < 0.0068$; *b*-waves, $p < 0.0098$, n.s. not significant.), in particular at higher light intensities ($-40, -30, -20, -10, 0$ dB). Six-month-old *Myo1c*-knockout mice had lower dark-adapted *a*- and *b*-wave amplitudes compared with those of controls (post-hoc ANOVA: *a*-waves, ** $p < 0.005$; *b*-waves, ** $p < 0.005$), in particular at higher light intensities ($-40, -30, -20, -10, 0$ dB). Photoreceptor cell responses (*a*-waves), which drive the *b*-waves, were equally affected in 6-month-old *Myo1c*-KO animals (both reduced on average between 38 and 45% of WT animals). Data are expressed as mean ± S.E. (*Myo1c*-KO mice and WT mice, *n* = 8 per genotype and age-group; 50:50 male-female ratio).

3.3. Localization of Rod and Cone Visual Pigments in Myo1c-KO Mice

Since the phototransduction protein rhodopsin constitutes 85–90% of photoreceptor OS protein content [48,49], and as the ERG responses were impaired in *Myo1c*-KO mice, we hypothesized that the loss of MYO1C might have affected proper opsin localization to the photoreceptor OS. To test this hypothesis, we analysed retinal sections from WT and *Myo1c*-KO mice (at 2 and 6 months of age; 5–7 retinal sections per eye from *n* = 8 mice per genotype and age-group; 50:50 male–female ratio), probing for rhodopsin, two types of cone opsins, medium wavelength R/G opsin (M-opsin) and short wavelength S-opsin, rod-specific phosphodiesterase 6b (Pde6b), rod-specific CNGA1, rod arrestin (ARR1), rod transducin (G-protein), and the general cone marker PNA lectin. In WT mice at 2 and 6 months of age, rhodopsin localized exclusively to the rod OS (Figure 3A). While the majority of rhodopsin trafficked to the OS in two-month-old *Myo1c*-KO mouse retinas, some mislocalization to the base of the rod IS and the cell bodies in the outer nuclear layer (ONL) was noted (Figure 3A; white arrows; rhodopsin levels within individual retinal layers were quantified and shown in Figure S3A–C). This suggested incomplete opsin localization to photoreceptor OS in the absence of MYO1C. An even more severe mislocalization of rhodopsin to the rod IS and within the ONL was observed in the 6-month-old *Myo1c*-KO mice, suggesting a progressive retinal phenotype in the absence of MYO1C (Figure 3A; rhodopsin expression within individual retinal layers were quantified and shown in Figure S3D,E). Staining for the two cone opsins showed that the cone OS were shorter and mis-shaped by two months, and this abnormality increased by six months of age (Figure 3B,C). Retinas stained for PNA lectin showed progressively shorter and mis-shaped cone OS, indicating that the cone OS structure was compromised in the absence of MYO1C as these mice aged (Figure 3D). Cone visual arrestin in WT mice retina typically outlines the entire cell, OS, IS, cell body, axon, and cone pedicle. Staining for cone arrestin in *Myo1c*-KO animals (2 months of age) confirmed the short and mis-shaped appearance of the cone OS compared to WT retinas at similar ages (Figure 4A, white arrows). In our case, this antibody did not reveal staining to other cone structures, except for the cone OS, so we could not distinguish changes (if any) in cone IS, cell body, axon, or pedicle among WT and Myo1c-KO animals. By contrast, staining for Pde6b, a lipidated rod-specific protein that traffics to the OS independently of rhodopsin [2], showed normal localization to the rod OS in both WT and *Myo1c*-KO retinas at 2 months of age (Figure 4B).

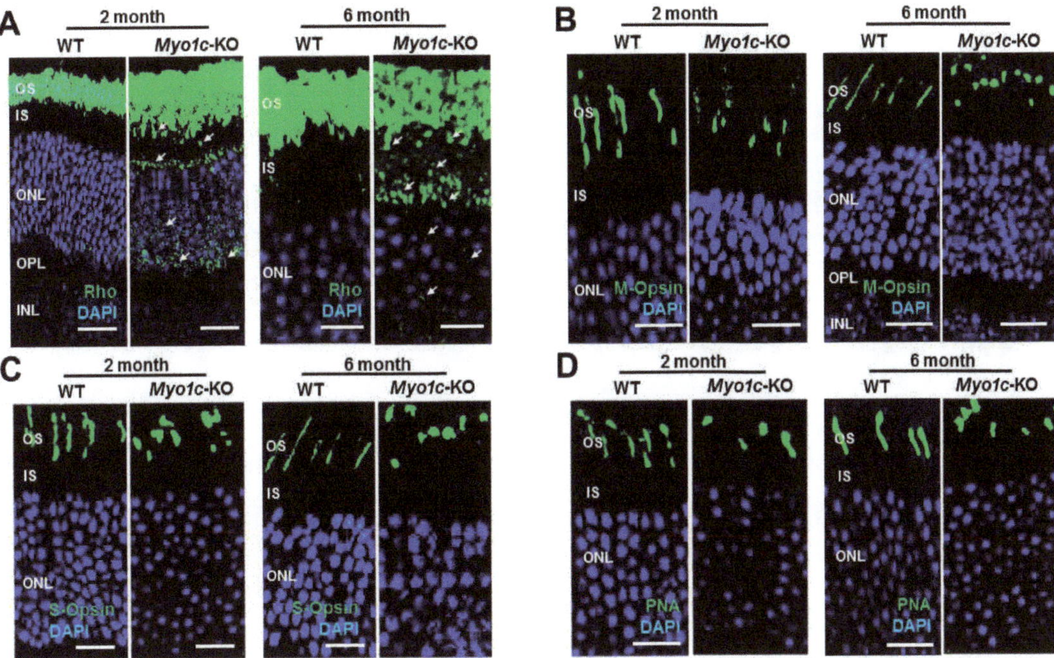

Figure 3. Immunohistochemical analysis of wild-type (WT) and *Myo1c*-knockout mice retinas shows rhodopsin localization defects: (**A**) Levels and localization of rhodopsin (Rho); (**B**) red/green medium wavelength cone opsin (M-opsin); (**C**) short wavelength cone opsin (S-opsin); (**D**) PNA-488 analysed in two- and six-month-old WT and *Myo1c*-KO mice retinas. *Arrows* in panel *A* highlight rhodopsin mislocalization to IS and cell bodies in *Myo1c*-knockout mouse retinas. Images in panels (**A–D**) are representative of immunostained retinal sections (*n* = 5–7 sections per eye) imaged from *n* = 8 animals per genotype and age group (50:50 male–female ratio). Scale bars = 75 μm and 25 μm (**A**, two months old and six months old, respectively); scale bar = 50 μm (**B–D**). *OS*, outer segments; *IS*, inner segments; *ONL*, outer nuclear layer; *OPL*, outer plexiform layer; *INL*, inner nuclear layer.

The CNG channels are also important mediators in the photoreceptor transduction pathways, and they require proper localization to the OS for normal photoreceptor cell function [5,49]. Additionally, the absence of CNGA1 or CNGB1 in mice led to decreased ERG responses and progressive rod and cone photoreceptor cell death [5]. Therefore, to rule out alternate mechanisms for the observed functional phenotypes in *Myo1c*-KO retinas, the retinas of WT and *Myo1c*-KO mice (3–4 months of age; 5–7 retinal sections per eye from $n = 8$ mice per genotype; 50:50 male–female ratio) were stained with the CNGA1 antibody. This analysis showed that even in the absence of MYO1C, both young and adult mice retinas showed no defects in the proper localization of CNGA1 protein to OS (Figure 4C; CNGA1 protein distribution in photoreceptor layer quantified and shown in Figure 4F).

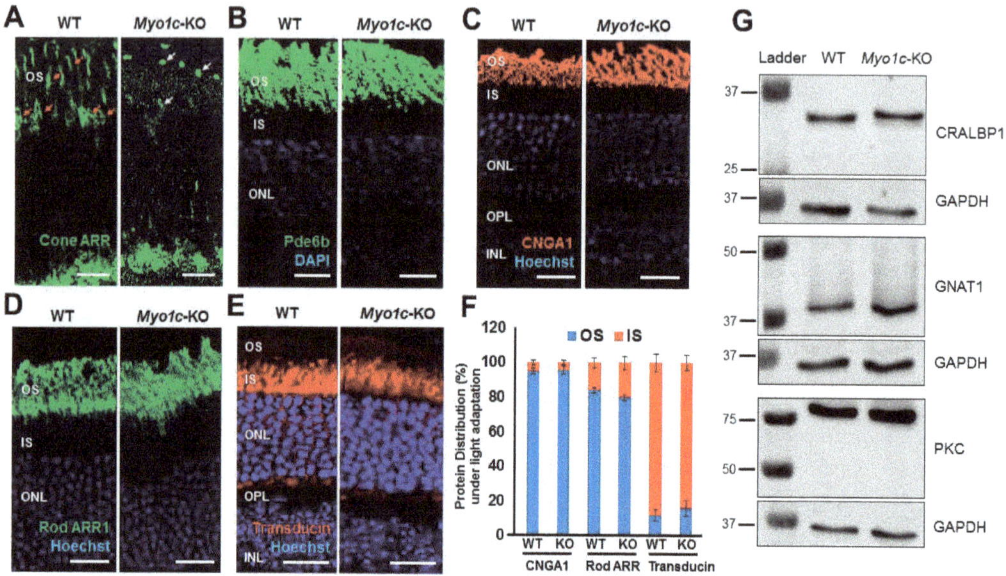

Figure 4. Immunohistochemical analysis of protein localization in photoreceptors of wild-type (WT) and *Myo1c*-knockout mice retinas: Levels and localization of (**A**) cone arrestin (ARR), (**B**) Pde6b; (**C**) CNGA1; (**D**) rod Arrestin (ARR1); and (**E**) *G-protein (transducin)* were analysed in WT and *Myo1c*-KO mice retinas to evaluate proper protein localization to photoreceptor OS. Red Arrows in panel **A** highlight cone photoreceptor nuclei and OS in WT mouse retinas that were significantly reduced or shorter, respectively, in *Myo1c*-KO animals (white arrows in **A**). Images in panels **A**–**E** are representative of immunostained retinal sections ($n = 5$–7 sections per eye) imaged from $n = 8$ animals per genotype and age group (50:50 male–female ratio). Panels (**A**,**B**), mice were 2–3 months of age. Panels **C**–**E**, mice were 3–4 months of age. (**F**) Protein distribution (in %) of CNGA1, rod ARR1, and transducin within the photoreceptor OS and IS in light-adapted mice. For quantification of protein distribution within retinal layers, 5–7 retinal sections from each eye ($n = 8$ animals for each genotype) were analysed using Image J. (**G**) Representative Western blot ($n = 3$ repeats) images of retinal proteins from 3–4-month-old WT and *Myo1c*-KO mice ($n = 2$ animals per genotype) showed no significant differences in protein expression of key retinal genes among genotypes. *OS*, outer segments; *IS*, inner segments; *ONL*, outer nuclear layer; *INL*, inner nuclear layer; *OPL*, outer plexiform layer; *IPL*, inner plexiform layer.

The soluble proteins, arrestin and transducin, exhibit light-dependent localization, where in response to light, arrestin migrates to rod OS and transducin translocates to rod IS [50]. To test whether the loss of MYO1C affected rod arrestin (ARR1) and rod G-protein (transducin) localization, we performed IHC staining for these proteins in retinas of light-adapted WT and *Myo1c*-KO mice (3–4 months of age; 5–7 retinal sections per eye from $n = 8$ mice per genotype; 50:50 male-female ratio). These analyses showed that in the presence of light, genetic loss of MYO1C had no negative effect on the localization of rod arrestin to the OS and G-protein to the IS and cell bodies in retinas of *Myo1c*-KO mice

(Figure 4D,E; rod ARR1 and transducin protein distribution in photoreceptor layer quantified and shown in Figure 4F). Using total protein lysates from retinas of WT and *Myo1c*-KO mice (3–4 months of age; four pooled retinas from $n = 2$ mice per genotype), we analysed protein expression of key retinal proteins in specific retinal cells: CRABLP1 (expressed in Müller cells), GNAT1 (expressed in photoreceptors), and PKCα (expressed in retinal bipolar cells). These analyses showed no significant differences in the expression of these genes in the inner or outer retinal layers of *Myo1c*-KO mice when compared to those of WT mice at 3–4 months of age (Figure 4G). Although MYO1C could not be detected by immunohistochemical analysis in mouse RPE, functional MYO1C and *Myo1C* mRNA were reported in human RPE cells [43] and mouse RPE [51], respectively. Since the elimination of the motor protein, *Myo7a*, in mice leads to alterations in protein localization in the RPE (RPE65) [52], we stained retinas of young and adult WT and *Myo1c*-KO mice (5–7 retinal sections per eye from $n = 8$ mice per genotype) with an anti-STRA6 antibody, another RPE-specific protein. This analysis showed that STRA6 expression and localization in the RPE was not affected in the absence of MYO1C (Figure S4). Since the motor protein MYO1C is proposed to have various functions, such as in protein trafficking, organization of F-actin, mitotic spindle regulation, and gene transcription [22,40], based on our observations above, we further investigated one of its roles in photoreceptor homeostasis. Our hypothesis was that its absence in photoreceptors of *Myo1c*-KO animals might contribute specifically to defective rhodopsin localization to the photoreceptor OS, which might result in retinal phenotypes.

3.4. Native Cre+ Mice Showed No Retinal Phenotypes

To rule out any Cre+-mediated effects on retinal phenotypes observed in the *Myo1c*-KO; Cre+ animals, the eyes from native Cre+ mice (3–4 months old; $n = 3$ animals) were harvested and subjected to similar histological and immunofluorescence analysis. As compared to age-matched WT mice retinas ($n = 3$ animals), the retinas of Cre+ mice showed no retinal pathology or mislocalization of opsins (Figure S5A vs. Figure S5B). These analyses support the view that genetic loss of MYO1C affects key components of phototransduction specifically, and this is further manifested in defects in visual function.

3.5. Myo1c-KO Mice Demonstrated Photoreceptor OS Loss

To further evaluate if rhodopsin mislocalization was associated with structural changes to the retina, histological and transmission electron microscopy (TEM) analyses of retinal sections of young and adult WT and *Myo1c*-KO mice were performed. In histological sections of retinas (5–7 retinal sections per eye from $n = 8$ mice per genotype and age), the progressive shortening of rod photoreceptor OS was observed. The OS of adult *Myo1c*-KO mice at 6 months of age were shorter than the OS of *Myo1c*-KO mice at 2 months of age, which in turn were shorter than those in WT mice at similar ages (Figure 5A,B; OS lengths quantified from H&E sections and represented using spider-plots in Figure 5C,D; ** $p < 0.05$). In comparison to WT mice, the photoreceptors in *Myo1c*-KO mice were less organized, especially in the 6-month-old mice (Figure 5B), suggesting that loss of MYO1C may progressively affect photoreceptor homeostasis. The retina outer nuclear layer (ONL) thickness between genotypes at both ages revealed no significant reduction in nuclear layers in *Myo1c*-KO animals compared to WT mice (ONL thickness quantified from H&E stained sections and represented using spider-plots in Figure 5E,F).

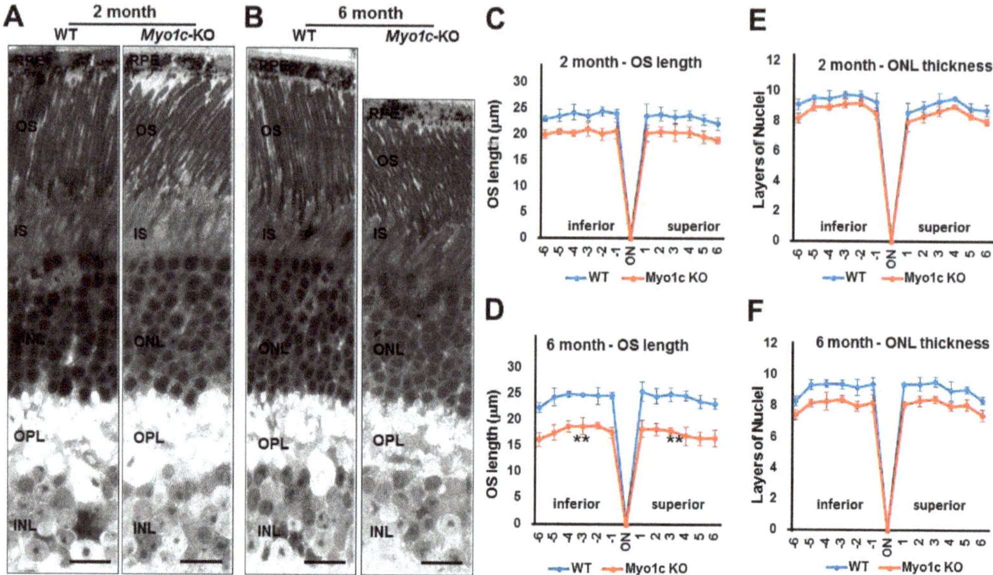

Figure 5. Histological analysis shows reduced photoreceptor OS lengths in *Myo1c*-KO mice retinas: (**A,B**) Retinas from 2- and 6-month-old WT and *Myo1c*-KO mice were sectioned, using an ultra-microtome, and semi-thin plastic sections were obtained to evaluate pathological consequences of MYO1C loss. Quantification of OS lengths from H and E sections (**C**), two- month-old mice; (**D**), six-month-old mice) and ONL thickness (**E**), two-month-old mice; (**F**), six month-old mice, using "spider graph" morphometry. The OS lengths and total number of layers of nuclei in the ONL from H and E sections through the optic nerve (ON; 0 μm distance from optic nerve and starting point) were measured at 12 locations around the retina, six each in the superior and inferior hemispheres, each equally at 150 μm distances. *RPE*, retinal pigmented epithelium; *OS*, outer segments; *IS*, inner segments; *ONL*, outer nuclear layer; *INL*, inner nuclear layer; *OPL*, outer plexiform layer. Retinal sections (*n* = 5–7 sections per eye) from *n* = 8 mice for each genotype and time point (50:50 male-female ratio) were analysed. Two-way ANOVA with Bonferroni post-tests compared *Myo1c*-KO mice with WT in all segments. ** $p < 0.005$, for OS length in only 6-month-old *Myo1c*-KO mice, compared to WT mice; and n.s. (not significant) for ONL thickness in both 2-month and 6-month-old *Myo1c*-KO animals, compared to WT mice). (**A,B**) Scale bar = 100 μm.

3.6. Ultrastructural TEM Analysis Showed Shorter Photoreceptor OS in Myo1c-KO Mice

To evaluate the structure of rod photoreceptors, ultrastructural analysis, using TEM, was performed (*n* = 6 retinal sections per eye from *n* = 8 mice per genotype and age). While the rod photoreceptor OS in the WT mice showed normal elongated morphology, they appeared slightly shorter in *Myo1c*-KO mice at two months of age (* $p < 0.05$; Figure 6A; rod OS lengths quantified in Figure 6E). Specifically, comparing *Myo1c*-KO with WT mouse rod OS lengths at six months of age demonstrated that OS segment lengths in *Myo1c* retinas were significantly (36–45%) shorter than those of WT mice (** $p < 0.005$; Figure 6B; rod OS lengths quantified in Figure 6E). Ultrastructurally, the cone OS in the *Myo1c*-KO mouse retina were shorter and had lost their typical cone shape (Figure 6C vs. Figure 6D; cone OS lengths quantified in Figure 6F), confirming the mis-shaped cone OS phenotype identified by immunohistochemistry (Figure 3B–D). These results suggest that the lack of MYO1C resulted in progressively severe opsin mislocalization (Figure 3A–D) and shorter photoreceptor OS (Figures 5 and 6), thus supporting the observed decrease in visual function by ERG (Figure 2).

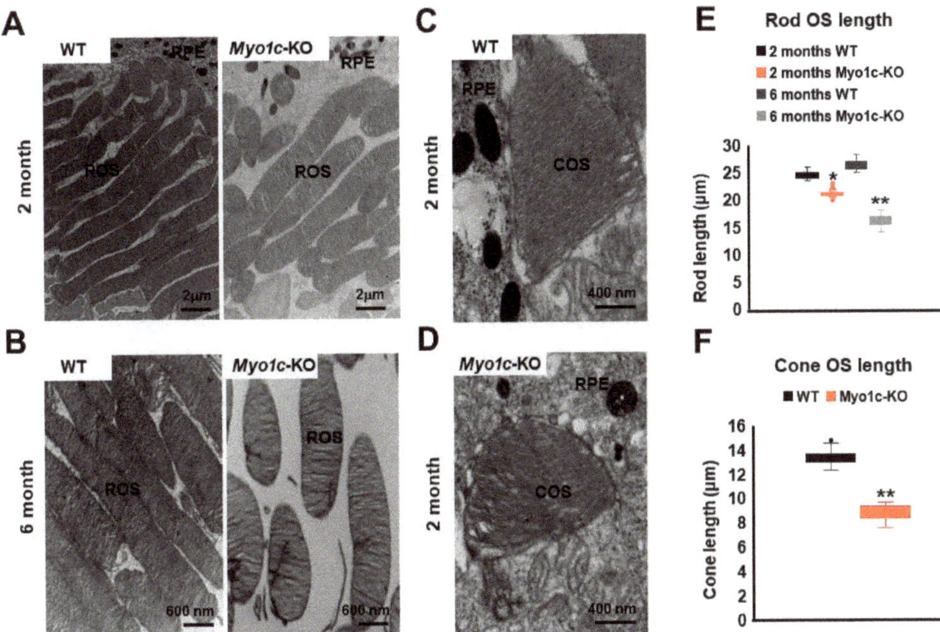

Figure 6. Ultrastructural analysis of rods and cone photoreceptors using transmission electron microscopy (TEM): Representative TEM images of rod photoreceptors from two month (**A**) and six month (**B**) old WT and *Myo1c*-KO mice are presented. Representative images of cone photoreceptors from 2-month-old WT (**C**) and *Myo1c*-KO (**D**) mice. (**A**) Scale bar = 2 μm (**B**) Scale bar = 600 nm (**C,D**) Scale bar = 400 nm. Data are representative of *n* = 6 retinal sections per eye from *n* = 8 mice per genotype and timepoint. (**E**) Rod OS (*ROS*) lengths in WT animals were measured and compared to those of *Myo1c*-KO animals. (**F**) Cone OS (*COS*) lengths in WT animals were measured and compared to those of *Myo1c*-KO animals. * $p < 0.05$; ** $p < 0.005$. RPE, retinal pigmented epithelium.

3.7. MYO1C Directly Interacted with Rhodopsin

Since the loss of MYO1C resulted in retinal function defects with significant alterations in the localization of opsins, we next evaluated whether MYO1C exerted this effect through a physical interaction with rhodopsin. Immunoprecipitation analysis, using WT and *Myo1c*-KO mice retinas (*n* = 6 retinas pooled from *n* = 3 animals per genotype, respectively), demonstrated that MYO1C was pulled down, using a rhodopsin antibody (Figure 7A; Co-IP). Using a baculovirus-produced purified recombinant mouse MYO1C protein in an overlay assay, we demonstrated that MYO1C directly interacted with rhodopsin, where opsin was overexpressed in HEK293 cells (transfected with pCDNA3 rod opsin). The cell lysate with overexpressed rhodopsin and control (cells transfected with empty pCDNA3 vector) was subjected to SDS PAGE and immobilized on nitrocellulose membrane, and probed with or without purified recombinant full-length MYO1C protein (Figure S6A schematic and Figure 7B) [13]. Post-incubation, the interaction of immobilized rhodopsin with MYO1C was probed using a MYO1C antibody. The immunoblot analysis of the over-layered MYO1C showed significant binding of MYO1C protein at the rhodopsin band, indicating a direct interaction between the two proteins (Figure 7B). Additionally, the direct interaction was also confirmed by ELISA, where mammalian-expressed and purified Flag rhodopsin was immobilized on individual wells of ELISA plate. The immobilized rhodopsin was then incubated with purified MYO1C protein, and the bound MYO1C was probed using MYO1C antibody (Figure S6A schematic and Figure S6B). These observations suggest that opsin is a cargo for MYO1C (arrows in Figure 7A,B).

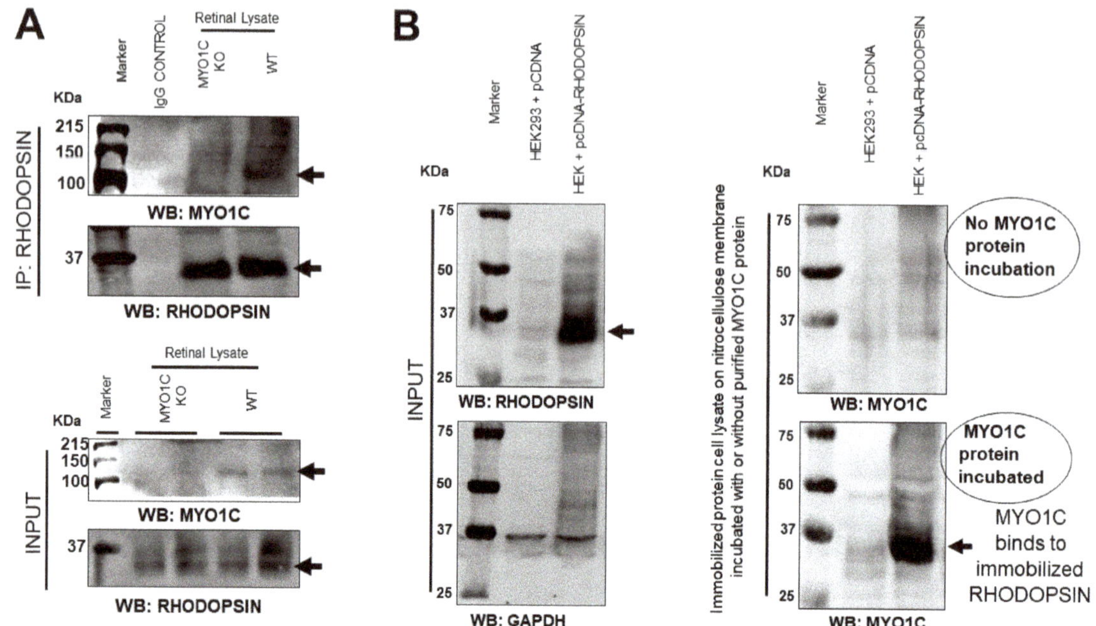

Figure 7. Rhodopsin is a direct cargo for MYO1C: (**A**) Mice retinal protein lysates were isolated from *Myo1c*-KO and wild-type (WT) mice (6 retinas pooled from *n* = 3 mice per genotype) and subjected to co-immunoprecipitation analysis. MYO1C was co-immunoprecipitated with a rhodopsin antibody. (**B**) Lysate from HEK293 cells transfected with pCDNA and pCDNA rhodopsin plasmid was separated using SDS-PAGE and transferred to nitrocellulose membranes. The rhodopsin bound to nitrocellulose membrane was then incubated with 5 ug of purified recombinant active full-length MYO1C generated from a baculovirus expression system. To analyse whether MYO1C binds to immobilized rhodopsin, blots were washed and Western blotted with MYO1C antibody. A positive signal with MYO1C showed direct binding of MYO1C to rhodopsin.

3.8. Genetic Deletion of Myo1c Did Not Affect Systemic Organs in Mice

Finally, to determine if the global deletion of *Myo1c* affected other organs, we harvested major systemic organs, including the liver, heart, and kidney of 2-month-old *Myo1c*-KO and WT mice (*n* = 4 per genotype), and performed histological analyses. Notably, *Myo1c*-KO mice developed and reproduced normally with no observable histological differences between the control and *Myo1c*-KO genotypes (Figure S7A–C). To further confirm that there were no functional defects in these systemic organs, we performed an echocardiogram (heart function), quantified protein/albumin levels in urine (kidney function), and measured alanine aminotransferase (ALT) enzyme levels (liver function) in *Myo1c*-KO mice (*n* = 4 mice per individual functional analysis) and compared these values to their WT littermates (*n* = 4 mice per individual functional analysis). All of these analyses showed no pathological defects in systemic organs of *Myo1c*-KO animals when compared to the age-matched WT littermates (Figures S7A'–C' and S8). Overall, these results indicate that except for the retinal phenotypes, *Myo1c*-KO animals retained normal physiology of the systemic organs examined.

4. Discussion

The proper localization of the G-protein coupled receptor (GPCR) type II opsins to the photoreceptor OS represents a critical event in the initiation of phototransduction for visual function in vertebrates [1–9]. Our work identified for the first time an unconventional motor protein, MYO1C, as a novel regulator of both rod and cone opsin localization to the photoreceptor OS in mice. In this study, based on MYO1C localization within the IS and OS of photoreceptors, and using a whole-body *Myo1c*-KO mouse model, we functionally identified MYO1C as a novel component of retinal physiology, which was specifically found to be involved in photoreceptor cell function. Retinal analysis of *Myo1c*-KO mice identified opsins as novel cargo for MYO1C. In the absence of MYO1C, both young and adult *Myo1c*-KO mice showed impaired opsin localization, where rhodopsin was retained in the photoreceptor IS and the cell bodies. In contrast, cone opsins showed no retention in the cell body or mislocalization to other retinal cell layers, although staining patterns revealed deformed cone OS shapes. These two phenotypes manifested as a progressive decline in visual responses in the rod ERGs and shorter photoreceptor OS lengths as *Myo1c*-KO animals aged, indicating a progressive retinal phenotype. Interestingly, localization of other OS proteins (CNGA1, arrestin, and transducin) was largely unaffected in the absence of MYO1C. The genetic deletion of *Myo1c* only affected retina, and the other systemic organs examined, including the heart, liver, and kidney, remained unaffected. Overall, our data point to a novel mechanism by which MYO1C regulates opsin localization to the photoreceptor OS, a critical event for photoreceptor function and long-term photoreceptor cell homeostasis. Our study identifies an unconventional motor protein, MYO1C, as an essential component of mammalian photoreceptors, where it plays a canonical role in promoting opsin localization and maintaining normal visual function.

4.1. MYO1C and Opsin Localization to Photoreceptor OS

Myo1c-KO mice exhibited rhodopsin mislocalization defects similar to those of $Rpgr^{-/-}$, $Myo7a^{Sh1}$, $Rp1^{-/-}$, $Kinesin\ II^{-/-}$, and $Tulp1^{-/-}$ mutant mice [1–9]. Since MYO1C, primarily localized to photoreceptor IS and OS, is proposed to be involved in protein trafficking (among other functions in different cell types), and uses actin as a track [32,40], we hypothesized that MYO1C likely participates in proper localization of opsins to the OS of photoreceptors. This hypothesis was supported by the observation that the rod opsins were mislocalized to IS and cell bodies. Defective assembly of cone OS in *Myo1c*-KO mice suggests that this phenotype is caused by an aberrant protein localization with OS degeneration as a secondary event. The normal ultrastructure of photoreceptors in our *Myo1c*-KO mice suggests that the retinal abnormalities in these animals were not due to structural defects in photoreceptors per se, but instead were induced by aberrant motor function leading to opsin mislocalization.

4.2. MYO1C Contributed to Phototransduction and Retinal Homeostasis

The opsin molecules and other phototransduction proteins are synthesized in the cell body of the photoreceptor [53,54]. They are then transported to the distal IS [55,56] and subsequently to the OS. Little is known about these transport processes and the molecular components involved in this process [1–9]. The localization of MYO1C in the rod photoreceptors' IS and OS, and in cone OS, suggested that opsins may utilize this molecular motor for transport to the OS. The immunohistochemical analysis of *Myo1c*-KO animals indicated that while rod and cone opsins trafficked to the OS, significant mislocalization was noted for rhodopsin in the IS and cell bodies in the ONL (Figure 2). Since they represent plasma membrane structural proteins, cone opsins presumably contribute to the cone OS stability and rhodopsin to the rod OS formation and stability [7]. Hence, photoreceptor OS shortening/degeneration in *Myo1c*-KO mice may be attributed, in a large part, to the mislocalization of opsins to the IS, or a progressive reduction of opsins in the OS membrane. Notably, the pattern of opsin mislocalization observed in *Myo1c*-KO mice closely resembled the retinal phenotype observed in our previously reported *Tulp1*-KO mice [4,56], $Cnga3^{-/-}$ mice [5], $Lrat^{-/-}$ and $Rpe65^{-/-}$ mice [9,57,58], and GC1-KO mice [1,9]. Importantly, in all of these studies, photoreceptor OS were unstable, and significant degeneration was noted. However, because 85–90% of OS protein is rhodopsin [59–61], the mislocalization of other less abundant proteins cannot be ruled out in the photoreceptors of *Myo1c*-KO mice.

4.3. Contributions from Other Motor Proteins in Proper Opsin Localization

Although this study demonstrates mislocalization of opsins due to a loss of MYO1C, the majority of opsin was still correctly localized, suggesting that contribution or compensation from other myosins cannot be ruled out. Nevertheless, the contributions from MYO1C were highly significant as its genetic deletion showed specific physiological defects in mouse retinas. It is likely that some redundancy exists among molecular motors, and several known candidates might compensate for the lack of MYO1C in photoreceptor function. However, the qPCR analysis of the retinas from WT and *Myo1c*-KO mice did not suggest compensation from other family myosin 1 members (Figure S9). Interestingly, the upregulation of Myo1f in our study was unable to rescue the Myo1c retinal phenotype, suggesting that Myo1f is unable to compensate for the functional loss of Myo1c in the retina (Figure S9). However, compensation by other motor proteins, including the members of kinesin superfamily [62,63], myosin VIIa, and conventional myosin (myosin II) [64,65], which have also been detected in the RPE and retina, cannot be ruled out and need further investigation. Additionally, Myo1C has been shown to be involved in several processes involving actin, such as actin–membrane interaction (by its PIP2 binding domain), in endocytosis, and in autophagosome–lysosome fusion. Therefore, the phenotypes observed upon the loss of Myo1c could also be caused by interfering with any of these processes, either in photoreceptors or in RPE or Mueller glia cells. To further understand the involvement of MYO1C in these retinal cell types, we are currently generating conditional knockout mice, using *Best1*-Cre+ (RPE), *Rho*-Cre+ (rod photoreceptors), and *HGRP*-Cre+ (cone photoreceptors) mice. Our findings have potential clinical implications for degenerative rod and cone diseases, as mutations in MYO1C, or its interacting partners, are predicted to affect retinal health and visual function by altering opsin localization to the photoreceptor OS, a fundamental step for maintaining visual function in humans. Overall, these results support a role for MYO1C in opsin localization in the photoreceptor OS and provide evidence that defective protein transport pathways are a pathologic mechanism, responsible for OS degeneration and decreased visual function in these mice.

Supplementary Materials: The following are available online at https://www.mdpi.com/article/10.3390/cells10061322/s1.

Author Contributions: G.P.L. and D.N. designed the research studies and wrote the manuscript. G.P.L., D.N., M.R.B., S.H., H.-J.K., and A.A.K. edited the manuscript. G.P.L., A.K.S., M.R.B., D.N., E.A., B.R., S.W., S.H., and S.R.M. conducted experiments and acquired data. A.K.S., G.P.L., A.A.K., M.R.B., S.W., S.H., and D.N. analyzed and interpreted the data. M.R.B. and S.H. performed ERG and interpreted the data. R.M., H.-J.K., M.R.B., S.R.M., and S.H. supplied reagents and software or provided equipment for data analysis. All authors have read and agreed to the published version of the manuscript.

Funding: This work was supported by the National Institute of Health (NIH) grants, R21EY025034 and R01EY030889 to G.P.L.; 2R01DK087956-06A1, R56-DK116887-01A1, and 1R03TR003038-01 to D.N.; EY027013-02 to M.R.B.; and R01EY027355 to S.H. This project was also supported in part by a SCTR-NIH/NCATS grant (5UL1TR001450) to G.P.L.

Institutional Review Board Statement: All mice experiments were approved by the Institutional Animal Care and Use Committee (IACUC protocol #00780; G.P.L.) of the Medical University of South Carolina and performed in compliance with ARVO Statement for the use of Animals in Ophthalmic and Vision Research.

Informed Consent Statement: Not applicable.

Data Availability Statement: Not applicable.

Acknowledgments: This work was supported by the National Institute of Health (NIH) grants R21EY025034 and R01EY030889 to G.P.L.; 2R01DK087956-06A1, R56-DK116887-01A1, and 1R03TR003038-01 to D.N.; EY027013-02 to M.R.B.; and R01EY027355 to S.H. This project was also supported in part by a DCI research grant (019898-001) and by a SCTR-NIH/NCATS grant (5UL1TR001450) to G.P.L. The pCDNA3 rod opsin construct was a gift from Robert Lucas (Addgene plasmid #109361; http://n2t.net/addgene:109361, accessed on 25 May 2021; RRID:Addgene_109361). The authors thank George Robertson (Keyence Microscopes) for the use of the Keyence BZ-X800 scope for imaging of semi-thin plastic and H&E sections and immunofluorescence-stained slides. The authors thank Don Rockey (MUSC) and Seok-Hyung Kim (MUSC) for recommending suitable liver function tests and/or for providing the ALT liver function kit. We also thank Rupak D. Mukherjee (MUSC) and Russell A. Norris (MUSC) for performing the ECHO tests. We also thank Linda McLoon (Department of Ophthalmology, University of Minnesota) and Barb Rohrer (Department of Ophthalmology, Medical University of South Carolina) for critical review of the manuscript.

Conflicts of Interest: All authors declare no conflict of interest.

References

1. Baehr, W.; Karan, S.; Maeda, T.; Luo, D.G.; Li, S.; Bronson, J.D.; Watt, C.B.; Yau, K.W.; Frederick, J.M.; Palczewski, K. The function of guanylate cyclase 1 and guanylate cyclase 2 in rod and cone photoreceptors. *J. Biol. Chem.* **2007**, *282*, 8837–8847. [CrossRef]
2. Concepcion, F.; Chen, J. Q344ter mutation causes mislocalization of rhodopsin molecules that are catalytically active: A mouse model of Q344ter-induced retinal degeneration. *PLoS ONE* **2010**, *5*, e10904. [CrossRef]
3. Hollingsworth, T.J.; Gross, A.K. Defective trafficking of rhodopsin and its role in retinal degenerations. *Int. Rev. Cell Mol. Biol.* **2012**, *293*, 1–44.
4. Grossman, G.H.; Watson, R.F.; Pauer, G.J.; Bollinger, K.; Hagstrom, S.A. Immunocytochemical evidence of Tulp1-dependent outer segment protein transport pathways in photoreceptor cells. *Exp. Eye Res.* **2011**, *93*, 658–668. [CrossRef]
5. Hüttl, S.; Michalakis, S.; Seeliger, M.; Luo, D.G.; Acar, N.; Geiger, H.; Hudl, K.; Mader, R.; Haverkamp, S.; Moser, M.; et al. Impaired channel targeting and retinal degeneration in mice lacking the cyclic nucleotide-gated channel subunit CNGB1. *J. Neurosci.* **2005**, *25*, 130–138. [CrossRef]
6. Nishimura, D.Y.; Fath, M.; Mullins, R.F.; Searby, C.; Andrews, M.; Davis, R.; Andorf, J.L.; Mykytyn, K.; Swiderski, R.E.; Yang, B.; et al. Bbs2-null mice have neurosensory deficits, a defect in social dominance, and retinopathy associated with mislocalization of rhodopsin. *Proc. Natl. Acad. Sci. USA* **2004**, *101*, 16588–16593. [CrossRef]
7. Pearring, J.N.; Salinas, R.Y.; Baker, S.A.; Arshavsky, V.Y. Protein sorting, targeting and trafficking in photoreceptor cells. *Prog. Retin. Eye Res.* **2013**, *36*, 24–51. [CrossRef] [PubMed]
8. Mallory, D.P.; Gutierrez, E.; Pinkevitch, M.; Klinginsmith, C.; Comar, W.D.; Roushar, F.J.; Schlebach, J.P.; Smith, A.W.; Jastrzebska, B. The Retinitis Pigmentosa-Linked Mutations in Transmembrane Helix 5 of Rhodopsin Disrupt Cellular Trafficking Regardless of Oligomerization State. *Biochemistry* **2018**, *57*, 5188–5201. [CrossRef]
9. Karan, S.; Zhang, H.; Li, S.; Frederick, J.M.; Baehr, W. A model for transport of membrane-associated phototransduction polypeptides in rod and cone photoreceptor inner segments. *Vis. Res.* **2008**, *48*, 442–452. [CrossRef] [PubMed]
10. Abdelkader, E.; Enani, L.; Schatz, P.; Safieh, L. Severe retinal degeneration at an early age in Usher syndrome type 1B associated with homozygous splice site mutations in MYO7A gene. *Saudi J. Ophthalmol.* **2018**, *32*, 119–125. [CrossRef]
11. Cheng, L.; Yu, H.; Jiang, Y.; He, J.; Pu, S.; Li, X.; Zhang, L. Identification of a novel MYO7A mutation in Usher syndrome type 1. *Oncotarget* **2018**, *9*, 2295–2303. [CrossRef]
12. Arif, E.; Wagner, M.C.; Johnstone, D.B.; Wong, H.N.; George, B.; Pruthi, P.A.; Lazzara, M.J.; Nihalani, D. Motor protein Myo1c is a podocyte protein that facilitates the transport of slit diaphragm protein Neph1 to the podocyte membrane. *Mol. Cell. Biol.* **2011**, *31*, 2134–2150. [CrossRef]
13. Woolner, S.; Bement, W.M. Unconventional myosins acting unconventionally. *Trends Cell Biol.* **2009**, *19*, 245–252. [CrossRef]
14. Bownds, D. Site of attachment of retinal in rhodopsin. *Nature* **1967**, *216*, 1178–1181. [CrossRef] [PubMed]
15. Stenkamp, R.E.; Teller, D.C.; Palczewski, K. Crystal structure of rhodopsin: A G protein-coupled receptor. *Science* **2000**, *289*, 739–745. [CrossRef]
16. Lee, E.S.; Flannery, J.G. Transport of truncated rhodopsin and its effects on rod function and degeneration. *Investig. Ophthalmol. Vis. Sci.* **2007**, *48*, 2868–2876. [CrossRef] [PubMed]
17. Wald, G.; Hubbard, R. The Synthesis of Rhodopsin from Vitamin A(1). *Proc. Natl. Acad. Sci. USA* **1950**, *36*, 92–102. [CrossRef]
18. Wang, J.K.; McDowell, J.H.; Hargrave, P.A. Site of attachment of 11-cis-retinal in bovine rhodopsin. *Biochemistry* **1980**, *19*, 5111–5117. [CrossRef]
19. McConnell, R.E.; Tyska, M.J. Leveraging the membrane—Cytoskeleton interface with myosin-1. *Trends Cell Biol.* **2010**, *20*, 418–426. [CrossRef]
20. Navinés-Ferrer, A.; Martín, M. Long-Tailed Unconventional Class I Myosins in Health and Disease. *Int. J. Mol. Sci.* **2020**, *21*, 2555. [CrossRef]

21. Hirokawa, N.; Takemura, R. Biochemical and molecular characterization of diseases linked to motor proteins. *Trends Biochem. Sci.* **2003**, *28*, 558–565. [CrossRef]
22. Tsang, S.H.; Sharma, T. Retinitis Pigmentosa (Non-syndromic). *Adv. Exp. Med. Biol.* **2018**, *1085*, 125–130.
23. Donato, L.; Scimone, C.; Alibrandi, S.; Pitruzzella, A.; Scalia, F.; D'Angelo, R.; Sidoti, A. Possible A2E Mutagenic Effects on RPE Mitochondrial DNA from Innovative RNA-Seq Bioinformatics Pipeline. *Antioxidants* **2020**, *9*, 1158. [CrossRef]
24. Jiang, M.; Volland, S.; Paniagua, A.E.; Wang, H.; Balaji, A.; Li, D.G.; Lopes, V.S.; Burgess, B.L.; Williams, D.S. Microtubule motor transport in the delivery of melanosomes to the actin-rich, apical domain of in the retinal pigment epithelium. *J. Cell Sci.* **2020**, *133*, 1233–1249. [CrossRef]
25. Scimone, C.; Alibrandi, S.; Scalinci, S.Z.; Trovato Battagliola, E.; D'Angelo, R.; Sidoti, A.; Donato, L. Expression of Pro-Angiogenic Markers Is Enhanced by Blue Light in Human RPE Cells. *Antioxidants* **2020**, *9*, 1154. [CrossRef]
26. Sripathi, S.R.; He, W.; Sylvester, O.; Neksumi, M.; Um, J.Y.; Dluya, T.; Bernstein, P.S.; Jahng, W.J. Altered Cytoskeleton as a Mitochondrial Decay Signature in the Retinal Pigment Epithelium. *Protein J.* **2016**, *35*, 179–192. [CrossRef]
27. Stürmer, K.; Baumann, O. Immunolocalization of a putative unconventional myosin on the surface of motile mitochondria in locust photoreceptors. *Cell Tissue Res.* **1998**, *292*, 219–227. [CrossRef]
28. Gibbs, D.; Kitamoto, J.; Williams, D.S. Abnormal phagocytosis by retinal pigmented epithelium that lacks myosin VIIa, the Usher syndrome 1B protein. *Proc. Natl. Acad. Sci. USA* **2003**, *100*, 6481–6486. [CrossRef]
29. Walsh, T.; Walsh, V.; Vreugde, S.; Hertzano, R.; Shahin, H.; Haika, S.; Lee, M.K.; Kanaan, M.; King, M.C.; Avraham, K.B. From flies' eyes to our ears: Mutations in a human class III myosin cause progressive nonsyndromic hearing loss DFNB30. *Proc. Natl. Acad. Sci. USA* **2002**, *99*, 7518–7523. [CrossRef]
30. Rehman, A.U.; Bird, J.E.; Faridi, R.; Shahzad, M.; Shah, S.; Lee, K.; Khan, S.N.; Imtiaz, A.; Ahmed, Z.M.; Riazuddin, S.; et al. Mutational Spectrum of MYO15A and the Molecular Mechanisms of DFNB3 Human Deafness. *Hum. Mutat.* **2016**, *37*, 991–1003. [CrossRef]
31. Samuels, I.S.; Bell, B.A.; Sturgill-Short, G.; Ebke, L.A.; Rayborn, M.; Shi, L.; Nishina, P.M.; Peachey, N.S. Myosin 6 is required for iris development and normal function of the outer retina. *Investig. Ophthalmol. Vis. Sci.* **2013**, *54*, 7223–7233. [CrossRef]
32. Adamek, N.; Geeves, M.A.; Coluccio, L.M. Myo1c mutations associated with hearing loss cause defects in the interaction with nucleotide and actin. *Cell. Mol. Life Sci.* **2011**, *68*, 139–150. [CrossRef]
33. Lin, T.; Greenberg, M.J.; Moore, J.R.; Ostap, E.M. A hearing loss-associated myo1c mutation (R156W) decreases the myosin duty ratio and force sensitivity. *Biochemistry* **2011**, *50*, 1831–1838. [CrossRef]
34. Skeie, J.M.; Mahajan, V.B. Proteomic interactions in the mouse vitreous-retina complex. *PLoS ONE* **2013**, *8*, e82140. [CrossRef]
35. Boguslavsky, S.; Chiu, T.; Foley, K.P.; Osorio-Fuentealba, C.; Antonescu, C.N.; Bayer, K.U.; Bilan, P.J.; Klip, A. Myo1c binding to submembrane actin mediates insulin-induced tethering of GLUT4 vesicles. *Mol. Biol. Cell* **2012**, *23*, 4065–4078. [CrossRef]
36. Bose, A.; Robida, S.; Furcinitti, P.S.; Chawla, A.; Fogarty, K.; Corvera, S.; Czech, M.P. Unconventional myosin Myo1c promotes membrane fusion in a regulated exocytic pathway. *Mol. Cell. Biol.* **2004**, *24*, 5447–5458. [CrossRef] [PubMed]
37. Fan, Y.; Eswarappa, S.M.; Hitomi, M.; Fox, P.L. Myo1c facilitates G-actin transport to the leading edge of migrating endothelial cells. *J. Cell Biol.* **2012**, *198*, 47–55. [CrossRef]
38. Mattapallil, M.J.; Wawrousek, E.F.; Chan, C.C.; Zhao, H.; Roychoudhury, J.; Ferguson, T.A.; Caspi, R.R. The Rd8 mutation of the Crb1 gene is present in vendor lines of C57BL/6N mice and embryonic stem cells, and confounds ocular induced mutant phenotypes. *Investig. Ophthalmol. Vis. Sci.* **2012**, *53*, 2921–2927. [CrossRef]
39. Lobo, G.P.; Fulmer, D.; Guo, L.; Zuo, X.; Dang, Y.; Kim, S.H.; Su, Y.; George, K.; Obert, E.; Fogelgren, B.; et al. The exocyst is required for photoreceptor ciliogenesis and retinal development. *J. Biol. Chem.* **2017**, *292*, 14814–14826. [CrossRef]
40. Wagner, M.C.; Barylko, B.; Albanesi, J.P. Tissue distribution and subcellular localization of mammalian myosin I. *J. Cell Biol.* **1992**, *119*, 163–170. [CrossRef]
41. Husain, S. Delta Opioids: Neuroprotective Roles in Preclinical Studies. *J. Ocul. Pharmacol. Ther.* **2018**, *34*, 119–128. [CrossRef]
42. Husain, S.; Ahmad, A.; Singh, S.; Peterseim, C.; Abdul, Y.; Nutaitis, M.J. PI3K/Akt Pathway: A Role in delta-Opioid Receptor-Mediated RGC Neuroprotection. *Investig. Ophthalmol. Vis. Sci.* **2017**, *58*, 6489–6499. [CrossRef]
43. Brandstaetter, H.; Kishi-Itakura, C.; Tumbarello, D.A.; Manstein, D.J.; Buss, F. Loss of functional MYO1C/myosin 1c, a motor protein involved in lipid raft trafficking, disrupts autophagosome-lysosome fusion. *Autophagy* **2014**, *10*, 2310–2323. [CrossRef]
44. Solanki, A.K.; Rathore, Y.S.; Badmalia, M.D.; Dhoke, R.R.; Nath, S.K.; Nihalani, D. Global shape and ligand binding efficiency of the HIV-1-neutralizing antibodies differ from those of antibodies that cannot neutralize HIV-1. *J. Biol. Chem.* **2014**, *289*, 34780–34800. [CrossRef] [PubMed]
45. Arif, E.; Solanki, A.K.; Srivastava, P.; Rahman, B.; Tash, B.R.; Holzman, L.B.; Janech, M.G.; Martin, R.; Knölker, H.J.; Fitzgibbon, W.R.; et al. The motor protein Myo1c regulates transforming growth factor-beta-signaling and fibrosis in podocytes. *Kidney Int.* **2019**, *96*, 139–158. [CrossRef] [PubMed]
46. Crosson, C.E.; Mani, S.K.; Husain, S.; Alsarraf, O.; Menick, D.R. Inhibition of histone deacetylase protects the retina from ischemic injury. *Investig. Ophthalmol. Vis. Sci.* **2010**, *51*, 3639–3645. [CrossRef]
47. Husain, S.; Potter, D.E.; Crosson, C.E. Opioid receptor-activation: Retina protected from ischemic injury. *Investig. Ophthalmol. Vis. Sci.* **2009**, *50*, 3853–3859. [CrossRef]
48. Krebs, W.; Kühn, H. Structure of isolated bovine rod outer segment membranes. *Exp. Eye Res.* **1977**, *25*, 511–526. [CrossRef]

49. Bales, K.L.; Gross, A.K. Aberrant protein trafficking in retinal degenerations: The initial phase of retinal remodeling. *Exp. Eye Res.* **2016**, *150*, 71–80. [CrossRef]

50. Calvert, P.D.; Strissel, K.J.; Schiesser, W.E.; Pugh, E.N., Jr.; Arshavsky, V.Y. Light-driven translocation of signaling proteins in vertebrate photoreceptors. *Trends Cell Biol.* **2006**, *16*, 560–568. [CrossRef]

51. Mustafi, D.; Kevany, B.M.; Genoud, C.; Bai, X.; Palczewski, K. Photoreceptor phagocytosis is mediated by phosphoinositide signaling. *FASEB J.* **2013**, *27*, 4585–4595. [CrossRef]

52. Lopes, V.S.; Gibbs, D.; Libby, R.T.; Aleman, T.S.; Welch, D.L.; Lillo, C.; Jacobson, S.G.; Radu, R.A.; Steel, K.P.; Williams, D.S. The Usher 1B protein, MYO7A, is required for normal localization and function of the visual retinoid cycle enzyme, RPE65. *Hum. Mol. Genet.* **2011**, *20*, 2560–2570. [CrossRef]

53. Young, R.W. The renewal of photoreceptor cell outer segments. *J. Cell Biol.* **1967**, *33*, 61–72. [CrossRef] [PubMed]

54. Young, R.W.; Droz, B. The renewal of protein in retinal rods and cones. *J. Cell Biol.* **1968**, *39*, 169–184. [CrossRef] [PubMed]

55. Findlay, J.B.; Pappin, D.J. The opsin family of proteins. *Biochem. J.* **1986**, *238*, 625–642. [CrossRef] [PubMed]

56. Lobo, G.P.; Au, A.; Kiser, P.D.; Hagstrom, S.A. Involvement of Endoplasmic Reticulum Stress in TULP1 Induced Retinal Degeneration. *PLoS ONE* **2016**, *11*, e0151806. [CrossRef]

57. Fan, J.; Rohrer, B.; Frederick, J.M.; Baehr, W.; Crouch, R.K. Rpe65−/− and Lrat−/− mice: Comparable models of leber congenital amaurosis. *Investig. Ophthalmol. Vis. Sci.* **2008**, *49*, 2384–2389. [CrossRef]

58. den Hollander, A.I.; Roepman, R.; Koenekoop, R.K.; Cremers, F.P. Leber congenital amaurosis: Genes, proteins and disease mechanisms. *Prog. Retin. Eye Res.* **2008**, *27*, 391–419. [CrossRef]

59. Shi, Y.; Obert, E.; Rahman, B.; Rohrer, B.; Lobo, G.P. The Retinol Binding Protein Receptor 2 (Rbpr2) is required for Photoreceptor Outer Segment Morphogenesis and Visual Function in Zebrafish. *Sci. Rep.* **2017**, *7*, 16207. [CrossRef] [PubMed]

60. Solanki, A.; Kondkar, A.A.; Fogerty, J.; Su, Y.; Kim, S.H.; Lipschutz, J.H.; Nihalani, D.; Perkins, B.D.; Lobo, G.P. A Functional Binding Domain in the Rbpr2 Receptor Is Required for Vitamin A Transport, Ocular Retinoid Homeostasis, and Photoreceptor Cell Survival in Zebrafish. *Cells* **2020**, *9*, 1099. [CrossRef] [PubMed]

61. Lobo, G.P.; Pauer, G.; Lipschutz, J.H.; Hagstrom, S.A. The Retinol-Binding Protein Receptor 2 (Rbpr2) Is Required for Photoreceptor Survival and Visual Function in the Zebrafish. *Adv. Exp. Med. Biol.* **2018**, *1074*, 569–576. [PubMed]

62. Beech, P.L.; Pagh-Roehl, K.; Noda, Y.; Hirokawa, N.; Burnside, B.; Rosenbaum, J.L. Localization of kinesin superfamily proteins to the connecting cilium of fish photoreceptors. *J. Cell Sci.* **1996**, *109*, 889–897. [CrossRef] [PubMed]

63. Muresan, V.; Bendala-Tufanisco, E.; Hollander, B.A.; Besharse, J.C. Evidence for kinesin-related proteins associated with the axoneme of retinal photoreceptors. *Exp. Eye Res.* **1997**, *64*, 895–903. [CrossRef] [PubMed]

64. Chaitin, M.H.; Coelho, N. Immunogold localization of myosin in the photoreceptor cilium. *Investig. Ophthalmol. Vis. Sci.* **1992**, *33*, 3103–3108. [PubMed]

65. Williams, D.S.; Hallett, M.A.; Arikawa, K. Association of myosin with the connecting cilium of rod photoreceptors. *J. Cell Sci.* **1992**, *103*, 183–190. [CrossRef] [PubMed]

Article

High-Fat Diet Alters the Retinal Transcriptome in the Absence of Gut Microbiota

David Dao [1], Bingqing Xie [2,3], Urooba Nadeem [4], Jason Xiao [1], Asad Movahedan [5], Mark D'Souza [2], Vanessa Leone [6,7], Seenu M. Hariprasad [1], Eugene B. Chang [7], Dinanath Sulakhe [3] and Dimitra Skondra [1,*]

[1] Department of Ophthalmology and Visual Science, University of Chicago, Chicago, IL 60637, USA; David.Dao@uchospitals.edu (D.D.); jason.xiao@uchospitals.edu (J.X.); sharipra@bsd.uchicago.edu (S.M.H.)
[2] Center for Research Informatics, University of Chicago, Chicago, IL 60637, USA; bxie@medicine.bsd.uchicago.edu (B.X.); dsouza@bsd.uchicago.edu (M.D.)
[3] Department of Medicine, University of Chicago, Chicago, IL 60637, USA; sulakhe@uchicago.edu
[4] Department of Pathology, University of Chicago, Chicago, IL 60637, USA; Urooba.Nadeem@uchospitals.edu
[5] Department of Ophthalmology and Visual Science, Yale University School of Medicine, New Haven, CT 06437, USA; asadolah.movahedan@yale.edu
[6] Department of Animal Biologics and Metabolism, University of Wisconsin, Madison, WI 53706, USA; valeone@wisc.edu
[7] Knapp Center for Biomedical Discovery, Department of Medicine, Microbiome Medicine Program, University of Chicago, Chicago, IL 60637, USA; echang@medicine.bsd.uchicago.edu
* Correspondence: dskondra@bsd.uchicago.edu

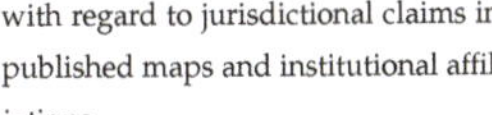

Citation: Dao, D.; Xie, B.; Nadeem, U.; Xiao, J.; Movahedan, A.; D'Souza, M.; Leone, V.; Hariprasad, S.M.; Chang, E.B.; Sulakhe, D.; et al. High-Fat Diet Alters the Retinal Transcriptome in the Absence of Gut Microbiota. *Cells* **2021**, *10*, 2119. https://doi.org/10.3390/cells10082119

Academic Editors: Maurice Ptito and Joseph Bouskila

Received: 12 July 2021
Accepted: 11 August 2021
Published: 18 August 2021

Publisher's Note: MDPI stays neutral with regard to jurisdictional claims in published maps and institutional affiliations.

Abstract: The relationship between retinal disease, diet, and the gut microbiome has shown increasing importance over recent years. In particular, high-fat diets (HFDs) are associated with development and progression of several retinal diseases, including age-related macular degeneration (AMD) and diabetic retinopathy. However, the complex, overlapping interactions between diet, gut microbiome, and retinal homeostasis are poorly understood. Using high-throughput RNA-sequencing (RNA-seq) of whole retinas, we compare the retinal transcriptome from germ-free (GF) mice on a regular diet (ND) and HFD to investigate transcriptomic changes without influence of gut microbiome. After correction of raw data, 53 differentially expressed genes (DEGs) were identified, of which 19 were upregulated and 34 were downregulated in GF-HFD mice. Key genes involved in retinal inflammation, angiogenesis, and RPE function were identified. Enrichment analysis revealed that the top 3 biological processes affected were regulation of blood vessel diameter, inflammatory response, and negative regulation of endopeptidase. Molecular functions altered include endopeptidase inhibitor activity, protease binding, and cysteine-type endopeptidase inhibitor activity. Human and mouse pathway analysis revealed that the complement and coagulation cascades are significantly affected by HFD. This study demonstrates novel data that diet can directly modulate the retinal transcriptome independently of the gut microbiome.

Keywords: age-related macular degeneration; high-fat diet; gut microbiome; gut-retina axis; RNA sequencing; germ-free mice; complement cascade; angiogenesis; retinal inflammation

1. Introduction

Over the last several decades, there is increasing evidence that diet and nutrient intake contribute to the pathophysiology of retinal diseases, including age-related macular degeneration (AMD), diabetic retinopathy (DR), and glaucoma [1–4]. The retina is one of the most metabolically active tissues in the body, and with its rich store of polyunsaturated fats, is vulnerable to oxidative, metabolic, and fatty acid perturbances [5,6]. In particular, multiple research groups have linked high-fat diets (HFDs) and fat-specific intake with increased prevalence of intermediate or advanced AMD, the leading cause of blindness in the developed world [7–10]. HFDs have been shown to replicate or exacerbate features of retinal disease through various proposed mechanisms: lipid signaling, metabolic dysfunction, vascularization, and inflammatory regulation [11]. Compared to mice fed on conventional or low-fat diets, HFD-fed mice exhibit impaired retinal sensitivity, greater macrophage/microglial cell activation, altered retinal fatty acid composition, and hallmark features of AMD such as choroidal neovascularization (CNV) and sub-retinal pigment epithelium (RPE) deposits [12–15]. HFDs also can lead to systemic changes such as hypercholesteremia and hyperinsulinemia, paving the way for retinal disease risk factors like obesity and diabetes [16]. HFD-induced vascular changes in the eye have also been reported, including changes in permeability and formation of acellular capillaries, though these effects are not consistent across studies [16,17]. Use of electroretinograms (ERGs) support the notion that HFDs can negatively affect function of several retinal cell types, including photoreceptors, bipolar cells, and retinal ganglion cells [16]. Altogether, a small but growing body of literature suggests that HFDs alter the retina and its microenvironment.

Recent studies suggest that the effects of HFDs in retinal diseases may be in part mediated through the gut microbiome [18]. The human gut microbiome is comprised of trillions of microbiota that live within our gastrointestinal tracts. These microbes can maintain and alter health homeostasis, playing diverse roles in immune regulation, metabolism, drug processing, and intercellular signaling [19]. This rapidly growing body of literature has linked the gut microbiome with anatomically distant sites, including the heart, liver, brain, and lungs [20–23]. Research regarding the gut microbiome's role in ocular tissues, particularly in the retina, has only recently begun [24,25]. These pioneering studies have revealed functional and compositional differences in the gut microbiome in patients with retinal diseases such as primary open-angle glaucoma (POAG), neovascular AMD, retinal artery occlusion (RAO), and retinopathy of prematurity (ROP) [26–29]. Our team has previously shown that modulating the gut microbiome impacts the retinal transcriptome across many biological pathways implicated in retinal disease [30]. Other investigators have found that gut microbiota can alter retinal lipid composition, as well as systemic factors such as endotoxemia and immune response that may set the stage for retinal pathophysiology [31,32]. Due to their location in the gastrointestinal system and intimate communication with human cell types, gut microbiota may be responsible for mediating many of the observed effects of HFD on the retina. For example, Andriessen et al. recently showed that a HFD modifies gut microbiota composition, which consequently exacerbates laser-induced CNV [18]. Other studies have also shown that HFD-fed mice exhibit altered gut microbiomes, such as reduced diversity and a decreased Bacteroidetes-to-Firmicutes ratio [33–35].

Taken together, there seems to be a connection between HFD, the gut microbiome, and retinal homeostasis. Nevertheless, whether and how HFD affects the retina in the absence of gut microbiome remain unknown. In this study, we sought to distinguish these overlapping components, and investigate whether HFD directly alters the retinal transcriptome in the absence of gut microbiome by using germ-free mice and high-throughput RNA-sequencing. Given the complexity of genetic and environmental factors that contribute to retinal disease progression, understanding how diet impacts retinal biology at the transcriptional level could help identify underlying mechanisms of how diet can affect retinal diseases, which ultimately could lead to discovery of novel targets for preventative or therapeutic interventions to treat ocular diseases.

2. Materials and Methods

2.1. Animals and Diets

Mouse experiments were approved by the University of Chicago Institutional Animal Care and Use Committee and conducted according to ophthalmic and vision research guidelines set by the Association for Research in Vision and Ophthalmology (ARVO). This study used germ-free (GF) C57B1/6 adult male mice, which were housed in the Gnotobiotic Research Animal Facility at the University of Chicago. At 7 weeks of age, GF mice were fed ad libitum either a normal diet (ND) or high-fat diet (HFD) for 8 consecutive weeks.

The HFD consisted of 23% saturated fat. Environmental conditions, including humidity and temperature, adhered to The Guide for the Care and Use of Laboratory Animals, 8th edition, and mice were subjected to a standard 12-h light cycle. At 15 weeks of age, the mice were euthanized by carbon dioxide and subsequent cervical dislocation. Samples were immediately placed on ice and processed for RNA-sequencing.

2.2. Sterility Monitoring

In order to provide a sterile environment, GF mice were housed in positive-pressure incubators at the University of Chicago's Gnotobiotic Research Animal Facility and fed diets which had been irradiated and autoclaved at 250 °F for 30 min. Germ-free status was assessed as described previously [36]. Briefly, fecal samples were collected weekly, and cultured aerobically at 37 °C and 42 °C and anaerobically at 37 °C. Cultures were checked after 1, 2, 3, and 5 days had passed—no positive cultures were identified during the study. Additionally, fecal samples were screened for contamination by DNA extraction and quantitative real-time polymerase chain reaction (PCR) using universal bacterial primers for the 16 S RNA-encoding gene (IDT, 8 F was 5′-AGA GTT TGA TCC TGG CTC AG-3′, and 338 R was 5′-TGC TGC CTC CCG TAG GAG T-3′).

2.3. RNA Extraction

Whole mouse retinas were isolated on ice from freshly enucleated eyes, with all equipment, surfaces, and tubes treated with RNase decontamination solution (Thermo Fisher Scientific, Waltham, MA, USA) prior to use. Dissected retinas were stored in RNAlater solution (Thermo Fisher Scientific, Waltham, MA, USA) at −80 °C until RNA extraction using the RNeasy kit from Qiagen (Qiagen, Hilden, Germany). Concentrations were quantified using a Nanodrop (Thermo Fisher Scientific, Waltham, MA, USA) before sequencing.

2.4. RNA Sequencing

RNA from eight samples was used for analysis (four per diet group). The quality was evaluated using a Bioanalyzer at the University of Chicago Genomics Core and was confirmed to meet appropriate RNA integrity numbers (RIN). Next, cDNA libraries were constructed using TruSeq RNA Sample Prep kits (Illumina, San Diego, CA, USA) to generate 100-bp paired-end reads, which were indexed for multiplexing and then sequenced using PE100bp on the NovaSeq 6000 System (Illumina, San Diego, CA, USA). Data was provided in FASTQ format and analyzed in R.

2.5. Statistical Analysis

The secondary analysis of sequence data was performed on Globus Genomics, an enhanced, cloud-based analytical platform that provides access to different versions of Next-Generation Sequence analysis tools and workflow capabilities. Tools such as STAR, featureCounts, and Limma were run from within the Globus Genomics platform. We used STAR (version 2.4.2 a, Stanford University, Stanford, CA, USA) aligner default parameters to align the RNA-seq reads to the reference mouse genome (GRCm38) for all eight samples. The raw gene expression count matrix was then generated by featureCounts (version subread-1.4.6-p1). The gene annotation was obtained from the Gencode vM23. STAR default parameter for the maximum mismatches is 10 which is optimized based on mammalian genomes and recent RNA-seq data.

Significant DEGs with a *p*-value < 0.01 and LogFC > 1 were extracted for further downstream analysis. Filtering for DEGs with low expression (count-per-million < 10) was performed using edgeR [37,38]. The enrichment analysis in EnrichR suite took both the upregulated and downregulated DEGs in GF and extracted the over-represented gene ontology functional classification (molecular functions, biological processes, and cellular component). The significance of the association between the datasets and bio functions were measured using a ratio of the number of genes from the dataset that map to the pathway divided by the total number of genes in that pathway. This enrichment analysis was based on mouse-to-human orthologs. A list of all DEGs and their *p*-values is available in Tables 1 and 2.

Table 1. Differentially expressed genes upregulated by high-fat diet (HFD).

Gene	LogFC	*p*-Value	Protein
Fat2	4.084	3.13×10^{-5}	FAT atypical cadherin 2
Npy4r	3.518	1.10×10^{-4}	Neuropeptide Y receptor Y4-2
C1qtnf2	3.248	9.57×10^{-4}	C1q And TNF-related 2
Deup1	3.133	1.81×10^{-4}	Deuterostome assembly protein 1
Ifi204	2.881	5.62×10^{-3}	Interferon gamma inducible protein
Siglec1	2.845	3.57×10^{-4}	Sialic acid-binding Ig-like lectin 1
Mrap	2.724	1.02×10^{-3}	Melanocortin 2 receptor accessory protein
Dmgdh	2.672	2.89×10^{-3}	Dimethylglycine dehydrogenase
Maats1	2.433	8.70×10^{-3}	Cilia and flagella-associated protein 91
Nppb	2.345	4.83×10^{-3}	Natriuretic-peptide B
Klf1	1.889	2.77×10^{-4}	Kruppel-like factor 1
Hba-a2	1.867	7.19×10^{-3}	Hemoglobin subunit alpha 2
Ppp1r3g	1.735	8.87×10^{-3}	Protein phosphatase 1 regulatory subunit 3G
Hbb-bs	1.378	4.73×10^{-3}	Hemoglobin subunit beta
Hba-a1	1.372	1.82×10^{-3}	Hemoglobin subunit alpha 1
Ms4a6b	1.336	4.78×10^{-3}	Membrane spanning 4-domains A6A
Gdf2	1.133	9.95×10^{-3}	Growth differentiation factor 2
Tspan11	1.124	7.60×10^{-3}	Tetraspanin 11
Nlrp10	1.103	2.63×10^{-3}	NLR family pyrin domain containing 10

Table 2. Differentially expressed genes downregulated by HFD.

Gene	LogFC	*p*-Value	Protein
Olfr690	−3.712	8.96×10^{-6}	Olfactory receptor family
Gtsf1	−3.503	1.05×10^{-4}	Gametocyte specific factor 1
Tcp10a	−3.121	1.77×10^{-3}	T-Complex 10-like 3, pseudogene
Olfr460	−3.077	7.60×10^{-4}	Olfactory receptor family
Cuzd1	−2.902	4.51×10^{-3}	CUB and zona pellucida-like domains 1
Serpinc1	−2.721	6.30×10^{-4}	Serpin family C member 1
Olfr691	−2.713	1.26×10^{-3}	Olfactory receptor family
Opn5	−2.708	8.80×10^{-3}	Opsin 5
Hist1h3i	−2.561	5.37×10^{-3}	H3 clustered histone 11
Rmi2	−2.546	6.18×10^{-3}	RecQ mediated genome instability 2
Rnf222	−2.370	6.91×10^{-3}	Ring finger protein 222
Serpinf2	−2.309	1.17×10^{-3}	Serpin family F member 2
Cstl1	−2.302	6.08×10^{-3}	Cystatin-like 1
Npb	−2.000	6.35×10^{-3}	Neuropeptide B
Anxa9	−1.746	3.87×10^{-4}	Annexin A9
Nanos2	−1.670	2.85×10^{-3}	Nanos C2HC-type zinc finger 2
Tssk4	−1.603	8.17×10^{-3}	Testis specific serine kinase 4
Eqtn	−1.510	3.93×10^{-3}	Equatorin
Klrg2	−1.436	6.80×10^{-5}	Killer-cell lectin-like receptor G2
Mcemp1	−1.419	3.67×10^{-3}	Mast cell expressed membrane protein 1
Lat	−1.339	2.98×10^{-3}	Linker for activation of T-cells
Plekha4	−1.334	7.90×10^{-3}	Pleckstrin homology domain containing A4
Misp	−1.270	7.52×10^{-3}	Mitotic spindle positioning
P4ha3	−1.176	4.91×10^{-3}	Prolyl 4-hydroxylase subunit alpha 3
Gpr84	−1.175	9.61×10^{-4}	G protein-coupled receptor 84
Pi16	−1.160	5.25×10^{-3}	Peptidase inhibitor 16
Slc10a5	−1.134	4.79×10^{-3}	Solute carrier family 10 member 5

Table 2. *Cont.*

Gene	LogFC	*p*-Value	Protein
Ankrd7	−1.118	7.66×10^{-3}	Ankyrin repeat domain 7
Tmprss5	−1.109	2.09×10^{-3}	Transmembrane serine protease 5
Cdhr3	−1.091	5.62×10^{-3}	Cadherin-related family member 3
Hmga2	−1.061	1.63×10^{-3}	High mobility group AT-hook 2
Sycp2	−1.061	4.34×10^{-3}	Synaptonemal complex protein 2
Cd27	−1.039	5.80×10^{-3}	T-cell activation antigen CD27
Mypn	−1.036	6.94×10^{-3}	Myopalladin

3. Results

3.1. HFD Is Associated with Differential Retinal Gene Expression in the Absence of the Microbiome

To compare the effect of a high-fat diet on the retinal transcriptome, we performed high-throughput RNA-seq analysis of mouse retinas from the GF-ND and GF-HFD. We sequenced four whole retinas from both experimental groups (n = 4 eyes from 4 different mice, controlled for age and sex). After the correction of the raw data to remove background noise, 19,681 genes were selected for differential gene analysis (Supplementary Table S1). DEGs were selected based on a stringent p-value cutoff < 0.01 and logFC > 1. Comparison between the two groups identifies 53 DEGs, of which 19 are upregulated and 34 are downregulated in the GF-HFD mice group. The National Center for Biotechnology Information (NCBI) gene database was used to filter pseudogenes and uncharacterized cDNA to compile a list of protein-coding genes only. A heatmap was plotted to show the hierarchical clustering of the DEGs (Figure 1). The sequencing data suggests that HFD is associated with changes in the retinal transcriptome in the absence of the microbiome. Detailed list and statistics of the upregulated and downregulated DEGs are available in Tables 1 and 2.

3.2. Significant Biologic Functions and Processes Are Overrepresented by Functional Enrichment Analysis

The enrichment analysis for gene ontology and pathways was performed using EnrichR [39–41]. Enrichment analysis was done to identify over-represented biological functions and classes from statistically significant differentially expressed genes. Human and mouse pathway analysis revealed that complement and coagulation cascades were significantly affected by HFD (Figures 2 and 3). The analysis also shows that the top 3 biological processes are regulation of blood vessel diameter, inflammatory response, and negative regulation of endopeptidase (Figure 4). Molecular functions altered include endopeptidase inhibitor activity, protease binding, and cysteine-type endopeptidase inhibitor activity (Figure 5).

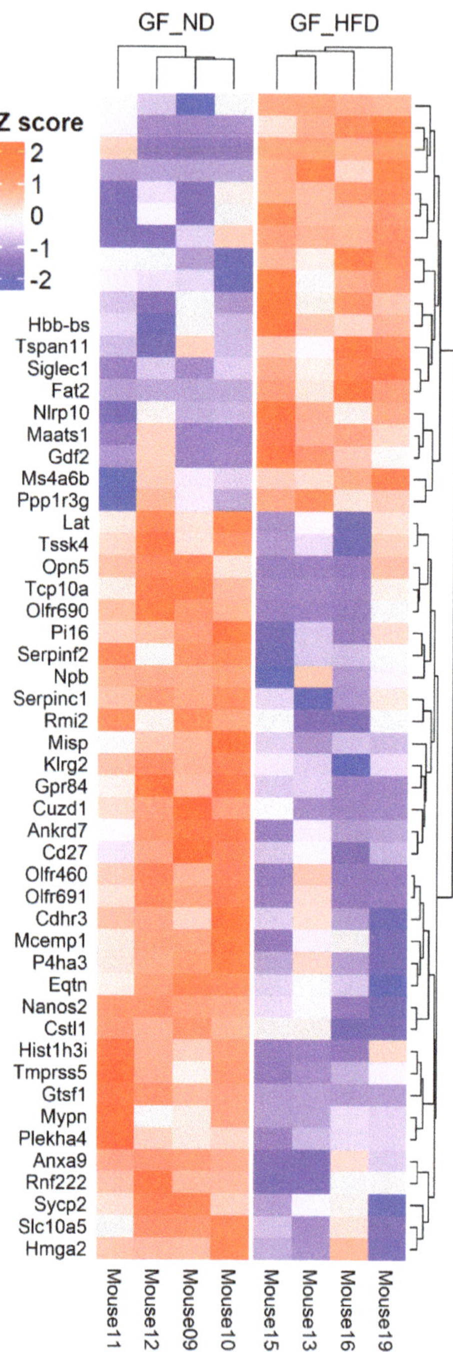

Figure 1. Heatmap with hierarchal clustering demonstrating 53 differentially expressed genes (DEGs) with a logFC greater than 1 and *p*-value < 0.01 between germ-free mice on normal diet (GF-ND, *n* = 4) and germ-free mice on high-fat diet (GF-HFD, *n* = 4). Z-score is calculated from LogFC with red and blue indicating upregulated and downregulated genes, respectively.

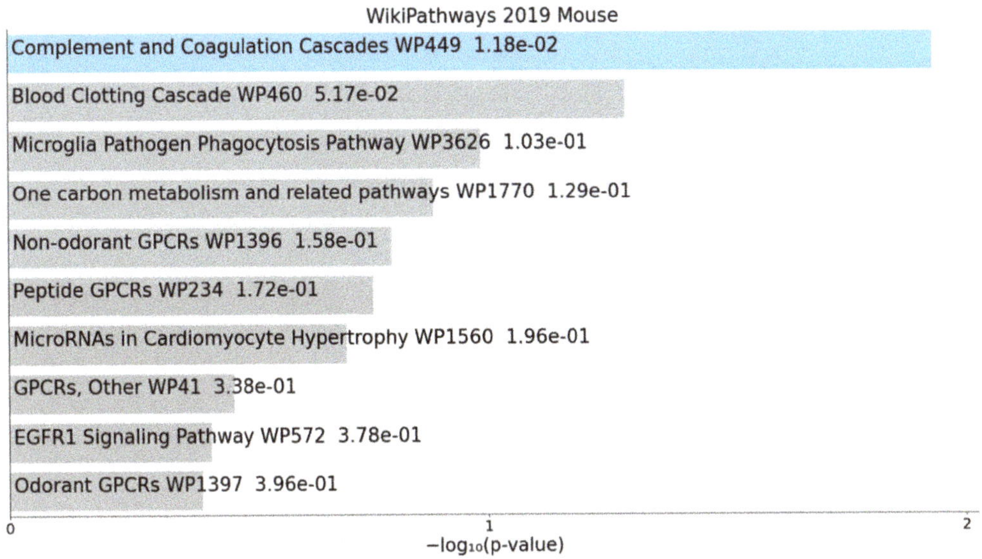

WikiPathways 2019 Mouse

Complement and Coagulation Cascades WP449	1.18e-02
Blood Clotting Cascade WP460	5.17e-02
Microglia Pathogen Phagocytosis Pathway WP3626	1.03e-01
One carbon metabolism and related pathways WP1770	1.29e-01
Non-odorant GPCRs WP1396	1.58e-01
Peptide GPCRs WP234	1.72e-01
MicroRNAs in Cardiomyocyte Hypertrophy WP1560	1.96e-01
GPCRs, Other WP41	3.38e-01
EGFR1 Signaling Pathway WP572	3.78e-01
Odorant GPCRs WP1397	3.96e-01

$-\log_{10}$(p-value)

WikiPathways 2019 Mouse	p-Value	Adjusted p-Value	Odds Ratio	Combined Score	Genes
* Complement and coagulation cascades WP449	1.18×10^{-2}	0.141	13	57.74	SERPINC1; SERPINF2
Blood clotting cascade WP460	5.17×10^{-2}	0.31	20.17	59.75	SERPINF2
Microglia pathogen phagocytosis pathway WP3626	1.03×10^{-1}	0.336	9.57	21.74	LAT
One carbon metabolism and related pathways WP1770	1.29×10^{-1}	0.336	7.5	15.36	DMGDH
Non-odorant GPCRs WP1396	1.58×10^{-1}	0.336	2.91	5.38	NPY4R; OPN5
Peptide GPCRs WP234	1.72×10^{-1}	0.336	5.46	9.61	NPY4R
MicroRNAs in cardiomyocyte hypertrophy WP1560	1.96×10^{-1}	0.336	4.72	7.69	NPPB
GPCRs, other WP41	3.38×10^{-1}	0.475	2.47	2.68	GPR84
EGFR1 signaling pathway WP572	3.78×10^{-1}	0.475	2.15	2.09	HIST1H3I
Odorant GPCRs WP1397	3.96×10^{-1}	0.475	2.02	1.87	GPR84

Figure 2. Gene list enrichment analysis of significant DEGs between GF-HFD mice ($n = 4$) and GF-ND mice ($n = 4$) using EnrichR. The bar graph shows a ranked list by p-value of the top 10 over-represented mouse pathways with significant pathways indicated in blue (p-value < 0.05). The corresponding table demonstrates detailed statistics and involved genes with significant pathways indicated by an asterisk (p-value < 0.05).

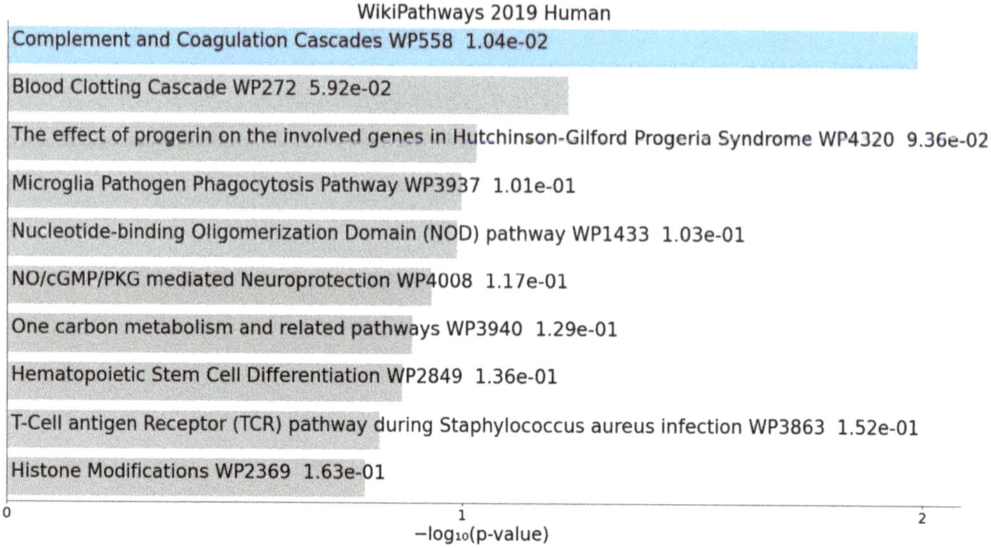

WikiPathways 2019 Human

Complement and Coagulation Cascades WP558 1.04e-02

Blood Clotting Cascade WP272 5.92e-02

The effect of progerin on the involved genes in Hutchinson-Gilford Progeria Syndrome WP4320 9.36e-02

Microglia Pathogen Phagocytosis Pathway WP3937 1.01e-01

Nucleotide-binding Oligomerization Domain (NOD) pathway WP1433 1.03e-01

NO/cGMP/PKG mediated Neuroprotection WP4008 1.17e-01

One carbon metabolism and related pathways WP3940 1.29e-01

Hematopoietic Stem Cell Differentiation WP2849 1.36e-01

T-Cell antigen Receptor (TCR) pathway during Staphylococcus aureus infection WP3863 1.52e-01

Histone Modifications WP2369 1.63e-01

$-\log_{10}(p\text{-value})$

WikiPathways 2019 Human	p-Value	Adjusted p-Value	Odds Ratio	Combined Score	Genes
* Complement and coagulation cascades WP558	1.04×10^{-2}	0.186	13.93	63.66	SERPINC1; SERPINF2
Blood clotting cascade WP272	5.92×10^{-2}	0.276	17.42	49.22	SERPINF2
The effect of progerin on the involved genes in Hutchinson–Gilford progeria syndrome WP4320	9.36×10^{-2}	0.276	10.64	25.19	HIST1H3I
Microglia pathogen phagocytosis pathway WP3937	1.01×10^{-1}	0.276	9.82	22.53	LAT
Nucleotide-binding oligomerization Domain (NOD) pathway WP1433	1.03×10^{-1}	0.276	9.57	21.74	NLRP10
NO/cGMP/PKG-mediated neuroprotection WP4008	1.17×10^{-1}	0.276	8.32	17.82	NPPB
One carbon metabolism and related pathways WP3940	1.29×10^{-1}	0.276	7.5	15.36	DMGDH
Hematopoietic stem cell differentiation WP2849	1.36×10^{-1}	0.276	7.08	14.14	KLF1
T-Cell antigen receptor (TCR) pathway during Staphylococcus aureus infection WP3863	1.52×10^{-1}	0.276	6.27	11.81	LAT
Histone modifications WP2369	1.63×10^{-1}	0.276	5.79	10.5	HIST1H3I

Figure 3. Gene list enrichment analysis of significant DEGs between GF-HFD mice ($n = 4$) and GF-ND mice ($n = 4$) using EnrichR. The bar graph shows a ranked list by p-value of the top 10 over-represented human pathways with significant pathways indicated in blue (p-value < 0.05). The corresponding table demonstrates detailed statistics and involved genes with significant pathways indicated by an asterisk (p-value < 0.05).

GO Biological Process 2018

regulation of blood vessel diameter (GO:0097746) *4.91e-05

inflammatory response (GO:0006954) 5.35e-04

negative regulation of endopeptidase activity (GO:0010951) 1.39e-03

positive regulation of multi-organism process (GO:0043902) 4.32e-03

positive regulation of interleukin-8 production (GO:0032757) 6.07e-03

cartilage development (GO:0051216) 6.62e-03

regulation of angiogenesis (GO:0045765) 1.15e-02

second-messenger-mediated signaling (GO:0019932) 1.18e-02

negative regulation of peptidase activity (GO:0010466) 1.18e-02

negative regulation of blood vessel morphogenesis (GO:2000181) 1.29e-02

$-\log_{10}$(p-value)

GO Biological Process 2018	*p*-Value	Adjusted *p*-Value	Odds Ratio	Combined Score	Genes
** Regulation of blood vessel diameter (GO:0097746)*	4.91×10^{-5}	0.019	49.808	494.138	NPPB; NANOS2; SERPINF2
** Inflammatory response (GO:0006954)*	5.35×10^{-4}	0.101	8.308	62.589	NANOS2; NLRP10; CD27; SIGLEC1; LAT
** Negative regulation of endopeptidase activity (GO:0010951)*	1.39×10^{-3}	0.146	14.9	98.019	SERPINC1; SERPINF2; CSTL1
** Positive regulation of multi-organism process (GO:0043902)*	4.32×10^{-3}	0.146	22.31	121.446	NANOS2; NLRP10
** Positive regulation of interleukin-8 production (GO:0032757)*	6.07×10^{-3}	0.146	18.585	94.867	NLRP10; GDF2
** Cartilage development (GO:0051216)*	6.62×10^{-3}	0.146	17.739	89.011	GDF2; HMGA2
** Regulation of angiogenesis (GO:0045765)*	1.15×10^{-2}	0.146	6.818	30.419	NPPB; GDF2; HMGA2
** Second messenger-mediated signaling (GO:0019932)*	1.18×10^{-2}	0.146	12.998	57.739	NANOS2; LAT
** Negative regulation of peptidase activity (GO:0010466)*	1.18×10^{-2}	0.146	12.998	57.739	SERPINC1; SERPINF2
** Negative regulation of blood vessel morphogenesis (GO:2000181)*	1.29×10^{-2}	0.146	12.377	53.864	NPPB; GDF2

Figure 4. Gene list enrichment analysis of significant DEGs between GF-HFD mice ($n = 4$) and GF-ND mice ($n = 4$) using EnrichR. The bar graph shows a ranked list by *p*-value of the top 10 over-represented biological processes with significant processes indicated in blue (*p*-value < 0.05). The corresponding table demonstrates detailed statistics and involved genes with significant processes indicated by an asterisk (*p*-value < 0.05).

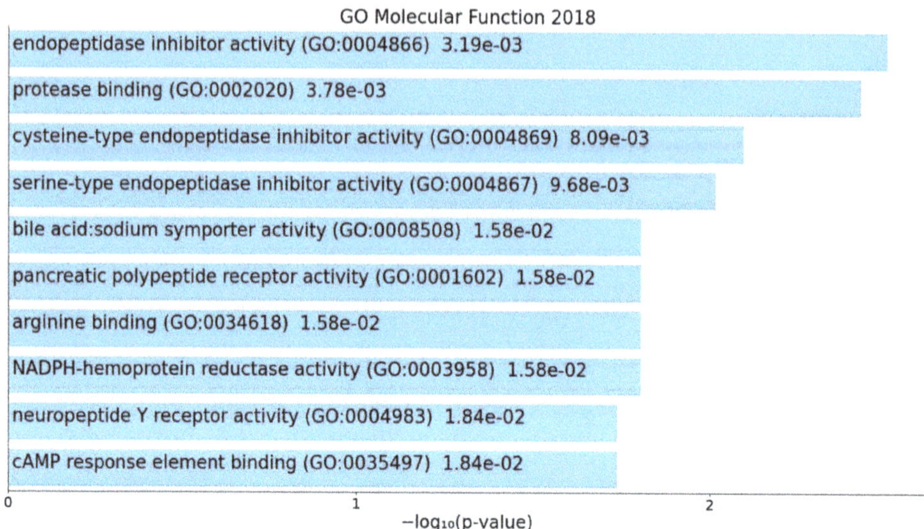

GO Molecular Function 2018	*p*-Value	Adjusted *p*-Value	Odds Ratio	Combined Score	Genes
** Endopeptidase inhibitor activity (GO:0004866)*	3.19×10^{-3}	0.132	11.02	63.37	SERPINC1; SERPINF2; CSTL1
** Protease binding (GO:0002020)*	3.78×10^{-3}	0.132	10.35	57.71	SERPINC1; SERPINF2; CSTL1
** Cysteine-type endopeptidase inhibitor activity (GO:0004869)*	8.09×10^{-3}	0.132	15.92	76.72	CD27; CSTL1
** Serine-type endopeptidase inhibitor activity (GO:0004867)*	9.68×10^{-3}	0.132	14.45	66.99	SERPINC1; SERPINF2
** Bile acid:sodium symporter activity (GO:0008508)*	1.58×10^{-2}	0.132	76.7	318.15	SLC10A5
** Pancreatic polypeptide receptor activity (GO:0001602)*	1.58×10^{-2}	0.132	76.7	318.15	NPY4R
** Arginine binding (GO:0034618)*	1.58×10^{-2}	0.132	76.7	318.15	NANOS2
** NADPH-hemoprotein reductase activity (GO:0003958)*	1.58×10^{-2}	0.132	76.7	318.15	NANOS2
** Neuropeptide Y receptor activity (GO:0004983)*	1.84×10^{-2}	0.132	63.91	255.34	NPY4R
** cAMP response element binding (GO:0035497)*	1.84×10^{-2}	0.132	63.91	255.34	HMGA2

Figure 5. Gene list enrichment analysis of significant DEGs between GF-HFD mice ($n = 4$) and GF-ND mice ($n = 4$) using EnrichR. The bar graph shows a ranked list by *p*-value of the top 10 over-represented molecular functions with significant functions indicated in blue (*p*-value < 0.05). The corresponding table demonstrates detailed statistics and involved genes with significant functions indicated by an asterisk (*p*-value < 0.05).

4. Discussion

To our knowledge, this is the first study to use high-throughput RNA sequencing of whole retinas from GF mice to demonstrate that high-fat diet alone is associated with changes in the retinal transcriptome in the absence of gut microbiome. Diet is a highly modifiable risk factor for development of vision threatening diseases, and understanding the relationship between diet and ocular pathology is a promising avenue for intervention. However, the biological pathways connecting diet to ocular disease are poorly understood and there is limited literature investigating the pathways involved.

Multiple clinical studies have demonstrated that diet plays a critical role in retinal health and contributes to diseases including age-related macular degeneration, diabetic retinopathy, and primary open angle glaucoma [42–44]. For example, recently published data from the AREDS study group reported that higher intake of saturated fatty acids, monounsaturated fatty acids, and oleic acid were associated with significant increased risk of progression to late AMD [45]. Supporting this, our team has previously published data showing that high-fat diet increased lesion size, vascular leakage, and sub-RPE deposits of laser-induced choroidal neovascularization in both wild-type and apolipoprotein E-deficient mice [15]. Recent evidence has suggested that the effects of high-fat diet on retinal disease are mediated by the gut microbiome. High-fat diets can cause gut microbial dysbiosis altering intestinal permeability and leading to low-grade inflammation with release of pro-angiogenic factors which may exacerbate ocular diseases such as proliferative diabetic retinopathy and neovascular AMD [18].

To further elucidate the biological connections between the diet–gut–retina axis, we aimed to investigate how diet affects the retinal transcriptome independently of the gut microbiome [30]. Germ-free mice, raised without exposure to any microbes, provide an ideal model to investigate this hypothesis [46]. In this study, we used GF mice fed a high-fat diet compared to a normal diet to explore retinal transcriptome changes induced by diet alone. After analysis of 19,681 total DEGs with removal of pseudogenes, 53 significant DEGs with LogFC > 1 were identified between groups (Figure 1). Enrichment analysis shows pathways involved in complement and coagulation cascades, inflammatory response, regulation of angiogenesis and blood vessel morphology, and endopeptidase inhibitor activity (Figure 2) were significantly affected by high-fat diet in germ-free mice.

4.1. High-Fat Diet May Affect Expression of Genes Involved in Inflammatory Pathways in Germ-Free Mice

Enrichment analysis of significant DEGs demonstrated that complement and coagulation cascades were significantly affected by high-fat diet (Figure 2). The complement and coagulation cascades are activated in response to retinal inflammation and vascular injury and have been highly implicated in retinal disease, especially in development of age-related macular degeneration, with multiple ongoing clinical trials currently being investigated [47–49]. Additional biological pathways identified were involved in inflammatory response, positive regulation of interleukin-8, protease binding, and regulation of endopeptidase activity (Figures 4 and 5). Our results demonstrate that DEGs in pathways involved in retinal inflammation were significantly affected by high-fat diet (Tables 1 and 2). *C1qtnf2* is a member of the C1q and tumor necrosis factor related-protein (CTRP) superfamily reported to be involved in retinal inflammation and associated with late-onset retinal degeneration [50,51]. High expression of CTRPs has been reported in the drusen of human donor eyes with AMD [52]. Additionally, the CTRP family has reported to mediate glucose-induced oxidative stress and apoptosis in RPE cells [53]. *Ifi204* (interferon gamma inducible protein) is a cytosolic DNA sensor involved in initiation of a type 1 interferon response and activation of the inflammasome pathway in response to bacterial or viral infection [54,55]. Multiple genes involved in activation of local ocular inflammatory response, including *Ifi204*, have been identified as mediators of retinal aging [56,57]. The H3 family of histones (including *Hist1h3i*) may be important in epigenetic modifications that promote a persistent pro-inflammatory state in diabetic retinopathy [58,59]. Multiple classes of histone genes are involved in regulation of the nucleosome and have been shown to be actively transcribed in both developing and aging retinal neurons [60]. *Serpinc1* and *Serpinf2* are members of the serine protease inhibitor (serpin) family, which were also found to be downregulated in our study. Proteins in the serpin family include endopeptidases that have been reported to be important in inhibiting angiogenesis and retinal cell death [61,62]. Proteomics analysis has identified multiple proteins in the serpin family including both *Serpinc1* and *Serpinf2* as potential serum biomarkers of retinal inflammation in diabetic retinopathy [63]. Additionally, VEGF is involved in the negative regulation of cysteine-type endopeptidase activity required for the apoptotic process [64].

4.2. High-Fat Diet May Influence Genes and Pathways Involved in Angiogenesis in Germ-Free Mice

Enrichment pathway analysis of the significant DEGs showed that pathways involved in regulation of angiogenesis, blood vessel diameter, and blood vessel morphogenesis (Figure 4) were affected by high-fat diet in germ-free mice. Bioactive lipids have been shown to be involved in regulation of pathologic retinal angiogenesis [65]. Our results identified several DEGs involved in regulation of angiogenesis (Tables 1 and 2). *Fat2* (FAT-like cadherin 2) was the most highly upregulated gene identified and has not previously been described in the retina [66]. The cadherin superfamily is involved in maintaining the blood–retinal barrier and cell migration during angiogenesis [67,68]. *Neuropsin* (Opn5) is expressed in retinal ganglion cells and has to been reported to mediate light-dependent retinal vascular development and mediate photoentrainment of circadian rhythm [69,70]. *Nppb* (Natriuretic peptide B) is involved in retinal response to hypoxia and may be an important regular of retinal vascular permeability [71,72].

4.3. High-Fat Diet May Affect Expression of Genes Involved in RPE Function and Ciliogenesis in Germ-Free Mice

Our data also suggest that high-fat diet may regulate expression of several genes involved RPE function and ciliogenesis in germ-free mice. Multiple DEGs related to olfactory receptor expression in the mouse retina (*Olfr460*, *Olfr690*, and *Olfr691*) were found to be affected by high-fat diet. Recent literature using RNA-sequencing of human retina has demonstrated that olfactory receptors are expressed in human retina in the retinal pigment epithelium, photoreceptor inner segments, ganglion cell layer, bipolar cells, and horizontal cells [73]. Retinal olfactory receptors may be important in retinal repair involving retinal pigment epithelium and retinal neurons [74,75]. Olfactory receptor expression is hypothesized to induce RPE proliferation and migration [76]. *Gtsf1* (gametocyte specific factor 1) was also identified as highly downregulated by high-fat diet in the retina of germ-free mice and has not previously been reported to be expressed in the retina. *Gtsf1* is involved in retrotransposon suppression in germ cells to prevent genomic instability [77]. Retrotransposons are also reported to be involved in propagation of *Alu* retroelements which may contribute to RPE cell death in age-related macular degeneration [78]. *Cstl1* (cystatin-like 1) was also identified to be downregulated by high-fat diet; however, the specific function of *Cstl1* is currently unknown. Other members of the cystatin superfamily, notably cystatin C, are highly expressed in the RPE and are associated with increased risk of AMD and Alzheimer's disease [79–82]. *Deup1* (deuterostome assembly protein 1) is an important component of the deuterosome involved in ciliogenesis [83]. While the exact role of Deup1 in the retina has not been determined, defects in primary cilium function in photoreceptors and the RPE leads to retinal degeneration as part of several syndromic ciliopathies like Bardet-Beidl syndrome and Alstrom syndrome [84]. *Maats1* (Cilia and Flagella associated protein 91) has been identified as an important component of sperm flagellum structure and has not previously been described in the retina [85]. Mutations in the cilia and flagella-associated protein family have been linked with retinitis pigmentosa in familial amyotrophic lateral sclerosis [86].

4.4. Additional Genes and Pathways of Retinal Transcriptome Affected by High-Fat Diet in Germ-Free Mice

Several neuroendocrine related pathways including pancreatic polypeptide receptor activity and neuropeptide Y receptor activity were found to be affected by high-fat diet in germ-free mice.

Npy4r (neuropeptide Y receptor) is expressed in human retinal RPE and glial cells, and it is involved in neuronal calcium release, neuroprotection, and proliferation of glial cells [87]. Clinically, polymorphisms in NPY have been associated with increased risk of type 2 diabetes and development of diabetic retinopathy [88,89]. Neuropeptide b (*Npb*) is a relatively novel neuropeptide associated with regulation of the neuroendocrine system, pain processing, stress, and feeding behaviors [90]. *Npb* is widely expressed in the central nervous system, but expression has not previously been described in the retina [91].

Several identified significant DEGs have not previously been reported to be expressed in the retina. *Rmi2* is involved in genome stability and has been reported to be associated with development of multiple types of cancer [92–94]. The physiologic role of *Rnf222* has not been described currently; however, other members of the ring finger protein family have been associated with cerebral vascular diseases like Moyamoya disease and atherosclerotic stroke [95]. *Cuzd1* (CUB and zona pellucida-like domains 1) has been reported to mediate epithelial proliferation of the mammary gland during pregnancy [96]. *Cuzd1* has also been identified in human embryonic stems cells [97]. Single nucleotide polymorphisms (SNPs) in *Cuzd1* have been associated with risk of age-related macular degeneration [98]. *Dmgdh* (dimethylglycine dehydrogenase) is involved in choline metabolism important in neurotransmitter and phospholipid biosynthesis [99]. *Dmgdh* was identified as part of a set of differentially expressed genes in the mitochondrial transcriptome human retinas with diabetic retinopathy [100]. *Nanos2* (nanos C2HC-type zinc finger 2) is involved in germ cell differentiation and was also identified as differentially expressed in the retinal transcriptome of a mouse model of diabetic retinopathy [101].

5. Conclusions and Limitations

This study demonstrates novel data that suggest diet may modulate the retinal transcriptome in the absence of the gut microbiome. Unbiased analysis of the retinal transcriptome using high-throughput RNA-sequencing identified genes and pathways involved in retinal inflammation, angiogenesis, and RPE function, whose expression was influenced by HFD in the absence of gut microbiome. These genes and pathways may be involved in complex diet-microbiome-retina axis interactions that have only recently been recognized to play roles in retinal physiology and retinal disease pathogenesis.

Our study is limited to RNA-sequencing alone, and confirmatory studies of protein expression or activity were outside the scope of this investigation. The goal of our study was to use germ-free mice and RNA-sequencing technology to provide an unbiased characterization of the effects of HFD on global retinal gene expression in the absence of gut microbiome, as well as to identify potential novel targets within the retinal transcriptome that may guide future investigation on the diet-microbiome-retina axis.

Future studies with quantitative RT-PCR, proteomics, or functional assays are needed to further investigate potential functional pathways affected by HFD. In addition, studies using animal models of retinal diseases should include protein markers of angiogenesis and retinal apoptosis using multiplex assays, ELISA, Western blotting, or flow cytometry to better characterize how the genes and pathways revealed by high-throughput RNA-sequencing may be modulated by HFD. Applying a multiomics approach towards investigating the diet-microbiome-retina axis will be critical to delineate the effects of HFD on protein function, retinal cell physiology, and retinal disease pathogenesis [102].

While the germ-free mouse model is considered the gold-standard for microbiome studies, our conclusions are limited due to changes in baseline physiologic parameters that were altered by lack of microbiome in these mice. Retinal transcriptome changes identified in germ-free mice may be influenced by changes in immune development, metabolism, and digestion affected by the absence of microbiome.

Dietary modification is an easily modifiable risk factor, and understanding the interaction between diet, gut microbiome, and retinal disease has the potential to advance our understanding of vision-threatening diseases. Delineating these complex interactions could lead to the discovery of novel targets for intervention. While much of the focus has been on alterations to the gut microbiome as a key effector in disease pathogenesis, we present novel data suggesting that diet may affect retinal gene transcription when the microbiome is absent.

However, the microbiome-dependent and microbiome-independent effects of HFD on the retinal transcriptome remain unclear. The gut microbiome is an important mediator of the effects of diet on the retinal transcriptome, and it is currently unclear if these effects are overall protective or deleterious. Pathways in the retinal transcriptome affected by high-fat diet could be both attenuated or exacerbated by the presence of the gut microbiome, and these interactions are still poorly understood.

Despite the limitations, our study provides novel insight about potential pathways that could be involved in the diet–microbiome–retina axis and furthers our understanding of how diet may regulate disease pathogenesis and severity. Future studies are needed to define the precise role of diet in retinal diseases and to elucidate the complex, overlapping relationships in the diet–microbiome–retina axis and its involvement in retinal disease pathobiology.

Supplementary Materials: The following are available online at https://www.mdpi.com/article/10.3390/cells10082119/s1, Table S1 provides all 19,681 genes selected for DEG analysis.

Author Contributions: Conceptualization and methodology was carried out by D.S. (Dimitra Skondra) and E.B.C. Bioinformatics and formal validation of data was performed by B.X., M.D., D.D., U.N. and D.S. (Dinanath Sulakhe). Experimental procedures were carried out by U.N., A.M., with assistance by V.L. Original manuscript draft preparation was performed by D.D., U.N. and J.X. with review and editing by D.S. (Dimitra Skondra). Funding acquisition was from D.S. (Dimitra Skondra) with contributions by E.B.C., S.M.H. and J.X. All authors have read and agreed to the published version of the manuscript.

Funding: This research was funded by BrightFocus Foundation "Role of high fat diet and gut microbiome in macular degeneration" (Dimitra Skondra, M2018042), NIDDK P30 (E.B.C., DK42086), The University of Chicago Women's Board (Dimitra Skondra), and the Illinois Society for the Prevention of Blindness (Dimitra Skondra, FP067271-01-PR and J.X., FP105447).

Institutional Review Board Statement: All animal experiments were performed in strict accordance with the recommendations in the Guide for the Care and Use of Laboratory Animals of the National Institutes of Health and approved by University of Chicago IACUC (#72557).

Informed Consent Statement: Not applicable.

Data Availability Statement: Complete dataset of identified DEGs is available in Supplementary Table S1.

Acknowledgments: We thank University of Chicago Core Genomics team, Peter Fabier and Mikayla Marchuk, for helping with RNA sequencing. Gnotobiotic Research Animal Facility (GRAF) staff (Betty Theriault and Melanie Spedale) and Donna Arvans for care of the germ-free animals and Chang lab members Candace Cham, Xiaorong Zhu, Jason Koval, and Amal Kambal.

Conflicts of Interest: The authors declare no conflict of interest. The funders had no role in the design of the study; in the collection, analyses, or interpretation of data; in the writing of the manuscript; or in the decision to publish the results.

References

1. Kang, J.H.; Willett, W.C.; Rosner, B.A.; Buys, E.; Wiggs, J.L.; Pasquale, L.R. Association of Dietary Nitrate Intake With Primary Open-Angle Glaucoma: A Prospective Analysis From the Nurses' Health Study and Health Professionals Follow-up Study. *JAMA Ophthalmol.* **2016**, *134*, 294–303. [CrossRef] [PubMed]
2. Kang, J.H.; Pasquale, L.R.; Willett, W.C.; Rosner, B.A.; Egan, K.M.; Faberowski, N.; Hankinson, S.E. Dietary Fat Consumption and Primary Open-Angle Glaucoma. *Am. J. Clin. Nutr.* **2004**, *79*, 755–764. [CrossRef] [PubMed]
3. Weikel, K.A.; Chiu, C.-J.; Taylor, A. Nutritional Modulation of Age-Related Macular Degeneration. *Mol. Asp. Med.* **2012**, *33*, 318–375. [CrossRef] [PubMed]
4. Rinninella, E.; Mele, M.C.; Merendino, N.; Cintoni, M.; Anselmi, G.; Caporossi, A.; Gasbarrini, A.; Minnella, A.M. The Role of Diet, Micronutrients and the Gut Microbiota in Age-Related Macular Degeneration: New Perspectives from the Gut–Retina Axis. *Nutrients* **2018**, *10*, 1677. [CrossRef]
5. Zemski Berry, K.A.; Gordon, W.C.; Murphy, R.C.; Bazan, N.G. Spatial Organization of Lipids in the Human Retina and Optic Nerve by MALDI Imaging Mass Spectrometry. *J. Lipid Res.* **2014**, *55*, 504–515. [CrossRef]
6. Wong-Riley, M.T.T. Energy Metabolism of the Visual System. *Eye Brain* **2010**, *2*, 99–116. [CrossRef]
7. Chiu, C.-J. The Relationship of Major American Dietary Patterns to Age-Related Macular Degeneration. *Am. J. Ophthalmol.* **2014**, *158*, 118–127. [CrossRef] [PubMed]
8. Parekh, N.; Voland, R.P.; Moeller, S.M.; Blodi, B.A.; Ritenbaugh, C.; Chappell, R.J.; Wallace, R.B.; Mares, J.A.; CAREDS Research Study Group. Association between Dietary Fat Intake and Age-Related Macular Degeneration in the Carotenoids in Age-Related Eye Disease Study (CAREDS): An Ancillary Study of the Women's Health Initiative. *Arch. Ophthalmol.* **2009**, *127*, 1483–1493. [CrossRef] [PubMed]
9. Bressler, N.M. Age-Related Macular Degeneration Is the Leading Cause of Blindness. *JAMA* **2004**, *291*, 1900–1901. [CrossRef] [PubMed]
10. Chong, E.W.-T.; Robman, L.D.; Simpson, J.A.; Hodge, A.M.; Aung, K.Z.; Dolphin, T.K.; English, D.R.; Giles, G.G.; Guymer, R.H. Fat Consumption and Its Association with Age-Related Macular Degeneration. *Arch. Ophthalmol.* **2009**, *127*, 674–680. [CrossRef] [PubMed]
11. Clarkson-Townsend, D.A.; Douglass, A.J.; Singh, A.; Allen, R.S.; Uwaifo, I.N.; Pardue, M.T. Impacts of High Fat Diet on Ocular Outcomes in Rodent Models of Visual Disease. *Exp. Eye Res.* **2021**, *204*, 108440. [CrossRef]
12. Chang, R.C.-A.; Shi, L.; Huang, C.C.-Y.; Kim, A.J.; Ko, M.L.; Zhou, B.; Ko, G.Y.-P. High-Fat Diet-Induced Retinal Dysfunction. *Investig. Ophthalmol. Vis. Sci.* **2015**, *56*, 2367–2380. [CrossRef]
13. Lee, J.-J.; Wang, P.-W.; Yang, I.-H.; Huang, H.-M.; Chang, C.-S.; Wu, C.-L.; Chuang, J.-H. High-Fat Diet Induces Toll-like Receptor 4-Dependent Macrophage/Microglial Cell Activation and Retinal Impairment. *Investig. Ophthalmol. Vis. Sci.* **2015**, *56*, 3041–3050. [CrossRef]
14. Albouery, M.; Buteau, B.; Grégoire, S.; Martine, L.; Gambert, S.; Bron, A.M.; Acar, N.; Chassaing, B.; Bringer, M.-A. Impact of a High-Fat Diet on the Fatty Acid Composition of the Retina. *Exp. Eye Res.* **2020**, *196*, 108059. [CrossRef] [PubMed]
15. Skondra, D.; She, H.; Zambarakji, H.J.; Connolly, E.; Michaud, N.; Chan, P.; Kim, I.K.; Gragoudas, E.S.; Miller, J.W.; Hafezi-Moghadam, A. Effects of ApoE Deficiency, Aging and High Fat Diet on Laser-Induced Choroidal Neovascularization and Bruch's Membrane-RPE Interface Morphology. *Investig. Ophthalmol. Vis. Sci.* **2007**, *48*, 1768.
16. Asare-Bediako, B.; Noothi, S.K.; Li Calzi, S.; Athmanathan, B.; Vieira, C.P.; Adu-Agyeiwaah, Y.; Dupont, M.; Jones, B.A.; Wang, X.X.; Chakraborty, D.; et al. Characterizing the Retinal Phenotype in the High-Fat Diet and Western Diet Mouse Models of Prediabetes. *Cells* **2020**, *9*, 464. [CrossRef] [PubMed]
17. Mohamed, I.N.; Hafez, S.S.; Fairaq, A.; Ergul, A.; Imig, J.D.; El-Remessy, A.B. Thioredoxin-Interacting Protein Is Required for Endothelial NLRP3 Inflammasome Activation and Cell Death in a Rat Model of High-Fat Diet. *Diabetologia* **2014**, *57*, 413–423. [CrossRef] [PubMed]
18. Andriessen, E.M.; Wilson, A.M.; Mawambo, G.; Dejda, A.; Miloudi, K.; Sennlaub, F.; Sapieha, P. Gut Microbiota Influences Pathological Angiogenesis in Obesity-driven Choroidal Neovascularization. *EMBO Mol. Med.* **2016**, *8*, 1366–1379. [CrossRef]
19. Jandhyala, S.M.; Talukdar, R.; Subramanyam, C.; Vuyyuru, H.; Sasikala, M.; Reddy, D.N. Role of the Normal Gut Microbiota. *World J. Gastroenterol. WJG* **2015**, *21*, 8787–8803. [CrossRef] [PubMed]
20. Fujimura, K.E.; Lynch, S.V. Microbiota in Allergy and Asthma and the Emerging Relationship with the Gut Microbiome. *Cell Host Microbe* **2015**, *17*, 592–602. [CrossRef]
21. Mohajeri, M.H.; Fata, G.L.; Steinert, R.E.; Weber, P. Relationship between the Gut Microbiome and Brain Function. *Nutr. Rev.* **2018**, *76*, 481–496. [CrossRef]
22. Tang, W.H.W.; Kitai, T.; Hazen, S.L. Gut Microbiota in Cardiovascular Health and Disease. *Circ. Res.* **2017**, *120*, 1183–1196. [CrossRef] [PubMed]
23. Ma, J.; Zhou, Q.; Li, H. Gut Microbiota and Nonalcoholic Fatty Liver Disease: Insights on Mechanisms and Therapy. *Nutrients* **2017**, *9*, 1124. [CrossRef] [PubMed]
24. Rowan, S.; Taylor, A. The Role of Microbiota in Retinal Disease. *Adv. Exp. Med. Biol.* **2018**, *1074*, 429–435. [PubMed]
25. Rowan, S.; Jiang, S.; Korem, T.; Szymanski, J.; Chang, M.-L.; Szelog, J.; Cassalman, C.; Dasuri, K.; McGuire, C.; Nagai, R.; et al. Involvement of a Gut-Retina Axis in Protection against Dietary Glycemia-Induced Age-Related Macular Degeneration. *Proc. Natl. Acad. Sci. USA* **2017**, *114*, E4472–E4481. [CrossRef] [PubMed]

26. Gong, H.; Zhang, S.; Li, Q.; Zuo, C.; Gao, X.; Zheng, B.; Lin, M. Gut Microbiota Compositional Profile and Serum Metabolic Phenotype in Patients with Primary Open-Angle Glaucoma. *Exp. Eye Res.* **2020**, *191*, 107921. [CrossRef] [PubMed]
27. Zinkernagel, M.S.; Zysset-Burri, D.C.; Keller, I.; Berger, L.E.; Leichtle, A.B.; Largiadèr, C.R.; Fiedler, G.M.; Wolf, S. Association of the Intestinal Microbiome with the Development of Neovascular Age-Related Macular Degeneration. *Sci. Rep.* **2017**, *7*, 40826. [CrossRef]
28. Zysset-Burri, D.C.; Keller, I.; Berger, L.E.; Neyer, P.J.; Steuer, C.; Wolf, S.; Zinkernagel, M.S. Retinal Artery Occlusion Is Associated with Compositional and Functional Shifts in the Gut Microbiome and Altered Trimethylamine-N-Oxide Levels. *Sci. Rep.* **2019**, *9*, 15303. [CrossRef]
29. Skondra, D.; Rodriguez, S.H.; Sharma, A.; Gilbert, J.; Andrews, B.; Claud, E.C. The Early Gut Microbiome Could Protect against Severe Retinopathy of Prematurity. *J. AAPOS Off. Publ. Am. Assoc. Pediatr. Ophthalmol. Strabismus* **2020**, *24*, 236–238. [CrossRef]
30. Nadeem, U.; Xie, B.; Movahedan, A.; D'Souza, M.; Barba, H.; Deng, N.; Leone, V.A.; Chang, E.; Sulakhe, D.; Skondra, D. High Throughput RNA Sequencing of Mice Retina Reveals Metabolic Pathways Involved in the Gut-Retina Axis. *bioRxiv* **2020**. [CrossRef]
31. Orešič, M.; Seppänen-Laakso, T.; Yetukuri, L.; Bäckhed, F.; Hänninen, V. Gut Microbiota Affects Lens and Retinal Lipid Composition. *Exp. Eye Res.* **2009**, *89*, 604–607. [CrossRef]
32. Belkaid, Y.; Hand, T.W. Role of the Microbiota in Immunity and Inflammation. *Cell* **2014**, *157*, 121–141. [CrossRef]
33. Do, M.H.; Lee, E.; Oh, M.-J.; Kim, Y.; Park, H.-Y. High-Glucose or -Fructose Diet Cause Changes of the Gut Microbiota and Metabolic Disorders in Mice without Body Weight Change. *Nutrients* **2018**, *10*, 761. [CrossRef]
34. Murphy, E.A.; Velazquez, K.T.; Herbert, K.M. Influence of High-Fat Diet on Gut Microbiota: A Driving Force for Chronic Disease Risk. *Curr. Opin. Clin. Nutr. Metab. Care* **2015**, *18*, 515–520. [CrossRef]
35. David, L.A.; Maurice, C.F.; Carmody, R.N.; Gootenberg, D.B.; Button, J.E.; Wolfe, B.E.; Ling, A.V.; Devlin, A.S.; Varma, Y.; Fischbach, M.A.; et al. Diet Rapidly and Reproducibly Alters the Human Gut Microbiome. *Nature* **2014**, *505*, 559–563. [CrossRef] [PubMed]
36. Theriault, B.; Wang, Y.; Chen, L.; Vest, A.; Bartman, C.; Alegre, M.-L. Long-Term Maintenance of Sterility After Skin Transplantation in Germ-Free Mice. *Transplant. Direct* **2015**, *1*, e28. [CrossRef] [PubMed]
37. Chen, Y.; Lun, A.T.L.; Smyth, G.K. From Reads to Genes to Pathways: Differential Expression Analysis of RNA-Seq Experiments Using Rsubread and the EdgeR Quasi-Likelihood Pipeline. *F1000Research* **2016**, *5*, 1438. [PubMed]
38. Robinson, M.D.; McCarthy, D.J.; Smyth, G.K. edgeR: A Bioconductor Package for Differential Expression Analysis of Digital Gene Expression Data. *Bioinformatics* **2010**, *26*, 139–140. [CrossRef]
39. Xie, Z.; Bailey, A.; Kuleshov, M.V.; Clarke, D.J.B.; Evangelista, J.E.; Jenkins, S.L.; Lachmann, A.; Wojciechowicz, M.L.; Kropiwnicki, E.; Jagodnik, K.M.; et al. Gene Set Knowledge Discovery with Enrichr. *Curr. Protoc.* **2021**, *1*, e90. [CrossRef]
40. Kuleshov, M.V. Enrichr: A comprehensive gene set enrichment analysis web server 2016 update. *Nucleic Acids Res.* **2016**, *44*, 90–97. [CrossRef]
41. Chen, E.Y.; Tan, C.M.; Kou, Y.; Duan, Q.; Wang, Z.; Meirelles, G.V.; Clark, N.R.; Ma'ayan, A. Enrichr: Interactive and Collaborative HTML5 Gene List Enrichment Analysis Tool. *BMC Bioinform.* **2013**, *14*, 128. [CrossRef] [PubMed]
42. Keenan, T.D.; Agrón, E.; Mares, J.; Clemons, T.E.; van Asten, F.; Swaroop, A.; Chew, E.Y.; Age-Related Eye Disease Studies (AREDS) 1 and 2 Research Groups. Adherence to the Mediterranean Diet and Progression to Late Age-Related Macular Degeneration in the Age-Related Eye Disease Studies 1 and 2. *Ophthalmology* **2020**, *127*, 1515–1528. [CrossRef]
43. Dow, C.; Mancini, F.; Rajaobelina, K.; Boutron-Ruault, M.-C.; Balkau, B.; Bonnet, F.; Fagherazzi, G. Diet and Risk of Diabetic Retinopathy: A Systematic Review. *Eur. J. Epidemiol.* **2018**, *33*, 141–156. [CrossRef] [PubMed]
44. Wang, Y.E.; Tseng, V.L.; Yu, F.; Caprioli, J.; Coleman, A.L. Association of Dietary Fatty Acid Intake With Glaucoma in the United States. *JAMA Ophthalmol.* **2018**, *136*, 141–147. [CrossRef] [PubMed]
45. Agrón, E.; Mares, J.; Clemons, T.E.; Swaroop, A.; Chew, E.Y.; Keenan, T.D.L.; AREDS and AREDS2 Research Groups. Dietary Nutrient Intake and Progression to Late Age-Related Macular Degeneration in the Age-Related Eye Disease Studies 1 and 2. *Ophthalmology* **2021**, *128*, 425–442. [CrossRef] [PubMed]
46. Kennedy, E.A.; King, K.Y.; Baldridge, M.T. Mouse Microbiota Models: Comparing Germ-Free Mice and Antibiotics Treatment as Tools for Modifying Gut Bacteria. *Front. Physiol.* **2018**, *9*, 1534. [CrossRef] [PubMed]
47. Boyer, D.S.; Schmidt-Erfurth, U.; Lookeren Campagne, M.; Henry, E.C.; Brittain, C. The Pathophysiology of Geographic Atrophy Secondary to Age-Related Macular Degeneration and the Complement Pathway as a Therapeutic Target. *Retina* **2017**, *37*, 819–835. [CrossRef] [PubMed]
48. Lorés-Motta, L.; Paun, C.C.; Corominas, J.; Pauper, M.; Geerlings, M.J.; Altay, L.; Schick, T.; Daha, M.R.; Fauser, S.; Hoyng, C.B.; et al. Genome-Wide Association Study Reveals Variants in CFH and CFHR4 Associated with Systemic Complement Activation: Implications in Age-Related Macular Degeneration. *Ophthalmology* **2018**, *125*, 1064–1074. [CrossRef]
49. Kassa, E.; Ciulla, T.A.; Hussain, R.M.; Dugel, P.U. Complement Inhibition as a Therapeutic Strategy in Retinal Disorders. *Expert Opin. Biol. Ther.* **2019**, *19*, 335–342. [CrossRef]
50. Lee, J.; Yoo, J.H.; Kim, H.S.; Cho, Y.K.; Lee, Y.L.; Lee, W.J.; Park, J.-Y.; Jung, C.H. C1q/TNF-Related Protein-9 Attenuates Palmitic Acid-Induced Endothelial Cell Senescence via Increasing Autophagy. *Mol. Cell. Endocrinol.* **2021**, *521*, 111114. [CrossRef]

51. Chavali, V.R.M.; Sommer, J.R.; Petters, R.M.; Ayyagari, R. Identification of a Promoter for the Human C1q-Tumor Necrosis Factor–Related Protein-5 Gene Associated with Late-Onset Retinal Degeneration. *Investig. Ophthalmol. Vis. Sci.* **2010**, *51*, 5499–5507. [CrossRef] [PubMed]

52. Shinomiya, K.; Mukai, A.; Yoneda, K.; Ueno, M.; Sotozono, C.; Kinoshita, S.; Hamuro, J. The High Expression of C1q and Tumor Necrosis Factor- Related Protein (CTRP) 6, a New Complement Regulatory Factor, in the Drusen of Age-Related Macular Degeneration Eyes. *Investig. Ophthalmol. Vis. Sci.* **2016**, *57*, 5013.

53. Cheng, Y.; Qi, Y.; Liu, S.; Di, R.; Shi, Q.; Li, J.; Pei, C. C1q/TNF-Related Protein 9 Inhibits High Glucose-Induced Oxidative Stress and Apoptosis in Retinal Pigment Epithelial Cells Through the Activation of AMPK/Nrf2 Signaling Pathway. *Cell Transplant.* **2020**, *29*, 963689720962052. [CrossRef] [PubMed]

54. Unterholzner, L.; Keating, S.E.; Baran, M.; Horan, K.A.; Jensen, S.B.; Sharma, S.; Sirois, C.M.; Jin, T.; Latz, E.; Xiao, T.S.; et al. IFI16 Is an Innate Immune Sensor for Intracellular DNA. *Nat. Immunol.* **2010**, *11*, 997–1004. [CrossRef]

55. Chen, W.; Yu, S.-X.; Zhou, F.-H.; Zhang, X.-J.; Gao, W.-Y.; Li, K.-Y.; Liu, Z.-Z.; Han, W.-Y.; Yang, Y.-J. DNA Sensor IFI204 Contributes to Host Defense Against Staphylococcus Aureus Infection in Mice. *Front. Immunol.* **2019**, *10*, 474. [CrossRef] [PubMed]

56. Chen, M.; Muckersie, E.; Forrester, J.V.; Xu, H. Immune Activation in Retinal Aging: A Gene Expression Study. *Investig. Ophthalmol. Vis. Sci.* **2010**, *51*, 5888–5896. [CrossRef] [PubMed]

57. Chen, M.; Luo, C.; Zhao, J.; Devarajan, G.; Xu, H. Immune Regulation in the Aging Retina. *Prog. Retin. Eye Res.* **2019**, *69*, 159–172. [CrossRef]

58. Zhong, Q.; Kowluru, R.A. Role of Histone Acetylation in the Development of Diabetic Retinopathy and the Metabolic Memory Phenomenon. *J Cell Biochem.* **2010**, *110*, 1306–1313. [CrossRef]

59. Villeneuve, L.M.; Reddy, M.A.; Lanting, L.L.; Wang, M.; Meng, L.; Natarajan, R. Epigenetic Histone H3 Lysine 9 Methylation in Metabolic Memory and Inflammatory Phenotype of Vascular Smooth Muscle Cells in Diabetes. *Proc. Natl. Acad. Sci. USA* **2008**, *105*, 9047–9052. [CrossRef]

60. Banday, A.R.; Baumgartner, M.; Al Seesi, S.; Karunakaran, D.K.P.; Venkatesh, A.; Congdon, S.; Lemoine, C.; Kilcollins, A.M.; Mandoiu, I.; Punzo, C.; et al. Replication-Dependent Histone Genes Are Actively Transcribed in Differentiating and Aging Retinal Neurons. *Cell Cycle* **2014**, *13*, 2526–2541. [CrossRef] [PubMed]

61. Steele, F.R.; Chader, G.J.; Johnson, L.V.; Tombran-Tink, J. Pigment Epithelium-Derived Factor: Neurotrophic Activity and Identification as a Member of the Serine Protease Inhibitor Gene Family. *Proc. Natl. Acad. Sci. USA* **1993**, *90*, 1526–1530. [CrossRef]

62. He, X.; Cheng, R.; Benyajati, S.; Ma, J.-X. PEDF and Its Roles in Physiological and Pathological Conditions: Implication in Diabetic and Hypoxia-Induced Angiogenic Diseases. *Clin. Sci.* **2015**, *128*, 805–823. [CrossRef] [PubMed]

63. Youngblood, H.; Robinson, R.; Sharma, A.; Sharma, S. Proteomic Biomarkers of Retinal Inflammation in Diabetic Retinopathy. *Int. J. Mol. Sci.* **2019**, *20*, 4755. [CrossRef] [PubMed]

64. Kou, B.; Ni, J.; Vatish, M.; Singer, D.R.J. Xanthine Oxidase Interaction with Vascular Endothelial Growth Factor in Human Endothelial Cell Angiogenesis. *Microcirculation* **2008**, *15*, 251–267. [CrossRef] [PubMed]

65. Elmasry, K.; Ibrahim, A.S.; Abdulmoneim, S.; Al-Shabrawey, M. Bioactive Lipids and Pathological Retinal Angiogenesis. *Br. J. Pharmacol.* **2019**, *176*, 93–109. [CrossRef] [PubMed]

66. Sadeqzadeh, E.; Bock, C.E.; Thorne, R.F. Sleeping Giants: Emerging Roles for the Fat Cadherins in Health and Disease. *Med. Res. Rev.* **2014**, *34*, 190–221. [CrossRef] [PubMed]

67. Horne-Badovinac, S. Fat-like Cadherins in Cell Migration—Leading from Both the Front and the Back. *Curr. Opin. Cell Biol.* **2017**, *48*, 26–32. [CrossRef]

68. Cao, J.; Ehling, M.; März, S.; Seebach, J.; Tarbashevich, K.; Sixta, T.; Pitulescu, M.E.; Werner, A.-C.; Flach, B.; Montanez, E.; et al. Polarized Actin and VE-Cadherin Dynamics Regulate Junctional Remodelling and Cell Migration during Sprouting Angiogenesis. *Nat. Commun.* **2017**, *8*, 2210. [CrossRef] [PubMed]

69. Nguyen, M.-T.T. An Opsin 5–Dopamine Pathway Mediates Light-Dependent Vascular Development in the Eye. *Nat. Cell Biol.* **2019**, *21*, 420–429. [CrossRef]

70. Buhr, E.D.; Yue, W.W.S.; Ren, X.; Jiang, Z.; Liao, H.-W.R.; Mei, X.; Vemaraju, S.; Nguyen, M.-T.; Reed, R.R.; Lang, R.A.; et al. Neuropsin (OPN5)-Mediated Photoentrainment of Local Circadian Oscillators in Mammalian Retina and Cornea. *Proc. Natl. Acad. Sci. USA* **2015**, *112*, 13093–13098. [CrossRef]

71. Aaltonen, V.; Kinnunen, K.; Jouhilahti, E.-M.; Peltonen, J.; Nikinmaa, M.; Kaarniranta, K.; Arjamaa, O. Hypoxic Conditions Stimulate the Release of B-Type Natriuretic Peptide from Human Retinal Pigment Epithelium Cell Culture. *Acta Ophthalmol. (Copenh.)* **2014**, *92*, 740–744. [CrossRef] [PubMed]

72. Rollín, R.; Mediero, A.; Roldán-Pallarés, M.; Fernández-Cruz, A.; Fernández-Durango, R. Natriuretic Peptide System in the Human Retina. *Mol. Vis.* **2004**, *10*, 15–22.

73. Jovancevic, N.; Wunderlich, K.A.; Haering, C.; Flegel, C.; Maßberg, D.; Weinrich, M.; Weber, L.; Tebbe, L.; Kampik, A.; Gisselmann, G.; et al. Deep Sequencing of the Human Retinae Reveals the Expression of Odorant Receptors. *Front. Cell. Neurosci.* **2017**, *11*, 3. [CrossRef]

74. Pavan, B.; Capuzzo, A.; Dalpiaz, A. Potential Therapeutic Effects of Odorants through Their Ectopic Receptors in Pigmented Cells. *Drug Discov. Today* **2017**, *22*, 1123–1130. [CrossRef]

75. Pavan, B.; Dalpiaz, A. Retinal Pigment Epithelial Cells as a Therapeutic Tool and Target against Retinopathies. *Drug Discov. Today* **2018**, *23*, 1672–1679. [CrossRef]

76. Jovancevic, N.; Khalfaoui, S.; Weinrich, M.; Weidinger, D.; Simon, A.; Kalbe, B.; Kernt, M.; Kampik, A.; Gisselmann, G.; Gelis, L.; et al. Odorant Receptor 51E2 Agonist β-Ionone Regulates RPE Cell Migration and Proliferation. *Front. Physiol.* **2017**, *8*, 888. [CrossRef]

77. Yoshimura, T.; Watanabe, T.; Kuramochi-Miyagawa, S.; Takemoto, N.; Shiromoto, Y.; Kudo, A.; Kanai-Azuma, M.; Tashiro, F.; Miyazaki, S.; Katanaya, A.; et al. Mouse GTSF1 Is an Essential Factor for Secondary PiRNA Biogenesis. *EMBO Rep.* **2018**, *19*, e42054. [CrossRef]

78. Fukuda, S.; Varshney, A.; Fowler, B.J.; Wang, S.-B.; Narendran, S.; Ambati, K.; Yasuma, T.; Magagnoli, J.; Leung, H.; Hirahara, S.; et al. Cytoplasmic Synthesis of Endogenous Alu Complementary DNA via Reverse Transcription and Implications in Age-Related Macular Degeneration. *Proc. Natl. Acad. Sci. USA* **2021**, *118*, e2022751118. [CrossRef] [PubMed]

79. Zurdel, J.; Finckh, U.; Menzer, G.; Nitsch, R.M.; Richard, G. CST3 Genotype Associated with Exudative Age Related Macular Degeneration. *Br. J. Ophthalmol.* **2002**, *86*, 214–219. [CrossRef] [PubMed]

80. Crawford, F.C.; Freeman, M.J.; Schinka, J.A.; Abdullah, L.I.; Gold, M.; Hartman, R.; Krivian, K.; Morris, M.D.; Richards, D.; Duara, R.; et al. A Polymorphism in the Cystatin C Gene Is a Novel Risk Factor for Late-Onset Alzheimer's Disease. *Neurology* **2000**, *55*, 763–768. [CrossRef] [PubMed]

81. Butler, J.M.; Sharif, U.; Ali, M.; McKibbin, M.; Thompson, J.P.; Gale, R.; Yang, Y.C.; Inglehearn, C.; Paraoan, L. A Missense Variant in CST3 Exerts a Recessive Effect on Susceptibility to Age-Related Macular Degeneration Resembling Its Association with Alzheimer's Disease. *Hum. Genet.* **2015**, *134*, 705–715. [CrossRef]

82. Carlsson, E.; Supharattanasitthi, W.; Jackson, M.; Paraoan, L. Increased Rate of Retinal Pigment Epithelial Cell Migration and Pro-Angiogenic Potential Ensuing From Reduced Cystatin C Expression. *Investig. Ophthalmol. Vis. Sci.* **2020**, *61*, 9. [CrossRef]

83. Yamamoto, S.; Yabuki, R.; Kitagawa, D. Biophysical and Biochemical Properties of Deup1 Self-Assemblies: A Potential Driver for Deuterosome Formation during Multiciliogenesis. *Biol. Open* **2021**, *10*, bio056432. [CrossRef]

84. Bujakowska, K.M.; Liu, Q.; Pierce, E.A. Photoreceptor Cilia and Retinal Ciliopathies. *Cold Spring Harb. Perspect. Biol.* **2017**, *9*, a028274. [CrossRef]

85. Martinez, G.; Beurois, J.; Dacheux, D.; Cazin, C.; Bidart, M.; Kherraf, Z.-E.; Robinson, D.R.; Satre, V.; Le Gac, G.; Ka, C.; et al. Biallelic Variants in MAATS1 Encoding CFAP91, a Calmodulin-Associated and Spoke-Associated Complex Protein, Cause Severe Astheno-Teratozoospermia and Male Infertility. *J. Med. Genet.* **2020**, *57*, 708–716. [CrossRef] [PubMed]

86. Kurashige, T.; Morino, H.; Matsuda, Y.; Mukai, T.; Murao, T.; Toko, M.; Kume, K.; Ohsawa, R.; Torii, T.; Tokinobu, H.; et al. Retinitis Pigmentosa Prior to Familial ALS Caused by a Homozygous Cilia and Flagella-Associated Protein 410 Mutation. *J. Neurol. Neurosurg. Psychiatry* **2020**, *91*, 220–222. [CrossRef] [PubMed]

87. Santos-Carvalho, A.; Ambrósio, A.F.; Cavadas, C. Neuropeptide Y System in the Retina: From Localization to Function. *Prog. Retin. Eye Res.* **2015**, *47*, 19–37. [CrossRef]

88. Jaakkola, U.; Pesonen, U.; Vainio-Jylhä, E.; Koulu, M.; Pöllönen, M.; Kallio, J. The Leu7Pro Polymorphism of Neuropeptide Y Is Associated with Younger Age of Onset of Type 2 Diabetes Mellitus and Increased Risk for Nephropathy in Subjects with Diabetic Retinopathy. *Exp. Clin. Endocrinol. Diabetes Off. J. Ger. Soc. Endocrinol. Ger. Diabetes Assoc.* **2006**, *114*, 147–152. [CrossRef]

89. Koulu, M.; Movafagh, S.; Tuohimaa, J.; Jaakkola, U.; Kallio, J.; Pesonen, U.; Geng, Y.; Karvonen, M.K.; Vainio-Jylhä, E.; Pöllönen, M.; et al. Neuropeptide Y and Y2-Receptor Are Involved in Development of Diabetic Retinopathy and Retinal Neovascularization. *Ann. Med.* **2004**, *36*, 232–240. [CrossRef] [PubMed]

90. Dvorakova, M.C. Distribution and Function of Neuropeptides W/B Signaling System. *Front. Physiol.* **2018**, *9*, 981. [CrossRef]

91. Singh, G.; Davenport, A.P. Neuropeptide B and W: Neurotransmitters in an Emerging G-Protein-Coupled Receptor System. *Br. J. Pharmacol.* **2006**, *148*, 1033–1041. [CrossRef] [PubMed]

92. Zhan, W.; Liu, Y.; Gao, Y.; Gong, R.; Wang, W.; Zhang, R.; Wu, Y.; Kang, T.; Wei, D. RMI2 Plays Crucial Roles in Growth and Metastasis of Lung Cancer. *Signal Transduct. Target. Ther.* **2020**, *5*, 1–3. [CrossRef] [PubMed]

93. Liu, J.; Nie, S.; Gao, M.; Jiang, Y.; Wan, Y.; Ma, X.; Zhou, S.; Cheng, W. Identification of EPHX2 and RMI2 as Two Novel Key Genes in Cervical Squamous Cell Carcinoma by an Integrated Bioinformatic Analysis. *J. Cell. Physiol.* **2019**, *234*, 21260–21273. [CrossRef]

94. Hudson, D.F.; Amor, D.J.; Boys, A.; Butler, K.; Williams, L.; Zhang, T.; Kalitsis, P. Loss of RMI2 Increases Genome Instability and Causes a Bloom-Like Syndrome. *PLoS Genet.* **2016**, *12*, e1006483. [CrossRef] [PubMed]

95. Choi, E.; Lee, H.; Chung, J.; Seo, W.; Kim, G.; Ki, C.; Kim, Y.; Bang, O.Y. Ring Finger Protein 213 Variant and Plaque Characteristics, Vascular Remodeling, and Hemodynamics in Patients With Intracranial Atherosclerotic Stroke: A High-Resolution Magnetic Resonance Imaging and Hemodynamic Study. *J. Am. Heart Assoc.* **2019**, *8*, e011996. [CrossRef]

96. Mapes, J.; Li, Q.; Kannan, A.; Anandan, L.; Laws, M.; Lydon, J.P.; Bagchi, I.C.; Bagchi, M.K. CUZD1 Is a Critical Mediator of the JAK/STAT5 Signaling Pathway That Controls Mammary Gland Development during Pregnancy. *PLoS Genet.* **2017**, *13*, e1006654. [CrossRef] [PubMed]

97. Brink, T.C.; Sudheer, S.; Janke, D.; Jagodzinska, J.; Jung, M.; Adjaye, J. The Origins of Human Embryonic Stem Cells: A Biological Conundrum. *Cells Tissues Organs* **2008**, *188*, 9–22. [CrossRef] [PubMed]

98. Schmidt, S.; Hauser, M.A.; Scott, W.K.; Postel, E.A.; Agarwal, A.; Gallins, P.; Wong, F.; Chen, Y.S.; Spencer, K.; Schnetz-Boutaud, N.; et al. Cigarette Smoking Strongly Modifies the Association of LOC387715 and Age-Related Macular Degeneration. *Am. J. Hum. Genet.* **2006**, *78*, 852–864. [CrossRef]

99. Wortmann, S.B.; Mayr, J.A. Choline-related-inherited metabolic diseases—A mini review. *J. Inherit. Metab. Dis.* **2019**, *42*, 237–242. [CrossRef]

100. Govindarajan, G.; Mathews, S.; Srinivasan, K.; Ramasamy, K.; Periasamy, S. Establishment of Human Retinal Mitoscriptome Gene Expression Signature for Diabetic Retinopathy Using Cadaver Eyes. *Mitochondrion* **2017**, *36*, 150–181. [CrossRef]
101. Kandpal, R.P.; Rajasimha, H.K.; Brooks, M.J.; Nellissery, J.; Wan, J.; Qian, J.; Kern, T.S.; Swaroop, A. Transcriptome Analysis Using next Generation Sequencing Reveals Molecular Signatures of Diabetic Retinopathy and Efficacy of Candidate Drugs. *Mol. Vis.* **2012**, *18*, 1123–1146.
102. Manzoni, C.; Kia, D.A.; Vandrovcova, J.; Hardy, J.; Wood, N.W.; Lewis, P.A.; Ferrari, R. Genome, Transcriptome and Proteome: The Rise of Omics Data and Their Integration in Biomedical Sciences. *Brief. Bioinform.* **2018**, *19*, 286–302. [CrossRef]

cells

Article

Regional Variation of Gap Junctional Connections in the Mammalian Inner Retina

Katalin Fusz [1,2,3], Tamás Kovács-Öller [2,3,4], Péter Kóbor [1,2,3], Edina Szabó-Meleg [2,5], Béla Völgyi [2,3,4,6], Péter Buzás [1,2,3,*] and Ildikó Telkes [1,2,3]

1 Institute of Physiology, Medical School, University of Pécs, 7624 Pécs, Hungary; katalin.fusz@aok.pte.hu (K.F.); peter.kobor@aok.pte.hu (P.K.); ildiko.telkes@aok.pte.hu (I.T.)
2 Szentágothai Research Centre, University of Pécs, 7624 Pécs, Hungary; kovacs-oller.tamas@pte.hu (T.K.-Ö.); edina.meleg@aok.pte.hu (E.S.-M.); volgyi01@gamma.ttk.pte.hu (B.V.)
3 Centre for Neuroscience, University of Pécs, 7624 Pécs, Hungary
4 MTA-PTE NAP-2 Retinal Electrical Synapses Research Group, 7624 Pécs, Hungary
5 Institute of Biophysics, Medical School, University of Pécs, 7624 Pécs, Hungary
6 Department of Experimental Zoology and Neurobiology, University of Pécs, 7624 Pécs, Hungary
* Correspondence: peter.buzas@aok.pte.hu; Tel.: +36-72-536000

Abstract: The retinas of many species show regional specialisations that are evident in the differences in the processing of visual input from different parts of the visual field. Regional specialisation is thought to retflect an adaptation to the natural visual environment, optical constraints, and lifestyle of the species. Yet, little is known about regional differences in synaptic circuitry. Here, we were interested in the topographical distribution of connexin-36 (Cx36), the major constituent of electrical synapses in the retina. We compared the retinas of mice, rats, and cats to include species with different patterns of regional specialisations in the analysis. First, we used the density of Prox1-immunoreactive amacrine cells as a marker of any regional specialisation, with higher cell density signifying more central regions. Double-labelling experiments showed that Prox1 is expressed in AII amacrine cells in all three species. Interestingly, large Cx36 plaques were attached to about 8–10% of Prox1-positive amacrine cell somata, suggesting the strong electrical coupling of pairs or small clusters of cell bodies. When analysing the regional changes in the volumetric density of Cx36-immunoreactive plaques, we found a tight correlation with the density of Prox1-expressing amacrine cells in the ON, but not in the OFF sublamina in all three species. The results suggest that the relative contribution of electrical synapses to the ON- and OFF-pathways of the retina changes with retinal location, which may contribute to functional ON/OFF asymmetries across the visual field.

Keywords: AII amacrine cell; Prox1; parvalbumin; gap junction; eccentricity; ON/OFF asymmetry

Citation: Fusz, K.; Kovács-Öller, T.; Kóbor, P.; Szabó-Meleg, E.; Völgyi, B.; Buzás, P.; Telkes, I. Regional Variation of Gap Junctional Connections in the Mammalian Inner Retina. *Cells* **2021**, *10*, 2396. https://doi.org/10.3390/cells10092396

Academic Editors: Maurice Ptito and Joseph Bouskila

Received: 6 August 2021
Accepted: 7 September 2021
Published: 12 September 2021

1. Introduction

The retinas of most species show some type of regional specialisation that is evident in the distribution of photoreceptors and other cell types. Regional specialisations are thought to reflect an adaptation to the natural visual environment, optical constraints, and lifestyle of the species [1,2]. The best-known example is the peaking density of photoreceptors around the optical centre, such as in the fovea of primates [3] or area centralis of carnivores [4,5]. In the central regions, the neural circuitry is specialised for high-acuity vision, which in primates is also reflected by the number of cones converging on bipolar cells and eventually, on ganglion cells [6,7]. A number of perceptual functions also vary with visual field eccentricity, including not only visual acuity, but also contrast sensitivity, colour sensitivity, critical fusion frequency, motion perception, reaction time and crowding, many of which can be traced back to regional differences across the retina (see [8,9] for a review).

Electrical synapses contribute to diverse microcircuits in the retina that underlie a variety of functions, including the transmission of rod signals to ganglion cells [10–14], surround suppression [15,16], the synchronization of ganglion cell firing [17–20] and dynamic adaptation of these functions to changes in light level or circadian rhythms [21]. Connexin-36 (Cx36), the major constituent of mammalian retinal gap junctions, is positioned in key signal pathways [22]. For example, in the outer retina, gap junctions formed by Cx36 connect cone photoreceptors [13,23–26] with each other. Cx36 gap junctions connecting rods to cones form the secondary rod pathway [10,11], which is thought to be responsible for light detection at mid-scotopic intensity levels [14,27].

In the inner retina, gap junctions are found on all major cell types but interestingly, their involvement in ON and OFF pathways appears to be asymmetric. First, among the ON and OFF-sublaminae of the inner plexiform layer (IPL), the ON-sublamina contains more gap junctions [28]. There are two gap junction pathways that are formed exclusively in the ON sublamina, both involving AII amacrine cells. One of them is between AII cells and ON BCs and serves the signalling through the primary rod pathway. The second one connects AII cells into a dense homologous network [21,22,29–33]. There are four subpopulations of gap junctions maintained in the OFF sublamina, including two formed by retinal ganglion cells (RGCs) either with neighbouring RGCs or amacrine cells [34–37], a third population that connects amacrine cells to one another and a fourth population that exists between bipolar cells [38–40]. Three types out of these gap junctional subpopulations are abundant in the ON sublamina as well, including bipolar-bipolar [41], amacrine–amacrine [42–46] and RGC–amacrine [18,21,22,36] contacts. Direct RGC-RGC gap junctions, however, seem to be considerably less frequent here than in the OFF sublamina [36]. These microcircuits have been studied extensively at the local level, but their relationship to the topography of the retina is essentially unknown. Therefore, we were interested in obtaining a large-scale view of how the contribution of gap junctional connections to ON- and OFF-pathways changes across the retina.

Previously, we have observed a centre-periphery gradient in the density of Cx36 immunolabelled puncta in the ON-sublamina of the inner plexiform layer of cats [47]. In the present study, we extend these observations to any eccentricity-driven variation of inner retinal gap junction distribution in cats, rats and mice that represent species with different patterns of retinal specialisations. In addition, we compared the density of electrical contacts between the ON and OFF sublayers of the inner plexiform layer of all three animal models. In order to do this, plaque distributions were correlated with the AII amacrine cell coverage, as AII cells contribute considerably to gap junctional contacts in the ON, but not in the OFF sublaminae.

2. Materials and Methods

2.1. Animals and Sample Preparation

Retinas of 4 adult cats (*Felis catus*, 1 male aged 0.85 year, 3 females, aged 2.75, 8 and 8 years, respectively), retinas of 4 adult rats (*Rattus norvegicus*, Wistar, male, body weight 300–400 g) and retinas of 5 mice (*Mus musculus*), were used (males, 1–12 months old). Wild-type mice were from strain C57BL/6J ($n = 3$). The genetically modified animals referred to as PV-tdT (parvalbumin-tdTomato, $n = 2$) were cross-bred from PV-Cre (JAX #017320) and tdTomato (JAX #007909) at the animal house of the Szentágothai Research Centre, University of Pécs, Hungary. The sources of animals are listed in Table 1. The animals were kept, and the experiments were performed in accordance with Hungarian and European legislation. All procedures were approved by the Directorate for Food Chain Safety and Animal Health of the Baranya County Government Office, Hungary.

Table 1. List of key resources.

Reagent or Resource	Source	Identifier
Antibodies		
rabbit polyclonal anti-calretinin	Merck Hungary, Budapest, Hungary	AB 5054
mouse monoclonal anti-connexin-36	Merck Hungary, Budapest, Hungary	MAB 3045
rabbit polyclonal anti-Prox1	Merck Hungary, Budapest, Hungary	ABN 278
mouse monoclonal anti-parvalbumin	Swant, Burgdorf, Switzerland	PV 235
mouse monoclonal anti-calbindin	Synaptic Systems, Göttingen, German	214011
biotinylated anti-mouse IgG (H + L)	Vector Laboratories, Burlingame, CA, USA	BA 2001
streptavidin-Alexa Fluor 488 conjugate	Invitrogen, Waltham, MA, USA	S 32354
donkey anti-mouse IgG (H + L) Alexa Fluor 488 conjugate	Jackson ImmunoResearch, West Grove, PA, USA	715-545-151
donkey anti-rabbit IgG (H + L) Alexa Fluor 594 conjugate	Jackson ImmunoResearch, West Grove, PA, USA	711-585-152
goat anti-rabbit IgG (H + L) Texas Red conjugate	Jackson ImmunoResearch, West Grove, PA, USA	111-075-003
Animals		
domestic cat (*Felis catus*)	Animal house of the Institute of Physiology, Medical School, University of Pécs, Hungary	N/A
rat (*Rattus norvegicus*), Wistar strain	Animal house of the Institute of Physiology, Medical School, University of Pécs, Hungary	N/A
mouse (*Mus musculus*), C57BL/6J strain	The Jackson Laboratory, Bar Harbor, USA	000664
mouse (*Mus musculus*), PV-Cre line	The Jackson Laboratory, Bar Harbor, USA	B6.129P2-Pvalb$^{tm1(cre)Arbr}$/J, JAX #017320
mouse (*Mus musculus*), tdTomato line	The Jackson Laboratory, Bar Harbor, USA	B6.Cg-Gt(ROSA)26Sor$^{tm9(CAG-tdTomato)Hze}$/J, JAX #007909
Equipment		
Zeiss LSM 710 confocal laser scanning microscope	Carl Zeiss, Jena, Germany	
Software		
Fiji	https://imagej.net/software/fiji/, accessed on 27 July 2018	N/A
Imaris 8.12	Oxford Instruments, Zurich, Switzerland	N/A
SPSS 24.0	IBM Corporation, Armonk, NY, USA	N/A
Excel for Office 365	Microsoft, Redmond, WA, USA	N/A

Cats were overdosed with 5% isoflurane followed by lethal injection of T61 (embutramide 250 mg/kg, tetracaine HCl 6.25 mg/kg, mebezonium iodide 63 mg/kg, Intervet, Boxmeer, The Netherlands) following unrelated physiological experiments. One animal was perfused intracardially with 4% paraformaldehyde in PBS (0.1 M phosphate-buffered saline, pH 7.5), the others were enucleated immediately after anaesthetic overdose. The eyes were cut along the *ora serrata* and the vitreous body was removed. The posterior eyecups were postfixed overnight at +4 °C and transferred into cold PBS. The retinas were prepared and cut into upper, lower, nasal, and temporal quadrants using the optic disk as the centre.

Rats were overdosed with 5% isoflurane and perfused intracardially with 4% paraform aldehyde in PBS. The eyes were cut along the *ora serrata* and the lens and vitreous body were removed. The posterior eyecups were postfixed overnight at +4 °C in 4% paraformaldehyde in PBS and the retinas were prepared in cold PBS.

Mice were sacrificed after isoflurane anaesthesia (0.2 mL/l) by cervical dislocation. The eyes were removed immediately after termination. Eyeballs were cut at the *ora serrata*, then the lens and vitreous body were removed. Retinas were fixed in 4% paraformaldehyde in PBS at room temperature for 15 min.

2.2. Immunohistochemistry and Confocal Microscopy

Free-floating retinal quadrants or whole retinas were first incubated with a blocking solution composed of 10% normal goat serum in antibody diluting solution (0.25% bovine serum albumin, 0.001% sodium azide, and 0.2% Triton X-100 in 0.1 M PBS) for 2 days. The same solution was used for all further antibodies unless stated otherwise. Tissue samples were then incubated with the primary antibodies at +4 °C for 4 days using the following dilutions: polyclonal anti-calretinin (CaR) produced in rabbit, 1:2000; monoclonal anti-Cx36 produced in mouse, 1:1000; polyclonal anti-Prox1 produced in rabbit, 1:250; monoclonal anti-parvalbumin produced in mouse, 1:600; monoclonal anti-calbindin produced in mouse, 1:100.

The following steps were done at +4 °C overnight. Cx36 immunoreactivity was visualized using biotinylated anti-mouse IgG (H + L) (1:100 dilution) followed by streptavidin-Alexa Fluor 488 conjugate (1:200 dilution in 0.1 M PBS). CaR immunoreactivity was visualised with goat anti-rabbit antibody Texas Red conjugate (1:100 dilution). Prox1 immunoreactivity was visualised with donkey anti-rabbit Alexa Fluor 594 (1:500 dilution), parvalbumin (PV) and calbindin (CaB) immunoreactivity were visualized with donkey anti-mouse Alexa Fluor 488 (1:500 dilution). The sources of antibodies are listed in Table 1.

We washed the retinal pieces between the incubations five times for 10 min in PBS. Retinal pieces were mounted in Aqua-PolyMount (rats, cats; Polysciences, Warrington, PA, USA) or VectaShield (mice; Vector Labs., Burlingame, California, USA) media with the ganglion cell layer facing the coverslip.

We inspected the flat-mounted retinas using a Zeiss LSM 710 confocal laser scanning microscope (Carl Zeiss, Jena, Germany) through a Plan-Apochromat 63× objective lens (NA 1.4), following a tile-scan at 10× for localisation. We took confocal stacks at selected regions of interest (ROIs, see below); the horizontal size of the ROIs was 135 × 135 μm and the z-stacks spanned depth from the outer nuclear layer to the optic fibres (for Cx36 density measurements) or a narrower range as needed for inspection of histological structures. Voxel size was 0.132 μm × 0.132 μm × 0.381 μm or 0.132 μm × 0.132 μm × 0.5 μm.

2.3. Measurement of Retinal Eccentricity and Feature Density

Regions of interest (ROIs) were selected randomly to cover eccentricities as equally as possible. To calculate eccentricity in cat retinas, first the location of each ROI was measured in polar coordinates with the optic disk as the centre (range of distances from the optic disk 0.16–14.2 mm; mean ± SD 8.14 ± 3.74 mm). These coordinates were then converted into Cartesian coordinates of the entire flattened retina, taking advantage of the fact that neighbouring quadrants had common edges. The eccentricity of each ROI was calculated from the position 3 mm lateral from the optic disk; this position is in good agreement with the location of the area centralis in the cat retina [48,49]. The eccentricities of ROIs were statistically not different in the four quadrants (Kruskal–Wallis test, $p > 0.05$). For mouse retinas, we only determined eccentricity categories of the ROIs (centre, mid-centre, periphery, Supplementary Figure S1).

To determine cell densities, we used the Fiji distribution of the ImageJ software package [50]. Cell bodies of the cell type of interest were identified based on their immuno-histochemical labelling and laminar position and each member of the cell type found in a z-stack was marked using the Cell Counter plugin. Their number was divided by the area of the ROI to obtain an areal density in mm^{-2}.

Cx36 plaques were identified using the 3D View-Surfaces module of the Imaris image processing software (Oxford Instruments, Zurich, Switzerland). This module fitted irregular solid shapes to aggregates of voxels showing increased Cx36 immunolabel. We used the following settings: Thresholding method, Background Subtraction; Diameter of the largest sphere that fits into the object: 0.1 μm. Based on their 3D coordinates, we assigned each Cx36 plaque to one of the 5 strata of the IPL. The Z coordinate ranges of the strata were determined by dividing the Z-distance between the limits of the IPL into five equal parts. Strata 1–2 were assigned to the OFF- sublamina and strata 3–5 were assigned to the ON-sublamina. This also allowed us to calculate the volume of each sublamina for each ROI. We then calculated the volumetric density of the Cx36 plaques by dividing their number found in a ROI by the volume of the IPL sublamina in that ROI. Only Cx36 plaques with a volume between 0.02 μm^3 and 2.5 μm^3 were counted to reduce the effect of pixel noise and to exclude objects too large for a Cx36 plaque.

2.4. 3D Colocalization Analysis

To show parvalbumin-tdTomato (PV-tdT) and Prox1 colocalization, a selected area was scanned by the confocal microscope with a 63x-oil objective at high resolution (voxel size of 0.15 μm × 0.15 μm × 0.4 μm). A side-view rotation was then rendered from this Z-stack using Fiji's '3D viewer' module. Four smaller ROIs (20 μm × 20 μm) containing selected amacrine cells were cut (three with evident colocalization and one with no PV-tdT label only) for subsequential rotations to show dendritic morphology. Intensity profiles comparing Prox1, Cx36 and tdT label along line segments were also produced in Fiji.

2.5. Data Analysis

ANOVA, correlation analysis, linear regression and statistical tests were done in IBM SPSS (IBM Corporation, Armonk, NY, USA). Significance level was set to $p < 0.05$.

3. Results

3.1. Prox1 Immunoreactive Cell Types in Cat, Rat and Mouse Retinas

The homeodomain protein Prox1 is expressed in horizontal cells, bipolar cells and AII amacrine cells of the adult mouse retina [51,52]. Here, we begin with testing the hypothesis that Prox1 is present in the same cell populations of rats as well as cats.

Figure 1 shows retinal whole mounts immunostained for Prox1 with the focal plane on the outer aspect of the inner nuclear layer (INL). Two types of cell nuclei expressing Prox1 could be readily distinguished. One population was relatively sparse (cell densities in cat: 485 ± 80 mm^{-2}, $n = 9$ z-stacks; rat: 838 ± 314 mm^{-2}, $n = 15$; mouse: 510 ± 102 mm^{-2}, $n = 20$), had larger nuclei (horizontal diameters for cat: 9.65 ± 0.29 μm, $n = 9$; rat: 8.27 ± 0.14 μm, $n = 15$; mouse: 9.33 ± 0.19 μm, $n = 21$) and was more intensely stained. In cat retinas, these cells also contained parvalbumin (PV, Figure 1A), a marker of A- and B-type horizontal cells [53,54]. PV immunoreactivity also revealed the dendritic trees so that type-A (larger somata, stouter proximal dendrites) and type-B (smaller somata, thinner proximal dendrites) cells could be distinguished confirming that both horizontal cell types are Prox1 positive in the cat retina. In rat and mouse retinas, there is a single type of horizontal cell [55], which can be identified through calbindin (CaB) immunoreactivity [56] and by their laminar position. We found that all CaB-immunoreactive horizontal cells also contain Prox1 in both rodent species (Figure 1B,C).

Prox1 immunostaining also revealed numerous smaller (cat: 6.09 ± 0.09 μm, $n = 238$, $p < 0.001$, rat: 5.33 ± 0.14 μm, $n = 344$, $p < 0.001$, mouse: 6.30 ± 0.11 μm, $n = 379$, $p < 0.001$ in t-test against horizontal cell profiles) and somewhat less strongly labelled cell nuclei, which did not contain PV or CaB (Figure 1). Their laminar location, high density (cat: 9573 ± 1276 mm^{-2}, $n = 3$ z-stacks; rat: $14,222 \pm 1299$ mm^{-2}, $n = 3$; mouse: 9340 ± 335 mm^{-2}, $n = 3$), and size were similar in all three species, suggesting that many bipolar cells are Prox1-positive in cats and rats just like in the mouse retina [52].

The proximal INL also contained Prox1 immunopositive cells in each species. These cells appeared more heterogeneous in size and staining intensity than the horizontal cells or bipolar cells described above. According to previous reports on mouse retina, the majority or maybe all of the Prox1 positive cells in this layer are AII amacrine cells [52]. In cat and rat retinas, we used double labelling with an anti-PV (Figure 2) and anti-calretinin (CaR, Figure 3) antibodies to further identify Prox1 expressing amacrine cells.

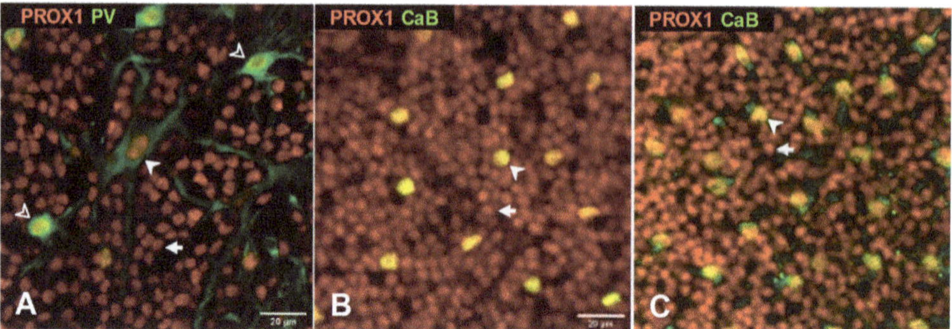

Figure 1. Prox1 immunoreactive cell nuclei (red channel) in the inner nuclear layer of cat (**A**), rat (**B**) and mouse (**C**) retina. Images show a horizontal view of whole-mount preparations at the level of horizontal cells. Nuclei of two types of cells are labelled in each species; larger and sparser horizontal cells (arrowheads) and numerous small bipolar cells (arrows). **A,** In the cat, Prox1 colocalizes with parvalbumin (green channel), which labels somata and dendrites of both A- (filled arrowhead) and B-type (open arrowheads) horizontal cells. In rat (**B**) and mouse (**C**) retina, Prox1 colocalizes with calbindin in the somata of horizontal cells. Scale bar 20 µm.

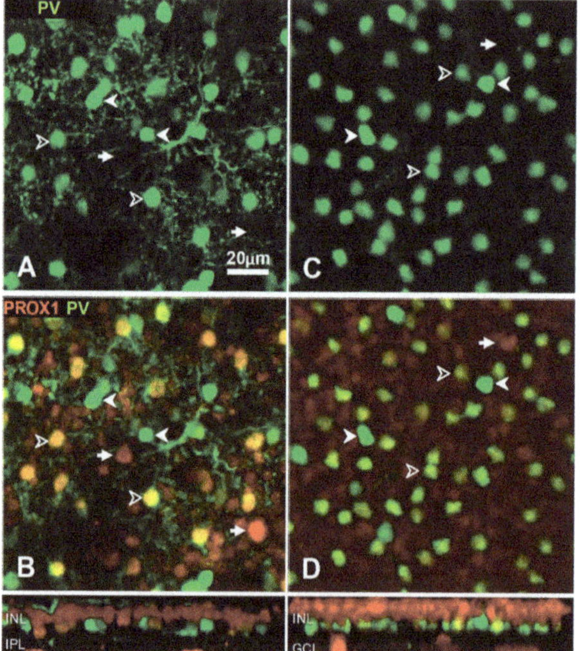

Figure 2. Typing of amacrine cells based on double labelling with parvalbumin (green channel) and Prox1 (red channel) immunoreactivity in cat (**A,B**) and rat (**C,D**) retina. Microscopic images show horizontal views of whole-mount preparations at the level of amacrine cells. Side-view reconstructions of representative sub-volumes spanning the INL (top) IPL and ganglion cell layer (GCL) are shown below (**B,D**). (**A,C**) Anti-parvalbumin antibody labelled two populations of amacrine cells in both species. The weakly parvalbumin-positive neurons (open arrowheads) include (in cats, A) or are exclusively (in rats C) AII amacrine cells. B, D. The combination of parvalbumin and Prox1 immunolabels reveals three amacrine cell types in both species. Amacrine cells with a strong PV expression are Prox1-negative. Weakly PV-positive amacrine cells are Prox-1 positive. A small population of Prox1-positive amacrine cells is PV-negative (arrows). Scale, 20 µm.

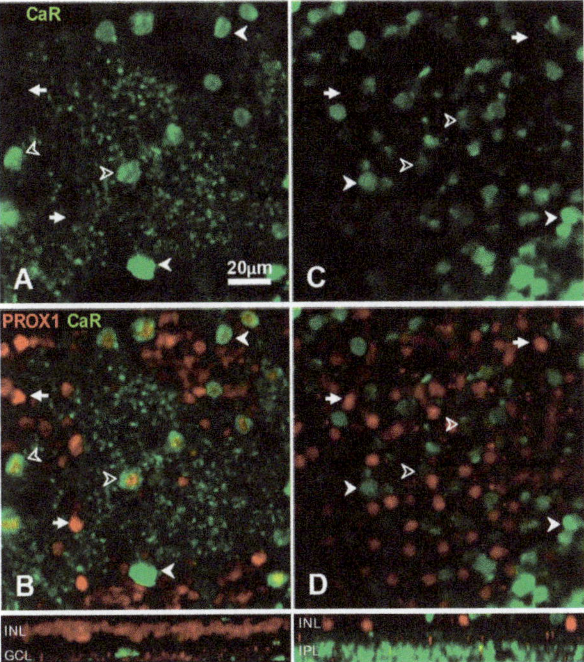

Figure 3. Typing of amacrine cells based on double labelling with calretinin (CaR, green channel) and Prox1 (red channel) immunoreactivity in cat (**A,B**) and rat (**C,D**) retina. Microscopic images show horizontal views of whole-mount preparations at the level of amacrine cells. Side-view reconstructions of representative sub-volumes spanning the INL (top) IPL and ganglion cell layer (GCL) are shown below B and D. In cat retina (**A,B**), the strongest CaR-immunoreactivity (**A**) is seen in sparsely distributed amacrine cells (solid arrowheads), which are Prox1-negative (**B**). AII amacrine cells correspond to the weakly CaR-immunoreactive cells (open arrowheads), which are more numerous and always Prox1 immunoreactive (**B**). Some amacrine cell bodies expressed only Prox1 and no CaR in the cat (arrows). The cell-free regions filled with PV-labelled varicosities belong to the IPL. In rat retina (**C,D**), the anti-CaR antibody did not differentiate amacrine cell types clearly (**C**). A combination with Prox1 immunolabel (**D**) revealed that the two markers label largely non-overlapping amacrine cell populations, with the majority showing either CaR (solid arrowheads) or Prox1 immunoreactivity (arrows) and only a few are double-labelled (open arrowheads).

In both species, PV labelling revealed two types of amacrine cells, which corresponded to earlier descriptions [54,57,58]. Strongly PV-immunoreactive amacrines were sparser (8.9% out of 325 labelled amacrine cells belonged to this group in cat retinas, and the same ratio was 6.2% out of 2544 cells in rat retinas). Their cell bodies emitted one or two main dendrites whose branches could be followed for at least 50 μm (Figure 2A); thus, they fall in the category of small-field or larger amacrine cells [59]. Typical members of this group did not express Prox1 (Figure 2B,D).

The second PV-immunopositive amacrine cell group was less intensely labelled and more numerous (Figure 2A,C; frequency 24.3% and 90.5% of all labelled cells in cat and rat retinas, respectively). In rats, these cells have been identified as AII amacrine cells [57]; in cat retina, the majority of them are also AII cells [54,58]. These amacrine cells also contained Prox1 immunoreactivity (Figure 2B,D).

Calretinin is known to be a marker of largely different amacrine cell types in cats and rats. In cats, most of the CaR-containing amacrines are of type AII along with a sparser, more strongly labelled population [28,47,56,58,60]. Our current experiments showed that CaR-positive AII amacrine cells are always Prox1-immunoreactive (Figure 3A,B). In rat retina, CaR is a marker of non-AII (mainly starburst) amacrine cells [61,62]. In our CaR and Prox1 double-labelled rat retinas, the two markers revealed in essence, complementary populations with only a few double labelled cells present (Figure 3C,D), which is compatible with our assumption that Prox1 labels AII amacrines in the rat.

In both PV- and CaR-labelled retinas, some of the amacrine cell bodies were revealed by the anti-Prox1 antibody alone (Figure 2B,D and Figure 3B,D). Without further neurochemical markers and any labelling in their dendrites, these cells could not be assigned to known subtypes. This population was more abundant in cat retinas (e.g., 66.8% of all labelled cells in the PV + Prox1 labelled material vs. 3.3% in rodent retinas), and included many that were more lightly labelled and seemed difficult to differentiate from Prox1-immunoreactive bipolar cells (Figure 3B and Figure 6A). Consequently, the density estimates of Prox1-positive amacrine cells were more uncertain and therefore were not used as a proxy for retinal location in our analysis of cat retinas (see below).

Whereas the majority of Prox1-expressing neurons were in the INL, some strongly Prox1-positive cell bodies were also found in the ganglion cell layer in cats as well as rodents (Figure 4). Their size was similar to the Prox1+ regular amacrine cells (cat: $7.37 \pm 0.16\ \mu m$, $n = 8$, $p = 0.13$, rat: $8.86 \pm 0.33\ \mu m$, $n = 6$, $p = 0.08$, mouse: $7.97 \pm 0.22\ \mu m$, $n = 6$, $p = 0.81$ in t-test against Prox1-positive regular amacrine cells), suggesting they are displaced amacrine cells. Only in the cat did some ganglion cells show Prox1-immunoreactivity, which was rather weak and unusually appeared also in the cytoplasm (Figure 4A).

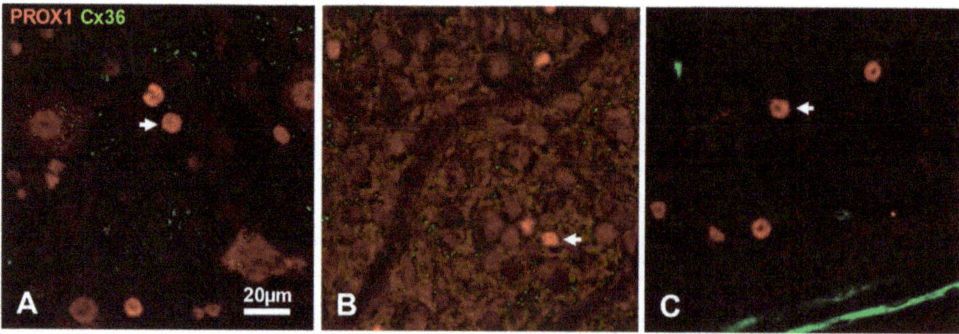

Figure 4. Prox1 immunoreactive cell nuclei (red channel) in the ganglion cell layer of flat mounted cat (**A**), rat (**B**) and mouse (**C**) retina. Smaller nuclei (arrows) are putative displaced amacrine cells (see details in the text). In the cat retina (**A**), some ganglion cell somata are also Prox1-immunoreactive. Connexin-36 puncta appear in irregular patches where the neuropil of the inner plexiform layer intrudes between cell bodies, but no somatic plaques can be observed on the Prox1-positive cells. Non-specific labelling by the anti-Cx36 antibody is seen in a bundle of optic fibers in the mouse retina (**C**). Scale bar 20 µm.

3.2. Identification of Individual AII Amacrine Cells in the PV-tdT Mouse Line

To further validate the Prox1 immunoreactive cell's background in the INL, we used the PV-tdT mouse line to show the morphology of these cells. Although PV expression is known to be limited to the GCs (and an unidentified group of amacrine cells) in mice [63,64], in this mouse line, tdT red-fluorescent reporter is expressed sporadically in some isolated amacrine cells, probably due to embryonal expression of Cre-recombinase. Expression of the fluorescent tdT protein allowed us to observe the morphology of these amacrine cells in fine detail.

Following immunolabelling of PV-tdT mouse retina with anti-Prox1 antibody (Figure 5a,b), we found that double-labelled neurons were exclusively amacrine cells, but only a small fraction of Prox1 immunoreactive amacrine cells (3.17% measured from six $224 \times 224\ \mu m$ regions on one retina)expressed tdT. They showed consistent morphology with axially ovoid somata, short, lobular dendrites in the OFF-sublamina and longer, thinner transversal dendrites projecting to the ON sublamina of the IPL, characteristics of AII cells (Figure 5b,c; [65–67]). Many of the tdT-labelled amacrine cells, however, were negative for Prox1 (29.67 cells per region on average). An example is shown in Figure 5c(4); this neuron had long, monostratified dendrites, suggesting that it was a wide-field amacrine cell. Taken together, these results corroborate earlier findings that Prox1 is a specific marker of AII cells [52].

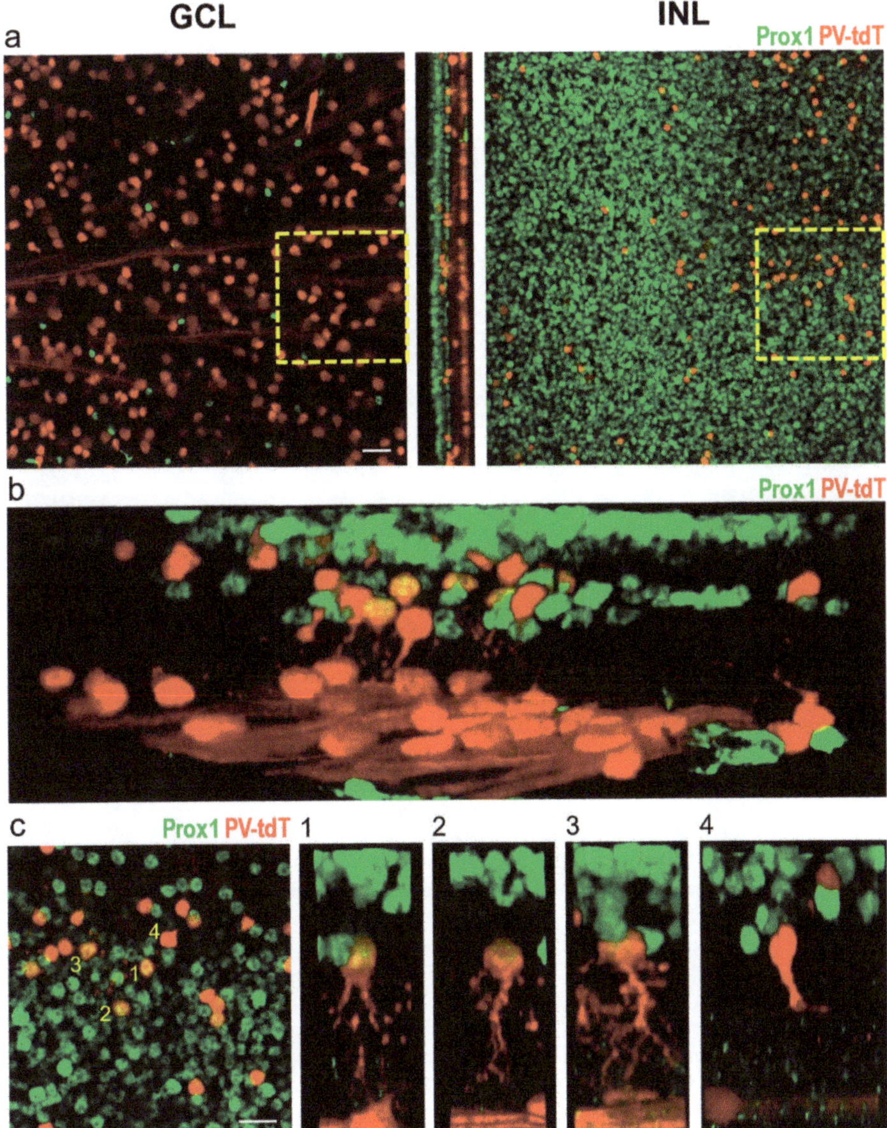

Figure 5. Colocalization of tdT and Prox1 in whole-mount preparation of the PV-tdT mouse retina. Overview images (**a**) show no colocalization of Prox1 with tdT in the ganglion cell layer (GCL), but in the INL, patches of amacrine cell bodies were double-labelled. The box outlined in yellow broken lines is shown at higher magnification in **c**. Three-dimensionally rendered and rotated views (**b,c**) of double labelled cells revealed their typical AII-like dendritic morphology (**c(1–3)**). Prox1-negative amacrine cells showed sparser, monostratified dendritic arbours reminiscent of wide-field amacrine cells (**c(4)**). Numbers in the left panel of **c** identify cells shown on side-views numbered 1 through 4, Scale = 25 μm.

3.3. Somatic Cx36 Plaques on Amacrine Cells of the Cat, Rat and Mouse Retina

In the following section, we turn our attention to the relationship of Prox1 immunoreactive cells to Cx36 immunoreactive punctate structures, the light-microscopic correlates of gap junctions. We begin our analysis with the inner nuclear layer. The INL is of interest because in previous studies, we saw large Cx36 plaques on cell bodies of some AII amacrine cells of cat [47] or mouse [68] retina. In our current material, we surveyed Prox1-positive amacrine cells systematically for the presence of Cx36 plaques (Figure 6). For this, we used optical sections focussed on the inner aspect of the INL and excluded regions where the neuropil of the IPL intruded between cell bodies.

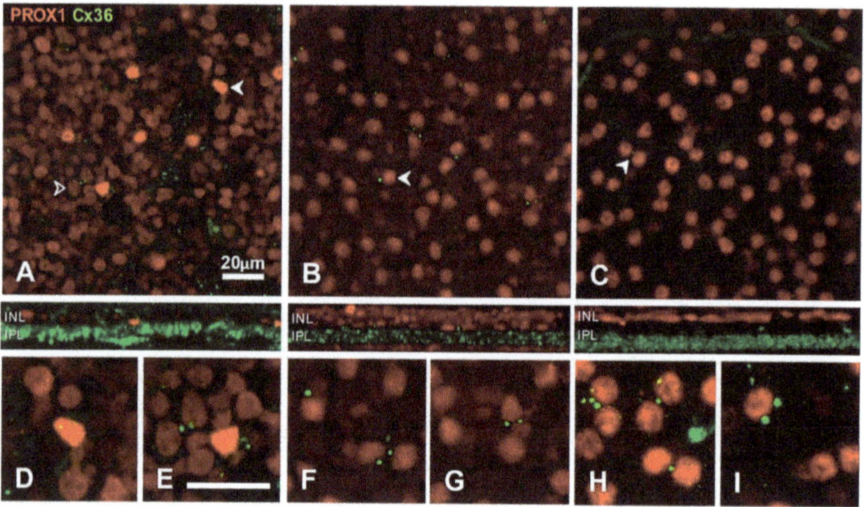

Figure 6. Somatic Cx36 plaques on Prox1 immunoreactive amacrine cells in whole-mount preparations of cat (**A,D,E**), rat (**B,F,G**) and mouse (**C,H,I**) retinas. Red channel, Prox1; green channel, Cx36. Side-view reconstructions of representative sub-volumes spanning the proximal INL (top) and IPL are shown below (**A–C**). Somatic Cx36 plaques were found in close apposition to a subset of Prox1-immunoreactive cell bodies. Examples are marked by arrowheads or shown at higher magnification in (**D–I**). In the cat retina (**A**), somatic plaques occurred in apposition to both strongly Prox1 positive (solid arrowheads) and lightly Prox1-positive (open arrowhead) cells.

Although Prox1 immunoreactivity is limited to the nuclei, and cell boundaries are not observable, plaques in the INL were almost exclusively in close apposition to Prox1-positive profiles, a few of the plaques sometimes surrounded the labelled nucleus (Figure 6D–I). The proportion of somatic Cx36 plaques and Prox1 immunoreactive amacrine cells was 8% in cat (n = 163 cells), 10% in rat (n = 103 cells) and 11% in mouse retina (n = 131 cells). In the cat retina (Figure 6A), somatic plaques were three times more frequent on strongly Prox1-immunoreactive amacrine cells (18% out of 470 cells) than they were on their weakly labelled counterparts (6% of cells). No somatic plaques could be observed on Prox1-positive neurons of the ganglion cell layer in either species (Figure 4).

The presence of such Cx36 plaques suggests a route for electrical coupling through the cell bodies of AII (and likely other Prox1 expressing) amacrine cells. The synaptic partners can be other amacrine cells of the same kind, since the somatic plaques were sometimes seen at the confluence of Prox1 positive cell bodies (Figure 6E–H). However, the neighbouring cell next to the plaque was quite often unstained (Figure 6A–C,I), raising the possibility of somatic coupling to other cell types.

We have addressed the question of heterocellular coupling in transgenic PV-tdT mouse retinas. Here, we identified Cx36 plaques between Prox1 positive (AII) and Prox1 negative (non-AII) amacrine cells that were revealed by tdT fluorescence (Figure 7). The detailed dendritic morphology revealed by tdT expression allowed us to confirm the existence of gap junctions between the cell bodies of different amacrine cell types.

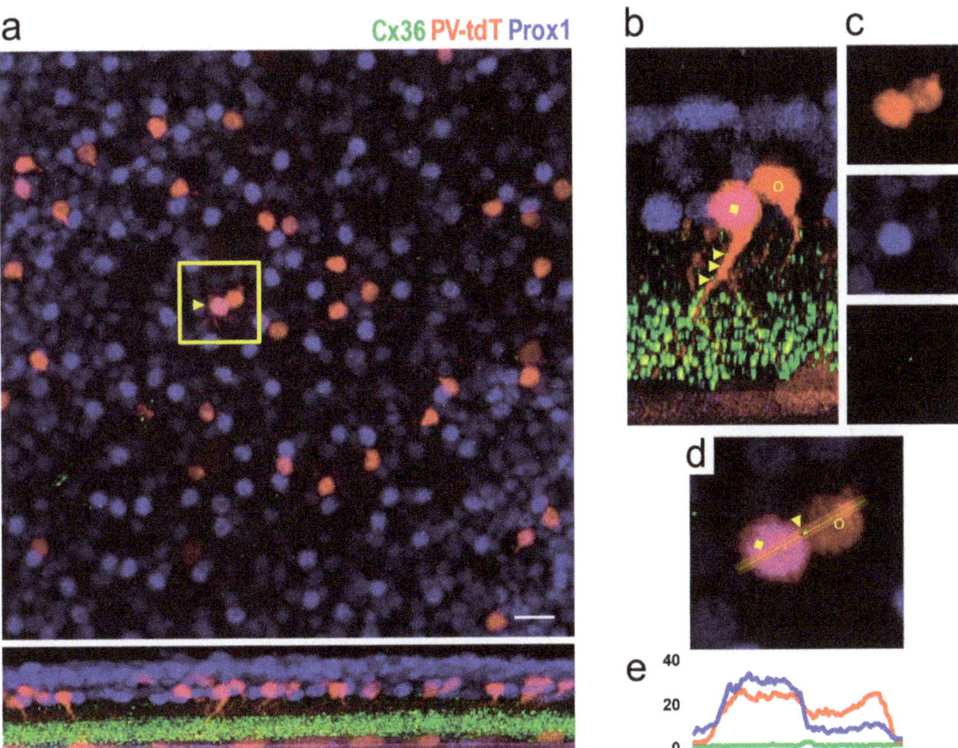

Figure 7. Colocalization of tdT, Prox1 and Cx36 immunofluorescence in amacrine cell somata of PV-tdT mouse retina shown in a tangential view at the level of the amacrine cell bodies (top of panel (**a**)) and as an orthogonal section along the line indicated by the yellow arrowhead (bottom of panel (**a**)). The region enclosed by the yellow box contains an AII amacrine cell (yellow arrowhead, magenta color due to double labelling with Prox1 and tdT) along with a tdT-positive neighbouring non-AII amacrine cell (red). The corresponding dendritic morphology of these cells is readily seen on the side-view in (**b**) (yellow square, AII amacrine cell; yellow circle, non-AII amacrine cell). Panel (**c**) shows the boxed region from **a** with the three channels separated (PV-tdT, Prox1 and Cx36 from top to bottom). Note the dot on the green channel, which indicates a Cx36 plaque where the two cell bodies touch each other. A close-up of this region (**d**) and an intensity profile, measured across the Cx36 plaque (**e**) confirm the close apposition of the plaque to both cell bodies. Some Cx36-puncta are also present on the proximal dendrite of the AII cell (**b**, yellow arrowheads). Markers as in (**b**). Scale, 25 μm.

3.4. Regional Variation of Connexin-36 Density in the OFF- and ON-Sublaminae of the Inner Plexiform Layer

In this part of the study, we sought to detect large scale variations in gap junction density with retinal location and compare this variation between the ON and OFF sublaminae of the IPL in cats, rats and mice. Regional specialisation is evident in the distribution of photoreceptors and several neuron types, but the topography of the cell distributions is different in cats and rodents (cat [4,69]; mouse [70]), it may have complicated shapes and also vary by individual [71–73]. Therefore, we decided to use cell density as a general measure of regional variation in rodents, instead of distance from some retinal landmark. In the following, high cell density has thus the same meaning as central location of a concentrically organised retina, and conversely, low cell density is equivalent to a peripheral location.

The densities of many cell types might be used as a surrogate for retinal location. Regarding Prox1-immunoreactive amacrine cells and horizontal cells, we found that their densities change with location, and they are positively correlated (rat, r = 0.87, $p < 0.001$; mice, r = 0.54, $p = 0.015$; Figure 8). It is important to note that mean densities for Prox1-positive amacrine cells (5085 ± 1139 for rats, 3687 ± 904 and 3520 ± 836 for wild type and PV-tdT mice, Table 2) were very much in line with the reported densities of AII amacrine cells in these species [52,57]. This supports the notion that the population of Prox1-positive amacrine cells largely overlaps with AII cells.

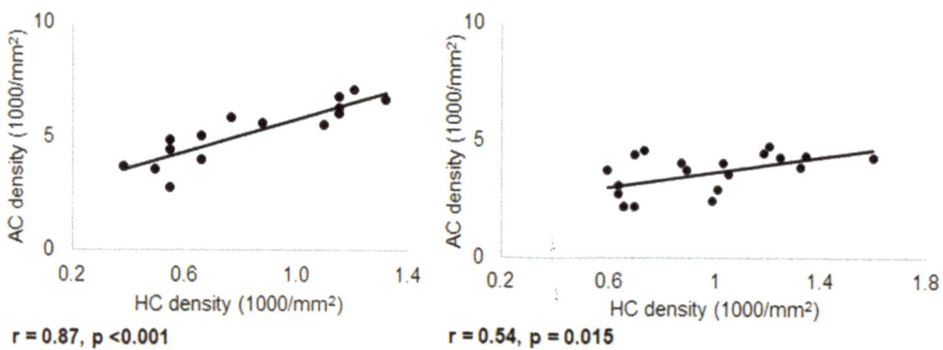

Figure 8. Correlation of density of Prox1 immunoreactive amacrine and horizontal cells in rat ($n = 15$) and mouse retina ($n = 20$). A strong positive correlation was found in both cases (r = 0.87, $p < 0.001$ and r = 0.54, $p = 0.015$).

Table 2. Relationship of connexin-36 plaque density and retinal position in mammalian retinas.

Ample (Number of Retinal Locations)			Cat ($n = 20$)	Rat 1 ($n = 5$)	Rat 2 ($n = 15$)	Mouse 1 and 2, Wild Type ($n = 8$)	Mouse 3 and 4, PV-tdT ($n = 12$)
Marker used for measurement of cell density			none	CaR	Prox1	Prox1	Prox1
Measure of retinal position			Eccentricity (mm)	Amacrine cell density (mm^{-2})			
Mean ± SD			7.34 ± 4.01	5766 ± 598	5085 ± 1139	3687 ± 904	3520 ± 836
IPL sublamina	OFF	Cx36-density (1/mm^{-3})	22.8 × 10^6 ±10.5 × 10^6	32.3 × 10^6 ±17.4 × 10^6	38.7 × 10^6 ±24.5 × 10^6	4.84 × 10^6 ±2.29 × 10^6	7.55 × 10^6 ±5.53 × 10^6
		Correlation with retinal position	r = −0.21 $p = 0.364$	r = 0.20 $p = 0.754$	r = −0.29 $p = 0.299$	r = 0.34 $p = 0.404$	r = 0.05 $p = 0.989$
	ON	Cx36-density (mm^{-3})	47.7 × 10^6 ±11.8 × 10^6	46.2 × 10^6 ±5.55 × 10^6	74.3 × 10^6 ±19.4 × 10^6	32.3 × 10^6 ±13.2 × 10^6	35.0 × 10^6 ±15.5 × 10^6
		Correlation with retinal position	r = −0.66 $p = 0.001$	r = 0.95 $p = 0.013$	r = 0.52 $p = 0.046$	r = 0.88 $p = 0.004$	r = 0.76 $p = 0.004$
Difference of Cx36-density between IPL OFF and IPL ON			t = −10.05 $p < 0.001$	t = −2.44 $p = 0.024$	t = −5.62 $p < 0.001$	t = −47.95 $p < 0.001$	t = −24.33 $p < 0.001$
Correlation of Cx36-density between IPL OFF and IPL ON			r = 0.11 $p = 0.325$	r = 0.33 $p = 0.587$	r = 0.42 $p = 0.121$	r = 0.54 $p = 0.463$	r = 0.77 $p = 0.075$

In the following analysis, we used either Prox1-positive amacrine cell density, or in one rat retina, CaR-positive amacrine cell density as a measure of retinal location. As described above, CaR is the marker of a large population of non-AII amacrine cells in the rat retina that includes the cholinergic starburst cells [61,62]. The densities of starburst and AII amacrine cells do, however, follow similar regional distribution patterns to each other [57,74].

Our previous analysis showed that Cx36 plaque density correlates with the density of AII amacrine cells so that central regions show a higher density of both features compared with peripheral regions [47], but this analysis was limited to the ON IPL of the cat retina. The OFF sublamina contains generally fewer gap junctions in the mammalian retinas studied so far [28]. This was confirmed in our material where the OFF sublamina contained significantly fewer Cx36 plaques per mm^3 than the ON sublamina ($p < 0.05$, Table 2). Here, we calculated plaque density per unit volume to eliminate the effect of the different thicknesses of the layers, which exist between the ON and OFF sublaminae, between species or retinal locations.

Figure 9 compares the densities of Cx36 plaques in the ON and OFF sublaminae of the IPL at various retinal locations of cat, rat and mouse retinas. In the ON sublamina of cat retina, we found a strong, significant negative correlation with eccentricity ($r = -0.66$, $p = 0.001$), confirming our earlier observation [47]. However, plaque density in the OFF sublamina was uncorrelated to eccentricity ($p = 0.364$, Table 2). Interestingly, data from rodent retinas suggested the same relationship. In their ON sublaminae, Cx36 plaque density and the density of immunolabeled amacrine cells were positively correlated ($p < 0.05$, Table 2), implying that density decreased from central to peripheral regions. In contrast, no significant correlation was found in the OFF sublamina in either sample ($p = 0.299$, Table 2).

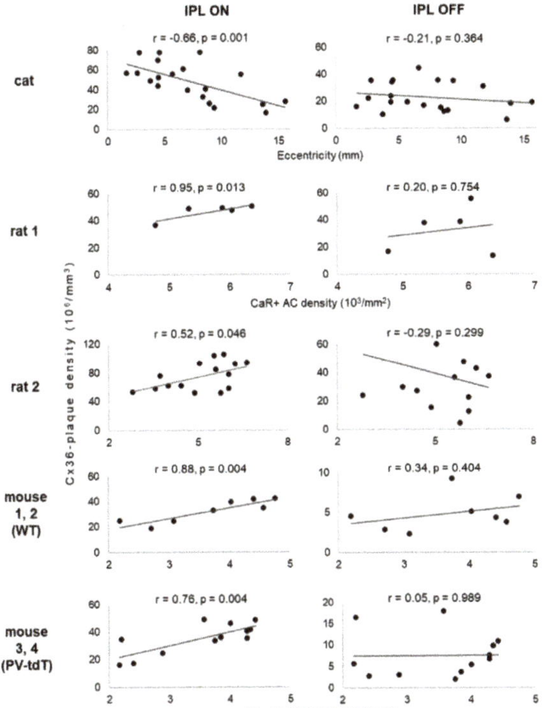

Figure 9. Relationship of volumetric connexin-36 plaque density to retinal position in the ON- and OFF-sublaminae of the inner plexiform layer of mammalian retinas. Each scatterplot shows data from several locations within one or more retinas treated the same way. Retinal location is indicated by the distance from the area centralis for cat retina (**top** row). For rodent retinas, retinal location is indicated by the areal densities of amacrine cells labelled by CaR or Prox1. Higher cell density indicates a higher sampling density of the retinal image; these regions must therefore be functionally more "central".

When data of the wild-type and PV-tdT mouse strains were pooled (Supplementary Figure S2), the same trends were evident ($r = 0.77$, $p < 0.001$ and $r = 0.04$, $p = 0.856$ for the ON and OFF sublaminae, respectively). This suggests that the Cx36 expression pattern was not affected by the Cre-mediated introduction of the tdT transgene. Another representation of the same data is when connexin density data are pooled by eccentricity category (centre, mid-centre, periphery, Supplementary Figure S1).

Finally, the independence of gap junctional connections in the OFF and ON sublaminae is further supported by the lack of significant correlation between the Cx36 plaque densities ($p > 0.05$, Table 2) of these sublaminae. In conclusion, our data suggest a fundamental difference in the organization of gap junctional connections between the ON and OFF sublaminae of the IPL.

4. Discussion

The main goal of the present study was to compare the regional variation in the density of punctate connexin-36 immunoreactive structures between the ON and OFF sublaminae of the inner plexiform layer in the retinae of multiple species. The central result is that regional variation in gap junction density follows different rules in the two sublaminae; whereas gap junctions become less frequent towards the periphery in the ON sublamina, there is no such correlation in the OFF sublamina. Since we calculated densities per unit volume of tissue, our results cannot be ascribed to systematic variation of the thickness of the laminae.

Two auxiliary results will also be discussed here. One is the regular occurrence of large Cx36 plaques on certain cell bodies of the amacrine cell layer. An additional result concerns the identity of Prox1 immunoreactive neurons in the retina, which we will discuss first.

4.1. Conservative Expression of Prox1 in Major Cell Types of the Inner Nuclear Layer

The Prox1 homeodomain transcription factor is expressed during development in various tissues including the brain and retina [75,76]. In adult animals, the retinal localization of Prox1 is best known for mice. Dyer et al. [51] found that Prox1 is expressed in horizontal cells and certain amacrine cells, that Pérez De Sevilla Müller et al. [52] later demonstrated to be AII amacrines. The same authors have shown using G0α immunohistochemistry that most of the large population of Prox1-positive bipolar cells are of the ON type. Similar immunohistochemical data from adult specimens of other species are not known to us except for those of Dyer et al. [51], also stating that Prox1 is localized to bipolar cells and parvalbumin-containing AII amacrine cells in rat retina. Although the density range of Prox1 immunoreactive amacrine cells matches the densities of AII cells identified by other methods in rats and in cats [57,77], further investigation is required to identify the population of Prox1 immunoreactive amacrines that do not express typical neurochemical markers of AII amacrine cells.

Our data provide evidence that the expression pattern of Prox1 is similar in carnivores and rodents as far as three major neuron types of the INL contain this protein. This corroborates the idea that the Prox1-dependent mechanism of retinal cell differentiation is conserved across mammalian species [51,78]. Moreover, the presence of Prox1 in both A- and B-type horizontal cells of cats [79] taken together with their neurochemical similarity [54,56,58,80,81] implies that despite their morphological differences, horizontal cell types differentiate late in development [82].

4.2. Somatic Gap Junctions of Retinal Amacrine Cells

Somato–somatic gap junctions are rather the exception than the rule in nervous tissue [83,84]. One notable example is the mesencephalic nucleus of the trigeminal nerve, where perikarya of primary afferent neurons form functional pairs or small clusters by way of electrical synapses [85–87].

In the neuropils of retina, gap junctions are most often localized on the dendritic or axonal processes of neurons [84,88] and they have also been discovered in the optic nerve head where they interconnect certain axons [89]. Somato–somatic gap junctions on AII amacrine cell bodies have been known for a while, although they have received little attention. They were first described in an electron microscopic study of the AII amacrine network of the cat retina [90], where large gap junctions were seen between AII cell bodies, sometimes connected by long appendages only a few hundred nanometres thin. It is reasonable to assume that the prominent, solitary Cx36 plaques seen on AII cell bodies in our material (Figures 6 and 7) are likely the light-microscopic correlates of these gap junctions. The fact that we see them in multiple species taken together with earlier observations in rat [28,68], human [28,39] and feline [47] retinas suggest that they are an ubiquitous feature among mammals.

Somatic Cx36 plaques were almost exclusively associated with Prox1 immunoreactive amacrine cells, although in some cases, the neighbouring soma (and putative synaptic partner) was not labelled. One simple explanation is that the putative synaptic partner is from a different cell type, an example of which is shown in Figure 7. Another possibility is, without excluding the previous one, that the synaptic partner is located further away, attached through a process emanating from the cell bodies; such processes may be up to 10–15 µm long [90,91], which amounts to the width of multiple amacrine cell bodies and is in the order of the mean nearest neighbour distance of AII cells [77,92]. It is, however, unclear at present whether the processes seen in the electron microscope to connect AII somata through gap junctions [90] are identical to those thin, short dendrites that are seen at the light-microscopic level to attach directly to the cell bodies [91].

The frequency of occurrence of somatic gap junctions is another difference between the electron microscopic reconstructions of Vardi and Smith [90] and what is observed by immunofluorescence microscopy. We counted that the average ratio of somatic Cx36 plaques and Prox1 positive amacrine cells is around 8–10% with little variation between species. Vardi and Smith [90], on the other hand, reported that "wherever AII somas abutted, they formed large gap junctions" and even those further apart were sometimes linked (although see [93]). Both techniques revealed that the somatic gap junctions of AII cells are particularly large and therefore, the reason for the difference in occurrence is unlikely to be the lower resolution of the light microscope. There seem to be two possibilities worth pursuing. First, that a significant proportion of the somatic AII gap junctions failed to react with the anti-Cx36 antibody used. We are, however, not aware that any other connexin isoform has been observed to form plaques at this site so far. Alternatively, the number of somatic gap junctions or their detectability as plaques may vary with the functional state of the retina. Light adaptation [94] or injury [95,96] are known to affect functional coupling, which may be accompanied by detectable structural changes given that the half-life of connexin protein is in the order of a few hours [97,98].

AII amacrine cells are coupled extensively through Cx36-containing gap junctions [99,100] located on their arboreal dendrites in the ON sublamina of the IPL [101,102]. Somatic coupling could simply add another route to this connectivity without conferring an additional function. However, morphologically accurate modelling suggests that AII cells are electrotonically not compact and thus, somatic and dendritic inputs may be processed differently [103]. It is therefore worthwhile considering high-conductivity somatic gap junctions when modelling the biophysical properties of AII cells.

4.3. Different Scaling Principles of Connexin-36 Gap Junction Density in ON and OFF Circuits with Retinal Position

The present work builds on previous research on the distribution of electrical synapses in the cat retina [47]. Although cats and rodents have adapted to quite different lifestyles and, accordingly, their retinas and visual systems are differently specialised, the similarities revealed here suggest that we have encountered regularities generalizable for mammalian retinas.

The role of gap junctional connections in the retina is generally characterized by the formation of functional syncytia in which electrotonic membrane potential changes and potentially other signal carriers can propagate in the lateral direction. In this light, the density of gap junctions can be interpreted as indicating the strength of interconnections in these syncytia. As gap junctions are involved in several parallel processing pathways [22], multiple electrically coupled networks can coexist and be superimposed. The anatomical separation of the ON and OFF sublaminae of the INL allowed us to observe the large-scale differences between gap junctions contributing to the ON- and OFF-pathways. Our results confirmed that in mammalian species, the ON sublamina contains significantly more Cx36 gap junctions than the OFF sublamina [28]. It is widely believed that this difference is mainly due to the gap junctions formed by AII amacrine cells with ON-cone bipolar cells and among themselves [13,99,100,104]. Although the estimated number of gap junctions formed by each AII cell varies between a few dozen (estimates from electron microscopy [93,105,106]) and over a hundred (estimates using Cx36 immunofluorescence [47,104,107]), the fact that AII cells are the most frequent type of amacrine cell [108] makes it likely that they are the main contributors to the high density of Cx36 plaques detected in the ON sublamina. Furthermore, calculations based on comparison of AII amacrine cell and Cx36 plaque densities in the cat retina [47] have suggested that at least half of the Cx36 plaques in the ON-sublamina belong to other cell types [18,21,22,36,41–46].

The positive correlation of Cx36 plaque density with AII amacrine cell density may not come as a surprise because of the large contribution of this cell type to gap junctional connections. It is, however, worth considering that for most retinal cell types, lower cell density is compensated by a larger dendritic (or axonal) field diameter, so that the potential for synaptic connections is maintained. Why would then the density of synaptic contacts diminish with cell density? A plausible explanation may be gained from the observation that larger dendritic fields are typically sparser [109], a regularity that also applies for AII cells [91]. As a result, the meshwork of dendrites of a given type of neuron tends to be more loosely knit in the periphery. Importantly, it is also known for several retinal cell types that the number of synapses formed by a cell scales linearly with the available cell membrane area (and not with the area covered by the dendritic field) [110–113]. As a result, synapses of smaller (more central) cells are expected to be present at higher volumetric density in the neuropil than the synapses of larger (more peripheral) cells.

If the above line of reasoning is accepted, it is more difficult to explain why gap junction density does not decrease towards the periphery in the OFF-sublamina. An important contributing factor may be the diversity of the Cx36 expressing cell population. Indeed, gap junctions of the OFF-sublamina fall into four categories including those formed by ganglion cells with either RGC neighbours or nearby amacrine cells [34–37]— the population that connects amacrine cells to one another and a fourth population that exists between bipolar cells [38–40]. We can safely assume that at large, the density and arborization of the contributing cell types underlies the same scaling principles as outlined for the ON-sublamina. However, the dendritic fields of different cell types may well scale by different factors with eccentricity and the structure of the dendritic trees may also change differently with size [109,112]. Ultimately, the superposition of a variety of such systems would obscure eccentricity-dependent regularities that may be present in the connectivity of a single cell type.

An alternative hypothesis for the different scaling of Cx36 density in the OFF sublamina may be based on a recent study [114]. Here, the authors tested the presence of functional electrical synapses on ON- and OFF-bipolar cells by using in vitro electrophysiology and the gap junction blocker meclofenic acid in rat retinas. Surprisingly, the measurements suggest that the gap junctions of OFF-bipolar cells may not be functional electrical synapses. Instead, they could, for instance, serve metabolic functions or pass signal molecules between coupled cells. In any case, the distribution of non-synaptic gap junctions on the neurites may underlie different scaling rules compared to those that apply for synaptic contacts. If a sufficiently large proportion of Cx36 gap junctions in the OFF sublamina are of this kind, our measurements of the gross gap junction density may also be affected.

Finally, one might contemplate the idea that the regularities seen in the ON sublamina are generalized across species. This may suggest some general mechanism that scales gap junction density to cell density. To test this idea, we pooled the Cx36 and Prox1-positive amacrine cell density data of mice and rats. Supplementary Figure S2 shows that indeed, the correlation for IPL-ON observed within species is retained ($r = 0.77$, $p < 0.0005$), and there is no significant correlation in the OFF sublamina. It is, however, worth noting that cross-species differences still exist because in the region where AC densities of the two species overlapped, the Cx36 densities were lower on average in mouse than in rat samples.

4.4. Concluding Remarks

Although the ON and OFF responses of visual neurons are conceptually symmetrical, the pathways that implement them in the retina are fundamentally different (see e.g., [115] for review). There is abundant evidence that the visual response properties of ON and OFF neurones are also different in many ways [116–118]. Typically, these so-called ON/OFF asymmetries are investigated at a given, limited retinal location, and results obtained from central and peripheral parts of the retina could sometimes not be reconciled [116,119,120]. This points to the existence of region-specific ON/OFF asymmetries in the retina [121], which must eventually be rooted in regional differences of synaptic circuitry. Our study demonstrates large-scale differences between gap junctions involved in the ON and OFF networks of the retina potentially contributing to region-specific functional ON/OFF asymmetries.

Supplementary Materials: The following are available online at https://www.mdpi.com/article/10.3390/cells10092396/s1, Figure S1: Data of Prox1 and Cx36 immunolabelled mouse retinas from Figure 9 plotted with eccentricity categories (centre, mid-centre, periphery) on the abscissa, Figure S2: Data of Prox1 and Cx36 immunolabelled rat and mouse retinas from Figure 9 plotted together for comparison.

Author Contributions: Conceptualization, T.K.-Ö., B.V. and P.B.; Data curation, K.F.; Formal analysis, K.F. and P.B.; Funding acquisition, B.V. and P.B.; Investigation, K.F., T.K.-Ö., P.K. and I.T.; Methodology, T.K.-Ö., E.S.-M. and I.T.; Project administration, P.K. and P.B.; Resources, T.K.-Ö., P.K., E.S.-M. and B.V.; Supervision, E.S.-M., B.V., P.B. and I.T.; Visualization, K.F., T.K.-Ö. and I.T.; Writing—original draft, K.F., T.K.-Ö., P.B. and I.T.; Writing—review and editing, B.V., P.B. and I.T. All authors have read and agreed to the published version of the manuscript.

Funding: This research was funded by the Hungarian Brain Research Program 2 (2017-1.2.1.-NKP-2017) and the Higher Education Institutional Excellence Programme (FIKP) of the Ministry for Innovation and Technology in Hungary, within the framework of the 5th thematic program of the University of Pécs to PB and BV. This research was also supported by OTKA NN128293 and the European Union and the State of Hungary, co-financed by the European Social Fund in the framework of TÁMOP-4.2.4.A/2-11/1-2012-0001 'National Excellence Program' to BV as well as by ÚNKP-17-3-I-PTE-155 and ÚNKP-20-4-II-PTE-683 New National Excellence Program of the Ministry of Human Capacities. The research was performed in collaboration with the Histology and Light Microscopy core facility at the Szentágothai Research Centre of the University of Pécs with support from GINOP-2.3.2-15-2016-00036.

Institutional Review Board Statement: The study was conducted according to the guidelines of the Declaration of Helsinki and approved by the Directorate for Food Chain Safety and Animal Health of the Baranya County Government Office, Hungary (license numbers BA02/2000-11/2015 and BA/73/00504-5/2021).

Informed Consent Statement: Not applicable.

Data Availability Statement: Raw data that are not reported in the paper are available on request from the corresponding author, P.B.

Acknowledgments: We are grateful to Erzsébet Korona, Rita Illés and Varna Gomes da Silveira for technical assistance.

Conflicts of Interest: The authors declare no conflict of interest.

References

1. Moore, B.A.; Tyrrell, L.P.; Kamilar, J.M.; Collin, S.P.; Dominy, N.J.; Hall, M.I.; Heesy, C.P.; Lisney, T.J.; Loew, E.R.; Moritz, G.L.; et al. Structure and Function of Regional Specializations in the Vertebrate Retina. In *Evolution of Nervous Systems*; Kaas, J.H., Ed.; Elsevier: Amsterdam, The Netherlands, 2017; Volume 1, pp. 351–372. ISBN 978-0-12-804096-6.
2. Ahnelt, P.K.; Kolb, H. The Mammalian Photoreceptor Mosaic-Adaptive Design. *Prog. Retin. Eye Res.* **2000**, *19*, 711–777. [CrossRef]
3. Curcio, C.A.; Sloan, K.R.; Kalina, R.E.; Hendrickson, A.E. Human Photoreceptor Topography. *J. Comp. Neurol.* **1990**, *292*, 497–523. [CrossRef]

4. Steinberg, R.H.; Reid, M.; Lacy, P.L. The Distribution of Rods and Cones in the Retina of the Cat (Felis Domesticus). *J. Comp. Neurol.* **1973**, *148*, 229–248. [CrossRef] [PubMed]
5. Linberg, K.A.; Lewis, G.P.; Shaaw, C.; Rex, T.S.; Fisher, S.K. Distribution of S- and M-Cones in Normal and Experimentally Detached Cat Retina. *J Comp Neurol* **2001**, *430*, 343–356. [CrossRef]
6. Curcio, C.A.; Allen, K.A. Topography of Ganglion Cells in Human Retina. *J. Comp. Neurol.* **1990**, *300*, 5–25. [CrossRef]
7. Telkes, I.; Lee, S.C.; Jusuf, P.R.; Grünert, U. The Midget-Parvocellular Pathway of Marmoset Retina: A Quantitative Light Microscopic Study. *J Comp Neurol* **2008**, *510*, 539–549. [CrossRef] [PubMed]
8. Stewart, E.E.M.; Valsecchi, M.; Schütz, A.C. A Review of Interactions between Peripheral and Foveal Vision. *J. Vis.* **2020**, *20*, 1–35. [CrossRef] [PubMed]
9. Strasburger, H.; Rentschler, I.; Jüttner, M. Peripheral Vision and Pattern Recognition: A Review. *J. Vis.* **2011**, *11*, 13. [CrossRef] [PubMed]
10. Nelson, R. Cat Cones Have Rod Input: A Comparison of the Response Properties of Cones and Horizontal Cell Bodies in the Retina of the Cat. *J. Comp. Neurol.* **1977**, *172*, 109–135. [CrossRef]
11. Schneeweis, D.M.; Schnapf, J.L. Photovoltage of Rods and Cones in the Macaque Retina. *Science* **1995**, *268*, 1053–1056. [CrossRef]
12. Deans, M.R.; Völgyi, B.; Goodenough, D.A.; Bloomfield, S.A.; Paul, D.L. Connexin36 Is Essential for Transmission of Rod-Mediated Visual Signals in the Mammalian Retina. *Neuron* **2002**, *36*, 703–712. [CrossRef]
13. Güldenagel, M.; Ammermüller, J.; Feigenspan, A.; Teubner, B.; Degen, J.; Söhl, G.; Willecke, K.; Weiler, R.; Sohl, G.; Willecke, K.; et al. Visual Transmission Deficits in Mice with Targeted Disruption of the Gap Junction Gene Connexin. *J. Neurosci.* **2001**, *21*, 6036–6044. [CrossRef]
14. Völgyi, B.; Deans, M.R.; Paul, D.L.; Bloomfield, S.A. Convergence and Segregation of the Multiple Rod Pathways in Mammalian Retina. *J. Neurosci.* **2004**, *24*, 11182–11192. [CrossRef] [PubMed]
15. Naka, K.I.; Nye, P.W. Role of Horizontal Cells in Organization of the Catfish Retinal Receptive Field. *J. Neurophysiol.* **1971**, *34*, 785–801. [CrossRef] [PubMed]
16. Shelley, J.; Dedek, K.; Schubert, T.; Feigenspan, A.; Schultz, K.; Hombach, S.; Willecke, K.; Weiler, R. Horizontal Cell Receptive Fields Are Reduced in Connexin57-Deficient Mice. *Eur. J. Neurosci.* **2006**, *23*, 3176–3186. [CrossRef] [PubMed]
17. Hu, E.H.; Bloomfield, S.A. Gap Junctional Coupling Underlies the Short-Latency Spike Synchrony of Retinal α Ganglion Cells. *J. Neurosci.* **2003**, *23*, 6768–6777. [CrossRef] [PubMed]
18. Völgyi, B.; Pan, F.; Paul, D.L.; Wang, J.T.; Huberman, A.D.; Bloomfield, S.A. Gap Junctions Are Essential for Generating the Correlated Spike Activity of Neighboring Retinal Ganglion Cells. *PLoS ONE* **2013**, *8*, e69426. [CrossRef]
19. Mastronarde, D.N. Interactions between Ganglion Cells in Cat Retina. *J. Neurophysiol.* **1983**, *49*, 350–365. [CrossRef]
20. Brivanlou, I.H.; Warland, D.K.; Meister, M. Mechanisms of Concerted Firing among Retinal Ganglion Cells. *Neuron* **1998**, *20*, 527–539. [CrossRef]
21. Bloomfield, S.A.; Völgyi, B. The Diverse Functional Roles and Regulation of Neuronal Gap Junctions in the Retina. *Nat. Rev. Neurosci.* **2009**, *10*, 495–506. [CrossRef]
22. Völgyi, B.; Kovács-Öller, T.; Atlasz, T.; Wilhelm, M.; Gábriel, R. Gap Junctional Coupling in the Vertebrate Retina: Variations on One Theme? *Prog. Retin. Eye Res.* **2013**, *34*, 1–18. [CrossRef]
23. O'Brien, J.J.; Chen, X.; Macleish, P.R.; O'Brien, J.J.; Massey, S.C. Photoreceptor Coupling Mediated by Connexin36 in the Primate Retina. *J. Neurosci.* **2012**, *32*, 4675–4687. [CrossRef] [PubMed]
24. Feigenspan, A.; Janssen-Bienhold, U.; Hormuzdi, S.; Monyer, H.; Degen, J.; Sohl, G.; Willecke, K.; Ammermuller, J.; Weiler, R. Expression of Connexin36 in Cone Pedicles and OFF-Cone Bipolar Cells of the Mouse Retina. *J. Neurosci.* **2004**, *24*, 3325–3334. [CrossRef] [PubMed]
25. Lee, E.J.; Han, J.W.; Kim, H.J.; Kim, I.B.; Lee, M.Y.; Oh, S.J.; Chung, J.W.; Chun, M.H. The Immunocytochemical Localization of Connexin 36 at Rod and Cone Gap Junctions in the Guinea Pig Retina. *Eur J Neurosci* **2003**, *18*, 2925–2934. [CrossRef]
26. Asteriti, S.; Gargini, C.; Cangiano, L. Connexin 36 Expression Is Required for Electrical Coupling between Mouse Rods and Cones. *Vis. Neurosci.* **2017**, *34*, E006. [CrossRef]
27. Smith, R.G.; Freed, M.A.; Sterling, P. Microcircuitry of the Dark-Adapted Cat Retina: Functional Architecture of the Rod-Cone Network. *J. Neurosci.* **1986**, *6*, 3505–3517. [CrossRef] [PubMed]
28. Kovács-Öller, T.; Debertin, G.; Balogh, M.; Ganczer, A.; Orbán, J.; Nyitrai, M.; Balogh, L.; Kántor, O.; Völgyi, B. Connexin36 Expression in the Mammalian Retina: A Multiple-Species Comparison. *Front. Cell. Neurosci.* **2017**, *11*, 65. [CrossRef]
29. Mills, S.L.; Massey, S.C. Differential Properties of Two Gap Junctional Pathways Made by AII Amacrine Cells. *Nature* **1995**, *377*, 734–737. [CrossRef]
30. Bloomfield, S.A.; Xin, D.; Osborne, T. Light-Induced Modulation of Coupling between AII Amacrine Cells in the Rabbit Retina. *Vis. Neurosci.* **1997**, *14*, 565–576. [CrossRef]
31. Massey, S.C.; Mills, S.L. Gap Junctions between AII Amacrine Cells and Calbindin-Positive Bipolar Cells in the Rabbit Retina. *Vis. Neurosci.* **1999**, *16*, 1181–1189. [CrossRef] [PubMed]
32. Bloomfield, S.A.; Dacheux, R.F. Rod Vision: Pathways and Processing in the Mammalian Retina. *Prog. Retin. Eye Res.* **2001**, *20*, 351–384. [CrossRef]
33. Hartveit, E.; Veruki, M.L. Electrical Synapses between AII Amacrine Cells in the Retina: Function and Modulation. *Brain Res.* **2012**, *1487*, 160–172. [CrossRef]

34. Schubert, T.; Degen, J.; Willecke, K.; Hormuzdi, S.G.; Monyer, H.; Weiler, R. Connexin36 Mediates Gap Junctional Coupling of Alpha-Ganglion Cells in Mouse Retina. *J. Comp. Neurol.* **2005**, *485*, 191–201. [CrossRef]

35. Völgyi, B.; Abrams, J.; Paul, D.L.; Bloomfield, S.A. Morphology and Tracer Coupling Pattern of Alpha Ganglion Cells in the Mouse Retina. *J. Comp. Neurol.* **2005**, *492*, 66–77. [CrossRef] [PubMed]

36. Völgyi, B.; Chheda, S.; Bloomfield, S.A. Tracer Coupling Patterns of the Ganglion Cell Subtypes in the Mouse Retina. *J. Comp. Neurol.* **2009**, *512*, 664–687. [CrossRef] [PubMed]

37. Pan, F.; Paul, D.L.; Bloomfield, S.A.; Völgyi, B. Connexin36 Is Required for Gap Junctional Coupling of Most Ganglion Cell Subtypes in the Mouse Retina. *J. Comp. Neurol.* **2010**, *518*, 911–927. [CrossRef] [PubMed]

38. Mills, S.L. Unusual Coupling Patterns of a Cone Bipolar Cell in the Rabbit Retina. *Vis. Neurosci.* **1999**, *16*, 1029–1035. [CrossRef]

39. Kántor, O.; Varga, A.; Nitschke, R.; Naumann, A.; Énzsöly, A.; Lukáts, A.; Szabó, A.; Németh, J.; Völgyi, B. Bipolar Cell Gap Junctions Serve Major Signaling Pathways in the Human Retina. *Brain Struct Funct* **2017**, *222*, 2603–2624. [CrossRef] [PubMed]

40. Manookin, M.B.; Patterson, S.S.; Linehan, C.M. Neural Mechanisms Mediating Motion Sensitivity in Parasol Ganglion Cells of the Primate Retina. *Neuron* **2018**, *97*, 1327–1340.e4. [CrossRef]

41. Sigulinsky, C.L.; Anderson, J.R.; Kerzner, E.; Rapp, C.N.; Pfeiffer, R.L.; Rodman, T.M.; Emrich, D.P.; Rapp, K.D.; Nelson, N.T.; Lauritzen, J.S.; et al. Network Architecture of Gap Junctional Coupling among Parallel Processing Channels in the Mammalian Retina. *J. Neurosci.* **2020**, *40*, 4483–4511. [CrossRef]

42. Xin, D.; Bloomfield, S.A. Tracer Coupling Pattern of Amacrine and Ganglion Cells in the Rabbit Retina. *J. Comp. Neurol.* **1997**, *383*, 512–528. [CrossRef]

43. Völgyi, B.; Xin, D.; Amarillo, Y.; Bloomfield, S.A. Morphology and Physiology of the Polyaxonal Amacrine Cells in the Rabbit Retina. *J. Comp. Neurol.* **2001**, *440*, 109–125. [CrossRef] [PubMed]

44. Li, W.; Zhang, J.; Massey, S.C. Coupling Pattern of S1 and S2 Amacrine Cells in the Rabbit Retina. *Vis. Neurosci.* **2002**, *19*, 119–131. [CrossRef]

45. Bloomfield, S.A.; Völgyi, B. Response Properties of a Unique Subtype of Wide-Field Amacrine Cell in the Rabbit Retina. *Vis. Neurosci.* **2007**, *24*, 459–469. [CrossRef]

46. Jacoby, J.; Nath, A.; Jessen, Z.F.; Schwartz, G.W. A Self-Regulating Gap Junction Network of Amacrine Cells Controls Nitric Oxide Release in the Retina. *Neuron* **2018**, *100*, 1149–1162.e5. [CrossRef]

47. Telkes, I.; Kóbor, P.; Orbán, J.; Kovács-Öller, T.; Völgyi, B.; Buzás, P. Connexin-36 Distribution and Layer-Specific Topography in the Cat Retina. *Brain Struct. Funct.* **2019**, *224*, 2183–2197. [CrossRef]

48. Bishop, P.O.; Kozak, W.; Vakkur, G.J. Some Quantitative Aspects of the Cat's Eye: Axis and Plane of Reference, Visual Field Co-Ordinates and Optics. *J. Physiol.* **1962**, *163*, 466–502. [CrossRef] [PubMed]

49. Hughes, A. A Quantitative Analysis of the Cat Retinal Ganglion Cell Topography. *J Comp Neurol* **1975**, *163*, 107–128. [CrossRef] [PubMed]

50. Schindelin, J.; Arganda-Carreras, I.; Frise, E.; Kaynig, V.; Longair, M.; Pietzsch, T.; Preibisch, S.; Rueden, C.; Saalfeld, S.; Schmid, B.; et al. Fiji: An Open-Source Platform for Biological-Image Analysis. *Nat Methods* **2012**, *9*, 676–682. [CrossRef]

51. Dyer, M.A.; Livesey, F.J.; Cepko, C.L.; Oliver, G. Prox1 Function Controls Progenitor Cell Proliferation and Horizontal Cell Genesis in the Mammalian Retina. *Nat. Genet.* **2003**, *34*, 53–58. [CrossRef]

52. Pérez De Sevilla Müller, L.; Azar, S.S.; de los Santos, J.; Brecha, N.C. Prox1 Is a Marker for Aii Amacrine Cells in the Mouse Retina. *Front. Neuroanat.* **2017**, *11*, 1–12. [CrossRef]

53. Röhrenbeck, J.; Wässle, H.; Heizmann, C.W. Immunocytochemical Labelling of Horizontal Cells in Mammalian Retina Using Antibodies against Calcium-Binding Proteins. *Neurosci Lett* **1987**, *77*, 255–260. [CrossRef]

54. Gábriel, R.; Straznicky, C. Immunocytochemical Localization of Parvalbumin- and Neurofilament Triplet Protein Immunoreactivity in the Cat Retina: Colocalization in a Subpopulation of AII Amacrine Cells. *Brain Res.* **1992**, *595*, 133–136. [CrossRef]

55. Peichl, L.; González-Soriano, J. Morphological Types of Horizontal Cell in Rodent Retinae: A Comparison of Rat, Mouse, Gerbil, and Guinea Pig. *Vis. Neurosci.* **1994**, *11*, 501–517. [CrossRef]

56. Pasteels, B.; Rogers, J.; Blachier, F.; Pochet, R. Calbindin and Calretinin Localization in Retina from Different Species. *Vis. Neurosci.* **1990**, *5*, 1–16. [CrossRef]

57. Wässle, H.; Grünert, U.; Röhrenbeck, J. Immunocytochemical Staining of AII-Amacrine Cells in the Rat Retina with Antibodies against Parvalbumin. *J Comp Neurol* **1993**, *332*, 407–420. [CrossRef] [PubMed]

58. Goebel, D.J.; Pourcho, R.G. Calretinin in the Cat Retina: Colocalizations with Other Calcium-Binding Proteins, GABA and Glycine. *Vis. Neurosci.* **1997**, *14*, 311–322. [CrossRef] [PubMed]

59. Kolb, H.; Nelson, R.; Mariani, A. Amacrine Cells, Bipolar Cells and Ganglion Cells of the Cat Retina: A Golgi Study. *Vision Res.* **1981**, *21*, 1081–1114. [CrossRef]

60. Jeon, M.H.; Jeon, C.J. Immunocytochemical Localization of Calretinin Containing Neurons in Retina from Rabbit, Cat, and Dog. *Neurosci Res* **1998**, *32*, 75–84. [CrossRef]

61. Gábriel, R.; Witkovsky, P. Cholinergic, but Not the Rod Pathway-Related Glycinergic (AII), Amacrine Cells Contain Calretinin in the Rat Retina. *Neurosci. Lett.* **1998**, *247*, 179–182. [CrossRef]

62. Hwang, I.K.; Yoo, K.Y.; Kim, D.S.; Jung, J.Y.; Shin, M.C.; Seo, K.; Kim, K.S.; Kang, T.C.; Won, M.H. Comparative Study on Calretinin Immunoreactivity in Gerbil and Rat Retina. *Anat Histol Embryol* **2005**, *34*, 129–131. [CrossRef]

63. Kovács-Öller, T.; Szarka, G.; Ganczer, A.; Tengölics, Á.; Balogh, B.; Völgyi, B. Expression of Ca2+-Binding Buffer Proteins in the Human and Mouse Retinal Neurons. *Int. J. Mol. Sci.* **2019**, *20*, 2229. [CrossRef]

64. Haverkamp, S.; Wässle, H. Immunocytochemical Analysis of the Mouse Retina. *J. Comp. Neurol.* **2000**, *424*, 1–23. [CrossRef]

65. Kolb, H.; Famiglietti, E.V. Rod and Cone Pathways in the Inner Plexiform Layer of Cat Retina. *Science* **1974**, *186*, 47–49. [CrossRef]

66. Famiglietti, E.V.; Kolb, H. A Bistratified Amacrine Cell and Synaptic Circuitry in the Inner Plexiform Layer of the Retina. *Brain Res.* **1975**, *84*, 293–300. [CrossRef]

67. Nelson, R. AII Amacrine Cells Quicken Time Course of Rod Signals in the Cat Retina. *J. Neurophysiol.* **1982**, *47*, 928–947. [CrossRef]

68. Kovács-Öller, T.; Raics, K.; Orbán, J.; Nyitrai, M.; Völgyi, B. Developmental Changes in the Expression Level of Connexin36 in the Rat Retina. *Cell Tissue Res.* **2014**, *358*, 289–302. [CrossRef]

69. Stone, J. A Quantitative Analysis of the Distribution of Ganglion Cells in the Cat's Retina. *J. Comp. Neurol.* **1965**, *124*, 337–352. [CrossRef]

70. Dräger, U.C.; Olsen, J.F. Ganglion Cell Distribution in the Retina of the Mouse. *Investig. Ophthalmol. Vis. Sci.* **1981**, *20*, 285–293.

71. Salinas-Navarro, M.; Jiménez-López, M.; Valiente-Soriano, F.J.; Alarcón-Martínez, L.; Avilés-Trigueros, M.; Mayor, S.; Holmes, T.; Lund, R.D.; Villegas-Pérez, M.P.; Vidal-Sanz, M. Retinal Ganglion Cell Population in Adult Albino and Pigmented Mice: A Computerized Analysis of the Entire Population and Its Spatial Distribution. *Vision Res.* **2009**, *49*, 637–647. [CrossRef]

72. Salinas-Navarro, M.; Mayor-Torroglosa, S.; Jiménez-López, M.; Avilés-Trigueros, M.; Holmes, T.M.; Lund, R.D.; Villegas-Pérez, M.P.; Vidal-Sanz, M. A Computerized Analysis of the Entire Retinal Ganglion Cell Population and Its Spatial Distribution in Adult Rats. *Vision Res.* **2009**, *49*, 115–126. [CrossRef]

73. Ortín-Martínez, A.; Jiménez-López, M.; Nadal-Nicolás, F.M.; Salinas-Navarro, M.; Alarcón-Martínez, L.; Sauvé, Y.; Villegas-Pérez, M.P.; Vidal-Sanz, M.; Agudo-Barriuso, M. Automated Quantification and Topographical Distribution of the Whole Population of S- and L-Cones in Adult Albino and Pigmented Rats. *Investig. Opthalmology Vis. Sci.* **2010**, *51*, 3171. [CrossRef] [PubMed]

74. Voigt, T. Cholinergic Amacrine Cells in the Rat Retina. *J. Comp. Neurol.* **1986**, *248*, 19–35. [CrossRef]

75. Oliver, G.; Sosa-Pineda, B.; Geisendorf, S.; Spana, E.P.; Doe, C.Q.; Gruss, P. Prox 1, a Prospero-Related Homeobox Gene Expressed during Mouse Development. *Mech. Dev.* **1993**, *44*, 3–16. [CrossRef]

76. Lavado, A.; Oliver, G. Prox1 Expression Patterns in the Developing and Adult Murine Brain. *Dev. Dyn.* **2007**, *236*, 518–524. [CrossRef]

77. Vaney, D.I. The Morphology and Topographic Distribution of AII Amacrine Cells in the Cat Retina. *Proc. R. Soc. London - Biol. Sci.* **1985**, *224*, 475–488. [CrossRef]

78. Cook, T. Cell Diversity in the Retina: More than Meets the Eye. *BioEssays* **2003**, *25*, 921–925. [CrossRef]

79. Boycott, B.B.; Peichl, L.; Wassle, H. Morphological Types of Horizontal Cell in the Retina of the Domestic Cat. *Proc. R. Soc. London - Biol. Sci.* **1978**, *203*, 229–245. [CrossRef]

80. Röhrenbeck, J.; Wässle, H.; Boycott, B.B. Horizontal Cells in the Monkey Retina: Immunocytochemical Staining with Antibodies against Calcium Binding Proteins. *Eur J Neurosci* **1989**, *1*, 407–420. [CrossRef] [PubMed]

81. Fyk-Kolodziej, B.; Qin, P.; Dzhagaryan, A.; Pourcho, R.G. Differential Cellular and Subcellular Distribution of Glutamate Transporters in the Cat Retina. *Vis. Neurosci.* **2004**, *21*, 551–565. [CrossRef]

82. Lyser, K.M.; Li, A.I.; Nuñez, M. Horizontal Cells in the Rabbit Retina: Differentiation of Subtypes at Neonatal and Postnatal Stages. *Int. J. Dev. Neurosci.* **1994**, *12*, 673–682. [CrossRef]

83. Sotelo, C.; Llinás, R. Specialized Membrane Junctions between Neurons in the Vertebrate Cerebellar Cortex. *J. Cell Biol.* **1972**, *53*, 271–289. [CrossRef]

84. Galarreta, M.; Hestrin, S. Electrical Synapses between Gaba-Releasing Interneurons. *Nat. Rev. Neurosci.* **2001**, *2*, 425–433. [CrossRef]

85. Hinrichsen, C.F.L.; Larramendi, L.M.H. Synapses and Cluster Formation of the Mouse Mesencephalic Fifth Nucleus. *Brain Res.* **1968**, *7*, 296–299. [CrossRef]

86. Pedroarena, C.M.; Pose, I.E.; Yamuy, J.; Chase, M.H.; Morales, F.R. Oscillatory Membrane Potential Activity in the Soma of a Primary Afferent Neuron. *J. Neurophysiol.* **1999**, *82*, 1465–1476. [CrossRef]

87. Curti, S.; Hoge, G.; Nagy, J.I.; Pereda, A.E. Synergy between Electrical Coupling and Membrane Properties Promotes Strong Synchronization of Neurons of the Mesencephalic Trigeminal Nucleus. *J. Neurosci.* **2012**, *32*, 4341–4359. [CrossRef]

88. Connors, B.W.; Long, M.A. Electrical Synapses in the Mammalian Brain. *Annu. Rev. Neurosci.* **2004**, *27*, 393–418. [CrossRef] [PubMed]

89. Smedowski, A.; Akhtar, S.; Liu, X.; Pietrucha-Dutczak, M.; Podracka, L.; Toropainen, E.; Alkanaan, A.; Ruponen, M.; Urtti, A.; Varjosalo, M.; et al. Electrical Synapses Interconnecting Axons Revealed in the Optic Nerve Head – a Novel Model of Gap Junctions' Involvement in Optic Nerve Function. *Acta Ophthalmol.* **2020**, *98*, 408–417. [CrossRef]

90. Vardi, N.; Smith, R.G. The AII Amacrine Network: Coupling Can Increase Correlated Activity. *Vis. Res* **1996**, *36*, 3743–3757. [CrossRef]

91. Zandt, B.-J.; Liu, J.H.; Veruki, M.L.; Hartveit, E. AII Amacrine Cells: Quantitative Reconstruction and Morphometric Analysis of Electrophysiologically Identified Cells in Live Rat Retinal Slices Imaged with Multi-Photon Excitation Microscopy. *Brain Struct. Funct.* **2017**, *222*, 151–182. [CrossRef]

92. Sterling, P. Microcircuitry of the Cat Retina. *Annu. Rev. Neurosci.* **1983**, *6*, 149–185. [CrossRef] [PubMed]

93. Tsukamoto, Y.; Omi, N. Functional Allocation of Synaptic Contacts in Microcircuits from Rods via Rod Bipolar to AII Amacrine Cells in the Mouse Retina. *J. Comp. Neurol.* **2013**, *521*, 3541–3555. [CrossRef]

94. Bloomfield, S.A.; Völgyi, B. Function and Plasticity of Homologous Coupling between AII Amacrine Cells. *Vision Res.* **2004**, *44*, 3297–3306. [CrossRef] [PubMed]

95. Wang, Y.; Song, J.-H.; Denisova, J.V.; Park, W.-M.; Fontes, J.D.; Belousov, A.B. Neuronal Gap Junction Coupling Is Regulated by Glutamate and Plays Critical Role in Cell Death during Neuronal Injury. *J. Neurosci.* **2012**, *32*, 713–725. [CrossRef]

96. Szarka, G.; Balogh, M.; Tengölics, Á.; Ganczer, A.; Völgyi, B.; Kovács-Öller, T. The Role of Gap Junctions in Cell Death and Neuromodulation in the Retina. *Neural Regen. Res.* **2021**, *16*, 1911. [CrossRef]

97. Wang, H.Y.; Lin, Y.-P.P.; Mitchell, C.K.; Ram, S.; O'Brien, J. Two-Color Fluorescent Analysis of Connexin 36 Turnover: Relationship to Functional Plasticity. *J Cell Sci* **2015**, *128*, 3888–3897. [CrossRef] [PubMed]

98. Hervé, J.-C.; Derangeon, M.; Bahbouhi, B.; Mesnil, M.; Sarrouilhe, D. The Connexin Turnover, an Important Modulating Factor of the Level of Cell-to-Cell Junctional Communication: Comparison with Other Integral Membrane Proteins. *J. Membr. Biol. 2007* *2171* **2007**, *217*, 21–33. [CrossRef] [PubMed]

99. Feigenspan, A.; Teubner, B.; Willecke, K.; Weiler, R. Expression of Neuronal Connexin36 in AII Amacrine Cells of the Mammalian Retina. *J. Neurosci.* **2001**, *21*, 230–239. [CrossRef]

100. Veruki, M.L.; Hartveit, E. AII (Rod) Amacrine Cells Form a Network of Electrically Coupled Interneurons in the Mammalian Retina. *Neuron* **2002**, *33*, 935–946. [CrossRef]

101. Vaney, D.I. Many Diverse Types of Retinal Neurons Show Tracer Coupling When Injected with Biocytin or Neurobiotin. *Neurosci. Lett.* **1991**, *125*, 187–190. [CrossRef]

102. Strettoi, E.; Raviola, E.; Dacheux, R.F. Synaptic Connections of the Narrow-Field, Bistratified Rod Amacrine Cell (AII) in the Rabbit Retina. *J Comp Neurol* **1992**, *325*, 152–168. [CrossRef]

103. Zandt, B.-J.; Veruki, M.L.; Hartveit, E. Electrotonic Signal Processing in AII Amacrine Cells: Compartmental Models and Passive Membrane Properties for a Gap Junction-Coupled Retinal Neuron. *Brain Struct. Funct.* **2018**, *223*, 3383–3410. [CrossRef] [PubMed]

104. Mills, S.L.; O'Brien, J.J.; Li, W.; O'Brien, J.; Massey, S.C. Rod Pathways in the Mammalian Retina Use Connexin. *J. Comp. Neurol.* **2001**, *436*, 336–350. [CrossRef]

105. Sterling, P.; Freed, M.A.; Smith, R.G. Architecture of Rod and Cone Circuits to the On-Beta Ganglion Cell. *J. Neurosci.* **1988**, *8*, 623–642. [CrossRef]

106. Marc, R.E.; Anderson, J.R.; Jones, B.W.; Sigulinsky, C.L.; Lauritzen, J.S. The AII Amacrine Cell Connectome: A Dense Network Hub. *Front Neural Circuits* **2014**, *8*, 104. [CrossRef] [PubMed]

107. Meyer, A.; Hilgen, G.; Dorgau, B.; Sammler, E.M.; Weiler, R.; Monyer, H.; Dedek, K.; Hormuzdi, S.G. AII Amacrine Cells Discriminate between Heterocellular and Homocellular Locations When Assembling Connexin36-Containing Gap Junctions. *J. Cell Sci.* **2014**, *127*, 1190–1202. [CrossRef]

108. Macneil, M.A.; Purrier, S.; Rushmore, R.J. The Composition of the Inner Nuclear Layer of the Cat Retina. *Vis. Neurosci.* **2009**, *26*, 365–374. [CrossRef]

109. Teeter, C.M.; Stevens, C.F. A General Principle of Neural Arbor Branch Density. *Curr. Biol.* **2011**, *21*, 2105–2108. [CrossRef] [PubMed]

110. Kier, C.; Buchsbaum, G.; Sterling, P. How Retinal Microcircuits Scale for Ganglion Cells of Different Size. *J. Neurosci.* **1995**, *15*, 7673–7683. [CrossRef]

111. Freed, M.A.; Smith, R.G.; Sterling, P. Computational Model of the On-Alpha Ganglion Cell Receptive Field Based on Bipolar Cell Circuitry. *Proc. Natl. Acad. Sci. USA* **1992**, *89*, 236–240. [CrossRef]

112. Xu, Y.; Vasudeva, V.; Vardi, N.; Sterling, P.; Freed, M.A. Different Types of Ganglion Cell Share a Synaptic Pattern. *J. Comp. Neurol.* **2008**, *507*, 1871–1878. [CrossRef]

113. Percival, K.A.; Jusuf, P.R.; Martin, P.R.; Grünert, U. Synaptic Inputs onto Small Bistratified (Blue-ON/Yellow-OFF) Ganglion Cells in Marmoset Retina. *J Comp Neurol* **2009**, *517*, 655–669. [CrossRef] [PubMed]

114. Fournel, R.; Hartveit, E.; Veruki, M.L. Differential Contribution of Gap Junctions to the Membrane Properties of ON- and OFF-Bipolar Cells of the Rat Retina. *Cell. Mol. Neurobiol.* **2021**, *41*, 229–245. [CrossRef] [PubMed]

115. Wässle, H. Parallel Processing in the Mammalian Retina. *Nat. Rev. Neurosci.* **2004**, *5*, 747–757. [CrossRef] [PubMed]

116. Chichilnisky, E.J.; Kalmar, R.S. Functional Asymmetries in ON and OFF Ganglion Cells of Primate Retina. *J. Neurosci.* **2002**, *22*, 2737–2747. [CrossRef] [PubMed]

117. Pandarinath, C.; Victor, J.D.; Nirenberg, S. Symmetry Breakdown in the ON and OFF Pathways of the Retina at Night: Functional Implications. *J. Neurosci.* **2010**, *30*, 10006–10014. [CrossRef]

118. Ravi, S.; Ahn, D.; Greschner, M.; Chichilnisky, E.J.; Field, G.D. Pathway-Specific Asymmetries between ON and OFF Visual Signals. *J. Neurosci.* **2018**, *38*, 9728–9740. [CrossRef]

119. Jin, J.; Wang, Y.; Lashgari, R.; Swadlow, H.A.; Alonso, J.M. Faster Thalamocortical Processing for Dark than Light Visual Targets. *J. Neurosci.* **2011**, *31*, 17471–17479. [CrossRef]

120. Jiang, Y.; Purushothaman, G.; Casagrande, V.A. The Functional Asymmetry of ON and OFF Channels in the Perception of Contrast. *J. Neurophysiol.* **2015**, *114*, 2816–2829. [CrossRef]

121. Escobar, M.J.; Reyes, C.; Herzog, R.; Araya, J.; Otero, M.; Ibaceta, C.; Palacios, A.G. Characterization of Retinal Functionality at Different Eccentricities in a Diurnal Rodent. *Front. Cell. Neurosci.* **2018**, *12*, 444. [CrossRef] [PubMed]

Review

The Vertical and Horizontal Pathways in the Monkey Retina Are Modulated by Typical and Atypical Cannabinoid Receptors

Joseph Bouskila [1], Maxime Bleau [1], Catarina Micaelo-Fernandes [1], Jean-François Bouchard [1] and Maurice Ptito [1,2,3,4,*]

1 School of Optometry, University of Montreal, Montreal, QC H3Y 1P1, Canada; joseph.bouskila@umontreal.ca (J.B.); maxime.bleau.1@umontreal.ca (M.B.); catarina.fernandes@umontreal.ca (C.M.-F.); jean-francois.bouchard@umontreal.ca (J.-F.B.)
2 Behavioral Science Foundations, Eastern Caribbean, Estridge KN0101, Saint Kitts and Nevis
3 Department of Neuroscience, Copenhagen University, 2200 Copenhagen, Denmark
4 Department of Neurology and Neurosurgery, Montreal Neurological Institute, McGill University, Montreal, QC H3A 2B4, Canada
* Correspondence: maurice.ptito@umontreal.ca; Tel.: +1-514-995-2824

Abstract: The endocannabinoid (eCB) system has been found in all visual parts of the central ner-vous system and plays a role in the processing of visual information in many species, including monkeys and humans. Using anatomical methods, cannabinoid receptors are present in the monkey retina, particularly in the vertical glutamatergic pathway, and also in the horizontal GABAergic pathway. Modulating the eCB system regulates normal retinal function as demonstrated by electrophysiological recordings. The characterization of the expression patterns of all types of cannabinoid receptors in the retina is progressing, and further research is needed to elucidate their exact role in processing visual information. Typical cannabinoid receptors include G-protein coupled receptor CB1R and CB2R, and atypical cannabinoid receptors include the G-protein coupled receptor 55 (GPR55) and the ion channel transient receptor potential vanilloid 1 (TRPV1). This review focuses on the expression and localization studies carried out in monkeys, but some data on other animal species and humans will also be reported. Furthermore, the role of the endogenous cannabinoid receptors in retinal function will also be presented using intraocular injections of known modulators (agonists and antagonists) on electroretinographic patterns in monkeys. The effects of the natural bioactive lipid lysophosphatidylglucoside and synthetic FAAH inhibitor URB597 on retinal function, will also be described. Finally, the potential of typical and atypical cannabinoid receptor acti-vity regulation in retinal diseases, such as age-related macular degeneration, diabetic retinopathy, glaucoma, and retinitis pigmentosa will be briefly explored.

Keywords: retina; typical cannabinoid receptors; atypical cannabinoid receptors; immunohistochemistry; electroretinography; monkeys; visual system

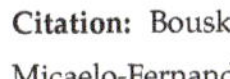
check for
updates

Citation: Bouskila, J.; Bleau, M.; Micaelo-Fernandes, C.; Bouchard, J.-F.; Ptito, M. The Vertical and Horizontal Pathways in the Monkey Retina Are Modulated by Typical and Atypical Cannabinoid Receptors. *Cells* **2021**, *10*, 3160. https://doi.org/10.3390/cells10113160

Academic Editors: Yvan Arsenijevic and Alexander E. Kalyuzhny

Received: 5 October 2021
Accepted: 11 November 2021
Published: 13 November 2021

Publisher's Note: MDPI stays neutral with regard to jurisdictional claims in published maps and institutional affil-iations.

1. Introduction

The organization of the normal retinal mosaic is presented in the introductory chapter (Ptito et al., 2021, this volume) and we will concentrate here on the general components of the endocannabinoid (eCB) system. The eCB system is comprised of various components. There are 2 main types of receptors, cannabinoid receptors type 1 (CB1R) and type 2 (CB2R); endogenous ligands acting upon cannabinoid receptors, termed endocannabinoids (eCBs), mainly anandamide (AEA) and 2-arachidonylglycerol (2-AG); and enzymes metabolizing eCBs, particularly N-acyl phosphatidylethanolamine phospholipase D (NAPE-PLD) and diacylglycerol lipase (DAGL) for the synthesis of eCBs and fatty acid amide hydrolase (FAAH) and monoacylglycerol lipase (MAGL) for the degradation of eCBs [1] (Figure 1).

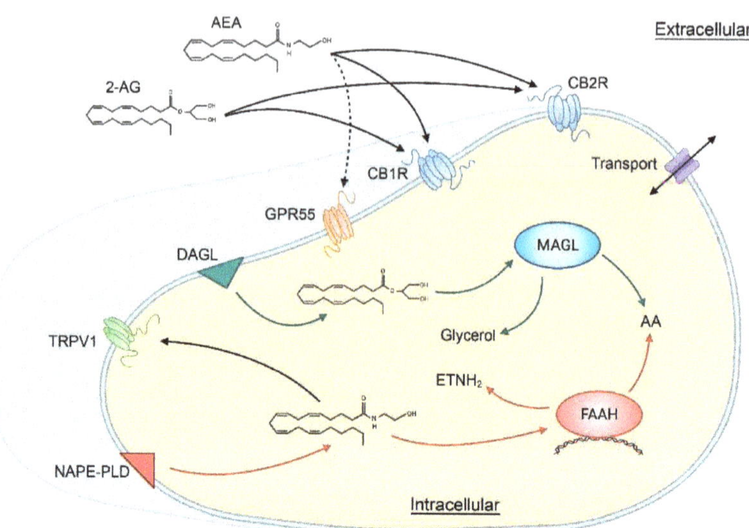

Figure 1. Schematic illustration showing the main components of the classical endocannabinoid system. 2-AG, 2-arachidonylglycerol; AA, arachidonic acid; AEA, anandamide or N-arachidonoylethanolamine; CB1R, cannabinoid receptor type 1; CB2R, cannabinoid receptor type 2; DAGL, Diacylglycerol lipase; ETNH$_2$, ethanolamine; FAAH, fatty acid amide hydrolase; GPR55, G protein-coupled receptor 55; MAGL, monoacylglycerol lipase; NAPE-PLD, N-acyl phosphatidylethanolamine phospholipase D; TRPV1, transient receptor potential vanilloid type 1. Created with BioRender.com and based on [1].

ECBs are derived from long-chain polyunsaturated fatty acids (amides, esters and ethers), mainly arachidonic acid, found in the central and peripheral nervous systems [2,3]. While FAAH converts AEA to ethanolamine and arachidonic acid, MAGL converts 2-AG to glycerol and arachidonic acid. AEA and 2-AG are the most studied eCBs, but others are still being unveiled [4]. Since THC, the main psychoactive compound of cannabis, binds to the same CB1R and CB2R as eCBs, the latter mimic most of the central and peripheral effects of cannabis [5,6].

CB1R and CB2R are the two classical cannabinoid receptors with seven transmembrane passages and coupled to the G proteins. They are most often coupled to the G_i/G_o proteins, inhibit adenylate cyclase, and regulate the calcium and potassium-type ion channels [7] (Figure 2). In 1990, CB1R was identified and cloned [8]. In 1993, CB2R was discovered [9] and, two years later, cloned [10]. Given their differential expression within the body, they have usually been studied independently. CB1R is abundantly expressed in the CNS of all vertebrates [11] and have been ascribed multiple roles, particularly, in the development of the brain (e.g., axon growth) [12–15] and visual functions [16–18].

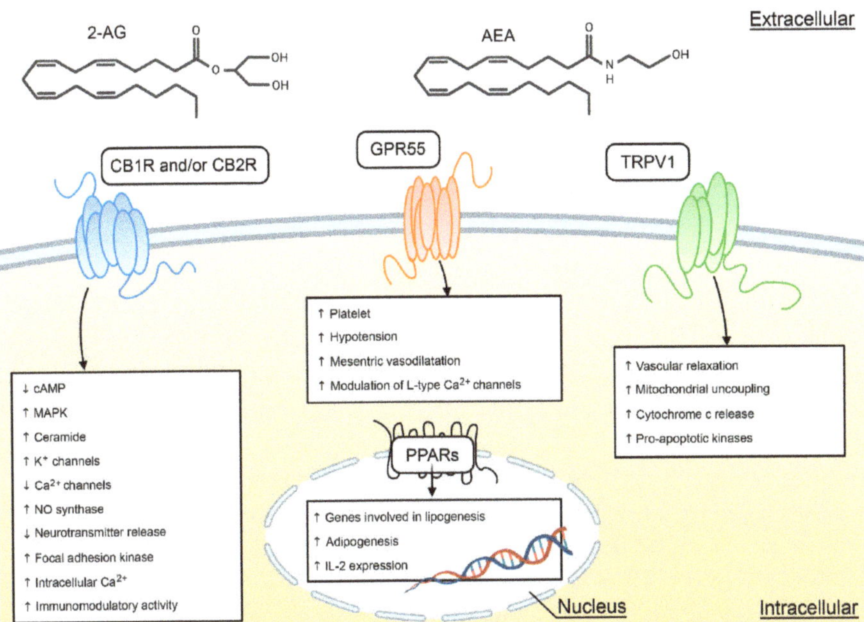

Figure 2. Cellular targets and signaling pathways of AEA and 2-AG. This figure illustrates the main cellular effects for CB1R/CB2R, GPR55, and TRPV1. Redrawn from [19] and created with BioRender.com.

Recently, several studies showed that there may be other cannabinoid receptors, such as the G protein-coupled receptor 55 (GPR55), the ionotropic transient receptor potential vanilloid 1 (TRPV1) and the nuclear peroxisome proliferator-activated receptor (PPAR) (see chapter 3 of [19]). Accordingly, cannabinoid receptors have been divided into typical ad atypical receptors. The typical cannabinoid receptors are CB1R and CB2R, and non-CB1R/CB2R cannabinoid effects define the atypical cannabinoid receptors, mainly GPR55 and TRPV1. Figure 2 shows the main signaling channels for CB1R, CB2R, GPR55, TRPV1, but also PPAR, a receptor that has not been extensively studied in the retina. Cannabinoids, including phytocannabinoids, endocannabinoids, and exogenous synthetic cannabinoid modulators, bind to the traditional two types of cannabinoid receptors CB1 and CB2, but also, to other GPCR and ion channels. In the past decade, it has been reported that both, typical and atypical cannabinoid receptors, are expressed in the visual system of rodents, monkeys, and humans [18,20,21].

2. The Retinal Endocannabinoid System

Expression studies have shown that CB1R, CB2R, GPR55, and TRPV1 are present in the retina of multiple species [13,14,17,18,21–26], but their differential expression has been found solely in the monkey retina [16,27–30]. More precisely, our studies demonstrate that these receptors are distributed in the vertical and horizontal pathways of the vervet monkey retina. It is important to stress out that good, specific, validated antibodies are important for proper characterization of cannabinoid expression patterns, e.g., targeting CB2R using immunohistochemistry is controversial in the literature [31,32]. Similarly, commercially available TRPV1 antibodies generate non-specific labeling in the retina of some species [33]. Using the non-human primate model and antibodies directed against human epitopes for cannabinoid receptors, in our experience, was found adequate [30,34]. In line with this, we compared the alignment of the human sequence for these 4 receptors (Figure 3).

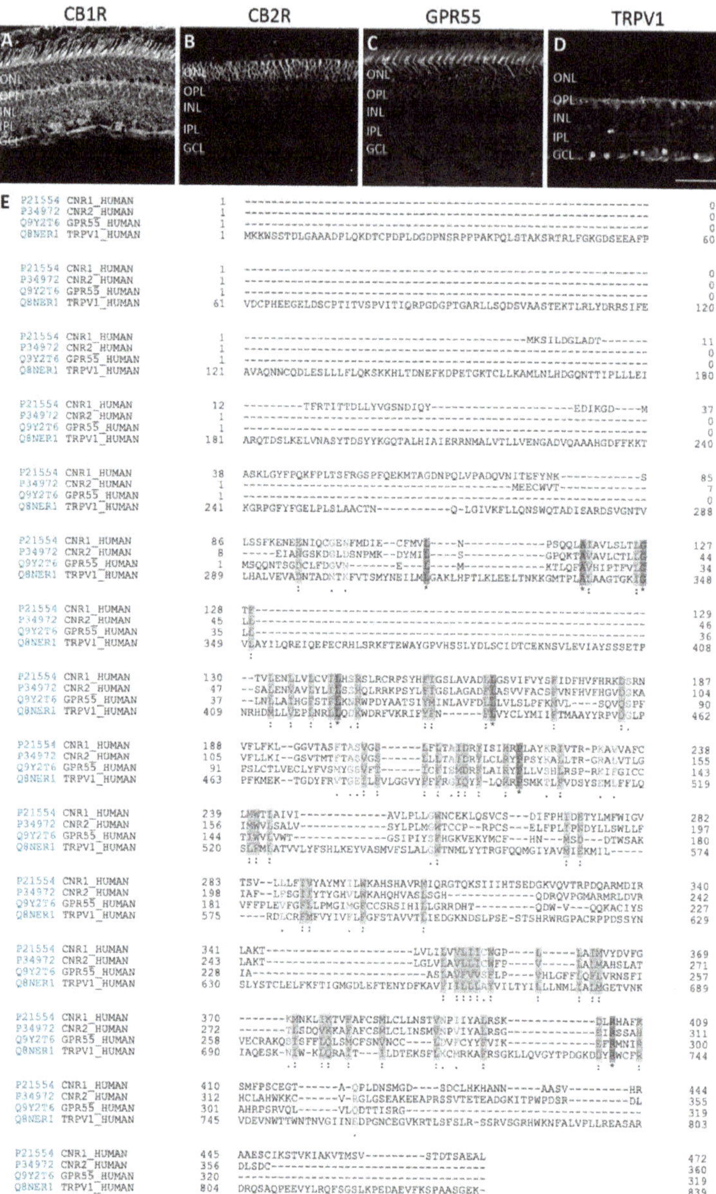

Figure 3. Comparison of typical and atypical cannabinoid receptors. Confocal immunofluorescence images of vervet monkey retinal sections (**A–D**) and human protein sequence alignment (**E**) of CB1R, CB2R, GPR55, and TRPV1. CB1R is localized in neural components, with very weak (albeit absence) in rods [16] (**A**). CB2R is strictly expressed in the glial components, the Müller cells [29] (**B**). GPR55 is found exclusively in rods, with the most prominent staining in the inner segments [28] (**C**). TRPV1 is expressed in the lateral pathway, particularly horizontal cells and amacrine cells [27] (**D**). Protein alignment show conserved regions that may be important for the structural and functional effects of cannabinoids. Scale bar = 75 μm.

2.1. Typical Cannabinoid Receptors CB1R and CB2R

2.1.1. Anatomical Localization and Function

Traditionally, CB1R has been associated with the CNS and CB2R with the immune system. This idea was then developed into CB1R in neurons to modulate neurotransmitter release, and CB2R in glial cells to modulate cytokine release. While this differential expression is not present in the rodent retina, it appears to be present in the vervet monkey retina. In healthy monkeys, the presence of CB1R in the neuroretina has been well characterized [16]. CB1R is highly expressed in the foveal region, owing to its abundant expression in cone photoreceptors. It is also present throughout the vertical pathway consisting of rod and cone photoreceptors, bipolar cells, and ganglion cells, and at a lesser degree in the horizontal pathway consisting of horizontal cells and amacrine cells. CB2R is exclusively expressed in the major glial element of the retina (Müller cells), extending from the outer limiting membrane to the inner limiting membrane [29]. Furthermore, localization of their associated metabolic enzymes suggests that eCBs are synthesized and released in the synapse surrounding the neurons from which they are released [13,18]. This dual CB1R/CB2R expression pattern already suggests differential retinal functions (Figure 4).

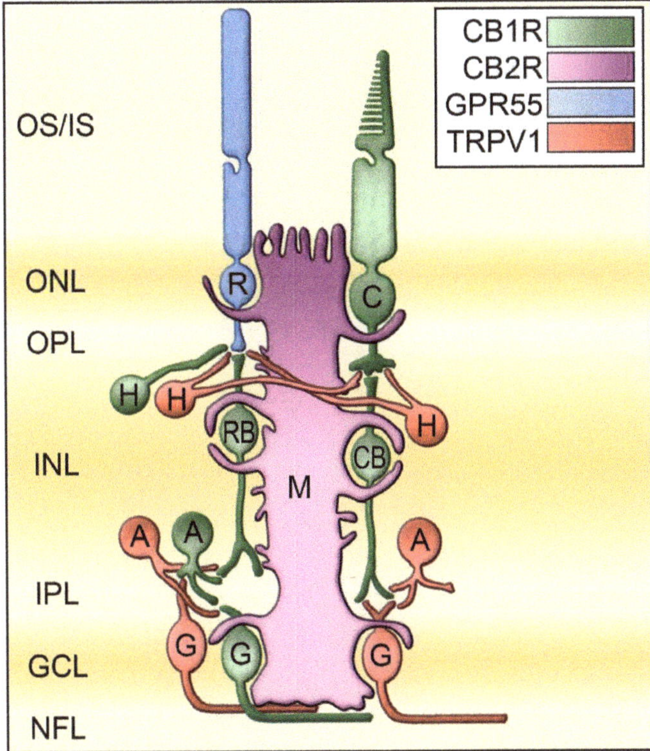

Figure 4. Mapping of the receptors CB1R, CB2R, GPR55, and TRPV1 in the monkey retina. These receptors are differently expressed in the retina of monkeys. These results are compiled from several published articles [16,27–29]. CB1R is represented in green, CB2R in magenta, GPR55 in blue, and TRPV1 in red. A, amacrine cells; C, cone photoreceptors; CB, cone bipolar cells; G, ganglion cells; GCL, ganglion cell layer; H, horizontal cells; INL, inner nuclear layer; IPL, inner plexiform layer; IS, photoreceptors inner segments; M, Müller cells; NFL, nerve fiber layer; ONL, outer nuclear layer; ONL, outer plexiform layer; OS, photoreceptors outer segments; R, rod photoreceptors; RB, rod bipolar cells. Redrawn from [27].

The hypothetical function of cannabinoid receptors in the monkey retina could be extended to the human retina given the strong similarities in its anatomical organization. Intraocular injections of specific blockers for CB1R and CB2R in the monkey retina lead to an increase of photopic and scotopic responses (Figure 5; [20]).

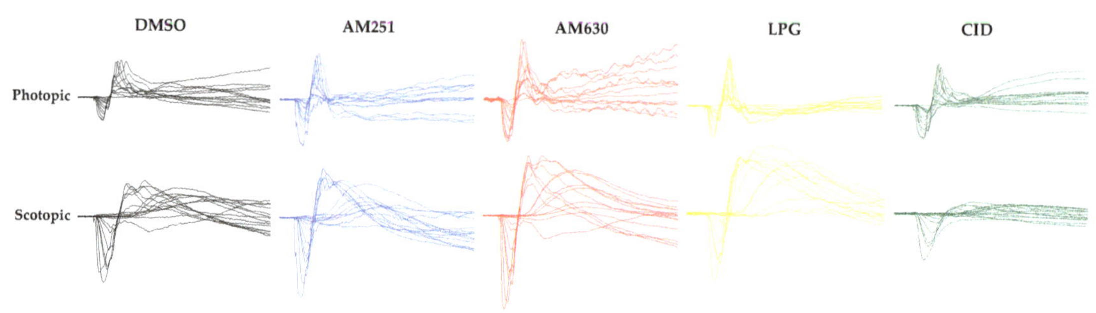

Figure 5. Raw photopic and scotopic ERGs in the different drug injection groups. The effect of modulating CB1R, CB2R, or GPR55 in the monkey retina. The intravitreal injection of AM251 (blue), an inverse agonist of CB1R, or AM630 (red), an inverse agonist of CB2R, causes an increase of the ERG photopic (under a 30 cd/m^2 white background-adapting field) and scotopic (recorded under dark-adapted conditions in a light-tight room) responses, compared to the vehicle, dimethyl sulfoxide (DMSO, in black) [20]. The intravitreal injection of lysophosphatidylglucoside (LPG), an agonist of GPR55 (yellow), causes an increase of the scotopic response, but not of the photopic response. Conversely, the intravitreal injection of CID16020046 (CID), an antagonist of GPR55, causes a decrease of the scotopic response, but not of the photopic response (green).

2.1.2. Interspecies CB1R/CB2R Comparison

Several pieces of evidence lead us to believe that these receptors are not expressed in the retina in a similar way across species, e.g. the protein sequence of mouse CB2R is different from that of the primate. There are some major differences in protein sequences as one goes up the animal ladder. CB1R and CB2R are unique to Chordata (phylum of animals that has a notochord, a cartilaginous lamella of mesodermal origin located on the dorsal side of the animal). Furthermore, synthetic and degradative enzymes are present in several species of the animal kingdom [35]. It is possible that receptor-acting proteins appeared much later than eCBs. Although the expression patterns of some eCBs components like CB1R and FAAH are similar in different species, this is not the case for CB2R and NAPE-PLD [34]. As the expression of CB2R in the mouse retina [22] is different from the monkey [29], we studied several components of the eCB system known to date in the mouse retina, monkeys and treeshrews (an intermediate species in the phylogenetic tree). In addition, the expression of this system has been compared in the retina of two types of monkeys, namely vervets and macaques. In the retina of mice, treeshrews, macaques, and vervets, CB1R, and its associated eCB degrading enzyme FAAH, have an overlapping expression pattern [34]. This suggests that degradation of eCBs may occur in the same CB1R positive retinal cell. For CB2R, some important expression differences are present. In the mouse rod-dominated retina, CB2R is expressed in the neuro-retina, in photoreceptors, horizontal cells, bipolar cells, amacrine cells, and ganglion cells, but not in Müller cells [22]. In the cone-dominated retina of treeshrews, CB2R is expressed in neuroretina and also in Müller cells. In the duplex retina of vervet and macaque monkeys, CB2R is found exclusively in Müller cells. One possible explanation for this transition of CB2R distribution (diffuse neuronal expression vs. specific expression in the retinal glia) is that this receptor has assumed a strategic place to exert the role of potassium modulator during the evolution of the retina. Interestingly, the synthetic enzyme NAPE-PLD also has an expression pattern specific to each of these species. In mice, NAPE-PLD is expressed in the neuroretina [34]. In the Tree shrew, NAPE-PLD is expressed strongly in the external retina and weakly in the internal retina [34]. In monkeys, NAPE-PLD is expressed only at the level of photoreceptors, cones and rods [34]. The bioactive lipid molecules, N-acyl ethanolamines (NAE), are synthesized by NAPE-PLD from cell membrane phospholipids. This interspecies retinal expression difference is noted despite the conserved protein sequence of NAPE-PLD, unlike CB2R. This variation may be of importance to NAPE-PLD constructs other than eCBs. This enzyme synthesizes several molecules such as anandamide (an eCB), but also N-palmitoylethanolamine (an anti-inflammatory agent; [36]) and Noleoylethanolamine (an anorexic agent; [37]). NAPE-PLD can even have proapoptotic effects [38]. In addition, the products of this enzyme, NAE, are present in axons and regulate postsynaptic neuronal activity by acting as anterograde synaptic signaling molecules ([39]. Finally, the expression of NAPE-PLD in monkey photoreceptors suggests a direct role of NAEs in primate phototransduction. But this remains to be verified experimentally. The expression patterns of DAGL and MAGL in the retina of mice, treeshrews and monkeys are similar, but have different signal intensities [34,40]. As eCBs are synthesized and degraded around their receptors, expression of DAGL and MAGL must be found in the vicinity of CB2R. In the mouse retina, these two enzymes in addition to CB1R and CB2R are expressed in an overlapping manner and therefore may exert a self-regulatory role in each of the cells of the retina. In the retina of tree shrews and monkeys, the expression of enzymes of the eCB system is found in a complementary manner. The eCB system may have specialized in parallel with the increasing complexity of the visual system to adopt a strategic position for the modulation of visual activity

The molecular basis of CB1R/CB2R action in retinal function can be explained by a model recently published [30]. In photopic conditions, when cones are stimulated by light, the ionic channels are inhibited, a process known as the "inhibition of the retinal dark currents." The resulting phototransduction reduces the glutamate release in the synapse and propagates an evoked potential to bipolar cells. Given the localization of the metabolic enzymes in monkeys, the same bipolar cells may be the main source of eCB production that will act in a retrograde manner and activate CB1R located in cone pedicles, thus regulating glutamate release. This eCB production will also synthesize 2-AG that will activate CB2R in Müller cells, thus modulating potassium spatial buffering throughout the retina. The activation of CB2R coupled to $G_{i/o}$ will reduce the levels of cyclic AMP and PKA activity. Given that PKA activates $KIR_{4.1}$ channels in Müller cells, CB2R will play a role by negatively modulating potassium. In scotopic conditions, the synaptic terminals of rods release a large quantity of glutamate. This glutamate binds to mGluR6 receptors located on the dendrites of ON rod bipolar cells. See Figure 6 for details.

2.2. Atypical Putative Cannabinoid Receptors GPR55 and TRPV1

To date, there are two atypical cannabinoid receptors that have been described in the monkey retina: (1) GPR55 has been found mainly in rods of the vertical pathway, and (2) TRPV1 in horizontal and amacrine cells of the horizontal pathway.

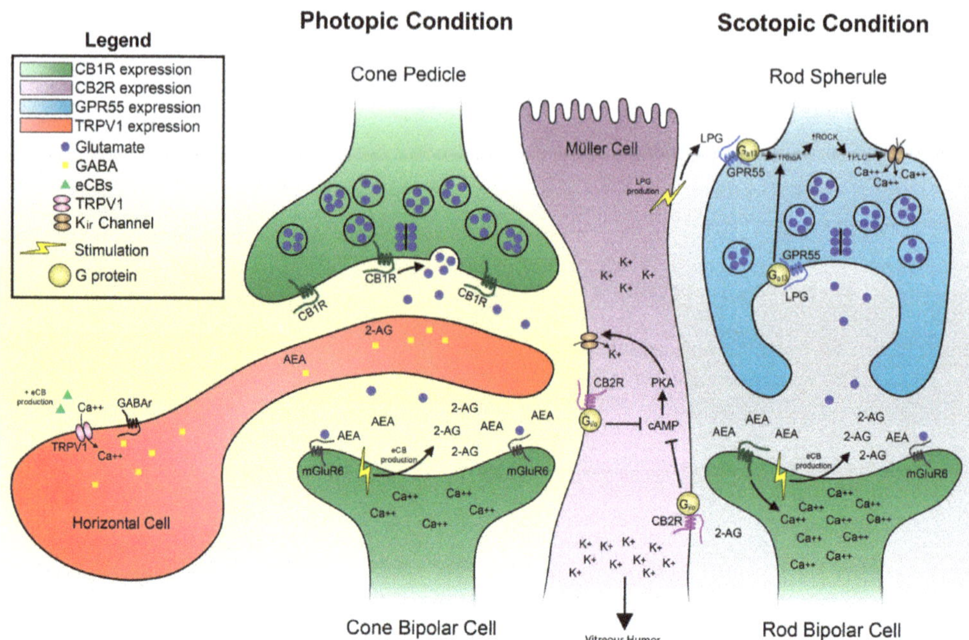

Figure 6. The hypothetical molecular mechanisms for typical and atypical cannabinoid receptors action in the retina is illustrated at the photoreceptor-bipolar cell synapse. This may also exist in several other synapses in the visual where this differential expression pattern exists. LPG, as suggested in previous models [30,41], could be produced by Müller cells. Adapted from [30].

2.2.1. Anatomical Localization and Function of GPR55

GPR55 is a GPCR that is naturally activated by LPI or LPG [42–44]. While the endogenous cannabinoid AEA and phytocannabinoid THC can also activate this receptor, the non-psychoactive compound CBD is an antagonist of this receptor [45]. GPR55 is an important modulator of retinal development in mice and hamsters [23]. GPR55 expression is widely expressed in the adult rodent retina [23], and it is exclusively expressed in rod photoreceptors, most prominently in inner segments of the vervet monkey retina [28] (Figure 4). Modulation of GPR55 by either blocking GPR55 (impaired retinal function, e.g., nyctalopia) or by activating it (increased retinal function, e.g., hyper-scotopia) impacts solely scotopic retinal function [41] (Figure 5).

2.2.2. Anatomical Localization and Function of TRPV1

Transient receptor potential vanilloid type 1 (TRPV1) is a cannabinoid-like non-selective cation channel receptor that is the main target of the pungent compound found in hot chili peppers capsaicin [46]. In neurons, endovannilloids and endocannabinoids, like anandamide, 2-arachidonoylglycerol and N-arachidonoyl dopamine bind to TRPV1 [47]. Evidence for TRPV1 expression and function in the retina of multiple species has also been put forward [48]. In the goldfish and zebrafish retinas, TRPV1 was found in amacrine cells, as well as the synaptic ribbons of photoreceptors [49,50]. In the rat retina, TRPV1 was found in microglial cells, blood vessels, and astrocytes and in neuronal structures such as synaptic boutons of both plexiform layers as well as in cell bodies of the INL and GCL [51]. TRPV1 mRNA was detected in ganglion and Müller cells in the rat retina [26,52]. In the rabbit and human retina, TRPV1 is intensely expressed in the RPE [53]. TRPV1 is also present in the outer nuclear layer (ONL) and INL and at the end of the nerve fiber layer as well as in Müller cells of the rabbit retina [53]. In the vervet monkey retina, TRPV1 was found in the

OPL and IPL, and in RGCL with a higher density in the periphery [27]. Co-immunolabeling of TRPV1 with parvalbumin, a primate horizontal cell marker, revealed a clear overlap of expression throughout the entire cell structure with most prominent staining in the cell body membrane and synaptic terminals. Furthermore, double labeling of TRPV1 and syntaxin was found throughout amacrine cells in the inner plexiform layer. Finally, double staining of TRPV1 and Brn3a allowed us to confirm its previously reported expression in the cell bodies and dendrites of RGCs [26]. The presence of TRPV1 in the horizontal pathway suggests a function of this receptor in lateral inhibition between photoreceptors through the horizontal cells, and between bipolar cells through amacrine cells (Figure 4). A role for TRPV1 channels in physiopathological retinal processes has been first ascribed in the rat retina [54]. Activation of TRPV1 has been then proposed to modify retinal protein expression patterns contributing to calcium-dependent signaling that maintains excitatory signaling in RGCs using TRPV1 knockout mice [26]. Inhibition of TRPV1 prevented retinal angiogenesis in a mouse model of oxygen-induced retinopathy [55,56]. We have not studied the role of TRPV1 in the retina of vervet monkeys. A good way to test this would be to perform intraocular injections of AM404 acting as a neuronal TRPV1 agonist, and SB-705498 acting as an antagonist in vervet monkeys, since the retina has no nociceptors [57].

2.2.3. Interspecies GPR55/TRPV1 Comparison

Anatomical data on GPR55 retinal expression stems largely from two studies on mice and vervet monkeys [23,28]. To our knowledge, there has not been in other immunohistochemical study reported since the available and specific GPR55 markers are scarce. TRPV1 expression pattern is different in the retina of mice, macaque monkeys, and humans [26].

3. Cannabinoid Receptors in Retinal Diseases

Eye-related vision loss can occur because of damage of the optical system (e.g., corneal and lens anomalies due to multiple factors such as age, genetics, and intrusion of a foreign compound) or pathology of the neuro-retina (e.g., photoreceptors, bipolar cells, ganglion cells physiological dysfunction). While surgical interventions can treat pathologies of the optical system (cataracts, corneal opacifications …), there are little approved clinical treatments for retinal diseases like age-related macular degeneration (AMD), diabetic retinopathy (DR), glaucoma, and retinitis pigmentosa (RP). However, intense research is being carried out targeting these pathologies using a wide variety of approaches [58–63]. Current strategies focus on slowing down or stop the initial triggers (e.g., AREDS supplementation, exercise, eating well, and smoking cessation in dry AMD, and anti-VEGF intraocular injections in wet AMD) [64,65], or focus on relieving the symptoms (e.g., lowering intraocular pressure (IOP) with prostaglandin analogues or beta blockers in glaucoma) [66]. Given the wide expression profile of the eCB system in the neuro-retina, the use of this system as a pharmacological target, particularly the typical and atypical cannabinoid receptors, in the management of these retinal diseases is of great interest. Concomitantly, data obtained from preclinical studies shows that cannabis and cannabinoid molecules used as neuroprotective agents may have potential benefits to prevent glaucoma and other retinal neurodegenerative diseases [67]. The purpose of this review was not to exhaustively list the evidence supporting the idea that cannabinoids are useful in the treatment of retinal diseases, rather, to present how typical and atypical cannabinoid receptors can be used as novel pharmacological targets. Indeed, the potential of cannabinoids as anti-inflammatory, neuroprotective, or IOP lowering agents in the management of some retinal diseases has been reviewed elsewhere [68]. Given that cannabinoids can activate typical CB1R and CB2R, as well as atypical GPR55 and TRPV1, these studies can definitely serve as the basis of using for further drug development. A current clinical trial is studying if cannabinoids may be beneficial in certain degenerative diseases of the retina by determining whether cannabis derivatives affect the visual functions in healthy adults, and examining the effect of cannabis derivatives on the retina of retinitis pigmentosa patients using the ERG as the primary outcome measure, as well as other optometric tests (https://clinicaltrials.gov/ct2/show/NCT03078309, accessed on 11 November 2021).

3.1. Age-Related Macular Degeneration (AMD)

Macular degeneration is an eye disease typically due to normal aging. It is estimated that 9% of the worldwide population has AMD with a projected number of people with the disease around 196 million in 2020, and increasing to 288 million in 2040 [69]. More than 1 million Canadians are affected and is the leading cause of visual deficiency in the elderly. The macula is located in the center of the retina and allows for central vision and detection of fine details. AMD is thus the degeneration of the macula. Blurred central vision makes daily activities such as reading and driving difficult. Peripheral vision is often preserved as AMD affects primarily central vision. There are two forms of this disease, the dry and wet AMD. Dry AMD, the most frequent form, is caracterized by thinning or pigmentation/coloration of the macula. Wet AMD has a rapid and severe onset and is caused by abnormal forming blood vessels under the macula. The resulting flow of liquid/blood destroys the macular neuro-retina. THC acting on typical and atypical receptors has been shown to inhibit VEGF pathways, slowing down retinal angiogenesis in mice [70]. Cannabidiol by activating A_{2A} adenosine receptors in retinal microglial cells can as an anti-inflammatory agent in rats [71]. Finally, cannabidiol acting upon GPR55 could act as an inhibitor of cytokine production and inflammation, thus decreasing the thinning and degeneration of the macular region in monkeys [72].

3.2. Diabetic Retinopathy (DR)

Diabetes is a hormonal disease primarily affecting the control of sugar in the blood. High glucose blood levels reach the retinal circulatory system and can make tiny blood vessels swell and cause retinal detachment. This is called non-proliferative DR. New blood vessels can also start forming and compromise normal vision. This is known as proliferative DR. The resulting inflammation and oxidation, as well as degeneration of the nearby neuro-retina can worsen the prognostic of this disease. Anti-VEGF intraocular injections can slow down and stop newly forming blood vessels. Yet, there is no treatment targeting the resulting inflammation and neurodegeneration process occurring in DR. Modulating the eCB system (e.g., cannabidiol activating typical and atypical cannabinoid receptors) is a promising strategy for the treatment of DR by increasing anti-inflammatory processes and neuroprotection [73].

3.3. Glaucoma

Glaucoma is an eye disease that results in the destruction of the optic nerve neurons. Controlling the main risk factor that is IOP is an important strategy to slow down or prevent the disease. Cannabinoids consumed *orally* can lower IOP by activating cannabinoid receptors and probably the trabecular meshwork that controls the production and secretion of aqueous humor, but result in unwanted systemic side effects [74]. Specific typical and atypical cannabinoid receptor modulation in many delivery methods could be used as therapeutics to decrease inflammation and the resulting cell death [75].

3.4. Retinitis Pigmentosa

RP is an eye disease characterized by a slow degeneration of the pigmented epithelium in both eyes mainly caused by genetic factors. It is first manifested by night vision blindness and narrowing of the visual field. Central vision may also be affected tardively in some patients. Of importance is the role exerted by atypical GPR55 on improving scotopic vision. Given that RP affects night vision by altering rod photoreceptor function, exploiting agonists of GPR55 could have a tremendous impact on bettering good vision at an early stage of the disease, and possibly, slowing down the disease [28,41].

4. Prospects and Limitations

The progress in cannabinoid research might have an impact in the treatment of various retinal diseases. Indeed, recent studies have shown that the expression and localization of cannabinoid receptors in specific retinal components led to relate cannabinoid receptors to specific visual functions (photopic and scotopic vision). This prospect is highlighted by a study that clearly showed that an agonist of GPR55 (LPG) increased significantly the ERG response in scotopic conditions [41]. Given that night vison is impacted by various eye diseases, this compound offers a nice potetial therapeutic venue to restore normal night vision. Moreover, we are seeing now an increase in the discovery of new atypical cannabinoid receptors and ligands that will increase the number of potential therapeutic targets [76]. However, extending this research to humans proves to be difficult for methodological and ethical reasons. Nonetheless, given the strong similarity between the organisation of the retina and the visual system of primates and humans, data obtained from monkeys could be easily applied to humans.

5. Conclusions

We have clearly demonstrated that there is an anatomical segregation of the four different endocannabinoid receptors, CB1R, CB2R, GPR55, and TRPV1 in the retina of the vervet monkey. The development of new therapeutic agents that interact with typical and atypical cannabinoid receptors is crucial for potential treatment of retinal disorders (Figure 7). The four cannabinoid receptors CB1R, CB2R, GPR55, and TRPV1 are ideally positioned in the primate retina to play important functions. Modulation of CB1R in the neuro-retina can serve as neuroprotection. By activating CB1R in cones, bipolar cells, horizontal cells, amacrine cells, and ganglion cells, these retinal cells could be protected from AMPA excitotoxicity [77]. Modulation of CB2R in glial Müller cells can control many cytokine production, and serve as an anti-inflammatory component. By activating CB2R in Müller cells, the production of cytokines, nitric oxide, and/or reactive oxygen species might be controled [78,79]. GPR55 modulation can enhance the activity of rods, and increase night vision. By specifically activating GPR55 in rod photoreceptors, agonists of this receptor could act as photosensitizers to induce night vision enhancement, similarly to what was observed with porphyrins [80]. The activation could enhance TRPV1 modulation as a way to preserve contrast perception. By increasing lateral inhibition in horizontal cells, visual perception could improve by boosting the image contrasts [81].

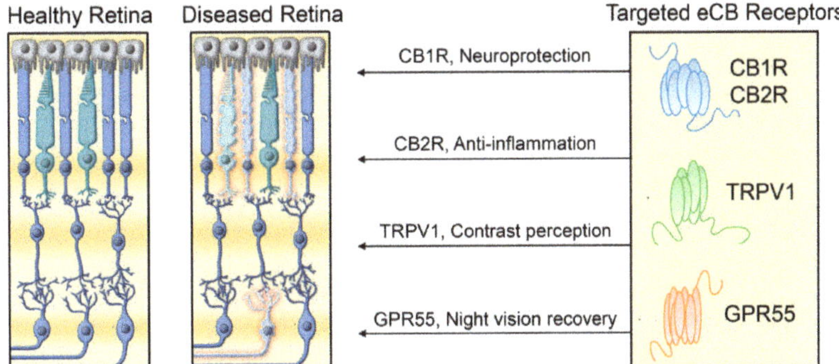

Figure 7. The main proposed mechanisms for the therapeutic effects of typical and atypical cannabinoid receptors in retinal diseases. Cannabinoid receptors can possibly contribute to treatment of retinal disorders by different mechanisms involving cannabinoid receptor type 1, cannabinoid receptor type 2, GPR55, and TRPV1 receptors.

Author Contributions: Conceptualization, M.P. and J.B.; writing—original draft preparation, J.B. and M.B.; writing—review and editing, M.P., J.B., M.B., C.M.-F. and J.-F.B.; supervision, M.P. and J.-F.B.; project administration, M.P. and J.B.; funding acquisition, M.P. and J.-F.B. All authors have read and agreed to the published version of the manuscript.

Funding: This research was funded by the National Science and Engineering Research Council of Canada (6362-2017: M.P.; RGPIN- 2020-05739: J.-F.B.), and the Canadian Institutes of Health Research (163014-2019: M.P. and J.-F.B., PJT-156029: J.-F.B.).

Institutional Review Board Statement: All experiments were performed according to the guidelines of the Canadian Council on Animal Care (CCAC) and the Association for Research in Vision and Ophthalmology (ARVO) Statement for the Use of Animals in Ophthalmic and Vision Research. The experimental protocol was also reviewed and approved by the local Animal Care and Use Committee (University of Montreal, Protocol # 19-115) and the Institutional Review Board of the Behavioral Science Foundation. None of the animals were euthanatized for this study.

Informed Consent Statement: Not applicable.

Data Availability Statement: Data supporting reported results can be requested from J.B. (joseph.bouskila@umontreal.ca).

Acknowledgments: The authors would like to thank Amy Beierschmitt and Maurice Matthew, M.Sc. for their expert technical assistance in handling the monkeys. We are very grateful to the late Frank Ervin, Roberta Palmour, and the staff of the Behavioral Sciences Foundation Laboratories located in St Kitts, West Indies, for their continued support to our monkey work.

Conflicts of Interest: The authors declare no conflict of interest. The funders had no role in the design of the study; in the collection, analyses or interpretation of data; in the writing of the manuscript or in the decision to publish the results.

References

1. Hoover, H.S.; Blankman, J.L.; Niessen, S.; Cravatt, B.F. Selectivity of inhibitors of endocannabinoid biosynthesis evaluated by activity-based protein profiling. *Bioorg. Med. Chem. Lett.* **2008**, *18*, 5838–5841. [CrossRef]
2. Bari, M.; Battista, N.; Fezza, F.; Gasperi, V.; Maccarrone, M. New insights into endocannabinoid degradation and its therapeutic potential. *Mini-Rev. Med. Chem.* **2006**, *6*, 257–268. [CrossRef]
3. De Petrocellis, L.; Cascio, M.G.; Di Marzo, V. The endocannabinoid system: A general view and latest additions. *Br. J. Pharmacol.* **2004**, *141*, 765–774. [CrossRef]
4. Hillard, C.J. Circulating endocannabinoids: From whence do they come and where are they going? *Neuropsychopharmacology* **2018**, *43*, 155–172. [CrossRef]
5. Howlett, A.C.; Breivogel, C.S.; Childers, S.R.; Deadwyler, S.A.; Hampson, R.E.; Porrino, L.J. Cannabinoid physiology and pharmacology: 30 years of progress. *Neuropharmacology* **2004**, *47* (Suppl. 1), 345–358. [CrossRef] [PubMed]
6. Mechoulam, R. Discovery of endocannabinoids and some random thoughts on their possible roles in neuroprotection and aggression. *Prostaglandins Leukot. Essent. Fatty Acids* **2002**, *66*, 93–99. [CrossRef]
7. Piomelli, D. The molecular logic of endocannabinoid signalling. *Nat. Rev. Neurosci.* **2003**, *4*, 873–884. [CrossRef]
8. Matsuda, L.A.; Lolait, S.J.; Brownstein, M.J.; Young, A.C.; Bonner, T.I. Structure of a cannabinoid receptor and functional expression of the cloned cDNA. *Nature* **1990**, *346*, 561–564. [CrossRef] [PubMed]
9. Munro, S.; Thomas, K.L.; Abu-Shaar, M. Molecular characterization of a peripheral receptor for cannabinoids. *Nature* **1993**, *365*, 61–65. [CrossRef] [PubMed]
10. Facci, L.; Dal Toso, R.; Romanello, S.; Buriani, A.; Skaper, S.D.; Leon, A. Mast cells express a peripheral cannabinoid receptor with differential sensitivity to anandamide and palmitoylethanolamide. *Proc. Natl. Acad. Sci. USA.* **1995**, *92*, 3376–3380. [CrossRef] [PubMed]
11. Eggan, S.M.; Lewis, D.A. Immunocytochemical distribution of the cannabinoid CB1 receptor in the primate neocortex: A regional and laminar analysis. *Cereb. Cortex* **2007**, *17*, 175–191. [CrossRef]
12. Argaw, A.; Duff, G.; Zabouri, N.; Cécyre, B.; Chaine, N.; Cherif, H.; Tea, N.; Lutz, B.; Ptito, M.; Bouchard, J.F. Concerted action of CB1 cannabinoid receptor and deleted in colorectal cancer in axon guidance. *J. Neurosci.* **2011**, *31*, 1489–1499. [CrossRef]
13. Bouchard, J.F.; Casanova, C.; Cecyre, B.; Redmond, W.J. Expression and Function of the Endocannabinoid System in the Retina and the Visual Brain. *Neural Plast.* **2016**, *2016*, 9247057. [CrossRef]
14. Duff, G.; Argaw, A.; Cécyre, B.; Cherif, H.; Tea, N.; Zabouri, N.; Casanova, C.; Ptito, M.; Bouchard, J.F. Cannabinoid receptor CB2 modulates axon guidance. *PLoS ONE* **2013**, *8*, e70849. [CrossRef]
15. Harkany, T.; Guzman, M.; Galve-Roperh, I.; Berghuis, P.; Devi, L.A.; Mackie, K. The emerging functions of endocannabinoid signaling during CNS development. *Trends Pharmacol. Sci.* **2007**, *28*, 83–92. [CrossRef]

16. Bouskila, J.; Burke, M.W.; Zabouri, N.; Casanova, C.; Ptito, M.; Bouchard, J.F. Expression and localization of the cannabinoid receptor type 1 and the enzyme fatty acid amide hydrolase in the retina of vervet monkeys. *Neuroscience* **2012**, *202*, 117–130. [CrossRef] [PubMed]

17. Schwitzer, T.; Schwan, R.; Angioi-Duprez, K.; Giersch, A.; Laprevote, V. The endocannabinoid system in the retina: From physiology to practical and therapeutic applications. *Neural Plast.* **2016**, *2016*, 2916732. [CrossRef] [PubMed]

18. Yazulla, S. Endocannabinoids in the retina: From marijuana to neuroprotection. *Prog. Retin. Eye Res.* **2008**, *27*, 501–526. [CrossRef] [PubMed]

19. Di Marzo, V. *Cannabinoids*; Wiley: Hoboken, NJ, USA, 2014.

20. Bouskila, J.; Harrar, V.; Javadi, P.; Beierschmitt, A.; Palmour, R.; Casanova, C.; Bouchard, J.-F.; Ptito, M. Cannabinoid receptors CB1 and CB2 modulate the Electroretinographic waves in Vervet monkeys. *Neural Plast.* **2016**, *2016*. [CrossRef] [PubMed]

21. Straiker, A.; Stella, N.; Piomelli, D.; Mackie, K.; Karten, H.J.; Maguire, G. Cannabinoid CB1 receptors and ligands in vertebrate retina: Localization and function of an endogenous signaling system. *Proc. Natl. Acad. Sci. USA.* **1999**, *96*, 14565–14570. [CrossRef]

22. Cécyre, B.; Zabouri, N.; Huppé-Gourgues, F.; Bouchard, J.F.; Casanova, C. Roles of cannabinoid receptors type 1 and 2 on the retinal function of adult mice. *Invest. Ophthalmol. Vis. Sci.* **2013**, *54*, 8079–8090. [CrossRef]

23. Cherif, H.; Argaw, A.; Cécyre, B.; Bouchard, A.; Gagnon, J.; Javadi, P.; Desgent, S.; Mackie, K.; Bouchard, J.-F. Role of GPR55 during Axon Growth and Target Innervation. *Eneuro* **2015**, *2*. [CrossRef]

24. Straiker, A.J.; Maguire, G.; Mackie, K.; Lindsey, J. Localization of cannabinoid CB1 receptors in the human anterior eye and retina. *Invest. Ophthalmol. Vis. Sci.* **1999**, *40*, 2442–2448.

25. Zabouri, N.; Bouchard, J.F.; Casanova, C. Cannabinoid receptor type 1 expression during postnatal development of the rat retina. *J. Comp. Neurol.* **2011**, *519*, 1258–1280. [CrossRef] [PubMed]

26. Sappington, R.M.; Sidorova, T.; Ward, N.J.; Chakravarthy, R.; Ho, K.W.; Calkins, D.J. Activation of transient receptor potential vanilloid-1 (TRPV1) influences how retinal ganglion cell neurons respond to pressure-related stress. *Channels* **2015**, *9*, 102–113. [CrossRef]

27. Bouskila, J.; Micaelo-Fernandes, C.; Palmour, R.M.; Bouchard, J.F.; Ptito, M. Transient receptor potential vanilloid type 1 is expressed in the horizontal pathway of the vervet monkey retina. *Sci. Rep.* **2020**, *10*, 12116. [CrossRef] [PubMed]

28. Bouskila, J.; Javadi, P.; Casanova, C.; Ptito, M.; Bouchard, J.-F. Rod photoreceptors express GPR55 in the adult vervet monkey retina. *PLoS ONE* **2013**, *8*, e81080. [CrossRef] [PubMed]

29. Bouskila, J.; Javadi, P.; Casanova, C.; Ptito, M.; Bouchard, J.F. Müller cells express the cannabinoid CB2 receptor in the vervet monkey retina. *J. Comp. Neurol.* **2013**, *521*, 2399–2415. [CrossRef] [PubMed]

30. Bouskila, J.; Palmour, R.; Bouchard, J.-F.; Ptito, M. The Endocannabinoid System in the Vervet Monkey Retina. *Primates* **2018**, 145–162. [CrossRef]

31. Ashton, J.C.; Wright, J.L.; McPartland, J.M.; Tyndall, J.D. Cannabinoid CB1 and CB2 receptor ligand specificity and the development of CB2-selective agonists. *Curr. Med. Chem.* **2008**, *15*, 1428–1443. [CrossRef]

32. Cécyre, B.; Thomas, S.; Ptito, M.; Casanova, C.; Bouchard, J.F. Evaluation of the specificity of antibodies raised against cannabinoid receptor type 2 in the mouse retina. *Naunyn Schmiedebergs Arch. Pharmacol.* **2014**, *387*, 175–184. [CrossRef]

33. Jo, A.O.; Noel, J.M.; Lakk, M.; Yarishkin, O.; Ryskamp, D.A.; Shibasaki, K.; McCall, M.A.; Križaj, D. Mouse retinal ganglion cell signalling is dynamically modulated through parallel anterograde activation of cannabinoid and vanilloid pathways. *J. Physiol.* **2017**, *595*, 6499–6516. [CrossRef]

34. Bouskila, J.; Javadi, P.; Elkrief, L.; Casanova, C.; Bouchard, J.F.; Ptito, M. A Comparative Analysis of the Endocannabinoid System in the Retina of Mice, Tree Shrews, and Monkeys. *Neural Plast.* **2016**, *2016*, 3127658. [CrossRef] [PubMed]

35. Elphick, M.R. The evolution and comparative neurobiology of endocannabinoid signalling. *Philos. Trans. R. Soc. Lond. B Biol. Sci.* **2012**, *367*, 3201–3215. [CrossRef] [PubMed]

36. Lambert, D.M.; Vandevoorde, S.; Jonsson, K.O.; Fowler, C.J. The palmitoylethanolamide family: A new class of anti-inflammatory agents? *Curr. Med. Chem.* **2002**, *9*, 663–674.

37. Rodríguez de Fonseca, F.; Navarro, M.; Gómez, R.; Escuredo, L.; Nava, F.; Fu, J.; Murillo Rodríguez, E.; Giuffrida, A.; LoVerme, J.; Gaetani, S.; et al. An anorexic lipid mediator regulated by feeding. *Nature* **2001**, *414*, 209–212. [CrossRef]

38. Maccarrone, M.; Pauselli, R.; Di Rienzo, M.; Finazzi-Agro, A. Binding, degradation and apoptotic activity of stearoylethanolamide in rat C6 glioma cells. *Biochem. J.* **2002**, *366*, 137–144. [CrossRef]

39. Egertová, M.; Simon, G.M.; Cravatt, B.F.; Elphick, M.R. Localization of N-acyl phosphatidylethanolamine phospholipase D (NAPE-PLD) expression in mouse brain: A new perspective on N-acylethanolamines as neural signaling molecules. *J. Comp. Neurol.* **2008**, *506*, 604–615. [CrossRef]

40. Cécyre, B.; Monette, M.; Beudjekian, L.; Casanova, C.; Bouchard, J.F. Localization of diacylglycerol lipase alpha and monoacylglycerol lipase during postnatal development of the rat retina. *Front. Neuroanat.* **2014**, *8*, 150. [CrossRef]

41. Bouskila, J.; Harrar, V.; Javadi, P.; Casanova, C.; Hirabayashi, Y.; Matsuo, I.; Ohyama, J.; Bouchard, J.-F.; Ptito, M. Scotopic vision in the monkey is modulated by the G protein-coupled receptor 55. *Vis. Neurosci.* **2016**, *33*, 1–8. [CrossRef] [PubMed]

42. Alhouayek, M.; Masquelier, J.; Muccioli, G.G. Lysophosphatidylinositols, from cell membrane constituents to GPR55 ligands. *Trends Pharmacol. Sci.* **2018**, *39*, 586–604. [CrossRef]

43. Guy, A.T.; Nagatsuka, Y.; Ooashi, N.; Inoue, M.; Nakata, A.; Greimel, P.; Inoue, A.; Nabetani, T.; Murayama, A.; Ohta, K.; et al. Glycerophospholipid regulation of modality-specific sensory axon guidance in the spinal cord. *Science* **2015**, *349*, 974–977. [CrossRef]

44. Sampath, A.P.; Rieke, F. Selective transmission of single photon responses by saturation at the rod-to-rod bipolar synapse. *Neuron* **2004**, *41*, 431–443. [CrossRef]

45. Morales, P.; Jagerovic, N. Advances towards the discovery of GPR55 ligands. *Curr. Med. Chem.* **2016**, *23*, 2087–2100. [CrossRef] [PubMed]

46. Gladkikh, I.N.; Sintsova, O.V.; Leychenko, E.V.; Kozlov, S.A. TRPV1 Ion Channel: Structural Features, Activity Modulators, and Therapeutic Potential. *Biochemistry (Moscow)* **2021**, *86*, S50–S70. [CrossRef]

47. Zygmunt, P.M.; Petersson, J.; Andersson, D.A.; Chuang, H.-h.; Sørgård, M.; Di Marzo, V.; Julius, D.; Högestätt, E.D. Vanilloid receptors on sensory nerves mediate the vasodilator action of anandamide. *Nature* **1999**, *400*, 452–457. [CrossRef]

48. Ryskamp, D.A.; Redmon, S.; Jo, A.O.; Križaj, D. TRPV1 and endocannabinoids: Emerging molecular signals that modulate mammalian vision. *Cells* **2014**, *3*, 914–938. [CrossRef]

49. Zimov, S.; Yazulla, S. Localization of vanilloid receptor 1 (TRPV1/VR1)-like immunoreactivity in goldfish and zebrafish retinas: Restriction to photoreceptor synaptic ribbons. *J. Neurocytol.* **2004**, *33*, 441–452. [CrossRef]

50. Zimov, S.; Yazulla, S. Vanilloid receptor 1 (TRPV1/VR1) co-localizes with fatty acid amide hydrolase (FAAH) in retinal amacrine cells. *Vis. Neurosci.* **2007**, *24*, 581–591. [CrossRef] [PubMed]

51. Leonelli, M.; Martins, D.O.; Kihara, A.H.; Britto, L.R. Ontogenetic expression of the vanilloid receptors TRPV1 and TRPV2 in the rat retina. *Int. J. Dev. Neurosci.* **2009**, *27*, 709–718. [CrossRef]

52. Lakk, M.; Young, D.; Baumann, J.M.; Jo, A.O.; Hu, H.; Križaj, D. Polymodal TRPV1 and TRPV4 sensors colocalize but do not functionally interact in a subpopulation of mouse retinal ganglion cells. *Front. Cell. Neurosci.* **2018**, *12*, 353. [CrossRef] [PubMed]

53. Martínez-García, M.C.; Martínez, T.; Pañeda, C.; Gallego, P.; Jimenez, A.I.; Merayo, J. Differential expression and localization of transient receptor potential vanilloid 1 in rabbit and human eyes. *Histol. Histopathol.* **2013**, *28*, 1507–1516.

54. Leonelli, M.; Martins, D.O.; Britto, L.R. Retinal cell death induced by TRPV1 activation involves NMDA signaling and upregulation of nitric oxide synthases. *Cell. Mol. Neurobiol.* **2013**, *33*, 379–392. [CrossRef]

55. O'Leary, C. *The Role of TRPV1 and TRPV4 Channels in Retinal Angiogenesis*; Queen's University Belfast: Belfast, Ireland, 2016.

56. O'Leary, C.; McGahon, M.K.; Ashraf, S.; McNaughten, J.; Friedel, T.; Cincola, P.; Barabas, P.; Fernandez, J.A.; Stitt, A.W.; McGeown, J.G. Involvement of TRPV1 and TRPV4 channels in retinal angiogenesis. *Invest. Ophthalmol. Vis. Sci.* **2019**, *60*, 3297–3309. [CrossRef]

57. Belmonte, C. Ocular Nociceptors. In *Encyclopedia of Pain*; Gebhart, G.F., Schmidt, R.F., Eds.; Springer: Berlin, Heidelberg, 2013; pp. 2378–2382.

58. Choi, S.; Guo, L.; Cordeiro, M.F. Retinal and Brain Microglia in Multiple Sclerosis and Neurodegeneration. *Cells* **2021**, *10*, 1507. [CrossRef] [PubMed]

59. Hamid, M.A.; Moustafa, M.T.; Nashine, S.; Costa, R.D.; Schneider, K.; Atilano, S.R.; Kuppermann, B.D.; Kenney, M.C. Anti-VEGF Drugs Influence Epigenetic Regulation and AMD-Specific Molecular Markers in ARPE-19 Cells. *Cells* **2021**, *10*, 878. [CrossRef] [PubMed]

60. Othman, R.; Cagnone, G.; Joyal, J.-S.; Vaucher, E.; Couture, R. Kinins and Their Receptors as Potential Therapeutic Targets in Retinal Pathologies. *Cells* **2021**, *10*, 1913. [CrossRef]

61. Holan, V.; Palacka, K.; Hermankova, B. Mesenchymal Stem Cell-Based Therapy for Retinal Degenerative Diseases: Experimental Models and Clinical Trials. *Cells* **2021**, *10*, 588. [CrossRef]

62. Yoo, H.-S.; Shanmugalingam, U.; Smith, P.D. Harnessing Astrocytes and Müller Glial Cells in the Retina for Survival and Regeneration of Retinal Ganglion Cells. *Cells* **2021**, *10*, 1339. [CrossRef]

63. Seah, I.; Zhao, X.; Lin, Q.; Liu, Z.; Su, S.Z.Z.; Yuen, Y.S.; Hunziker, W.; Lingam, G.; Loh, X.J.; Su, X. Use of biomaterials for sustained delivery of anti-VEGF to treat retinal diseases. *Eye* **2020**, *34*, 1341–1356. [CrossRef]

64. Mukhtar, S.; Ambati, B.K. The value of nutritional supplements in treating Age-Related Macular Degeneration: A review of the literature. *Int. Ophthalmol.* **2019**, *39*, 2975–2983. [CrossRef]

65. Wallsh, J.O.; Gallemore, R.P. Anti-VEGF-Resistant Retinal Diseases: A Review of the Latest Treatment Options. *Cells* **2021**, *10*, 1049. [CrossRef]

66. Johny, M.E.; Suresh, S.; Begum, R.F.; Bharathi, S.P.; Shankari, P.K.; Kumar, V.S.; Velmurugan, R.; Shanmugasundaram, P. A comparative study of prostaglandin analogs and ocular beta-blockers on intraocular pressure reduction. *Drug Invent. Today* **2020**, *14*, 456–461.

67. Rapino, C.; Tortolani, D.; Scipioni, L.; Maccarrone, M. Neuroprotection by (endo) cannabinoids in glaucoma and retinal neurodegenerative diseases. *Curr. Neuropharmacol.* **2018**, *16*, 959–970. [CrossRef]

68. Xu, H.; Azuara-Blanco, A. Phytocannabinoids in Degenerative and Inflammatory Retinal Diseases: Glaucoma, Age-Related Macular Degeneration, Diabetic Retinopathy and Uveoretinitis. In *Handbook of Cannabis*; Oxford University Press: Oxford, UK, 2014; pp. 601–618.

69. Wong, W.L.; Su, X.; Li, X.; Cheung, C.M.G.; Klein, R.; Cheng, C.-Y.; Wong, T.Y. Global prevalence of age-related macular degeneration and disease burden projection for 2020 and 2040: A systematic review and meta-analysis. *Lancet Glob. Health* **2014**, *2*, e106–e116. [CrossRef]

70. Blázquez, C.; González-Feria, L.; Alvarez, L.; Haro, A.; Casanova, M.L.; Guzmán, M. Cannabinoids inhibit the vascular endothelial growth factor pathway in gliomas. *Cancer Res.* **2004**, *64*, 5617–5623. [CrossRef]
71. Liou, G.I.; Auchampach, J.A.; Hillard, C.J.; Zhu, G.; Yousufzai, B.; Mian, S.; Khan, S.; Khalifa, Y. Mediation of cannabidiol anti-inflammation in the retina by equilibrative nucleoside transporter and A2A adenosine receptor. *Invest. Ophthalmol. Vis. Sci.* **2008**, *49*, 5526–5531. [CrossRef] [PubMed]
72. Nagarkatti, P.; Pandey, R.; Rieder, S.A.; Hegde, V.L.; Nagarkatti, M. Cannabinoids as novel anti-inflammatory drugs. *Future Med. Chem.* **2009**, *1*, 1333–1349. [CrossRef] [PubMed]
73. Liou, G.I.; El-Remessy, A.; Ibrahim, A.; Caldwell, R.B.; Khalifa, Y.; Gunes, A.; Nussbaum, J.J. Cannabidiol as a putative novel therapy for diabetic retinopathy: A postulated mechanism of action as an entry point for biomarker-guided clinical development. *Curr. Pharm. Pers. Med. (Former. Curr. Pharm.)* **2009**, *7*, 215–222. [CrossRef]
74. Passani, A.; Posarelli, C.; Sframeli, A.T.; Perciballi, L.; Pellegrini, M.; Guidi, G.; Figus, M. Cannabinoids in Glaucoma Patients: The Never-Ending Story. *J. Clin. Med.* **2020**, *9*, 3978. [CrossRef]
75. Vernazza, S.; Oddone, F.; Tirendi, S.; Bassi, A.M. Risk Factors for Retinal Ganglion Cell Distress in Glaucoma and Neuroprotective Potential Intervention. *Int. J. Mol. Sci.* **2021**, *22*, 7994. [CrossRef] [PubMed]
76. Shahbazi, F.; Grandi, V.; Banerjee, A.; Trant, J.F. Cannabinoids and cannabinoid receptors: The story so far. *Iscience* **2020**, *23*, 101301. [CrossRef]
77. Kokona, D.; Thermos, K. Synthetic and endogenous cannabinoids protect retinal neurons from AMPA excitotoxicity in vivo, via activation of CB1 receptors: Involvement of PI3K/Akt and MEK/ERK signaling pathways. *Exp. Eye Res.* **2015**, *136*, 45–58. [CrossRef] [PubMed]
78. Eastlake, K.; Banerjee, P.J.; Angbohang, A.; Charteris, D.G.; Khaw, P.T.; Limb, G.A. Müller glia as an important source of cytokines and inflammatory factors present in the gliotic retina during proliferative vitreoretinopathy. *Glia* **2016**, *64*, 495–506. [CrossRef]
79. Fernández-Ruiz, J.; Romero, J.; Velasco, G.; Tolón, R.M.; Ramos, J.A.; Guzmán, M. Cannabinoid CB2 receptor: A new target for controlling neural cell survival? *Trends Pharmacol. Sci.* **2007**, *28*, 39–45. [CrossRef] [PubMed]
80. Washington, I.; Brooks, C.; Turro, N.J.; Nakanishi, K. Porphyrins as photosensitizers to enhance night vision. *J. Am. Chem. Soc.* **2004**, *126*, 9892–9893. [CrossRef]
81. Kramer, R.H.; Davenport, C.M. Lateral Inhibition in the Vertebrate Retina: The Case of the Missing Neurotransmitter. *PLoS Biol.* **2015**, *13*, e1002322. [CrossRef]

Article

High-Density EEG in a Charles Bonnet Syndrome Patient during and without Visual Hallucinations: A Case-Report Study

Andrea Piarulli [1,2], Jitka Annen [2,3], Ron Kupers [4], Steven Laureys [2,3] and Charlotte Martial [2,3,*]

1   Department of Surgical, Medical, Molecular Pathology and Critical Care Medicine, University of Pisa, 56126 Pisa, Italy; andrea.piarulli@uliege.be
2   Coma Science Group, GIGA-Consciousness, University of Liège, 4000 Liège, Belgium; jitka.annen@uliege.be (J.A.); steven.laureys@uliege.be (S.L.)
3   Centre du Cerveau, University Hospital of Liège, 4000 Liège, Belgium
4   Department of Neuroscience, University of Copenhagen, 1050 Copenhagen, Denmark; endogonidia@gmail.com
*   Correspondence: cmartial@uliege.be; Tel.: +32-428-43612

Abstract: Charles Bonnet syndrome (CBS) is a rare clinical condition characterized by complex visual hallucinations in people with loss of vision. So far, the neurobiological mechanisms underlying the hallucinations remain elusive. This case-report study aims at investigating electrical activity changes in a CBS patient during visual hallucinations, as compared to a resting-state period (without hallucinations). Prior to the EEG, the patient underwent neuropsychological, ophthalmologic, and neurological examinations. Spectral and connectivity, graph analyses and signal diversity were applied to high-density EEG data. Visual hallucinations (as compared to resting-state) were characterized by a significant reduction of power in the frontal areas, paralleled by an increase in the midline posterior regions in delta and theta bands and by an increase of alpha power in the occipital and midline posterior regions. We next observed a reduction of theta connectivity in the frontal and right posterior areas, which at a network level was complemented by a disruption of small-worldness (lower local and global efficiency) and by an increase of network modularity. Finally, we found a higher signal complexity especially when considering the frontal areas in the alpha band. The emergence of hallucinations may stem from these changes in the visual cortex and in core cortical regions encompassing both the default mode and the fronto-parietal attentional networks.

Keywords: Charles Bonnet syndrome; EEG; visual hallucination; resting state

Citation: Piarulli, A.; Annen, J.; Kupers, R.; Laureys, S.; Martial, C. High-Density EEG in a Charles Bonnet Syndrome Patient during and without Visual Hallucinations: A Case-Report Study. *Cells* **2021**, *10*, 1991. https://doi.org/10.3390/cells10081991

Academic Editors: Maurice Ptito and Joseph Bouskila

Received: 7 May 2021
Accepted: 31 July 2021
Published: 5 August 2021

Publisher's Note: MDPI stays neutral with regard to jurisdictional claims in published maps and institutional affiliations.

1. Introduction

Charles Bonnet syndrome (CBS) is a rare syndrome characterized by the appearance of recurrent and complex visual hallucinations in elderly people without mental disorders, who generally recognize their hallucinations as unreal [1]. The syndrome is typically reported by people who have a severe decrease or complete loss of vision, often caused by retinitis pigmentosa, cataracts, macular degeneration, or optic neuritis [2].

Though CBS was first described in the 18th century by the biologist who gave his name to the syndrome [3], the neurobiological mechanisms underlying the hallucinations remain elusive, with no satisfactory response as to why some blind patients hallucinate and many do not. Nonetheless, some attempts have been made to find plausible explanations. Currently, research has shown that CBS may be associated with lesions in the visual system, ranging from the retina to the occipital cortex [4–9], likely causing the visual hallucinations. So far, the most widely accepted mechanism of the emergence of such visual hallucinations refers to the "release phenomenon" suggesting that deafferentation of the visual cortex might lead to cortical hyperexcitability in this area [10,11]. While a few studies, including electroencephalography (EEG) ones, suggest that hyperexcitability in the visual cortex of partially blind CBS patients is associated with external visual stimuli (e.g., [12,13]), other studies demonstrate that CBS visual hallucinations can rather arise in the absence of external visual stimuli and, thus, suggest the possibility that spontaneous brain activity could generate conscious percepts (e.g., [14]). Since CBS is relatively rare [15], its investigation is only in the preliminary stages. Only a few studies have investigated the electrophysiological correlates in CBS and none of these investigations has ever used high-density EEG.

We here describe the case of an 85-year-old man with late-onset visual impairment and CBS. First, the patient underwent neuropsychological, ophthalmologic, and neurological examinations. Next, high-density EEG was performed as long as necessary to record data during rest and while vivid visual hallucinations were present. Investigating and comparing both conditions may provide novel insights into the neural pathways that underlie the emergence of visual hallucinations in CBS.

2. Materials and Methods

2.1. Case Report

An 85-year-old man with visual impairment and no psychiatric history presented to the Centre du Cerveau[2] (University Hospital of Liège, Belgium). The patient suffered from retinitis pigmentosa, a degenerative eye disease characterized by progressive degeneration of the rod photoreceptors in the retina, which typically leads to severe vision impairments [16]. He was three when he began experiencing a progressive loss of vision. Some years later, during the teenage years, he experienced a progressive development of "tunnel vision", i.e., his peripheral field of vision progressively narrowed. Concurrently, the patient developed hemeralopia, a night vision deterioration caused by the abolition of rod cells. At the age of 70 years, he lost his central vision, causing complete blindness at the age of 80 years. He also reported a positive family history of CBS.

The patient described a five-year history of increasing frequency of visual hallucinations. His visual hallucinatory experiences started at the age of 80, together with the occurrence of his complete blindness. Overall, he was able to report a coherent and very detailed description of his hallucinations, as well as its occurrence context. The patient reported that his visual hallucinations gradually became more frequent and could occur many times during the day. He described his hallucinations as, in general, well formed, ranging from simple flashes or colored backgrounds to more complex with the appearance of faces, objects, people, or landscapes. They were usually binocular, covering the entire visual field and could vary in size and color. He nevertheless specified that animations (i.e., scenes in motion) were only present in his right visual field. The visual hallucinations generally occurred when he has his eyes open, and they did not disappear when he closed his eyes. They were never accompanied by abnormal perception in any other sensory modality. He reported his hallucinations as pleasant. The patient is apparently not able to consciously control the occurrence and content of the visual hallucinations. He was fully aware of their unreal nature.

2.2. Procedure

The patient went through a consultation with a neurologist, as well as with a neuropsychologist. The latter examination included the administration of the Mattis Dementia Rating Scale [17] and of the Montreal Cognitive Assessment (MoCA, [18]) using the version adapted for blind individuals (MoCA-BLIND, [19]). Given his visual impairments, these cognitive tests were administered verbally, thereby omitting all vision-specific items. The patient was then submitted to an ophthalmologic examination, notably including measurements of visual acuity and visually evoked potentials.

The patient finally underwent a high-density EEG recording session. The EEG was recorded using a Net Amps 300 system (GES300) with a 256-electrode Hydrocel Geodesic Sensor Net (Electrical Geodesic Inc., Eugene, OR, USA). EEG channels were referenced to the vertex, electrode impedances were kept below 50 kΩ throughout the recording, and signals were acquired at a sampling rate of 250 Hz, using Electrical Geodesic Net Station software version 4.5.4 (Electrical Geodesic Inc., Eugene, OR, USA). During the EEG acquisition, which lasted about an hour, the patient was comfortably seated in a quiet room at the hospital. During the first 17 min, the patient was in an eyes-closed resting state condition, thus providing a baseline measure of brain electrical activity. After the resting state period, he started having vivid visual hallucinations, which lasted for about 20 min. During the hallucinatory period, he described the visual hallucinations he was experiencing, giving additional details about the hallucinations at the end of the EEG acquisition.

An MRI exam was also performed; the results are reported in [20].

2.3. EEG Pre-Processing

EEG recordings were analyzed using tailored codes written in MATLAB (MathWorks, Natick, MA, USA) and taking advantage of EEGLAB toolbox functions [21].

We analyzed about 37 min of the one hour recording due to the presence of massive movement and muscular artifacts especially during the second part of the recording (visual hallucination (VH) period): 17 min in the resting state (RS) condition, and 20 min in the VH condition were retained for further analyses, thus discarding the last 23 min. Signals were filtered between 0.5 and 30 Hz. Channels on the forehead, neck, and cheeks, which mostly contribute to movement-related noise were discarded [22], thus, retaining 183 channels out of 256. For each condition (RS and VH conditions), epochs with signals exceeding 100 μV were automatically discarded. Retained signals were visually inspected for the removal of large non-stationary artifacts and noisy channels. The retained epochs of each phase were then concatenated and submitted to independent component analysis (ICA) to remove ocular and/or muscular artifacts [23]. Noisy signals were then substituted with signals obtained via spline-interpolation [24], and signals were average-referenced. At the end of the artifact's rejection procedure, 7.5 min of the RS condition and 8 min of the VH condition, were kept for further analyses. Even after the artifact removal procedure, the EEG visual inspection highlighted the presence of residual high levels of electromyographic noise caused by involuntary muscle movements both during RS and VS periods. On this basis, we restricted our analyses to delta (1–4 Hz), theta (4–8 Hz), and alpha (8–13 Hz) bands, where the influence of EMG noise is negligible [25]. As a final pre-processing step, both RS and VS conditions were divided into 15 s non-overlapping epochs (thus obtaining 30 and 32 epochs, respectively).

2.4. Spectral and Connectivity Analyses

A spectral analysis was conducted for each condition. The EEG power spectral density (PSD) was obtained for the bands of interest: delta (1–4 Hz), theta (4–8 Hz), and alpha (8–13 Hz). Power density distributions were estimated by applying a Hamming-windowed FFT on 15 s consecutive epochs. For each epoch and electrode, the mean band PSD was estimated as the average over its frequency bins. For each condition, the PSD of each band and electrode was finally obtained by averaging among epochs and then log-transformed.

Both for RS and VH, we estimated the connectivity between each couple of electrodes using the debiased weighted phase lag index (dwPLI, [26]). The dwPLI at frequency bins of 0.5 Hz was estimated for each 15 s epoch (for the purpose, each epoch was segmented into 2 s periods with a 50% overlap between contiguous ones). For each epoch and couple of electrodes, the dwPLI in each band of interest was estimated by averaging over its frequency bins. The dwPLI of each condition, band, epoch, and couple of electrodes was finally calculated by averaging over the time samples belonging to the epoch itself. For each condition and epoch, the dwPLI values across all channel pairs were used to construct 183 × 183 connectivity matrices for delta, theta, and alpha bands. Scalp power maps were visualized using EEGLAB functions [21], while connectivity maps were displayed taking advantage of the Brain Net toolbox [27].

2.5. Graph Analysis

Connectivity matrices were thresholded by varying the connection density to retain between 90% and 10% of the higher connectivity values in steps of 2.5% [28]. At each density, the matrix was represented as a weighted graph, with channels as nodes and non-zero connectivities as links between the nodes. Each graph was characterized by using graph-theoretical parameters implemented in the Brain Connectivity Toolbox [29]. We adopted the following of metrics:

- Strength: The strength of a node is defined as the sum of its edges. The mean graph strength is, thus, estimated as the average over nodal strengths.
- Local efficiency provides a measure of the degree of information integration between the immediate neighbors of a given network node. The mean local efficiency thus reflects the degree of local connectivity within a graph [30].
- Global efficiency provides a measure of network integration and is defined as the average inverse shortest path length [30].
- Modular structure and modularity: The modular structure of a graph is estimated by subdividing the network into groups of nodes (maximizing the number of within-group links and minimizing the number of between-group links). Modularity indicates the degree of reliability of a given modular structure [31].
- Participation coefficient provides an estimate of the degree to which an included node of a given module is linked with other modules. Nodes with a high participation coefficient promote inter-modular integration and, as such, a network with a high participation coefficient is likely to also be globally interconnected.

Finally, for each condition and epoch, each metric was averaged over the considered connection densities.

2.6. Lempel–Ziv Complexity

For each condition, epoch and electrode, Lempel–Ziv Complexity (LZC) was estimated for each band of interest. As a first step, the EEG of both the RS and of the VH conditions were band-pass filtered (Chebyshev II filters) to obtain signals in delta, theta, and alpha bands. For each band, condition, epoch, and channel, the EEG signal was converted into a binary sequence. The threshold used to binarize the signal for each band, epoch, and electrode was computed in line with the following [32]:

- The average signal value (over the single epoch) was estimated and subtracted from the signal's original time series (s_{je}); the resulting signal was then linearly detrended (s_{je}^*). Note that j identifies the jth channel and e the eth epoch.
- The analytic signal of s_{je}^* was estimated using the Hilbert transform.
- The binarization threshold (th_{je}) for each channel and epoch was obtained as the average over the epoch of the analytic signal absolute value.
- For each band, the EEG signal at each electrode and epoch (s_{je}) was binarized based on the estimated threshold (th_{je}). If $s_{jek} \geq th_{je}$, $sb_{jek} = 1$, otherwise $sb_{jek} = 0$. Note that k identifies the kth time sample, and sb is the resulting binarized signal.

After binarizing the EEG signals related to each band, epoch, and channel, each resulting binary time series was submitted to the LZC algorithm [33]. At the end of the procedure, we obtained for each band, condition, and electrode a series of LZC values (one for each 15 s epoch).

2.7. Statistical Procedures

2.7.1. Spectral Power, Lempel–Ziv Complexity and Classical Connectivity Analyses

For each band and electrode (or couples of electrodes when considering connectivity), comparisons between VH (32 samples, each corresponding to a 15 s epoch) and RS conditions (30 samples) were performed using a single threshold permutation test for the maximum t-statistics (two-samples test, 5000 permutations; [34]; for details see Supplementary Material, SM).

2.7.2. Graph Metrics

For each band and graph parameter (graph strength, local efficiency, global efficiency, modularity, and participation coefficient), between-condition differences (VH-32 samples and RS-30 samples) were assessed using a permutation test on the t-statistic (two-samples test, [35]) based on 5000 randomizations. For each band, p-values (one for each parameter) were then adjusted for multiple testing using the false discovery rate (FDR) procedure [36]. Significance level was set at $p = 0.05$.

3. Results

3.1. Neuropsychological Examination

The Mattis Dementia Rating Scale and the MoCA-BLIND scores were 79/79 and 22/22, respectively. The patient's cognitive and global functioning can be considered as normal.

3.2. Ophthalmological Examination

An ophthalmologic disease was diagnosed. The examination revealed retinal atrophy in both eyes, as well as a decompensated corneal grafting in his left eye. Flash visual evoked potentials were flat. During the examination, he could perceive light in both eyes. He underwent a cataract operation of both eyes.

3.3. Neurological Examination

A diagnosis of CBS was made by the neurologist, based on his clinical history and the results of the diagnostic and clinical assessments the patient underwent. The neurologist confirmed that the patient fulfilled the four diagnostic criteria for CBS: presence of (i) complex, repetitive, and persistent hallucinations and (ii) awareness that the hallucinations are not real; and absence of (iii) additional delusions and (iv) additional hallucinations in the other senses [1].

3.4. EEG

3.4.1. Content of the Visual Hallucinations Experienced during the VH Condition

Vivid visual hallucinations were reported by the patient during the VH condition (during 20 min). First, he described the vision of a huge stone cathedral, with a ~100-m-high ceiling. The stones were terracotta. He described the hallucinations as really beautiful, joyful, and truly magic. The colors that he saw inspired joy and were reported as having a granulated texture. There was no animation in the scene. Then, the orange shade was changing and became darker to finally have a mysterious appearance. The patient reported what he saw at this moment as very powerful and strong. He also reported some lines of light. He stated that it was the first time he saw a cathedral in his hallucinations.

The visual hallucinations he had were binocular, covering the entire visual field. He reported not knowing why the visual hallucinations started to arise, he was not able to control their appearance or their visual content. They were not accompanied by abnormal perception in any other sensory modality, and he was fully aware of their unreal nature.

3.4.2. EEG Analyses

Since the high-density EEG recording was extremely noisy, 7.5 min of RS and 8 min of VH were retained for the analyses. Furthermore, we analyzed only delta, theta, and alpha bands due to movement and muscular artifacts at higher frequencies, which were present also after the cleaning procedure both during RS and VH periods (see [25]).

Power Spectral Density

For each band and condition, we first estimated the average scalp map (see Figure 1, columns 1–2) and then for each band, we evaluated putative between-condition differences (Figure 1, column 3).

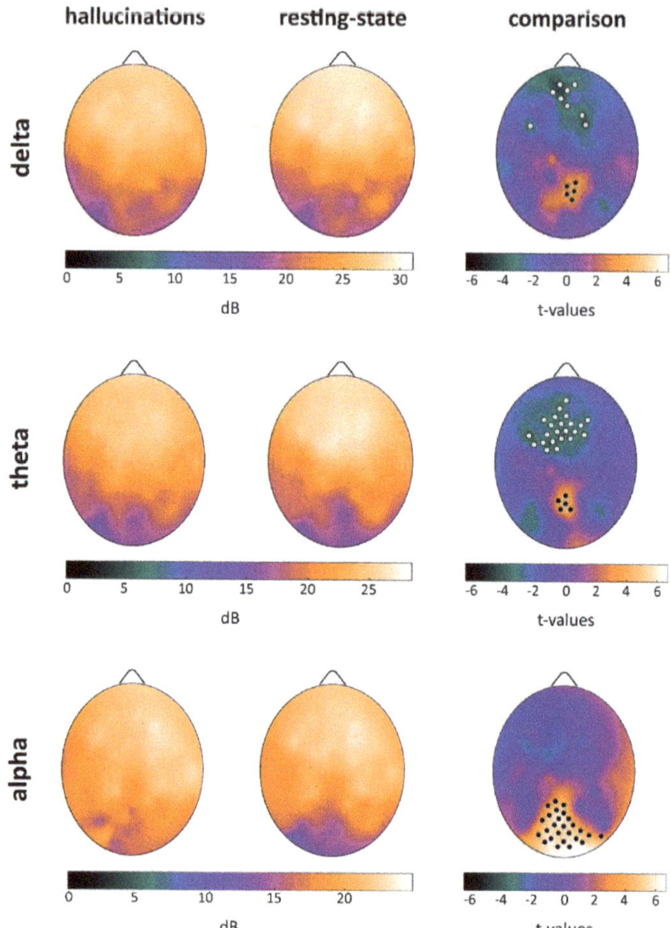

Figure 1. Average scalp maps for each band and condition are presented in the first two columns. Band-wise comparisons are presented in the third column. Black dots indicate electrodes showing a significantly higher PSD during the VH condition as compared to RS, white dots the opposite relationship. Critical t-values (in absolute value) for significance at $p = 0.05$ are, respectively, $|t| = 3.22$ for delta, $|t| = 3.13$ for theta and $|t| = 3.12$ for alpha band (single-threshold tests for the maximum t-statistics).

When considering the delta band, we observed a significant PSD decrease in the midline frontal areas during the VH condition as compared to the RS one. An analogous decrease, but one involving virtually the whole frontal area, was found when considering the theta band. Both for the delta and the theta band, the frontal decrease was paralleled by an increase of PSD in the midline posterior areas. The alpha band was instead characterized by an increase of PSD in occipital areas covering the whole visual cortex and extending toward midline posterior areas (for statistical details, see SM, Figure S1).

Based on whole scalp comparisons, we selected four representative electrodes (one for each area showing significant between-condition differences, and one "control" electrode): Fz, Pz, Oz, and Cz. As apparent from Figure 2, both VH and RS conditions were characterized by a peak of theta activity in the frontal areas (midline frontal theta, see Fz subplot). In line with whole scalp comparisons, both delta and theta PSD were higher during the RS condition as compared to those in VH. The same theta peak was observed when considering the Cz electrode, although in this case, no significant PSD difference was found between VH and RS, in line with findings about whole scalp comparisons (see Figure 1).

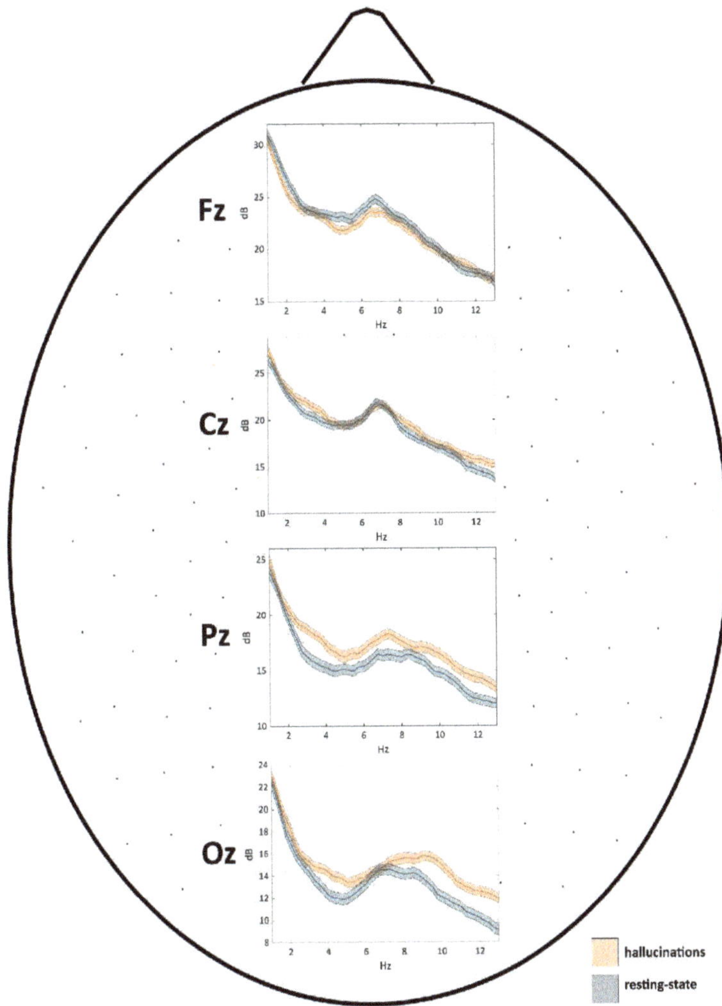

Figure 2. The PSD of VH and RS conditions are presented as a function of frequency for four representative electrodes. In each subplot, the solid line represents, for each condition, the mean PSD value, whereas the colored areas highlight the values enclosed between the mean-standard error and mean + standard error interval.

When considering the Pz derivation, we observed a local increase of PSD encompassing the theta and alpha bands, coherently for the two conditions. Of note, PSD during VH was significantly higher than that observed during RS coherently for the three bands of interest (in line with whole scalp analyses).

Finally, regarding occipital areas (Oz subplot), we found a local increase of PSD (theta, but mostly alpha band) analogous to that observed for Pz when considering the VH condition.

Classical Connectivity Analysis (dwPLI)

We next verified whether the VH condition induced significant changes in connectivity as compared to RS. When considering either the delta or the alpha band, we did not observe any between-condition difference (see SM, Figure S2). At variance with the latter bands, we found a widespread decrease of connectivity within the theta band during the VH condition (Figure 3 and SM, Figure S2). Significant decreases involved the whole frontal area and posterior-central areas of the right hemisphere.

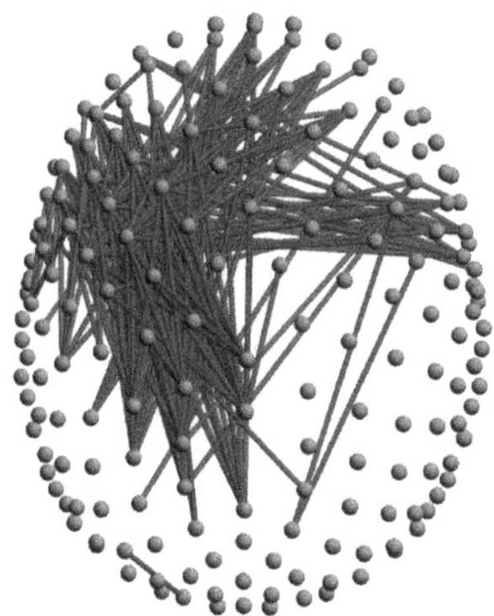

Figure 3. Significant decreases in theta band connectivity are presented for the VH versus RS comparisons (blue lines). The critical t-value (in absolute value) for significance at $p = 0.05$ is $|t| = 3.12$ (single threshold tests for the maximum t-statistics).

Graph Theoretic Metrics

For each subject, condition, and epoch, connectivity values across all channel pairs were used to construct symmetric 183×183 connectivity matrices for each considered band. Connectivity matrices were then thresholded by varying the connection density to retain between 90% and 10% of the highest dwPLI values (steps of 2.5%). At each connection density, we characterized the weighted network's topological features using the following metrics: (a) network strength, (b) local efficiency, (c) global efficiency, (d) modularity, and (e) participation coefficient. Each metric was averaged across connection densities (see Figure 4 and SM, Table S1). The estimated metrics as a function of connection density are presented in SM (Figures S3–S5).

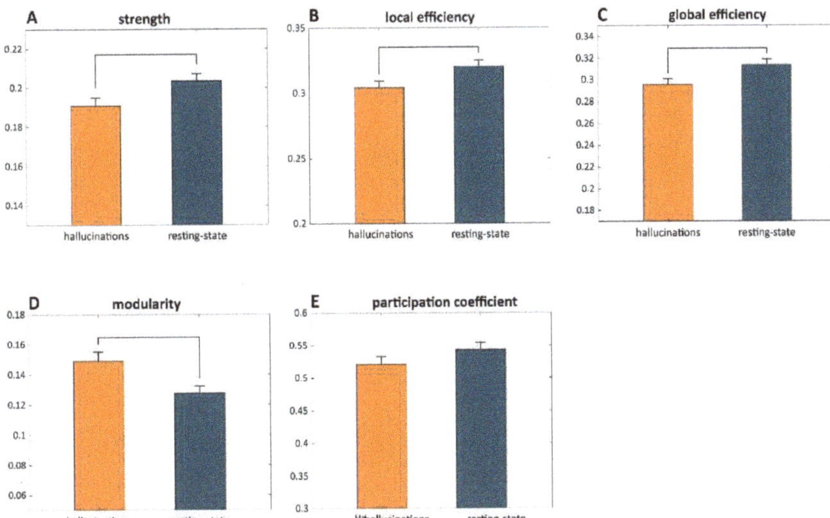

Figure 4. For each graph metric in the theta band, descriptive statistics for the two conditions are presented (mean + standard error). In each subplot, the orange bar represents the considered metric estimated during the VH condition, while the blue bar represents the metric related to the RS. Significant between-condition differences ($p < 0.05$ after FDR correction) are highlighted by a black line connecting the two bars. Graph strength is presented in (**A**), local efficiency in (**B**), global efficiency in (**C**), modularity in (**D**) and participation coefficient in (**E**).

No significant difference was found for any graph metric when considering either the delta or alpha networks (see SM, Tables S1 and S3). On the other side, we observed a disruption of small-world attributes typical of complex networks [37] during VH as compared to RS. Indeed, theta networks showed a lower network strength, paralleled by a decrease of both local and global efficiency. On the other side, the whole scalp networks showed a higher modularity complemented by a lower participation coefficient although the latter comparison was not significant (see Figure 4). Of note, findings about theta networks were consistent across connection densities (SM, Table S2).

Lempel–Ziv Complexity

We next estimated band-wise LZC. No significant between-condition difference was found either when considering delta or theta bands (see SM, Figure S6). A significant enhancement of alpha band complexity over large scalp areas was observed during the VH period as compared to that in the RS condition (Figure 5). Significant increases were found in the midline frontal areas, posterior areas of the left hemisphere, and centro-posterior areas of the right hemisphere.

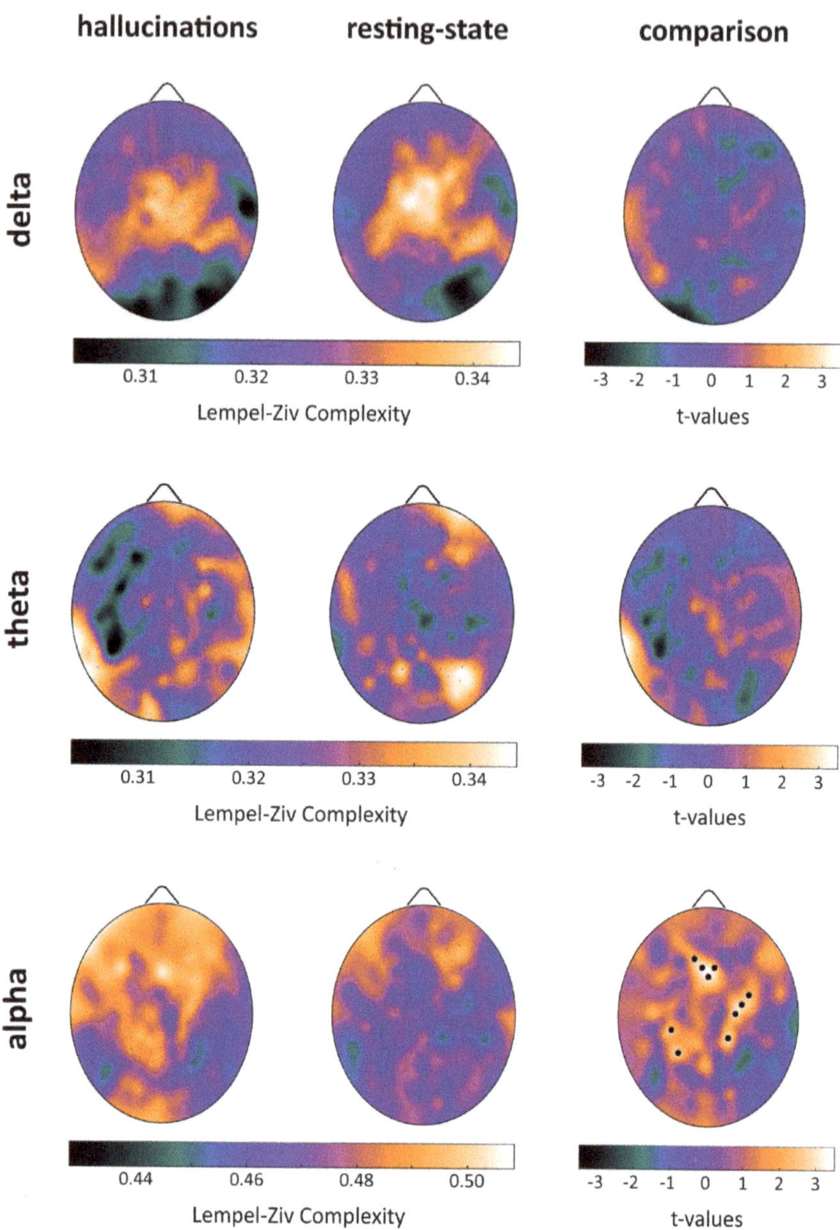

Figure 5. Average Lempel-Ziv complexity (LZC) scalp maps for each band and condition are presented in the first two columns. Band-wise comparisons are presented in the third column. Black dots indicate electrodes showing a significantly higher LZC during the VH condition as compared to the RS one, white dots (if present) indicate the opposite relationship. The critical t-value (in absolute value) for significance at $p = 0.05$ is $|t| = 2.41$ (single threshold tests for the maximum t-statistics).

4. Discussion

In the present study, the electrical brain activity of an 85-year-old patient with a late-onset visual impairment and CBS was investigated by comparing two conditions:

(1) resting state (RS, without any hallucinatory experience, 7.5 min of cleaned data) and (2) visual hallucinations (VH, 8 min of cleaned data).

When looking at PSD, during the RS condition, we observed the emergence of theta activity in the frontal areas, and of theta–alpha in the posterior and occipital ones (Figures 1 and 2). These findings seem at odds with those observed in healthy subjects, as during relaxed wakefulness the human brain activity is characterized by a marked rhythmic electrical activity in the alpha band (8–13 Hz; [38]), which in an eyes-closed condition shows an occipital prevalence. However, the results herein described come as no surprise, as previous studies highlighted how blind people show a reduction of alpha activity over parieto-occipital areas [39], paralleled by a reduced gray matter volume in these very same areas as previously observed in this patient [20]. Moreover, the onset and duration of blindness are predictors of the decline of alpha activity in blind people [40]. At variance with resting state phase, during the VH period, we observed the reinstatement of a "pure" low-alpha activity in occipital areas, see Figure 1. Of note, the presence of alpha activity has recently been proposed as an active mechanism for visual processing [41].

In both conditions, the patient showed a topologically widespread enhancement of slow activity (delta–theta bands) paralleled by a reduced alpha activity. A cofactor contributing to the emergence of this EEG pattern could be related to aging: indeed, as observed by Ishii and colleagues [42], physiological aging is characterized by (i) a reduction in the amplitude of alpha activity (8–13 Hz), (ii) a slowing of the background activity, and (iii) a global increase of delta (1–4 Hz) and theta (4–8 Hz) power.

As compared to RS, the VH condition was characterized by PSD decreases in the midline frontal areas for delta and theta bands. Both decreases were paralleled by PSD increases in the midline posterior areas corresponding at a cortical level to posteromedial cortical regions, thus including the precuneus. This latter structure has been associated with self and visual awareness [43], and its activity seems to correlate with self-reflection processes [44,45], including mental imagery [46]. Based on these considerations, we speculate that the activity of the posteromedial cortex could play a crucial role in the patient's awareness of perceiving visual hallucinations instead of real scenarios. During the VH condition, coherently with what always happens when the patient is hallucinating outside (as reported during the anamnesis with the neurologist), the subject was aware of the unreality of his hallucinations, suggesting a successful and preserved reality monitoring. Interestingly, the precuneus has been identified as one of the regions of interest supporting successful reality monitoring in humans [47]. In general, CBS patients do recognize the unreality of their visual hallucination [1]: this ability may be related to the fact that they were sighted people becoming blind very late in their life. The unreality of the hallucinations is experienced as pleasant by our CBS patient, but this is not the case of all CBS patients who may, in some cases, suffer from anxiety [48].

When considering the alpha band, we observed a PSD increase during the VH condition (as compared to RS), encompassing both the visual cortex and midline posterior areas. Interestingly, the patient showed increased functional connectivity (estimated by fMRI) in the same posterior midline regions, which are included both in the secondary visual and salience networks, as compared to a sample of late-blind subjects [20]. The observed functional reorganization involving regions known to be crucial for self-awareness and for the processing of visual, salient stimuli could be related to the emergence of visual hallucinations in CBS. On the other side, frontal midline theta has been associated with cognitive control [49]: indeed, the significant decrease of this activity during the VH phase could be at the basis of an individual's inability to control either the emergence of the hallucinations or their visual content. Finally, the significant increase of alpha activity in occipital areas is plausibly related to the appearance and development of the visual hallucinations, which have been linked to the endogenous activation of large sections of the visual cortex [6]. In CBS patients, neural populations within the visual system are deprived of external inputs due to visual loss, a condition that may lead to a hyper-sensitization to spontaneous activity fluctuations, which may in turn result in the emergence of visual hallucinations [50]. We make the hypothesis that the increase of occipital alpha activity could be the neural correlate of such a hyper-sensitization.

We have also revealed a widespread decrease of connectivity within theta band including the whole frontal area and posterior-central areas of the right hemisphere during the emergence of visual hallucinations. This widespread decrease was accompanied at a network level, by a concurrent decrease of small-worldness (lower local and global efficiency) and an increase of the network's modularity. Interestingly, similar disruptions of small-worldness (i.e., decreased local efficiency and longer path length/decreased global efficiency) are consistently reported in studies on schizophrenia, whose core symptoms are hallucinatory experiences (mainly auditory, 75% of patients) and delusions [51,52], although about 35% of patients also experience visual hallucinations [53].

While a comparison between hallucinations in two such different pathological conditions seems at least hazardous, this parallelism is motivated by the hypothesis of common neurobiological mechanisms subtending all hallucinatory experiences be they auditory or visual [53,54]. The proposed model posits that the emergence of hallucinations stems from widespread impairments in attention and/or perception and stands on evidence from resting-state studies showing impaired connectivity within the Default Mode Network (DMN) and between brain attentional networks (i.e., ventral and dorsal attention networks; [54]). This hypothesis and the "release phenomenon" hypothesis [10,11] are not mutually exclusive. Our findings may support both, considering, respectively, the decrease of connectivity in the frontal areas and the heightened alpha activity in occipital regions during visual hallucinations.

Finally, we observed a significant enhancement of alpha band diversity (LZC) over large scalp areas including midline frontal areas, centro-posterior areas of the left hemisphere, and posterior areas of the right hemisphere during the VH period as compared to RS. Interestingly, an increase of alpha diversity was observed following the administration of N, N-dimethyltryptamine (DMT), a psychedelic drug known to induce visual hallucinations [55], as compared to the administration of a placebo, and during stroboscopically induced visual hallucinations [56]. In line with Timmermann and colleagues' work [55], the increase in signal diversity here described, can be interpreted within the "entropic brain hypothesis" framework, which posits that the quality of any conscious state depends on the system entropy [57], and as such, signal diversity measures such as LZC are reliable indices of the richness of contents of any conscious experience be it "real" or "hallucinatory".

We have described significant electrical brain changes in the visual cortex and in core cortical regions encompassing both the default mode and fronto-parietal attentional networks, which could contribute to the occurrence of visual hallucinations in CBS. We believe that this study provides a substantial contribution to the investigation of the electrophysiological signatures of CB syndrome at large and on the emergence of visual hallucinations linked to CBS. That said, we must acknowledge that this work is not free of limitations, and thus, our results should be interpreted cautiously as this is a case report study, which limits the generalizability of the presented results. Further studies should focus on electrical brain activity changes in a higher number of patients with CBS, in comparison with control groups such as for example matched late-blind people who do not suffer from hallucinations. Nevertheless, the CBS syndrome is quite rare. Furthermore, the spatial resolution of EEG is relatively low, and therefore, we do not have compelling evidence for the involvement of specific brain regions in the hallucinations. Finally, due to the presence of muscular artifacts, we limited our analyses to slow frequency bands (i.e., delta, theta, and alpha), preventing a thorough characterization of CBS electrophysiological correlates both during the resting state and during visual hallucinations.

5. Conclusions

As a concluding remark we would like to stress that, as this syndrome is relatively rare, both CBS trait and state (i.e., visual hallucinations) findings herein described, provide a first and up-to-now unique electrophysiological window on CBS and visual hallucinations at large. Indeed, we believe that the CBS can provide a useful clinical model for advancing our understanding of the electrophysiological mechanisms underlying the emergence of visual hallucinations. Future high-density EEG studies including larger samples of CBS patients are needed to confirm and extend the findings of this study.

Supplementary Materials: The following are available online at https://www.mdpi.com/article/10.3390/cells10081991/s1, Figure S1: Statistics of the PSD analysis, Figure S2: Statistics of the connectivity analysis, Figure S3: Graph theoretical metrics: Delta, Figure S4: Graph theoretical metrics: Theta, Figure S5: Graph theoretical metrics: Alpha, Figure S6: Statistics of the Lempel–Ziv complexity, Table S1: Delta networks, Table S2: Theta networks, Table S3: Alpha networks.

Author Contributions: C.M. and S.L. designed the study. C.M. handled data acquisitions. A.P., J.A. and C.M. analyzed the data. R.K. supervised the revision. All authors have designed the study, interpreted results, and drafted and revised the manuscript. All authors have read and agreed to the published version of the manuscript.

Funding: This research was funded by the BIAL Foundation, the Belgian National Funds for Scientific Research (FRS-FNRS), the University and University Hospital of Liège, the fund Léon Fredericq, the Fund Generet, the Mind Care International Foundation, the King Baudouin Foundation, DOCMA project (EU-H2020-MSCA–RISE–778234), the AstraZeneca Foundation, the European Union's Horizon 2020 Framework Program for Research and Innovation under the Specific Grant Agreement No. 945539 (Human Brain Project SGA3), the European Space Agency (ESA), and the Belgian Federal Science Policy Office (BELSPO) in the framework of the PRODEX Program, the Center-TBI project (FP7-HEALTH- 602150), the Public Utility Foundation "Université Européenne du Travail", "Fondazione Europea di Ricerca Biomedica", the Mind Science Foundation, and the European Commission. S.L. is research director at the F.R.S-FNRS.

Institutional Review Board Statement: The study was conducted according to the guidelines of the Declaration of Helsinki (1964) and its later amendments and approved by the Ethics Committee of the Faculty of Medicine of the University of Liège.

Informed Consent Statement: Informed consent was obtained from all subjects involved in the study.

Data Availability Statement: Some or all data used during the study are available from the corresponding author by request.

Conflicts of Interest: The authors declare no conflict of interest.

References

1. Teunisse, R.J.; Cruysberg, J.R.M.; Verbeek, A.; Zitman, F.G. The Charles Bonnet syndrome: A large prospective study in the Netherlands. *Br. J. Psychiatry* **1995**, *166*, 254–257. [CrossRef] [PubMed]
2. Damas-Mora, J.; Skelton-Robinson, M.; Jenner, F.A. The Charles Bonnet syndrome in perspective. *Psychol. Med.* **1982**, *12*, 251–261. [CrossRef]
3. Bonnet, C. *Essai Analytique Sur Les Facultés de L'âme*; Philibert: Copenhagen, Denmark, 1760; pp. 426–428.
4. Adachi, N.; Nagayama, M.; Anami, K.; Arima, K.; Matsudal, H. Asymmetrical blood flow in the temporal lobe in the Charles Bonnet syndrome: Serial neuroimaging study. *Behav. Neurol* **1994**, *7*, 97–99. [CrossRef] [PubMed]
5. Chen, J.J. Diagnosis and treatment of psychiatric commordity in a patient with Charles Bonnet syndrome. *Case Rep. Psychiatry* **2014**, *2014*, 195847. [CrossRef]
6. Kazui, H.; Ishii, R.; Yoshida, T.; Ikezawa, K.; Takaya, M.; Tokunaga, H.; Tanaka, T.; Takeda, M. Neuroimaging studies in patients with Charles Bonnet Syndrome. *Psychogeriatrics* **2009**, *9*, 77–84. [CrossRef] [PubMed]
7. Kishi, T.; Uegaki, J.; Kitani, M.; Fujimoto, A.; Naganuma, R. The usefulness of single photon emission computed tomography in Charles Bonnet syndrome: A case with occipital lobe involvement. *Gen. Hosp. Psychiatry* **2000**, *22*, 132–135. [CrossRef]
8. Rovner, B.W. The Charles Bonnet syndrome: Visual hallucinations caused by vision impairment. *Geriatrics* **2002**, *57*, 45–46.
9. Terao, T. The present state of Charles Bonnet syndrome. *Psychogeriatrics* **2002**, *2*, 6–14. [CrossRef]
10. Cogan, D. Visual hallucinations as release phenomena. *Graefes Arch. Clin. Exp. Ophtahlmol.* **1973**, *188*, 139–150. [CrossRef] [PubMed]
11. Schultz, G.; Melzack, R. The Charles Bonnet syndrome: 'Phantom visual images'. *Perception* **1991**, *20*, 809–825. [CrossRef]
12. Painter, D.R.; Dwyer, M.F.; Kamke, M.R.; Mattingley, J.B. Stimulus-driven cortical hyperexcitability in individuals with Charles Bonnet hallucinations. *Curr. Biol.* **2018**, *28*, 3475–3480. [CrossRef]
13. Vacchiano, V.; Tonon, C.; Mitolo, M.; Evangelisti, S.; Carbonelli, M.; Liguori, R.; Lodi, R.; Carelli, V.; la Morgia, C. Functional MRI study in a case of Charles Bonnet syndrome related to LHON. *BMC Neurol.* **2019**, *19*, 350. [CrossRef]

14. Hahamy, A.; Wilf, M.; Rosin, B.; Behrmann, M.; Malach, R. How do the blind 'see'? The role of spontaneous brain activity in self-generated perception. *Brain* **2021**, *144*, 340–353. [CrossRef] [PubMed]

15. Shiraishi, Y.; Terao, T.; Ibi, K.; Nakamura, J.; Tawara, A. Charles Bonnet syndrome and visual acuity—The involvement of dynamic or acute sensory deprivation. *Eur. Arch. Psychiatry Clin. Neurosci.* **2004**, *254*, 362–364. [CrossRef] [PubMed]

16. Caras, I.; Littman, N.; Abo, A. Proceedings: Debilitating Eye Diseases. *Stem Cells Transl. Med.* **2014**, *20*, 1393–1400. [CrossRef]

17. Mattis, S. *Dementia Rating Scale: Professional Manual*; Psychological Assessment Resources: Odessa, FL, USA, 1988.

18. Sokal, R.R.; Rohlf, F.J. *Biometry: The Principles and Practice of Statistics in Biological Research*; Freeman: New York, NY, USA, 1995.

19. Nasreddine, Z.S.; Phillips, N.A.; Bedirian, V.; Charbonneau, S.; Whitehead, V.; Collin, I.; Cummings, J.L.; Chertkow, H. The Montreal Cognitive Assessment, MoCA: A brief screening tool for mild cognitive impairment. *J. Am. Geriatr. Soc.* **2005**, *53*, 695–699. [CrossRef]

20. Martial, C.; Larroque, S.K.; Cavaliere, C.; Wannez, S.; Annen, J.; Kupers, R.; Laureys, S.; Di Perri, C. Resting-state functional connectivity and cortical thickness characterization of a patient with Charles Bonnet Syndrome. *PLoS ONE* **2019**, *14*, e0219656.

21. Delorme, A.; Makeig, S. EEGLAB: An open source toolbox for analysis of single-trial EEG dynamics including independent component analysis. *J. Neurosci. Methods* **2004**, *134*, 9–21. [CrossRef]

22. Piarulli, A.; Zaccaro, A.; Laurino, M.; Menicucci, D.; De Vito, A.; Bruschini, L.; Berrettini, S.; Bergamasco, M.; Laureys, S.; Gemignani, A. Ultra-slow mechanical stimulation of olfactory epithelium modulates consciousness by slowing cerebral rhythms in humans. *Sci. Rep.* **2018**, *8*, 6581. [CrossRef] [PubMed]

23. Makeig, S.; Bell, A.J.; Jung, T.P.; Sejnowski, T.J. Independent component analysis of electroencephalographic data. In *Advances in Neural Information Processing Systems 8*; Touretzky, D.S., Hasselmo, M.E., Eds.; Neural Information Processing Systems, NIPS Foundation; MIT Press: Cambridge, MA, USA, 1996; pp. 145–151.

24. Junghofer, M.; Elbert, T.; Tucker, D.M.; Rockstroh, B. Statistical control of artifacts in dense array EEG/MEG studies. *Psychophysiology* **2000**, *37*, 523–532. [CrossRef]

25. Chennu, S.; Finoia, P.; Kamau, E.; Allanson, J.; Williams, G.B.; Monti, M.M.; Noreika, V.; Arnatkeviciute, A.; Canales-Johnson, A.; Olivares, F.; et al. Spectral signatures of reorganised brain networks in disorders of consciousness. *PLoS Comput. Biol.* **2004**, *10*, e1003887. [CrossRef]

26. Vinck, M.; Oostenveld, R.; van Wingerden, M.; Battaglia, F.; Pennartz, C.M. An improved index of phase-synchronization for electrophysiological data in the presence of volume-conduction, noise and sample-size bias. *Neuroimage* **2011**, *55*, 1548–1565. [CrossRef] [PubMed]

27. Xia, M.; Wang, J.; He, Y. BrainNet Viewer: A network visualization tool for human brain connectomics. *PLoS ONE* **2013**, *8*, e68910. [CrossRef]

28. Chennu, S.; Annen, J.; Wannez, S.; Thibaut, A.; Chatelle, C.; Cassol, H.; Martens, G.; Schnakers, C.; Gosseries, O.; Menon, D.; et al. Brain networks predict metabolism, diagnosis and prognosis at the bedside in disorders of consciousness. *Brain* **2017**, *140*, 2120–2132. [CrossRef] [PubMed]

29. Rubinov, M.; Kötter, R.; Hagmann, P.; Sporns, O. Brain connectivity toolbox: A collection of complex network measurements and brain connectivity datasets. *NeuroImage* **2009**, *47*, S169. [CrossRef]

30. Latora, V.; Marchiori, M. Efficient behavior of small-world networks. *Phys. Rev. Lett.* **2001**, *87*, 198701. [CrossRef]

31. Newman, M.E. Analysis of weighted networks. *Phys. Rev. E Stat. Nonlin Soft Matter Phys.* **2004**, *70*, 056131. [CrossRef] [PubMed]

32. Schartner, M.; Seth, A.; Noirhomme, Q.; Boly, M.; Bruno, M.-A.; Laureys, S.; Barrett, A. Complexity of Multi-Dimensional Spontaneous EEG Decreases during Propofol Induced General Anaesthesia. *PLoS ONE* **2015**, *10*, e0133532. [CrossRef] [PubMed]

33. Lempel, A.; Ziv, J. On the Complexity of Finite Sequences. *IEEE Trans. Inf. Theory* **1976**, *22*, 75–81. [CrossRef]

34. Nichols, T.E.; Holmes, A.P. Nonparametric permutation tests for functional neuroimaging: A primer with examples. *Hum. Brain Mapp.* **2001**, *15*, 1–25. [CrossRef]

35. Ludbrook, J.; Dudley, H. Why permutation tests are superior to t and f tests in biomedical research. *Am. Stat.* **1998**, *52*, 127–132. [CrossRef]

36. Benjamini, Y.; Hochberg, Y. Controlling the False Discovery Rate: A practical and powerful approach to multiple testing. *J. R Stat. Soc.* **1995**, *57*, 289–300. [CrossRef]

37. Bullmore, E.; Sporns, O. Complex brain networks: Graph theoretical analysis of structural and functional systems. *Nat. Rev. Neurosci.* **2009**, *10*, 186–198. [CrossRef]

38. Hughes, S.W.; Crunelli, V. Thalamic Mechanisms of EEG Alpha Rhythms and Their Pathological Implications. *Neuroscientist* **2005**, *11*, 357–372. [CrossRef] [PubMed]

39. Krigseis, A.; Henninghausen, E.; Rösler, F.; Röder, B. Reduced EEG alpha activity over parieto-occipital brain areas in congenitally blind adults. *Clin. Neuropysiol.* **2006**, *117*, 1560–1573. [CrossRef]

40. Novikova, L.A. *Blindness and the Electrical Activity of the Brain: Electroencephalographic Studies of the Effects of Sensory Impairment*, 23rd ed.; American Foundation for the Blind: Arlington, VA, USA, 1974.

41. Bacigalupo, F.; Luck, S.J. Lateralized suppression of alpha-band EEG activity as a mechanism of target processing. *J. Neurosci.* **2019**, *39*, 900–917. [CrossRef]

42. Ishii, R.; Canuet, L.; Aoki, Y.; Hata, M.; Iwase, M.; Ikeda, S.; Nishida, K.; Ikeda, M. Healthy and Pathological Brain Aging: From the Perspective of Oscillations, Functional Connectivity, and Signal Complexity. *Neuropsychobiology* **2017**, *75*, 151–161. [CrossRef]

43. Cavanna, A.E.; Trimble, M.R. The precuneus: A review of its functional anatomy and behavioural correlates. *Brain* **2006**, *129*, 564–583. [CrossRef] [PubMed]
44. Gusnard, D.A. Being a self: Considerations from functional imaging. *Conscious. Cogn.* **2005**, *14*, 679–697. [CrossRef]
45. Northoff, G.; Heinzel, A.; de Greck, M.; Bermpohl, F.; Dobrowolny, H.; Panksepp, J. Self-referential processing in our brain-A meta-analysis of imaging studies on the self. *NeuroImage* **2006**, *31*, 440–457. [CrossRef]
46. Cavanna, A.E. The precuneus and consciousness. *CNS Spectr.* **2014**, *12*, 545–552. [CrossRef]
47. Garrison, J.R.; Saviola, F.; Morgenroth, E.; Barker, H.; Lurhs, M.; Simons, J.S.; Fernyhough, C.; Allen, P. Did I imagine that? The functional role of paracingulate cortex in reality monitoring. *BioRxiv* **2020**. [CrossRef]
48. Geueke, A.; Morley, M.G.; Morley, K.; Lorch, A.; Jackson, M.; Lambrou, A.; Oteng-Amoako, A. Anxiety and Charles Bonnet Syndrome. *J. Vis. Impair. Blind.* **2012**, *106*, 145–153. [CrossRef]
49. Cavanagh, J.F.; Frank, M.J. Frontal theta as a mechanism for cognitive control. *Trends Cogn. Sci.* **2014**, *18*, 414–421. [CrossRef] [PubMed]
50. Ffytche, D.H.; Howard, R.J.; Brammer, M.J.; David, A.; Woodruff, P.; Williams, S. The anatomy of conscious vision: An fMRI study of visual hallucinations. *Nat. Neurosc.* **1998**, *1*, 738–742. [CrossRef] [PubMed]
51. Liu, Y.; Liang, M.; Zhou, Y.; He, Y.; Hao, Y.; Song, M.; Yu, C.; Liu, H.; Jiang, T. Disrupted small-world networks in schizophrenia. *Brain* **2008**, *131*, 945–961. [CrossRef]
52. Ottet, M.-C.; Schaer, M.; Debbané, M.; Cammoun, L.; Thiran, J.-P.; Eliez, S. Graph theory reveals disconnected hubs in 22q11DS and altered nodal efficiency in patients with hallucinations. *Front. Hum. Neurosci.* **2013**, *7*, 402. [CrossRef] [PubMed]
53. Alderson-Day, B.; Diederen, K.; Fernyhough, C.; Ford, J.M.; Horga, G.; Margulies, D.S.; McCarthy-Jones, S.; Northoff, G.; Shine, J.M.; Turner, J.; et al. Auditory Hallucinations and the Brain's Resting-State Networks: Findings and Methodological Observations. *Schizophr. Bull.* **2016**, *42*, 1110–1123. [CrossRef] [PubMed]
54. Shine, J.M.; O'Callaghan, C.; Halliday, G.M.; Lewis, S.J.G. Tricks of the mind: Visual hallucinations as disorders of attention. *Prog Neurobiol.* **2014**, *116*, 58–65. [CrossRef] [PubMed]
55. Timmermann, C.; Roseman, L.; Schartner, M.; Milliere, R.; Williams, L.T.J.; Erritzoe, D.; Muthukumaraswamy, S.; Ashton, M.; Bendrioua, A.; Kaur, O.; et al. Neural correlates of the DMT experience assessed with multivariate EEG. *Sci. Rep.* **2019**, *9*, 16324. [CrossRef] [PubMed]
56. Schwartzman, D.J.; Schartner, M.; Ador, B.B.; Simonelli, F.; Chang, A.Y.-C.; Seth, A.K. Increased spontaneous EEG signal diversity during stroboscopically-induced altered states of consciousness. *BioRxiv* **2019**, 511766. [CrossRef]
57. Carhart-Harris, R. The entropic brain-revisited. *Neuropharmacology* **2018**, *142*, 167–178. [CrossRef] [PubMed]

Article

Retinal Venular Tortuosity Jointly with Retinal Amyloid Burden Correlates with Verbal Memory Loss: A Pilot Study

Oana M. Dumitrascu [1,*], Ryan Rosenberry [2], Dale S. Sherman [3], Maziyar M. Khansari [4], Julia Sheyn [5], Tania Torbati [5], Ayesha Sherzai [6], Dean Sherzai [6], Kenneth O. Johnson [7], Alan D. Czeszynski [7], Steven Verdooner [7], Keith L. Black [5], Sally Frautschy [8], Patrick D. Lyden [9], Yonggang Shi [4], Susan Cheng [2], Yosef Koronyo [5] and Maya Koronyo-Hamaoui [5,10,*]

1. Department of Neurology, Mayo Clinic, Scottsdale, AZ 85251, USA
2. Department of Cardiology, Cedars-Sinai Medical Center, Los Angeles, CA 90048, USA; ryan.rosenberry@cshs.org (R.R.); susan.cheng@cshs.org (S.C.)
3. Department of Neuropsychology, Cedars-Sinai Medical Center, Los Angeles, CA 90048, USA; dale.sherman@cshs.org
4. USC Stevens Neuroimaging and Informatics Institute, Keck School of Medicine of University of Southern California, Los Angeles, CA 90007, USA; Maziyar.Khansari@loni.usc.edu (M.M.K.); Yonggang.Shi@loni.usc.edu (Y.S.)
5. Department of Neurosurgery, Cedars-Sinai Medical Center, Los Angeles, CA 90048, USA; julia.sheyn@cshs.org (J.S.); tania.torbati@westernu.edu (T.T.); keith.black@cshs.org (K.L.B.); yosef.koronyo@cshs.org (Y.K.)
6. Department of Neurology, Loma Linda University, Loma Linda, CA 92350, USA; Asherzai@llu.edu (A.S.); dsherzai@llu.edu (D.S.)
7. NeuroVision Imaging Inc., Sacramento, CA 95833, USA; kjohnson@neurovision.com (K.O.J.); aczeszynski@neurovision.com (A.D.C.); sverdooner@neurovision.com (S.V.)
8. Department of Neurology, University of California Los Angeles, Los Angeles, CA 90095, USA; sfrautschy@mednet.ucla.edu
9. Department of Neurology, Cedars-Sinai Medical Center, Los Angeles, CA 90048, USA; plyden@usc.edu
10. Department of Biomedical Sciences, Cedars-Sinai Medical Center, Los Angeles, CA 90048, USA
* Correspondence: dumitrascu.oana@mayo.edu (O.M.D.); Maya.koronyo@csmc.edu (M.K.-H.); Tel.: +480-301-8100 (O.M.D.); Fax: +480-301-9494 (O.M.D.)

check for updates

Citation: Dumitrascu, O.M.; Rosenberry, R.; Sherman, D.S.; Khansari, M.M.; Sheyn, J.; Torbati, T.; Sherzai, A.; Sherzai, D.; Johnson, K.O.; Czeszynski, A.D.; et al. Retinal Venular Tortuosity Jointly with Retinal Amyloid Burden Correlates with Verbal Memory Loss: A Pilot Study. *Cells* **2021**, *10*, 2926. https://doi.org/10.3390/cells10112926

Academic Editors: Maurice Ptito and Joseph Bouskila

Received: 29 June 2021
Accepted: 25 October 2021
Published: 28 October 2021

Publisher's Note: MDPI stays neutral with regard to jurisdictional claims in published maps and institutional affiliations.

Abstract: Introduction: Retinal imaging is a non-invasive tool to study both retinal vasculature and neurodegeneration. In this exploratory retinal curcumin-fluorescence imaging (RFI) study, we sought to determine whether retinal vascular features combined with retinal amyloid burden correlate with the neurocognitive status. Methods: We used quantitative RFI in a cohort of patients with cognitive impairment to automatically compute retinal amyloid burden. Retinal blood vessels were segmented, and the vessel tortuosity index (VTI), inflection index, and branching angle were quantified. We assessed the correlations between retinal vascular and amyloid parameters, and cognitive domain Z-scores using linear regression models. Results: Thirty-four subjects were enrolled and twenty-nine (55% female, mean age 64 ± 6 years) were included in the combined retinal amyloid and vascular analysis. Eleven subjects had normal cognition and 18 had impaired cognition. Retinal VTI was discriminated among cognitive scores. The combined proximal mid-periphery amyloid count and venous VTI index exhibited significant differences between cognitively impaired and cognitively normal subjects (0.49 ± 1.1 vs. 0.91 ± 1.4, $p = 0.006$), and correlated with both the Wechsler Memory Scale-IV and SF-36 mental component score Z-scores ($p < 0.05$). Conclusion: This pilot study showed that retinal venular VTI combined with the proximal mid-periphery amyloid count could predict verbal memory loss. Future research is needed to finesse the clinical application of this retinal imaging-based technology.

Keywords: retinopathy; retinal vessels; retinal fluorescence imaging; amyloid; cognitive decline; Alzheimer's disease

1. Introduction

By 2025, the number of people aged 65 years and older with Alzheimer's dementia (AD) is projected to reach 7.1 million, which is almost a 22% increase from 2020 [1,2]. The contribution of vascular disease to cognitive performance is increasingly recognized, as the mechanisms linking vascular dysfunction and neurodegeneration are better characterized [3–7]. Recent reports implicate cerebral vascular pathology as an early and core contributor to the development of not only vascular dementia but also AD [7–11], a neurodegenerative condition and looming public health threat [2,12]. Considering the emerging vascular hypothesis [8,13–15], there is a critical need to incorporate both vascular and AD biomarkers [16–20] into predictive models to allow for early and sensitive detection of AD and mixed dementias. Yet, imaging of the skull-shielded brain poses various limitations for widespread screening in the clinical setting. The retina is a central nervous system organ that exhibits Aβ deposition and vascular changes [21–33] and is far more accessible for repeated and high-resolution imaging [34–42]. Dysfunctional pericytes in the blood-brain barrier (BBB) are significant contributors to the pathogenesis of vascular cognitive impairment, including cerebral small vessel and cerebral large vessel disease, as well as AD [28,43]. BBB pericyte injury is a predictor of apolipoprotein E (APOE) ε4-associated cognitive decline [4]. In contrast, BBB dysfunction mediates cerebral Aβ deposition, the retinal–blood barrier mirrors the BBB, and its disruption in the form of retinopathy was shown to predict cognitive decline [28,44–49]. Post-mortem retinal vessels derived from patients with mild cognitive impairment (MCI) and AD exhibited early and progressive pericyte loss as well as Aβ accumulation inside retinal pericytes, correlating with similar AD pathology in the brain [28]. Several studies demonstrated the linkage between retinal vascular fractal dimensions, caliber, and both tortuosity and cognitive deterioration [45–51]. The retinal arteriolar central reflex to vessel width ratio in digital retinal photographs was significantly higher in APOE ε4 allele carriers [48], hence the retina may allow for non-invasive monitoring of the effects of APOE ε4 on the cerebrovascular disease. Similarly, as targeting vascular risk factors is being considered in dementia prevention trials [52], retinal vascular assessments could offer a window for assessing the response to various interventions.

Recent work has highlighted the promising utility of retinal fluorescence imaging, an emerging technique capable of non-invasively imaging and quantifying the retinal amyloid, which is the pathological marker of AD [22,23,26,34,53–55]. Using this technique, our group previously identified a significant association between retinal amyloid count, especially in the proximal mid-periphery area, and the severity of cognitive impairment as well as hippocampal volumes [34,35]. As the same retinal imaging modality also allows for retinal vasculature analysis, we aimed to quantitatively examine both retinal vascular and retinal amyloid biomarkers in a cohort of subjects with cognitive decline. In this proof-of-concept exploratory study, we sought to examine the relationship between retinal microvascular features and retinal amyloid burden, with global and domain-specific cognitive scores.

2. Materials and Methods

2.1. Participants

This pilot study was approved by the Cedars-Sinai Institutional Review Board. All subjects older than 40 years of age presenting to our Neurology clinic with subjective cognitive decline and interest in undergoing retinal fluorescence imaging were included in this cohort. All subjects underwent a neurological examination, a standard battery of neuropsychological tests, and standard-of-care 3 Tesla non-contrast structural brain magnetic resonance imaging (MRI). No exclusion criteria were prespecified, except for a history of glaucoma, allergy to mydriatic eye drops, curcumin, or vitamin E. All subjects provided written informed consent prior to the commencement of the study.

2.2. Retinal Imaging

After ocular dilation, the retinal imaging was performed with a confocal scanning ophthalmoscope (Retia™, CenterVue SpA) that utilizes blue light for the excitation of curcumin emission to obtain fluorescent images of the retina, following a study design described in prior reports (Figure 1A) [34,35]. Curcumin has high affinity and specificity for the β-pleated sheets of Aβ, specifically for Aβ42, oligomers, and fibrils, which are linked to AD [56–61]. The researchers conducting the retinal image processing and quantifications were blinded to the patients' clinical characteristics.

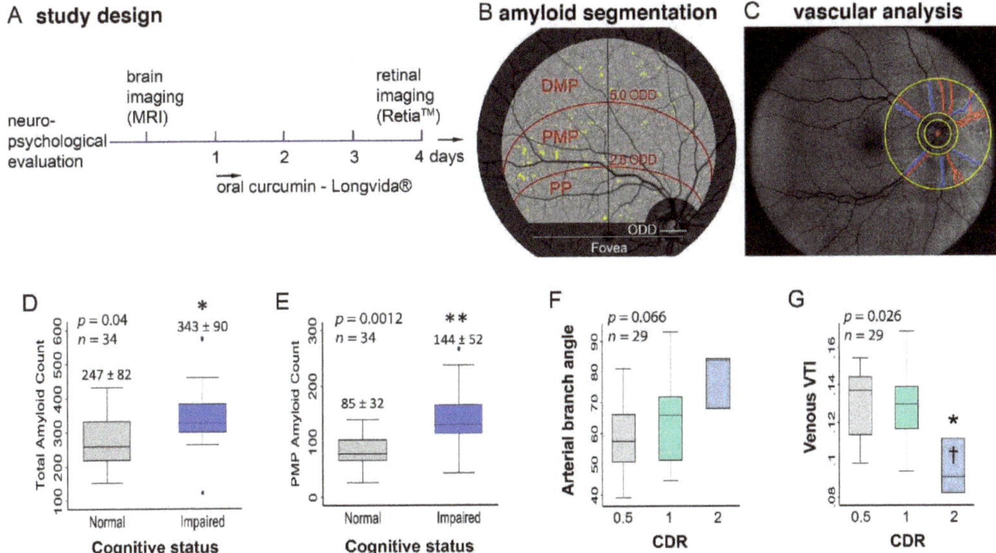

Figure 1. Study timeline of brain and retinal imaging followed by sectoral amyloid and vascular analysis. Study design scheme illustrating that subjects underwent baseline brain imaging and neuropsychological evaluation, followed by retinal fluorescence imaging after 4 days of daily oral curcumin intake (**A**). Illustration of the region of interest in the right eye supero-temporal retinal quadrant and its three subregions, which were used for quantifying retinal amyloid counts (**B**). Illustration of the region of interest used for the retinal vascular analysis. The red circle indicates the center of the optic nerve-head and the smallest yellow circle shows the optic nerve-head area. The two larger circles indicate the region of interest for the vascular analysis, which were 1.5 and 4 times the diameter of the optic disc. The branching angle and tortuosity of vessels within the region of interest were calculated. Arteries and veins are outlined by red and blue lines, respectively (**C**). Graphs illustrating differences in total amyloid (**D**) and proximal amyloid counts (**E**) when stratified by cognitive status. Graphs illustrating the differences between arterial branching angle (**F**) and the venous tortuosity index (**G**) when stratified by CDR. * $p < 0.05$; ** $p < 0.01$, by two-tailed unpaired student *t*-test or one-way ANOVA and Bonferroni's post-hoc test. Abbreviations: MRI, magnetic resonance imaging; PP, posterior pole; PMP, proximal mid-periphery; DMP, distal mid-periphery; ODD, optic disc diameter; CDR, Clinical Dementia Rating; and VTI, vessel tortuosity index.

2.3. Retinal Amyloid Quantification

The set of retinal images were processed using an automated retinal fluorescence measurement software system (NeuroVision Imaging, Inc., Sacramento, CA, USA). A combination of algorithms, including background correction, followed by characterization of the corrected retina using a mixture model, were used to identify pixels that were abnormally bright. Specifically, the primary factor of the variation in pixel intensity is illumination variability across the entire field of view (e.g., edges of the image become dark). This variability is addressed by estimating the background level and correcting it. The secondary factor in pixel variability is the structure to which it depends on. Vessels appear dark or hypofluorescent, while the amyloid appears identically as bright or hyperfluorescent. The background correction produces all vessels at a more consistent pixel value. In a similar manner, this occurs for the retina and amyloid spots. Choosing the appropriate threshold is possible using the mixture model, which characterizes hypofluorescent, isofluorescent, and hyperfluorescent pixels appropriately. A common region of interest (ROI) in the supero-temporal quadrant was applied with a field of view of 50 degrees, positioned on the image center using fovea and optic nerve-head centers as reference points to correct for eye rotation, with a zone around the fovea and optic nerve-head masked, as previously reported [35]. The ROI was further divided into three subregions: posterior pole, proximal mid-periphery, and distal mid-periphery (Figure 1B). Retinal amyloid count was quantified in the target ROI and three specified subregions.

2.4. Retinal Vascular Quantification

From the same retinal fundus images, an ROI was defined within a circumpapillary region centered on the optic nerve-head (ONH) and extending between 1.5 and 4 ONH radii (Figure 1C) [62]. Before the analysis, retinal images were visually inspected to ensure vessels were visible and that there was no reflectivity that could influence the result. The major vessels were detected after intensity normalization to minimize the effect of other influencing factors. Retinal vessels within the ROI were segmented using the Frangi vesselness filter to generate a binary image [63]. The vessels were classified into arteries and veins by a human observer based on the facts that retinal arteries are brighter in color and thinner in width compared to veins [64]. For each vessel segment on the binary image, vessel endpoints were selected, and distance transformation was used to extract the vessel centerline. The extracted centerlines were smoothed using a cubic spline with a regularization parameter of 3×10^{-5}. For each centerline, several geometric features, including the vessel tortuosity index (VTI), vessel inflection index, and branching angle, were non-automatically quantified. The VTI was calculated for each centerline based on a combination of local and global centerline geometric variables, as explained previously, that can detect alterations in the retinal vessels' curvature with pixel-level accuracy [65]. Equation (1) shows the formula for the VTI.

$$\text{VTI} = 0.1 \times \left(\text{SD}_{\theta}.\text{N.M.}\frac{\text{L}_\text{A}}{\text{L}_\text{C}} \right) \tag{1}$$

where SD_{θ} the is standard deviation of the angle difference between lines tangent to each centerline pixel and a reference axis (i.e., x-axis), and M is the average ratio of the centerline length to its chord length between pairs of inflection points, including centerline endpoints. N is number of critical points where the first derivative of the centerline vanishes, while L_A and Lc are the length of the vessel and its chord length, respectively. The VTI is shown to provide good correspondence with human perception of tortuosity and is invariant to rigid transformations. Similar to other measures of tortuosity, VTI is unitless. Its minimum value is zero, while it has no theoretical maximum as it can increase with the twistedness of a vessel. The vessel inflexion index was determined based on a number of inflection points along the vessels. Mathematically, these were pixels where the second derivative of the centerline vanishes. The vessel inflexion index represents local changes in the tortuosity of vessels and was found to be robust for ranking the tortuosity of vessels with similar lengths [66]. The branching angle of the vessels was calculated interactively using the open-source tool GIMP 2.8.

2.5. Cognitive Evaluation

All participants underwent a standard battery of neuropsychometric testing performed by a licensed neuropsychologist (DS). Standard neuropsychological testing included the Montreal Cognitive Assessment (MOCA), global Clinical Dementia Rating (CDR), as well as general cognitive (ACS-test of Premorbid Functioning) and specific cognitive domain assessments: attention and concentration (Wechsler Adult Intelligence Scale (WAIS)-IV); verbal memory (California Verbal Learning Test (CVLT) II, Wechsler Memory

Scale (WMS)-IV, and Logical Memory II); non-verbal memory (Rey Complex Figure Test and Recall (RCFT) 30 min, and Brief Visuo-Spatial Memory Test Revised (BVMT-R) Delayed Recall); language (Fluency-Letter (FAS) and Fluency-category (animals)); visuo-spatial ability (Rey Complex Figure Test and Recognition Trial (RCFT) Copy); speed of information processing (Trails A and B); and symptom validity and functional status (SF-36 Physical Component Score (PCS) and Mental Component Score (MCS)). We also evaluated the subject's emotional status using the Beck Depression Inventory II, Geriatric Depression Scale, and Profile of Mood State/Total Mood Disturbance.

2.6. Statistical Analysis

Descriptive statistics were calculated for patient demographics and clinical character-istics. Unless otherwise stated, data are expressed as mean $\pm$ standard deviation. Subjects were partitioned into three groups according to the Clinical Dementia Rating (CDR) (0.5, questionable impairment; 1, mild cognitive impairment; and 2, moderate cognitive impairment) [67] and dichotomized using MOCA, which demonstrates excellent sensitivity and specificity for both mild cognitive impairment (MCI) and AD. Using the cutoff score of <26, the MOCA has excellent sensitivity for MCI (90%) and AD (100%), as well as for the specificity for normal controls (87%). Positive (PPA) and negative predictive accuracy (NPA) were also reported to be excellent with a PPA of 89% and NPA of 91% for MCI, and a PPA of 89% and NPA of 100% for AD [68]. The subjects were also partitioned into groups according to the neuropsychometric diagnosis (normal cognition versus impaired cognition).

To produce combined indices of retinal vascular and amyloid measures, each variable was first inspected for normality; any non-normal variables were then log-transformed to produce a normal distribution. Each normalized variable was then standardized to a mean of 0 and a standard deviation set equal to 1. While higher amyloid count was associated with worse cognitive function, higher venous vascular tortuosity index (VTI) values were associated with better cognitive function. To account for this inverted scale, the standardized values of venous VTI were multiplied by -1. Standardized variables were then summed to produce exploratory, combined index measures of retinal amyloid and retinal vascular features.

Differences in continuous variables between levels of CDR were assessed through one-way analysis of variance (ANOVA), with Bonferroni's post-hoc test for the correction of multiple comparisons. Differences in the continuous variables between diagnostic scores were assessed using Student's *t*-test. Linear regression was performed to assess the relationship between retinal vascular and retinal amyloid measures, as well as to assess the relationship between combined retinal vascular and amyloid counts, and cognitive parameters. All statistical analyses were performed using STATA v15.1 (StataCorp, College Station, TX, USA) with an a priori significance level of 0.05.

3. Results

Our study included a total of 34 subjects that presented to our Neurology clinic with cognitive concerns. Out of those 34, 29 had retinal images of sufficient quality to undergo both retinal amyloid and vascular analysis; their demographics and preexisting conditions are shown in Table 1. Mean MOCA was 26 (range of 4–32) and median MOCA was 27. Eleven subjects had a CDR of 0.5, 15 had a CDR of 1, and 3 had a CDR of 2. Regarding the formal neuropsychometric cognitive evaluation, 11 (37.93%) patients had normal cognition and 18 (62.06%) had impaired cognition (six with amnestic MCI, nine with multidomain MCI, two probable AD cases, and one with possible fronto-temporal lobar degeneration).

Linear regression analyses revealed that the venous branching angle correlated with the distal mid-periphery amyloid count ($p = 0.03$) and the arterial inflexion index correlated with the posterior pole amyloid count ($p = 0.02$). There were no associations between retinal vascular parameters and amyloid count in the proximal mid-periphery (Table S1).

Table 1. Demographics and medical history of subjects in the combined retinal vascular and retinal amyloid analysis.

N (% female)	29 (55)
Age (years)	64 ± 6
Preexisting health conditions, N (%)	
Hypertension	11 (38)
Hyperlipidemia	15 (52)
Diabetes	3 (10)
Hyperthyroidism	8 (28)
Stroke/TIA	1 (3)
Heart disease/CAD/CHF	1 (3)
Smoking h/o	2 (7)

The analysis of retinal vascular and amyloid measures according to strata of cognitive function showed that the retinal PMP amyloid count and total amyloid count were significantly higher in the cognitively impaired compared to normal cognition participants (PMP: 144 ± 52 vs. 85 ± 32, p = 0.0012; total: 343 ± 90 vs. 247 ± 82, p = 0.04; Figure 1D,E and Table 2). There was no significant difference in the venous branching angle (p = 0.98) or arterial VTI (p = 0.53) across levels of CDR, whereas the arterial branching angle reached near significance (p = 0.066; Figure 1F). Venous VTI was significantly different across levels of CDR (mean ± SD of venous VTI values across increasing CDR categories: 0.13 ± 0.02, 0.13 ± 0.02, and 0.09 ± 0.02; p = 0.026; Figure 1G). Given these group differences and because of the independence of retinal vascular and retinal amyloid measures, the following combined amyloid-vascular indexes were calculated as exploratory variables: proximal mid-periphery amyloid count-venous VTI, total amyloid count-venous VTI, proximal mid-periphery amyloid count-arterial branching angle, and total amyloid count-arterial branching angle. One-way ANOVA revealed significant group differences in the VTI indices when compared according to the CDR level (Figure 2A–D). The combined proximal mid-periphery amyloid-venous VTI index was the only combined index measure exhibiting significant group differences when the cognitively impaired were compared to the cognitively normal subjects (0.49 ± 1.1 vs. −0.91 ± 1.4, p = 0.006; Figure 2E and Table 2).

Table 2. Vascular and amyloid parameters stratified by the cognitive status.

Variable	Normal Cognition (n = 11)	Impaired Cognition (n = 18)	p
Age (years; mean ± SD)	66.92 ± 7.7	67.06 ± 7.41	0.954
Years of education (mean ± SD)	16.05 ± 1.43	15.59 ± 2.76	0.432
Arterial hypertension (%)	6 (54.4)	10 (55.55)	0.633
Dyslipidemia (%)	6 (54.4)	11 (61.11)	0.924
Diabetes Mellitus (%)	0 (0)	3 (16.66)	0.563
Hippocampal volume (cm^3; mean ± SD)	7.82 ± 0.78	6.12 (±0.87)	0.051
Arterial branching angle (mean ± SD)	66.45 ± 13.0	63.3 ± 13.9	0.55
Venous branching angle (mean ± SD)	58.5 ± 6.1	59.0 ± 12.3	0.94
Arterial vessel tortuosity Index (mean ± SD)	0.15 ± 0.05	0.14 ± 0.06	0.42
Venous vessel tortuosity index (mean ± SD)	0.13 ± 0.02	0.12 ± 0.02	0.40
Arterial length (mean ± SD)	2687 ± 288	2706 ± 297	0.87
Venous length (mean ± SD)	2614 ± 270	2689 ± 276	0.50
Arterial vessel inflexion index (mean ± SD)	5.6 ± 0.80	5.5 ± 0.81	0.70

Table 2. *Cont.*

Variable	Normal Cognition ($n = 11$)	Impaired Cognition ($n = 18$)	p
Venous vessel inflexion index (mean ± SD)	5.2 ± 0.91	5.3 ± 0.78	0.83
Proximal mid-periphery amyloid count (mean ± SD)	**85 ± 32**	**144 ± 52**	**0.0012**
Distal mid-periphery amyloid count (mean ± SD)	91.3 ± 63	93.1 ± 45	0.92
Posterior pole amyloid count (mean ± SD)	98 ± 60	106 ± 46	0.66
Total amyloid count (mean ± SD)	**247 ± 82**	**343 ± 90**	**0.04**
Combined proximal mid-periphery amyloid count–arterial branching angle index (mean ± SD)	−0.56 ± 1.3	0.27 ± 1.4	0.11
Combined proximal mid-periphery amyloid count–venous tortuosity index (mean ± SD)	**−0.91 ± 1.4**	**0.49 ± 1.1**	**0.0068**
Combined total amyloid count–arterial branching angle index (mean ± SD)	−0.27 ± 1.3	0.08 ± 1.4	0.51
Combined total amyloid count–venous tortuosity index (mean ± SD)	−0.62 ± 1.5	0.29 ± 1.3	0.09

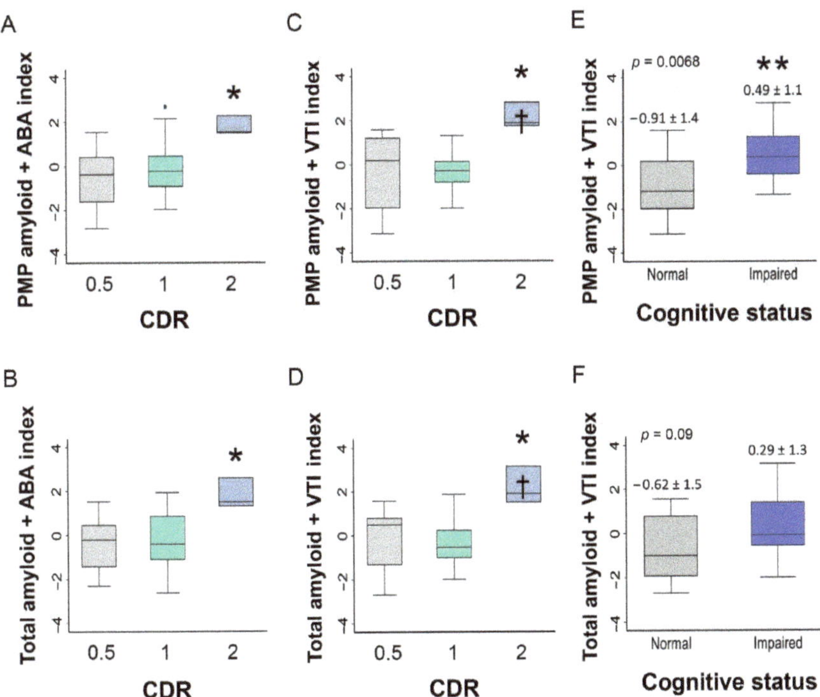

Figure 2. Combined retinal amyloid and vascular parameters in patients stratified by cognitive scores. Graphs illustrating the differences in the combined proximal mid-periphery amyloid-arterial branching angle index (**A**), total amyloid-arterial branching angle index (**B**), proximal mid-periphery amyloid-venous tortuosity index (**C**), and total amyloid-venous tortuosity index (**D**) when stratified by CDR score. Graphs illustrating the differences between the combined proximal mid periphery amyloid-venous tortuosity index (**E**) and total amyloid-venous tortuosity index (**F**) when stratified by the cognitive status. Bar graphs show the mean and deviation (* $p < 0.05$ and ** $p < 0.01$ by two-tailed paired Student's *t*-test). Abbreviations: VTI, vessel tortuosity index; ABA, arterial branching angle; PMP, proximal mid-periphery; and CDR, Clinical Dementia Rating.

We performed regression analyses to evaluate the correlations between retinal vascular geometric parameters and retinal amyloid counts with cognitive domain Z-scores. We found that the venous branching angle correlated with the WAIS-IV-digit span Z-score (Beta −0.045 (SE 0.015), $p = 008$). The total amyloid count correlated with the SF-36-MCS Z-score (Beta −0.004 (SE 0.002), $p = 0.046$), whereas the proximal mid-periphery amyloid count correlated with two verbal memory measures, namely CVLT-II Long Delay (Beta −0.009 (SE 0.003), $p = 0.027$) and WMS-IV LM-II (Beta −0.007 (SE 0.003), $p = 0.028$). The distal mid-periphery amyloid count correlated with non-verbal memory, RCFT Delayed Recall (Beta −0.01 (SE 0.005), $p = 0.04$), and SF-36-MCS (Beta −0.014 (SE 0.004), $p = 0.004$; Table 3).

Table 3. Retinal vascular and amyloid parameter predictors of cognitive domain measures.

Retinal Parameter	Cognitive Measures' Z-Score	Beta (Std. Err)	p
Venous branching angle	WAIS-IV	−0.045 (0.015)	0.008
PMP amyloid-venous VTI	CVLT Long Delay	−0.37 (0.17)	0.04
Total amyloid-venous VTI	WMS LM-II	−0.44 (0.13)	0.03
PMP amyloid-arterial branching angle	WMS LM-II	−0.35 (0.16)	0.04
PMP amyloid-venous VTI	WMS LM-II	−0.53 (0.13)	0.001
Total amyloid-venous VTI	MCS	−0.30 (0.14)	0.04
PMP amyloid-venous VTI	MCS	−0.33 (0.15)	0.03
PMP amyloid count	CVLT Long Delay	−0.009 (0.003)	0.02
PMP amyloid count	WMS LM-II	−0.007 (0.03)	0.02
DMP amyloid count	RCFT 30 min Recall	−0.010 (0.005)	0.04
DMP amyloid count	MCS	−0.014 (0.004)	0.004
Total amyloid count	MCS	−0.004 (0.002)	0.04

Abbreviations: VTI, vessel tortuosity index; PMP, proximal mid-periphery; DMP, distal mid-periphery; WAIS, Wechsler Adult Intelligence Scale; CVLT, California Verbal Learning Test; WMS LM-II, Wechsler Memory Scale Logical Memory II; RCFT, Rey Complex Figure Test and Recall; MCS, Mental Component Score; and Std. Err, standard error.

The combined proximal mid-periphery amyloid-venous VTI index correlated with both verbal memory performance Z-scores (WMS-IV LM-II (Beta −0.537 (SE 0.138), $p = 0.001$) and CVLT-II Long Delay (Beta −0.370 (SE 0.176), $p = 0.046$)), as well as with the mental component of the cognitive-related quality-of-life score (SF-36-MCS (Beta −0.338 (SE 0.153), $p = 0.039$); Figure 3C,D). The combined total amyloid-venous VTI index correlated with WMS-IV LM-II (Beta −0.440 (SE 0.132), $p = 0.003$) and SF-36-MCS (Beta −0.302 (SE 0.141), $p = 0.045$; Figure 3A,B and Table 3).

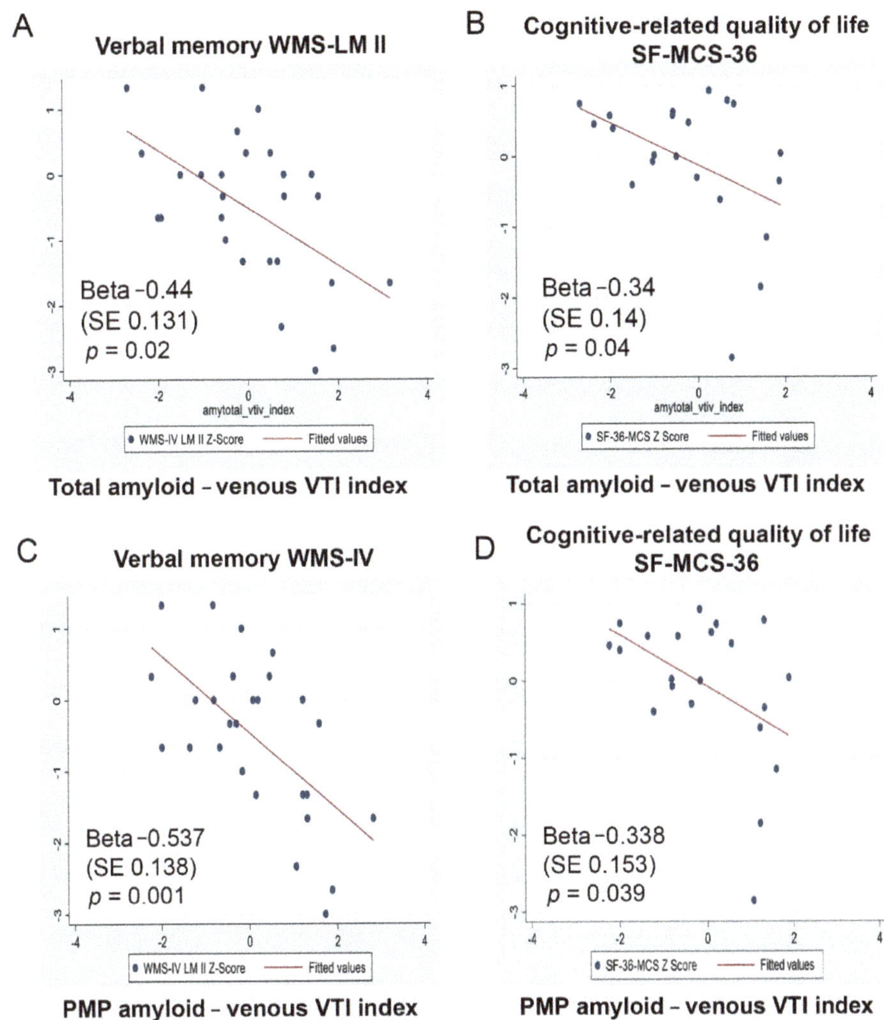

Figure 3. Retinal amyloid count combined with retinal venous VTI correlated with verbal memory and cognitive-related quality-of-life measures. Graphs illustrating the correlations between the combined total amyloid-venous tortuosity index and verbal memory (**A**) and cognitive-related quality-of-life Z-scores (**B**), and the correlations between the combined proximal mid-periphery amyloid-venous tortuosity index and verbal memory (**C**) and cognitive-related quality-of-life Z-scores (**D**). Abbreviations: WMS-IV LM II, Wechsler Memory Scale IV Logical Memory II; WMS-IV, Wechsler Memory Scale IV; SF-36 MCS, SF-36 Mental Component Score; PMP, proximal mid-periphery; and VTI, vessel tortuosity index.

4. Discussion

The main findings from this exploratory investigation of retinal fluorescence imaging are that retinal vascular features do not significantly correlate with retinal amyloid deposition in the proximal mid-periphery area; proximal mid-periphery retinal amyloid count correlates with verbal memory; and the combination of the retinal amyloid and venous tortuosity index into standardized index scores can provide a more comprehensive indicator of cognitive performance.

Microvascular damage is increasingly recognized as a critical initiator of vascular cognitive impairment and AD pathology [5,9,69]. Vascular dysregulation, leading to cerebral amyloid accumulation, and the link between cerebrovascular disease and dementia are explained by several mechanisms [4,5,70]. Pericyte loss and deficient vascular platelet-derived growth factor receptor-β signaling were identified in both the retinal and cerebral vasculature in subjects with MCI and AD [4,28]. Prior reports demonstrated that retinal vasculature may be used as a biomarker of early or preclinical dementia [71], and retinal microvascular abnormalities in MCI and dementia have been demonstrated using various retinal vasculature imaging modalities (e.g., retinal fundus photography [48,72,73], optical coherence tomography angiography [74,75], high-frequency flicker-light stimulation [51,76], and the retinal function imager [74]). Conversely, den Haan et al. [77] showed that retinal vascular measures did not differ between patients with AD and control participants, and venular tortuosity was smaller in subjects with greater white matter disease burden. Previous investigations have also shown a strong relationship between retinal vasculopathy and brain amyloid deposition [30]. Sharafi et al. [30] evaluated the relationship between retinal vascular statuses (vessel diameter and both tortuosity and spatial-spectral texture measures) using hyperspectral retinal imaging and CNS amyloid status (assessed with (18) F-florbetaben positron-emission tomography). They found that retinal venules of amyloid-positive subjects showed a higher mean tortuosity compared with the amyloid-negative subjects. This study suggested that the inclusion of metrics related to retinal vasculature and the surrounding tissue-related texture could improve the discrimination performance of the cerebral amyloid status [30].

As both retinal amyloid accumulation and retinal vascular pathology [28,34–37] are reported in patients with MCI, we explored the interplay between retinal vascular geometric measures and retinal amyloid burden using retinal fluorescence imaging. Prior studies showed that the retinal proximal mid-periphery area may be the target of amyloid quantification to reflect cerebral AD pathology, as it correlates with cognitive performance and hippocampal volume [35,37]. In this pilot cohort, we found that retinal vascular features correlated with amyloid deposition in the posterior pole and retinal distal mid-periphery area, but not with the proximal mid-periphery area. A possible explanation is that this investigation measured physical features of the retinal vasculature (e.g., branching angle and tortuosity index) and did not assess functional endpoints. Additionally, our quantitative vascular analysis could not target the smaller retinal blood vessels. The mechanisms driving vascular remodeling and amyloid deposition may occur at different rates, leading to the appearance of these clinical signs at different stages in disease progression. The investigation of subjects with mainly mild cognitive impairment in our cohort may explain why venous VTI was lower in subjects with worse cognition and why the other arterial or venous vascular parameters did not show any significant differences across cognitive strata. This hypothesis is supported by the lack of association between retinal vascular features and most of the neuropsychometric cognitive scores in our cohort. Conversely, retinal vascular measures did not correlate with any cognitive measures except for attention and concentration. The total and proximal mid-periphery retinal amyloid count correlated with verbal memory measures, while the distal mid-periphery amyloid count correlated with non-verbal memory measures. Interestingly, in this cohort with early cognitive disorders, subjects with higher amyloid counts and worse cognition levels had lower retinal venular tortuosity. The only combined index that discriminated between individuals with impaired cognition and normal cognition was the proximal mid-periphery amyloid count-venular VTI. This combined index was also associated with verbal memory and the 'mental component' summary of psychological functioning (SF-36 Mental Component Score). This latter finding reflects the association with cognitive-related psychological and emotional functioning. This appears to represent an exclusive contribution, as physical functioning status, as demonstrated by the SF-36 Physical Component Score, was not associated with amyloid or vascular retinal markers.

Two or more retinal vascular abnormalities were associated in a dose-response manner with an increased risk of disabling dementia in a prior study [49]. It is possible that combined amyloid–vascular indexes are better discriminators of cognitive function, with the potential for use as outcome measures in AD and mixed dementia trials. Our study is limited by a small sample size, heterogeneity, and the absence of genetic and CSF or brain amyloid biomarkers. Similarly, our patients were not evaluated for all ocular conditions other than a history of glaucoma. Due to the limited sample size, we could not adjust for the presence of traditional vascular risk factors or the presence of retinopathy, which are known contributors to retinal vascular geometric changes. Given the heterogeneity in the sample size across study groups, further confirmation of these preliminary results will be necessary in the future for specific groups of early AD and vascular and mixed dementias.

Our findings underscore the potential value of the exploratory amyloid-vascular indexes presented herein. Future investigations are warranted to explore the clinical utility of retinal fluorescence imaging in concert with combined amyloid-vascular index measures. More comprehensive cohort studies including a larger sample size and a greater range of disease severity among the participants could help to elucidate the stage at which retinal amyloid and/or retinal venular versus arterial impairments begin to develop in cognitive disorders. Given the cost and technical requirements of gold-standard methods for assessing cerebral amyloid deposition and vascular pathology, further validation of these retinal imaging methods could potentially yield greater accessibility to testing, thus facilitating more extensive clinical trials as well as improving the detection of early dementia.

Supplementary Materials: The following are available online at https://www.mdpi.com/article/10.3390/cells10112926/s1, Table S1: Retinal vascular parameter correlation with retinal amyloid measures.

Author Contributions: Conceptualization, O.M.D., R.R., S.F., P.D.L., S.C., Y.K. and M.K.-H.; methodology, O.M.D., R.R., M.M.K., A.S., D.S., D.S.S., Y.K., P.D.L., S.C. and M.K.-H.; software, R.R., J.S., T.T., K.O.J., A.D.C., S.V., K.L.B., S.F. and M.K.-H.; validation, D.S.S, J.S., T.T., S.C.; formal analysis, R.R., M.M.K., Y.S., S.C. and M.K.-H.; investigation, D.S.S., A.S., D.S., K.L.B., P.D.L. and M.K.-H.; resources, K.L.B., P.D.L. and M.K.-H.; data curation, O.M.D., R.R., J.S., T.T., Y.K., S.C. and M.K.-H.; writing—original draft preparation, O.M.D.; writing—review and editing, O.M.D., R.R., D.S.S., M.K.-H., K.O.J., A.D.C., S.V., K.L.B., S.F., Y.K., S.F., P.D.L., Y.S., S.C. and M.K.-H.; visualization, S.C., P.D.L. and M.K.-H.; supervision, P.D.L., Y.K., M.K.-H. and K.L.B.; project administration, O.M.D. and M.K.-H.; funding acquisition, K.L.B. and M.K.-H. All authors have read and agreed to the published version of the manuscript.

Funding: We received support from a National Institute on Aging award (AG044897, Koronyo-Hamaoui, PI) and from the Saban, Gordon, and Marciano Private Foundations (Koronyo-Hamaoui, PI). The funders had no role in the design or conduct of this research.

Institutional Review Board Statement: The study was conducted according to the guidelines of the Declaration of Helsinki and was approved by the Institutional Review Board (or Ethics Committee) of Cedars-Sinai Medical Center (protocol code 00052349, approved in 2018).

Informed Consent Statement: Informed consent was obtained from all subjects involved in the study.

Data Availability Statement: Data available upon request due to restrictions, e.g., privacy or ethical restrictions.

Acknowledgments: We thank Mia Oviatt for editing the manuscript. The authors dedicate the manuscript to the memory of Salomon Moni Hamaoui and Lillian Jones Black, who died of Alzheimer's disease.

Conflicts of Interest: Black, Verdooner, Koronyo, and Koronyo-Hamaoui are founding members of NeuroVision Imaging Inc., 1395 Garden Highway, Suite 250, Sacramento, CA 95833, USA. Dr. Frautschy is co-inventor of the US patent US9192644B2 for a curcumin formulation. Johnson, Czeszynski, and Verdooner are currently employed by NeuroVision Imaging Inc. The remaining authors declare that the research study was conducted in the absence of any commercial or financial relationships that could be construed as a potential conflict of interest.

References

1. Hebert, L.E.; Weuve, J.; Scherr, P.A.; Evans, D.A. Alzheimer disease in the United States (2010–2050) estimated using the 2010 census. *Neurology* **2013**, *80*, 1778–1783. [CrossRef] [PubMed]
2. 2020 Alzheimer's disease facts and figures. *Alzheimer's Dement.* **2020**, *16*, 391–460. [CrossRef] [PubMed]
3. Bell, R.D.; Zlokovic, B.V. Neurovascular mechanisms and blood-brain barrier disorder in Alzheimer's disease. *Acta Neuropathol.* **2009**, *118*, 103–113. [CrossRef] [PubMed]
4. Montagne, A.; Nation, D.A.; Sagare, A.P.; Barisano, G.; Sweeney, M.D.; Chakhoyan, A.; Pachicano, M.; Joe, E.; Nelson, A.R.; D'Orazio, L.M.; et al. APOE4 leads to blood-brain barrier dysfunction predicting cognitive decline. *Nature* **2020**, *581*, 71–76. [CrossRef]
5. He, J.T.; Zhao, X.; Xu, L.; Mao, C.Y. Vascular Risk Factors and Alzheimer's Disease: Blood-Brain Barrier Disruption, Metabolic Syndromes, and Molecular Links. *J. Alzheimer's Dis.* **2020**, *73*, 39–58. [CrossRef]
6. Riphagen, J.M.; Ramakers, I.; Freeze, W.M.; Pagen, L.H.G.; Hanseeuw, B.J.; Verbeek, M.M.; Verhey, F.R.J.; Jacobs, H.I.L. Linking APOE-epsilon4, blood-brain barrier dysfunction, and inflammation to Alzheimer's pathology. *Neurobiol. Aging* **2020**, *85*, 96–103. [CrossRef]
7. Nikolakopoulou, A.M.; Montagne, A.; Kisler, K.; Dai, Z.; Wang, Y.; Huuskonen, M.T.; Sagare, A.P.; Lazic, D.; Sweeney, M.D.; Kong, P.; et al. Pericyte loss leads to circulatory failure and pleiotrophin depletion causing neuron loss. *Nat. Neurosci.* **2019**, *22*, 1089–1098. [CrossRef]
8. Solis, E.; Hascup, K.N.; Hascup, E.R. Alzheimer's Disease: The Link Between Amyloid-beta and Neurovascular Dysfunction. *J. Alzheimer's Dis.* **2020**, *76*, 1179–1198. [CrossRef] [PubMed]
9. Sweeney, M.D.; Montagne, A.; Sagare, A.P.; Nation, D.A.; Schneider, L.S.; Chui, H.C.; Harrington, M.G.; Pa, J.; Law, M.; Wang, D.J.J.; et al. Vascular dysfunction-The disregarded partner of Alzheimer's disease. *Alzheimer's Dement.* **2019**, *15*, 158–167. [CrossRef] [PubMed]
10. Frost, S.; Kanagasingam, Y.; Sohrabi, H.; Vignarajan, J.; Bourgeat, P.; Salvado, O.; Villemagne, V.; Rowe, C.C.; Lance Macaulay, S.; Szoeke, C.; et al. Retinal vascular biomarkers for early detection and monitoring of Alzheimer's disease. *Transl. Psychiatry* **2013**, *3*, e233. [CrossRef] [PubMed]
11. Williams, M.A.; McGowan, A.J.; Cardwell, C.R.; Cheung, C.Y.; Craig, D.; Passmore, P.; Silvestri, G.; Maxwell, A.P.; McKay, G.J. Retinal microvascular network attenuation in Alzheimer's disease. *Alzheimer's Dement.* **2015**, *1*, 229–235. [CrossRef]
12. Quinones, A.R.; Kaye, J.; Allore, H.G.; Botoseneanu, A.; Thielke, S.M. An Agenda for Addressing Multimorbidity and Racial and Ethnic Disparities in Alzheimer's Disease and Related Dementia. *Am. J. Alzheimer's Dis. Other Dement.* **2020**, *35*, 1533317520960874. [CrossRef]
13. Sweeney, M.D.; Sagare, A.P.; Zlokovic, B.V. Blood-brain barrier breakdown in Alzheimer disease and other neurodegenerative disorders. *Nat. Rev. Neurol.* **2018**, *14*, 133–150. [CrossRef]
14. Kisler, K.; Nelson, A.R.; Montagne, A.; Zlokovic, B.V. Cerebral blood flow regulation and neurovascular dysfunction in Alzheimer disease. *Nat. Rev. Neurosci.* **2017**, *18*, 419–434. [CrossRef] [PubMed]
15. Shi, H.; Koronyo, Y.; Fuchs, D.T.; Sheyn, J.; Wawrowsky, K.; Lahiri, S.; Black, K.L.; Koronyo-Hamaoui, M. Retinal capillary degeneration and blood-retinal barrier disruption in murine models of Alzheimer's disease. *Acta Neuropathol. Commun.* **2020**, *8*, 202. [CrossRef]
16. Jack, C.R., Jr.; Bennett, D.A.; Blennow, K.; Carrillo, M.C.; Dunn, B.; Haeberlein, S.B.; Holtzman, D.M.; Jagust, W.; Jessen, F.; Karlawish, J.; et al. NIA-AA Research Framework: Toward a biological definition of Alzheimer's disease. *Alzheimer's Dement.* **2018**, *14*, 535–562. [CrossRef] [PubMed]
17. Mantzavinos, V.; Alexiou, A. Biomarkers for Alzheimer's Disease Diagnosis. *Curr. Alzheimer Res.* **2017**, *14*, 1149–1154. [CrossRef] [PubMed]
18. Olsson, B.; Lautner, R.; Andreasson, U.; Ohrfelt, A.; Portelius, E.; Bjerke, M.; Holtta, M.; Rosen, C.; Olsson, C.; Strobel, G.; et al. CSF and blood biomarkers for the diagnosis of Alzheimer's disease: A systematic review and meta-analysis. *Lancet Neurol.* **2016**, *15*, 673–684. [CrossRef]
19. Allegri, R.F.; Chrem Mendez, P.; Calandri, I.; Cohen, G.; Martin, M.E.; Russo, M.J.; Crivelli, L.; Pertierra, L.; Tapajoz, F.; Clarens, M.F.; et al. Prognostic value of ATN Alzheimer biomarkers: 60-month follow-up results from the Argentine Alzheimer's Disease Neuroimaging Initiative. *Alzheimer's Dement.* **2020**, *12*, e12026. [CrossRef] [PubMed]
20. Baldacci, F.; Mazzucchi, S.; Della Vecchia, A.; Giampietri, L.; Giannini, N.; Koronyo-Hamaoui, M.; Ceravolo, R.; Siciliano, G.; Bonuccelli, U.; Elahi, F.M.; et al. The path to biomarker-based diagnostic criteria for the spectrum of neurodegenerative diseases. *Expert Rev. Mol. Diagn.* **2020**, *20*, 421–441. [CrossRef] [PubMed]
21. Hart, N.J.; Koronyo, Y.; Black, K.L.; Koronyo-Hamaoui, M. Ocular indicators of Alzheimer's: Exploring disease in the retina. *Acta Neuropathol.* **2016**, *132*, 767–787. [CrossRef]
22. Koronyo-Hamaoui, M.; Koronyo, Y.; Ljubimov, A.V.; Miller, C.A.; Ko, M.K.; Black, K.L.; Schwartz, M.; Farkas, D.L. Identification of amyloid plaques in retinas from Alzheimer's patients and noninvasive in vivo optical imaging of retinal plaques in a mouse model. *Neuroimage* **2011**, *54* (Suppl. 1), S204–S217. [CrossRef]
23. Koronyo, Y.; Salumbides, B.C.; Black, K.L.; Koronyo-Hamaoui, M. Alzheimer's disease in the retina: Imaging retinal abeta plaques for early diagnosis and therapy assessment. *Neurodegener. Dis.* **2012**, *10*, 285–293. [CrossRef]

24. Tsai, Y.; Lu, B.; Ljubimov, A.V.; Girman, S.; Ross-Cisneros, F.N.; Sadun, A.A.; Svendsen, C.N.; Cohen, R.M.; Wang, S. Ocular changes in TgF344-AD rat model of Alzheimer's disease. *Invest. Ophthalmol. Vis. Sci.* **2014**, *55*, 523–534. [CrossRef] [PubMed]

25. La Morgia, C.; Ross-Cisneros, F.N.; Koronyo, Y.; Hannibal, J.; Gallassi, R.; Cantalupo, G.; Sambati, L.; Pan, B.X.; Tozer, K.R.; Barboni, P.; et al. Melanopsin retinal ganglion cell loss in Alzheimer disease. *Ann. Neurol.* **2016**, *79*, 90–109. [CrossRef] [PubMed]

26. den Haan, J.; Morrema, T.H.J.; Verbraak, F.D.; de Boer, J.F.; Scheltens, P.; Rozemuller, A.J.; Bergen, A.A.B.; Bouwman, F.H.; Hoozemans, J.J. Amyloid-beta and phosphorylated tau in post-mortem Alzheimer's disease retinas. *Acta Neuropathol. Commun.* **2018**, *6*, 147. [CrossRef]

27. Grimaldi, A.; Pediconi, N.; Oieni, F.; Pizzarelli, R.; Rosito, M.; Giubettini, M.; Santini, T.; Limatola, C.; Ruocco, G.; Ragozzino, D.; et al. Neuroinflammatory Processes, A1 Astrocyte Activation and Protein Aggregation in the Retina of Alzheimer's Disease Patients, Possible Biomarkers for Early Diagnosis. *Front. Neurosci.* **2019**, *13*, 925. [CrossRef] [PubMed]

28. Shi, H.; Koronyo, Y.; Rentsendorj, A.; Regis, G.C.; Sheyn, J.; Fuchs, D.T.; Kramerov, A.A.; Ljubimov, A.V.; Dumitrascu, O.M.; Rodriguez, A.R. Identification of early pericyte loss and vascular amyloidosis in Alzheimer's disease retina. *Acta Neuropathol.* **2020**, *139*, 813–836. [CrossRef]

29. Habiba, U.; Merlin, S.; Lim, J.K.H.; Wong, V.H.Y.; Nguyen, C.T.O.; Morley, J.W.; Bui, B.V.; Tayebi, M. Age-Specific Retinal and Cerebral Immunodetection of Amyloid-beta Plaques and Oligomers in a Rodent Model of Alzheimer's Disease. *J. Alzheimer's Dis.* **2020**, *76*, 1135–1150. [CrossRef]

30. Sharafi, S.M.; Sylvestre, J.P.; Chevrefils, C.; Soucy, J.P.; Beaulieu, S.; Pascoal, T.A.; Arbour, J.D.; Rheaume, M.A.; Robillard, A.; Chayer, C.; et al. Vascular retinal biomarkers improves the detection of the likely cerebral amyloid status from hyperspectral retinal images. *Alzheimer's Dement.* **2019**, *5*, 610–617. [CrossRef]

31. Vit, J.P.; Fuchs, D.T.; Angel, A.; Levy, A.; Lamensdorf, I.; Black, K.L.; Koronyo, Y.; Koronyo-hamaoui, M. Color and contrast vision in mouse models of aging and Alzheimer's disease using a novel visual-stimuli four-arm maze. *Sci. Rep.* **2021**, *11*, 1255. [CrossRef]

32. Mirzaei, N.; Shi, H.; Oviatt, M.; Doustar, J.; Rentsendorj, A.; Fuchs, D.T.; Sheyn, J.; Black, K.L.; Koronyo, Y.; Koronyo-Hamaoui, M. Alzheimer's Retinopathy: Seeing Disease in the Eyes. *Front. Neurosci.* **2020**, *14*, 921. [CrossRef] [PubMed]

33. Snyder, P.J.; Alber, J.; Alt, C.; Bain, L.J.; Bouma, B.E.; Bouwman, F.H.; DeBuc, D.C.; Campbell, M.C.W.; Carrillo, M.C.; Chew, E.Y.; et al. Retinal imaging in Alzheimer's and neurodegenerative diseases. *Alzheimer's Dement.* **2021**, *17*, 103–111. [CrossRef] [PubMed]

34. Koronyo, Y.; Biggs, D.; Barron, E.; Boyer, D.S.; Pearlman, J.A.; Au, W.J.; Kile, S.J.; Blanco, A.; Fuchs, D.T.; Ashfaq, A.; et al. Retinal amyloid pathology and proof-of-concept imaging trial in Alzheimer's disease. *JCI Insight* **2017**, *2*, e93621. [CrossRef] [PubMed]

35. Dumitrascu, O.M.; Lyden, P.D.; Torbati, T.; Sheyn, J.; Sherzai, A.; Sherzai, D.; Sherman, D.S.; Rosenberry, R.; Cheng, S.; Johnson, K.O.; et al. Sectoral segmentation of retinal amyloid imaging in subjects with cognitive decline. *Alzheimer's Dement.* **2020**, *12*, e12109. [CrossRef] [PubMed]

36. Schultz, N.; Byman, E.; Netherlands Brain, B.; Wennstrom, M. Levels of Retinal Amyloid-beta Correlate with Levels of Retinal IAPP and Hippocampal Amyloid-beta in Neuropathologically Evaluated Individuals. *J. Alzheimer's Dis.* **2020**, *73*, 1201–1209. [CrossRef] [PubMed]

37. Lee, S.; Jiang, K.; McIlmoyle, B.; To, E.; Xu, Q.A.; Hirsch-Reinshagen, V.; Mackenzie, I.R.; Hsiung, G.R.; Eadie, B.D.; Sarunic, M.V.; et al. Amyloid Beta Immunoreactivity in the Retinal Ganglion Cell Layer of the Alzheimer's Eye. *Front. Neurosci.* **2020**, *14*, 758. [CrossRef] [PubMed]

38. Doustar, J.; Torbati, T.; Black, K.L.; Koronyo, Y.; Koronyo-Hamaoui, M. Optical Coherence Tomography in Alzheimer's Disease and Other Neurodegenerative Diseases. *Front. Neurol.* **2017**, *8*, 701. [CrossRef] [PubMed]

39. Hampel, H.; Toschi, N.; Babiloni, C.; Baldacci, F.; Black, K.L.; Bokde, A.L.W.; Bun, R.S.; Cacciola, F.; Cavedo, E.; Chiesa, P.A.; et al. Revolution of Alzheimer Precision Neurology. Passageway of Systems Biology and Neurophysiology. *J. Alzheimer's Dis.* **2018**, *64*, S47–S105. [CrossRef]

40. Carare, R.O.; Aldea, R.; Agarwal, N.; Bacskai, B.J.; Bechman, I.; Boche, D.; Bu, G.; Bulters, D.; Clemens, A.; Counts, S.E.; et al. Clearance of interstitial fluid (ISF) and CSF (CLIC) group-part of Vascular Professional Interest Area (PIA): Cerebrovascular disease and the failure of elimination of Amyloid-beta from the brain and retina with age and Alzheimer's disease-Opportunities for Therapy. *Alzheimer's Dement.* **2020**, *12*, e12053.

41. Shi, H.; Koronyo, Y.; Rentsendorj, A.; Fuchs, D.T.; Sheyn, J.; Black, K.L.; Mirzaei, N.; Koronyo-Hamaoui, M. Retinal Vasculopathy in Alzheimer's Disease. *Front. Neurosci.* **2021**, *15*, 1211. [CrossRef]

42. Cabrera DeBuc, D.; Somfai, G.M.; Arthur, E.; Kostic, M.; Oropesa, S.; Mendoza Santiesteban, C. Investigating Multimodal Diagnostic Eye Biomarkers of Cognitive Impairment by Measuring Vascular and Neurogenic Changes in the Retina. *Front. Physiol.* **2018**, *9*, 1721. [CrossRef]

43. Uemura, M.T.; Maki, T.; Ihara, M.; Lee, V.M.Y.; Trojanowski, J.Q. Brain Microvascular Pericytes in Vascular Cognitive Impairment and Dementia. *Front. Aging Neurosci.* **2020**, *12*, 80. [CrossRef]

44. McGrory, S.; Cameron, J.R.; Pellegrini, E.; Warren, C.; Doubal, F.N.; Deary, I.J.; Dhillon, B.; Wardlaw, J.M.; Trucco, E.; MacGillivray, T.J. The application of retinal fundus camera imaging in dementia: A systematic review. *Alzheimer's Dement.* **2017**, *6*, 91–107. [CrossRef]

45. Deal, J.A.; Sharrett, A.R.; Rawlings, A.M.; Gottesman, R.F.; Bandeen-Roche, K.; Albert, M.; Knopman, D.; Selvin, E.; Wasserman, B.A.; Klein, B.; et al. Retinal signs and 20-year cognitive decline in the Atherosclerosis Risk in Communities Study. *Neurology* **2018**, *90*, e1158–e1166. [CrossRef]

46. Deal, J.A.; Sharrett, A.R.; Albert, M.; Bandeen-Roche, K.; Burgard, S.; Thomas, S.D.; Gottesman, R.F.; Knopman, D.; Mosley, T.; Klein, B.; et al. Retinal signs and risk of incident dementia in the Atherosclerosis Risk in Communities study. *Alzheimer's Dement.* **2019**, *15*, 477–486. [CrossRef]

47. Jung, N.Y.; Han, J.C.; Ong, Y.T.; Cheung, C.Y.; Chen, C.P.; Wong, T.Y.; Kim, H.J.; Kim, Y.J.; Lee, J.; Lee, J.S.; et al. Retinal microvasculature changes in amyloid-negative subcortical vascular cognitive impairment compared to amyloid-positive Alzheimer's disease. *J. Neurol. Sci.* **2019**, *396*, 94–101. [CrossRef]

48. Frost, S.; Bhuiyan, A.; Offerman, D.; Doecke, J.D.; Macaulay, S.L.; Sohrabi, H.R.; Ames, D.; Masters, C.; Martins, R.N.; Kanagasingam, Y.; et al. Modulation of Retinal Arteriolar Central Reflection by APOE Genotype. *Curr. Alzheimer Res.* **2017**, *14*, 916–923. [CrossRef] [PubMed]

49. Jinnouchi, H.; Kitamura, A.; Yamagishi, K.; Kiyama, M.; Imano, H.; Okada, T.; Cui, R.; Umesawa, M.; Muraki, I.; Hayama-Terada, M.; et al. Retinal Vascular Changes and Prospective Risk of Disabling Dementia: The Circulatory Risk in Communities Study (CIRCS). *J. Atheroscler. Thromb.* **2017**, *24*, 687–695. [CrossRef] [PubMed]

50. Dumitrascu, O.M.; Demaerschalk, B.M.; Valencia Sanchez, C.; Almader-Douglas, D.; O'Carroll, C.B.; Aguilar, M.I.; Lyden, P.D.; Kumar, G. Retinal Microvascular Abnormalities as Surrogate Markers of Cerebrovascular Ischemic Disease: A Meta-Analysis. *J. Stroke Cerebrovasc. Dis.* **2018**, *27*, 1960–1968. [CrossRef] [PubMed]

51. Golzan, S.M.; Goozee, K.; Georgevsky, D.; Avolio, A.; Chatterjee, P.; Shen, K.; Gupta, V.; Chung, R.; Savage, G.; Orr, C.F.; et al. Retinal vascular and structural changes are associated with amyloid burden in the elderly: Ophthalmic biomarkers of preclinical Alzheimer's disease. *Alzheimer's Res. Ther.* **2017**, *9*, 13. [CrossRef]

52. Kobe, T.; Gonneaud, J.; Pichet Binette, A.; Meyer, P.F.; McSweeney, M.; Rosa-Neto, P.; Breitner, J.C.S.; Poirier, J.; Villeneuve, S. Association of Vascular Risk Factors With beta-Amyloid Peptide and Tau Burdens in Cognitively Unimpaired Individuals and Its Interaction With Vascular Medication Use. *JAMA Netw. Open* **2020**, *3*, e1920780. [CrossRef]

53. Ngolab, J.; Donohue, M.; Belsha, A.; Salazar, J.; Cohen, P.; Jaiswal, S.; Tan, V.; Gessert, D.; Korouri, S.; Aggarwal, N.T.; et al. Feasibility study for detection of retinal amyloid in clinical trials: The Anti–Amyloid Treatment in Asymptomatic Alzheimer's Disease (A4) trial. *Alzheimer's Dement. Diagn. Assess. Dis. Monit.* **2021**, *13*, e12199.

54. Sidiqi, A.; Wahl, D.; Lee, S.; Ma, D.; To, E.; Cui, J.; To, E.; Faisal Beg, M.; Sarunic, M.; Matsubara, J.A. In vivo Retinal Fluorescence Imaging with Curcumin in an Alzheimer Mouse Model. *Front. Neurosci.* **2020**, *14*, 713. [CrossRef] [PubMed]

55. Tadokoro, K.; Yamashita, T.; Kimura, S.; Nomura, E.; Ohta, Y.; Omote, Y.; Takemoto, M.; Hishikaw, N.; Morihara, R.; Morizane, Y.; et al. Retinal Amyloid Imaging for Screening Alzheimer's Disease. *J. Alzheimer's Dis.* **2021**, *83*, 927–934. [CrossRef] [PubMed]

56. Ryu, E.K.; Choe, Y.S.; Lee, K.H.; Choi, Y.; Kim, B.T. Curcumin and Dehydrozingerone Derivatives: Synthesis, Radiolabeling, and Evaluation for β-Amyloid Plaque Imaging. *J. Med. Chem.* **2006**, *49*, 6111–6119. [CrossRef]

57. Mutsuga, M.; Chambers, J.K.; Uchida, K.; Tei, M.; Makibuchi, T.; Mizorogi, T.; Hishikawa, N.; Morihara, R.; Morizane, Y.; Abe, K. Binding of curcumin to senile plaques and cerebral amyloid angiopathy in the aged brain of various animals and to neurofibrillary tangles in Alzheimer's brain. *J. Vet. Med Sci.* **2012**, *74*, 51–57. [CrossRef]

58. Masuda, Y.; Fukuchi, M.; Yatagawa, T.; Tada, M.; Takeda, K.; Irie, K.; Akagi, K.; Monobe, Y.; Imazawa, T.; Takegoshi, K. Solid-state NMR analysis of interaction sites of curcumin and 42-residue amyloid beta-protein fibrils. *Bioorg. Med. Chem.* **2011**, *19*, 5967–5974. [CrossRef]

59. Yanagisawa, D.; Taguchi, H.; Yamamoto, A.; Shirai, N.; Hirao, K.; Tooyama, I. Curcuminoid binds to amyloid-beta1-42 oligomer and fibril. *J. Alzheimer's Dis.* **2011**, *24* (Suppl. 2), 33–42. [CrossRef] [PubMed]

60. Yang, F.; Lim, G.P.; Begum, A.N.; Ubeda, O.J.; Simmons, M.R.; Ambegaokar, S.S.; Chen, P.P.; Kayed, R.; Glabe, C.G.; Frautschy, S.A.; et al. Curcumin inhibits formation of amyloid beta oligomers and fibrils, binds plaques, and reduces amyloid in vivo. *J. Biol. Chem.* **2005**, *280*, 5892–5901. [CrossRef]

61. Kumaraswamy, P.; Sethuraman, S.; Krishnan, U.M. Mechanistic insights of curcumin interactions with the core-recognition motif of beta-amyloid peptide. *J. Agric. Food Chem.* **2013**, *61*, 3278–3285. [CrossRef]

62. Khansari, M.M.; Garvey, S.L.; Farzad, S.; Shi, Y.; Shahidi, M. Relationship between retinal vessel tortuosity and oxygenation in sickle cell retinopathy. *Int. J. Retin. Vitr.* **2019**, *5*, 47. [CrossRef]

63. Frangi, F.A.N.W.; Vincken, K.L.; Viergever, M.A. Multiscale vessel enhancement filtering. In *International Conference on Medical Image Computing and Computer-Assisted Intervention*; Springer: Berlin/Heidelberg, Germany, 1998; pp. 130–137.

64. Ayub, L.K.A.; Ayub, J.; Ayub, S.; Akram, S.; Irshad, S. Differentiation of blood vessels in retina into arteries and veins using neural network. In Proceedings of the 2016 International Conference on Computing, Electronic and Electrical Engineering (ICE Cube), Quetta, Pakistan, 11–12 April 2016; pp. 301–306.

65. Khansari, M.M.; O'Neill, W.; Lim, J.; Shahidi, M. Method for quantitative assessment of retinal vessel tortuosity in optical coherence tomography angiography applied to sickle cell retinopathy. *Biomed. Opt. Express* **2017**, *8*, 3796–3806. [CrossRef] [PubMed]

66. Bullitt, E.; Gerig, G.; Pizer, S.M.; Lin, W.; Aylward, S.R. Measuring tortuosity of the intracerebral vasculature from MRA images. *IEEE Trans. Med. Imaging* **2003**, *22*, 1163–1171. [CrossRef]

67. Morris, J.C. The Clinical Dementia Rating (CDR): Current version and scoring rules. *Neurology* **1993**, *43*, 2412–2414. [CrossRef] [PubMed]
68. Nasreddine, Z.S.; Phillips, N.A.; Bedirian, V.; Charbonneau, S.; Whitehead, V.; Collin, I.; Cummings, J.L.; Chertkow, H. The Montreal Cognitive Assessment, MoCA: A brief screening tool for mild cognitive impairment. *J. Am. Geriatr. Soc.* **2005**, *53*, 695–699. [CrossRef]
69. Nation, D.A.; Sweeney, M.D.; Montagne, A.; Sagare, A.P.; D'Orazio, L.M.; Pachicano, M.; Sepehrband, F.; Nelson, A.R.; Buennagel, D.P.; Harrington, M.G.; et al. Blood-brain barrier breakdown is an early biomarker of human cognitive dysfunction. *Nat. Med.* **2019**, *25*, 270–276. [CrossRef]
70. Goulay, R.; Mena Romo, L.; Hol, E.M.; Dijkhuizen, R.M. From Stroke to Dementia: A Comprehensive Review Exposing Tight Interactions between Stroke and Amyloid-beta Formation. *Transl. Stroke Res.* **2020**, *11*, 601–614. [CrossRef] [PubMed]
71. O'Bryhim, B.E.; Apte, R.S.; Kung, N.; Coble, D.; Van Stavern, G.P. Association of Preclinical Alzheimer Disease With Optical Coherence Tomographic Angiography Findings. *JAMA Ophthalmol.* **2018**, *136*, 1242–1248. [CrossRef]
72. Cheung, C.Y.; Ong, Y.T.; Ikram, M.K.; Ong, S.Y.; Li, X.; Hilal, S.; Catindig, J.A.; Venketasubramanian, N.; Yap, P.; Seow, D.; et al. Microvascular network alterations in the retina of patients with Alzheimer's disease. *Alzheimer's Dement.* **2014**, *10*, 135–142. [CrossRef]
73. Csincsik, L.; MacGillivray, T.J.; Flynn, E.; Pellegrini, E.; Papanastasiou, G.; Barzegar-Befroei, N.; Csutak, A.; Bird, A.C.; Ritchie, C.W.; Peto, T.; et al. Peripheral Retinal Imaging Biomarkers for Alzheimer's Disease: A Pilot Study. *Ophthalmic. Res.* **2018**, *59*, 182–192. [CrossRef] [PubMed]
74. Jiang, H.; Liu, Y.; Wei, Y.; Shi, Y.; Wright, C.B.; Sun, X.; Rundek, T.; Baumel, B.S.; Landman, J.; Wang, J. Impaired retinal microcirculation in patients with Alzheimer's disease. *PLoS ONE* **2018**, *13*, e0192154. [CrossRef]
75. van de Kreeke, J.A.; Nguyen, H.T.; Konijnenberg, E.; Tomassen, J.; den Braber, A.; Ten Kate, M.; Yaqub, M.; van Berckel, B.; Lammertsma, A.A.; Boomsma, D.I.; et al. Optical coherence tomography angiography in preclinical Alzheimer's disease. *Br. J. Ophthalmol.* **2020**, *104*, 157–161. [CrossRef] [PubMed]
76. Bettermann, K.; Slocomb, J.E.; Shivkumar, V.; Lott, M.E. Retinal vasoreactivity as a marker for chronic ischemic white matter disease? *J. Neurol. Sci.* **2012**, *322*, 206–210. [CrossRef] [PubMed]
77. den Haan, J.; van de Kreeke, J.A.; van Berckel, B.N.; Barkhof, F.; Teunissen, C.E.; Scheltens, P.; Verbraak, F.D.; Bouwman, F.H. Is retinal vasculature a biomarker in amyloid proven Alzheimer's disease? *Alzheimers Dement.* **2019**, *11*, 383–391.

cells

Article

Retinal Protection from LED-Backlit Screen Lights by Short Wavelength Absorption Filters

Celia Sanchez-Ramos [1], Cristina Bonnin-Arias [1], Vanesa Blázquez-Sánchez [1], Victoria Aguirre-Vilacoro [1], Teresa Cobo [2], Olivia García-Suarez [3], María Jesús Perez-Carrasco [4], Cristina Alvarez-Peregrina [5,*] and José A. Vega [3,6]

[1] Grupo de Neuro-Computación y Neuro-Robótica, Facultad de Óptica y Optometría, Universidad Complutense de Madrid, 28037 Madrid, Spain; celiasr@opt.ucm.es (C.S.-R.); cbonnina@ucm.es (C.B.-A.); vblazquez@ucm.es (V.B.-S.); victoriaguirre@telefonica.net (V.A.-V.)
[2] Departamento de Cirugía y Especialidades Médico-Quirúrgicas, Universidad de Oviedo, 33006 Oviedo, Spain; teresacobo@uniovi.es
[3] Grupo SINPOS, Departamento de Morfología y Biología Celular, Universidad de Oviedo, 33006 Oviedo, Spain; garciaolivia@uniovi.es (O.G.-S.); javega@uniovi.es (J.A.V.)
[4] Departamento de Óptica II: Optometría y Visión, Facultad de Óptica y Optometría, Universidad Complutense de Madrid, 28037 Madrid, Spain; mjpc@opt.ucm.es
[5] Faculty of Biomedical and Health Sciences, Universidad Europea de Madrid, 28670 Madrid, Spain
[6] Facultad de Ciencias de la Salud, Universidad Autónoma de Chile, Santiago 7500912, Chile
* Correspondence: cristina.alvarez@universidadeuropea.es

check for
updates

Citation: Sanchez-Ramos, C.; Bonnin-Arias, C.; Blázquez-Sánchez, V.; Aguirre-Vilacoro, V.; Cobo, T.; García-Suarez, O.; Perez-Carrasco, M.J.; Alvarez-Peregrina, C.; Vega, J.A. Retinal Protection from LED-Backlit Screen Lights by Short Wavelength Absorption Filters. *Cells* **2021**, *10*, 3248. https://doi.org/10.3390/cells10113248

Academic Editors: Maurice Ptito and Joseph Bouskila

Received: 29 September 2021
Accepted: 16 November 2021
Published: 19 November 2021

Publisher's Note: MDPI stays neutral with regard to jurisdictional claims in published maps and institutional affiliations.

Abstract: (1) Background: Ocular exposure to intense light or long-time exposure to low-intensity short-wavelength lights may cause eye injury. Excessive levels of blue light induce photochemical damage to the retinal pigment and degeneration of photoreceptors of the outer segments. Currently, people spend a lot of time watching LED screens that emit high proportions of blue light. This study aims to assess the effects of light emitted by LED tablet screens on pigmented rat retinas with and without optical filters. (2) Methods: Commercially available tablets were used for exposure experiments on three groups of rats. One was exposed to tablet screens, the other was exposed to the tablet screens with a selective filter and the other was a control group. Structure, gene expression (including life/death, extracellular matrix degradation, growth factors, and oxidative stress related genes), and immunohistochemistry in the retina were compared among groups. (3) Results: There was a reduction of the thickness of the external nuclear layer and changes in the genes involved in cell survival and death, extracellular matrix turnover, growth factors, inflammation, and oxidative stress, leading decrease in cell density and retinal damage in the first group. Modulation of gene changes was observed when the LED light of screens was modified with an optical filter. (4) Conclusions: The use of short-wavelength selective filters on the screens contribute to reduce LED light-induced damage in the rat retina.

Keywords: retinal light injury; LED screen; optical filter; retinal protection

1. Introduction

As early as 1966 it was suggested that exposure to low-intensity short-wavelength light for a long time can induce retinal damage [1], with the action spectrum (400–440 nm) of blue light the most dangerous [2,3], able to trigger or exacerbate macular and retinal damage [4,5]. Both human and animal studies suggest that excessive levels of blue light induces immediate photochemical damage to the retinal pigment epithelial cells (RPE), photoreceptors, and ganglion cells [2,3,6–10]. Thus, the phototoxicity of blue light may contribute to the progression and severity of age-related macular degeneration (AMD) and vision loss.

Nowadays, light-emitting diodes (LEDs) are gradually becoming the majority of the domestic light sources, replacing conventional ones [2,11,12]. The most commonly used method to produce white LED lights is the use of blue diodes covered with a phosphor compound; thus, one of the biggest risks of LEDs is its emission spectrum, since it contains a large blue light component [2,12,13]. Additionally, there is time-related degradation of phosphorus which leads to a progressively increasing release of short wavelengths (blue light) [9]. Since LEDs do not directly emit ultraviolet (UV) and infrared (IR) wavelengths, the blue light risk is the main ones to focus on when considering LEDs and LED systems [12].

Recent studies assessing LED effects on the retina using ambient LED lighting have demonstrated LED phototoxicity [2,3,9,11,12,14,15]. To our knowledge, no data are available regarding backlit LED screens. People spend increasing amounts of time viewing backlit LED screens, such as personal computers, tablets, smartphones, etc., at short distances. Thus, this raised a range of public concerns about their potential risks as retinal hazards [16–18]. In 2017, Lin et al. investigated the effects of periodic exposure to the blue LED in rats showing marked retinal damage [19] characterized by an accumulation of macrophages (identified because express ionized calcium binding adapter molecule-1), drusen-like materials around the outer segments of the photoreceptors, and finally degeneration of the photoreceptors [7]. Additionally, LED induce strong damage in rat retinal pigment epithelium (RPE) characterized by the breakdown of the blood–retinal barrier and the induction of necrotic cell death [20]. Furthermore, Balb/c mice exposed to LED light show a reduction of outer nuclear layer (ONL), increase in TUNEL-positive apoptotic nuclei, changes in several differentially expressed genes, and downregulation of ubiquitin and autophagy function [21]. At the basis of those changes is the oxidative stress induced by LED [22] which affects mitochondria and triggers mitochondrial death signaling pathways [17,23], as well as accumulation of damaged lysosomes and subsequent lysosomal cell death [24].

On the other hand, the deleterious effects of LED lights on the retina can be prevented partially by blue-light-blocking films [25], or more effectively by lenses with a brown or gray tint [26].

Previous studies from our research group have demonstrated that exposure to blue light reduces the number of retinal cells, upregulates genes related to cell death, downregulates genes involved in cell survival, increases the activity of genes involved in extracellular matrix degradation, and alters the expression of growth factor that participates in cell maintenance and survival. These effects can be prevented or reversed by filtering blue light [27,28]. Therefore, this study aims to evaluate the effects of light emitted by back-lit LED tablet screens on the retina of pigmented rats and determine whether or not they can be modified by partially filtering out the emitted short-wavelengths of the visible spectrum. Based on the multiple pathways that can participate in LED-induced retinal damage, and our previous experience and studies, the following genes were analyzed: genes involved in cell survival and death (Bcl-2; Bcl-XL; Bax; Bak; Bcl-XS, Caspases-3 and 9), genes related to the extracellular matrix (ECM) turnover (MMPs-2 and 9, ADAMTS-12 and 14, TIMPs-1 and 2), genes related to growth factors (BDNF-Trk-B system, VEGF-VEGFr-2 system), and inflammation (TNF-α, SODs-1 and 2). Furthermore, immunohistochemistry was used to study determined the protein product of the genes showing greater variation after exposure to LED. The results could support the effects of short-wavelengths emitted by back-lit LED screens on gene regulation and assess the efficacy of the filter in removing excessive blue light radiation, thus potentially providing a retinal photoprotective effect.

2. Materials and Methods

2.1. Animals and Rearing Conditions

Male Lister-Hooded rats obtained from Harlan Laboratories Models, S.L. (Barcelona, Spain), were housed at the bioterium of the Medical School of the Universidad Complutense de Madrid (UCM, Madrid, Spain). The animals were kept in a dark environment for 14 days to remove the effect of light exposure from their previous breeding location, with access to food and water ad libitum. The use of rats complied with the Statement for the Use of

Animals in Ophthalmic and Vision Research (ARVO 2013). Animals were treated humanely and with regard to the alleviation of suffering. This study has the approval of the Animal Experimentation Committee of the UCM and the Department of Health of the Comunidad de Madrid, Spain (Reference PROEX 310/15).

Rats were divided into three groups, with 12 animals per group. Group 1 was exposed to the light emitted by the LED-backlit tablet screens; group 2 was exposed to the light emitted by the LED-backlit tablet screens with a selective short-wavelength absorption filter adhered to the screen; the control group was unexposed to LED-backlit tablet screens. Animals were housed in clear Makrolon Polycarbonate cages, 58 × 38 × 18 cm in size, with two animals per cage.

2.2. Light Source and Exposure

Light emitted by blank LED-backlit tablet screens, with size 231 × 147.2 × 8.7 mm (Cristal IPS multitouch 8.9", 1920 × 1200 px Full HD), set at full brightness was used. The selective short-wavelength absorption filter used for group 2 exposure conditions was Reticare® Intensive (Tecnología Sostenible y Responsable S.L., Madrid, Spain). Before starting the study, the filter was adhered to each tablet screen, following the manufacturer's instructions.

The photo exposure process was designed aiming at simulating the conditions of use of touch screens by children. The screens were placed surrounding the cages, at a distance between 4.72 and 5.90 inches of the animal's eyes, leaving the upper and lower areas without screens. As for the devices used for photo exposure, tablets were used instead of smartphones for two reasons: firstly, because the time of use of tablets, in general, is longer than smartphones; and secondly, because the tablets emit with less intensity (roughly half) than mobile phones.

Tablet screens were set at 10 cm from each of the four cage walls. For groups 1 and 2, each cage contained six screens (without or with filter, respectively), one screen in each of the two short walls, and two screens in each of the two long walls. No screens were attached to cage ceilings. Figure 1 shows the characteristic of the light LED screen emission and the illuminance (lux) measured inside the study cages, as well as the transmittance curve of selective short-wavelength absorption filter (Reticare® Intensive). LED-backlit screen tablet light emission (with and without filter) was measured with an Ocean Optics USB2000 + Spectrometer.

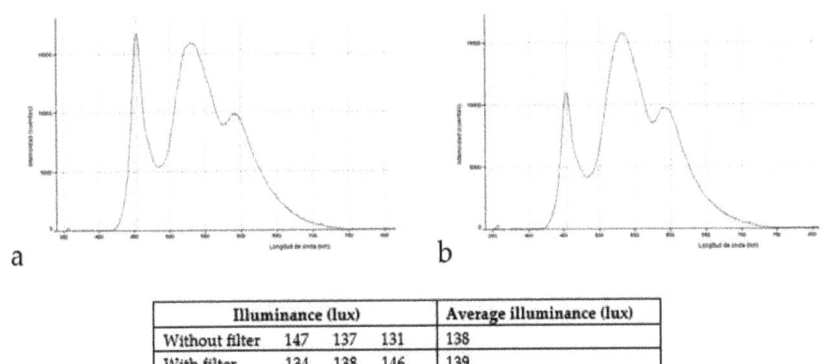

Illuminance (lux)			Average illuminance (lux)
Without filter	147	137	131	138
With filter	134	138	146	139

Figure 1. Characteristic of the light LED screen emission and the illuminance (lux) measured inside the study cages; (**a**) without optical filter; (**b**) with the optical filter.

Animals of groups 1 and 2 were exposed to 8-hr dark/16-hr LED-backlit screen tablet light cycles for three months. All animals, including a control group, were sacrificed at the end of the exposure period.

2.3. Tissue Collection, RNA Extraction, and qPCR

The animals were sacrificed with an overdose of pentobarbital sodium (200 mg/kg, ip) and eyes were removed. The left eyes of each animal were frozen at $-80\,^{\circ}$C and maintained until quantitative polymerase chain reaction (qPCR) analysis. Right eyes were fixed in 4% formaldehyde for 24 h, then washed in tap water, and routinely processed for paraffin inclusion, and used for structural and immunohistochemical studies.

Total RNA was extracted from the whole retina using a commercial kit (Trizol Reagent, Invitrogen, Carlsbad, CA, USA), following the manufacturer's instructions. After precipitation and cold ethanol washing, RNA was dried and dissolved in an appropriate volume of Tris–EDTA buffer (10 mM Tris–HCl pH 8.0 and 1mM EDTA-Na2). Each sample was treated with 1U of DNase I for 1-h at 37 $^{\circ}$C to digest genomic DNA. RNA was precipitated, washed, and dissolved again in the same buffer. RNA solution was quantified at 260 nm (Biomate 3, Thermo Electron Corporation, Waltham, MA, USA) and its purity was assessed by the ratio of 260/280 nm readings. We used the High Capacity cDNA Archive kit (Applied Biosystems, Foster City, CA, USA), random hexamers, and 10 μg of total RNA to make cDNA following the manufacturer's instructions. Then, 1 μg of cDNA was used to detect the expression of the genes analyzed in this qPCR study.

Quantitative PCR was performed using 1 μg of the cDNA, and the primers used to detect the investigated genes, and the β-actin were designed based upon the published mRNA sequences for Rattus norvegicus (Table 1; GenBank accession numbers are included).

Table 1. Genes sequences analyzed in this study.

	Accession Number	Forward	Reverse
Life/Death			
Bcl-2	NM_016993	5′-ATCTTCTCCTTCCAGCCTGA3′	5′-CTGGACATCTCTGCAAAGTC3′
Bcl-xl	BC094213	5′-AAGAAACTGAACCAGAAAGG3′	5′-TAGATCACTGAATGCTCTCC3′
Bax	NM_017059	5′-CTACAGGGTTTCATCCAGGA3′	5′-ATCCACATCAGCAATCATCC3′
Bak	AF259504	5′-CTATTTAAGAGCGGCATCAG3′	5′-ATTACCACTACACTCAGGAT3′
Bcl-xs	AF279286	5′-CGGAGAGCATTCAGTGAT 3′	5′-CCAGCAGAACTACACCAG 3′
Caspase 3	NM_012922	5′-GTCATGGAGATGAAGGAGTA3′	5′-AGAGTAAGCATACAGGAAGT3′
Caspase 9	NM_031632	5′-GTCTGTACTCCAGGGAAGAT3′	5′-TTAGCAGTCAGGTCGTTCTT3′
ECM degradation			
MMP2	NM_031054	5′-GCAATACCTGAACACTTTCT3′	5′-ATCTGATTCTTGTCCCACTT3′
MMP9	NM_031055	5′-GACTACGACACAGACAGAAA3′	5′-GAGTAGGACAGAAGCCATAC3′
ADAMTS-12	NM_001106420	5′-CTGCCAGAATACCACATAGT3′	5′-TATCTCCTCTCCACGACATA3′
ADAMTS-14	NM_001107636	5′-GCTACCTCCTATCCTACAAT3′	5′-CTTGGTCTTGCAGAAGTATG3′
TIMP1	U06179	5′-CCACCTTATACCAGCGTTAT3′	5′-CTGGGACTTGTGGACATATC3′
TIMP22	NM_021989	5′-AGATGTTCAAAGGACCTGAC3′	5′-CTTCTTCTGGGTGATGCTAA3′
Growth factors			
BDNF	NM_001270630	5′-GTGACAACAATGTGACTCCA3′	5′-CATTCACGCTCTCCAGAGTC3′
TrkB	AY265419	5′-CCAGAGAACATCACCGAAAT3′	5′-ATCAGGTCAGACAAGTCAAG3′
VEG	NM_001110333	5′-GTATATCTTCAAGCCGTCCT3′	5′-CATTCACATCTGCTATGCTG3′
VEGFR2	NM_013062	5′-GGCAAATACAACCCTTCAGA3′	5′-CCGATAGAAGCACTTGTAGG3′
TNFα	NM_012675	5′-GCTCTTCTGTCTACTGAACT3′	5′-CTTTGAGATCCATGCCATTG3′
Oxidative stress			
SOD1	NM_017050	5′-CTTTGAGATCCATGCCATT3′	5′-ACACGATCTTCAATGGACAC3′
SOD2	NM_017051	5′-GAGAACCCAAAGGAGAGTTG3′	5′-CTGAAGATAGTAAGCGTGCT3′
β-actin	NM_031144	5′-ATCGTGCGTGACATTAAAGA3′	5′-GATGCCACAGGATTCCATAC3′

Homemade TaqMan probes were labeled at the 50 with 60 FAM fluorochromes for all investigated sequences, and VIC fluorochrome for β-actin, while the 3′ ends were labeled with the Minor Groove Binder (MGB) quencher. The PCR reactions were performed using the TaqMan Universal PCR Master Mix (Applied Biosystems, Foster City, CA, USA) using 5 pmol of each primer and 9 pmol of both target and β-actin probe. The assays were performed in triplicate in independent experiments using a 7500 PCR real-time system (Applied Biosystems), and quantification was calculated using the 2-ΔΔCt algorithm against β-actin and expressed as the n-fold difference compared to an arbitrary calibrator, chosen as a higher value than ΔΔCt.

Average values obtained in control group animals were compared to those exposed to LED-backlit tablet screen light, without and with filter (group 1 and 2, respectively), and the results were expressed as fold difference compared to control (relative expression).

2.4. Structural and Immunohistochemical Studies

The structure of the retinal in all animal groups evaluated by staining representative formol fixed, paraffin embedded sections (10 per eye and animal) with hematoxylin & eosin. The sections were then scanned by an SCN400F scanner (Leica Biosystems™, Newcastle, UK), and the scans were computerized using SlidePath Gateway LAN software (Leica, Leica Biosystems™). Then, the whole thickness of the retina was measured, evaluating in 10 points the distance from the inner to the other surface of the retina.

Deparaffinized and rehydrated sections were processed for detection SYN using the EnVision antibody complex detection kit (DakoCytomation, Copenhagen, Denmark), following supplier's instructions. Briefly, the endogenous peroxidase activity and non-specific binding were blocked, and the sections were then incubated overnight at 4 °C with the primary antibodies included in the Table 2. Subsequently, sections were rinsed incubated with Dako EnVision System labeled polymer-HR anti-mouse IgG (DakoCytomation) for 30 min at room temperature. Finally, sections were washed, and immunoreaction visualized using 3-3′-diaminobenzidine as chromogen. To ascertain structural details some sections were counterstained with Mayer's hematoxylin, dehydrated, and mounted with Entellan® (Merk, Dramstadt, Germany).

Table 2. Antibodies used in the study.

Antigen (Clone)	Origin	Dilution	Supplier
Life/Death			
Bcl-2 **Cat #** PA5-11379	Rabbit	1:200	ThermoFisher [1]
Bcl-xl **Cat #** PA5-17805	Rabbit	1:100	ThermoFisher [1]
Caspase 3 **Cat #** PA5-77887	Rabbit	1:400	ThermoFisher [1]
Caspase 9 ab25758	Rabbit	1:200	Abcam [2]
ECM degradation			
MMP2 (clone MMP2/2C1) LS-C2814-100	Mouse	1:100	LifeSpan Biosciences [3]
MMP9 ab38898	Rabbit	1:200	Abcam [4]
Growth factors/Cytokines			
TrkB catalog # sc-12	Rabbit	1:100	Santa Cruz Biotechnol [5]
VEGFR2 PA5-16487	Rabbit	1:100	Invitrogen [6]
Oxidative stress			
SOD1 Ab13498	Rabbit	1:200	Abcam [4]
SOD2 ab13533	Rabbit	1:100	Abcam [4]

[1] Watham, MA, USA; [2] Abcam: Cambridge, UK; [3] Seattle, WA, USA; [4] Santa Cruz, CA, USA; [5] Dallas, TX, USA; [6] Waltman, MA, USA.

The variations in the intensity of immunostaining developed for each investigated antibody was evaluated in five sections per animal separated by 100 μm. Each section was scanned with an SCN400F scanner (Leica Biosystems™) and annotated using SlidePath Gateway LAN software (Leica Biosystems™). A 1 mm^2 grid was then applied randomly onto 2 × 500 μm enlarged images in 4 non-overlapping fields (4 mm^2 per section; 20 mm^2 per subject), and the free nerve endings and sensory corpuscles within the grid were counted by two independent observers. The results are expressed as absolute numbers for the densities of the sensory corpuscles and free nerve endings per cm^2. The intensity of immunostaining developed with the different retinal layers was evaluated in arbitrary units of grey levels ranging from 1 (black) to 256 (white) using an image analysis system (MIP System, Servicio de Análisis de Imágenes, Universidad de Oviedo, Spain). The intensities were therefore divided into four groups (64 units of grey level each), referred in the text and tables as strong (1–64, ++++), high (65–128, +++), intermediate (129–192), and low (193 to level of the background in control sections, +). Intensities of grey higher than those of background of the corresponding control sections were considered unreactive. Although we attempted to process all samples identically, there may have been variations in the intensity of the final immunoreaction due to technical aspects, i.e., differences in penetration of the fixative. On the other hand, for the immunohistochemical study, the retinal segment and orientation of the sections were not taken into account, and therefore the thickness of the retina may appear in the images with greater thickness in the experimental animals than in the controls.

2.5. Statistical Analysis

Significant differences among the three groups were assessed with the Kruskal–Wallis H test, and p-values < 0.05 were considered statistically significant (marked in the figures as * p < 0.05, ** p < 0.01).

3. Results

The study was carried out with 36 male Lister-Hooded rats obtained from Harlan Laboratories Models, S.L. (Barcelona, Spain), housed at the bioterium of the Medical School of the Universidad Complutense de Madrid (UCM, Madrid, Spain), with food and water ad libitum, kept in a dark environment for 14 days to remove the effect of light exposure from their previous breeding location. Rats were divided into 3 groups, with 12 animals per group. Group 1 was exposed to the light emitted by the LED-backlit tablet screens; Group 2 was exposed to the light emitted by the LED-backlit tablet screens with a selective short-wavelength absorption filter adhered to the screen; the control group was unexposed to LED-backlit tablet screens.

3.1. Structural Study

The retinal structure was studied in the three groups of animals. In those exposed to LED-backlit tablet screen light, there was a significant decrease in the whole retinal thickness (reduction of 23.82 ± 6.21%), apparently due to a reduction in the number of cells in both the outer and inner nuclear layers. The whole thickness of the retina was measured by evaluating in 10 points the distance from the inner to the other surface of the retina. These structural changes were almost reversed by the selective short-wavelength absorption filter (Figure 2).

3.2. Gene Expression Study

3.2.1. Life/Death Cell-Related Genes

Different genes related to cell survival and cell death were analyzed. The eyes of the animals exposed to LED-backlit tablet screen light without filter (group 1) showed a down-regulation of the genes encoding anti-apoptotic proteins, especially Bcl-2 (−13.2-fold), and to a lesser extent Bcl-XL (−3.5-fold), as Figure 3a shows. Selective short-wavelength absorption almost eliminated these responses. Regarding genes directly or indirectly involved in cell death, exposition to the LED-backlit tablet screen light without filter resulted in a moderate up-regulation of Bax and Bak (5- and 3.3-fold, respectively) and a marked up-regulation of Bcl-XS (16.3 folds), as Figure 3b shows. Similarly, in this group (group 1), there was an up-regulation of both caspase-3 (6-fold) and caspase-9 (11.8-fold) genes. This scenario changes after the use of the selective short-wavelength absorption filter. In fact, in this experimental situation, Bak and caspase-3 were undetectable, and the expression of Bax, Bcl-XL, and caspase-9 remained up-regulated for the controls, but the expression was significantly reduced for the group of animals exposed to LED-backlit screens, as Figure 3c shows.

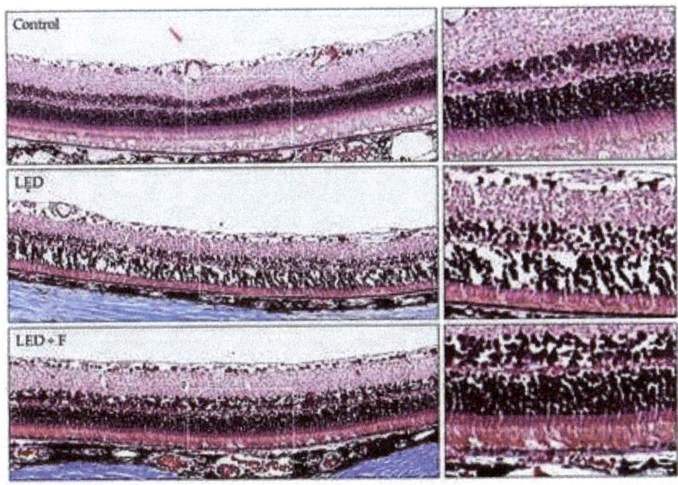

Figure 2. Retinal structure from control rats, rats exposed to LED Screen (LED-S), and rats exposed to LED screen with a protective filter (LED-S-P).

3.2.2. Extracellular Matrix Degradation: MMP/TIMPs System and ADAMTS Genes

Figure 4 shows that exposure to LED-backlit tablet screen light without filter (group 1) entails an increase in MMP-2 and MMP-9 mRNA expression, suggesting an increase in the extracellular matrix (ECM) turnover mediated by these proteases. As for changes in TIMPs expression, exposure to light emitted by LED-backlit tablet screens caused a decrease of both TIMP-1 and TIMP-2. The use of the selective short-wavelength absorption filter almost reversed the decrease, although in levels below control group values, as Figure 4c shows. Regarding ADAMTS-12, no differences were found between the control group versus either group 1 or group 2. On the other hand, exposure to LED-backlit tablet screen light without filter (group 1) increased the levels of ADAMTS-14 expression, whereas exposure effects were reverted with the filter (group 2), as Figure 4b shows.

3.2.3. BDNF/TrkB and VEGF/VEGFR-2 System, and TNF-α

Figure 5a shows how exposure to LED-backlit tablet screen light did not modify expression levels of BDNF either with or without the selective short-wavelength absorption filter, while TrkB gene expression increased significantly. It is important to highlight that, with the filter, TrkB gene expression persisted at a high level, although below the level observed in group 1 (without filter).

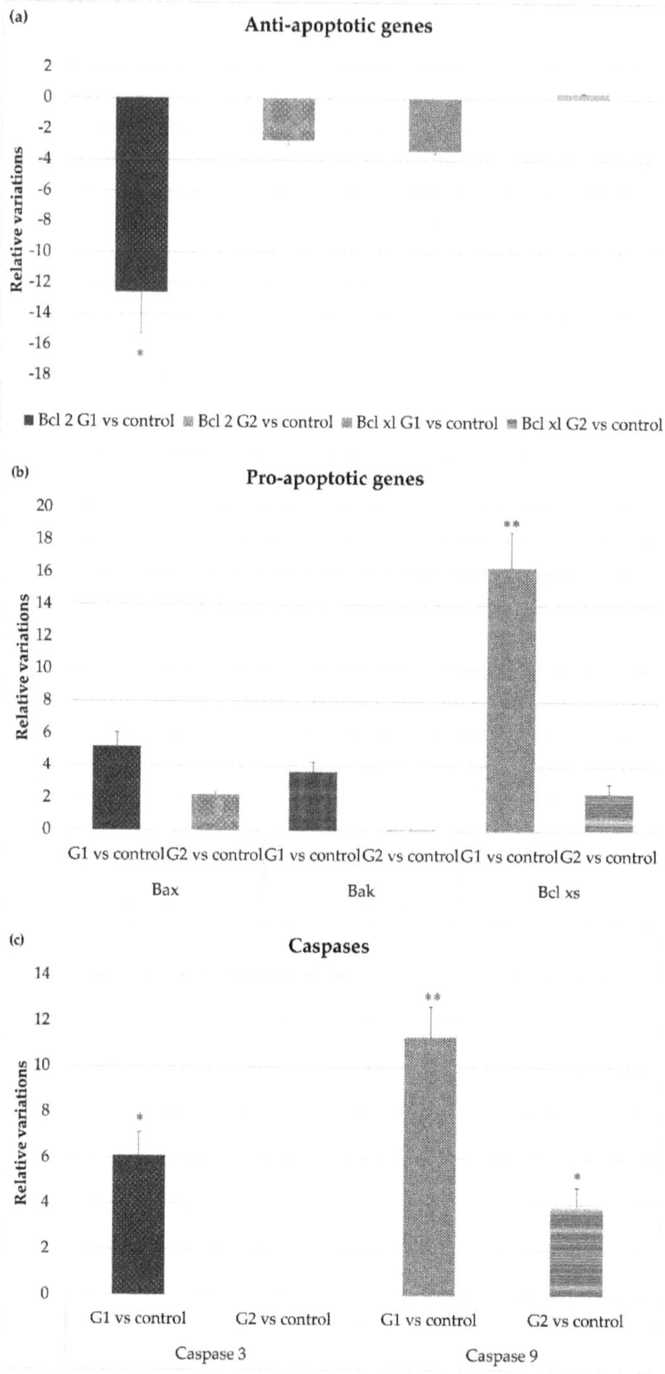

Figure 3. Gene expression comparisons between group 1 vs. control and group 2 vs. control. (**a**) Anti-apoptotic gene expression, (**b**) pro-apoptotic gene expression, and (**c**) gene expression of apoptotic-related enzymes. * $p < 0.05$; ** $p < 0.01$.

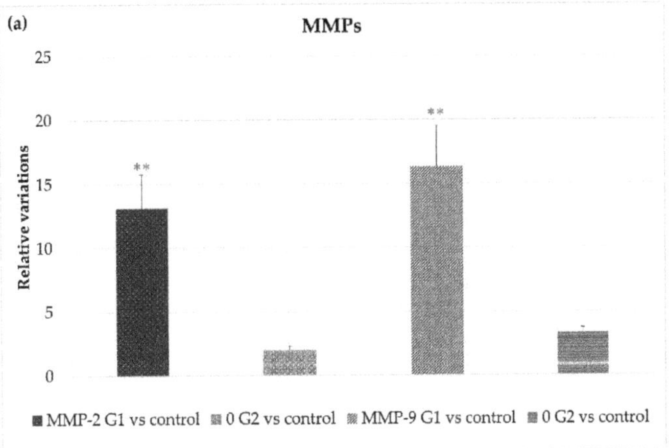

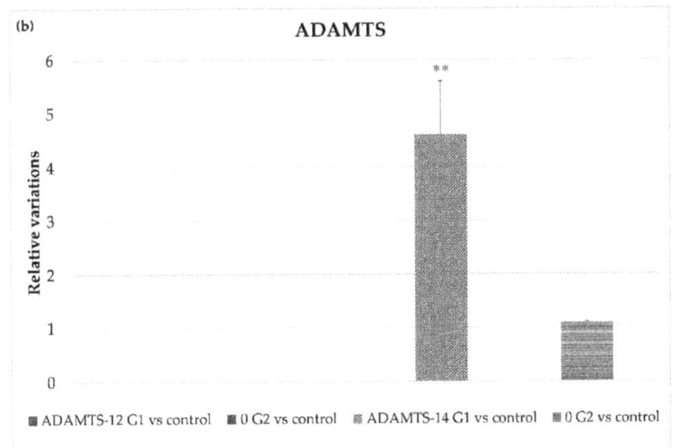

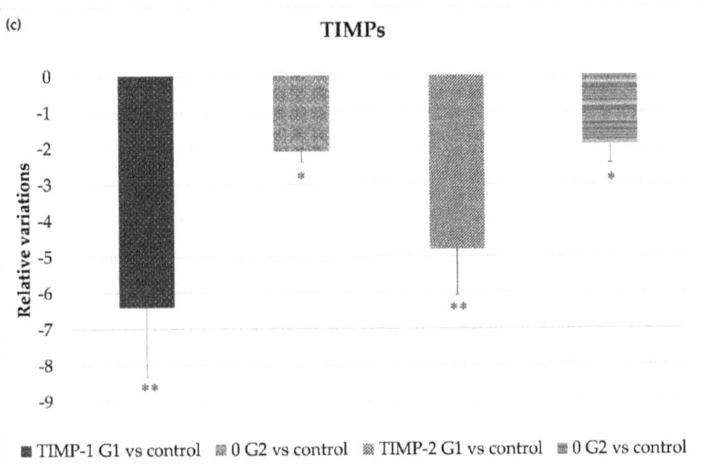

Figure 4. Variations in gene expression in both, experimental vs. control groups of (**a**) MMP2 and MMP9; (**b**) ADAMTS-12 and ADMSTS-14; and (**c**) TIMP1 and TIMP-2, * $p < 0.05$; ** $p < 0.01$.

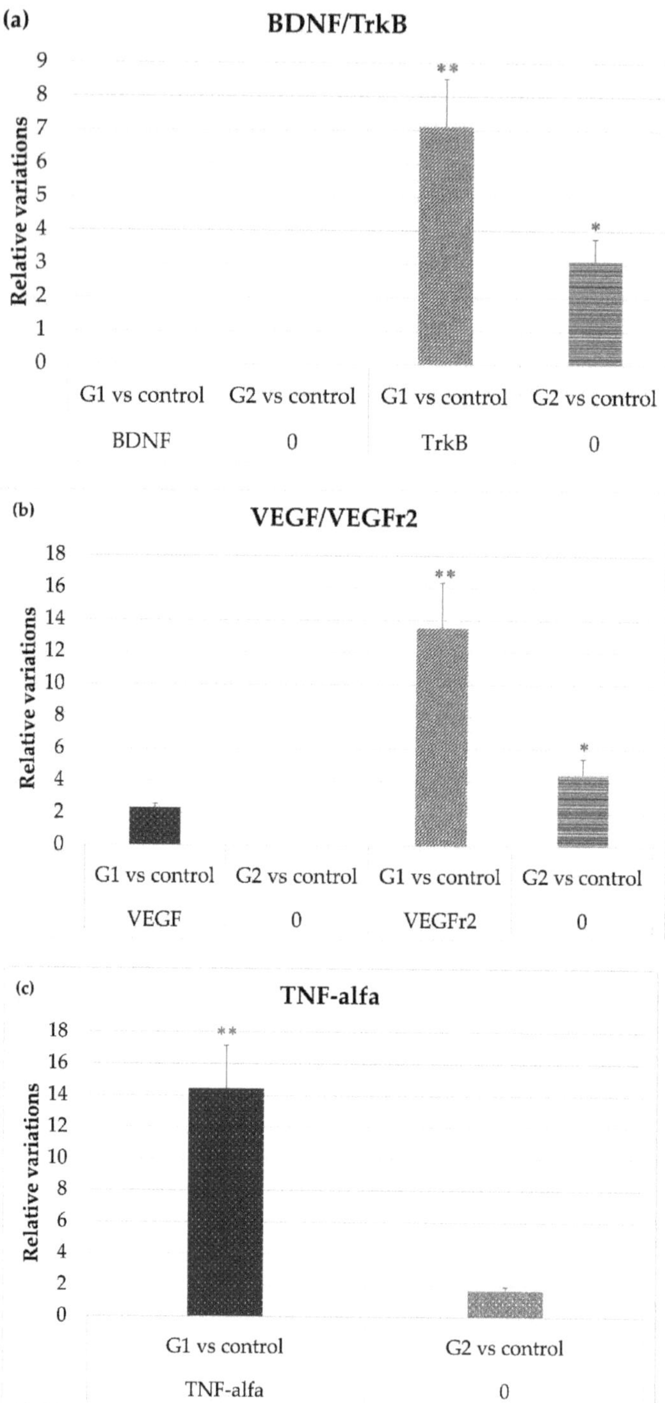

Figure 5. Variation of the expression in both experimental vs. control groups, of (**a**) TrKB; (**b**) VEHG and VEGFr2; and (**c**) and TNF-α. * $p < 0.05$; ** $p < 0.01$.

Figure 5b shows that after exposure to LED-backlit tablet screen light (group 1), levels of VEGF mRNA increase moderately, while levels of its receptor increased by nearly 14-fold. These effects were highly attenuated using the selective short-wavelength absorption filter (group 2): VEGF levels were practically identical to controls, and VEGFr2 were lower, although above controls.

Finally, Figure 5c shows that TNF-α mRNA expression levels increased up to 14-fold in group 1 rats (without filter) but were like controls in group 2 (with filter).

3.2.4. Oxidative Stress

Figure 6 shows how the expression of superoxide dismutase-1 and -2 were strongly up-regulated after exposure to LED-backlit tablet screen light (group 1). These effects are highly attenuated, but increased for the controls, using the selective short-wavelength absorption filter.

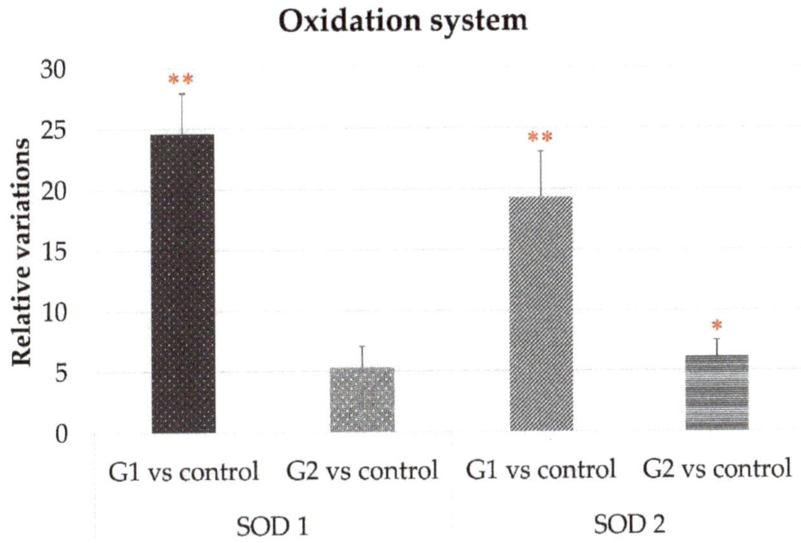

Figure 6. Variation in the expression of SOD1 and SOD2 in both experimental vs. control groups. * $p < 0.05$; ** $p < 0.01$.

Table 3 summarizes the results of the variations in the expression of the genes analyzed, as a complement of the figures.

3.3. Immunohistochemistry

Using immunohistochemistry, we analyzed the detection in retinal sections of the protein products of the genes whose expression were most affected by the experimental conditions to which the two groups of animals. The increases or decreases in the intensity of immunostaining paralleled that observed in the gene expression. The immunolabelling for Bcl-2 was greatly reduced in animals exposed to LEDs and partially recovered when light was filtered (Figure 7), whereas no notable variations were observed for Bcl-X (Figure 7). The detection of caspase-3 showed an increase in immunostaining in the group subjected to LED light and distribution and intensity similar to the controls in animals subjected to filtered light (Figure 7).

Table 3. Changes in the gene expression in the two established experimental groups in relation to the controls.

Gene Expression	Group 1 Average vs. Control	Group 2 Average vs. Control
Life/Death		
Bcl-2	-12.6 ± 2.71	-2.7 ± 0.3
Bcl-xl	-3.4 ± 0.4	0.4 ± 0.01
Bax	5.2 ± 0.86	2.2 ± 0.2
Bak	3.6 ± 0.63	0.1 ± 0.0
Bcl-xs	1623 ± 2.22	2.3 ± 0.61
Caspase 3	6.1 ± 1.03	0.0 ± 0.0
Caspase 9	11.3 ± 1.37	3.9 ± 0.83
ECM degradation		
MMP2	13.1 ± 2.65	2 ± 0.33
MMP9	16.3 ± 3.21	$3.3. \pm 0.39$
ADAMTS-12	0.0 ± 0.0	0.0 ± 0.0
ADAMTS-14	4.6 ± 0.99	1.1 ± 0.0
TIMP1	-6.4 ± 1.91	-2.1 ± 0.3
TIMP22	-4.8 ± 1.27	-1.9 ± 0.52
Growth factors		
BDNF	0.0 ± 0.0	0.0 ± 0.0
TrkB	7.1 ± 1.42	3.1 ± 0.66
VEG	2.3 ± 0.25	0.0 ± 0.0
VEGFR2	11.5 ± 2.84	4.4 ± 0.98
TNFα	14.4 ± 2.72	1.7 ± 0.23
Oxidative stress		
SOD1	24.6 ± 6.33	5.3 ± 1.75
SOD2	19.3 ± 3.77	6.1 ± 1.42

Regarding the two metalloproteases investigated, MMP2 did not show positive immunoreaction in the control animals, while showing a slight immunopositivity in the animals subjected to LEDs that was reduced in the group of filtered LED light (Figure 8). On the other hand, the immunoreaction for MMP9 was very similar in the three groups (Figure 8).

The immunoreactivity for TrkB, the high affinity receptor for the neurotrophin BDNF, was increased in animals exposed to LED, especially in the photoreceptors, but also in the inner nuclear layer and ganglionic cells layer, with levels similar to those of the animal controls when filtering LEDs (Figure 9).

The immunoreactivity for SOD1 increased in the photoreceptors layer of LED exposed rats and returned to similar levels of the control after LED filtering. However, SOD2 immunoreactivity increased in the photoreceptors, inner nuclear and ganglionic cell layers in LED-exposed animals, and the pattern of expression was not reverted by filtering (Figure 10).

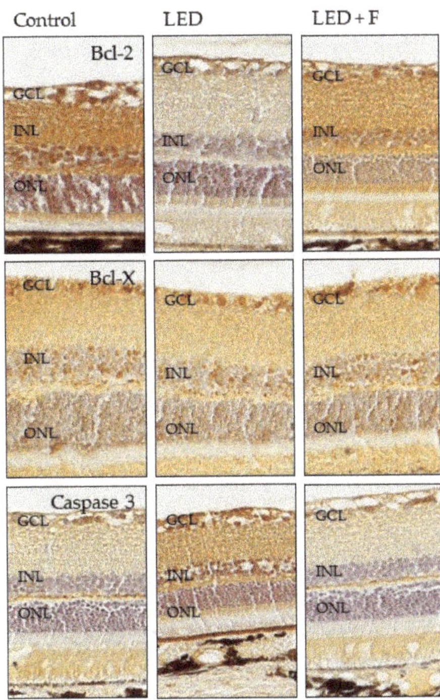

Figure 7. Immunohistochemical detection of Bcl-2 (**upper panel**), Bcl-X (**central panel**) and Caspase-3 (**lower panel**) in the three groups of rats investigated. LED: animals exposed to LED-backlit screen for 3 months; LED + F: animals exposed to filtered LED-backlit screen for 3 months. GCL: ganglionic cells layer; INL: inner nuclear layer; ONL: outer nuclear layer.

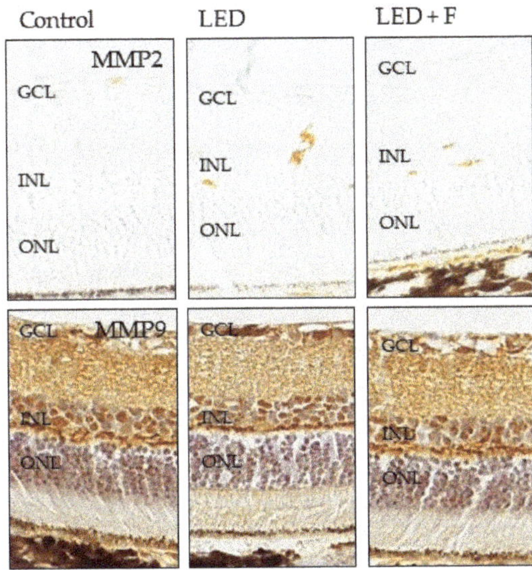

Figure 8. Immunohistochemical detection of MMP2 (**upper panel**), and MMP9 (**lower panel**) in the three groups of rats investigated. LED: animals exposed to LED-backlit screen for 3 months; LED + F: animals exposed to filtered LED-backlit screen for 3 months. GCL: ganglionic cells layer; INL: inner nuclear layer; ONL: outer nuclear layer.

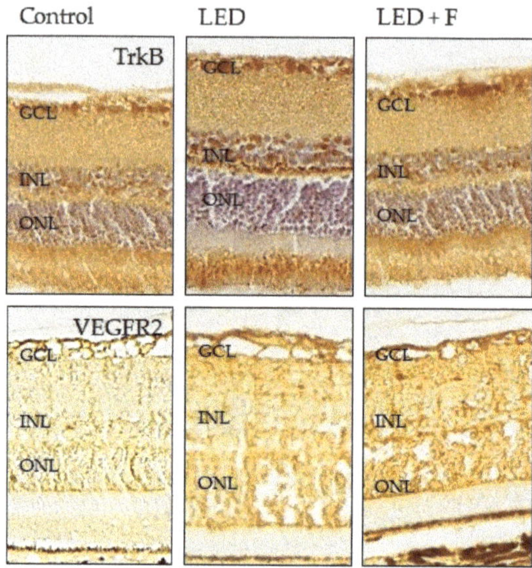

Figure 9. Immunohistochemical detection of TrkB (**upper panel**), and VEGFR2 (**lower panel**) in the three groups of rats investigated. LED: animals exposed to LED-backlit screen for 3 months; LED + F: animals exposed to filtered LED-backlit screen for 3 months. GCL: ganglionic cells layer; INL: inner nuclear layer; ONL: outer nuclear layer.

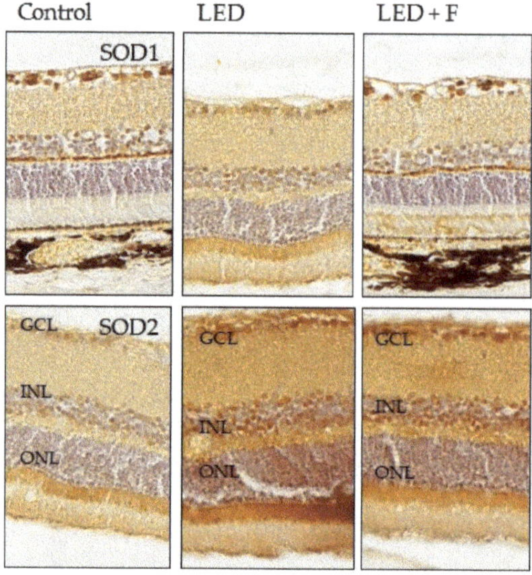

Figure 10. Immunohistochemical detection of SOD1 (**upper panel**), and SOD2 (**lower panel**) in the three groups of rats investigated. LED: animals exposed to LED-backlit screen for 3 months; LED + F: animals exposed to filtered LED-backlit screen for 3 months. GCL: ganglionic cells layer; INL: inner nuclear layer; ONL: outer nuclear layer.

The results of the quantitative study performed on immunohistochemical sections are in Table 4.

Table 4. Variations in the intensity of immunostaining in the control and the two experimental groups.

Retinal Layer	Protein	Control	LED	LED + F
Photoreceptors	Blc2	+/++	−/+	+
	BclX	−	−	−
	Caspase 3	+	+/++	+
	MMP9	−	−/+	−
	TrkB	+/++	+++	++
	VEGFR2	−	−	−
	SOD1	−/+	++	+
	SOD2	+	+++	+/++
ONL	Blc2	++	−	−/+
	BclX	+/++	+/++	+/++
	Caspase 3	−/+	+/++	−/+
	MMP9	+/++	+/++	+/++
	TrkB	+/++	+/++	+/++
	VEGFR2	+	++	++
	SOD1	+	+	+
	SOD2	+	++/+++	++
INL	Blc2	++	−	−
	BclX	+/++	+/++	+/++
	Caspase 3	−	+/++	−
	MMP9	++	++/+++	++/+++
	TrkB	+/++	+++	++
	VEGFR2	+	++	+/++
	SOD1	−/+	−/+	−/+
	SOD2			
Ganglionic cells	Blc2	+++	+	++
	BclX	++	++	++
	Caspase 3	+/++	++/+++	−/+
	MMP9	++	+++	+++
	TrkB	++/+++	+++	+++
	VEGFR2	++	++	++
	SOD1	++/+++	+	++/+++
	SOD2	+/+	+++	+++

4. Discussion

The present study was designed to investigate the effects of the exposure to LED-backlit tablet screen light on the structure, gene expression, and protein localization of the retina of a rat model. Moreover, we have analyzed whether or not those effects could be reversed partial or using a selective short-wavelength absorption filter. The genes and proteins investigated were related to cell survival and cell death, as well as some proteases, and protease-blockers involved in the turnover of the ECM, growth factors previously known to be affected by light exposure, one involved in inflammation, and others related to oxidative stress. This approach is necessary because of the multiple factors involved in phototoxicity [15]. Because the results of the gene expression and immunohistochemistry were in parallel, they are discussed together.

In 2001, Dawson showed that blue LEDs (460 nm) and argon lasers (458 nm) induced retinal damage with corneal irradiances of 10 J/cm^2 [29]. Ueda et al. also observed detrimental effects in the macula with blue LEDs (465 nm) [30]; besides, Shang et al. showed that the spectral range distribution of blue-white LEDs contains a significant fraction of short-wavelengths that induced irreversible retinal neuronal cell death in rats; they exposed Sprague-Dawley rats to white and blue LED light (750 lux) for 28 days, finding an increase in free radical production in the LED-exposed group [9]. Finally, Krigel et al. used albino Wistar and pigmented Long Evans rats exposed for 1–28 days to 500–6000 lux LEDs (cold white, blue, and green) [15]. The authors found that the blue component of the white-LED reduced photoreceptor layer thickness and induced retinal toxicity. In previous in vitro studies, Chamorro et al. exposed human retina cells to LED lighting in 12-h dark/12-h light cycles affected RPE cell growth and induced cellular stress, increasing levels of reactive oxygen species, DNA damage, and apoptotic cells [3]. Additionally, when a selective blue-light absorption filter was added, phototoxic damage in human RPE cells was reduced, providing a retinal photoprotector effect. However, few studies have investigated long-term and low luminance light-induced retinal phototoxicity and, to our knowledge, all have been carried out using different ambient LED lights. In contrast to previous studies, we investigated the induced retinopathy in a rat model due with long-term, low-intensity exposure to commercially available LED-backlit tablet light. Thus, the main objective of the current study was to assess the long-term effects on the retina of pigmented rats by light emitted by a backlit commercially-available digital device LED screen (tablet), with and without a short-wavelength selective absorption filter, in 16-h light/8-h dark cycles.

Although albino rats are commonly used for studies about retinal damage [9], in our study we have used pigmented rats. Compared to albino strains, the age of the animals at the onset of the light exposure, and not ocular pigmentation, is the most important factor in regulating the severity of light-induced retinal damage [31]. However, following the Shang et al. protocol, the conditions of the animals analyzed included a 14-day pre-study housing period in a dark environment to clear the light exposure effect prior to breeding conditions. Experimental animal retinas were baseline evaluated after this pre-study wash-out period. The experimental conditions of the present study differ substantially and significantly in various aspects from that of Shang et al. study [9]. Firstly, in the light source used: those authors used single-wavelength blue LEDs (460 $\pm$ 10 nm) and custom-made PC white LEDs at domestic lighting levels, while we have used commercially available LED-backlit tablet screens, to replicate real-life usage. Secondly, the exposure conditions: in our study, back-lit LED tablet screens exposed to rats with and without a selective short-wavelength absorption filter, aimed to show that, by the use of the filter, the transmission reduction of the most energetic LED light emitted band (blue light) resulted in reduced cell damage, as shown in studies conducted by Sparrow et al. and Nagai et al., who used yellow optic filtered intraocular lenses [32,33]. Thirdly, the mean luminance used by Shang et al. was 750 lux, the usual domestic luminance level, while in our study mean luminance was 140 lux, which corresponds to the light emitted by a commercially available LED-backlit tablet screen-blank screen set at full brightness. Finally, in the exposure time: whereas Shang et al. used light exposure times of 3, 9, and 28 days under 12-h dark/12-h light cycles, we extended the study to 3 months under more aggressive cycles: 8-h dark/16-h light.

To our knowledge, this is the first study regarding the effects of LED-backlit screens on the mammalian retina, and it is relevant because the time we spend watching LED-backlit screens of domestic devices (as personal computers, tablets, smartphones) at short distances has increased progressively in recent years, and this also includes infants and young people.

The retinal structure in the group of animals exposed to long periods of LED-backlit tablets light was altered, showing a significant decreases in the number of cells in the outer nuclear layer. No further significant changes were observed neither in the inner nuclear layer nor in the ganglion cell layer. These results agree with those obtained by Shang et al. in 2014 [9]. Lin et al. in 2017 reported marked retinal damage by regular exposure to a blue light-emitting diode in rats [19]. However, interestingly, these structural changes were prevented by filtering light specifically, as previously reported by different authors, including our research group.

The reduction in the number of retinal cells due to blue light exposure might be due to multiple factors, such as apoptosis of photoreceptors and RPE reported earlier [2,3,9,11,12,14,15]. Our results on the life/death-related genes further argue for the causes of cell loss after back-lit LED screens exposure. We observed that long periods of light exposure increase the expression of some genes related to cell death, such as Bax, Bak, and Bcl-XS as well as caspase 3 and caspase 9; conversely, the genes related to cell survival like Bcl-2 and Bcl-xl were down-regulated. Supporting our results, Lin et al. also observed increased expression of Bax and caspase-3, decreased expression of inhibited Bcl-2 and Bcl-xL, and inhibition of Bcl-2/Bax association in the RPE after regular exposure to blue light-emitting diode [19].

The exposition to LED-backlit screens strongly up-regulates MMP-2 and MMP-9, to a lesser extent ADAMTS-14, and was without effect on ADAMTS-12, and down-regulates both TIMP-1 and TIMP-2. These changes could have deleterious effects on the retina due to the extracellular matrix (ECM). Increased MMP expression could induce a faster ECM turnover to elude matrix deposit genesis [34]. Our experiments indicate upregulation in MMP-2 and MMP-9 expression, especially in rats exposed to the unfiltered LED tablet screen, whereas Sanchez-Ramos et al. in 2010 did not find any change in MMP-2 expression in other animal model trials [27]. Plantner et al. detected an increase in this MMP [35]. As for MMP-9 expression, our results are consistent with those obtained by Papp et al. and by Sanchez-Ramos [27,36]. In general, these results cannot support the hypothesis that the formation of deposits that give rise to retinal drusen is due to a decrease in pigmentary epithelium MMPs [37], as there are two possible interpretations. On one hand, long-term exposure to light causes an increase in the expression of some MMPs and ADAMTS-14 concurrently by a decrease of TIMP-1 and TIMP-2. These facts could affect the retina due to ECM damage. On the other, the increase in MMPs and ADMATS-14 expression could be related to increased ECM turnover to avoid the appearance of deposits that causes drusen [27].

As for the analysis of the results of BDNF and its receptor TrkB, exposure to light in our study conditions, whether with or without filter, had no effects on BDNF gene expression, which is consistent with findings by Wen et al., who reported that continuous exposure to white light had no significant effects on BDNF expression [38]. Asai et al. also obtained similar results by Western blot densitometry [39]. On the contrary, TrkB receptor expression levels were affected by LED-backlit screen tablet light in rat retinas, both with and without the filter. However, compared to filtered conditions, gene change was greater when exposure was without the filter. Given TrkB essentially has a survival and protector role, the increased levels of TrkB could be related to retinal protection against phototoxicity [40]. Interestingly, maximum TrkB increases were obtained in rats exposed to light in unfiltered conditions, with a higher proportion of short wavelengths, the most deleterious for the retina [41]. Therefore, an increase in TrkB expression, after LED-light exposure, could be considered as a neuroprotection response. On the other hand, photoreceptor integrity and survival could be irreversibly compromised by the short-term stress of the RPE. Exposure to intense visible light activates the VEGF signal and, thus, the breakdown of the outer blood–retinal barrier. The resulting permeability increase seems to induce photoreceptor apoptosis due to light exposure. The breakdown and successive apoptosis effects can be prevented by inhibiting VEGF [42]. Results obtained in our study show an increase in VEGF expression in the retinas of rats exposed to LED-backlit tablet screen light. However, results obtained with the absorbing filter, which partially filters short wavelengths, do not show VEGF expression differences compared to control group values. However, it is important to point out that the toxic outcome of VEGF on photoreceptors is not a direct effect, but secondary to VEGF-induced RPE permeability. It seems plausible that after exposure to high levels of light, the breakdown of the epithelium integrity alters the metabolite exchange between retina and choroid, which, in turn, may further affect photoreceptors [42].

We also found after up-regulation of TNFα, SOD1, and SOD2 after exposition to LED tablet screen light. As far as we know, the regulation of the expression of TNFα by light has not been previously reported; nevertheless, this could be in agreement with the inflammatory mechanism that underlines the age-related macular degeneration induced or exacerbated by blue light [43]. Finally, the increase in SOD1 and SOD2, two enzymes intimately related to oxidative stress after light exposure, suggests a contribution to the whole retinal damage. Shang et al. observed free radical production in the retina after LED-exposition [9]. Similarly, Jaadane et al. and Nakamura et al. indicated that oxidative stress was partially involved in blue LED light-induced retinal damage [2,7].

Several studies have speculated on the damaging effects that short wavelength radiation can cause on the eye and the possible protective effect of optical filters that selectively absorb this light. In 2004, Sparrow et al. already suggested that a blue light partial absorption filter reduces approximately 80% and 78% in the death of the RPE exposed to blue (430 nm) and white light (5400 °K) [32]. Other studies, such as those published by Yanagi et al. in 2006 and by Hui et al. in 2009, demonstrated that the blue light partial absorption filter protects the RPE from the damage produced by short wave visible radiation, increasing cell viability by 42% and 79.5%, respectively [44,45]. On the other hand, in 2011, Zhou et al. observed that the viability of cells exposed to 430 ± 20 nm was reduced by 40% [46]. To evaluate possible artificial protection mechanisms, they interposed optical filters with different levels of a pigment that absorbs blue light and verified that the protection provided by the filter was a function of its absorbance. In 2013, Chamorro et al. published a study whose results showed that the absorption of blue light decreases apoptotic cell death by 50–89% and inhibits DNA damage by 57–81% [3].

Although all these studies have results consistent with ours, the research carried out by Chamorro et al. stands out, since, in their experiment, they used LED light sources, just like us. However, our study presents two important differences with theirs when analyzing the results. First, their light exposure system consisted of an experimental device with direct emission LEDs, while our lighting system consisted of LED-backlit screens, currently commercialized, which emit white light diffusely. It should be noted that, under normal conditions, a person can spend long hours watching screens and not specific sources of LEDs, which gives our study greater similarities with real experiences. Second, Chamorro et al. performed an in vitro experiment exposing an RPE cell culture to LED light. However, our study of animal experimentation has allowed us to expose the eyes to the LED radiation, keeping intact the ocular optical systems of protection, such as, for example, the filtering of light provided by the cornea, the lens, and the intraocular media. In addition, in our study, the retinal tissue analyzed was contained in the eye of a living animal during exposure to light, therefore, it kept intact the regeneration mechanisms of photopigments and cells that the eye presents under normal conditions of life during the exposure period. This gives our results greater consistency with the effects that blue light and selective absorbance filters for this radiation can produce on living retinal tissue.

In this study, we assessed genes involved in retinal damage in the classic model of pigmented rats. Although extrapolation of the results obtained in rats to humans is questionable, it is useful to gain knowledge on the type of detrimental effects that the new LED-backlit screens can generate. Additionally, it is important to consider that results obtained in in-vitro studies with human retinal cells are similar to those obtained in our study. Along this same line, Nagai et al. have shown that blue-light filtering intraocular lenses that absorb a percentage of short-wavelengths that reach the human retina, decrease the incidence of retinal degeneration [33].

Although further investigations are still needed, there is evidence about how LED light damage depends on the exposure time and the wavelength, and involves different pathways related to oxidative stress, inflammation, ECM degradation, regulation of apoptotic genes, and growth factors. As LED-backlit screens are watched at a short distance, we stress the importance of controlling excessive exposure to LED light currently imposed by the widespread use of digital devices such as computers, laptops, tablets, smartphones, etc. The use of short-wavelength selective filters on screens may contribute to reducing potentially irreversible damage to the human retina, by partially cutting off the excessive and most energetic blue-light emissions.

Author Contributions: Conceptualization, C.S.-R., V.A.-V. and J.A.V.; Data curation, T.C. and O.G.-S.; Formal analysis, V.B.-S., T.C. and O.G.-S.; Investigation, C.B.-A., T.C., O.G.-S. and C.A.-P.; Methodology, C.B.-A. and M.J.P.-C.; Project administration, C.S.-R. and J.A.V.; Resources, T.C. and M.J.P.-C.; Software, V.B.-S. and O.G.-S.; Supervision, C.S.-R., C.B.-A., V.A.-V. and J.A.V.; Validation, C.S.-R. and C.A.-P.; Visualization, M.J.P.-C.; Writing–original draft, C.A.-P.; Writing–review & editing, C.S.-R., C.B.-A., V.B.-S., V.A.-V., T.C., M.J.P.-C., O.G.-S. and J.A.V. All authors have read and agreed to the published version of the manuscript.

Funding: This research received no external funding.

Institutional Review Board Statement: This study has the approval of the Animal Experimentation Committee of the UCM and of the Department of Health of the Comunidad de Madrid, Spain (Reference PROEX 310/15).

Informed Consent Statement: Not applicable.

Data Availability Statement: The data that support the findings of this study are available from corresponding author upon reasonable request.

Conflicts of Interest: The authors declare no conflict of interest.

References

1. Noell, W.K.; Walker, V.S.; Kang, B.S.; Berman, S. Retinal damage by light in rats. *Investig. Ophthalmol.* **1966**, *5*, 450–473.
2. Jaadane, I.; Boulenguez, P.; Chahory, S.; Carré, S.; Savoldelli, M.; Jonet, L.; Behar-Cohen, F.; Martinsons, C.; Torriglia, A. Retinal damage induced by commercial light emitting diodes (LEDs). *Free Radic. Biol. Med.* **2015**, *84*, 373–384. [CrossRef]
3. Chamorro, E.; Bonnin-Arias, C.; Pérez-Carrasco, M.J.; Muñoz de Luna, J.; Vázquez, D.; Sánchez-Ramos, C. Effects of light-emitting diode radiations on human retinal pigment epithelial cells in vitro. *Photochem. Photobiol.* **2013**, *89*, 468–473. [CrossRef]
4. Roberts, J.E. Ultraviolet radiation as a risk factor for cataract and macular degeneration. *Eye Contact Lens* **2011**, *37*, 246–249. [CrossRef] [PubMed]
5. Wielgus, A.R.; Roberts, J.E. Retinal photodamage by endogenous and xenobiotic agents. *Photochem. Photobiol.* **2012**, *88*, 1320–1345. [CrossRef]
6. Mohan, K.; Kecova, H.; Hernandez-Merino, E.; Kardon, R.H.; Harper, M.M. Retinal ganglion cell damage in an experimental rodent model of blast-mediated traumatic brain injury. *Investig. Ophthalmol. Vis. Sci.* **2013**, *54*, 3440–3450. [CrossRef]
7. Nakamura, M.; Yako, T.; Kuse, Y.; Inoue, Y.; Nishinaka, A.; Nakamura, S.; Shimazawa, M.; Hara, H. Exposure to excessive blue LED light damages retinal pigment epithelium and photoreceptors of pigmented mice. *Exp. Eye Res.* **2018**, *177*, 1–11. [CrossRef] [PubMed]
8. Shang, Y.-M.; Wang, G.-S.; Sliney, D.H.; Yang, C.-H.; Lee, L.-L. Light-emitting-diode induced retinal damage and its wavelength dependency in vivo. *Int. J. Ophthalmol.* **2017**, *10*, 191–202. [CrossRef]
9. Shang, Y.-M.; Wang, G.-S.; Sliney, D.; Yang, C.-H.; Lee, L.-L. White light-emitting diodes (LEDs) at domestic lighting levels and retinal injury in a rat model. *Environ. Health Perspect.* **2014**, *122*, 269–276. [CrossRef]
10. Zhang, P.; Huang, C.; Wang, W.; Wang, M. Early changes in staurosporine-induced differentiated RGC-5 cells indicate cellular injury response to nonlethal blue light exposure. *Photochem. Photobiol. Sci. Off. J. Eur. Photochem. Assoc. Eur. Soc. Photobiol.* **2015**, *14*, 1093–1099. [CrossRef]
11. Kuse, Y.; Ogawa, K.; Tsuruma, K.; Shimazawa, M.; Hara, H. Damage of photoreceptor-derived cells in culture induced by light emitting diode-derived blue light. *Sci. Rep.* **2014**, *4*, 5223. [CrossRef]
12. Behar-Cohen, F.; Martinsons, C.; Viénot, F.; Zissis, G.; Barlier-Salsi, A.; Cesarini, J.P.; Enouf, O.; Garcia, M.; Picaud, S.; Attia, D. Light-emitting diodes (LED) for domestic lighting: Any risks for the eye? *Prog. Retin. Eye Res.* **2011**, *30*, 239–257. [CrossRef] [PubMed]
13. Martinsons, C. *Potential Health Issues of Solid-State Lighting*; IEA 4E Final Report; IEA: Paris, France, 2014.
14. Contín, M.A.; Arietti, M.M.; Benedetto, M.M.; Bussi, C.; Guido, M.E. Photoreceptor damage induced by low-intensity light: Model of retinal degeneration in mammals. *Mol. Vis.* **2013**, *19*, 1614–1625.
15. Krigel, A.; Berdugo, M.; Picard, E.; Levy-Boukris, R.; Jaadane, I.; Jonet, L.; Dernigoghossian, M.; Andrieu-Soler, C.; Torriglia, A.; Behar-Cohen, F. Light-induced retinal damage using different light sources, protocols and rat strains reveals LED phototoxicity. *Neuroscience* **2016**, *339*, 296–307. [CrossRef] [PubMed]
16. Landry, R.J.; James, R.H.; Miller, S.A.; Ilev, I.K. Evaluation of Potential Optical Radiation Hazards from LED Flashlights. *Health Phys.* **2021**, *120*, 56–61. [CrossRef]
17. Tao, J.-X.; Zhou, W.-C.; Zhu, X.-G. Mitochondria as potential targets and initiators of the blue light hazard to the retina. *Oxid. Med. Cell. Longev.* **2019**, *2019*, 6435364. [CrossRef]
18. Touitou, Y.; Point, S. Effects and mechanisms of action of light-emitting diodes on the human retina and internal clock. *Environ. Res.* **2020**, *190*, 109942. [CrossRef] [PubMed]

19. Lin, C.-H.; Wu, M.-R.; Li, C.-H.; Cheng, H.-W.; Huang, S.-H.; Tsai, C.-H.; Lin, F.-L.; Ho, J.-D.; Kang, J.-J.; Hsiao, G.; et al. Editor's Highlight: Periodic Exposure to Smartphone-Mimic Low-Luminance Blue Light Induces Retina Damage Through Bcl-2/BAX-Dependent Apoptosis. *Toxicol. Sci.* **2017**, *157*, 196–210. [CrossRef]

20. Jaadane, I.; Villalpando Rodriguez, G.E.; Boulenguez, P.; Chahory, S.; Carré, S.; Savoldelli, M.; Jonet, L.; Behar-Cohen, F.; Martinsons, C.; Torriglia, A. Effects of white light-emitting diode (LED) exposure on retinal pigment epithelium in vivo. *J. Cell. Mol. Med.* **2017**, *21*, 3453–3466. [CrossRef]

21. Xie, C.; Zhu, H.; Chen, S.; Wen, Y.; Jin, L.; Zhang, L.; Tong, J.; Shen, Y. Chronic retinal injury induced by white LED light with different correlated color temperatures as determined by microarray analyses of genome-wide expression patterns in mice. *J. Photochem. Photobiol. B Biol.* **2020**, *210*, 111977. [CrossRef]

22. Song, J.; Li, D.; Shan, Z.; Kurskaya, O.; Sharshov, K.; Gao, T.; Bi, H. Photocytotoxicity of white light-emitting diode irradiation on human lens epithelium and retinal pigment epithelium via the JNK and p38 MAPK signaling pathways. *J. Photochem. Photobiol. B Biol.* **2020**, *213*, 112058. [CrossRef]

23. Benedetto, M.M.; Contin, M.A. Oxidative stress in retinal degeneration promoted by constant LED light. *Front. Cell. Neurosci.* **2019**, *13*, 139. [CrossRef]

24. Otsu, W.; Ishida, K.; Nakamura, S.; Shimazawa, M.; Tsusaki, H.; Hara, H. Blue light-emitting diode irradiation promotes transcription factor EB-mediated lysosome biogenesis and lysosomal cell death in murine photoreceptor-derived cells. *Biochem. Biophys. Res. Commun.* **2020**, *526*, 479–484. [CrossRef]

25. Liu, X.; Zhou, Q.; Lin, H.; Wu, J.; Wu, Z.; Qu, S.; Bi, Y. The Protective Effects of Blue Light-Blocking Films With Different Shielding Rates: A Rat Model Study. *Transl. Vis. Sci. Technol.* **2019**, *8*, 19. [CrossRef] [PubMed]

26. Park, S.-I.; Jang, Y.P. The Protective Effect of Brown-, Gray-, and Blue-Tinted Lenses against Blue LED Light-Induced Cell Death in A2E-Laden Human Retinal Pigment Epithelial Cells. *Ophthalmic Res.* **2017**, *57*, 118–124. [CrossRef]

27. Sanchez-Ramos, C.; Vega, J.A.; del Valle, M.E.; Fernandez-Balbuena, A.; Bonnin-Arias, C.; Benitez-del Castillo, J.M. Role of metalloproteases in retinal degeneration induced by violet and blue light. *Adv. Exp. Med. Biol.* **2010**, *664*, 159–164. [CrossRef] [PubMed]

28. Sánchez-Ramos, C.; Bonnin-Arias, C.; Guerrera, M.C.; Calavia, M.G.; Chamorro, E.; Montalbano, G.; López-Velasco, S.; López-Muñiz, A.; Germanà, A.; Vega, J.A. Light regulates the expression of the BDNF/TrkB system in the adult zebrafish retina. *Microsc. Res. Tech.* **2013**, *76*, 42–49. [CrossRef] [PubMed]

29. Dawson, W.; Nakanishi-Ueda, T.; Armstrong, D.; Reitze, D.; Samuelson, D.; Hope, M.; Fukuda, S.; Matsuishi, M.; Ozawa, T.; Ueda, T.; et al. Local fundus response to blue (LED and laser) and infrared (LED and laser) sources. *Exp. Eye Res.* **2001**, *73*, 137–147. [CrossRef]

30. Ueda, T.; Nakanishi-Ueda, T.; Yasuhara, H.; Koide, R.; Dawson, W.W. Eye damage control by reduced blue illumination. *Exp. Eye Res.* **2009**, *89*, 863–868. [CrossRef]

31. Polosa, A.; Bessaklia, H.; Lachapelle, P. Strain Differences in Light-Induced Retinopathy. *PLoS ONE* **2016**, *11*, e0158082. [CrossRef]

32. Sparrow, J.R.; Miller, A.S.; Zhou, J. Blue light-absorbing intraocular lens and retinal pigment epithelium protection in vitro. *J. Cataract Refract. Surg.* **2004**, *30*, 873–878. [CrossRef] [PubMed]

33. Nagai, H.; Hirano, Y.; Yasukawa, T.; Morita, H.; Nozaki, M.; Wolf-Schnurrbusch, U.; Wolf, S.; Ogura, Y. Prevention of increased abnormal fundus autofluorescence with blue light-filtering intraocular lenses. *J. Cataract Refract. Surg.* **2015**, *41*, 1855–1859. [CrossRef]

34. Singh, M.; Tyagi, S.C. Metalloproteinases as mediators of inflammation and the eyes: Molecular genetic underpinnings governing ocular pathophysiology. *Int. J. Ophthalmol.* **2017**, *10*, 1308–1318. [CrossRef]

35. Plantner, J.J.; Smine, A.; Quinn, T.A. Matrix metalloproteinases and metalloproteinase inhibitors in human interphotoreceptor matrix and vitreous. *Curr. Eye Res.* **1998**, *17*, 132–140. [CrossRef] [PubMed]

36. Papp, A.M.; Nyilas, R.; Szepesi, Z.; Lorincz, M.L.; Takács, E.; Abrahám, I.; Szilágyi, N.; Tóth, J.; Medveczky, P.; Szilágyi, L.; et al. Visible light induces matrix metalloproteinase-9 expression in rat eye. *J. Neurochem.* **2007**, *103*, 2224–2233. [CrossRef] [PubMed]

37. Elliot, S.; Catanuto, P.; Stetler-Stevenson, W.; Cousins, S.W. Retinal pigment epithelium protection from oxidant-mediated loss of MMP-2 activation requires both MMP-14 and TIMP-2. *Investig. Ophthalmol. Vis. Sci.* **2006**, *47*, 1696–1702. [CrossRef] [PubMed]

38. Wen, R.; Cheng, T.; Song, Y.; Matthes, M.T.; Yasumura, D.; LaVail, M.M.; Steinberg, R.H. Continuous exposure to bright light upregulates bFGF and CNTF expression in the rat retina. *Curr. Eye Res.* **1998**, *17*, 494–500. [CrossRef]

39. Asai, N.; Abe, T.; Saito, T.; Sato, H.; Ishiguro, S.-I.; Nishida, K. Temporal and spatial differences in expression of TrkB isoforms in rat retina during constant light exposure. *Exp. Eye Res.* **2007**, *85*, 346–355. [CrossRef]

40. Nikoletopoulou, V.; Lickert, H.; Frade, J.M.; Rencurel, C.; Giallonardo, P.; Zhang, L.; Bibel, M.; Barde, Y.-A. Neurotrophin receptors TrkA and TrkC cause neuronal death whereas TrkB does not. *Nature* **2010**, *467*, 59–63. [CrossRef]

41. Wu, J.; Seregard, S.; Algvere, P. V Photochemical damage of the retina. *Surv. Ophthalmol.* **2006**, *51*, 461–481. [CrossRef]

42. Cachafeiro, M.; Bemelmans, A.P.; Samardzija, M.; Afanasieva, T.; Pournaras, J.A.; Grimm, C.; Kostic, C.; Philippe, S.; Wenzel, A.; Arsenijevic, Y. Hyperactivation of retina by light in mice leads to photoreceptor cell death mediated by VEGF and retinal pigment epithelium permeability. *Cell Death Dis.* **2013**, *4*, e781. [CrossRef]

43. Algvere, P.V.; Kvanta, A.; Seregard, S. Drusen maculopathy: A risk factor for visual deterioration. *Acta Ophthalmol.* **2016**, *94*, 427–433. [CrossRef] [PubMed]

44. Yanagi, Y.; Inoue, Y.; Iriyama, A.; Jang, W.-D. Effects of yellow intraocular lenses on light-induced upregulation of vascular endothelial growth factor. *J. Cataract Refract. Surg.* **2006**, *32*, 1540–1544. [CrossRef] [PubMed]
45. Hui, S.; Yi, L.; Fengling, Q.L. Effects of light exposure and use of intraocular lens on retinal pigment epithelial cells in vitro. *Photochem. Photobiol.* **2009**, *85*, 966–969. [CrossRef] [PubMed]
46. Zhou, J.; Sparrow, J.R. Light filtering in a retinal pigment epithelial cell culture model. *Optom. Vis. Sci. Off. Publ. Am. Acad. Optom.* **2011**, *88*, 759–765. [CrossRef] [PubMed]

MDPI

St. Alban-Anlage 66

4052 Basel

Switzerland

Tel. +41 61 683 77 34

Fax +41 61 302 89 18

www.mdpi.com

Cells Editorial Office

E-mail: cells@mdpi.com

www.mdpi.com/journal/cells